Springer Texts in Statistics

Springer Texts in Statistics

Athreya/Lahiri: Measure Theory and Probability Theory
Bilodeau/Brenner: Theory of Multivariate Statistics
Brockwell/Davis: An Introduction to Time Series and Forecasting
Carmona: Statistical Analysis of Financial Data in S-PLUS
Casella: Statistical Design
Chow/Teicher: Probability Theory: Independence, Interchangeability, Martingales, Third Edition
Christensen: Advanced Linear Modeling: Multivariate, Time Series, and Spatial Data; Nonparametric Regression and Response Surface Maximization, Second Edition
Christensen: Log-Linear Models and Logistic Regression, Second Edition
Christensen: Plane Answers to Complex Questions: The Theory of Linear Models, Second Edition
Cryer/Chan: Time Series Analysis, Second Edition
DasGupta: Asymptotic Theory of Statistics and Probability
Davis: Statistical Methods for the Analysis of Repeated Measurements
Dean/Voss: Design and Analysis of Experiments
Dekking/Kraaikamp/Lopuhaä/Meester: A Modern Introduction to Probability and Statistics
Durrett: Essentials of Stochastic Processes
Edwards: Introduction to Graphical Modeling, Second Edition
Everitt: An R and S-PLUS Companion to Multivariate Analysis
Gentle: Matrix Algebra: Theory, Computations, and Applications in Statistics
Ghosh/Delampady/Samanta: An Introduction to Bayesian Analysis
Gut: Probability: A Graduate Course
Heiberger/Holland: Statistical Analysis and Data Display; An Intermediate Course with Examples in S-PLUS, R, and SAS
Jobson: Applied Multivariate Data Analysis, Volume I: Regression and Experimental Design
Jobson: Applied Multivariate Data Analysis, Volume II: Categorical and Multivariate Methods
Karr: Probability
Kulkarni: Modeling, Analysis, Design, and Control of Stochastic Systems
Lange: Applied Probability
Lange: Optimization
Lehmann: Elements of Large Sample Theory
Lehmann/Romano: Testing Statistical Hypotheses, Third Edition
Lehmann/Casella: Theory of Point Estimation, Second Edition
Longford: Studying Human Populations: An Advanced Course in Statistics
Marin/Robert: Bayesian Core: A Practical Approach to Computational Bayesian Statistics
Nolan/Speed: Stat Labs: Mathematical Statistics Through Applications
Pitman: Probability
Rawlings/Pantula/Dickey: Applied Regression Analysis

(continued after index)

Anirban DasGupta

Asymptotic Theory of Statistics and Probability

Anirban DasGupta
Purdue University
Department of Statistics
150 North University Street
West Lafayette, IN 47907
dasgupta@stat.purdue.edu

ISBN 978-0-387-75970-8 e-ISBN 978-0-387-75971-5
DOI: 10.1007/978-0-387-75971-5

Library of Congress Control Number: 2008921241

Printed on acid-free paper.

9 8 7 6 5 4 3 2 1

springer.com

To my mother, and to the loving memories of my father

Preface

This book developed out of my year-long course on asymptotic theory at Purdue University. To some extent, the topics coincide with what I cover in that course. There are already a number of well-known books on asymptotics. This book is quite different. It covers more topics in one source than are available in any other single book on asymptotic theory. Numerous topics covered in this book are available in the literature in a scattered manner, and they are brought together under one umbrella in this book. Asymptotic theory is a central unifying theme in probability and statistics. My main goal in writing this book is to give its readers a feel for the incredible scope and reach of asymptotics. I have tried to write this book in a way that is accessible and to make the reader appreciate the beauty of theory and the insights that only theory can provide.

Essentially every theorem in the book comes with at least one reference, preceding or following the statement of the theorem. In addition, I have provided a separate theorem-by-theorem reference as an entry on its own in the front of the book to make it extremely convenient for the reader to find a proof that was not provided in the text. Also particularly worth mentioning is a collection of nearly 300 practically useful inequalities that I have collected together from numerous sources. This is appended at the very end of the book. Almost every inequality in this collection comes with at least one reference. I have often preferred to cite a book rather than an original publication for these inequalities, particularly if the book contained many of the inequalities that I present. I also emphasize in this book conceptual discussion of issues, working out many examples and providing a good collection of unusual exercises. Another feature of this book is the guidance to the literature for someone who wishes to dig deeper into the topic of a particular chapter. I have tried to make the chapter-by-chapter bibliographies both modern and representative. The book has 574 exercises and 293 worked-out examples. I have marked the more nonroutine exercises with an asterisk.

I hope that this book is useful as a graduate text, for independent reading, and as a general and nearly encyclopedic research reference on asymptotic theory. It should be possible to design graduate-level courses using this book with emphasis on parametric methods or nonparametric methods, on classic topics or more current topics, on frequentist topics or Bayesian topics, or even on probability theory. For the benefit of instructors, I have provided recommended chapters for ten different one-semester courses, with emphasis on different themes. I hope that this provides some useful guidance toward designing courses based on this book.

Because the book covers a very broad range of topics, I do not have a uniform answer for what background I assume for a reader of this book. For most chapters, a knowledge of advanced calculus and linear algebra is enough to enable the reader to follow the material. However, some chapters require some use of measure theory and advanced analysis and some exposure to stochastic processes. One course on statistical theory at the level of Bickel and Doksum (cited in Chapter 3 of this volume) or Casella and Berger (1990) and one on probability at the level of Hoel, Port, and Stone (1971) or Durrett (1994) are certainly needed to follow the discussion in this book. Chapter 1 is essentially a review of somewhat more advanced probability should one need it. The more advanced chapters in this book can be much better appreciated if one has had courses on the two books of Erich Lehmann (Lehmann and Casella (cited in Chapter 16), Lehmann and Romano (cited in Chapter 24)) and a course based on Breiman (1992), Durrett (2004) or Billingsley (see Chapter 1).

My greatest thanks are due to Peter Hall for being an inspirational and caring advisor, reader, and intellectual filter over the last several years as I was writing drafts of this book. Peter has deeply influenced my understanding, appreciation, and taste for probability and statistics, and I have felt grateful that I have had access to him at all times and with unlimited patience. I have received much more from Peter than I could wish or expect. I could not have written this book without Peter's exemplary warmth and mentorship. However, all mistakes and ambiguities in the book are exclusively my responsibility. I would love to know of all serious mistakes that a reader finds in this book, and there must be mistakes in a book of this length.

I also want to express my very special thanks to John Marden and Larry Wasserman for repeatedly offering their friendly and thoughtful suggestions on various decisions I had to make on this book. I want to mention the generous help and support from Erich Lehmann, Peter Bickel, Rabi Bhattacharya, and Jon Wellner on specific chapters in the book. Numerous colleagues, and in particular C. R. Rao, Arup Bose, Persi Diaconis, Joe Eaton, Jianqing Fan, Iain Johnstone, T. Krishnan, Bruce Lindsay, Wei-Liem Loh, Peter McCullagh, Dimitris Politis, B. V. Rao, Bob Serfling, J. Sethuraman, Kesar Singh, and Michael Woodroofe, made helpful comments on parts of earlier drafts of the book. Chun Han, Martina Muehlbach, and Surya Tokdar helped me graciously with putting together TeX files of the chapers. John Kimmel and Jeffrey Taub at Springer were extremely helpful and professional, and I enjoyed working with them very much. I will work with John and Jeff at any time with pleasure. Several anonymous referees did unbelievably helpful and constructive readings of many parts of the book. The Springer series editors gave me gracious input whenever needed. The copyeditor Hal Henglein and the typesetters – Integra India of Springer did a magnificent job. I am immensely thankful to all of them. I am also thankful to Purdue University for providing me with computing and secretarial assistance. Doug and Cheryl Crabill, in particular, assisted me numerous times with a smile.

I was an impressionable fifteen-year-old when I entered the Indian Statistical Institute (ISI) as a first-year student. I had heard that statisticians do boring calculations with large numbers using clumsy calculating machines. Dev Basu entered the lecture room on my first day at the ISI and instantly changed my perception of statistics. No one I met could explain so effortlessly the study of randomness and how to use what we learn about it to make useful conclusions. There was not one person at the ISI who didn't regard him as an incomparable role model, a personification of scholarship, and an angelic personality. I am fortunate that I had him as my foremost teacher. I am grateful to C. R. Rao for the golden days of the ISI and for making all of us feel that even as students we were equals in his eyes.

At a personal level, I am profoundly grateful to Jenifer Brown for the uniqueness and constancy of her treasured support, counsel, well wishes, and endearing camaraderie for many years, all of which have enriched me at my most difficult times and have helped me become a better human being. I will always remain much indebted to Jenifer for the positive, encouraging, and crystallizing influence she has been at all times. I have considered Jenifer to be an impeccable role model.

I am also thankful to Supriyo and Anuradha Datta, Julie Marshburn, Teena Seele, Gail Hytner, Norma Lucas, Deb Green, Tanya Winder, Hira Koul, Rajeeva Karandikar, Wei-Liem Loh, Dimitris Politis, and Larry Shepp for their loyalty, friendship and warmth. Jim and Ann Berger, Herman Rubin, B.V. Rao, T. Krishnan, Larry Brown, Len Haff, Jianqing Fan, and Bill Strawderman have mentored, supported and cared for me for more than a decade. I appreciate all of them. But most of all, I appreciate the love and warmth of my family. I dedicate this book to the cherished memories of my father, and to my mother on her eightieth birthday.

Anirban DasGupta
Purdue University, West Lafayette, IN

Recommended Chapter Selections

Course Type	Chapters
Semester I, Classical Asymptotics	1, 2, 3, 4, 7, 8, 11, 13, 15, 17, 21, 26, 27
Semester II, Classical Asymptotics	9, 14, 16, 22, 24, 25, 28, 29, 30, 31, 32
Semester I, Inference	1, 2, 3, 4, 7, 14, 16, 17, 19, 20, 21, 26, 27
Semester II, Inference	8, 11, 12, 13, 22, 24, 25, 29, 30, 32, 33, 34
Semester I, Emphasis on Probability	1, 2, 3, 4, 5, 6, 8, 9, 10, 11, 12, 23
Semester I, Contemporary Topics	1, 2, 3, 8, 10, 12, 14, 29, 30, 32, 33, 34
Semester I, Nonparametrics	1, 3, 5, 7, 11, 13, 15, 18, 24, 26, 29, 30, 32
Semester I, Modeling and Data Analysis	1, 3, 4, 8, 9, 10, 16, 19, 26, 27, 29, 32, 33
My Usual Course, Semester I	1, 2, 3, 4, 6, 7, 8, 11, 13, 14, 15, 16, 20
My Usual Course, Semester II	5, 9, 12, 17, 21, 22, 24, 26, 28, 29, 30, 32, 34

Key Theorems and References

Theorem Number	A Source for Proof
1.2	Breiman (1968, 1992)
1.4	Billingsley (1995)
1.5	Ferguson (1996)
1.6	Breiman (1968, 1992)
1.7	Ash (1973), Billingsley (1995)
1.11	Breiman (1968, 1992)
1.13	Billingsley (1995)
1.15	Breiman (1968, 1992)
2.1	Reiss (1989)
2.2	Reiss (1989), Billingsley (1995)
2.7	Johnson (2004)
2.12	Le Cam (1960)
3.1	Feller (1966), Ferguson (1996)
3.4	Breiman (1968, 1992)
3.5	Breiman (1968, 1992)
3.6	Proved in the text
3.9	Bickel and Doksum (2001)
4.1, 4.3	Brown, Cai, and DasGupta (2006)
5.1	Billingsley (1995)
5.9	Anscombe (1952)
5.10, 5.11	Breiman (1968, 1992), Ash (1973)
5.13, 5.16, 5.18, 5.19	Breiman (1968, 1992)
6.1, 6.2	Billingsley (1995)
6.6	Shiryaev (1980)
7.1, 7.2, 7.3	Serfling (1980)
7.5	Koenker and Bassett (1978)
8.4, 8.5, 8.6	Galambos (1977)
8.12, 8.13, 8.14	Galambos (1977)
8.16, 8.17	Leadbetter et al. (1983)

9.1, 9.2	Lehmann (1999)
9.3, 9.4, 9.5	Hall and Heyde (1980)
10.1	Gudynas (1991)
10.5	Diaconis and Stroock (1991), Fill (1991)
11.1–11.4	Petrov (1975)
11.5	Bentkus (2003)
12.1	Durrett (1996), Csörgo (2003)
12.3–12.5	Csörgo (2003)
12.8	Dudley (1984), Gineé (1996)
12.10	Vapnik and Chervonenkis (1971), Sauer (1972)
12.11, 12.13, 12.15	van der Vaart and Wellner (1996)
12.19–12.21	Csörgo (2003)
13.1	Lahiri (2003)
13.3	Hall (1987)
13.4	Barndorff-Nielsen and Cox (1991)
14.1, 14.2	Olver (1997)
14.3	Daniels (1954), Section 14.2.1 in the text
14.4	Corollary of 14.3
14.5	Lugannani and Rice (1980)
15.1–15.3	Serfling (1980)
16.1	Proved in the text
16.2	Lehmann and Casella (1998), Lehmann (1999)
16.3	Lehmann and Casella (1998), Lehmann (1999)
16.4, 16.5	Pfanzagl (1973)
16.6, 16.7	van der Vaart (1998)
16.8	Efron (1975)
17.1, 17.2	Serfling (1980)
18.1	Lehmann (1983)
19.1	Hettmansperger and McKean (1998)
20.1, 20.2	Bickel and Doksum (2001)
20.3	Le Cam (1986)
20.4, 20.6	Johnson (1967, 1970)
20.8, 20.9	Bickel and Doksum (2001)
Prop. 20.1	Proved in the text
20.11	Proved in the text
20.13	Diaconis and Freedman (1986)
21.1	Ferguson (1996)
21.2	Sen and Singer (1993)
21.4	Serfling (1980), van der Vaart (1998), Sen and Singer (1993)
22.1–22.4	Serfling (1980)

22.6	Serfling (1980)
23.1–23.3	Bucklew (2004)
23.4	Chaganty and Sethuraman (1993)
24.2	Proved in the text
24.3, 24.5, 24.6, 24.7	Hettmansperger (1984)
24.9	Basu and DasGupta (1995)
24.11	Proved in the text
24.12	Bahadur and Savage (1956)
24.14	Lehmann and Romano (2005)
25.1	Hettmansperger (1984)
25.4	Proved in the text
26.1	del Barrio et al. (2007), van der Vaart (1998), Chapter 12 in the text
26.3	del Barrio et al. (2007)
26.4, 26.5	Jager and Wellner (2006)
26.6	Proved in the text
26.10	Brown, DasGupta, Marden, and Politis (2004)
26.11	de Wet and Ventner (1972), Sarkadi (1985)
27.1, 27.2	Proved in the text
27.3	Sen and Singer (1993)
27.5, 27.6	Kallenberg et al. (1985)
28.1	del Barrio et al. (2007)
29.1	Proved in the text, Bickel and Freedman (1981), Singh (1981)
29.2	Hall (1990)
29.3, 29.4	Shao and Tu (1995)
29.5	Lahiri (2003)
29.8, 29.10–29.12	Shao and Tu (1995)
29.13	Hall (1988)
29.15	Freedman (1981)
29.16	Hall (1989)
29.17	Freedman (1981)
29.18	Bose (1988)
29.19, 29.20	Lahiri (1999)
30.1	Proved in the text
30.2–30.7	Shao and Tu (1995)
31.2	Proved in the text
31.3	Bickel and van Zwet (1978)
Prop. 32.1	Proved in the text
Thm. 32.1	Proved in the text
32.4	Devroye and Penrod (1984), Chow et al. (1983)

32.5	Hall (1983)
32.6	Stone (1984)
32.7	Hall and Wand (1988)
32.8	Devroye and Penrod (1984)
32.9, 32.10	Bickel and Rosenblatt (1975)
32.12	DasGupta (2000)
32.13	Scott (1992)
32.14	Groeneboom (1985), Birge (1989)
32.15	Birge (1989)
32.17, 32.18	Parzen (1962), Chernoff (1964), Venter (1967)
33.1, 33.2	Cheney and Light (1999)
33.3	Feller (1966), Jewel (1982)
33.6	Redner and Walker (1984)
33.7	Beran (1977)
33.8	Tamura and Boos (1986)
33.9	Lindsay (1983a,b)
33.10	Kiefer and Wolfowitz (1956)
33.13	Chen (1995)
33.14	Fan (1991), Carroll and Hall (1988)
34.1–34.3	Holst (1972), Morris (1975)
34.4, 34.5	Portnoy (1984, 1985)
34.6	Brown (1979, 1984), Hayter (1984)
34.7	See the text
34.8	Finner and Roters (2002)
34.10	Storey, Taylor, and Siegmund (2004)
34.11, 34.12	Donoho and Jin (2004)
34.13, 34.14	Genovese and Wasserman (2002)
34.15, 34.16	DasGupta and Zhang (2006)
34.18	Meinshausen and Rice (2006)
34.19, 34.21	Sarkar (2006)
34.20	Karlin and Rinott (1980)
34.22	Hall and Jin (2007)

Contents

Chapter 1
Basic Convergence Concepts and Theorems

The fundamental concepts of convergence and the most important theorems related to convergence are presented in the form of a review in this chapter with illustrative examples. Although some use of measure theory terminology has been made in this chapter, it is possible to understand the meanings of most of the results without a formal background in measure theory. There are many excellent texts on the advanced theory of probability that treat the material presented here. Among these, we recommend Breiman (1968), Billingsley (1995), and Ash (1972). Discussions with a statistical orientation are given in Serfling (1980) and Ferguson (1996). More specific references for selected key theorems are included later.

1.1 Some Basic Notation and Convergence Theorems

Definition 1.1 Let $\{X_n, X\}$ be random variables defined on a common probability space. We say X_n converges to X in probability if, for any $\epsilon > 0$, $P(|X_n - X| > \epsilon) \to 0$ as $n \to \infty$. If a sequence X_n converges in probability to zero, then we write $X_n \stackrel{\mathcal{P}}{\Rightarrow} 0$ or $X_n \stackrel{P}{\to} 0$ and also $X_n = o_p(1)$. If, more generally, $a_n X_n \stackrel{\mathcal{P}}{\Rightarrow} 0$ for some sequence a_n, then we write $X_n = o_p(\frac{1}{a_n})$.

Definition 1.2 A sequence of random variables X_n is said to be bounded in proba-bility if, given $\epsilon > 0$, one can find a constant k such that $P(|X_n| > k) \leq \epsilon$ for all $n \geq n_0 = n_0(\epsilon)$. If X_n is bounded in probability, then we write $X_n = O_p(1)$. If $a_n X_n = O_p(1)$, we write $X_n = O_p(\frac{1}{a_n})$. If $X_n = o_p(1)$, then also $X_n = O_p(1)$.

Definition 1.3 Let $\{X_n, X\}$ be defined on the same probability space. We say that X_n converges almost surely to X (or X_n converges to X with probability 1) if $P(\omega : X_n(\omega) \to X(\omega)) = 1$. We write $X_n \stackrel{\text{a.s.}}{\to} X$ or $X_n \stackrel{\text{a.s.}}{\Rightarrow} X$.

A. DasGupta, *Asymptotic Theory of Statistics and Probability*,

Remark. Almost sure convergence is a stronger mode of convergence than convergence in probability. In fact, a characterization of almost sure convergence is that, for any given $\epsilon > 0$,

$$\lim_{m\to\infty} P(|X_n - X| \le \epsilon \quad \forall n \ge m) = 1.$$

It is clear from this characterization that almost sure convergence is stronger than convergence in probability.

Example 1.1 For iid Bernoulli trials with a success probability $p = \frac{1}{2}$, let T_n denote the number of times in the first n trials that a success is followed by a failure. Denoting $I_i = I\{i\text{th trial is a success and } (i+1)\text{st trial is a failure}\}$, $T_n = \sum_{i=1}^{n-1} I_i$, and therefore $E(T_n) = \frac{n-1}{4}$, and $\mathrm{Var}(T_n) = \sum_{i=1}^{n-1} \mathrm{Var}(I_i) + 2\sum_{i=1}^{n-2} \mathrm{cov}(I_i, I_{i+1}) = \frac{3(n-1)}{16} - \frac{2(n-2)}{16} = \frac{n+1}{16}$. It follows by an application of Chebyshev's inequality that $\frac{T_n}{n} \overset{P}{\Rightarrow} \frac{1}{4}$.

Example 1.2 Suppose $X_1, X_2, \ldots$ is an infinite sequence of iid $U[0,1]$ random variables, and let $X_{(n)} = \max\{X_1, \ldots, X_n\}$. Intuitively, $X_{(n)}$ should get closer and closer to 1 as n increases. In fact, $X_{(n)}$ converges almost surely to 1 for $P(|1 - X_{(n)}| \le \epsilon \quad \forall n \ge m) = P(1 - X_{(n)} \le \epsilon \quad \forall n \ge m) = P(X_{(n)} \ge 1 - \epsilon \quad \forall n \ge m) = P(X_{(m)} \ge 1 - \epsilon) = 1 - (1-\epsilon)^m \to 1$ as $m \to \infty$, and hence $X_{(n)} \overset{\text{a.s.}}{\Rightarrow} 1$.

Example 1.3 Suppose $X_n \sim N(\frac{1}{n}, \frac{1}{n})$ is a sequence of independent variables. Since the mean and the variance are both converging to zero, intuitively one would expect that the sequence X_n converges to zero in some sense. In fact, it converges almost surely to zero. Indeed, $P(|X_n| \le \epsilon \quad \forall n \ge m) = \prod_{n=m}^{\infty} P(|X_n| \le \epsilon) = \prod_{n=m}^{\infty} [\Phi(\epsilon\sqrt{n} - \frac{1}{\sqrt{n}}) + \Phi(\epsilon\sqrt{n} + \frac{1}{\sqrt{n}}) - 1] = \prod_{n=m}^{\infty}\left[1 + O\left(\frac{\phi(\epsilon\sqrt{n})}{\sqrt{n}}\right)\right] = 1 + O\left(\frac{e^{-\frac{m\epsilon^2}{2}}}{\sqrt{m}}\right) \to 1$ as $m \to \infty$, implying $X_n \overset{\text{a.s.}}{\Rightarrow} 0$. In the above, the next to last equality follows on using the tail property of the standard normal cumulative distribution function (CDF) that $\frac{1-\Phi(x)}{\phi(x)} = \frac{1}{x} + O(\frac{1}{x^3})$ as $x \to \infty$.

We next state two fundamental theorems known as the *laws of large numbers*. Breiman (1968) can be seen for the proofs.

Theorem 1.1 (Khintchine) *Suppose $X_1, X_2, \ldots$ are independent and identically distributed (iid) with a finite mean μ. Let $\bar{X}_n = \frac{1}{n}\sum_{i=1}^{n} X_i$. Then $\bar{X}_n \xrightarrow{\mathcal{P}} \mu$.*

This is known as the *weak law of large numbers*. The next result is known as the *strong law of large numbers*.

Theorem 1.2 (Kolmogorov) If X_i are iid, then $\bar{X}_n$ has an a.s. limit iff $E(|X_1|) < \infty$, in which case $\bar{X}_n \stackrel{\text{a.s.}}{\to} \mu = E(X_1)$.

Convergence in distribution, which is of great importance in applications, is defined next.

Definition 1.4 Let $\{X_n, X\}$ be real-valued random variables defined on a common probability space. We say that X_n converges in distribution (in law) to X and write $X_n \xrightarrow{\mathcal{L}} X$ or $X_n \stackrel{\mathcal{L}}{\Rightarrow} X$ if $P(X_n \leq x) \to P(X \leq x)$ as $n \to \infty$ at every point x that is a continuity point of the CDF of X.

If $\{X_n, X\}$ are random variables taking values in a multidimensional Euclidean space, the same definition applies, with the CDFs of X_n, X being defined in the usual way.

The following two results are frequently useful in calculations; neither of the two parts is hard to prove.

Theorem 1.3 (a) If $X_n \xrightarrow{\mathcal{L}} X$ for some X, then $X_n = O_p(1)$.

(b) **Polya's Theorem** If $F_n \xrightarrow{\mathcal{L}} F, F$ is a continuous CDF, then $\sup_{-\infty<x<\infty} |F_n(x) - F(x)| \to 0$ as $n \to \infty$.

A large number of equivalent definitions for convergence in distribution are known. Some of these are summarized in the following *Portmanteau theorem*. Billingsley (1995) can be seen for a proof of most of the parts of this theorem.

Theorem 1.4 The Portmanteau Theorem
Let $\{X_n, X\}$ be random variables taking values in a finite-dimensional Euclidean space. The following are characterizations of $X_n \xrightarrow{\mathcal{L}} X$:

(a) $E(g(X_n)) \to E(g(X))$ for all bounded continuous functions g.
(b) $E(g(X_n)) \to E(g(X))$ for all bounded Lipschitz functions g.
(c) $E(g(X_n)) \to E(g(X))$ for all functions in $C_0(\mathbb{R})$ if X_n are real valued.
(d) $P(X_n \in B) \to P(X \in B)$ for all Borel sets B that have a null boundary ∂B under X; i.e., $P(X \in \partial B) = 0$.
(e) $\liminf P(X_n \in G) \geq P(X \in G)$ for all open sets G.
(f) $\limsup P(X_n \in S) \leq P(X \in S)$ for all closed sets S.

Remark. Some of these characterizations are more useful for proving theorems rather than as easily verifiable tools in a practical application.

Example 1.4 Consider $X_n \sim \text{Uniform}\{\frac{1}{n}, \frac{2}{n}, \ldots, \frac{n-1}{n}, 1\}$. Then, it can be shown easily that the sequence X_n converges in law to the $U[0, 1]$ distribution. Consider now the function $g(x) = x^{10}, 0 \leq x \leq 1$. Note that g is continuous and bounded. Therefore, by part (a) of the Portmanteau theorem, $E(g(X_n)) = \sum_{k=1}^{n} \frac{k^{10}}{n^{11}} \to E(g(X)) = \int_0^1 x^{10} dx = \frac{1}{11}$. This is of course a simple fact from Riemann integration, but it is instructive that it can be proved by using the Portmanteau theorem.

Example 1.5 This is a classic example from analysis. For $n \geq 1, 0 \leq p \leq 1$, and a given continuous function $g : [0, 1] \to \mathcal{R}$, define the sequence of Bernstein polynomials $B_n(p) = \sum_{k=0}^{n} g(\frac{k}{n})\binom{n}{k} p^k(1 - p)^{n-k}$. Note that $B_n(p) = E[g(\frac{X}{n})|X \sim \text{Bin}(n, p)]$. As $n \to \infty$, $\frac{X}{n} \xrightarrow{P} p$, and it follows that $\frac{X}{n} \stackrel{\mathcal{L}}{\Rightarrow} \delta_p$, the point mass at p. Since g is continuous and hence bounded, it follows from the Portmanteau theorem that $B_n(p) \to g(p)$. Thus, pointwise convergence of the Bernstein polynomials to the function g follows very easily from the Portmanteau theorem. Even uniform convergence holds, but it requires more work to prove it probabilistically. The Bernstein polynomials give an example of a sequence of polynomials that uniformly approximate a given continuous function on the compact interval [0, 1].

The following theorem is simple but very useful; Serfling (1980) presents a proof.

Theorem 1.5 (Slutsky)

(a) If $X_n \xrightarrow{\mathcal{L}} X$ and $Y_n \xrightarrow{\mathcal{P}} c$, then $X_n \cdot Y_n \xrightarrow{\mathcal{L}} cX$.

(b) If $X_n \xrightarrow{\mathcal{L}} X$ and $Y_n \xrightarrow{\mathcal{P}} c \neq 0$, then $\frac{X_n}{Y_n} \xrightarrow{\mathcal{L}} \frac{X}{c}$.

(c) If $X_n \xrightarrow{\mathcal{L}} X$ and $Y_n \xrightarrow{\mathcal{P}} c$, then $X_n + Y_n \xrightarrow{\mathcal{L}} X + c$.

A very useful tool for establishing almost sure convergence is stated next.

Theorem 1.6 (Borel-Cantelli Lemma) Let $\{A_n\}$ be a sequence of events on a probability space. If

$$\sum_{n=1}^{\infty} P(A_n) < \infty,$$

then P(infinitely many A_n occur) = 0. If $\{A_n\}$ are pairwise independent and

$$\sum_{n=1}^{\infty} P(A_n) = \infty,$$

then P(infinitely many A_n occur) = 1. See Breiman (1968) for a proof.

Remark. Although pairwise independence suffices for the conclusion of the second part of the Borel-Cantelli lemma, common applications involve cases where the A_n are mutually independent.

Example 1.6 Consider iid $N(\mu, 1)$ random variables $X_1, X_2, \ldots$, and suppose $\overline{X}_n$ is the mean of the first n observations. Fix an $\epsilon > 0$ and consider $P(|\overline{X}_n| > \epsilon)$. By Markov's inequality, $P(|\overline{X}_n| > \epsilon) \le \frac{E(\overline{X}_n)^4}{\epsilon^4} = \frac{3}{\epsilon^4 n^2}$. Since $\sum_{n=1}^{\infty} \frac{1}{n^2} < \infty$, from the Borel-Cantelli lemma it follows that $P(|\overline{X}_n| > \epsilon$ infinitely often) $= 0$, and hence $\overline{X}_n \stackrel{\text{a.s.}}{\rightarrow} 0$. One obtains the Kolmogorov strong law's conclusion in this case by a direct application of the Borel-Cantelli lemma.

Example 1.7 Consider a sequence of independent Bernoulli trials in which success occurs with probability p and failure with probability $q = 1 - p$. Suppose $p > q$, so that successes are more likely than failures. Consider a hypothetical long uninterrupted run of m failures, say $FF \ldots F$, for some fixed m. Break up the Bernoulli trials into nonoverlapping blocks of m trials, and consider A_n to be the event that the nth block consists of only failures. The probability of each A_n is q^m, which is free of n. Therefore, $\sum_{n=1}^{\infty} P(A_n) = \infty$ and it follows from the second part of the Borel-Cantelli lemma that no matter how large p may be, as long as $p < 1$, a string of consecutive failures of any given arbitrary length reappears infinitely many times in the sequence of Bernoulli trials.

The key concept of an empirical CDF is defined next.

Definition 1.5 Let $X_1, \ldots, X_n$ be independent with a common CDF $F(x)$. Let $F_n(x) = \frac{\#\{i : X_i \le x\}}{n} = \frac{1}{n}\sum_{i=1}^{n} I_{X_i \le x}$. $F_n(x)$ is called the empirical CDF of the sample $X_1, \ldots, X_n$.

The next two results say that F_n is close to F for large n. Again, Serfling (1980) can be consulted for a proof.

Theorem 1.7 (Glivenko-Cantelli)

$$\sup_{-\infty < x < \infty} |F_n(x) - F(x)| \stackrel{\text{a.s.}}{\rightarrow} 0.$$

The next two results give useful explicit bounds on probabilities of large values for the deviation of F_n from F.

Theorem 1.8 (Dvoretzky-Kiefer-Wolfowitz) There exists a finite positive constant C (independent of F) such that $P(\sup_{-\infty<x<\infty} |F_n(x) - F(x)| \geq \lambda) \leq Ce^{-2n\lambda^2}$.

Massart (1990) gave an explicit value for the best possible value for the constant C. It is stated next.

Theorem 1.9 (Massart) If $n\lambda^2 \geq \frac{\log 2}{2}$, $P(\sup_{-\infty<x<\infty} |F_n(x) - F(x)| > \lambda) \leq 2e^{-2n\lambda^2}$.

1.2 Three Series Theorem and Kolmogorov's Zero-One Law

Although in statistics we often look at averages, sometimes the more relevant quantity is the sum itself, not the average. Thus, if $X_1, X_2, \cdots$ is a sequence of independent random variables and $S_n = \sum_{i=1}^n X_i$ denotes the nth partial sum, we are also often interested in the asymptotic behavior of S_n. The first question would be whether S_n converges. Kolmogorov proved a famous theorem, now known as *Kolmogorov's three series theorem*, that completely settles the convergence question. Interestingly, S_n either converges with probability one or diverges with probability one; it cannot sometimes converge and sometimes diverge. Such a make-or-break phenomenon is usually called a *zero-one law*. We first state an important zero-one law.

Theorem 1.10 (Kolmogorov's Zero-One Law) Let $X_1, X_2, \cdots$ be a sequence of independent random variables on a probability space $(\Omega, \mathcal{A}, P)$ and let $\mathcal{F}_k$ be the σ-field generated by $X_1, X_2, \cdots, X_k, 1 \leq k < \infty$. Let $E \in \mathcal{A}$ be independent of $\mathcal{F}_k$ for each $k < \infty$; i.e., let E be a *tail event*. Then $P(E) = 0$ or 1.

Remark. Since the event E that S_n converges is a tail event, this implies that for an independent sequence $X_1, X_2, \cdots, S_n$ converges with probability 1 or probability 0.

Here is Kolmogorov's three series theorem.

Theorem 1.11 (Three Series Theorem) Let $X_1, X_2, \cdots$ be a sequence of independent random variables and fix $c > 0$. Then, $S_n = \sum_{i=1}^n X_i$ converges iff

(a) $\sum_{n=1}^{\infty} P(|X_n| > c) < \infty$;
(b) $\sum_{n=1}^{\infty} E[|X_n| I_{|X_n| \le c}] < \infty$;
(c) $\sum_{n=1}^{\infty} \text{Var}[|X_n| I_{|X_n| \le c}] < \infty$.

These two theorems are proved in Chow and Teicher (1988).

Example 1.8 Suppose X_i are iid $U[-1, 1]$. Then, if we take c to be, for example, $\frac{1}{2}$, clearly the three series theorem fails and so $\sum_{n=1}^{\infty} X_n$ does not converge to a finite limit. Now suppose X_i are independent $U[-\sigma_i, \sigma_i]$. For simplicity, let us suppose that the σ_i are decreasing and $\sigma_1 = 1$. Then, taking $c = 1$ in the three series theorem, conditions (b) and (c) hold when $\sum_{i=1}^{\infty} \sigma_i < \infty$. So, in that case, $\sum_{n=1}^{\infty} X_n$ converges to a finite value with probability 1. Actually, it can be shown that $\sum_{n=1}^{\infty} X_n$ converges iff $\sum_{i=1}^{\infty} \sigma_i^2 < \infty$.

Example 1.9 Suppose X_i are independent Bernoulli (p_i) random variables. We know that if we add a large number of independent Bernoullis with small success probabilities, then the total number of successes is distributed approximately like a Poisson (see Chapter 2). So, we might expect that if p_i are small, then $\sum_{n=1}^{\infty} X_n$ converges even almost surely to a finite value. Indeed, by taking $c = 1$ in the three series theorem, we see that conditions (b) and (c) hold if $\sum_{i=1}^{\infty} p_i < \infty$.

For partial sums of bounded random variables, the following theorem holds; see Breiman (1968) for a proof.

Theorem 1.12 Suppose X_i are independent and uniformly bounded by some finite number M. Then, $\sum_{n=1}^{\infty} X_n$ converges to a finite value iff $\sum_{n=1}^{\infty} E(X_n)$ converges and $\sum_{n=1}^{\infty} \text{Var}(X_n) < \infty$.

1.3 Central Limit Theorem and Law of the Iterated Logarithm

The most fundamental result on convergence in law is the central limit theorem (CLT) for means of iid random variables. We state the finite-variance case in this section.

Theorem 1.13 (CLT) Let $X_i, i \ge 1$ be iid with $E(X_i) = \mu$ and $\text{Var}(X_i) = \sigma^2 < \infty$. Then

$$\frac{\sqrt{n}(\bar{X}-\mu)}{\sigma} \xrightarrow{\mathcal{L}} Z \sim N(0,1).$$

We also write

$$\frac{\sqrt{n}(\bar{X}-\mu)}{\sigma} \xrightarrow{\mathcal{L}} N(0,1).$$

See Billingsley (1995) or any text on probability for a proof.

Remark. For example, this theorem can be used to approximate $P(\bar{X} \leq \mu + \frac{k\sigma}{\sqrt{n}})$ by $\Phi(k)$, where Φ is the CDF of $Z \sim N(0,1)$. This is very useful because the sampling distribution of $\bar{X}$ is not available except for some special cases such as when X_i are iid normal, Poisson, Bernoulli, exponential, Cauchy, or uniform.

It turns out that continuous transformations preserve many types of convergence, and this fact is useful in many applications. We record it next.

Theorem 1.14 (Continuous Mapping Theorem) Suppose X_n converges to X in probability, almost surely, or in law. Let $g(\cdot)$ be a continuous function. Then $g(X_n)$ converges to $g(X)$ in probability, almost surely, or in law, respectively.

Example 1.10 Suppose $X_1, \ldots, X_n$ are iid with $E(|X_1|) < \infty$ and $E(X_1) = \mu$. By the Khinchine weak law, $\bar{X} \xrightarrow{\mathcal{P}} \mu$. Since $g(x) = x^2$ is a continuous function, it follows that $\bar{X}^2 \xrightarrow{\mathcal{P}} \mu^2$.

Example 1.11 Suppose $X_1, \ldots, X_n$ are iid with mean μ and variance $\sigma^2 < \infty$. Since $\frac{\sqrt{n}(\bar{X}-\mu)}{\sigma} \xrightarrow{\mathcal{L}} N(0,1)$, it follows that $\frac{n(\bar{X}-\mu)^2}{\sigma^2} \xrightarrow{\mathcal{L}} \chi^2(1)$.

Sometimes the following result seems contradictory to the CLT, but a little reflection shows that it is not.

Theorem 1.15 (Law of the Iterated Logarithm (LIL)) Suppose $X_1, X_2, \ldots$ are iid with mean μ and variance σ^2. Set $S_n = X_1 + \cdots + X_n$. Then

$$(a) \limsup_{n\to\infty} \frac{S_n - n\mu}{\sqrt{2n\log\log n}} = \sigma \text{ (a.s.)}$$

$$(b) \liminf_{n\to\infty} \frac{S_n - n\mu}{\sqrt{2n\log\log n}} = -\sigma \text{ (a.s.)}$$

(c) If $\text{Var}(X_1) = \infty$, then

$$\limsup_{n\to\infty} \frac{S_n - n\mu}{\sqrt{2n \log\log n}} = \infty \text{ (a.s.)}$$

(*d*) If finite constants a, τ satisfy $\limsup_{n\to\infty} \frac{S_n - na}{\sqrt{2n \log\log n}} = \tau$ (a.s.), then $a = E(X_1)$ and $\tau^2 = \text{Var}(X_1)$.

The LIL is proved in almost any text on probability; see Chow and Teicher (1988) as a specific reference.

As regards parts (a) and (b), what is going on is that, for a given n, there is some collection of sample points ω for which the partial sum $S_n - n\mu$ stays in a specific $\sqrt{n}$ -neighborhood of zero. But this collection keeps changing with changing n, and any particular ω is sometimes in the collection and at other times out of it. Such *unlucky* values of n are unbounded, giving rise to the LIL phenomenon. The exact rate $\sqrt{n \log\log n}$ is a technical aspect and cannot be explained intuitively.

The LIL also provides an example of almost sure convergence being truly stronger than convergence in probability.

Example 1.12 Let $X_1, X_2, \ldots$ be iid with a finite variance. Then,

$$\frac{S_n - n\mu}{\sqrt{2n \log\log n}} = \frac{S_n - n\mu}{\sqrt{n}} \frac{1}{\sqrt{2 \log\log n}} = O_p(1) \cdot o_p(1) \overset{\text{Slutsky}}{=} o_p(1).$$

But, by the LIL, $\frac{S_n - n\mu}{\sqrt{2n \log\log n}}$ does *not* converge a.s. to zero. Hence, convergence in probability is weaker than almost sure convergence, in general.

The next result gives a standard and useful method for establishing convergence in distribution of a sequence of vector-valued random variables; see Serfling (1980) for a proof.

Theorem 1.16 (**Cramér-Wold Theorem**) Let X_n be a sequence of p-dimensional random vectors. $X_n \overset{\mathcal{L}}{\Rightarrow} X$ iff for every p-dimensional vector $c, c'X_n \overset{\mathcal{L}}{\Rightarrow} c'X$.

The multivariate central limit theorem for iid random vectors with a finite covariance matrix is stated next; it follows from the Cramér-Wold theorem and the one-dimensional central limit theorem.

Theorem 1.17 (**Multivariate CLT**) Let X_i be iid p-dimensional random vectors with $E(X_1) = \mu$ and covariance matrix $\text{cov}(X_1) = \Sigma$. Then,

$$\sqrt{n}(\bar{X} - \mu) \stackrel{\mathcal{L}}{\Rightarrow} N_p(0, \Sigma).$$

1.4 Further Illustrative Examples

Given a sequence of random variables Y_n, often by itself Y_n converges in probability to some number (a constant) c. Thus, the limit distribution of Y_n itself is degenerate at c and is not of any interest. However, often it is also true that for some suitable norming sequence $b_n, b_n(Y_n - c)$ converges to some nondegenerate distribution (e.g., a normal distribution). Often, however, colloquially we refer to *the limiting distribution of* Y_n. It is always understood that suitable centering and norming may be needed to obtain a nondegenerate limit law. This colloquial usage is made throughout the book and should not cause any confusion.

Example 1.13 Let $X_1, \ldots, X_n$ be iid $N(\mu, \sigma^2)$ and let $T_n = \frac{\sqrt{n}(\bar{X}-\mu)}{s}$, where $s^2 = \frac{1}{n-1}\sum_{i=1}^{n}(X_i - \bar{X})^2$. T_n has the central $t_{(n-1)}$ distribution. Write $T_n = \frac{\sqrt{n}(\bar{X}-\mu)/\sigma}{s/\sigma}$. Now

$$s^2 = \frac{1}{n-1}\sum_{i=1}^{n}(X_i - \bar{X})^2 = \frac{1}{n-1}\left(\sum X_i^2 - n\bar{X}^2\right)$$
$$= \frac{n}{n-1}\left(\sum \frac{X_i^2}{n} - \bar{X}^2\right) \stackrel{\mathcal{P}}{\longrightarrow} 1 \cdot \left(E(X_1^2) - \mu^2\right) = \sigma^2$$

(in fact, $s^2 \stackrel{\text{a.s.}}{\rightarrow} \sigma^2$). Therefore, by Slutsky's theorem,

$$T_n = \frac{\sqrt{n}(\bar{X} - \mu)/\sigma}{s/\sigma} \stackrel{\mathcal{L}}{\longrightarrow} \frac{N(0,1)}{1};$$

i.e., $t_{(n-1)} \stackrel{\mathcal{L}}{\longrightarrow} N(0, 1)$. Actually, denoting $f_n(t)$ as the density of the $t_{(n)}$ distribution and $\phi(t)$ as the $N(0, 1)$ density, it is true that $f_n(t) \longrightarrow \phi(t)$ pointwise (one can use the first-order Stirling approximation to show this). This implies a much stronger convergence, known as convergence in total variation (see Chapter 2).

Definition 1.6 Let P_n, P be probability measures on some probability space. We say that P_n converges to P in total variation if $\sup_{\text{Borel} A} |P_n(A) - P(A)| \rightarrow 0$ as $n \rightarrow \infty$.

Example 1.14 Suppose X_n has the discrete uniform distribution on $\{0, \frac{1}{n}, \frac{2}{n}, \ldots, \frac{n-1}{n}\}$. If we plot the CDF of X_n for large n, it will look as though it has no jumps and will look like the function $F(x) = x$ on $[0, 1]$. Simple calculation gives that $X_n \xrightarrow{\mathcal{L}} U[0, 1]$.

Note that convergence in this example is *not* in total variation. Indeed, because F_n is finitely supported, it is clear that the total variation distance between
F_n and F is

$$\sup_{\text{Borel } A} |P_n(A) - P(A)| = 1 \quad \forall n \geq 1,$$

where P_n, P are the distributions corresponding to F_n, F.

Example 1.15 Consider $X_n \sim \text{Beta}(\frac{1}{n}, \frac{1}{n}), n \geq 1$. As n gets large, the density of X_n looks almost like a 50-50 mixture of point masses near 0 and 1. Indeed, a simple calculation shows that $X_n \xrightarrow{\mathcal{L}} \text{Bernoulli}(\frac{1}{2})$. Note that the CDF of X_n does not converge to the Bernoulli($\frac{1}{2}$) CDF at 0. However, this is not needed, as 0 is a jump point of the Bernoulli($\frac{1}{2}$) CDF. This is also an example where a sequence of continuous random variables converges in distribution to a discrete random variable.

Example 1.16 Let $X_n \sim \chi^2(n)$. For large n, the density of X_n has very little skewness. Indeed, for large n, X_n is almost normal. A precise statement is that

$$\frac{X_n - n}{\sqrt{2n}} \xrightarrow{\mathcal{L}} N(0, 1).$$

The simplest way to see this is to write $X_n = Y_1 + \cdots + Y_n$, where $Y_i \stackrel{\text{iid}}{\sim} \chi^2(1)$, and to use the CLT for the mean of iid random variables with a finite variance.

Example 1.17 Suppose $E(|X_n - c|) \to 0$ for some $-\infty < c < \infty$. By Markov's inequality, $P(|X_n - c| > \epsilon) \leq \frac{E(|X_n - c|)}{\epsilon} \to 0$ as $n \to \infty$. Thus $X_n \xrightarrow{\mathcal{P}} c$.

Example 1.18 Suppose $a_n(X_n - c_n) \xrightarrow{\mathcal{L}} X$ for some constants a_n, c_n, and suppose $c_n \to c$. Note that

$$X_n - c = X_n - c_n + c_n - c = \frac{1}{a_n} a_n(X_n - c_n) + (c_n - c)$$
$$= O_p\left(\frac{1}{a_n}\right) + o(1) = o_p(1)$$

if $a_n \to \infty$. Thus, $a_n \to \infty$ suffices for $X_n \xrightarrow{\mathcal{P}} c$.

Example 1.19 Suppose X_n, X are integer-valued random variables. Then, by a simple calculation, one can show that $X_n \xrightarrow{\mathcal{L}} X \Leftrightarrow P(X_n = k) \to P(X = k)$ for every integer k. This is a useful characterization of convergence in distribution for integer-valued random variables.

1.5 Exercises

Exercise 1.1 Consider testing the hypothesis that the mean of some distribution with a finite variance is a given number, say zero. If the distribution under consideration is normal, then the t-test has various finite sample optimality properties. Without the normal assumption, such properties are not true. But the t-test is nearly universally used in all problems (just pick up any popular text and verify this). Can we provide any justification for this?

Show that the t-test of nominal level α is asymptotically of level α if the distribution under consideration has finite variance.

Exercise 1.2 Consider the very basic problem of estimating a binomial proportion p. Suppose we wish to construct a $100(1 - \alpha)\%$ confidence interval for p. Due to the discreteness of the binomial distribution, an exact $100(1 - \alpha)\%$ confidence interval cannot (in general) be constructed.

Show how the central limit theorem for iid random variables may be used to construct an *approximate* $100(1 - \alpha)\%$ confidence interval for p. Explain what "approximate" means.

Exercise 1.3 Consider the purely mathematical problem of finding a definite integral $\int_a^b f(x)dx$ for some (possibly complicated) function $f(x)$. Show that the usual laws of large numbers provide a method for approximately finding the value of the integral by using appropriate averages $\frac{1}{n}\sum_{i=1}^n f(X_i)$. Numerical analysts call this *Monte Carlo integration*.

Remark. Providing an error bound is a subtler issue, as error formulas involve other integrals involving f.

Exercise 1.4 Elementary Facts on Various Convergences

(a) Show that if a sequence $X_n \xrightarrow{r} X$ for some $r > 0$ (this notation means convergence in the rth mean; i.e., $E|X_n - X|^r \to 0$), then convergence in probability follows.

(b) Show that $X_n \xrightarrow{2} c$ iff the bias and the variance each converge to zero.

(c) Prove that the following is indeed a characterization of almost sure convergence:

$$\forall \epsilon > 0, \lim_{n\to\infty} P(|X_m - X| \leq \epsilon \quad \forall m \geq n) = 1.$$

(d) Hence, almost sure convergence is stronger than convergence in probability.

(e) Show by an example (different from the text) that convergence in probability does not necessarily imply almost sure convergence.

(f) Show that if $X_n \xrightarrow{\mathcal{P}} X$, then a subsequence converges to X almost surely.

Exercise 1.5 (a) Consider the discrete uniform distribution on $\{0, \frac{1}{n}, \frac{2}{n}, \ldots, 1\}$. Show that these have a limit in the sense of convergence in law, and identify the limit.

(b) Suppose X_n has the Bin$(n, \frac{1}{n})$ distribution. Show that X_n converges in law to the Poisson(1) distribution.

(c) Consider the symmetric Beta distribution with each parameter equal to $\frac{1}{n}$. Show that these converge in law to a Bernoulli distribution, and identify the parameter of that Bernoulli distribution.

(d) Show by a simple example that it is possible for $X_n \xrightarrow{\mathcal{P}} c$ but; $E(X_n) \not\to c$.

(e) Suppose $U_1, \ldots, U_{2m+1}$ are iid $U[0, 1]$. Find the exact density of the median $U_{(m+1)}$ and hence find the limiting distribution of $U_{(m+1)}$.

(f) * Suppose $U_1, \ldots, U_n$ are iid $U[0, 1]$. Where does the correlation between $U_{(1)}$ and $U_{(n)}$ go when $n \to \infty$?

(g) * Suppose $U_1, \ldots, U_{2m+1}$ are iid $U[0, 1]$. Where does the correlation between $U_{(m)}$ and $U_{(m+1)}$ go as $m \to \infty$?

Exercise 1.6 Convergence of t to Normal

(a) Suppose that T_n has a $t_{(n)}$ distribution. Prove that $T_n \xrightarrow{\mathcal{L}} N(0, 1)$ by proving that the densities of T_n converge to the $N(0, 1)$ density.

(b) * Suppose T_n has a $t_{(n)}$ distribution. Do the percentiles of T_n converge to the percentiles of $N(0, 1)$ uniformly?

Exercise 1.7 (a) Suppose $E|Y_n - c| \to 0$. Does $Y_n \xrightarrow{\mathcal{P}} c$ necessarily?

(b) Suppose $a_n(X_n - \theta) \to N(0, \tau^2)$. What can be said about the limiting distribution of $|X_n|$ when $\theta \neq 0, \theta = 0$?

(c) Suppose X_i are iid $U[0, \theta], i = 1, 2, \ldots, n$. What can be said about the limiting distribution of $n(\theta - X_{(n-1)})$, where $X_{(n-1)}$ is the $(n-1)$st-order statistic?

(d) * Suppose X_i are iid Bernoulli(p). What can be said about the limiting distribution of the sample variance s^2 when $p = \frac{1}{2}; p \neq \frac{1}{2}$?

(e) Give an example of $X_n \xrightarrow{\mathcal{L}} X, Y_n \xrightarrow{\mathcal{L}} Y$, but jointly $(X_n, Y_n) \stackrel{\mathcal{L}}{\nrightarrow} (X, Y)$. Under what additional condition is it true?

(f) Give a necessary and sufficient condition in terms of the probability mass functions for $X_n \xrightarrow{\mathcal{L}} X$ if X_n, X are integer-valued random variables, and prove it.

(g) $o_p(1) + O_p(1) = ?$

(h) $o_p(1)O_p(1) = ?$

(i) $o_p(1) + o_p(1)O_p(1) = ?$

(j) Suppose $X_n \xrightarrow{\mathcal{L}} X$. Then $o_p(1)X_n =?$

Exercise 1.8 Suppose (X_n, Y_n) is bivariate normal and $(X_n, Y_n) \xrightarrow{\mathcal{P}} (X, Y)$. Show that (X, Y) is also bivariate normal.

Exercise 1.9 Suppose X_i are iid with density $\frac{c}{x^2 \log|x|}$ for $|x| \geq 2$. Does $E|X_1|$ exist? Does $\bar{X}_n$ have a limit in probability?

Exercise 1.10 Suppose $g : \mathbb{R} \to \mathbb{R}$ is continuous and bounded. Show that $e^{-n\lambda} \sum_{k=0}^{\infty} g(\frac{k}{n}) \frac{(n\lambda)^k}{k!} \to g(\lambda)$ as $n \to \infty$.

Exercise 1.11 (a) * Give an example of an iid sequence $X_1, \ldots, X_n$ with finite variance such that the CLT holds for the sum $S_n = X_1 + \cdots + X_n$ but the density of $\sqrt{n}(\bar{X} - \mu)/\sigma$ does not converge to the $N(0, 1)$ density pointwise.

Exercise 1.12 * For $p \geq 1$, let U_p be distributed uniformly on the boundary of the unit ball in $\mathbb{R}^p$. Show that for fixed $k \geq 1$, as $p \to \infty$,

$$\sqrt{p}(U_{p1}, \ldots, U_{pk}) \xrightarrow{\mathcal{L}} N_k(0, I).$$

Exercise 1.13 Let X_i be iid Exp(1) random variables. Define $Y_n = \frac{1}{n}\left(\sum_{i=1}^n X_i, \sum_{i=1}^n X_i^2, \sum_{i=1}^n X_i^3, \sum_{i=1}^n X_i^4\right)'$. Find, with appropriate centering and norming, the limiting distribution of Y_n.

Exercise 1.14 Let $MN(n, p_1, p_2, \ldots, p_k)$ be the multinomial distribution with k cells and cell probabilities p_i, and let $n_1, n_2, \ldots, n_k$ denote the cell frequencies. Show that the vector of cell frequencies converges in distribution to a suitable multivariate normal distribution on appropriate centering and norming. Identify the covariance matrix of the limiting multivariate normal distribution.

Exercise 1.15 * Let X_i be iid geometric with parameter p and N an independent Poisson variable with parameter λ. Let $S_n = \sum_{i=1}^n X_i$. Does S_N converge in distribution on any centering and norming as $\lambda \to \infty$?

Exercise 1.16 Suppose X_i are iid standard Cauchy. Show that

(a) $P(|X_n| > n$ infinitely often$) = 1$,
(b) * $P(|S_n| > n$ infinitely often$) = 1$.

Exercise 1.17 * Suppose X_i are iid standard exponential. Show that $\limsup_n \frac{X_n}{\log n} = 1$ with probability 1 (w.p. 1) (i.e., almost surely).

Exercise 1.18 Suppose $\{A_n\}$ is an infinite sequence of independent events. Show that $P($infinitely many A_n occur$) = 1 \Leftrightarrow P(\bigcup A_n) = 1$.

Exercise 1.19 * Suppose X_n is a sequence of random variables converging in probability to a random variable X; X is absolutely continuous with a strictly positive density. Show that the medians of X_n converge to the median of X.

Exercise 1.20 * Suppose m balls are distributed at random and independently into n urns. Let $W_{m,n}$ denote the number of empty urns.

(a) Find the mean and variance of $W_{m,n}$.
(b) Suppose, for some $0 < \lambda < 1$, $\frac{m}{n} = \lambda + o(1)$ as $n \to \infty$. Show that $\frac{W_{m,n}}{n} \stackrel{\mathcal{P}}{\Rightarrow} c$ and identify c.

Remark. For an extensive treatment of empty cells asymptotics, a classic reference is Kolchin, Sevastyanov, and Chirstyakov (1978).

Exercise 1.21 Suppose X_i are independent $N(0, \sigma_i^2)$ random variables. Characterize the sequences $\{\sigma_i\}$ for which $\sum_{n=1}^{\infty} X_n$ converges to a finite value.

Exercise 1.22 * Suppose X_i are independent $N(\mu_i, \sigma_i^2)$ random variables. Characterize the sequences $\{\mu_i\}, \{\sigma_i\}$ for which $\sum_{n=1}^{\infty} X_n$ converges to a finite value.

Exercise 1.23 * Suppose X_i are independent Beta(α_i, β_i) random variables. Characterize the sequences $\{\alpha_i\}, \{\beta_i\}$ for which $\sum_{n=1}^{\infty} X_n$ converges to a finite value.

Exercise 1.24 Which of the following events E are tail events? Below, you may take the X_i to be a sequence of nonnegative independent random variables.

(a) $E = \sum_{n=1}^{\infty} X_n$ converges to a finite number.
(b) $E = \sum_{n=1}^{\infty} X_n$ converges to a number < 5.
(c) $E = \prod_{n=1}^{\infty} X_n$ converges to a finite number.
(d) $E = \prod_{n=1}^{\infty} X_n$ converges to a rational number.

References

Ash, R. (1972). *Real Analysis and Probability*, Academic Press, New York.
Billingsley, P.(1995). *Probability and Measure*, 3rd edition, John Wiley, New York.
Breiman, L. (1968). *Probability*, Addison-Wesley, New York.

Chow, Y. and Teicher, H. (1988). *Probability Theory*, Springer, New York.

Ferguson, T.S. (1996). *A Course in Large Sample Theory*, Chapman and Hall, London.

Kolchin, V., Sevastyanov, B., and Chistyakov, V. (1978). *Random Allocations*, V.H. Winston & Sons, Washington, distributed by Halsted Press, New York.

Massart, P. (1990). The tight constant in the Dvoretzky-Kiefer-Wolfowitz inequality, Ann. Prob. 3, 1269–1283.

Serfling, R.(1980). *Approximation Theorems of Mathematical Statistics*, John Wiley, New York.

Chapter 2
Metrics, Information Theory, Convergence, and Poisson Approximations

Sometimes it is technically convenient to prove a certain type of convergence by proving that, for some suitable metric d on the set of CDFs, $d(F_n, F) \to 0$ instead of proving the required convergence directly from the definition. Here F_n, F are CDFs on some space, say the real line. Metrics are also useful as statistical tools to assess errors in distribution estimation and to study convergence properties in such statistical problems. The metric, of course, will depend on the type of convergence desired.

The central limit theorem justifiably occupies a prominent place in all of statistics and probability theory. Fourier methods are most commonly used to prove the central limit theorem. This is technically efficient but fails to supply any intuition as to *why* the result should be true. It is interesting that proofs of the central limit theorem have been obtained that avoid Fourier methods and use instead much more intuitive information-theoretic methods. These proofs use convergence of entropies and Fisher information in order to conclude convergence in law to normality. It was then realized that such information-theoretic methods are useful also to establish convergence to Poisson limits in suitable paradigms; for example, convergence of appropriate Bernoulli sums to a Poisson limit. In any case, Poisson approximations are extremely useful in numerous complicated problems in both probability theory and statistics. In this chapter, we give an introduction to the use of metrics and information-theoretic tools for establishing convergences and also give an introduction to Poisson approximations.

Good references on metrics on distributions are Dudley (1989), Rachev (1991), and Reiss (1989). The role of information theory in establishing central limit theorems can be seen, among many references, in Linnik (1959), Brown (1982), and Barron (1986). Poisson approximations have a long history. There are first-generation methods and then there are the modern methods, often called the *Stein-Chen methods*. The literature is huge. A few references are LeCam (1960), Sevas'tyanov (1972), Stein (1972, 1986),

A. DasGupta, *Asymptotic Theory of Statistics and Probability*,

Chen (1975), and Barbour, Holst and Janson (1992). Two other references where interesting applications are given in an easily readable style are Arratia, Goldstein, and Gordon (1990) and Diaconis and Holmes (2004).

2.1 Some Common Metrics and Their Usefulness

There are numerous metrics and distances on probability distributions on Euclidean spaces. The choice depends on the exact purpose and on technical feasibility. We mention a few important ones only and give some information about their interrelationships, primarily in the form of inequalities. The inequalities are good to know in any case.

(i) Metric for convergence in probability
$d_E(X, Y) = E\left(\frac{|X-Y|}{1+|X-Y|}\right)$. This extends to the multidimensional case in the obvious way by using the Euclidean norm $||X - Y||$.

(ii) Kolmogorov metric
$d_K(F, G) = \sup_x |F(x) - G(x)|$. This definition includes the multidimensional case.

(iii) Lévy metric
$d_L(F, G) = \inf\{\epsilon > 0 : F(x - \epsilon) - \epsilon \leq G(x) \leq F(x + \epsilon) + \epsilon \quad \forall x\}$.

(iv) Total variation metric
$d_{TV}(P, Q) = \sup_{\text{Borel } A} |P(A) - Q(A)|$. This also includes the multidimensional case. If P, Q are both absolutely continuous with respect to some measure μ, then $d_{\text{TV}}(P, Q) = \frac{1}{2}\int |f(x) - g(x)| d\mu(x)$, where f is the density of P with respect to μ and g is the density of Q with respect to μ.

(v) Kullback-Leibler distance
$K(P, Q) = -\int (\log \frac{q}{p}) dP = -\int (\log \frac{q}{p}) p d\mu$, where $p = \frac{dP}{d\mu}$ and $q = \frac{dQ}{d\mu}$ for some μ. Again, the multidimensional case is included. Note that K is not symmetric in its arguments P, Q.

(vi) Hellinger distance
$H(P, Q) = \left[\int (\sqrt{p} - \sqrt{q})^2 d\mu\right]^{1/2}$, where again $p = \frac{dP}{d\mu}$ and $q = \frac{dQ}{d\mu}$ for some μ, and the multidimensional case is included.

Theorem 2.1

(i) $X_n \xrightarrow{\mathcal{P}} X$ iff $d_E(X_n, X) \to 0$.

(ii) $X_n \xrightarrow{\mathcal{L}} X$ iff $d_L(F_n, F) \to 0$, where $X_n \sim F_n$ and $X \sim F$.

(iii) $X_n \xrightarrow{\mathcal{L}} X$ if $d_K(F_n, F) \to 0$, the reverse being true only under additional conditions.

(iv) If $X \sim F$, where F is continuous and $X_n \sim F_n$, then $X_n \xrightarrow{\mathcal{L}} X$ iff $d_K(F_n, F) \to 0$ (Polyá's theorem).

(v) $X_n \xrightarrow{\mathcal{L}} X$ if $d_{TV}(P_n, P) \to 0$, where $X_n \sim P_n$, $X \sim P$ (the converse is not necessarily true).

(vi) $H(P, Q) \leq \sqrt{K(P, Q)}$.

(vii) $H(P, Q) \geq d_{\text{TV}}(P, Q)$.

(viii) $H(P, Q)/\sqrt{2} \leq \sqrt{d_{\text{TV}}(P, Q)}$.

Proofs of parts of Theorem 2.1 are available in Reiss (1989).

Corollary 2.1 (a) The total variation distance and the Hellinger distance are equivalent in the sense $d_{TV}(P_n, P) \to 0 \Leftrightarrow H(P_n, P) \to 0$.

(b) If P_n, P are all absolutely continuous with unimodal densities, and if P_n converges to P in law, then $H(P_n, P) \to 0$.

(c) Convergence in Kullback-Leibler distance implies convergence in total variation and hence convergence in law.

Note that the proof of part (b) also uses Ibragimov's theorem stated below.

Remark. The Kullback-Leibler distance is very popular in statistics. Specifically, it is frequently used in problems of model selection, testing for goodness of fit, Bayesian modeling and Bayesian asymptotics, and in certain estimation methods known as minimum distance estimation. The Kolmogorov distance is one of the easier ones computationally and has been used in many problems, too, and notably so in the literature on robustness and Bayesian robustness. The Hellinger distance is a popular one in problems of density estimation and in time series problems. The Lévy metric is technically hard to work with but metrizes weak convergence, a very useful property. It, too, has been used in the robustness literature, but it is more common in probability theory. Convergence in total variation is extremely strong, and many statisticians seem to consider it unimportant. But it has a direct connection to $\mathcal{L}_1$ distance, which is intuitive. It has a transformation invariance property and, when it holds, convergence in total variation is extremely comforting.

Notice the last two parts in Theorem 2.1. We have inequalities in *both directions* relating the total variation distance to the Hellinger distance. Since computation of the total variation distance is usually difficult, Hellinger distances are useful in establishing useful bounds on total variation.

2.2 Convergence in Total Variation and Further Useful Formulas

Next, we state three important results on when convergence in total variation can be asserted; see Reiss (1989) for all three theorems and also almost any text on probability for a proof of Scheffé's theorem.

Theorem 2.2 (Scheffé) Let $f_n, n \geq 0$ be a sequence of densities with respect to some measure μ. If $f_n \to f_0$ a.e. (μ), then $d_{\text{TV}}(f_n, f_0) \to 0$.

Remark. Certain converses to Scheffé's theorem are available, and the most recent results are due to Sweeting (1986) and Boos (1985). As we remarked before, convergence in total variation is very strong, and even for the simplest weak convergence problems, convergence in total variation should not be expected without some additional structure. The following theorem exemplifies what kind of structure may be necessary. This is a general theorem (i.e., no assumptions are made on the structural forms of the statistics). In the Theorem 2.4 below, convergence in total variation is considered for sample means of iid random variables (i.e., there is a restriction on the structural form of the underlying statistics). It is not surprising that this theorem needs fewer conditions than Theorem 2.3 to assert convergence in total variation.

Theorem 2.3 (Ibragimov) Suppose P_0 and (for large n) P_n are unimodal, with densities $f_0 = \frac{dP_0}{d\lambda}$ and $f_n = \frac{dP_n}{d\lambda}$, where λ denotes Lebesgue measure. Then $P_n \xrightarrow{\mathcal{L}} P_0$ iff $d_{\text{TV}}(P_n, P_0) \to 0$.

Definition 2.1 A random variable X is said to have a lattice distribution if it is supported on a set of the form $\{a + nh : n \in \mathcal{Z}\}$, where a is a fixed real, h a fixed positive real, and $\mathcal{Z}$ the set of integers.

Theorem 2.4 Suppose $X_1, \ldots, X_n$ are iid nonlattice random variables with a finite variance and characteristic function $\psi(t)$. If, for some $p \geq 1$, $\psi \in L^p(\lambda)$, where λ denotes Lebesgue measure, then $\frac{\sqrt{n}(\bar{X}-\mu)}{\sigma}$ converges to $N(0, 1)$ in total variation.

Example 2.1 Suppose X_n is a sequence of random variables on $[0, 1]$ with density $f_n(x) = 1+\cos(2\pi nx)$. Then, $X_n \stackrel{\mathcal{L}}{\Rightarrow} U[0, 1]$ by a direct verification of the definition using CDFs. However, note that the densities f_n *do not* converge to the uniform density 1 as $n \to \infty$. The limit distribution P_0 is unimodal, but the distribution P_n of X_n is not unimodal. The example shows that the condition in Ibragimov's theorem above that the P_n need to be unimodal as well cannot be relaxed.

Example 2.2 Suppose $X_1, X_2, \ldots$ are iid $\chi^2(2)$ with density $\frac{1}{2}e^{-x/2}$. The characteristic function of X_1 is $\psi(t) = \frac{1}{1-2it}$, which is in $L^p(\lambda)$ for any $p > 1$. Hence, by Theorem 2.4, $\frac{\sqrt{n}(\bar{X}-2)}{2}$ converges in total variation to $N(0, 1)$. We now verify that in fact the density of $Z_n = \frac{\sqrt{n}(\bar{X}-2)}{2}$ converges pointwise to the $N(0, 1)$ density, which by Scheffé's theorem will also imply convergence in total variation. The pointwise convergence of the density is an interesting calculation.

Since $S_n = \sum_{i=1}^{n} X_i$ has the $\chi^2(2n)$ distribution with density $\frac{e^{-x/2}x^{n-1}}{2^n \Gamma(n)}$, Z_n has density $f_n(z) = \frac{e^{-(z\sqrt{n}+n)}(1+\frac{z}{\sqrt{n}})^{n-1}n^{n-\frac{1}{2}}}{\Gamma(n)}$. Hence, $\log f_n(z) = -z\sqrt{n} - n + (n-1)(\frac{z}{\sqrt{n}} - \frac{z^2}{2n} + O(n^{-3/2})) + (n - \frac{1}{2}) \log n - \log \Gamma(n) = -z\sqrt{n} - n + (n - 1)(\frac{z}{\sqrt{n}} - \frac{z^2}{2n} + O(n^{-3/2})) + (n - \frac{1}{2}) \log n - (n \log n - n - \frac{1}{2} \log n + \log \sqrt{2\pi} + O(n^{-1}))$ on using Stirling's approximation for $\log \Gamma(n)$.

On canceling terms, this gives $\log f_n(z) = -\frac{z}{\sqrt{n}} - \log \sqrt{2\pi} - \frac{(n-1)z^2}{2n} + O(n^{-1/2})$, implying that $\log f_n(z) \to -\log \sqrt{2\pi} - \frac{z^2}{2}$, and hence $f_n(z) \to \frac{1}{\sqrt{2\pi}} e^{-\frac{z^2}{2}}$, establishing the pointwise density convergence.

Example 2.3 The Hellinger and the Kullback-Leibler distances are generally easier to calculate than the total variation distance. The normal case itself is a good example. For instance, the Kullback-Leibler distance $K(N_p(\mu, \mathbf{I}), N_p(0, \mathbf{I})) = \frac{1}{2}||\mu||^2$.

Many bounds on the total variation distance between two multivariate normal distributions are known; we mention a few below that are relatively neat.

$$d_{\text{TV}}(N_p(\mu_1, \mathbf{I}), N_p(\mu_2, \mathbf{I})) \leq \frac{1}{\sqrt{2}}||\mu_1 - \mu_2||,$$

$$d_{\text{TV}}(N_p(0, \Sigma), N_p(0, \mathbf{I})) \leq \min \left\{ \begin{array}{l} \frac{1}{\sqrt{2}} \left(\sum_{i=1}^{p} (\sigma_i^2 - 1) - \log |\Sigma| \right)^{\frac{1}{2}} \\ p2^{p+1}||\Sigma - \mathbf{I}||_2 \end{array} \right. ,$$

where $||A||_2$ denotes the usual Euclidean matrix norm $(\sum_i \sum_j a_{ij}^2)^{1/2}$. These and other bounds can be seen in Reiss (1989).

Example 2.4 Suppose $X_n \sim N(\mu_n, \sigma_n^2)$ and $X_0 \sim N(\mu, \sigma^2)$. Then X_n converges to X_0 in total variation if and only if $\mu_n \to \mu$ and $\sigma_n^2 \to \sigma^2$. This can be proved directly by calculation.

Remark. There is some interest in finding projections in total variation of a fixed distribution to a given class of distributions. This is a good problem but usually very hard, and even in simple one-dimensional cases, the projection can only be found by numerical means. Here is an example; the exercises at the end of the chapter offer some more cases.

Example 2.5 If $X_n \sim \text{Bin}(n, p_n)$ and $np_n \to \lambda, 0 < \lambda < \infty$, then X_n converges in law to the Poi(λ) distribution. In practice, this result is used to approximate a Bin(n, p) distribution for large n and small p by a Poisson distribution with mean np. One can ask what is the best Poisson approximation for a given Bin(n, p) distribution (e.g., what is the total variation projection of a given Bin(n, p) distribution onto the class of all Poisson distributions). An explicit description would not be possible. However, the total variation projection can be numerically computed.

For instance, if $n = 50$, $p = .01$, then the total variation projection is the Poisson distribution with mean .5025. If $n = 100$, $p = .05$, then the total variation projection is the Poisson distribution with mean 5.015. The best Poisson approximation seems to have a mean slightly off from np. In fact, if the total variation projection has mean λ_n, then $|\lambda_n - \lambda| \to 0$. We will come back to Poisson approximations to binomials later in this chapter.

2.3 Information-Theoretic Distances, de Bruijn's Identity, and Relations to Convergence

Entropy and Fisher information are two principal information-theoretic quantities. Statisticians, by means of well-known connections to inference such as the Cramér-Rao inequality and maximum likelihood estimates, are very familiar with the Fisher information. Probabilists, on the other hand, are very familiar with entropy. We first define them formally.

Definition 2.2 Let f be a density in $\mathcal{R}^d$. The entropy of f, or synonymously of a random variable $X \sim f$, is $H(X) = -\int f(x) \log f(x) dx = -E_f[\log f(X)]$.

For integer-valued variables, the definition is similar.

Definition 2.3 Let X be integer valued with $P(X = j) = p_j$. Then, the entropy of X is $H(X) = -\sum_j p(j) \log p(j)$.

Fisher information is defined only for smooth densities. Here is the definition.

Definition 2.4 Let f be a density in $\mathcal{R}^d$. Suppose f has one partial derivative with respect to each coordinate everywhere in its support $\{x : f(x) > 0\}$. The Fisher information of f, or synonymously of a random variable $X \sim f$, is $I(X) = \int_{x:f(x)>0} \frac{||\nabla f(x)||^2}{f(x)} dx = E_f[|| \nabla \log f(X)||^2]$, where $\nabla(.)$ denotes the gradient vector.

Remark. The function $\nabla \log f(x)$ is called *the score function* of f.

Entropy and Fisher information each satisfy certain suitable subadditivity properties. We record their most basic properties below. Johnson (2004) can be consulted for proofs of the theorems in this section apart from the specific references given for particular theorems below.

Theorem 2.5 (a) For jointly distributed random variables X, Y, $H(X, Y) \leq H(X) + H(Y)$ with equality iff X, Y are independent:

(b) For any $\sigma > 0$, $H(\mu + \sigma X) = \log \sigma + H(X)$.

(c) For independent random variables X, Y, $H(X+Y) \geq \max\{H(X), H(Y)\}$.

(d) For jointly distributed random variables X, Y, $I(X, Y) \geq \max\{I(X), I(Y)\}$.

(e) For any σ, $I(\mu + \sigma X) = \frac{I(X)}{\sigma^2}$.

(f) For independent random variables X, Y, $I(X + Y) \leq \alpha^2 I(X) + (1 - \alpha)^2 I(Y) \, \forall 0 \leq \alpha \leq 1$ with equality iff X, Y are each normal.

(g) For independent random variables X, Y, $I(X + Y) \leq \left(\frac{1}{I(X)} + \frac{1}{I(Y)}\right)^{-1}$ with equality iff X, Y are each normal.

Example 2.6 For some common distributions, we give expressions for the entropy and Fisher information when available.

Distribution	$H(X)$	$I(X)$
Exponential(1)	1	1
$N(0, 1)$	$\frac{1}{2}\log(2\pi) + \frac{1}{2}$	1
Gamma$(\alpha, 1)$	$\alpha + \log \Gamma(\alpha) + (\alpha - 1)\psi(\alpha)$	$\frac{1}{\alpha-2}(\alpha > 2)$
$C(0, 1)$	—	$\frac{1}{2}$
$N_d(0, \Sigma)$	$\frac{d}{2}\log(2\pi) + \log|\Sigma| + \frac{d}{2}$	$\mathrm{tr}\Sigma^{-1}$

Remark. In the table above, ψ is the di-Gamma function (i.e., the derivative of $\log \Gamma$).

Entropy and Fisher information, interestingly, are connected to each other. They are connected by a link to the normal distribution and also through an algebraic relation known as *de Bruijn's identity*. We mention the link through the normal distribution first.

Theorem 2.6 Among all densities with mean 0 and variance $\sigma^2 < \infty$, the entropy is maximized by the $N(0, \sigma^2)$ density. On the other hand, among all densities with mean 0 and variance $\sigma^2 < \infty$ such that the Fisher information is defined, Fisher information is minimized by the $N(0, \sigma^2)$ density.

Remark. The theorem says that normal distributions are extremals in two optimization problems with a variance constraint, namely the maximum entropy and the minimum Fisher information problems. Actually, although we state the theorem for $N(0, \sigma^2)$, the mean is irrelevant. This theorem establishes an indirect connection between H and I inherited from a connection of each to normal distributions.

We can use H and I to define distances between two different distributions. These are defined as follows.

Definition 2.5 Let $X \sim f$, $Y \sim g$, and assume that $g(x) = 0 \Rightarrow f(x) = 0$. The *entropy divergence or differential entropy between f and g* is defined as

$$D(f||g) = \int f(x) \log\left(\frac{f(x)}{g(x)}\right).$$

The *Fisher information distance between f and g* is defined as

$$I(f||g) = I(X||Y) = \int [|| \nabla \log f - \nabla \log(g)||^2] f(x) dx.$$

Using the normal distribution as a benchmark, we can define a standardized Fisher information as follows.

Definition 2.6 Let $X \sim f$ have finite variance σ^2. The *standardized Fisher information of f* is defined as $I_s(f) = I_s(X) = \sigma^2 I(f||N(0, \sigma^2))$.

The advantage of the standardization is that $I_s(f)$ can be zero only when f itself is a normal density. Similarly, the entropy divergence of a density f with a normal density can be zero only if f is that same normal density.

We state the elegant algebraic connection between entropy divergence and standardized Fisher information next.

Theorem 2.7 (**De Bruijn's Identity**) Let $X \sim f$ have variance 1. Let Z be a standard normal variable independent of X. For $t > 0$, let f_t denote the density of $X + \sqrt{t} Z$. Then, $I(f_t) = 2\frac{d}{dt}[H(f_t)]$.

Remark. De Bruijn's identity (which extends to higher dimensions) is a consequence of the heat equation of partial differential equations; see Johnson (2004). A large number of such $\frac{d}{dt}$ identities of use in statistics (although not de Bruijn's identity itself) are proved in Brown et al. (2006). That such a neat algebraic identity links entropy with Fisher information is a pleasant surprise.

We now describe how convergence in entropy divergence is a very strong form of convergence.

Theorem 2.8 Let f_n, f be densities in $\mathcal{R}^d$. Suppose $D(f_n||f) \to 0$. Then f_n converges to f in total variation; in particular, convergence in distribution follows.

This theorem has completely general densities f_n, f. In statistics, often one is interested in densities of normalized convolutions. Calculating their entropies or entropy distances from the density of the limiting $N(0, 1)$ distribution could be hard because convolution densities are difficult to write. In a remarkable result, Barron (1986) proved the following.

Theorem 2.9 Let $X_1, X_2, \cdots$ be iid zero-mean, unit-variance random variables and let f_n denote the density (assuming it exists) of $\sqrt{n}\bar{X}$. If, for some m, $D(f_m||N(0, 1)) < \infty$, then $D(f_n||N(0, 1)) \to 0$.

Analogously, one can use Fisher information in order to establish weak convergence. The intuition is that if the Fisher information of $\sqrt{n}\bar{X}$ is converging to 1, which is the Fisher information of the $N(0, 1)$ distribution, then by virtue of the unique Fisher information minimizing property of the $N(0, 1)$ subject to a fixed variance of 1 (stated above), it ought to be the case that $\sqrt{n}\bar{X}$ is converging to $N(0, 1)$ in distribution. The intuition is pushed to a proof in Brown (1982), as stated below.

Theorem 2.10 Let $X_1, X_2, \cdots$ be iid zero-mean, unit-variance random variables and $Z_1, Z_2, \cdots$ be an iid $N(0, 1)$ sequence independent of the $\{X_i\}$. Let $v > 0$ and $Y_n(v) = \sqrt{n}\bar{X} + \sqrt{v}Z_n$. Then, for any v, $I_s(Y_n(v)) \to 0$ and hence $\sqrt{n}\bar{X} \stackrel{\mathcal{L}}{\Rightarrow} N(0, 1)$.

Remark. It had been suspected for a long time that there should be such a proof of the central limit theorem by using Fisher information. It was later found that Brown's technique was so powerful that it extended to central limit theorems for many kinds of non-iid variables. These results amounted to a triumph of information theory tools and provided much more intuitive proofs of the central limit results than proofs based on Fourier methods.

An interesting question to ask is what can be said about the rates of convergence of the entropy divergence and the standardized Fisher information in the canonical CLT situation (i.e., for $\sqrt{n}\bar{X}$ when X_i are iid with mean 0 and variance 1). This is a difficult question. In general, one can hope for convergence at the rate of $\frac{1}{n}$. The following is true.

Theorem 2.11 Let $X_1, X_2, \cdots$ be iid zero-mean, unit-variance random variables. Then, each of $D(\sqrt{n}\bar{X}||N(0,1))$ and $I_s(\sqrt{n}\bar{X})$ is $O(\frac{1}{n})$.

Remark. This is quite a bit weaker than the best results that are now known. In fact, one can get bounds valid for *all* n, although they involve constants that usually cannot be computed. Johnson and Barron (2003) may be consulted to see the details.

2.4 Poisson Approximations

Exercise 1.5 in Chapter 1 asks to show that the sequence of Bin$(n, \frac{1}{n})$ distributions converges in law to the Poisson distribution with mean 1. The Bin$(n, \frac{1}{n})$ is a sum of n independent Bernoullis but with a success probability that is small and also depends on n. The Bin$(n, \frac{1}{n})$ is a count of the total number of occurrences among n independent rare events. It turns out that convergence to a Poisson distribution can occur even if the individual success probabilities are small but not the same, and even if the Bernoulli variables are not independent. Indeed, approximations by Poisson distributions are extremely useful and accurate in many problems. The problems arise in diverse areas. Poisson approximation is a huge area, with an enormous body of literature, and there are many book-length treatments. We provide here a glimpse into the area with some examples.

Definition 2.7 Let p, q be two mass functions on the integers. The total variation distance between p and q is defined as $d_{\text{TV}}(p,q) = \sup_{A \subseteq \mathcal{Z}} |P_p(X \in A) - P_q(X \in A)|$, which equals $\frac{1}{2}\sum_j |p(j) - q(j)|$.

A simple and classic result is the following.

Theorem 2.12 (LeCam (1960)) (a) $d_{\text{TV}}(\text{Bin}(n, \frac{\lambda}{n}), \text{Poi}(\lambda)) \leq \frac{8\lambda}{n}$. (b) For $n \geq 1$, let $\{X_{in}\}_{i=1}^n$ be a triangular array of independent Ber(p_{in}) variables. Let $S_n = \sum_{i=1}^n X_{in}$ and $\lambda_n = \sum_{i=1}^n p_{in}$. Then, $d_{\text{TV}}(S_n, \text{Poi}(\lambda_n)) \leq \frac{8}{\lambda_n}\sum_{i=1}^n p_{in}^2$, if $\max\{p_{in}, 1 \leq i \leq n\} \leq \frac{1}{4}$.

A neat corollary of LeCam's theorem is the following.

Corollary 2.2 If X_{in} is a triangular array of independent Ber(p_{in}) variables such that $\max\{p_{in}, 1 \leq i \leq n\} \to 0$, and $\lambda_n = \sum_{i=1}^n p_{in} \to \lambda, 0 < \lambda < \infty$, then $d_{\text{TV}}(S_n, \text{Poi}(\lambda)) \to 0$ and hence $S_n \stackrel{\mathcal{L}}{\Rightarrow} \text{Poi}(\lambda)$.

The Poisson distribution has the property of having equal mean and variance, so intuition would suggest that if a sum of independent Bernoulli variables had, asymptotically, an equal mean and variance, then it should converge to a Poisson distribution. That, too, is true.

Corollary 2.3 If X_{in} is a triangular array of independent Ber(p_{in}) variables such that $\sum_{i=1}^{n} p_{in}$ and $\sum_{i=1}^{n} p_{in}(1 - p_{in})$ each converge to $\lambda, 0 < \lambda < \infty$, then $S_n \stackrel{\mathcal{L}}{\Rightarrow} \text{Poi}(\lambda)$.

It is a fact that, in many applications, although the variable can be represented as a sum of Bernoulli variables, they are not independent. The question arises if a Poisson limit can still be proved. The question is rather old. Techniques that we call *first-generation techniques*, using combinatorial methods, are successful in some interesting problems. These methods typically use generating functions or sharp Bonferroni inequalities. Two very good references for looking at those techniques are Kolchin, Sevas'tyanov, and Chistyakov (1978) and Galambos and Simonelli (1996). Here is perhaps the most basic result of that type.

Theorem 2.13 For $N \geq 1$, let $X_{in}, i = 1, 2, \cdots, n = n(N)$ be a triangular array of Bernoulli random variables, and let $A_i = A_{in}$ denote the event where $X_{in} = 1$. For a given k, let $M_k = M_{kn}$ be the kth binomial moment of S_n; i.e., $M_k = \sum_{j=k}^{n} \binom{j}{k} P(S_n = j)$. If there exists $0 < \lambda < \infty$ such that, for every fixed k, $M_k \to \frac{\lambda^k}{k!}$ as $N \to \infty$, then $S_n \stackrel{\mathcal{L}}{\Rightarrow} \text{Poi}(\lambda)$.

Remark. In some problems, typically of a combinatorial nature, careful counting lets one apply this basic theorem and establish convergence to a Poisson distribution.

Example 2.7 (The Matching Problem) Cards are drawn one at a time from a well-shuffled deck containing N cards, and a match occurs if the card bearing a number, say j, is drawn at precisely the jth draw from the deck. Let S_N be the total number of matches. Theorem 2.13 can be used in this example. The binomial moment M_k can be shown to be $M_k = \binom{N}{k} \frac{1}{N(N-1)\ldots(N-k+1)}$, and from here, by Stirling's approximation, for every fixed k, $M_k \to \frac{1}{k!}$, establishing that the total number of matches converges to a Poisson distribution with mean 1 as the deck size $N \to \infty$. Note that the mean value of S_N is exactly 1 for any N. Convergence to a Poisson distribution is extremely fast in this problem; even for $N = 5$, the Poisson approximation is quite good. For $N = 10$, it is almost exact!

For information, we note the following superexponential bound on the error of the Poisson approximation in this problem; this is proved in DasGupta (1999).

Theorem 2.14 $d_{\mathrm{TV}}(S_N, \mathrm{Poi}(1)) \leq \frac{2^N}{(N+1)!}\ \forall N.$

Example 2.8 (**The Committee Problem**) From n people, $N = N(n)$ committees are formed, each committee of a fixed size m. We let $N, n \to \infty$, holding m fixed. The Bernoulli variable X_{in} is the indicator of the event that the ith person is not included in any committee. Under the usual assumptions of independence and also the assumption of random selection, the binomial moment M_k can be shown to be $M_k = \binom{n}{k}[\frac{\binom{n-k}{m}}{\binom{n}{m}}]^N$.

Stirling's approximation shows that $M_k \sim \frac{n^k}{k!}e^{-kN(\frac{m}{n}+O(n^{-2}))}$ as $n \to \infty$. One now sees on inspection that if N, n are related as $N = \frac{n \log n}{m} - n \log \lambda + o(n^{-1})$ for some $0 < \lambda < \infty$, then $M_k \to \frac{\lambda^{km}}{k!}$ and so, from the basic convergence theorem above, the number of people who are left out of *all* committees converges to $\mathrm{Poi}(\lambda^m)$.

Example 2.9 (**The Birthday Problem**) This is one of the most colorful examples in probability theory. Suppose each person in a group of n people has, mutually independently, a probability $\frac{1}{N}$ of being born on any given day of a year with N calendar days. Let S_n be the total number of pairs of people (i, j) such that they have the same birthday. $P(S_n > 0)$ is the probability that there is at least one pair of people in the group who share the same birthday. It turns out that if n, N are related as $n^2 = 2N\lambda + o(N)$, for some $0 < \lambda < \infty$, then $S_n \stackrel{\mathcal{L}}{\Rightarrow} \mathrm{Poi}(\lambda)$. For example, if $N = 365, n = 30$, then S_n is roughly Poisson with mean 1.233.

A review of the birthday and matching problems is given in DasGupta (2005). Many of the references given at the beginning of this chapter also discuss Poisson approximation in these problems.

We earlier described the binomial moment method as a first-generation method for establishing Poisson convergence. The modern method, which has been fantastically successful in hard problems, is known as the *Stein-Chen method*. It has a very interesting history. In 1972, Stein gave a novel method of obtaining error bounds in the central limit theorem. Stein (1972) gave a technique that allowed him to have dependent summands and also allowed him to use non-Fourier methods, which are the classical methods in that problem. We go into those results, generally called Berry-Esseen bounds, later in the book (see Chapter 11). Stein's method was based on a very simple identity, now universally known as Stein's identity (published later in Stein (1981)), which says that if $Z \sim N(0, 1)$, then for *nice* functions f, $E[Zf(Z)] = E[f'(Z)]$. It was later found that if Stein's identity holds for *many* nice functions, then the underlying variable Z *must* be $N(0, 1)$. So, the intuition is that if for some random variable $Z = Z_n$, $E[Zf(Z) - f'(Z)] \approx 0$, then Z should be close to $N(0, 1)$ in distribution. In a manner that many still find mysterious, Stein reduced this to a comparison of the mean of a suitable function h, related to f by a differential equation, under the true distribution of Z and the $N(0, 1)$ distribution. From here, he was able to obtain non-Fourier bounds on errors in the CLT for dependent random variables. A Stein type identity was later found for the Poisson case in the decision theory literature; see Hwang (1982). Stein's method for the normal case was successfully adapted to the Poisson case in Chen (1975). The Stein-Chen method is now regarded as the principal tool in establishing Poisson limits for sums of dependent Bernoulli variables. Roughly speaking, the dependence should be weak, and for any single Bernoulli variable, the number of other Bernoulli variables with which it shares a dependence relation should not be very large. The Stein-Chen method has undergone a lot of evolution with increasing sophistication since Chen (1975). The references given in the first section of this chapter contain a wealth of techniques, results, and, most of all, numerous new applications. Specifically, we recommend Arratia, Goldstein, and Gordon (1990), Barbour, Holst and Janson (1992), Dembo and Rinott (1996), and the recent monograph by Diaconis and Holmes (2004). See Barbour, Chen, and Loh (1992) for use of the Stein-Chen technique for compound Poisson approximations.

2.5 Exercises

Exercise 2.1 * Let $X \sim F$ with density $\frac{1}{\pi(1+x^2)}, -\infty < x < \infty$. Find the total variation projection of F onto the family of all normal distributions.

Exercise 2.2 For each of the following cases, evaluate the indicated distances.

(i) $d_{\text{TV}}(P, Q)$ when $P = \text{Bin}(20, .05)$ and $Q = \text{Poisson}(1)$.
(ii) $d_K(F, G)$ when $F = N(0, \sigma^2)$ and $G = \text{Cauchy}(0, \tau^2)$.
(iii) $H(P, Q)$ when $P = N(\mu, \sigma^2)$ and $Q = N(\nu, \tau^2)$.

Exercise 2.3 * Write an expansion in powers of ϵ for $d_{\text{TV}}(P, Q)$ when $P = N(0, 1)$ and $Q = N(\epsilon, 1)$.

Exercise 2.4 Calculate and plot (as a function of μ) $H(P, Q)$ and $d_{TV}(P, Q)$ when $P = N(0, 1)$ and $Q = N(\mu, 1)$.

Exercise 2.5 * Suppose $P_n = \text{Bin}(n, p_n)$ and $P = \text{Poi}(\lambda)$. Give a sufficient condition for $d_{TV}(P_n, P) \to 0$. Can you give a nontrivial necessary condition?

Exercise 2.6 Show that if $X \sim P, Y \sim Q$, then $d_{TV}(P, Q) \leq P(X \neq Y)$.

Exercise 2.7 Suppose $X_i \overset{\text{indep.}}{\sim} P_i, Y_i \overset{\text{indep.}}{\sim} Q_i$. Then $d_{TV}(P_1 * P_2 * \cdots * P_n, Q_1 * Q_2 * \cdots * Q_n) \leq \sum_{i=1}^{n} d_{TV}(P_i, Q_i)$, where $*$ denotes convolution.

Exercise 2.8 Suppose X_n is a Poisson variable with mean $\frac{n}{n+1}$ and X is Poisson with mean 1.

(a) Show that the total variation distance between the distributions of X_n and X converges to zero.

(b) * (Harder) Find the rate of convergence to zero in part (a).

Exercise 2.9 * Let $P = N(0, 1)$ and $Q = N(\mu, \sigma^2)$. Plot the set $S = \{(\mu, \sigma) : d_{TV}(P, Q) \leq \epsilon\}$ for some selected values of ϵ.

Exercise 2.10 Suppose $X_1, X_2, \ldots$ are iid Exp(1). Does $\sqrt{n}(\bar{X} - 1)$ converge to standard normal in total variation?

Exercise 2.11 If X_i are iid, show that $\bar{X}_n \xrightarrow{\mathcal{P}} 0$ iff $E\left(\frac{\bar{X}_n^2}{1+\bar{X}_n^2}\right) \to 0$.

Exercise 2.12 * Let $X \sim U[-1, 1]$. Find the total variation projection of X onto the class of all normal distributions.

Exercise 2.13 * Consider the family of densities with mean equal to a specified μ. Find the density in this family that maximizes the entropy.

Exercise 2.14 * (**Projection in Entropy Distance**) Suppose X has a density with mean μ and variance σ^2. Show that the projection of X onto the class of all normal distributions has the same mean and variance as X.

Exercise 2.15 * (**Projection in Entropy Distance Continued**) Suppose X is an integer-valued random variable with mean μ. Show that the projection of X onto the class of all Poisson distributions has the same mean as X.

Exercise 2.16 * First write the exact formula for the entropy of a Poisson distribution, and then prove that the entropy grows at the rate of $\log \lambda$ as the mean $\lambda \to \infty$.

Exercise 2.17 What can you say about the existence of entropy and Fisher information for Beta densities? What about the double exponential density?

Exercise 2.18 Prove that the standardized Fisher information of a Gamma$(\alpha, 1)$ density converges to zero at the rate $\frac{1}{\alpha}$, α being the shape parameter.

Exercise 2.19 * Consider the Le Cam bound $d_{\mathrm{TV}}(\mathrm{Bin}(n, p), \mathrm{Poi}(np)) \leq 8p$. Compute the ratio $\frac{d_{\mathrm{TV}}(\mathrm{Bin}(n,p),\mathrm{Poi}(np))}{p}$ for a grid of (n, p) pairs and investigate the best constant in Le Cam's inequality.

Exercise 2.20 * For $N = 5, 10, 20, 30$, compute the distribution of the total number of matches in the matching problem, and verify that the distribution in each case is unimodal.

Exercise 2.21 Give an example of a sequence of binomial distributions that converge neither to a normal (on centering and norming) nor to a Poisson distribution.

References

Arratia, R., Goldstein, L., and Gordon, L. (1990). Poisson approximation and the Chen-Stein method, Stat. Sci., 5(4), 403–434.

Barbour, A., Chen, L., and Loh, W-L. (1992). Compound Poisson approximation for non-negative random variables via Stein's method, Ann. Prob., 20, 1843–1866.

Barbour, A., Holst, L., and Janson, S. (1992). *Poisson Approximation*, Clarendon Press, New York.

Barron, A. (1986). Entropy and the central limit theorem, Ann. Prob., 14(1), 336–342.

Boos, D. (1985). A converse to Scheffe's theorem, Ann. Stat., 1, 423–427.

Brown, L. (1982). A proof of the central limit theorem motivated by the Cramér-Rao inequality, G. Kallianpur, P.R. Krishnaiah, and J.K. Ghosh *Statistics and Probability, Essays in Honor of C.R. Rao*, North-Holland, Amsterdam, 141–148.

Brown, L., DasGupta, A., Haff, L.R., and Strawderman, W.E. (2006). The heat equation and Stein's identity: Connections, applications, J. Stat. Planning Infer, Special Issue in Memory of Shanti Gupta, 136, 2254–2278.

Chen, L.H.Y. (1975). Poisson approximation for dependent trials, Ann. Prob., 3, 534–545.

DasGupta, A. (1999). The matching problem and the Poisson approximation, Technical Report, Purdue University.

DasGupta, A. (2005). The matching, birthday, and the strong birthday problems: A contemporary review, J. Stat. Planning Infer, Special Issue in Honor of Herman Chernoff, 130, 377–389.

Dembo, A. and Rinott, Y. (1996). Some examples of normal approximations by Stein's method, in *Random Discrete Structures*, D. Aldous and Pemantle R. IMA Volumes in Mathematics and Its Applications, Vol. 76, Springer, New York, 25–44.

Diaconis, P. and Holmes, S. (2004). *Stein's Method: Expository Lectures and Applications*, IMS Lecture Notes Monograph Series, vol. 46, Institute of Mathematical Statistics, Beachwood, OH.

Dudley, R. (1989). *Real Analysis and Probability*, Wadsworth, Pacific Grove, CA.

Galambos, J. and Simonelli, I. (1996). *Bonferroni-Type Inequalities with Applications*, Springer, New York.

Hwang, J.T. (1982). Improving upon standard estimators in discrete exponential families with applications to Poisson and negative binomial cases, Ann. Stat., 10(3), 857–867.

Johnson, O. (2004). *Information Theory and the Central Limit Theorem*, Imperial College Press, Yale University London.

Johnson, O. and Barron, A. (2003). Fisher information inequalities and the central limit theorem, Technical Report.

Kolchin, V., Sevas'tyanov, B., and Chistyakov, V. (1978). *Random Allocations*, V.H. Winston & Sons, Washington, distributed by Halsted Press, New York.

LeCam, L. (1960). An approximation theorem for the Poisson binomial distribution, Pac. J. Math., 10, 1181–1197.

Linnik, Y. (1959). An information theoretic proof of the central limit theorem, Theory Prob. Appl., 4, 288–299.

Rachev, S. (1991). *Probability Metrics and the Stability of Stochastic Models*, John Wiley, Chichester.

Reiss, R. (1989). *Approximate Distributions of Order Statistics, with Applications to Nonparametric Statistics*, Springer-Verlag, New York.

Sevas'tyanov, B.A. (1972). A limiting Poisson law in a scheme of sums of dependent random variables, Teor. Veroyatni. Primen., 17, 733–738.

Stein, C. (1972). A bound for the error in the normal approximation to the distribution of a sum of dependent random variables, L. Le Cam, J. Neyman, and E. Scott in *Proceedings of the Sixth Berkeley Symposium*, Vol. 2, University of California Press, Berkeley, 583–602.

Stein, C. (1981). Estimation of the mean of a multivariate normal distribution, Ann. Stat., 9, 1135–1151.

Stein, C. (1986). *Approximate Computations of Expectations*, Institute of Mathematical Statistics, Hayward, CA.

Sweeting, T. (1986). On a converse to Scheffe's theorem, Ann. Stat., 3, 1252–1256.

Chapter 3
More General Weak and Strong Laws and the Delta Theorem

For sample means of iid random variables $X_1, \ldots, X_n$, the usual weak and strong laws say that $\bar{X}_n \xrightarrow{\mathcal{P}} \mu$ and $\bar{X}_n \xrightarrow{\text{a.s.}} \mu$ if $E(|X_1|) < \infty$. Weaker versions can be proved under more general conditions. The amount of literature on this is huge. A sample of a few results is stated below. See Feller (1966), Revesz (1968), Chung (2001), or Sen and Singer (1993) for a more detailed exposition of general laws of large numbers (LLNs). Schneider (1987) is an interesting reading on the history of laws of large numbers. We also present in this chapter the delta theorem, one of the most useful tools in asymptotic theory. For delta theorems, good general references are Bickel and Doksum (2001), Lehmann (1999), Serfling (1980), and van der Vaart (1998).

3.1 General LLNs and Uniform Strong Law

Theorem 3.1 (i) Suppose $X_1, X_2, \ldots$ are iid. The sequence $\{\bar{X}_n\}$ is unbounded with probability 1 if the distribution of X_1 does not have a finite mean.

(ii) If $\{X_n\}$ is any sequence of random variables (possibly vector valued), then there always exist constants $\{b_n\}$ such that $\frac{X_n}{b_n} \xrightarrow{\text{a.s.}} 0$.

(iii) If $X_1, X_2, \ldots$ are iid random variables with distribution F, then there exist constants μ_n such that $\bar{X}_n - \mu_n \xrightarrow{\mathcal{P}} 0$ if and only if

$$x(1 - F(x) + F(-x)) \xrightarrow{x \to \infty} 0.$$

In such a case, one may choose $\mu_n = \int_{[-n,n]} x dF(x)$.

(iv) If $X_1, X_2, \ldots$ are independent with $E(X_n) = \mu$, $\sigma_n^2 = \text{Var}(X_n) < \infty$, and $\sum_{n=1}^{\infty} \sigma_n^{-2} = \infty$, then

A. DasGupta, *Asymptotic Theory of Statistics and Probability*,

$$\frac{\sum_{i=1}^{n} \frac{X_i}{\sigma_i^2}}{\sum_{i=1}^{n} \frac{1}{\sigma_i^2}} \xrightarrow{\text{a.s.}} \mu.$$

(v) If $X_1, X_2, \ldots$ are independent with $E(X_i) = \mu_i$, $\sigma_i^2 = \text{Var}(X_i) < \infty$, $\sum_{i=1}^{\infty} \frac{\sigma_i^2}{c_i^2} < \infty$, $\{c_n\}$ ultimately monotone, and $c_n \to \infty$, then

$$\frac{\sum_{i=1}^{n}(X_i - \mu_i)}{c_n} \xrightarrow{\text{a.s.}} 0.$$

(vi) **Uniform Strong Law** Suppose $X_1, X_2, \ldots$ are iid and, for some auxiliary parameter θ, $g(x, \theta)$ is such that $E(|g(X_1, \theta)|) < \infty$ for each $\theta \in \Theta$. If g is continuous in θ for each x, $|g(x, \theta)| \leq M(x)$ for some $M(.)$ with $E(M(X_1)) < \infty$, and if Θ is a compact set, then $\sup_{\theta \in \Theta} |\frac{1}{n} \sum_{i=1}^{n} g(X_i, \theta) - \mu(\theta)| \xrightarrow{\text{a.s.}} 0$, where $\mu(\theta) = E(g(X_1, \theta))$.

There is no single source where all of the parts of Theorem 3.1 are proved. Feller (1966, Vol. II) can be consulted for several of the parts, while Ferguson (1996) proves the uniform strong law.

Example 3.1 Suppose $X_i \overset{\text{indep.}}{\sim} (\mu, \sigma_i^2)$; i.e., X_i have a common expectation μ and variance σ_i^2. Then, by simple calculus, the BLUE (best linear unbiased estimate) of μ is $\frac{\sum_{i=1}^{n} \frac{X_i}{\sigma_i^2}}{\sum_{i=1}^{n} \frac{1}{\sigma_i^2}}$. Suppose now that the σ_i^2 do not grow at a rate faster than i; i.e., for some finite constant K, $\sigma_i^2 \leq iK$. Then, $\sum_{i=1}^{n} \frac{1}{\sigma_i^2}$ clearly diverges as $n \to \infty$, and so by Theorem 3.1 the BLUE of μ is strongly consistent.

Example 3.2 As an example of an application of the uniform strong law, suppose $X_1, \ldots, X_n$ are iid $N(\theta, 1)$. Consider the statistic $T_n = \frac{1}{n} \sum_{i=1}^{n} | X_i - \bar{X}|$, known as the *mean absolute deviation*. Since $\bar{X}$ is close to the true value of θ, say θ_0, for large n, one expects that T_n should behave like $\frac{1}{n} \sum_{i=1}^{n} |X_i - \theta_0|$ and therefore ought to converge to $E|X_1 - \theta_0|$, perhaps a.s.

Toward this end, note first that $X_1 - \theta_0 \sim N(0, 1)$ and $X_1 - \bar{X} \sim N(0, 1 - \frac{1}{n})$. Hence, $E|X_1 - \theta_0| - E|X_1 - \bar{X}| = \sqrt{2\pi}(1 - \sqrt{1 - \frac{1}{n}}) \to 0$ as $n \to \infty$. Also note that, by the SLLN for $\bar{X}$, one may assume that $\bar{X}$ lies in a fixed compact neighborhood C of θ_0 for all large n. If we define $g(x,\theta) = |x-\theta|$, then the assumptions of the uniform strong law in Theorem 3.1 hold. Thus,

$$\left|\frac{1}{n}\sum_{i=1}^{n}|X_i-\bar{X}| - E|X_1-\theta_0|\right| = \left|\frac{1}{n}\sum_{i=1}^{n}|X_i-\bar{X}| - E|X_1-\bar{X}|\right.$$
$$\left. + E|X_1-\bar{X}| - E|X_1-\theta_0|\right|$$

$$\le \sup_{\theta\in C}\left|\frac{1}{n}\sum_{i=1}^{n}|X_i-\theta| - E|X_1-\theta|\right|$$
$$+|E|X_1-\bar{X}| - E|X_1-\theta_0||.$$

The first term goes to zero a.s. by the uniform strong law, and we noted above that the second term goes to zero. So, indeed, $T_n \stackrel{\text{a.s.}}{\Rightarrow} E|X_1-\theta_0|$.

It should be pointed out here that the conclusion of this example can be derived by more direct means and without using the uniform strong law.

Example 3.3 Suppose (X_i, Y_i), $i = 1,\ldots,n$ are iid bivariate samples from some distribution with $E(X_1) = \mu_1$, $E(Y_1) = \mu_2$, $\text{Var}(X_1) = \sigma_1^2$, $\text{Var}(Y_1) = \sigma_2^2$, and $\text{corr}(X_1, Y_1) = \rho$. Let

$$r_n = \frac{\sum(X_i-\bar{X})(Y_i-\bar{Y})}{\sqrt{\sum(X_i-\bar{X})^2\sum(Y_i-\bar{Y})^2}}$$

denote the sample correlation coefficient. If we write

$$r_n = \frac{\frac{1}{n}\sum X_iY_i - \bar{X}\bar{Y}}{\sqrt{\left(\sum\frac{X_i^2}{n}-\bar{X}^2\right)\left(\sum\frac{Y_i^2}{n}-\bar{Y}^2\right)}},$$

then from the strong law of large numbers for iid random variables,

$$r_n \xrightarrow{\text{a.s.}} \frac{E(X_1Y_1)-\mu_1\mu_2}{\sqrt{\sigma_1^2\sigma_2^2}} = \rho.$$

Remark. The almost sure convergence of r_n to ρ thus follows very easily. However, the asymptotic distribution of r is not trivial to derive. The tool required is known as the delta theorem, one of the most useful theorems in asymptotic theory. The delta theorem specifies the asymptotic distribution of a smooth transformation $g(T_n)$ if the asymptotic distribution of T_n is known. We present it in a later section in this chapter.

3.2 Median Centering and Kesten's Theorem

The strong law due to Kolmogorov says that, for iid random variables with a finite mean, if we normalize the sequence of partial sums $S_n = \sum_{i=1}^n X_i$ by n, then the normalized sequence converges almost surely to the mean. That is, the normalized sequence has, almost surely, only one limit point if $E(|X_1|) < \infty$. It is natural to ask what happens in general. The following theorem elegantly describes what happens for general iid sequences and with a general normalization. Theorems 3.2 and 3.3 below are stated mainly for the purpose of reference and are not illustrated with examples.

Theorem 3.2 Let $X_1, X_2, \cdots$ be iid with distribution F and γ_n a sequence converging to ∞. Let $Z_{n,\gamma} = \frac{S_n}{\gamma(n)}, n \geq 1$, and $B(F, \gamma)$ be the set of all limit points of $Z_{n,\gamma}$. Then, there is a nonrandom set $A(F, \gamma)$ such that, with probability one, $B(F, \gamma)$ coincides with $A(F, \gamma)$. In particular:

(a) If $0 < \alpha < \frac{1}{2}, \gamma_n = n^\alpha$, then for any F not degenerate at 0, $A(F, \gamma)$ equals the whole extended real line if it contains at least one finite real number.

(b) If $\gamma_n = n$ and if $A(F, \gamma)$ contains at least two finite reals, then it has to contain $\pm\infty$.

(c) If $\gamma_n = 1$, then $A(F, \gamma) \subseteq \{\pm\infty\}$ iff, for some $a > 0$, $\int_{-a}^{a} \text{Re}\{\frac{1}{1-\psi(t)}\}dt < \infty$, where $\psi(t)$ is the characteristic function of F.

(d) When $E(X_1) = 0$ and $\text{Var}(X_1) = \sigma^2 < \infty$, then, with $\gamma_n = \sqrt{2n \log \log n}$, $A(F, \gamma) = [-\sigma, \sigma]$, and when $E(X_1) = 0$ and $\text{Var}(X_1) = \infty$ with $\gamma_n = \sqrt{2n \log \log n}$, $A(F, \gamma)$ contains at least one of $\pm\infty$.

Remark. Parts (c) and (d) are due to Strassen (1964, 1966) and Chung and Fuchs (1951); Kesten (1972) can be seen for the other parts. Another interesting question is what can be said about the set of limit points if we center the partial sums at some sequence a_n. We take the specific sequence $a_n = \text{med}(S_n)$, any median of S_n. Choosing the centering sequence to be the median can give more interesting answers, even if the mean of F exists. Here is a result on the set of limit points by centering S_n at its median.

Theorem 3.3 Let m_n be such that $P(S_n \le m_n) \ge \frac{1}{2}$ and $P(S_n \ge m_n) \ge \frac{1}{2}$. Then, there exists a positive sequence γ_n satisfying $-\infty < \liminf \frac{S_n - m_n}{\gamma_n} < \limsup \frac{S_n - m_n}{\gamma_n} < \infty$ iff, for all $c \ge 1$, $\liminf_{x\to\infty} \frac{P(|X_1|>cx)}{P(|X_1|>x)} \le c^{-2}$.

Remark. To our knowledge, no general explicit choices for γ_n in Theorem 3.3 are known, although necessary conditions they have to satisfy are known. A reference for this entire section is Kesten (1972). The condition on F as we have stated it in Theorem 3.3 is due to Maller (1980).

3.3 The Ergodic Theorem

Imagine that corresponding to each individual x in an ensemble there is a value $f(x)$ of some function f. Suppose the ensemble average is some number c. Now suppose that you start with one individual x_0 and iteratively change the location from x_0 to x_1 to x_2, etc., by repeatedly applying a transformation T as the clock moves from time 0 to 1 to 2, etc. Consider now the time average of these function values. For well-behaved functions and certain types of transformations T, the ensemble average equals the time average of one individual. Phase transitions by repeated applications of T mix things up so well that the time average for one individual equals the ensemble average after infinite time has passed, roughly speaking. This is Birkhoff's ergodic theorem (Birkhoff (1931)); Breiman (1968) is a more modern reference. The ergodic theorem is immensely useful in probability and statistics. We need some definitions to give a statement of it.

Definition 3.1 Let $(\Omega, \mathcal{A}, P)$ be a probability space and $T : \Omega \to \Omega$ a (measurable) transformation. T is called *ergodic* if $A \in \mathcal{A}, T(A) = A \Rightarrow P(A)(1 - P(A)) = 0$. T is called *invariant* if it is measure-preserving; i.e., $P \circ T^{-1} = P$.

Theorem 3.4 Let $(\Omega, \mathcal{A}, P)$ be a probability space and $T : \Omega \to \Omega$ an ergodic transformation. Let $f \in L_1(P)$. Then, for almost all x with respect to P, $\lim_{n\to\infty} \frac{1}{n} \sum_{k=0}^{n-1} f(T^k(x)) = \int f dP$.

For statisticians, the best applications of the ergodic theorem are in studies of properties of time series and stochastic processes.

Definition 3.2 Let $\{X_n\}_{n=1}^{\infty}$ be a (discrete time) stochastic process. $\{X_n\}_{n=1}^{\infty}$ is called stationary if the shift operator T defined as $T(\mathbf{X})_\mathbf{n} = \mathbf{X}_{\mathbf{n+1}}$ keeps the distribution of the process the same; i.e., $(X_2, X_3, \cdots)$ has the same distribution as $(X_1, X_2, \cdots)$.

The definition of ergodicity of a process requires development of some measure-theoretic background. See Breiman (1968) for a rigorous definition. Roughly speaking, a stationary process is ergodic if the long-run behavior of its paths is not affected by the initial conditions. One useful corollary of the ergodic theorem for statisticians is the following result.

Theorem 3.5 Let $\{X_n\}_{n=1}^{\infty}$ be a stationary ergodic process, and suppose $E(|X_1|) < \infty$. Then $\frac{1}{n}\sum_{k=1}^{n} X_k \stackrel{a.s.}{\rightarrow} E(X_1)$.

For example, for a stationary Gaussian process with autocorrelation function ρ_k, the sample mean is a strongly consistent estimate of the process mean if $\rho_k \to 0$ as $k \to \infty$.

3.4 Delta Theorem and Examples

Distributions of transformations of a statistic are of importance in applications. The delta theorem says how to approximate the distribution of a transformation of a statistic in large samples if we can approximate the distribution of the statistic itself. Serfling (1980), Lehmann (1999), and Bickel and Doksum (2001) can be consulted for more examples and implications of the delta theorem in addition to the material that we present below.

Theorem 3.6 (Delta Theorem) Let T_n be a sequence of statistics such that

$$\sqrt{n}(T_n - \theta) \xrightarrow{\mathcal{L}} N(0, \sigma^2(\theta)), \ \sigma(\theta) > 0.$$

Let $g : \mathbb{R} \to \mathbb{R}$ be once differentiable at θ with $g'(\theta) \neq 0$. Then

$$\sqrt{n}\,(g(T_n) - g(\theta)) \xrightarrow{\mathcal{L}} N(0, [g'(\theta)]^2\sigma^2(\theta)).$$

Proof. First note that it follows from the assumed CLT for T_n that T_n converges in probability to θ and hence $T_n - \theta = o_p(1)$. The proof of the theorem now follows from a simple application of Taylor's theorem that says that

$$g(x_0 + h) = g(x_0) + hg'(x_0) + o(h)$$

if g is differentiable at x_0. Therefore

$$g(T_n) = g(\theta) + (T_n - \theta)g'(\theta) + o_p(T_n - \theta).$$

That the remainder term is $o_p(T_n - \theta)$ follows from our observation that $T_n - \theta = o_p(1)$. Taking $g(\theta)$ to the left and multiplying both sides by $\sqrt{n}$, we obtain

$$\sqrt{n}\,(g(T_n) - g(\theta)) = \sqrt{n}(T_n - \theta)g'(\theta) + \sqrt{n}\; o_p(T_n - \theta).$$

Observing that $\sqrt{n}(T_n - \theta) = O_p(1)$ by the assumption of the theorem, we see that the last term on the right-hand side is $\sqrt{n}\; o_p(T_n - \theta) = o_p(1)$. Hence, an application of Slutsky's theorem to the above gives

$$\sqrt{n}\,(g(T_n) - g(\theta)) \xrightarrow{\mathcal{L}} N(0, [g'(\theta)]^2\sigma^2(\theta))$$

Remark. There are instances in which $g'(\theta) = 0$ (at least for some special value of θ), in which case the limiting distribution of $g(T_n)$ is determined by the third term in the Taylor expansion. Thus, if $g'(\theta) = 0$, then

$$g(T_n) = g(\theta) + \frac{(T_n - \theta)^2}{2} g''(\theta) + o_p\left((T_n - \theta)^2\right)$$

and hence

$$n(g(T_n) - g(\theta)) = n\frac{(T_n - \theta)^2}{2} g''(\theta) + o_p\,(1) \xrightarrow{\mathcal{L}} \frac{g''(\theta)\sigma^2(\theta)}{2}\chi^2(1).$$

Next we state the multivariate version of the delta theorem, which is similar to the univariate case.

Theorem 3.7 Suppose $\{T_n\}$ is a sequence of k-dimensional random vectors such that $\sqrt{n}(T_n - \theta) \xrightarrow{\mathcal{L}} N_k(0, \Sigma(\theta))$. Let $g : \mathbb{R}^k \to \mathbb{R}^m$ be once differentiable at θ with the gradient matrix $\nabla g(\theta)$. Then

$$\sqrt{n}\,(g(T_n) - g(\theta)) \xrightarrow{\mathcal{L}} N_m\left(0, \nabla g(\theta)^T \Sigma(\theta) \nabla g(\theta)\right)$$

provided $\nabla g(\theta)^T \Sigma(\theta) \nabla g(\theta)$ is positive definite.

Example 3.4 Suppose $X_1, \ldots, X_n$ are independent and identically distributed with mean μ and variance σ^2. By taking $T_n = \overline{X}_n,\ \theta = \mu,\ \sigma^2(\theta) = \sigma^2$, and $g(x) = x^2$, one gets

$$\sqrt{n}(\overline{X}_n^2 - \mu^2) \xrightarrow{\mathcal{L}} N(0, 4\mu^2\sigma^2).$$

Example 3.5 Suppose X_i's are as above and $\mu > 0$. Then, taking $T_n = \overline{X}_n,\ \theta = \mu,\ \sigma^2(\theta) = \sigma^2$, and $g(x) = \log x$, one gets

$$\sqrt{n}(\log \overline{X}_n - \log \mu) \xrightarrow{\mathcal{L}} N\left(0, \frac{\sigma^2}{\mu^2}\right).$$

Example 3.6 Suppose X_i's are as above with $E(X^4) < \infty$. Then, by taking

$$T_n = \begin{pmatrix} \overline{X}_n \\ \frac{1}{n}\sum_{i=1}^n X_i^2 \end{pmatrix}, \ \theta = \begin{pmatrix} EX_1 \\ EX_1^2 \end{pmatrix}, \ \Sigma(\theta) = \begin{pmatrix} \mathrm{Var}(X_1) & \mathrm{cov}(X_1, X_1^2) \\ \mathrm{cov}(X_1, X_1^2) & \mathrm{Var}(X_1^2) \end{pmatrix}$$

and using $g((u, v)) = v - u^2$, it follows that

$$\sqrt{n}\left(\frac{1}{n}\sum_{i=1}^n (X_i - \overline{X}_n)^2 - \mathrm{Var}(X_1)\right) \xrightarrow{\mathcal{L}} N(0, \mu_4 - \sigma^4),$$

where $\mu_4 = E(X_1 - \mu)^4$ is the centered fourth moment of X_1.

If we choose $s_n^2 = \sum_{i=1}^n (X_i - \overline{X}_n)^2/(n-1)$, then

$$\sqrt{n}(s_n^2 - \sigma^2) = \frac{\sum_{i=1}^n (X_i - \overline{X}_n)^2}{(n-1)\sqrt{n}} + \sqrt{n}\left(\frac{1}{n}\sum_{i=1}^n (X_i - \overline{X}_n)^2 - \sigma^2\right),$$

which also converges in law to $N(0, \mu_4 - \sigma^4)$ by Slutsky's theorem. From here, by another use of the delta theorem, one sees that

$$\sqrt{n}(s_n - \sigma) \xrightarrow{\mathcal{L}} N\left(0, \frac{\mu_4 - \sigma^4}{4\sigma^2}\right).$$

The delta theorem is also useful in finding the limiting distribution of sample moments. We state it next.

Theorem 3.8 Let $\{X_i\}$ be an iid sequence with $E(X_1^{2k}) < \infty$. Let $m_k = \frac{1}{n}\sum_{i=1}^n (X_i - \overline{X}_n)^k$ and $\mu_k = E(X_1 - \mu)^k$. Then

$$\sqrt{n}(m_k - \mu_k) \xrightarrow{\mathcal{L}} N\left(0, V_k\right),$$

where $V_k = \mu_{2k} - \mu_k^2 - 2k\mu_{k-1}\mu_{k+1} + k^2\sigma^2\mu_{k-1}^2$.

Remark. In fact, a tedious calculation also shows that the joint limiting distribution of any finite number of central sample moments is a multivariate normal. See Serfling (1980) for explicit formulas of the covariance matrix for that limiting distribution.

Example 3.7 Let $\{X_i\}$ be iid with $E(X_1^4) < \infty$. Then, by using the multivariate delta theorem on some algebra,

$$\sqrt{n}\begin{pmatrix} \overline{X}_n - \mu \\ s_n^2 - \sigma^2 \end{pmatrix} \xrightarrow{\mathcal{L}} N_2\left(\begin{pmatrix} 0 \\ 0 \end{pmatrix}, \begin{pmatrix} \sigma^2 & \mu_3 \\ \mu_3 & \mu_4 - \sigma^4 \end{pmatrix}\right).$$

Thus $\overline{X}_n$ and s_n^2 are asymptotically independent if the population skewness is 0 (i.e., $\mu_3 = 0$).

Example 3.8 Another use of the multivariate delta theorem is the derivation of the limiting distribution of the samplecorrelation coefficient r_n for iid bivariate data (X_i, Y_i). We have

$$r_n = \frac{\frac{1}{n}\sum_{i=1}^{n} X_i Y_i - \overline{X}_n \overline{Y}_n}{\sqrt{\frac{1}{n}\sum_{i=1}^{n}(X_i - \overline{X}_n)^2 \frac{1}{n}\sum_{i=1}^{n}(Y_i - \overline{Y}_n)^2}}.$$

By taking

$$T_n = \left(\overline{X}_n, \overline{Y}_n, \frac{1}{n}\sum_{i=1}^{n} X_i^2, \frac{1}{n}\sum_{i=1}^{n} Y_i^2, \frac{1}{n}\sum_{i=1}^{n} X_i Y_i\right)^T,$$

$$\theta = \left(EX_1, EY_1, EX_1^2, EY_1^2, EX_1Y_1\right)^T,$$

$$\Sigma = \text{cov}(X_1, Y_1, X_1^2, Y_1^2, X_1Y_1),$$

and using the transformation $g(u_1, u_2, u_3, u_4, u_5) = (u_5 - u_1u_2)/\sqrt{(u_3 - u_1^2)(u_4 - u_2^2)}$, it follows that

$$\sqrt{n}(r_n - \rho) \xrightarrow{\mathcal{L}} N(0, v^2)$$

for some $v > 0$. It is not possible to write a clean formula for v in general. If (X_i, Y_i) are iid $N_2(\mu_X, \mu_Y, \sigma_X^2, \sigma_Y^2, \rho)$, then the calculation can be done in closed form and

$$\sqrt{n}(r_n - \rho) \xrightarrow{\mathcal{L}} N(0, (1 - \rho^2)^2).$$

However, convergence to normality is very slow. The fixed sample density of r_n in the bivariate normal case can be written as an infinite series or in terms of hypergeometric functions (see Tong (1990)).

3.5 Approximation of Moments

The delta theorem is proved by an ordinary Taylor expansion of T_n around θ. The same method also produces approximations, with error bounds, on the moments of $g(T_n)$. The order of the error can be made smaller the more moments T_n has. To keep notation simple, we give approximations to the mean and variance of a function $g(T_n)$ below when T_n is a sample mean. See Bickel and Doksum (2001) for proofs and more information.

Theorem 3.9 Suppose $X_1, X_2, \ldots$ are iid observations with a finite fourth moment. Let $E(X_1) = \mu$ and $\text{Var}(X_1) = \sigma^2$. Let g be a scalar function with four uniformly bounded derivatives. Then

$$(a)\ E[g(\overline{X})] = g(\mu) + \frac{g^{(2)}(\mu)\sigma^2}{2n} + O(n^{-2}),$$

$$(b)\ \text{Var}[g(\overline{X})] = \frac{(g'(\mu))^2\sigma^2}{n} + O(n^{-2}).$$

Remark. The variance approximation above is simply what the delta theorem says. With more derivatives of g that are uniformly bounded, higher-order approximations can be given.

Example 3.9 Suppose $X_1, X_2, \ldots$ are iid Poi(μ) and we wish to estimate $P(X_1 = 0) = e^{-\mu}$. The MLE is $e^{-\overline{X}}$, and suppose we want to find an approximation to the bias and variance of $e^{-\overline{X}}$.

We apply Theorem 3.9 with the function $g(x) = e^{-x}$ so that $g'(x) = -g^{(2)}(x) = -e^{-x}$. Plugging into the theorem, we get the approximations $\text{Bias}(e^{-\overline{X}}) = \frac{\mu e^{-\mu}}{2n} + O(n^{-2})$, and $\text{Var}(e^{-\overline{X}}) = \frac{\mu e^{-2\mu}}{n} + O(n^{-2})$.

Note that it is in fact possible to derive exact expressions for the mean and variance of $e^{-\overline{X}}$ in this case, as $\sum_{i=1}^n X_i$ has a Poi$(n\mu)$ distribution and therefore its mgf (moment generating function) equals $\psi_n(t) = E(e^{t\overline{X}}) = (e^{\mu(e^{t/n}-1)})^n$. In particular, the mean of $e^{-\overline{X}}$ is $(e^{\mu(e^{-1/n}-1)})^n$.

It is possible to recover the approximation for the bias given above from this exact expression. Indeed, $(e^{\mu(e^{-1/n}-1)})^n = e^{n\mu(e^{-1/n}-1)} = e^{n\mu(\sum_{k=1}^{\infty}\frac{(-1)^k}{k!n^k})} = e^{-\mu}\prod_{k=0}^{\infty} e^{\frac{(-1)^{k-1}\mu}{(k+1)!n^k}} = e^{-\mu}(1+\frac{\mu}{2n}+O(n^{-2}))$ on collecting the terms of the exponentials together. On subtracting $e^{-\mu}$, this reproduces the bias approximation given above. The delta theorem produces it more easily than the direct calculation. The direct calculation is enlightening, however.

3.6 Exercises

Exercise 3.1 For each F below, find the limit distributions of $\frac{\bar{X}}{s}$ and $\frac{s}{\bar{X}}$.

(i) $F = U[0, 1]$.
(ii) $F = \text{Exp}(\lambda)$.
(iii) $F = \chi^2(p)$.

Exercise 3.2 * Suppose $X_1, \ldots, X_n$ are iid $N(\mu, \mu^2)$, $\mu > 0$. Therefore, $\bar{X}$ and s are both reasonable estimates of μ. Find the limit of $P(|s-\mu| < |\bar{X}-\mu|)$.

Hint: Use the joint limiting distribution of $(\bar{X}, s^2)$ and the delta theorem.

Exercise 3.3 * Suppose $X_1, \ldots, X_n$ are Poi(μ). Find the limit distribution of $e^{-\bar{X}}$ and $(1-\frac{1}{n})^{n\bar{X}}$.

Remark. These are the MLE and UMVUE of $e^{-\mu}$.

Exercise 3.4 Suppose $X_1, \ldots, X_n$ are iid $N(\mu, \sigma^2)$. Let $b_1 = \frac{\frac{1}{n}\sum(X_i-\bar{X})^3}{[\frac{1}{n}\sum(X_i-\bar{X})^2]^{3/2}}$ and $b_2 = \frac{\frac{1}{n}\sum(X_i-\bar{X})^4}{[\frac{1}{n}\sum(X_i-\bar{X})^2]^2} - 3$ be the sample skewness and kurtosis coefficients. Find the joint limiting distribution of (b_1, b_2).

Exercise 3.5 Suppose $X_1, \ldots, X_n$ are iid Poi(μ). Find the limit distribution of $\frac{1}{\bar{X}+\bar{X}^2+\bar{X}^3}$.

Exercise 3.6 Let X_n, Y_m be independent Poisson variables with means n, m. Find the limiting distribution of $\frac{X_n-Y_m-(n-m)}{\sqrt{X_n+Y_m}}$ as $n, m \to \infty$.

Exercise 3.7 * Suppose X_i are iid Exp(1). Find the limiting distribution of $\sqrt{n}(\bar{X}^2 - X_{(1)} - 1)$, where $X_{(1)}$ is the sample minimum.

Exercise 3.8 Suppose X_i are iid $N(0, 1)$. Find the limit distribution of

$$\frac{n(X_1X_2 + X_3X_4 + \cdots + X_{2n-1}X_{2n})^2}{\left(\sum_{i=1}^{2n} X_i^2\right)^2}.$$

Exercise 3.9 Suppose $X \sim \text{Bin}(n, p)$. Derive an approximation to the bias of the MLE of $\frac{p}{1-p}$ up to the order $O(n^{-2})$.

Exercise 3.10 Suppose $X_1, X_2, \ldots, X_n$ are iid $N(\mu, 1)$.

(a) Find the UMVUE of $P(X_1 > 0)$ by applying the Rao-Blackwell theorem.
(b) Derive an approximation to its variance up to the order $O(n^{-2})$.

Exercise 3.11 Suppose $X_1, X_2, \ldots, X_n$ are iid with mean 0 and variance 1. Show that $\frac{\sum_{i=1}^n X_i}{\sqrt{n}\log n} \stackrel{\text{a.s.}}{\Rightarrow} 0$.

Exercise 3.12 * Cereal boxes contain independently and with equal probability exactly one of n different celebrity pictures. Someone having the entire set of n pictures can cash them in for money. Let W_n be the minimum number of cereal boxes one would need to purchase to own a complete set of the pictures. Find a_n such that $\frac{W_n}{a_n} \stackrel{P}{\Rightarrow} 1$.
Hint: Approximate the mean of W_n.

Exercise 3.13 Suppose $X_i \stackrel{\text{indep.}}{\sim} \text{Poi}(\lambda_i)$. Suppose $\sum_{i=1}^{\infty} \lambda_i = \infty$. Show that $\frac{\sum_{i=1}^n X_i}{\sum_{i=1}^n \lambda_i} \stackrel{P}{\Rightarrow} 1$. Is the convergence almost sure as well?

Exercise 3.14 * Suppose X_i is a sequence of iid $C(0, 1)$ random variables. Is there any sequence γ_n such that $\frac{S_n}{\gamma_n}$ has finite lim inf and lim sup but they are not equal?

Exercise 3.15 * Suppose X_i is a sequence of iid Exp(1) random variables. Give examples of sequences $\{a_n\}, \{\gamma_n\}$ such that $-\infty < \liminf \frac{S_n - a_n}{\gamma_n} \le \limsup \frac{S_n - a_n}{\gamma_n} < \infty$.

Exercise 3.16 * Suppose X_i is a sequence of iid $C(0, 1)$ random variables. Is the set of limit points of $S_n \subseteq \{\pm\infty\}$?

Exercise 3.17 Give a proof of Kolmogorov's SLLN by using the ergodic theorem.

References

Bickel, P. and Doksum, K. (2001). *Mathematical Statistics: Basic Ideas and Selected Topics*, Vol. I, Prentice-Hall, Englewood Cliffs, NJ.

Birkhoff, G. (1931). Proof of the Ergodic theorem, Proc. Nat. Acad. Sci. USA, 17, 656–660.

Breiman, L. (1968). *Probability*, Addison-Wesley, New York.

Chung, K.L. (2001). *A Course in Probability Theory*, 3rd ed., Academic Press, San Diego, CA.

Chung, K.L. and Fuchs, W. (1951). On the distribution of values of sums of random variables, Memoir 6, American Mathematical Society, Providence, RI.

Feller, W. (1966). *An Introduction to Probability Theory and Its Applications*, Vols. I, II, John Wiley, New York.

Ferguson, T. (1996). *A Course in Large Sample Theory*, Chapman and Hall, London.

Kesten, H. (1972). Sums of independent random variables—without moment conditions, the 1971 Rietz Lecture, Ann. Math. Stat., 43, 701–732.

Lehmann, E.L. (1999). *Elements of Large Sample Theory*, Springer, New York.

Maller, R. (1980). A note on domains of partial attraction, Ann. Prob., 8, 576–583.

Revesz, P. (1968). *The Laws of Large Numbers*, Academic Press, New York.

Schneider, I. (1987). *The Intellectual and Mathematical Background of the Law of Large Numbers and the Central Limit Theorem in the 18th and the 19th Centuries*, Cahiers Histoire et de Philosophie des Sciences Nowlle Serie Societe Francaise d' Histoire des Sciences et des Techniques, Paris.

Sen, P.K. and Singer, J. (1993). *Large Sample Methods in Statistics: An Introduction with Applications*, Chapman and Hall, New York.

Serfling, R. (1980). *Approximation Theorems of Mathematical Statistics*, John Wiley, New York.

Strassen, V. (1964). An invariance principle for the law of the iterated logarithm, Z. Wahr. Verw. Geb., 3, 211–226.

Strassen, V. (1966). A converse to the law of the iterated logarithm, Z. Wahr. Verw. Geb., 4, 265–268.

Tong, Y. (1990). *The Multivariate Normal Distribution*, Springer-Verlag, New York.

van der Vaart, A. (1998). *Asymptotic Statistics*, Cambridge University Press, Cambridge.

Chapter 4
Transformations

A principal use of parametric asymptotic theory is to construct asymptotically correct confidence intervals. More precisely, suppose $\hat{\theta}$ is a reasonable estimate of some parameter θ. Suppose it is consistent and even asymptotically normal; i.e., $\sqrt{n}(\hat{\theta} - \theta) \xrightarrow{\mathcal{L}} N(0, \sigma^2(\theta))$ for some function $\sigma(\theta) > 0$. Then, a simple calculation shows that the confidence interval $\hat{\theta} \pm z_{\alpha/2}\sigma(\hat{\theta})/\sqrt{n}$ is *asymptotically correct*; i.e., its limiting coverage is $1-\alpha$ under every θ.

A number of approximations have been made in using this interval. The exact distribution of $\hat{\theta}$ has been replaced by a normal, the correct standard deviation has been replaced by another plug-in estimate $\sigma(\hat{\theta})$, and the true mean of $\hat{\theta}$ has been replaced by θ. If n is not adequately large, the actual distribution of $\hat{\theta}$ may be far from being approximately normal; e.g., it could have a pronounced skewness. The plug-in standard deviation estimate is quite often an underestimate of the true standard deviation. And depending on the situation, $\hat{\theta}$ may have a nontrivial bias as an estimate of θ. Interest has centered on finding transformations, say $g(\hat{\theta})$, that (i) have an asymptotic variance function free of θ, eliminating the annoying need to use a plug-in estimate, (ii) have skewness ≈ 0 in some precise sense, and (iii) have bias ≈ 0 as an estimate of $g(\theta)$, again in some precise sense.

Transformations of the first type are known as *variance-stabilizing transformations* (VSTs), those of the second type are known as *symmetrizing transformations* (STs), and those of the third type are known as *bias-corrected transformations* (BCTs). Ideally, we would like to find one transformation that achieves all three goals. However, usually no transformation can even achieve any two of the three goals simultaneously. There is an inherent tension between the three goals. One can often achieve more than one goal through iterations; e.g., first obtain a VST and then find a bias-corrected adjustment of that. Or first obtain an ST and then find a bias-corrected adjustment of that. There is an enormous body of literature on transformations, going back to the early 1940s. We will limit ourselves

A. DasGupta, *Asymptotic Theory of Statistics and Probability*,

to the topics of VSTs, STs, and BCTs, with an explanation of some general techniques and illustrative examples. The leading early literature includes Curtiss (1943), Bartlett (1947), Anscombe (1948), Hotelling (1953), and Fisher (1954). More recent key references include Bickel and Doksum (1981), Efron (1982), and Hall (1992). Also see Sprott (1984), Bar-Lev and Enis (1990), and DiCiccio and Stern (1994). The material in this chapter is also taken from Brown, Cai, and DasGupta (2006).

4.1 Variance-Stabilizing Transformations

A major use of the delta theorem is to construct VSTs. Unfortunately, the concept does not generalize to multiparameter cases, i.e., it is generally infeasible to find a dispersion-stabilizing transformation. It is, however, a useful tool in one-parameter problems.

Suppose T_n is a sequence of statistics such that $\sqrt{n}\,(T_n - \theta) \xrightarrow{\mathcal{L}} N(0, \sigma^2(\theta)), \sigma^2(\theta) > 0$. By the delta theorem, if $g(\cdot)$ is once differentiable at θ with $g'(\theta) \neq 0$, then

$$\sqrt{n}\,(g(T_n) - g(\theta)) \xrightarrow{\mathcal{L}} N(0, [g'(\theta)]^2\sigma^2(\theta)).$$

Therefore, if we want the variance in the asymptotic distribution of $g(T_n)$ to be constant, we set

$$[g'(\theta)]^2\sigma^2(\theta) = k^2$$

for some constant k. Thus, a way of deriving g from σ is

$$g(\theta) = k \int \frac{1}{\sigma(\theta)} d\theta$$

if $\sigma(\theta)$ is continuous in θ. k can obviously be chosen as any nonzero real number. In the above, the integral is to be interpreted as a primitive.

For such a $g(\cdot)$, $g(T_n)$ has an asymptotic distribution with a variance that is free of θ. Such a statistic or transformation of T_n is called a variance-stabilizing transformation. Note that the transformation is monotone. So, if we use $g(T_n)$ to make an inference for $g(\theta)$, then we can automatically retransform to make an inference for θ, which is the parameter of interest.

As long as there is an analytical formula for the asymptotic variance function in the limiting normal distribution for T_n, and as long as the reciprocal of its square root can be integrated in closed form, a VST can be written down. For one-parameter problems, it is a fairly general tool, and there are no complicated ideas involved. It is important to remember, however, that the VST $g(T_n)$ may have large biases and its variance may not actually be nearly a constant unless n is adequately large. Consequently, we may wish to suitably adjust the VST in order to reduce its bias and/or make its variance more nearly a constant. We come to these later. First, we work out some examples of VSTs and show how they are used to construct asymptotically correct confidence intervals for an original parameter of interest.

4.2 Examples

Example 4.1 Suppose $X_1, X_2, \ldots$ are iid Poisson(θ). Then $\sqrt{n}(\overline{X}_n - \theta) \xrightarrow{\mathcal{L}} N(0, \theta)$. Thus $\sigma(\theta) = \sqrt{\theta}$ and so a variance-stabilizing transformation is

$$g(\theta) = \int \frac{k}{\sqrt{\theta}} d\theta = 2k\sqrt{\theta}.$$

Taking $k = 1/2$ gives that $g(\theta) = \sqrt{\theta}$ is a variance-stabilizing transformation for the Poisson case. Indeed $\sqrt{n}(\sqrt{\overline{X}_n} - \sqrt{\theta}) \xrightarrow{\mathcal{L}} N(0, \frac{1}{4})$. Thus, an asymptotically correct confidence interval for $\sqrt{\theta}$ is

$$\sqrt{\overline{X}_n} \pm \frac{z_{1-\alpha/2}}{2\sqrt{n}},$$

where z_α is the αth percentile of the standard normal distribution. This implies that an asymptotically correct confidence interval for θ is

$$\left(\left(\sqrt{\overline{X}_n} - \frac{z_{1-\alpha/2}}{2\sqrt{n}} \right)^2, \left(\sqrt{\overline{X}_n} + \frac{z_{1-\alpha/2}}{2\sqrt{n}} \right)^2 \right).$$

If $\sqrt{\overline{X}_n} - z_{1-\alpha/2}/(2\sqrt{n}) < 0$, that expression should be replaced by 0. This confidence interval is different from the more traditional interval, namely $\bar{X} \pm \frac{z_{1-\alpha/2}}{\sqrt{n}} \sqrt{\bar{X}}$, which goes by the name of the *Wald interval*. The actual coverage properties of the interval based on the VST are significantly better than those of the Wald interval, but we do not go into that issue in detail.

Example 4.2 Suppose $X_n \sim \text{Bin}(n, p)$. Then $\sqrt{n}(X_n/n - p) \xrightarrow{\mathcal{L}} N(0, p(1-p))$. So $\sigma(p) = \sqrt{p(1-p)}$ and consequently, on taking $k = \frac{1}{2}$,

$$g(p) = \int \frac{1/2}{\sqrt{p(1-p)}} dp \;=\; \arcsin(\sqrt{p}).$$

Hence, $g(X_n) = \arcsin(\sqrt{X_n/n})$ is a variance-stabilizing transformation and indeed

$$\sqrt{n}\left(\arcsin\left(\sqrt{\frac{X_n}{n}}\right) - \arcsin\left(\sqrt{p}\right)\right) \xrightarrow{\mathcal{L}} N\left(0, \frac{1}{4}\right).$$

Thus, an asymptotically correct confidence interval for p is

$$\sin^2\left(\arcsin\left(\sqrt{\frac{X_n}{n}}\right) \mp \frac{z_{1-\alpha/2}}{2\sqrt{n}}\right).$$

Again, this interval has significantly better coverage properties than the traditional Wald interval $\frac{X_n}{n} \pm \frac{z_{1-\alpha/2}}{\sqrt{n}}\sqrt{\frac{X_n}{n}(1-\frac{X_n}{n})}$. Brown, Cai, and DasGupta (2001) demonstrate this in detail.

Example 4.3 Suppose $X_1, \ldots, X_n$ are iid with a Gamma density $\Gamma(\alpha)^{-1}$ $\theta^{-\alpha}\exp(-x/\theta)$ with $\alpha > 0$ known. Then $\sqrt{n}(\overline{X}_n - \alpha\theta) \xrightarrow{\mathcal{L}} N(0, \alpha\theta^2)$. This gives

$$g(\theta) = \int \frac{1}{\sigma(\theta)} d\theta = \int \frac{1}{\sqrt{\alpha}\theta} d\theta = \frac{\log\theta}{\sqrt{\alpha}}.$$

Thus, the log transformation is variance stabilizing for the Gamma case and

$$\sqrt{n}(\log\overline{X}_n - \log\alpha\theta) \xrightarrow{\mathcal{L}} N\left(0, \frac{1}{\alpha}\right).$$

Example 4.4 Suppose (X_i, Y_i), $i = 1, \ldots, n$, are iid bivariate normal with parameters $\mu_X, \mu_Y, \sigma_X^2, \sigma_Y^2, \rho$. Then, as we saw before, $\sqrt{n}(r_n - \rho) \xrightarrow{\mathcal{L}} N(0, (1-\rho^2)^2)$, where r_n is the sample correlation coefficient. Thus

$$g(\rho) = \int \frac{1}{(1-\rho)^2} d\rho = \frac{1}{2}\log\frac{1+\rho}{1-\rho} = \operatorname{arctanh}(\rho)$$

is a variance-stabilizing transformation for r_n. This is the famous arctanh transformation of Fisher, popularly known as *Fisher's z*. By the delta theorem, $\sqrt{n}(\operatorname{arctanh}(r_n) - \operatorname{arctanh}(\rho))$ converges in law to the $N(0, 1)$

distribution. Confidence intervals for ρ are computed from the arctanh transformation as

$$\tanh\left(\operatorname{arctanh}(r_n) \pm \frac{z_\alpha}{\sqrt{n}}\right)$$

rather than by using the asymptotic distribution of r_n itself. The arctanh transformation of r_n attains normality much quicker than r_n itself.

Example 4.5 For estimating the variance of a normal distribution, there is an exact $100(1-\alpha)\%$ confidence interval based on the fact that $(n-1)s^2 \sim \chi^2_{n-1}$. The equal-tailed interval is given by $\frac{(n-1)s^2}{\chi^2_{\alpha/2,n-1}} \le \sigma^2 \le \frac{(n-1)s^2}{\chi^2_{1-\alpha/2,n-1}}$.

On the other hand, from the result that $\sqrt{n}(s^2-\sigma^2) \overset{\mathcal{L}}{\Rightarrow} N(0, 2\sigma^4)$, it follows by the delta theorem that $\sqrt{n}(\log s - \log\sigma) \overset{\mathcal{L}}{\Rightarrow} N(0, \frac{1}{2})$. Therefore, an asymptotically correct $100(1-\alpha)\%$ confidence interval for $\log\sigma$ is $\log s \pm z_{\alpha/2}\frac{1}{\sqrt{2n}}$, and hence an asymptotically correct $100(1-\alpha)\%$ confidence interval for $\log\sigma^2$ is $\log s^2 \pm z_{\alpha/2}\sqrt{\frac{2}{n}}$. Exponentiating, an asymptotically correct $100(1-\alpha)\%$ confidence interval for σ^2 is $s^2/e^{z_{\alpha/2}\sqrt{2/n}} \le \sigma^2 \le s^2 e^{z_{\alpha/2}\sqrt{2/n}}$.

The coverage probability of this interval is not exactly $(1-\alpha)$ for fixed n. However, in contrast, the interval based on the χ^2 distribution for $(n-1)s^2/\sigma^2$ is *exact*; i.e., for any n, and under all values of the parameters, it has exactly a $(1-\alpha)$ coverage probability. This example has a feature that was not shared by the preceding examples: there is an invariance structure in this example. In problems with such an invariance structure, there is no clear need for using VSTs to obtain confidence intervals, even though they are available.

Example 4.6 This is an uncommon example. The $N(\theta,\theta)$ distribution, the normal distribution with an equal mean and variance, is a continuous analog of the Poisson distribution. It is easily seen that $\sum X_i^2$ is the minimal sufficient statistic, and by a straightforward calculation, the MLE of θ is $\hat{\theta} = \frac{\sqrt{1+4c_2}-1}{2}$, where $c_2 = \frac{1}{n}\sum X_i^2$. From standard asymptotic theory for maximum likelihood estimates (Chapter 16 in this book), it follows that $\sqrt{n}(\hat{\theta}-\theta) \overset{\mathcal{L}}{\Rightarrow} N(0, \frac{2\theta^2}{2\theta+1})$. Therefore, the VST is given by the function $\xi = g_v(\theta) = \int \frac{\sqrt{2\theta+1}}{\theta} = 2[\sqrt{2\theta+1} - \operatorname{arccoth}(\sqrt{2\theta+1})] = 2[\sqrt{2\theta+1} - \operatorname{arctanh}((2\theta+1)^{-\frac{1}{2}})]$ on actually doing the primitive calculation and on using a well-known trigonometric identity between arctanh and arccoth. Finally, plugging in the MLE $\hat{\theta}$ for θ, the VST is $g_v(\hat{\theta}) = (1+4c_2)^{1/4} - \text{arctanh}((1+4c_2)^{-1/4})$, and $\sqrt{n}(g_v(\hat{\theta}) - \xi) \stackrel{\mathcal{L}}{\Rightarrow} N(0, 1/2)$.

Example 4.7 Here is a *nonregular* example on the idea of variance stabilization. Suppose we have iid observations $X_1, X_2, \ldots$ from the $U[0, \theta]$ distribution. Then, the MLE of θ is the sample maximum $X_{(n)}$, and $n(\theta - X_{(n)}) \stackrel{\mathcal{L}}{\Rightarrow}$ Exp(θ). The asymptotic variance function in the distribution of the MLE is therefore simply θ^2, and therefore the primitive of the reciprocal of the asymptotic standard deviation is $\int \frac{1}{\theta} d\theta = \log\theta$. So, a straight application of the VST methodology produces the transformation $g_v(X_{(n)}) = \log X_{(n)}$. It is easily verified that $n(\log\theta - \log X_{(n)}) \stackrel{\mathcal{L}}{\Rightarrow}$ Exp(1). However, the interesting fact is that, for every n, the distribution of $n(\log\theta - \log X_{(n)})$ is *exactly* an exponential with mean 1. There is no nontrivial example such as this in the regular cases (the $N(\theta, 1)$ is a trivial example).

4.3 Bias Correction of the VST

We remarked before that, in finite samples, the VST generally has a bias, and its variance is not a constant. Intuitively, correcting the VST for its bias, or correcting it to make it have a variance that is more approximately a constant, should lead to better inference, an idea that goes back to Anscombe (1948).

Let $X \sim$ Bin(n, p). We know that the transformation $\arcsin\sqrt{\frac{X}{n}}$ is a variance-stabilizing transformation in this case. Anscombe (1948) pointed out that the transformation $g(X) = \sqrt{\arcsin}\sqrt{\frac{X+3/8}{n+3/4}}$ is a *better* VST in the sense that an asymptotic expansion for its variance works out to Var$[g(X)] = 1/(4n) + O(n^{-3})$, although for the more traditional VST $\arcsin\sqrt{\frac{X}{n}}$, an asymptotic expansion for the variance would be of the form $1/(4n) + c_1(p)/n^2 + O(n^{-3})$; i.e., the first nontrivial term in the expansion for the variance is nonzero and actually depends on p. The traditional VST *does not* have the second-order variance stabilization property, but the new one suggested by Anscombe does.

Likewise, let X_i iid Poi(λ). We know that in this case $\sqrt{\bar{X}}$ is a variance-stabilizing transformation in the sense $\sqrt{n}(\sqrt{\bar{X}} - \sqrt{\lambda}) \stackrel{\mathcal{L}}{\Rightarrow} N(0, 1/4)$. Anscombe (1948) pointed out that $\sqrt{\bar{X} + 3/(8n)}$ is a better VST in the sense that an asymptotic expansion for its variance works out to Var$[g(X)] = 1/(4n) + O(n^{-3})$, although for the more traditional VST $\sqrt{\bar{X}}$, an asymptotic expansion for the variance would be of the form $1/(4n) + c_1(\lambda)/n^2 + O(n^{-3})$;

i.e., again, the first nontrivial term in the expansion for the variance is nonzero and actually depends on λ. The traditional VST does not, once again, have the second-order variance stabilization property, but the one suggested by Anscombe does.

Although Anscombe did not explicitly point it out, it is not very hard to show that the traditional VST in each of these two cases also has a second-order bias problem. In the binomial case, $E[\arcsin(\sqrt{\bar{X}})] = \arcsin(\sqrt{p}) + (2p-1)/(8n\sqrt{pq}) + O(n^{-2})$, while in the Poisson case, $E[\sqrt{\bar{X}}] = \sqrt{\lambda} - 1/(8n\sqrt{\lambda}) + O(n^{-2})$. For greater variance stabilization, Anscombe had used $\arcsin\sqrt{\frac{X+3/8}{n+3/4}}$ in the binomial case and $\sqrt{\bar{X}+3/(8n)}$ in the Poisson case. Brown, Cai, and DasGupta (2006) show that in the binomial case $E[\arcsin(\sqrt{\frac{X+1/4}{n+1/2}})] = \arcsin(\sqrt{p}) + O(n^{-2})$, and in the Poisson case, $E[\sqrt{\bar{X}+1/(4n)}] = \sqrt{\lambda} + O(n^{-2})$. Thus, by using sets of constants different from Anscombe's, second-order bias correction can also be achieved. However, there are no sets of constants that can simultaneously achieve second-order bias *and* variance correction.

The method that Anscombe used, namely perturbing by a constant, is usually not productive. A more natural method of bias correction is to use the following method, which we simply outline. Suppose that in a generic problem $\hat{\theta}$ is an estimate for a parameter θ and that one can write expansions for the mean and variance of $\hat{\theta}$ such as $E(\hat{\theta}) = \theta + b(\theta)/n + O(n^{-2})$, $\mathrm{Var}(\hat{\theta}) = \frac{\sigma^2(\theta)}{n}(1 + c(\theta)/n) + O(n^{-3})$. Then, by Taylor series expansions, it *may be* possible to write corresponding expansions for the mean and variance of a transformation $g(\hat{\theta})$ as

$$E[g(\hat{\theta})] = g(\theta) + (g'(\theta)b(\theta) + \sigma^2(\theta)g''(\theta)/2)/n + O(n^{-2})$$

and

$$\begin{aligned}\mathrm{Var}[g(\hat{\theta})] = & \frac{(g'(\theta))^2\sigma^2(\theta)}{n} + 1/n^2 \times \{(g'(\theta))^2\sigma^2(\theta)c(\theta) \\ & +(g''(\theta))^2\sigma^4(\theta)/2\} + O(n^{-3}).\end{aligned}$$

They are derived in Brown, Cai, and DasGupta (2006). These are used to find second-order bias-corrected and variance-stabilized transforms. We mention only the bias-correction result.

Theorem 4.1 Let $g_v(\theta) = \int \frac{1}{\sigma(\theta)} d\theta$. Then, under various conditions, a second-order bias-corrected VST is $g_{bcv}(\hat{\theta}) = g_v(\hat{\theta}) + \frac{\sigma'(\hat{\theta})}{2} - \frac{b(\hat{\theta})}{\sigma(\hat{\theta})}$.

We now give two examples of bias-corrected VSTs and discuss how they improve on the quality of inference in these two examples.

Example 4.8 The first example is on Fisher's z, given by $z = \frac{1}{2}\log[\frac{1+r}{1-r}]$. As we showed before, this is a VST for the correlation parameter in the bivariate normal case. However, it has a second-order bias; i.e., the $\frac{1}{n}$ term in the bias of z does not vanish.

To use Theorem 4.1, we need an expansion for the expectation of r. This turns out to be $E(r) = \rho - \frac{\rho(1-\rho^2)}{2n} + O(n^{-2})$. We will not provide a proof of this, but it is classic and is available, For example, in Hotelling (1953). From here, a straightforward application of Theorem 4.1 leads to the bias-corrected VST $z - \frac{r}{2n}$.

How does this improve on z itself? As an example, consider the testing problem $H_0 : \rho = 0$ vs. $H_1 : \rho > 0$. We can conduct the test by using z itself or the bias-corrected version $z - \frac{r}{2n}$. The critical region of each test based on their asymptotic normal distributions is of the form $r > r(\alpha, n)$, α being the nominal level; $r(\alpha, n)$ depends only on the null distribution and is easily computable. Therefore, one can find the actual type I error probability of each test and compare it with the nominal level as a measure of the level accuracy of the tests. For example, if the nominal level is 5%, then the actual type I error rate of the test that uses z itself is .068 for $n = 10$ and .062 for $n = 20$. If we use the bias-corrected VST $z - \frac{r}{2n}$, the corresponding type I error rates are .06 and .055 (Brown, Cai, and DasGupta (2006)). The bias-corrected VST offers some improvement in level accuracy.

Example 4.9 In this example, the bias-corrected VST gives a more striking improvement in level accuracy than in the previous example. Consider the $N(\theta, \theta)$ example. We saw earlier that the VST is given by $g_v(\hat{\theta}) = (1 + 4c_2)^{1/4} - \text{arctanh}((1 + 4c_2)^{-1/4})$, where $c_2 = \frac{1}{n}\sum X_i^2$. Again, this has a second-order bias. On using Theorem 4.1, the bias-corrected version works out to $g_{bcv}(\hat{\theta}) = g_v(\hat{\theta}) + \frac{1+3\hat{\theta}}{2(2\hat{\theta}+1)^{3/2}n}$ (see Brown, Cai, and DasGupta (2006)).

Once again, consider the testing problem $H_0 : \theta = \theta_0$ vs. $H_1 : \theta > \theta_0$ for comparing the VST with the bias-corrected VST. The critical region of the tests is of the form $c_2 > c(\theta_0, \alpha, n)$. The type I error rates can be found by using noncentral chi-square probabilities(Brown, Cai, and DasGupta (2006)). A table of the actual type I error rate of each test is given below for the 5% and the 1% cases for selected values of n; θ_0 is taken to be 1. The test based on the VST has a serious level inaccuracy, and it is striking how much improvement the bias-corrected version provides.

n	VST(5%)	Bias Corrected(5%)	VST(1%)	Bias Corrected(1%)
10	.028	.043	.003	.005
30	.036	.045	.005	.007
100	.042	.047	.007	.008
300	.045	.048	.008	.009
500	.046	.049	.009	.009
1000	.047	.049	.009	.009

Remark. It is usually the case that a bias correction of the VST leads to better inference than using the VST itself. Bias correction is a good general principle.

4.4 Symmetrizing Transformations

Often in statistical applications a statistic $\hat{\theta}$ has an asymptotically normal distribution. When we apply the limiting distribution as an approximation to the actual distribution of $\hat{\theta}$, except in very special cases, the skewness in the actual distribution of $\hat{\theta}$ is ignored. The seriousness of the omission depends on the amount of skewness truly present in the finite sample distribution. If there are transforms $g(\hat{\theta})$ that have, comparatively speaking, less skewness in their finite sample distributions, then the normal approximation should be better for $g(\hat{\theta})$ than for $\hat{\theta}$ (see Hall (1992)). Symmetrizing transformations are such that, in an asymptotic expansion for their skewness coefficient, the first nontrivial term disappears. We make it precise below. One feature of symmetrizing transforms is that they are not variance-stabilizing. The two goals are intrinsically contradictory. Also, symmetrizing transforms generally have a second-order bias. So, a bias correction of a symmetrizing transform may lead to further improvement in the quality of inference but at the expense of increasing formal complexity. We need some notation to define symmetrizing transformations.

Let T_n be a sequence of asymptotically normal statistics with $\sqrt{n}(T_n - \theta) \stackrel{\mathcal{L}}{\Rightarrow} N(0, \sigma^2(\theta))$. Let $g(.)$ be a (sufficiently smooth) function and let $Y_n = \sqrt{n}(g(T_n) - g(\theta))$. Suppose T_n admits the moment expansions

$$E(T_n - \theta) = b(\theta)/n + O(n^{-2}),$$
$$\text{Var}(T_n) = \sigma^2(\theta)/n \times (1 + c(\theta)/n) + O(n^{-3}),$$

and

$$E(T_n - \theta)^3 = d_{31}(\theta)/n^2 + d_{32}(\theta)/n^3 + O(n^{-4}).$$

Let $d(\theta) = \frac{d_{31}(\theta)}{\sigma^3(\theta)}$ and let $\beta_3(n, \theta) = \frac{E(Y_n - EY_n)^3}{(\text{Var}(Y_n))^{3/2}}$ denote the skewness coefficient of Y_n. The definition of a symmetrizing transform is based on the following asymptotic expansion for $\beta_3(n, \theta)$. The derivation is available in Brown, Cai, and DasGupta (2006).

Theorem 4.2 Under various conditions, $\beta_3(n, \theta) = \frac{1}{\sqrt{n}} \times \text{sgn}(g'(\theta)\sigma(\theta))$ $[d(\theta) - 3\frac{b(\theta)}{g'(\theta)\sigma(\theta)} + 3\sigma(\theta)\frac{g''(\theta)}{g'(\theta)}] + O(n^{-3/2})$.

If the transform $g(.)$ is such that the coefficient of the $\frac{1}{\sqrt{n}}$ term vanishes for all values of θ, then it is called a symmetrizing transform. Here then is the definition.

Definition 4.1 In the notation given above, $g(T_n)$ is a symmetrizing transform iff $d(\theta) - 3\frac{b(\theta)}{g'(\theta)\sigma(\theta)} + 3\sigma(\theta)\frac{g''(\theta)}{g'(\theta)} = 0 \, \forall \theta$.

It can be proved that symmetrizing transforms that would be monotone in T_n are characterized by solutions, if any exist, of a second-order nonhomogeneous linear differential equation. The general solution of this differential equation is available in Brown, Cai, and DasGupta (2006), but we will not present it here. We specialize to the case of the one-parameter exponential family, where, it turns out, there is essentially a unique symmetrizing transform, and it is relatively simple to describe it. Here is the result of Brown, Cai, and DasGupta (2006).

Theorem 4.3 Let $X \sim f(x|\theta) = e^{\theta x - \psi(\theta)} h(x)(d\nu)$ for some σ-finite measure ν. Let $\mu = \mu(\theta) = E_\theta(X) = \psi'(\theta)$. Let $k(\mu)$ be the function defined as $k(\mu) = (\psi')^{-1}(\mu)$. Then, for estimating μ, the symmetrizing transforms are of the form $g_s(\bar{X}) = C_1 \int (\psi''(\theta))^{2/3} d\theta|_{\theta = k(\bar{X})} + C_2$ for constants C_1, C_2.

Now that we have a general representation of the symmetrizing transform in the entire one-parameter exponential family, we can work out some examples and observe that the symmetrizing transforms are different from the VSTs.

Example 4.10 Let $X \sim$ Poisson(μ), where μ is the mean. In the general notation above, in this case $\theta = \log \mu$, $\psi(\theta) = e^\theta$, and $k(\mu) = \log \mu$. Therefore, essentially the only symmetrizing transform of $\bar{X}$ is $g_s(\bar{X}) = \bar{X}^{2/3}$. In contrast, as we saw before, the VST is $g_v(\bar{X}) = \bar{X}^{1/2}$. Both transforms are fractional powers of $\bar{X}$ but different powers.

Example 4.11 Let $X \sim$ Ber(μ). So μ stands for just the traditional parameter p. In this case, $\theta = \log(\frac{\mu}{1-\mu})$, $\psi(\theta) = \log(1 + e^\theta)$, and $k(\mu) = \log(\frac{\mu}{1-\mu})$.

On using the general form of the symmetrizing transform of $\bar{X}$ in the exponential family, one gets that essentially the only symmetrizing transform is $g(\bar{X}) = F(\frac{2}{3}, \frac{2}{3}, \bar{X})$, where $F(\alpha, \beta, t)$ denotes the CDF of a Beta(α, β) density. In contrast, the VST, namely the arcsine-square root transform, is also in fact the CDF of a Beta$(\frac{1}{2}, \frac{1}{2})$ density. Thus, the VST and the symmetrizing transform are both Beta CDFs but with different parameters for the Beta distribution. This is interesting.

Example 4.12 In the Gamma case, we saw earlier that the log transform stabilizes variance. An exercise asks to show that the symmetrizing transform is the cube root transform of $\bar{X}$. This is the well-known *Wilson-Hilferty transform* for the Gamma distribution.

4.5 VST or Symmetrizing Transform?

Since the goals of stabilizing the variance and reducing the skewness are intrinsically contradictory and the corresponding transforms are different, the question naturally arises if one is better than the other. No general statements can be made. In Brown, Cai, and DasGupta (2006), it is demonstrated that a plain VST usually is not a very good performer. It often has quite a bit of bias. A bias correction of the VST seems to be the least that one should do. In the Poisson case, a bias-corrected VST seems to produce more accurate inference than the symmetrizing transform. But in the Gamma case, the symmetrizing transform is slightly better than even the bias-corrected VST. No studies seem to have been made that compare bias-corrected VSTs to bias-corrected symmetrizing transforms.

4.6 Exercises

Exercise 4.1 * Derive an expansion with an error of $O(n^{-2})$ for Var$(\sqrt{\bar{X}})$ if $X_1, X_2, \ldots$ are iid Poi(θ).

Exercise 4.2 Write an expression for the true coverage of the confidence interval for θ constructed from the variance-stabilizing transformation $\sqrt{\bar{X}}$ under Poi(θ) and plot it for $n = 5, 15, 30$, and $\alpha = 0.05$.

Exercise 4.3 Let $X \sim$ Bin(n, p). Is $\arcsin\sqrt{\frac{X}{n}}$ unbiased for $\arcsin\sqrt{p}$?

Exercise 4.4 Suppose $X_1, \ldots, X_n$ are iid $N(\mu, \mu^2)$, $\mu > 0$. Find a variance-stabilizing transformation for μ.

Exercise 4.5 Suppose $X_1, X_2, \ldots$ are iid Exp(1). What would you guess $\lim E(\log \bar{X})$ to be? Find an exact expression for $E(\log \bar{X})$ and derive its limit. Was your guess right?

Exercise 4.6 * Derive an expansion up to $O(n^{-2})$ for the variance of the Anscombe transformation in the Poisson case. Compare it with Exercise 4.1.

Exercise 4.7 * Suppose X_i are iid with the Gamma density $\frac{e^{-x}x^{\alpha-1}}{\Gamma(\alpha)}, \alpha > 0$ being unknown. How would you go about constructing a variance-stabilizing transformation?

Exercise 4.8 * Suppose X_i are iid $N(\mu, \mu), \mu > 0$. In the text, a specific variance-stabilizing transformation was described. How would you go about constructing other variance-stabilizing transformations? More precisely, it is possible to start with various statistics $\hat{\mu}$ and transform them for variance-stabilization. Think of a few such statistics and investigate the corresponding variance-stabilizing transformations.

Exercise 4.9 Let $X_1, \ldots, X_n \overset{\text{iid}}{\sim} N(\mu, \sigma^2)$. Consider the equal-tailed exact confidence interval and the asymptotically correct confidence interval of Example 4.5. Compare their coverages and lengths at some selected values of n, σ^2 by means of a simulation.

Exercise 4.10 * Let $X_1, X_2, \ldots \overset{\text{iid}}{\sim}$ Poi(θ). Derive an expansion up to $O(n^{-1})$ for the mean and an expansion up to $O(n^{-2})$ for the variance of the symmetrizing transformation in the Poisson case. Compare it with Exercises 4.1 and 4.6.

Exercise 4.11 * Let $X_1, X_2, \ldots \overset{\text{iid}}{\sim}$ Poi(θ). Using Theorem 4.2, derive an expansion up to $O(n^{-\frac{1}{2}})$ for the skewness coefficient of the VST $\sqrt{\bar{X}}$. Verify that the $n^{-\frac{1}{2}}$ term does not vanish.

Exercise 4.12 * Let $X \sim$ Bin(n, p). Using Theorem 4.2, derive an expansion up to $O(n^{-\frac{1}{2}})$ for the skewness coefficient of the VST $\arcsin(\sqrt{\frac{X}{n}})$. Verify that the $n^{-\frac{1}{2}}$ term does not vanish.

References

Anscombe, F. (1948). Transformation of Poisson, Binomial, and Negative Binomial data, Biometrika, 35, 246–254.

Bar-Lev, S. and Enis, P. (1990). On the construction of classes of variance stabilizing transformations, Stat. Prob. Lett., 2, 95–100.

Bartlett, M.S. (1947). The use of transformations, Biometrics, 3, 39–52.

Bickel, P. and Doksum, K. (1981). An analysis of transformations revisited, J. Am. Stat. Assoc., 76(374), 296–311.

Brown, L., Cai, T. and DasGupta, A. (2001). Interval estimation for a binomial proportion, Statist. Sci., 16, 2, 101–133.

Brown, L., Cai, T., and DasGupta, A. (2006). On selecting an optimal transformation, preprint.

Curtiss, J.H. (1943). On transformations used in the analysis of variance, Ann. Math. Stat., 14, 107–122.

DiCiccio, T. and Stern, S.E. (1994). Constructing approximately standard normal pivots from signed roots of adjusted likelihood ratio statistics, Scand. J. Stat., 21(4), 447–460.

Efron, B. (1982). Transformation theory: How normal is a family of distributions?, Ann. Stat., 10(2), 323–339.

Fisher, R.A. (1954). The analysis of variance with various binomial transformations, Biometrics, 10, 130–151.

Hall, P. (1992). On the removal of skewness by transformation, J. R. Stat. Soc. Ser. B, 54(1), 221–228.

Hotelling, H. (1953). New light on the correlation coefficient and its transforms, J. R. Stat. Soc. Ser. B, 15, 193–232.

Sprott, D.A. (1984). Likelihood and maximum likelihood estimation, C. R. Math. Rep. Acad. Sci. Can., 6(5), 225–241.

Chapter 5
More General Central Limit Theorems

Theoretically, as well as for many important applications, it is useful to have CLTs for partial sums of random variables that are independent but not iid. We present a few key theorems in this chapter. We also address possibilities of CLTs without a finite variance and CLTs for *exchangeable sequences*. A nearly encyclopedic reference is Petrov (1975). Other useful references for this chapter are Feller (1966), Billingsley (1995), Lehmann (1999), Ferguson (1996), Sen and Singer (1993), and Port (1994). Other specific references are given later.

5.1 The Independent Not IID Case and a Key Example

Theorem 5.1 (Lindeberg-Feller) Suppose $\{X_n\}$ is a sequence of independent variables with $E(X_n) = \mu_n$ and Var $(X_n) = \sigma_n^2 < \infty$. Let $s_n = \sqrt{\sum_{i=1}^n \sigma_i^2}$. If for any $\epsilon > 0$

$$\frac{1}{s_n^2} \sum_{j=1}^{n} \int_{|x|>\epsilon s_n} x^2 dF_j(x) \to 0,$$

where F_j is the CDF of X_j, then

$$\frac{\sum_{i=1}^{n}(X_i - \mu_i)}{s_n} \xrightarrow{\mathcal{L}} N(0, 1).$$

Remark. A proof can be seen in Billingsley (1995). It can be shown that the condition

A. DasGupta, *Asymptotic Theory of Statistics and Probability*,

$$\frac{1}{s_n^2}\sum_{j=1}^{n}\int_{|x|>\epsilon s_n} x^2 dF_j(x) \to 0$$

is equivalent to

$$\frac{1}{s_n^2}\sum_{j=1}^{n}\int_{|x|>\epsilon s_j} x^2 dF_j(x) \to 0.$$

This condition is called the Lindeberg-Feller condition. The Lindeberg-Feller theorem is a landmark theorem in probability and statistics. Generally, it is hard to verify the Lindeberg-Feller condition. A simpler theorem is the following.

Theorem 5.2 (Liapounov) Suppose $\{X_n\}$ is a sequence of independent random variables and for some $\delta > 0$

$$\frac{1}{s_n^{2+\delta}}\sum_{j=1}^{n} E|X_j - \mu_j|^{2+\delta} \to 0 \text{ as } n \to \infty.$$

Then

$$\frac{\sum_{i=1}^{n}(X_i - \mu_i)}{s_n} \xrightarrow{\mathcal{L}} N(0, 1).$$

Remark. A proof is given in Sen and Singer (1993). If $s_n \to \infty$ and $\sup_{j\geq 1} E|X_j - \mu_j|^{2+\delta} < \infty$, then the condition of Liapounov's theorem is satisfied. In practice, usually one tries to work with $\delta = 1$ or 2 for algebraic convenience.

A consequence especially useful in regression is the following theorem, which is also proved in Sen and Singer (1993).

Theorem 5.3 (Hájek-Sidak) Suppose $X_1, X_2, \ldots$ are iid random variables with mean μ and variance $\sigma^2 < \infty$. Let $c_n = (c_{n1}, c_{n2}, \ldots, c_{nn})$ be a vector of constants such that

$$\max_{1\leq i\leq n} \frac{c_{ni}^2}{\sum_{j=1}^{n} c_{nj}^2} \to 0 \text{ as } n \to \infty.$$

Then

$$\frac{\sum_{i=1}^{n} c_{ni}(Y_i - \mu)}{\sigma \sqrt{\sum_{j=1}^{n} c_{nj}^2}} \xrightarrow{\mathcal{L}} N(0, 1).$$

Remark. The condition that

$$\max_{1 \le i \le n} \frac{c_{ni}^2}{\sum_{j=1}^{n} c_{nj}^2} \to 0 \text{ as } n \to \infty$$

is to ensure that no coefficient dominates the vector c_n. For example, if $c_n = (1, 0, \ldots, 0)$, then the condition would fail and so would the theorem.

The Hájek-Sidak theorem has many applications, including in the regression problem. Here is an important example.

Example 5.1 Consider the simple linear regression model $y_i = \beta_0 + \beta_1 x_i + \epsilon_i$, where ϵ_i's are iid with mean 0 and variance σ^2 but are not necessarily normally distributed. The least squares estimate of β_1 based on n observations is

$$\hat{\beta}_1 = \frac{\sum_{i=1}^{n} (y_i - \overline{y}_n)(x_i - \overline{x}_n)}{\sum_{i=1}^{n} (x_i - \overline{x}_n)^2} = \beta_1 + \frac{\sum_{i=1}^{n} \epsilon_i (x_i - \overline{x}_n)}{\sum_{i=1}^{n} (x_i - \overline{x}_n)^2}$$

so

$$\hat{\beta}_1 - \beta_1 = \frac{\sum_{i=1}^{n} \epsilon_i c_{ni}}{\sum_{j=1}^{n} c_{nj}^2},$$

where $c_{ni} = (x_i - \overline{x}_n)$. Hence, by the Hájek-Sidak theorem,

$$\sqrt{\sum_{i=1}^{n}(x_i - \overline{x}_n)^2}\frac{\hat{\beta}_1 - \beta_1}{\sigma} = \frac{\sum_{i=1}^{n} \epsilon_i c_{ni}}{\sigma\sqrt{\sum_{j=1}^{n} c_{nj}^2}} \xrightarrow{\mathcal{L}} N(0, 1),$$

provided

$$\max_{1 \leq i \leq n} \frac{(x_i - \overline{x}_n)^2}{\sum_{j=1}^{n}(x_j - \overline{x}_n)^2} \to 0 \text{ as } n \to \infty.$$

For most reasonable designs, this condition is satisfied. Thus, the asymptotic normality of the LSE (least squares estimate) is established under some conditions on the design variables, an important result.

5.2 CLT without a Variance

It is not as widely known that existence of a variance is not necessary for asymptotic normality of partial sums of iid random variables. A CLT without a finite variance can sometimes be useful. We present the general result below and then give an illustrative example. Feller (1966) contains detailed information on the availability of CLTs without the existence of a variance, along with proofs. First, we need a definition.

Definition 5.1 A function $L : \mathcal{R} \to \mathcal{R}$ is called slowly varying at ∞ if, for every $t > 0$, $\lim_{x\to\infty} \frac{L(tx)}{L(x)} = 1$.

Remark. Examples of slowly varying functions are $\log x$, $\frac{x}{1+x}$, and indeed any function with a finite limit as $x \to \infty$. But, for example, x or e^{-x} are not slowly varying.

Theorem 5.4 Let $X_1, X_2, \ldots$ be iid from a CDF F on $\mathcal{R}$. Let $v(x) = \int_{[-x,x]} y^2 dF(y)$. Then, there exist constants $\{a_n\}$, $\{b_n\}$ such that $\frac{\sum_{i=1}^{n} X_i - a_n}{b_n} \stackrel{\mathcal{L}}{\Rightarrow} N(0, 1)$ if and only if $v(x)$ is slowly varying at ∞.

Remark. If F has a finite second moment, then automatically $v(x)$ is slowly varying at ∞. We present an example below where asymptotic normality of the sample partial sums still holds, although the summands do not have a finite variance.

Example 5.2 Suppose $X_1, X_2, \ldots$ are iid from a t-distribution with 2 degrees of freedom that has a finite mean but not a finite variance. The density of the t-distribution with 2 degrees of freedom is $f(y) = \frac{c}{(2+y^2)^{\frac{3}{2}}}$ for some positive c. Hence, by a direct integration, for some other constant k, $v(x) = k\sqrt{\frac{1}{2+x^2}}(x - \sqrt{2+x^2}\text{arcsinh}(\frac{x}{\sqrt{2}}))$. Therefore, on using the fact that $\text{arcsinh}(x) = \log 2x + O(x^{-2})$ as $x \to \infty$, we get, for any $t > 0$, $\frac{v(tx)}{v(x)} \sim \frac{\sqrt{2+x^2}}{\sqrt{2+t^2x^2}} \frac{tx - \sqrt{2+t^2x^2}\log(\sqrt{2}tx)}{x - \sqrt{2+x^2}\log(\sqrt{2}x)} \to 1$ as $x \to \infty$ on some algebra. It follows that for iid observations from a t-distribution with 2 degrees of freedom, on suitable centering and norming, the partial sums $S_n = \sum_{i=1}^n X_i$ converge to a normal distribution, although the X_i do not have a finite variance. The centering can be taken to be zero for the centered t-distribution; it can be shown that the norming required is $b_n = \sqrt{n \log n}$.

5.3 Combinatorial CLT

The following result is useful in finite population sampling and in certain problems in nonparametric testing. It is also of basic interest to probabilists. A proof can be seen in Hoeffding (1951). Port (1994) is another useful reference for combinatorial CLTs.

Theorem 5.5 Let $\pi = \pi_n$ denote a random permutation of $(1, 2, \ldots, n)$. Let $a_n(i), b_n(i), i = 1, 2, \ldots, n$, be two double arrays of constants. If, for all $r > 2$,

$$n^{\frac{r}{2}-1} \frac{\sum(a_n(i) - \bar{a}_n)^r}{(\sum(a_n(i) - \bar{a}_n)^2)^{r/2}} = O(1),$$

and if

$$\frac{\max_i (b_n(i) - \bar{b}_n)^2}{\sum(b_n(i) - \bar{b}_n)^2} = o(1),$$

then $\frac{S_n - E(S_n)}{\sqrt{\text{Var}(S_n)}} \xrightarrow{\mathcal{L}} N(0, 1)$, where $S_n = \sum_{i=1}^n a_n(i) b_n(\pi(i))$. Furthermore, with $c_n(i, j) = a_n(i) b_n(j)$, $1 \le i, j \le n$,

$$E(S_n) = \frac{1}{n} \sum \sum c_n(i, j)$$

and

$$\text{Var}(S_n) = \frac{1}{n-1} \sum \sum d_n^2(i, j),$$

where

$$d_n(i, j) = c_n(i, j) - \frac{1}{n}\sum_k c_n(k, j) - \frac{1}{n}\sum_k c_n(i, k) + \frac{1}{n^2}\sum_k\sum_l c_n(k, l).$$

Remark. It follows from this theorem, for example, that for a random permutation the sum $\sum_{i=1}^n i\pi(i)$ is asymptotically normally distributed. See the exercises.

Example 5.3 In many applications of this theorem, the sequence $a_n(i)$ is zero-one valued. Consider, for example, the familiar hypergeometric distribution, wherein an urn has n balls, D of which are black, and m are sampled at random without replacement. Let X denote the number of black balls among those sampled. We can represent X in the form $S_n = \sum_{i=1}^n a_n(i)b_n(\pi(i))$ in the sense that X and S_n have the same distribution. What one needs is a clever embedding in the random permutation setup. Thus consider random variables $X_1, \ldots, X_n$, where $X_i = j \Leftrightarrow \pi(j) = i$. Then, if we choose the sequence $a_n(i)$ to be $a_n(i) = I_{i \le D}$, and $b_n(j) = I_{j \le m}$, one has distributional equivalence of X and S_n. The conditions of the combinatorial central limit theorem hold if, for example, $m, n - m \to \infty$ and $\frac{D}{n} \to c$ with $0 < c < 1$. One can thus expect a hypergeometric random variable to be approximately normally distributed when the population size and sample size are large, the unsampled number is large, too, and the proportion of black balls is neither too small nor very close to 1.

5.4 CLT for Exchangeable Sequences

A very important generalization of the concept of an iid sequence is that of exchangeability. In statistics, for example, the natural interpretation of a sequence of sample observations for a *Bayesian* would be that they are exchangeable, as opposed to the *frequentist* interpretation that they are iid. Central limit theorems for exchangeable sequences bear some similarity and also a lot of differences from the iid situation. Some key references on the central limit results are Blum et al. (1958), Chernoff and Teicher (1958), and Klass and Teicher (1987). For expositions on exchangeability, we recommend Aldous (1983) and Diaconis (1988). Interesting examples can be seen in Diaconis and Freedman (1987). We first define the notion of exchangeability.

Definition 5.2 An infinite sequence $\{X_1, X_2, \cdots\}$ of random variables (on some probability space) is called exchangeable if, for any finite $n \geq 1$ and any n distinct positive integers $i_1, \cdots, i_n$, the joint distributions of $(X_{i_1}, \cdots, X_{i_n})$ and $(X_1, \cdots, X_n)$ are the same.

It follows from the definition that if $\{X_1, X_2, \cdots\}$ is an iid sequence, then it is exchangeable. However, the converse is not true. A famous theorem of de Finetti says that an infinite sequence of exchangeable Bernoulli random variables must be a mixture of an iid sequence of Bernoulli variables. More precisely, the theorem (de Finetti (1931)) says the following.

Theorem 5.6 Let $\{X_1, X_2, \cdots\}$ be an exchangeable sequence of $\{0, 1\}$-valued random variables. Then there exists a unique probability measure μ on $[0, 1]$ such that, for any $n \geq 1$, $P(X_1 = x_1, \cdots, X_n = x_n) = \int_{[0,1]} p^{\sum_{i=1}^n x_i}(1-p)^{n-\sum_{i=1}^n x_i} d\mu(p), \forall x_1, \cdots, x_n \in \{0, 1\}$.

Very general versions of this famous theorem of de Finetti have been obtained assuming some suitable compactness structure for the state space of the random variables; an early reference is Hewitt and Savage (1955).

We only treat the case of a finite variance. Here are two main CLTs for exchangeable sequences with a finite variance. Consult Klass and Teicher (1987) and Chernoff and Teicher (1958) for their proofs.

Theorem 5.7 (a) Let $\{X_1, X_2, \cdots\}$ be an exchangeable sequence with $E(X_1) = 0, \text{Var}(X_1) = 1$. Then $\sqrt{n}\bar{X} \stackrel{\mathcal{L}}{\Rightarrow} N(0, 1)$ iff $\text{cov}(X_1, X_2) = \text{cov}(X_1^2, X_2^2) = 0$.

(b) Let $\{X_1, X_2, \cdots\}$ be an exchangeable sequence with $E(X_1) = 0, \text{cov}(X_1, X_2) = \tau > 0$. Suppose for all $k \geq 1$ that $E(X_1 \cdots X_k) = E(Z^k)$, where $Z \sim N(0, 1)$. Then $\frac{\sum_{i=1}^n X_i}{n\tau} \stackrel{\mathcal{L}}{\Rightarrow} N(0, 1)$.

We see from Theorem 5.7 that convergence to normality can occur with normalization by n in the exchangeable (but not independent) case. There are other such differences from the iid case in the central limit theorem for exchangeable sequences.

In some important applications, the correct structure is of a triangular array of finitely exchangeable sequences rather than just one infinite sequence. The next result is on such an exchangeable array; we first need a definition.

Definition 5.3 A finite collection of random variables is called *interchangeable* if every subset has a joint distribution that is invariant under permutations.

Theorem 5.8 For each $n \geq 1$, suppose $\{X_{n,1}, \cdots, X_{n,k_n}\}$ are interchangeable random variables satisfying $\sum_{i=1}^{k_n} X_{n,i} = 0, \sum_{i=1}^{k_n} X_{n,i}^2 = k_n, n \geq 1$. Suppose

$$k_n \to \infty, 0 < m_n < k_n, \frac{m_n}{k_n} \to \alpha, 0 < \alpha < 1,$$

and

$$\max_{1 \leq i \leq k_n} \frac{X_{n,i}}{\sqrt{k_n}} \stackrel{P}{\Rightarrow} 0.$$

Then, $\frac{\sum_{i=1}^{m_n} X_{n,i}}{\sqrt{m_n}} \stackrel{\mathcal{L}}{\Rightarrow} N(0, 1 - \alpha)$.

Here is an interesting application of this theorem.

Example 5.4 Let $(R_1, \cdots, R_{k_n})$ be a random permutation of $(1, \cdots, k_n)$; i.e., the joint distribution of $(R_1, \cdots, R_{k_n})$ is uniform on the set of $k_n!$ permutations of $(1, \cdots, k_n)$. $(R_1, \cdots, R_{k_n})$ arise as ranks of the observed data in some distribution-free testing problems; see Chapter 24. The uniform distribution applies when the appropriate null hypothesis is true.

Let $X_{n,i} = \frac{R_i - (k_n+1)/2}{\sqrt{(k_n^2-1)/12}}$. Then, by simple algebra, $\sum_{i=1}^{k_n} X_{n,i} = 0, \sum_{i=1}^{k_n} X_{n,i}^2 = k_n$. Furthermore, $\max_{1 \leq i \leq k_n} \frac{X_{n,i}}{\sqrt{k_n}} = \frac{k_n-1}{2} \frac{\sqrt{12}}{\sqrt{k_n}\sqrt{k_n^2-1}} = O(\frac{1}{\sqrt{k_n}}) \to 0$ as $n \to \infty$. Therefore, by Theorem 5.8, if $\frac{m_n}{k_n} \to \alpha, 0 < \alpha < 1$, then $\frac{1}{\sqrt{m_n}} \sum_{i=1}^{m_n} \frac{R_i - (k_n+1)/2}{k_n} \stackrel{\mathcal{L}}{\Rightarrow} N(0, \frac{1-\alpha}{12})$. This can be used to test if the joint distribution of $(R_1, \cdots, R_{k_n})$ is indeed uniform, which, in turn, is a test for whether the appropriate null hypothesis holds. Note the important point that the statistic *truncates* the sum at some m_n less than k_n because the sum over all k_n indices would be zero by assumption, and so there cannot be a nondegenerate distribution.

5.5 CLT for a Random Number of Summands

The canonical CLT for the iid case says that if $X_1, X_2, \ldots$ are iid with mean zero and a finite variance σ^2, then the sequence of partial sums $S_n = \sum_{i=1}^{n} X_i$ obeys the central limit theorem in the sense $\frac{S_n}{\sigma\sqrt{n}} \stackrel{\mathcal{L}}{\Rightarrow} N(0, 1)$. There are some practical problems that arise in applications, for example in sequential statistical analysis, where the number of terms present in a partial sum is a random variable. Precisely, $\{N(t)\}, t \geq 0$, is a family of (nonnegative) integer-valued random variables, and we want to approximate the distribution of $S_{N(t)}$, where for each fixed n, S_n is still the sum of n iid variables as above. The question is whether a CLT still holds under appropriate conditions. It was shown in Anscombe (1952) and Rényi (1957) that a CLT still holds when the underlying time parameter $t \to \infty$ if $N(t)$ converges to ∞ as $t \to \infty$ in a sense to be described below. If we have problems more general than the iid case with a finite variance, then the theorem does need another condition on the underlying random variables $X_1, X_2, \ldots$; here is the Anscombe-Rényi theorem.

Theorem 5.9 Let $X_1, X_2, \ldots$ be iid with mean zero and a finite variance σ^2, and let $\{N(t)\}, t \geq 0$, be a family of nonnegative integer-valued random variables all defined on a common probability space. Assume that $\frac{N(t)}{t} \stackrel{P}{\Rightarrow} c, 0 < c < \infty$, as $t \to \infty$. Then,

$$\frac{S_{N(t)}}{\sigma\sqrt{N(t)}} \stackrel{\mathcal{L}}{\Rightarrow} N(0,1)$$

as $t \to \infty$.

Example 5.5 Consider the so-called coupon collection problem in which a person keeps purchasing boxes of cereals until she obtains a full set of some n coupons. The assumptions are that the boxes have an equal probability of containing any of the n coupons mutually independently. Suppose that the costs of buying the cereal boxes are iid with some mean μ and some variance σ^2. If it takes N_n boxes to obtain the complete set of all n coupons, then $\frac{N_n}{n \log n} \stackrel{P}{\Rightarrow} 1$ as $n \to \infty$ (see Exercise 3.12). The total cost to the customer to obtain the complete set of coupons is $S_{N_n} = X_1 + \ldots + X_{N_n}$ with obvious notation. By the Anscombe-Rényi theorem and Slutsky's theorem, we have that $\frac{S_{N_n} - N_n\mu}{\sigma\sqrt{n \log n}}$ is approximately $N(0, 1)$.

5.6 Infinite Divisibility and Stable Laws

Infinitely divisible distributions were introduced by de Finetti in 1929, and the most fundamental results were developed by Kolmogorov, Lévy, and Khintchine in the 1930s. The area has since continued to flourish, and a huge body of deep and elegant results now exist in the literature. We provide a short account of infinitely divisible distributions on the real line. Infinitely divisible and stable distributions are extensively used in applications, but they are also fundamentally related to the question of convergence of distributions of partial sums of independent random variables. Three review papers on the material of this section are Fisz (1962), Steutel (1979), and Bose, DasGupta, and Rubin (2004). Feller (1966) is a classic reference on infinite divisibility and stable laws.

Definition 5.4 A real-valued random variable X with cumulative distribution function F and characteristic function ϕ is said to be infinitely divisible if, for each $n > 1$, there exist iid random variables $X_1, \ldots, X_n$ with CDF, say F_n, such that X has the same distribution as $X_1 + \ldots + X_n$.

The following important property of the class of infinitely divisible distributions describes the connection of infinite divisibility to possible weak limits of partial sums of independent random variables.

Theorem 5.10 For $n \geq 1$, let $\{X_{ni}\}, 1 \leq i \leq n$ be iid random variables with common CDF H_n. Let $S_n = \sum_{i=1}^{n} X_{ni} \stackrel{\mathcal{L}}{\Rightarrow} Z \sim F$. Then F is infinitely divisible.

The result above allows *triangular arrays* of independent random variables with possibly different common distributions H_n for the different rows. An important special case is that of just one iid sequence $X_1, X_2, \ldots$ with a common CDF H. If the partial sums $S_n = \sum_{i=1}^{n} X_i$, possibly after suitable centering and norming, converge in distribution to some random variable Z, then Z belongs to a subclass of the class of infinitely divisible distributions. This class is the so-called *stable family*. We first give a more direct definition of a stable distribution that better explains the reason for the name *stable*.

Definition 5.5 A CDF F is said to be stable if for every $n \geq 1$ there exist constants b_n and a_n such that $S_n = X_1 + X_2 + \ldots + X_n$ and $b_n X_1 + a_n$ have the same distribution, where $X_1, X_2, \ldots, X_n$ are iid with the given distribution F.

Remark. It turns out that b_n has to be $n^{1/\alpha}$ for some $0 < \alpha \leq 2$. The constant α is said to be the *index* of the stable distribution F. The case $\alpha = 2$ corresponds to the normal case, and $\alpha = 1$ corresponds to the Cauchy case, which is also obviously stable.

Theorem 5.11 Let $X_1, X_2, \ldots$ be iid with the common CDF H. Suppose for constant sequences $\{a_n\}, \{b_n\}$, $\frac{S_n - a_n}{b_n} \stackrel{\mathcal{L}}{\Rightarrow} Z \sim F$. Then F is stable.

Remark. This theorem says that sample means of iid random variables can have only stable limits. For example, if the underlying random variables have a finite variance, then we know from the CLT that sample means converge to a normal limit, which is stable with index $\alpha = 2$. But for other types of underlying distributions, the limit can be some other stable law. This suggests the following general definition.

Definition 5.6 Let $X_1, X_2, \ldots$ be iid with the common CDF H and let F be a stable distribution with index α, $0 < \alpha \leq 2$. We say that H belongs to the *domain of attraction* of F if there exist constant sequences $\{a_n\}$, $\{b_n\}$ such that $\frac{S_n - a_n}{b_n} \stackrel{\mathcal{L}}{\Rightarrow} Z \sim F$ and write it as $H \in \mathcal{D}(\alpha)$.

Remark. Complete characterizations for a distribution H to be in the domain of attraction of a specific stable law are known. In particular, one can characterize all distributions H for which sample means have a normal limit on appropriate centering and norming; we address this later in this section.

Random variables with a bounded support cannot be infinitely divisible. This is a well-known fact and is not hard to prove. Interestingly, however, most common distributions with an unbounded support *are* infinitely divisible. A few well-known ones among them are also stable.

Example 5.6 Let X be $N(\mu, \sigma^2)$. For any $n > 1$, let $X_1, \ldots, X_n$ be iid $N(\mu/n, \sigma^2/n)$. Then X has the same distribution as $X_1 + \ldots + X_n$. Thus, all normal distributions are infinitely divisible.

Example 5.7 Let X have a Poisson distribution with mean λ. For a given n, take $X_1, \ldots, X_n$ as iid Poisson variables with mean λ/n. Then X has the same distribution as $X_1 + \ldots + X_n$. Thus, all Poisson distributions are infinitely divisible.

Example 5.8 Let X have the continuous $U[0, 1]$ distribution. Then X is *not* infinitely divisible. If it is, then, for any n, there exist iid random variables $X_1, \ldots X_n$ with some distribution F_n such that X has the same distribution as $X_1 + \ldots + X_n$. Clearly, the supremum of the support of F_n is at most $1/n$. This implies $\text{Var}(X_1) \leq 1/n^2$ and hence $V(X) \leq 1/n$, an obvious contradiction.

The most common means of characterizing infinitely divisible (id) laws is by their characteristic functions. Several forms are available; we give two of these forms, namely Form A for the finite-variance case and Form B for the general case.

Theorem 5.12 Form A Let F be infinitely divisible with mean b and a finite variance and let $\phi(t)$ be its characteristic function. Then

$$\log \phi(t) = ibt + \int_{-\infty}^{\infty} (e^{itx} - 1 - itx) \frac{d\mu(x)}{x^2},$$

where μ is a finite measure on the real line. Furthermore, $\mu(\mathcal{R}) = \text{Var}_F(X)$.

Example 5.9 Suppose F is the normal distribution with mean 0 and variance σ^2. Then the measure μ is degenerate at 0 with mass σ^2 there and b is 0.

Example 5.10 Suppose Y has a Poisson distribution with mean λ, and let F be the distribution of $X = c(Y - \lambda)$. Then μ is degenerate with mass c^2 at c and $b = 0$. Here c is arbitrary.

Theorem 5.13 Form B Let F be an id law and let $\phi(t)$ denote its characteristic function. Then

$$\log \phi(t) = ibt - t^2\sigma^2/2 + \int_{-\infty}^{\infty} \left(e^{itx} - 1 - \frac{itx}{1+x^2} \right) \frac{1+x^2}{x^2} d\lambda(x),$$

where b is a real number and λ is a finite measure on the real line giving mass 0 to the value 0 and the integrand is defined to be $-t^2/2$ at the origin by continuity.

Remark. This is the original canonical representation given in Lévy (1937). For certain applications and special cases, Form A is more useful. The measures μ and λ in the two forms are both termed the *Lévy measure* of F. We now give the explicit identification of Lévy measures for some special important distributions. Below, Form A will be used for cases of a finite variance.

Example 5.11
1. Normal distribution with mean 0 and variance σ^2. Here $w(t) = \log \phi(t) = -\sigma^2 t^2/2$. Thus, $b = 0$ and μ is concentrated at 0 with mass σ^2.
2. Poisson distribution with mean λ. Here $w(t) = \lambda \exp\{it - 1\}, b = \lambda$, and μ is concentrated at 1 with mass λ.
3. For the exponential distribution with mean 1, $\phi(t) = (1 - it)^{-1}, b = 1$, and $d\mu(x) = xe^{-x}dx, x > 0$.
4. For the Gamma density $f(x) = \frac{\alpha^p}{\Gamma(p)} e^{-\alpha x} x^{p-1}$, $\phi(t) = (1 - \frac{it}{\alpha})^{-p}$ and

$$w(t) = i\frac{p}{\alpha} + p \int_0^{\infty} \{e^{itx} - 1 - itx\} \frac{e^{-\alpha x}}{x} dx.$$

Hence, $b = \frac{p}{\alpha}$ and $d\mu(x) = px \exp\{-\alpha x\} dx, \ x > 0$.
5. Log Beta distribution. Suppose Y is a Beta random variable with density $f_Y(x) = [\text{Beta}(\alpha, \beta)]^{-1} x^{\alpha-1}(1-x)^{\beta-1}, \ \ 0 < x < 1$. Let $X = -\log Y$. Then X is id and its Lévy measure concentrates on the positive real line with measure $\lambda(dx) = \frac{x^2}{1+x^2} x^{-1} e^{-\alpha x}(1 - e^{-\beta x})/(1 - e^{-x}) dx$.

6. Log Gamma distribution. Suppose Y has the Gamma $(1, p)$ distribution. Let $X = \log Y$. Then the distribution's Lévy measure is $\lambda(dx) = \frac{x^2}{1+x^2}\exp(px)/(1-\exp(x))dx,\, x < 0$.
7. Non-Gaussian stable distributions with index α, $0 < \alpha < 2$. In this case,

$$w(t) = ibt + \beta_1 \int_{-\infty}^{0} \frac{A(t,x)}{|x|^{1+\alpha}}dx + \beta_2 \int_{0}^{\infty} \frac{A(t,x)}{x^{1+\alpha}}dx,$$

where $A(t,x) = e^{itx} - 1 - \frac{itx}{1+x^2}$ and β_1 and β_2 are nonnegative. Hence $\sigma^2 = 0$, and $d\lambda(x) = \frac{x^2}{1+x^2}|x|^{-(1+\alpha)}dx, \; -\infty < x < \infty$.

Remark. Characteristic functions of id laws satisfy some interesting properties. Such properties are useful to exclude particular distributions from being id and to establish further properties of id laws as well. They generally do not provide much probabilistic insight but are quite valuable as analytical tools in studying id laws. A collection of properties is listed below.

Theorem 5.14 Let $\phi(t)$ be the characteristic function (cf) of an infinitely divisible distribution.

(a) ϕ has no real zeros; the converse is false.
(b) $\overline{\phi}$, the complex conjugate of ϕ, and $|\phi|^2$ are also cfs of some infinitely divisible distributions.
(c) Let $\phi_n(t)$ be a sequence of id cfs converging pointwise to another cf, $\phi(t)$. Then $\phi(t)$ is also the cf of an infinitely divisible distribution.

Large classes of positive continuous random variables can be shown to be infinitely divisible by using the following famous result.

Theorem 5.15 (The Goldie-Steutel Law) Let a positive random variable X have a density $f(x)$ that is *completely monotone*; i.e., f is infinitely differentiable and $(-1)^k f^{(k)}(x) \geq 0 \,\forall x > 0$. Then X is infinitely divisible.

Remark. It is well-known that a positive random variable X has a completely monotone density iff X has the same distribution as YZ, where Z is exponential with mean 1 and Y is nonnegative and independent of Z. That is, all scale mixtures of exponentials are infinitely divisible.

Example 5.12 Let X have the Pareto density $f(x) = \frac{\alpha}{\mu}(\frac{\mu}{x+\mu})^{\alpha+1}$, $\mu, \alpha > 0$. Then, easily, f is completely monotone, and so X is id. It can be verified that, in the representation $X = YZ$ as above, Y has a Gamma density.

We end this section with a result on characteristic functions of stable laws and a characterization of the domain of attraction of a stable law. Stable laws occupy a special position in the class of infinitely divisible distributions. They have found numerous applications in statistics. Starting from Form B of the characteristic function of infinitely divisible distributions, it is possible to derive the following characterization for characteristic functions of stable laws.

Theorem 5.16 $\phi(t)$ is the cf of a stable law F with index $\alpha \neq 1$ if and only if it has the representation

$$\log \phi(t) = ibt - \sigma^{\alpha}|t|^{\alpha}(1 - i\beta \text{sgn}(t) \tan(\pi\alpha/2)),$$

and if $\alpha = 1$,

$$\log \phi(t) = ibt - \sigma|t| \left(1 + i\beta \text{sgn}(t)\frac{2}{\pi} \log|t|\right).$$

Remark. The *scale parameter* $\sigma > 0$, the *location parameter* b, and the *skewness* parameter β of F above are *unique* (except that, if $\alpha = 2$, the value of β is irrelevant). The possible value of β ranges in the closed interval $[-1, 1]$. The possible values of b are the entire real line. It follows trivially from the characteristic function that F is *symmetric* (about b) if and only if $\beta = 0$. If $\alpha = 2$, then F is normal.

If $\alpha = 1$ and $\beta = 0$, then F is a Cauchy law with scale σ and location b.

Remark. Normal distributions are the *only* stable laws with a finite variance. In fact, this is one of the principal reasons that stable laws are used to model continuous random variables with densities having heavy tails. Here is an important result on moments of stable laws.

Theorem 5.17 If X is stable with an index $0 < \alpha < 2$, then, for any $p > 0$, $E|X|^p < \infty$ if and only if $0 < p < \alpha$.

Remark. This says that stable laws with an index ≤ 1 cannot have a mean.

Finally, we provide a characterization of all distributions H that belong to the domain of attraction of a normal or more generally to the domain of attraction of some other stable law.

Theorem 5.18 A CDF $H \in \mathcal{D}(2)$ if and only if

$$\lim_{x\to\infty} \frac{x^2 \int_{|z|>x} dH(z)}{\int_{|z|\le x} z^2 dH(z)} = 0.$$

Remark. If H has a finite second moment, then this limit can be shown to be necessarily zero. However, the limit can be zero without H having a finite second moment. The condition above is equivalent to $\int_{|z|\le x} z^2 dH(z)$ being slowly varying at ∞; see Theorem 5.4. We saw in Example 5.2 that the t_2-distribution is indeed in the domain of attraction of a normal, although it does not have a finite second moment.

The corresponding characterization result for a general stable law is the following.

Theorem 5.19 A CDF $H \in \mathcal{D}(\alpha)$ if and only if there exist nonnegative constants c_1, c_2, not both zero, such that

$$\lim_{x\to\infty} \frac{1 - H(x)}{H(-x)} = \frac{c_2}{c_1},$$

$$c_2 > 0 \Rightarrow \lim_{x\to\infty} \frac{1 - H(tx)}{1 - H(x)} = t^{-\alpha}\ \forall t > 0,$$

$$c_1 > 0 \Rightarrow \lim_{x\to\infty} \frac{H(-tx)}{H(-x)} = t^{-\alpha}\ \forall t > 0.$$

Example 5.13 Consider the symmetric density function $h(x) = \frac{3}{4}|x|^{-5/2}$, $|x| \ge 1$. Then, for any $u > 1, 1 - H(u) = \frac{1}{2}u^{-3/2}$. Plugging in, for any $t > 0$, $\lim_{x\to\infty} \frac{1-H(tx)}{1-H(x)} = t^{-3/2}$. Since the density h is symmetric, it follows that $H \in \mathcal{D}(\frac{3}{2})$.

5.7 Exercises

Exercise 5.1 Let $X_{ni} \overset{\text{indep.}}{\sim} \text{Bin}(1, \theta_{ni}), 1 \le i \le n$, and suppose $\sum_{i=1}^n \theta_{ni}(1 - \theta_{ni}) \to \infty$. Show that

$$\frac{\sum_{i=1}^{n} X_{ni} - \sum_{i=1}^{n} \theta_{ni}}{\sqrt{\sum_{i=1}^{n} \theta_{ni}(1-\theta_{ni})}} \xrightarrow{\mathcal{L}} N(0,1).$$

Exercise 5.2 * Let X_i be iid, having a t-distribution with 2 degrees of freedom. Show, by truncating X_i at $|X_i| > \sqrt{n}\log\log n$, that $b_n\bar{X} \xrightarrow{\mathcal{L}} N(0,1)$ for suitable b_n. Identify b_n.

Exercise 5.3 Let $X_i \overset{\text{indep.}}{\sim} \text{Exp}(\lambda_i), i \geq 1$, and suppose $\frac{\max_{1\leq i\leq n}\lambda_i}{\sum_{i=1}^{n}\lambda_i} \to 0$. Show that $\bar{X}$ is asymptotically normally distributed.

Exercise 5.4 Let $X_i \overset{\text{indep.}}{\sim} \text{Poi}(\lambda_i), i \geq 1$. Find a sufficient condition on $\{\lambda_i\}$ so that $\bar{X}$ is asymptotically normally distributed.

Exercise 5.5 * Let $\pi = \pi_n$ denote a random permutation of $(1, 2, \ldots, n)$. Show that $\sum_{i=1}^{n} i\pi(i)$ is asymptotically normally distributed. How about $\sum_{i=1}^{n} i^{\alpha}\pi(i)$ for general $\alpha > 0$?

Exercise 5.6 * Consider the setup of Exercise 5.1. Suppose $\sum_{i=1}^{n}\theta_{ni} \to 0 < \lambda < \infty$. Where does $\sum_{i=1}^{n} X_{ni}$ converge in law?

Exercise 5.7 * Suppose $X_i \overset{\text{iid}}{\sim} f(x) = \frac{1}{|x|^3}, |x| \geq 1$. Show that there exist constants a_n, b_n such that $\frac{\sum_{i=1}^{n} X_i - a_n}{b_n} \overset{\mathcal{L}}{\Rightarrow} N(0,1)$, although f does not have a finite variance.

Exercise 5.8 X_i are discrete random variables with the pmf $P(X = n) = c\frac{1}{n^3}, n \geq 1$.

(a) Find c and hence the mean of X.
Hint: The answer involves the Riemann zeta function.

(b) Does $\bar{X}$ converge to normal on any centering and norming? If so, establish what the centering and norming are.

Exercise 5.9 * **Records** Suppose X_i are iid random variables, $i \geq 1$. A *record* is said to occur at time n if $X_n = X_{n:n}$, the maximum of the first n observations. Let T_n be the number of records up to time n.

(a) Find the mean and variance of T_n.

(b) Show that, on suitable centering and norming, $T_n \stackrel{\mathcal{L}}{\Rightarrow} N(0, 1)$.

Exercise 5.10 * Suppose $X_i \stackrel{\text{indep.}}{\sim} U[-a_i, a_i], i \geq 1$.

(a) Give a condition on $\{a_i\}$ such that the Lindeberg-Feller condition holds.

(b) Give $\{a_i\}$ such that the Lindeberg-Feller condition does not hold.

Exercise 5.11 * **(Weyl Equidistribution Theorem)** Define $X_n = \{n\pi\}$, where $\{.\}$ denotes the fractional part. Show that the empirical distribution of the $X_i, i \leq n$ converges weakly to $U[0, 1]$ as $n \to \infty$.

Remark. This is a famous result and is not very easy to prove.

Exercise 5.12 Suppose $\{X_1, X_2, \cdots\}$ is an infinite exchangeable sequence with finite variance. Can the correlation between X_1 and X_2 be negative?

Exercise 5.13 Suppose $\{X_1, X_2, \cdots\}$ is an infinite exchangeable Bernoulli sequence. Can the correlation between X_1 and X_2 be zero? When?

Exercise 5.14 Write half a dozen infinite exchangeable Bernoulli sequences by using de Finetti's theorem.

Exercise 5.15 Suppose, given λ, $X_i \stackrel{\text{iid}}{\sim} \text{Poi}(\lambda)$ and $\lambda \sim \text{Exp}(1)$. Is the infinite sequence $\{X_1, X_2, \cdots\}$ exchangeable?

Exercise 5.16 Find the Lévy measure in Form A for a general geometric distribution starting at 1.

Exercise 5.17 * Find the Lévy measure in Form A for a general double exponential density.

Exercise 5.18 * Let Z_1, Z_2 be iid $N(0, 1)$. Show that $Z_1 Z_2$ is infinitely divisible.

Exercise 5.19 * Let Z_1, Z_2 be iid $C(0, 1)$. Show that $Z_1 Z_2$ is infinitely divisible.

Exercise 5.20 * Let $n > 1$, and $Z_1, \ldots, Z_n$ iid $N(0, 1)$. Show that $Z_1 Z_2 \ldots Z_n$ is infinitely divisible.

Exercise 5.21 * Let $n > 1$, and $Z_1, \ldots, Z_n$ iid $C(0, 1)$. Show that $Z_1 Z_2 \ldots Z_n$ is infinitely divisible.

Exercise 5.22 * Give an example of an infinitely divisible continuous random variable with a density that is not unimodal.

Hint: Look at convolutions of a normal with a Poisson distribution.

Exercise 5.23 * Give an example of a characteristic function without any zeros but where the distribution is not infinitely divisible.

Exercise 5.24 * Which of the following distributions belong to the domain of attraction of a normal?

(a) $P(X = k) = c/k^2$, where c is a normalizing constant and $k \geq 1$; (b) $P(X = k) = 1/2^k, k \geq 1$; (c) $P(X = k) = c \log k/k^3$, where c is a normalizing constant and $k \geq 1$; (d) $C(0, 1)$; (e) $f(x) = 2/x^3, x \geq 1$; (f) $f(x) = c(1 - \cos x)/x^3, x \geq 1$, where c is a normalizing constant.

References

Aldous, D. (1983). *Exchangeability and Related Topics*, Lecture Notes in Mathematics, Vol. 1117, Springer, Berlin.

Anscombe, F. (1952). Large sample theory of sequential estimation, Proc. Cambridge Philos. Soc., 48, 600–607.

Billingsley, P. (1995). *Probability and Measure*, 3rd ed., John Wiley, New York.

Blum, J., Chernoff, H., Rosenblatt, M., and Teicher, H. (1958). Central limit theorems for interchangeable processes, Can. J. Math., 10, 222–229.

Bose, A., DasGupta, A., and Rubin, H. (2004). A contemporary review of infinitely divisible distributions and processes, Sankhya Ser. A, 64(3), Part 2, 763–819.

Chernoff, H. and Teicher, H. (1958). A central limit theorem for sums of interchangeable random variables, Ann. Math. Stat., 29, 118–130.

de Finetti, B. (1931). Funzione caratteristica di un fenomeno allatorio. Atti R. Accad. Naz. Lincii Ser. 6, Mem., Cl. Sci. Fis. Mat. Nat., 4, 251–299.

Diaconis, P. (1988). *Recent Progress on de Finetti's Notions of Exchangeability*, Bayesian Statistics, Vol. 3, Oxford University Press, New York.

Diaconis, P. and Freedman, D. (1987). A dozen de Finetti style results in search of a theory, Ann. Inst. Henri. Poincaré Prob. Stat., 23(2), 397–423.

Feller, W. (1966). *An Introduction to Probability Theory and Its Applications*, Vol. II, John Wiley, New York.

Ferguson, T. (1996). *A Course in Large Sample Theory*, Chapman and Hall, London.

Fisz, M. (1962). Infinitely divisible distributions: recent results and applications, Ann. Math. Stat., 33, 68–84.

Hewitt, E. and Savage, L. (1955). Symmetric measures on Cartesian products, Trans. Am. Math. Soc., 80, 470–501.

Hoeffding, W. (1951). A combinatorial central limit theorem, Ann. Math. Stat., 22, 558–566.

Klass, M. and Teicher, H. (1987). The central limit theorem for exchangeable random variables without moments, Ann. Prob., 15, 138–153.

Lehmann, E.L. (1999). *Elements of Large Sample Theory*, Springer, New York.

Lévy, P. (1937). *Théorie de ĺAddition des Variables Aléatoires*, Gauthier-Villars, Paris.

Petrov, V. (1975). *Limit Theorems for Sums of Independent Random Variables* (translation from Russian), Springer-Verlag, New York.

Port, S. (1994). *Theoretical Probability for Applications*, John Wiley, New York.

Rényi, A. (1957). On the asymptotic distribution of the sum of a random number of independent random variables, Acta Math. Hung., 8, 193–199.

Sen, P.K. and Singer, J. (1993). *Large Sample Methods in Statistics: An Introduction with Applications*, Chapman and Hall, New York.

Steutel, F.W. (1979). Infinite divisibility in theory and practice, Scand. J. Stat., 6, 57–64.

Chapter 6
Moment Convergence and Uniform Integrability

Sometimes we need to establish that moments of some sequence $\{X_n\}$, or at least some lower-order moments, converge to moments of X when $X_n \xrightarrow{\mathcal{L}} X$. Convergence in distribution by itself simply cannot ensure convergence of any moments. An extra condition that ensures convergence of appropriate moments is uniform integrability. On the other hand, convergence of moments is useful for proving convergence in distribution itself. This is related to a classic and extremely deep problem in analysis and probability called the *moment problem*. We discuss uniform integrability and the moment problem briefly in this chapter. Billingsley (1995) is a good reference for convergence of moments and its relation to uniform integrability. Two other useful references are Chung (2001) and Shiryaev (1980).

6.1 Basic Results

Definition 6.1 Let $\{X_n\}$ be a sequence of random variables on some probability space $(\Omega, \mathbb{A}, \mathbb{P})$. The sequence $\{X_n\}$ is called uniformly integrable if $\sup_{n\geq 1} E(|X_n|) < \infty$ and if for any $\epsilon > 0$ there exists a sufficiently small $\delta > 0$ such that whenever $P(A) < \delta$, $\sup_{n\geq 1} \int_A |X_n| dP < \epsilon$.

Remark. Direct verification of the definition is usually not easy, so easier sufficient conditions are useful. Here is one such sufficient condition.

Theorem 6.1 Suppose, for some $\delta > 0$, that $\sup_n E|X_n|^{1+\delta} < \infty$. Then $\{X_n\}$ is uniformly integrable.

Remark. A proof can be seen in Billingsley (1995). Often, $\delta = 1$ would work in practical applications.

A. DasGupta, *Asymptotic Theory of Statistics and Probability*,

Example 6.1 An interesting and useful application is the uniform integrability of an order statistic of an iid sample under a moment condition on the population. Suppose X_i are iid with a common CDF F, and suppose $E(X^m) < \infty$ for some given $m > 0$. An elegant general inequality on expectations of functions of order statistics says that if $g(x_1, \ldots, x_k)$ is a function of k real variables and $X_1, X_2, \ldots, X_n$ is an iid sample from some CDF F, then for any selection of k order statistics $X_{r_1:n}, X_{r_2:n}, \ldots, X_{r_k:n}$,

$$E(g(X_{r_1:n}, X_{r_2:n}, \ldots, X_{r_k:n})) \leq \frac{n!}{\prod_{i=1}^{n}(r_i - r_{i-1} - 1)!} E(g(X_1, X_2, \ldots, X_k))$$

(see Reiss (1989)). Here, $r_{0:n} = 0$. As a very simple application, take $k = 1$, $g(x) = x$, and $r_1 = r$(fixed). The inequality says that if $E(X^m) < \infty$, then so is $E(X_{r:n})^m$ for any r. In particular, if X has a finite second moment, so will any order statistic, and hence, by Theorem 6.1, the sequence $\{X_{r:n}\}$ will be uniformly integrable.

The following theorem describes the link between uniform integrability and convergence of moments.

Theorem 6.2 Suppose $X_n \xrightarrow{\mathcal{L}} X$ for some X and $|X_n|^k$ is uniformly integrable. Then $EX_n^r \rightarrow EX^r$ for every $1 \leq r \leq k$. In particular, if $\sup_n E|X_n|^{k+\delta} < \infty$ for some $\delta > 0$, then $E(X_n^r) \rightarrow E(X^r)$ for every $1 \leq r \leq k$.

Remark. Billingsley (1995) gives a proof. There are some simple cases where uniform integrability follows easily. For example, the following are true.

Proposition 6.1 (a) If $|X_n| \leq M < \infty$, then $\{X_n\}$ are uniformly integrable.

(b) If $\{X_n\}$ and $\{Y_n\}$ are uniformly integrable, then so is $\{X_n + Y_n\}$.

(c) If $\{X_n\}$ is uniformly integrable and $|Y_n| \leq M < \infty$, then $\{X_n Y_n\}$ is uniformly integrable.

Convergence of moments in the canonical CLT for iid random variables is an interesting question. The next theorem addresses that question; see von Bahr (1965).

Theorem 6.3 (von Bahr) Suppose $X_1, \ldots, X_n$ are iid with mean μ and finite variance σ^2, and suppose that, for some specific k, $E|X_1|^k < \infty$.

Suppose $Z \sim N(0, 1)$. Then

$$E\left(\frac{\sqrt{n}(\overline{X}_n - \mu)}{\sigma}\right)^r = E(Z^r) + O\left(\frac{1}{\sqrt{n}}\right)$$

for every $r \leq k$.

The next result is for more general functions.

Theorem 6.4 (Cramér) Let $\{T_n\}$ be a sequence of statistics and $g(T_n)$ a function of T_n such that $|g(T_n)| \leq cn^p$ for some $0 < c, p < \infty$. Let $r > 0$ be fixed, and suppose that, for some $k \geq r(1 + p)$, $E[\sqrt{n}(T_n - \theta)]^{2k} < \infty$. Then

$$E[\sqrt{n}(g(T_n) - g(E(T_n)))]^r = \left[g'(E(T_n))\right]^r E(T_n - E(T_n))^r + O\left(n^{-\frac{r+1}{2}}\right),$$

provided $g(\cdot)$ is sufficiently smooth.

Remark. The Cramér theorem is very general in its probability structure but limited in its scope of applications as far as the function g is concerned. See Cramér (1946) and Sen and Singer (1993) for the smoothness conditions on g.

6.2 The Moment Problem

Sometimes one can establish that $X_n \xrightarrow{\mathcal{L}} X$ for some X by directly verifying $E(X_n^k) \to E(X^k)$ for all $k \geq 1$. A condition on X is needed. A good reference for this section is Shiryaev (1980). Diaconis (1987) gives a catalog of interesting problems where weak convergence is established by showing that the moments converge. A reference at a more advanced level is Dette and Studden (1997). Two recent references are Diaconis and Freedman (2004a,b). A number of pretty examples are worked out in Stoyanov (1987).

For Theorem 6.5 below, see Billingsley (1995), and for Theorems 6.6 and 6.7, see Shiryaev (1980). But first we need a definition.

Definition 6.2 A real-valued random variable X is said to be determined by its moments if $E(X^k) < \infty$ for all $k \geq 1$ and if $E(X^k) = E(Y^k)\,\forall k \geq 1$ implies that X and Y have the same distribution.

Theorem 6.5 Suppose for some sequence $\{X_n\}$ and a random variable X that $E(X_n^k) \rightarrow E(X^k)$ for every $k \geq 1$. If the limit random variable is determined by its moments, then it follows that $X_n \xrightarrow{\mathcal{L}} X$.

We mention two important sufficient conditions for a distribution to be determined by its moment sequence. The standard distributions that are in fact determined by their moments are all covered by Theorem 6.7. But Theorem 6.6 is an important one in its own right.

Theorem 6.6 (Carleman) Let X be a real-valued random variable with finite moments $c_n = E(X^n), n \geq 1$. Suppose $\sum_{n\geq 1} c_{2n}^{-\frac{1}{2n}} = \infty$. Then X is determined by its moment sequence.

Theorem 6.7 Suppose a CDF F has a finite mgf (moment generating function) in some neighborhood of zero. Then F is determined by its moment sequence.

Example 6.2 Suppose $X_1, X_2, \ldots$ are iid $U[0, 1]$, and let $X_{(n)} = \max\{X_1, \ldots, X_n\}$. Denote $Z_n = n(1 - X_{(n)})$. Since $X_{(n)}$ has density nx^{n-1} on $[0, 1]$, for any $k \geq 1$, $E(Z_n^k) = n^{k+1} \int_0^1 (1-x)^k x^{n-1} dx = n^{k+1} \frac{\Gamma(n)\Gamma(k+1)}{\Gamma(n+k+1)} \rightarrow k!$ as $n \rightarrow \infty$ by applying Stirling's approximation to $\Gamma(n)$ and $\Gamma(n + k + 1)$. Note now that if $W \sim \text{Exp}(1)$, then for any $k \geq 1$, $E(W^k) = k!$, and W is determined by its moment sequence, as it has a finite mgf on $(-\infty, 1)$. It follows that Z_n converges in distribution to an Exp(1) random variable. It can be proved otherwise as well, but the moment proof is neat.

Example 6.3 Let $X \sim \text{Poisson}(1)$. Then $E(X^k) = B_k$, the kth *Bell number*, defined as the total number of ways to partition a k-element set into nonempty partitioning classes.

It is known that $B_n = O(n^n)$. Denoting $E(X^n)$ as c_n, it follows that $c_{2n} \sim 2^{2n} n^{2n}$ and hence $c_{2n}^{-\frac{1}{2n}} \sim \frac{1}{n}$. It follows that $\sum_{n\geq 1} c_{2n}^{-\frac{1}{2n}} = \infty$. This implies that the Poisson(1) distribution is determined by its moment sequence by Theorem 6.6.

Remark. In fact, a Poisson distribution with any mean is determined by its moment sequence by Theorem 6.7.

Example 6.4 Let $Z \sim N(0, 1)$. Then

$$c_{2n} = E(Z^{2n}) = \frac{1}{\sqrt{2\pi}} \int_{-\infty}^{\infty} z^{2n} e^{-z^2/2} dz = \sqrt{\frac{2}{\pi}} \int_0^{\infty} z^{2n} e^{-z^2/2} dz$$

$$\stackrel{z=\sqrt{2x}}{=} \sqrt{\frac{1}{\pi}} \cdot 2^n \cdot \int_0^\infty x^{n-\frac{1}{2}} e^{-x} dx = \frac{2^n}{\sqrt{\pi}} \Gamma\left(n + \frac{1}{2}\right).$$

On applying Stirling's formula, it follows that $c_{2n}^{-\frac{1}{2n}} \sim \frac{1}{n}$ and hence, by the Carleman theorem, the $N(0, 1)$ distribution is determined by its moments.

Remark. Again, any normal distribution is determined by its moments due to the existence of an mgf everywhere.

Example 6.5 Let $X \sim U[0, 1]$. Then $c_n = \frac{1}{n+1}$. Therefore, immediately, by the Carleman result, X is determined by its moments.

Remark. In fact, the Weierstrass approximation theorem implies that any distribution on a compact set in $\mathbb{R}$ is determined by its moment sequence.

The following theorem gives a clean necessary and sufficient condition for an absolutely continuous distribution with an unbounded support $(-\infty, \infty)$, $(-\infty, b)$, or (a, ∞) to be determined by its moments.

Theorem 6.8 Let X have an absolutely continuous distribution with density $f(x)$ on an unbounded interval (a, b). Then, X is determined by its moments iff $\int_a^b \frac{\log f(x)}{1+x^2} dx = -\infty$.

Example 6.6 Stoyanov (1987) is a reference for Theorem 6.8. Of the named distributions, lognormal distributions are not determined by their moments. In fact, Theorem 6.8 would show this. We present only the case of the lognormal distribution with parameters 0 and 1. The case of general parameters is the same.

Let $X \sim$ lognormal(0, 1) with density $f(x) = \frac{1}{x\sqrt{2\pi}} e^{-\frac{(\log x)^2}{2}}, x > 0$. Therefore,

$$\int_0^\infty \frac{\log f(x)}{1 + x^2} dx$$

$=$ constant $- \int_0^\infty \frac{\log x}{1+x^2} dx - \frac{1}{2} \int_0^\infty \frac{(\log x)^2}{1+x^2} dx$. Clearly, each of $\frac{\log x}{1+x^2}$ and $\frac{(\log x)^2}{1+x^2}$ is integrable on the interval $[1, \infty)$. Furthermore, $\int_0^1 \frac{\log x}{1+x^2} dx \sim \int_0^1 (\log x) dx = \int_{-\infty}^0 z e^z dz > -\infty$, and similarly $\int_0^1 \frac{(\log x)^2}{1+x^2} dx < \infty$. Putting it all together, $\int_0^\infty \frac{\log f(x)}{1+x^2} dx > -\infty$, and therefore it follows from Theorem 6.8 that the lognormal (0,1) distribution is not determined by its moments. Note that it does not have a finite mgf in any two-sided neighborhood of zero.

Example 6.7 Consider the well-known matching problem where one considers a random permutation π of $\{1, 2, \cdots, n\}$. Let

$$T_n = \#\text{fixed points of } \pi = \#\{i : \pi(i) = i\}.$$

See Chapter 2 for some treatment of the matching problem. It is a remarkable fact that $E(T_n^k) = E(\text{Poisson}(1))^k, \forall k \leq n$. Since the Poisson(1) distribution is determined by its moments, it follows from this fact and Theorem 6.5 that $T_n \xrightarrow{\mathcal{L}} \text{Poisson}(1)$.

Remark. One can see one proof of the equality of the first n moments in DasGupta (1999).

6.3 Exercises

Exercise 6.1 Convergence of Binomial to Poisson * Let $X_n \sim \text{Bin}(n, 1/n)$. Then show that, for any k, the kth moment of X_n converges to the kth moment of Poisson(1). Hence, X_n converges in distribution to Poisson(1).

Exercise 6.2 Suppose $E(\sup_n |X_n|) < \infty$. Show that $\{X_n\}$ is uniformly integrable.

Exercise 6.3 * Suppose $g \geq 0$ is a monotone increasing function and $\lim_{x\to\infty} \frac{g(x)}{x} = \infty$. Show that if $\sup_n Eg(|X_n|) < \infty$, then $\{X_n\}$ is uniformly integrable.

Exercise 6.4 Suppose $X_n \xrightarrow{\mathcal{P}} X$ and $\{|X_n|\}$ is uniformly integrable. Show that $E|X_n - X| \to 0$.

Exercise 6.5 Suppose $X_i, i \geq 1$ are iid Poi(μ), and let $S_n = \sum_{i=1}^n X_i$. Suppose $T_n = (-1)^{S_n}$. What is the limit of $E(T_n)$?

Exercise 6.6 * Suppose $X_i, i \geq 1$ are iid standard Cauchy with CDF $F(x)$. What is the limit of $E(F(\bar{X}))$?

Exercise 6.7 * Suppose $X_i, i \geq 1$ are iid standard Cauchy. Let $X_{(k)} = X_{(k:n)}$ denote the kth order statistic of the first n observations. Is $X_{(n)}$ uniformly integrable? Is $X_{(n-4)}$ uniformly integrable? Is $X_{([\frac{n}{2}])}$ uniformly integrable?

Exercise 6.8 Suppose X has the double exponential distribution with density $\frac{1}{2}e^{-|x|}$. Show that X is determined by its moment sequence.

Exercise 6.9 * Suppose $Z \sim N(0, 1)$ and $X = Z^3$. Show that X does not satisfy the Carleman condition.

Remark. In fact, X is not determined by its moment sequence, a slightly surprising result.

Exercise 6.10 Find a distribution whose first ten moments coincide with that of the $N(0, 1)$.

Hint: Think of discrete symmetric distributions.

Exercise 6.11 Suppose Z_1, Z_2 are iid $N(0, 1)$. Show that $Z_1 Z_2$ is determined by its moments.

Exercise 6.12 * Suppose Z_1, Z_2, Z_3 are iid $N(0, 1)$. Show that $Z_1 Z_2 Z_3$ is not determined by its moments.

Hint: Use Theorem 6.8.

References

Billingsley, P. (1995). *Probability and Measure*, 3rd ed., John Wiley, New York.

Chung, K.L. (2001). *A Course in Probability Theory*, 3rd ed., Academic Press, New York.

Cramér, H. (1946). *Mathematical Methods of Statistics*, Princeton University Press, Princeton, NJ.

DasGupta, A. (1999). A differential equation and recursion for Poisson moments, Technical Report, Purdue University.

Dette, H. and Studden, W.J. (1997). *The Theory of Canonical Moments with Applications in Statistics, Probability and Analysis*, John Wiley, New York.

Diaconis, P. (1987). Application of the method of moments in probability and statistics, in *Moments in Mathematics*, American Mathematical Society, Providence, RI, 125–142.

Diaconis, P. and Freedman, D. (2004a). The Markov moment problem and de Finetti's theorem, I, Math. Z., 247(1), 183–199.

Diaconis, P. and Freedman, D. (2004b). The Markov moment problem and de Finetti's theorem, II, Math. Z., 247(1), 201–212.

Reiss, R. (1989). Approximate Distribution of Order Statistics, Springer-Verlag, New York.

Sen, P.K. and Singer, J. (1993). *Large Sample Methods in Statistics:An Introduction with Applications*, Chapman and Hall, New York.

Shiryaev, A. (1980). *Probability*, Nauka, Moscow.

Stoyanov, J. (1987). *Counterexamples in Probability*, John Wiley, New York.

von Bahr, B. (1965). On the convergence of moments in the central limit theorem, Ann. Math. Stat., 36, 808–818.

Chapter 7
Sample Percentiles and Order Statistics

Sample percentiles have long been regarded as useful summary and diagnostic statistics. The full empirical CDF approximates the underlying population CDF for iid data; a few selected sample percentiles provide useful diagnostic summaries of the full empirical CDF. For example, the three quartiles of the sample already provide some information about symmetry of the underlying population, and extreme percentiles give information about the tail. However, the exact sampling distributions of sample percentiles are complicated, except in the simplest cases. So asymptotic theory of sample percentiles is of intrinsic interest in statistics. In this chapter, we present a selection of the fundamental results on the asymptotic theory for percentiles. The iid case and then an extension to the regression setup are discussed.

Suppose $X_1, \ldots, X_n$ are iid real-valued random variables with CDF F. We denote the order statistics of $X_1, \ldots, X_n$ by $X_{1:n}, X_{2:n}, \ldots, X_{n:n}$, or sometimes simply as $X_{(1)}, X_{(2)}, \ldots, X_{(n)}$. The empirical CDF is denoted as $F_n(x) = \frac{\#\{i : X_i \leq x\}}{n}$, and $F_n^{-1}(p) = \inf\{x : F_n(x) \geq p\}$ will denote the empirical quantile function. The population quantile function is $F^{-1}(p) = \inf\{x : F(x) \geq p\}$. $F^{-1}(p)$ will also be denoted as ξ_p.

Consider the kth order statistic $X_{k:n}$, where $k = k_n$. Three distinguishing cases are:

(a) $k_n = [np_n]$, where for some $0 < p < 1$, $\sqrt{n}(p_n - p) \to 0$. This is called the *central case*. Here, [.] denotes the integer part function.
(b) $n - k_n \to \infty$ and $\frac{k_n}{n} \to 1$. This is called the *intermediate case*.
(c) $n - k_n = O(1)$. This is called the *extreme case*.

Different asymptotics apply to the three cases; the case of extremes is considered in the next chapter. Good general references for this chapter are Serfling (1980) and Ferguson (1996). More specific references are given within the following sections.

A. DasGupta, *Asymptotic Theory of Statistics and Probability*,

7.1 Asymptotic Distribution of One Order Statistic

The first three theorems in this chapter are proved in Serfling (1980). The proofs essentially use the convergence of binomials to the normal. Theorem 7.4, an important fact, seems not to have been proved anywhere explicitly but is straightforward to derive.

Theorem 7.1 (Central case) Let $X_1, \ldots, X_n$ be iid with CDF F. Let $\sqrt{n}(\frac{k_n}{n} - p) \to 0$ for some $0 < p < 1$ as $n \to \infty$. Then

(a)

$$\lim_{n\to\infty} P_F(\sqrt{n}(X_{k_n:n} - \xi_p) \le t) = P\left(N\left(0, \frac{p(1-p)}{(F'(\xi_p-))^2}\right) \le t\right)$$

for $t < 0$, provided the left derivative $F'(\xi_p-)$ exists and is > 0.
(b)

$$\lim_{n\to\infty} P_F(\sqrt{n}(X_{k_n:n} - \xi_p) \le t) = P\left(N\left(0, \frac{p(1-p)}{(F'(\xi_p+))^2}\right) \le t\right)$$

for $t > 0$, provided the right derivative $F'(\xi_p+)$ exists and is > 0.
(c) If $F'(\xi_p-) = F'(\xi_p+) \overset{\text{say}}{=} f(\xi_p) > 0$, then

$$\sqrt{n}(X_{k_n:n} - \xi_p) \xrightarrow{\mathcal{L}} N\left(0, \frac{p(1-p)}{f^2(\xi_p)}\right).$$

Remark. Part (c) of Theorem 7.1 is the most ubiquitous version and the most used in applications. The same results hold with $X_{k_n:n}$ replaced by $F_n^{-1}(p)$.

Example 7.1 Suppose $X_1, X_2, \ldots$ are iid $N(\mu, 1)$. Let $M_n = X_{[\frac{n}{2}]:n}$ denote the sample median. Since the standard normal density $f(0)$ at zero equals $\frac{1}{\sqrt{2\pi}}$, it follows from Theorem 7.1 that $\sqrt{n}(M_n - \mu) \overset{\mathcal{L}}{\Rightarrow} N(0, \frac{\pi}{2})$. On the other hand, $\sqrt{n}(\bar{X} - \mu) \overset{\mathcal{L}}{\Rightarrow} N(0, 1)$. The ratio of the variances in the two asymptotic distributions, $1/\frac{\pi}{2} = \frac{2}{\pi}$, is called the ARE (asymptotic relative efficiency) of M_n relative to $\bar{X}$. Thus, for normal data, M_n is less efficient than $\bar{X}$.

Example 7.2 This is an example where the CDF F possesses left and right derivatives at the median but they are unequal. Suppose

$F(x) = \begin{cases} x, & \text{for } 0 \le x \le \frac{1}{2}, \\ 2x - \frac{1}{2}, & \text{for } \frac{1}{2} \le x \le \frac{3}{4}. \end{cases}$ By Theorem 7.1, $P_F(\sqrt{n}(X_{[\frac{n}{2}]:n} - \frac{1}{2}) \le t)$ can be approximated by $P(N(0, \frac{1}{4}) \le t)$ when $t < 0$ and by $P(N(0, \frac{1}{16}) \le t)$ when $t > 0$. Separate approximations are necessary because F changes slope at $x = \frac{1}{2}$.

7.2 Joint Asymptotic Distribution of Several Order Statistics

Just as a single central order statistic is asymptotically normal under very mild conditions, any fixed number of central order statistics are jointly asymptotically normal under similar conditions. Furthermore, any two of them are positively correlated, and there is a simple explicit description of the asymptotic covariance matrix. We present that result next.

Theorem 7.2 Let $X_{k_i:n}$, $1 \le i \le r$, be r specified order statistics, and suppose, for some $0 < q_1 < q_2 < \cdots < q_r < 1$, that $\sqrt{n}(\frac{k_i}{n} - q_i) \to 0$ as $n \to \infty$. Then

$$\sqrt{n}[(X_{k_1:n}, X_{k_2:n}, \ldots, X_{k_r:n}) - (\xi_{q_1}, \xi_{q_2}, \ldots, \xi_{q_r})] \xrightarrow{\mathcal{L}} N_r(\mathbf{0}, \Sigma),$$

where for $i \le j$, $\sigma_{ij} = \frac{q_i(1-q_j)}{F'(\xi_{q_i})F'(\xi_{q_j})}$, provided $F'(\xi_{q_i})$ exists and is > 0 for each $i = 1, 2, \ldots, r$.

Remark. Note that $\sigma_{ij} > 0$ for $i \ne j$ in Theorem 7.2. In fact, what is true is that if $X_1, \ldots, X_n$ are iid from any CDF F, then, provided the covariance exists, $\text{cov}(X_{i:n}, X_{j:n}) \ge 0$ for any i, j and any $n \ge 2$. See Bickel (1967) and Tong (1990).

Example 7.3 Sometimes, for estimating the scale parameter σ of a location-scale density $\frac{1}{\sigma} f(\frac{x-\mu}{\sigma})$, an estimate based on the interquartile range IQR $= X_{[\frac{3n}{4}]:n} - X_{[\frac{n}{4}]:n}$ is used. Use of such an estimate is quite common when normality is suspect. Since, from Theorem 7.2, the joint asymptotic distribution of $(X_{[\frac{3n}{4}]:n}, X_{[\frac{n}{4}]:n})$ is bivariate normal, it follows by a simple calculation that

$$\sqrt{n}(\text{IQR} - (\xi_{\frac{3}{4}} - \xi_{\frac{1}{4}})) \xrightarrow{\mathcal{L}} N\left(0, \frac{3}{f^2(\xi_{\frac{3}{4}})} + \frac{3}{f^2(\xi_{\frac{1}{4}})} - \frac{2}{f(\xi_{\frac{1}{4}})f(\xi_{\frac{3}{4}})}\right).$$

In particular, if $X_1, \ldots, X_n$ are iid $N(\mu, \sigma^2)$, then, by using the general result above, on some algebra,

$$\sqrt{n}(\text{IQR} - 1.35\sigma) \xrightarrow{\mathcal{L}} N(0, 2.48\sigma^2).$$

Consequently, for normal data, $\frac{\text{IQR}}{1.35}$ is a consistent estimate of σ(the 1.35 value of course is an approximation!) with asymptotic variance $2.48/1.35^2 = 1.36$. On the other hand, $\sqrt{n}(s - \sigma) \stackrel{\mathcal{L}}{\Rightarrow} N(0, .5)$ (see Chapter 3). The ratio of the asymptotic variances, namely $\frac{.5}{1.36} = .37$, is the ARE of the IQR-based estimate relative to s. Thus, for normal data, one is better off using s. For populations with thicker tails, IQR-based estimates can be more efficient. DasGupta and Haff (2006) work out the general asymptotic theory of the ratio $\frac{\text{IQR}}{s}$ and present several other examples and discuss practical uses of the ratio $\frac{\text{IQR}}{s}$.

Example 7.4 Suppose $X_1, \ldots, X_n$ are continuous and distributed as iid $F(x - \mu)$, where $F(-x) = 1 - F(x)$ and we wish to estimate the location parameter μ. An obvious idea is to use a convex combination of order statistics $\sum_{i=1}^n c_{in} X_{i:n}$. Such statistics are called *L-statistics*. A particular L-statistic that was found to have attractive versatile performance is the Gastwirth estimate (see Gastwirth (1966))

$$\hat{\mu} = 0.3X_{[\frac{n}{3}]:n} + 0.4X_{[\frac{n}{2}]:n} + 0.3X_{[\frac{2n}{3}]:n}.$$

This estimate is asymptotically normal with an explicitly available variance formula since we know from our general theorem that $(X_{[\frac{n}{3}]:n}, X_{[\frac{n}{2}]:n}, X_{[\frac{2n}{3}]:n})$ is jointly asymptotically trivariate normal under mild conditions. An alternative called the 10-20-30-40 estimator is discussed in Chapter 18.

7.3 Bahadur Representations

The sample median of an iid sample from some CDF F is clearly not a *linear statistic*; i.e., it is not a function of the form $\sum_{i=1}^n h_i(X_i)$. In 1966, Raghu Raj Bahadur proved that the sample median, and more generally any fixed sample percentile, is *almost* a linear statistic. The result in Bahadur (1966) not only led to an understanding of the probabilistic structure of percentiles but also turned out to be an extremely useful technical tool. For example, as we shall shortly see, it follows from Bahadur's result that, for iid samples from a CDF F, under suitable conditions not only are $\bar{X}, X_{[\frac{n}{2}]:n}$ marginally asymptotically normal but they are jointly asymptotically bivariate normal.

The result derived in Bahadur (1966) is known as the *Bahadur representation* of percentiles. Similar representations are now available for much more general types of statistics, sometimes for non-iid sequences.

Theorem 7.3 (Bahadur Representation of Sample Percentiles) Let $X_1, \ldots, X_n$ be iid from some CDF F. Let $0 < p < 1$, and suppose F is once differentiable at ξ_p with $f(\xi_p) = F'(\xi_p) > 0$. Let $Y_i = \frac{p - I(X_i \leq \xi_p)}{f(\xi_p)}$. Then the pth sample percentile $F_n^{-1}(p)$ admits the representation

$$F_n^{-1}(p) - \xi_p = \frac{1}{n}\sum_{i=1}^{n} Y_i + R_n,$$

where $R_n = o_p(\frac{1}{\sqrt{n}})$.

Remark. Actually, Bahadur gave an a.s. order for R_n under the stronger assumption that F is twice differentiable at ξ_p with $F'(\xi_p) > 0$. The theorem stated here is in the form later given in Ghosh (1971). The exact a.s. order of R_n was shown to be $n^{-3/4}(\log\log n)^{3/4}$ by Kiefer (1967) in a landmark paper. However, the weaker version presented here suffices for proving CLTs.

The Bahadur representation easily leads to the joint asymptotic distribution of $\bar{X}$ and $F_n^{-1}(p)$ for any $0 < p < 1$.

Theorem 7.4 Let $X_1, \ldots, X_n$ be iid from a CDF F. Let $0 < p < 1$ and suppose $\mathrm{Var}_F(X_1) < \infty$. If F is differentiable at ξ_p with $F'(\xi_p) = f(\xi_p) > 0$, then

$$\sqrt{n}(\bar{X} - \mu, F_n^{-1}(p) - \xi_p) \xrightarrow{\mathcal{L}} N_2(\mathbf{0}, \Sigma),$$

where

$$\Sigma = \begin{pmatrix} \mathrm{Var}(X_1) & \frac{p}{f(\xi_p)} E_F(X_1) - \frac{1}{f(\xi_p)} \int_{x \leq \xi_p} x dF(x) \\ \frac{p}{f(\xi_p)} E_F(X_1) - \frac{1}{f(\xi_p)} \int_{x \leq \xi_p} x dF(x) & \frac{p(1-p)}{f^2(\xi_p)} \end{pmatrix}.$$

Example 7.5 As an application of this theorem, consider iid $N(\mu, 1)$ data. Take $p = \frac{1}{2}$ so that Theorem 7.4 gives the joint asymptotic distribution of the sample mean and the sample median. The covariance entry in the matrix Σ equals (assuming without any loss of generality that $\mu = 0$) $-\sqrt{2\pi} \int_{-\infty}^{0} x\phi(x)dx = 1$. Therefore, the asymptotic correlation between the sample mean and median in the normal case is $\sqrt{\frac{2}{\pi}} = .7979$, a fairly strong correlation.

7.4 Confidence Intervals for Quantiles

Since the population median and more generally population percentiles provide useful summaries of the population CDF, inference for them is of clear interest. Confidence intervals for population percentiles are therefore of interest in inference. Suppose $X_1, X_2, \ldots, X_n \stackrel{\text{iid}}{\sim} F$ and we wish to estimate $\xi_p = F^{-1}(p)$ for some $0 < p < 1$. The corresponding sample percentile $\hat{\xi}_p = F_n^{-1}(p)$ is typically a fine point estimate for ξ_p. But how does one find a confidence interval of guaranteed coverage?

One possibility is to use the quantile transformation and observe that

$$(F(X_{1:n}), F(X_{2:n}), \ldots, F(X_{n:n})) \stackrel{\mathcal{L}}{=} (U_{1:n}, U_{2:n}, \ldots, U_{n:n}),$$

where $U_{i:n}$ is the ith order statistic of a $U[0, 1]$ random sample, provided F is continuous. Therefore, for given $1 \leq i_1 < i_2 \leq n$,

$$\begin{aligned} P_F(X_{i_1:n} \leq \xi_p \leq X_{i_2:n}) &= P_F(F(X_{i_1:n}) \leq p \leq F(X_{i_2:n})) \\ &= P(U_{i_1:n} \leq p \leq U_{i_2:n}) \geq 1 - \alpha \end{aligned}$$

if i_1, i_2 are appropriately chosen. The pair (i_1, i_2) can be chosen by studying the joint density of $(U_{i_1:n}, U_{i_2:n})$, which has an explicit formula. However, the formula involves incomplete Beta functions, and for certain n and α, the actual coverage can be substantially larger than $1 - \alpha$. This is because no pair (i_1, i_2) may exist such that the event involving the two uniform order statistics has exactly or almost exactly $1 - \alpha$ probability. This will make the confidence interval $[X_{i_1:n}, X_{i_2:n}]$ larger than one wishes and therefore less useful.

Alternatively, under previously stated conditions,

$$\sqrt{n}(\hat{\xi}_p - \xi_p) \stackrel{\mathcal{L}}{\longrightarrow} N\left(0, \frac{p(1-p)}{(F'(\xi_p))^2}\right).$$

Hence, an asymptotically correct $100(1 - \alpha)\%$ confidence interval for ξ_p is $\hat{\xi}_p \pm \frac{z_{1-\alpha/2}}{\sqrt{n}} \frac{\sqrt{p(1-p)}}{F'(\xi_p)}$. This confidence interval typically will have a lower coverage than the nominal $1 - \alpha$ level. But the interval has a simplistic appeal and is computed much more easily than the interval based on order statistics.

7.5 Regression Quantiles

Least squares estimates in regression minimize the sum of squared deviations of the observed and the expected values of the dependent variable. In the location-parameter problem, this principle would result in the sample mean as the estimate. If instead one minimizes the sum of the absolute values of the deviations, one would obtain the median as the estimate. Likewise, one can estimate the regression parameters by minimizing the sum of the absolute deviations between the observed values and the regression function. For example, if the model says $y_i = \beta'\mathbf{x_i} + \epsilon_i$, then one can estimate the regression vector β by minimizing $\sum_{i=1}^{n} |y_i - \beta'\mathbf{x_i}|$, a very natural idea. This estimate is called the *least absolute deviation* (LAD) regression estimate. While it is not as good as the least squares estimate when the errors are exactly Gaussian, it outperforms the least squares estimate for a variety of error distributions that are either thicker tailed than normal or have a sharp peak at zero. Generalizations of the LAD estimate, analogous to sample percentiles, are called *regression quantiles*. Good references for the material in this section and proofs of Theorems 7.5 and 7.6 below are Koenker and Bassett (1978) and Ruppert and Carroll (1980).

Definition 7.1 For $0 < p < 1$, the pth regression quantile is defined as $\hat{\beta}_{\mathbf{n,p}} = \text{argmin}_\beta[\sum_{i:y_i \geq \beta'\mathbf{x_i}} p|y_i - \beta'\mathbf{x_i}| + \sum_{i:y_i < \beta'\mathbf{x_i}} (1-p)|y_i - \beta'\mathbf{x_i}|]$.

Note that neither existence nor uniqueness are obvious. Existence is discussed in Koenker and Bassett (1978). If the minimum is not unique, then the regression quantile is interpreted as any member of the set of all minima. The following theorems describe the limiting distribution of one or several regression quantiles. There are some neat analogies in these results to limiting distributions of sample percentiles for iid data.

Theorem 7.5 Let $y_i = \beta'\mathbf{x_i} + \epsilon_i$, where $\epsilon_i \stackrel{\text{iid}}{\sim} F$, with F having median zero. Let $0 < p < 1$, and let $\hat{\beta}_{\mathbf{n,p}}$ be any pth regression quantile. Suppose F has a strictly positive derivative $f(F^{-1}(p))$ at $F^{-1}(p)$. Then, $\sqrt{n}(\hat{\beta}_{\mathbf{n,p}} - \beta - F^{-1}(p)\mathbf{e_1}) \stackrel{\mathcal{L}}{\Rightarrow} \mathbf{N_p}(\mathbf{0}, \mathbf{vD^{-1}})$, where $\mathbf{e_1} = (1, 0, \ldots, 0)^T$, $D = \lim \frac{1}{n}\mathbf{X'X}$ (assumed to exist), and $v = \frac{p(1-p)}{f^2(F^{-1}(p))}$.

The analogy to the limiting distribution of a sample percentile for iid data from F is clear. Similar to the joint asymptotic distribution of several percentiles for iid data, the joint asymptotic distribution of several regression quantiles is also multivariate normal. The next theorem describes that result.

Theorem 7.6 Let $k \geq 1$ be fixed and let $0 < p_1 < p_2 < \ldots < p_k < 1$. Assume that F is differentiable with strictly positive derivatives at $F^{-1}(p_i)$ for each $i = 1, 2, \ldots, k$. Then, $\sqrt{n}(\hat{\beta}_{\mathbf{n},\mathbf{p_1}} - \beta - F^{-1}(p_1)\mathbf{e_1}, \ldots, \hat{\beta}_{\mathbf{n},\mathbf{p_k}} - \beta - F^{-1}(p_k)\mathbf{e_1}) \stackrel{\mathcal{L}}{\Rightarrow} N_{pk}(\mathbf{0}, V \bigotimes D^{-1})$, where V is the symmetric matrix with elements $v_{ij} = \frac{p_i(1-p_j)}{f(F^{-1}(p_i))f(F^{-1}(p_j))}, i \leq j$, D is as in Theorem 7.5, and $\bigotimes$ denotes the Kronecker product.

Remark. An interesting aspect of the LAD estimate of the regression vector is that it has a smaller asymptotic covariance matrix than the least squares estimate iff the median has a smaller asymptotic variance than the mean in the case of iid data for the corresponding F. This brings out a unifying comparison between $\mathcal{L}_1$ and $\mathcal{L}_2$ methods for iid data and the regression model. Computation of the regression quantiles would typically involve iterative methods or constrained linear programming and is not discussed here.

7.6 Exercises

Exercise 7.1 Find the limiting distribution of the interquartile range for sampling from the normal, double exponential, and Cauchy distributions.

Exercise 7.2 Find the limiting distribution of the quartile ratio defined as $X_{[\frac{3n}{4}]:n}/X_{[\frac{n}{4}]:n}$ for the exponential, Pareto, and uniform distributions.

Exercise 7.3 * On denoting the sample median by M, find the limit of $P(|\bar{X} - \mu| \leq |M - \mu|)$ when sampling is iid from the normal, double exponential, and t_m distributions, $m > 2$.

Exercise 7.4 * Find the limiting distribution of the studentized IQR defined as $\frac{\text{IQR}}{s}$ for the normal, double exponential, Beta, and exponential distributions. Hint: Use Bahadur representation of the sample quantiles and then the delta theorem. Many examples of this kind are worked out in DasGupta and Haff (2006).

Exercise 7.5 Suppose $X_1, X_2, \ldots$ are iid $f(x-\mu)$. For each of the following cases, find the estimate of the form

$$pX_{[n\alpha_1]:n} + pX_{[n-n\alpha_1]:n} + (1-2p)X_{[\frac{n}{2}]:n}$$

that has the smallest asymptotic variance:

(a) $f(x) = \frac{1}{\sqrt{2\pi}} e^{-\frac{x^2}{2}}$.
(b) $f(x) = \frac{1}{2} e^{-|x|}$.
(c) $f(x) = \frac{1}{\pi(1+x^2)}$.

Exercise 7.6 * There is a unique t-distribution such that

$$\frac{1}{3} X_{[\frac{n}{4}]:n} + \frac{1}{3} X_{[\frac{n}{2}]:n} + \frac{1}{3} X_{[\frac{3n}{4}]:n}$$

has the smallest asymptotic variance among all estimates that are convex combinations of three sample percentiles. Identify the degrees of freedom of that t-distribution (the degree of freedom eventually has to be found numerically).

Exercise 7.7 * Suppose $X_1, X_2, \ldots$ are iid $f(x - \mu)$. Suppose T_n is the proportion of the sample data that are less than or equal to the sample mean. For each of the following cases, find the limiting distribution of T_n, centering and norming it suitably:

(a) f = standard normal.
(b) f = standard Cauchy.

Hint: $T_n = F_n(\bar{X})$, where F_n is the empirical CDF. This exercise is hard without the hint.

Exercise 7.8 Find the asymptotic variance of the Gastwirth estimate $0.3F_n^{-1}(\frac{1}{3}) + 0.4F_n^{-1}(\frac{1}{2}) + 0.3F_n^{-1}(\frac{2}{3})$ for the normal, double exponential, and Cauchy location-parameter cases.

Exercise 7.9 For iid samples from a CDF F, consider the statistic $R_n = \frac{\bar{X} - X_{[\frac{n}{2}]:n}}{X_{[\frac{3n}{4}]:n} - X_{[\frac{n}{4}]:n}}$. For symmetric F, R_n is sometimes used as a measure of the tail. Find the limiting distribution of R_n for each of the following cases:

(a) $F = N(\mu, \sigma^2)$.
(b) $F = U[\theta_1, \theta_2]$.

Exercise 7.10 * Suppose X_i are iid from a Poisson distribution with mean 1. How would the sample median behave asymptotically? Specifically, does it converge in probability to some number? Does it converge in distribution on any centering and norming?

Exercise 7.11 Prove or disprove that the LAD estimate of the regression vector in the multiple linear regression model is unbiased.

Exercise 7.12 For the simple linear regression model, give a confidence interval for the slope parameter centered at the LAD estimate such that its coverage is asymptotically $1 - \alpha$ for a given error distribution.

References

Bahadur, R.R. (1966). A note on quantiles in large samples, Ann. Math. Stat., 37, 577–580.
Bickel, P. (1967). Some contributions to the theory of order statistics, in *Proceedings of the Fifth Berkeley Symposium on Mathematical Statistics and Probability*, L. Le Cam and J. Neyman, (eds.), Vol. I, University of California Press, Berkeley, 575–591.
DasGupta, A. and Haff, L. (2006). Asymptotic expansions for correlations between different measures of spread, J.Stat. Planning Infer., 136, 2197–2212.
Ferguson, T.S. (1996). *A Course in Large Sample Theory*, Chapman and Hall, London.
Gastwirth, J. (1966). On robust procedures, J. Am. Stat. Assoc., 61, 929–948.
Ghosh, J.K. (1971). A new proof of the Bahadur representation of quantiles and an application, Ann. Math. Stat., 42, 1957–1961.
Kiefer, J. (1967). On Bahadur's representation of sample quantiles, Ann. Math. Stat., 38, 1323–1342.
Koenker, R. and Bassett, G. (1978). Regression quantiles, Econometrica, 46(1), 33–50.
Ruppert, D. and Carroll, R. (1980). Trimmed least squares estimation in the linear model, J. Am. Stat. Assoc., 75(372), 828–838.
Serfling, R. (1980). *Approximation Theorems of Mathematical Statistics*, John Wiley, New York.
Tong, Y.L. (1990). *Probability Inequalities in Multivariate Distributions*, Academic Press, New York.

Chapter 8
Sample Extremes

For sample observations $X_1, \ldots, X_n$, order statistics $X_{k:n}$ or $X_{n-k:n}$ for fixed k are called sample extremes. The asymptotic theory of sample extremes is completely different from that of central order statistics. This asymptotic theory is useful, as extremes are heavily used in the study of weather, natural disasters, financial markets, and many other practical phenomena. Although we emphasize the iid case, due to the current interest in dependent data, we also discuss some key results for certain types of dependent sequences. Good general references for this chapter are Galambos (1977), Reiss (1989), Leadbetter, Lindgren, and Rootzén (1983), and Sen and Singer (1993); more specific references are given later.

8.1 Sufficient Conditions

We start with an easy example that illustrates the different kinds of asymptotics extremes have compared with central order statistics.

Example 8.1 Let $U_1, \ldots, U_n \stackrel{\text{iid}}{\sim} \mathcal{U}[0, 1]$. Then

$$P(n(1 - U_{n:n}) > t) = P\left(1 - U_{n:n} > \frac{t}{n}\right) = P\left(U_{n:n} < 1 - \frac{t}{n}\right) = \left(1 - \frac{t}{n}\right)^n$$

$$\text{if } 0 < t < n,$$

so $P(n(1 - U_{n:n}) > t) \to e^{-t}$ for all real t, which implies that $n(1 - U_{n:n}) \xrightarrow{\mathcal{L}} \text{Exp}(1)$. Notice that the limit is nonnormal and the norming constant is n, not $\sqrt{n}$. The norming constant in general depends on the tail of the underlying CDF F.

It turns out that if $X_1, X_2, \ldots$ are iid from some F, then the limit distribution of $X_{n:n}$, if it exists at all, can be only one of three types.

A. DasGupta, *Asymptotic Theory of Statistics and Probability*,

Characterizations are available and were obtained rigorously by Gnedenko (1943), although some of his results were previously known to Frechet, Fisher, Tippett, and von Mises. The characterizations are somewhat awkward to state and can be difficult to verify. Therefore, we will first present more easily verifiable sufficient conditions.

Definition 8.1 A CDF F on an interval with $\xi(F) = \sup\{x : F(x) < 1\} < \infty$ is said to have terminal contact of order m if $F^{(j)}(\xi(F)-) = 0$ for $j = 1, \ldots, m$ and $F^{(m+1)}(\xi(F)-) \neq 0$.

Example 8.2 Consider the Beta density $f(x) = (m+1)(1-x)^m, 0 < x < 1$. Then the CDF $F(x) = 1 - (1-x)^{m+1}$. For this distribution, $\xi(F) = 1$ and $F^{(j)}(1) = 0$ for $j = 1, \ldots, m$, while $F^{(m+1)}(1) = (m+1)!$. Thus, F has terminal contact of order m.

Definition 8.2 A CDF F with $\xi(F) = \infty$ is said to be of an exponential type if F is absolutely continuous and infinitely differentiable and if, for each fixed $j \geq 2$, $\frac{F^{(j)}(x)}{F^{(j-1)}(x)} \asymp (-1)^{j-1}\frac{f(x)}{1-F(x)}$ as $x \to \infty$, where $\asymp$ means that the ratio converges to 1.

Example 8.3 Suppose $F(x) = \Phi(x)$, the $N(0, 1)$ CDF. Then

$$F^{(j)}(x) = (-1)^{j-1}H_{j-1}(x)\phi(x),$$

where $H_j(x)$ is the jth Hermite polynomial and is of degree j. Therefore, for every j, $\frac{F^{(j)}(x)}{F^{(j-1)}(x)} \asymp -x$. Thus $F = \Phi$ is a CDF of the exponential type.

Definition 8.3 A CDF F with $\xi(F) = \infty$ is said to be of a polynomial type of order k if $x^k(1 - F(x)) \to c$ for some $0 < c < \infty$ as $x \to \infty$.

Example 8.4 All t-distributions, therefore including the Cauchy, are of polynomial type. Consider the t-distribution with α degrees of freedom and with median zero. Then, it is easily seen that $x^\alpha(1 - F(x))$ has a finite nonzero limit. Hence, a t_α-distribution is of the polynomial type of order α.

We can now present our sufficient conditions for weak convergence of the maximum to three different types of limit distributions. The first three theorems below are proved in Sen and Singer (1993). The first result handles cases such as the uniform distribution on a bounded interval.

Theorem 8.1 Suppose $X_1, X_2, \ldots$ are iid from a CDF with mth-order terminal contact at $\xi(F) < \infty$. Then, for suitable a_n, b_n,

$$\frac{X_{n:n} - a_n}{b_n} \xrightarrow{\mathcal{L}} G,$$

where $G(t) = \begin{cases} e^{-(-t)^{m+1}} & t \le 0, \\ 1 & t > 0. \end{cases}$ Moreover, a_n can be chosen to be $\xi(F)$, and one can choose $b_n = \left\{ \frac{(-1)^m (m+1)!}{nF^{(m+1)}(\xi(F)-)} \right\}^{\frac{1}{m+1}}$. Here it is assumed that $(-1)^m F^{(m+1)}(\xi(F)-) > 0$.

The second result handles cases such as the t-distribution.

Theorem 8.2 Suppose $X_1, X_2, \ldots$ are iid from a CDF F of a polynomial type of order k. Then, for suitable a_n, b_n,

$$\frac{X_{n:n} - a_n}{b_n} \xrightarrow{\mathcal{L}} G,$$

where $G(t) = \begin{cases} e^{-t^{-k}} & t \ge 0 \\ 0 & t < 0. \end{cases}$ Moreover, a_n can be chosen to be 0, and one can choose $b_n = F^{-1}(1 - \frac{1}{n})$.

The last result handles in particular the important normal case.

Theorem 8.3 Suppose $X_1, X_2, \ldots$ are iid from a CDF F of an exponential type. Then, for suitable a_n, b_n,

$$\frac{X_{n:n} - a_n}{b_n} \xrightarrow{\mathcal{L}} G,$$

where $G(t) = e^{-e^{-t}}, \ -\infty < t < \infty$.

Remark. The choice of a_n, b_n will be discussed later.

Example 8.5 Suppose $X_1, \ldots, X_n$ are iid from a triangular density on $[0, \theta]$. So,

$$F(x) = \begin{cases} \frac{2x^2}{\theta^2} & \text{if } 0 \le x \le \frac{\theta}{2}, \\ \frac{4x}{\theta} - \frac{2x^2}{\theta^2} - 1 & \text{if } \frac{\theta}{2} \le x \le \theta. \end{cases}$$

It follows that $1 - F(\theta) = 0$ and $F^{(1)}(\theta-) = 0$, $F^{(2)}(\theta-) = -\frac{4}{\theta^2} \neq 0$. Thus F has terminal contact of order $(m+1)$ at θ with $m = 1$. It follows that $\frac{\sqrt{2n}(X_{n:n}-\theta)}{\theta} \xrightarrow{\mathcal{L}} G(t)$, where $G(t) = e^{-t^2}$, $t \leq 0$.

Example 8.6 Suppose $X_1, \ldots, X_n$ are iid standard Cauchy. Then

$$1 - F(x) = \frac{1}{\pi}\int_x^\infty \frac{1}{1+t^2}\,dt = \frac{1}{2} - \frac{\arctan(x)}{\pi} \sim \frac{1}{\pi x}$$

as $x \to \infty$; i.e., $F(x) \sim 1 - \frac{1}{\pi x}$. Therefore, $F^{-1}(1 - \frac{1}{n}) \sim \frac{n}{\pi}$. Hence, it follows that $\frac{\pi X_{n:n}}{n} \xrightarrow{\mathcal{L}} G(t)$, where $G(t)$ is the CDF of the reciprocal of an Exponential(1) random variable.

Example 8.7 Suppose $X_1, X_2, \ldots$ are iid $N(0, 1)$. Then one can show with nontrivial calculations (see Galambos (1977)) that

$$\sqrt{2\log n}\Big(X_{n:n} - \sqrt{2\log n} + \frac{\log\log n + \log 4\pi}{2\sqrt{2\log n}}\Big) \xrightarrow{\mathcal{L}} G(t) = e^{-e^{-t}},$$

$-\infty < t < \infty$. This is the CDF of an *extreme value* distribution, also known as the *Gumbel law*.

Remark. It is often the case that the rates of convergence in these limit theorems are very slow. Establishing an exact rate of convergence in the CLT for extremes is a hard problem. Some results are known. The rate of convergence of the Kolmogorov distance between the CDF of $X_{n:n}$ (suitably centered and normed) and its limit in the normal case was proved to be $\frac{1}{\log n}$ in Hall (1979). Other results are also available. For example, the Hellinger distance between $\Phi^n(a_n + b_n x)$ and the limiting extreme value CDF is shown to be $O(\frac{1}{\log n})$ in Reiss (1989), with a_n, b_n as in Theorem 8.3.

There are also results available on convergence of sample extremes to their limit distribution in total variation. For example,

$$\sup_{\text{Borel } A}\left|P\left(\frac{\pi}{n}X_{n:n} \in A\right) - P\left(\frac{1}{\text{Exp}(1)} \in A\right)\right| = O\Big(\frac{1}{n}\Big)$$

if $X_1, X_2, \ldots$ are standard Cauchy; see Reiss (1989).

8.2 Characterizations

We now present the characterization results on weak convergence of sample extremes. The results say that the sequence of sample extremes of an iid sequence can converge in distribution to one of only three types of laws. This is arguably the most fundamental result in extreme value theory and is known as the *convergence of types theorem*. For convenience, we introduce the following notation. For $r > 0$, let

$$\left.\begin{aligned} G_{1,r}(x) &= e^{-x^{-r}}, && x > 0 \\ &= 0, && x \le 0 \\ G_{2,r}(x) &= e^{-(-x)^r}, && x \le 0 \\ &= 1, && x > 0 \\ G_3(x) &= e^{-e^{-x}}, && -\infty < x < \infty \end{aligned}\right\}.$$

We also need a few definitions.

Definition 8.4 A function g is called slowly varying at ∞ if, for each $t > 0$, $\lim_{x\to\infty} \frac{g(tx)}{g(x)} = 1$ and is said to be of regular variation of index r if, for each t > 0, $\lim_{x\to\infty} \frac{g(tx)}{g(x)} = t^{-r}$.

Definition 8.5 Suppose $X_1, X_2, \ldots$ are iid with CDF F. We say that F is in the domain of maximal attraction of a CDF G and write $F \in \mathcal{D}(G)$ if, for some a_n, b_n,

$$\frac{X_{n:n} - a_n}{b_n} \xrightarrow{\mathcal{L}} G.$$

We now state the three main theorems of this section. See Galambos (1977) for their proofs.

Theorem 8.4 $F \in \mathcal{D}(G_{1,r})$ iff $\xi(F) = \infty$ and $1 - F$ is of regular variation at $\xi(F) = \infty$.

Theorem 8.5 $F \in \mathcal{D}(G_{2,r})$ iff $\xi(F) < \infty$ and $\widetilde{F} \in \mathcal{D}(G_{1,r})$, where $\widetilde{F}(x)$ is the CDF $\widetilde{F}(x) = F(\xi(F) - \frac{1}{x})$, $x > 0$.

Theorem 8.6 $F \in \mathcal{D}(G_3)$ iff there is a function $u(t) > 0$ such that $\lim_{t\to\xi(F)} \frac{1-F(t+xu(t))}{1-F(t)} = e^{-x}$ for all x.

Remark. In this last case, a_n, b_n can be chosen as $a_n = F^{-1}(1-\frac{1}{n})$ and $b_n = u(a_n)$. However, the question of choosing such a function $u(t)$ is nontrivial. We will present several available options under different circumstances.

Definition 8.6 Given a CDF F on the real line that is differentiable with derivative $F'(t)$, the hazard rate of F is the function $r(t) = \frac{F'(t)}{1-F(t)}$.

Definition 8.7 The residual mean survival time at time t is the function $R(t) = E_F(X - t|X > t)$.

Remark. Clearly, $R(t)$ equals the expression $R(t) = \frac{1}{1-F(t)} \int_t^{\xi(F)} (1 - F(x))\, dx$. It is easily proved that the function $R(t)$ determines F; so does the function $r(t)$.

In many applications, the choice $u(t) = \frac{1}{r(t)}$ works. But this choice need not work universally. It is true under mild conditions that $R(t)r(t) \to 1$ as $t \to \infty$. One may therefore suspect that $R(t)$ may be a candidate for the $u(t)$ function. The following results describe sufficient conditions under which this is true.

Theorem 8.7 Suppose $\lim_{t\to\xi(F)} \left(\frac{1}{r(t)}\right)' = 0$. Then $F \in \mathcal{D}(G_3)$ with $u(t) = R(t)$.

An alternative sufficient condition is the following, covering the case of unbounded upper endpoints.

Theorem 8.8 Suppose $\xi(F) = \infty$, that $R(t)$ is of regular variation at $t = \infty$, and that $\lim_{t\to\infty} \frac{R(t)}{t} = 0$. Then $F \in \mathcal{D}(G_3)$.

Example 8.8 Let $X \sim \text{Exp}(1)$. Then, trivially, $R(t) = \text{constant}$. Any constant function is obviously slowly varying and so is of regular variation. Therefore, the exponential CDF $F \in \mathcal{D}(G_3)$.

Example 8.9 Let $X \sim N(0, 1)$. Then, by using the second order approximation to Mill's ratio $\frac{1-\Phi(x)}{\phi(x)} = \frac{1}{x} - \frac{1}{x^3} + O(x^{-5})$, it follows that $R(t) \sim \frac{t}{t^2-1}$ as $t \to \infty$, and therefore $R(t)$ is of regular variation. Again, therefore it follows that $\Phi \in \mathcal{D}(G_3)$.

We end with a result on when the choice $u(t) = \frac{1}{r(t)}$ itself works.

Theorem 8.9 If the hazard rate $r(t)$ is monotone increasing or decreasing, then $F \in \mathcal{D}(G_3)$ and one can choose $u(t) = \frac{1}{r(t)}$.

Example 8.10 Let $X \sim \text{Gamma}(\alpha, 1)$ with density $f(x) = \frac{e^{-x}x^{\alpha-1}}{\Gamma(\alpha)}$, and let $\alpha > 1$. Then the hazard rate $r(t)$ is increasing, and so one can choose $u(t)$ to be $u(t) = \frac{1}{r(t)} = \frac{\int_t^\infty e^{-x}x^{\alpha-1}dx}{e^{-t}t^{\alpha-1}}$; for integer α, this can be further simplified in closed form, and otherwise it involves the incomplete Gamma function.

8.3 Limiting Distribution of the Sample Range

Suppose $X_1, X_2, \ldots, X_n$ are iid observations from some CDF F. The extreme order statistics and the central order statistics have widely separated indices for large n. Therefore, one may suspect that central order statistics and extremes are approximately independent for large n. This is true. In fact, any linear statistic is also asymptotically independent of the sample extremes. A general result is Theorem 8.10; it is then generalized in Theorem 8.11. See Reiss (1989) for Theorem 8.10; Theorem 8.11 is proved in essentially the same manner. A major consequence of these asymptotic independence results is that they often lead to the asymptotic distribution of the sample range $X_{n:n} - X_{1:n}$.

Theorem 8.10 Suppose $X_1, X_2, \ldots \overset{\text{iid}}{\sim} F$ and $g : \mathbb{R} \to \mathbb{R}$ such that $E_F(g(X)) = 0$ and $\text{Var}_F(g(X)) = 1$. Suppose $X_{n:n}$ has a weak limit; i.e., for suitable a_n, b_n, $M_n = \frac{X_{n:n}-a_n}{b_n} \xrightarrow{\mathcal{L}} H$. Let $Z_n = \frac{\sum_{i=1}^n g(X_i)}{\sqrt{n}}$. Then $(M_n, Z_n) \xrightarrow{\mathcal{L}} H \otimes \Phi$, where $H \otimes \Phi$ is the product measure, with H and Φ as the marginals.

Theorem 8.11 Suppose $X_1, \ldots, X_n \overset{\text{iid}}{\sim} F$, and let $g : \mathbb{R} \to \mathbb{R}$ be such that $E_F(g(X)) = 0$ and $\text{Var}_F(g(X)) = 1$. Suppose $X_{1:n}, X_{n:n}$ each have a weak limit; i.e., for suitable a_n, b_n, c_n, d_n, $m_n = \frac{X_{1:n}-c_n}{d_n} \xrightarrow{\mathcal{L}} G$ and $M_n = \frac{X_{n:n}-a_n}{b_n} \xrightarrow{\mathcal{L}} H$. Also let $Z_n = \frac{\sum_{i=1}^n g(X_i)}{\sqrt{n}}$. Then

$$(m_n, M_n, Z_n) \xrightarrow{\mathcal{L}} G \otimes H \otimes \Phi.$$

Corollary 8.1 If m_n, M_n each have a weak limit, then they are asymptotically independent.

Here is a very interesting application.

Example 8.11 Suppose $X_1, X_2, \ldots \overset{\text{iid}}{\sim} N(0, 1)$. Then, as we saw earlier, for $b_n = \frac{1}{\sqrt{2\log n}}$ and $a_n = \sqrt{2\log n} - \frac{\log\log n + \log 4\pi}{2\sqrt{2\log n}}$,

$$\frac{X_{n:n} - a_n}{b_n} \xrightarrow{\mathcal{L}} G_3,$$

where $G_3(x) = e^{-e^{-x}}$, $-\infty < x < \infty$. By symmetry, $\frac{X_{n:n}-a_n}{b_n} \stackrel{\mathcal{L}}{=} \frac{-X_{1:n}-a_n}{b_n}$. Since $X_{1:n}$ and $X_{n:n}$ are asymptotically independent by Theorem 8.11, if $W_n = X_{n:n} - X_{1:n}$ denotes the sample range, then

$$\frac{W_n - 2a_n}{b_n} = \frac{X_{n:n} - a_n - X_{1:n} - a_n}{b_n} \xrightarrow{\mathcal{L}} G_3 * G_3,$$

where $G_3 * G_3$ denotes the convolution of G_3 with itself. Now, the density of $G_3 * G_3$ is

$$\int_{-\infty}^{\infty} e^{-e^{-(x-y)}} e^{-(x-y)} e^{-y} dy = 2e^{-x} K_0(2e^{-\frac{x}{2}}),$$

where K_0 is the Bessel function denoted by that notation.

Here is another important example on the limiting distribution of the sample range.

Example 8.12 Let $X_1, \ldots, X_n \stackrel{\text{iid}}{\sim} U[0, 1]$. Then the order statistics $X_{k:n}$, $1 \le k \le n$, have the following elegant representation. Let $Z_1, \ldots, Z_{n+1}$ be iid Exp(1) and $S_{n+1} = \sum_{i=1}^{n+1} Z_i$. Then

$$(X_{1:n}, X_{2:n}, \ldots, X_{n:n}) \stackrel{\mathcal{L}}{=} \left(\frac{Z_1}{S_{n+1}}, \frac{Z_1 + Z_2}{S_{n+1}}, \ldots, \frac{Z_1 + Z_2 + \ldots + Z_n}{S_{n+1}} \right);$$

see Reiss (1989). Therefore the sample range $W_n = X_{n:n} - X_{1:n} = \frac{Z_2+\cdots+Z_n}{S_{n+1}}$ $\Rightarrow 1 - W_n = \frac{Z_1+Z_{n+1}}{S_{n+1}} \Rightarrow n(1 - W_n) = (Z_1 + Z_{n+1})\frac{1}{\frac{S_{n+1}}{n}} \xrightarrow{\mathcal{L}} \frac{1}{2}\chi^2(4)$ since $\frac{S_{n+1}}{n} \stackrel{P}{\Rightarrow} 1$.

8.4 Multiplicative Strong Law

Besides weak convergence, another interesting question is whether extreme order statistics satisfy a strong law. The results here are mixed; sometimes they do, and sometimes they do not. We discuss the possibilities of strong laws for sample maxima in this section and the next.

Definition 8.8 The sample maximum $X_{n:n}$ of an iid sequence is said to admit a *multiplicative strong law* if there exists b_n such that $\frac{X_{n:n}}{b_n} \stackrel{\text{a.s.}}{\to} 1$.

There is an elegant characterization of when a multiplicative strong law holds.

Theorem 8.12 $\{X_{n:n}\}$ admits a multiplicative strong law iff, for all $k > 1$,

$$\sum_{n=1}^{\infty}\left[1 - F\left(kF^{-1}\left(1 - \frac{1}{n}\right)\right)\right] < \infty,$$

in which case $\frac{X_{n:n}}{F^{-1}(1-\frac{1}{n})} \stackrel{\text{a.s.}}{\rightarrow} .1$.

See Galambos (1977) for a proof.

Example 8.13 Let $X_1, X_2, \ldots \stackrel{\text{iid}}{\sim} N(0, 1)$. Then, using the fact that $1 - \Phi(t) \sim \frac{\phi(t)}{t}$ as $t \to \infty$ and hence $\Phi^{-1}(1 - \frac{1}{n}) \sim \sqrt{2\log n}$, one has

$$1 - \Phi\left(k\Phi^{-1}\left(1 - \frac{1}{n}\right)\right) = O\Big(\frac{1}{n^{k^2}\sqrt{\log n}}\Big).$$

Therefore, for all $k > 1$, $\sum_{n\geq 1}\left[1-\Phi\big(k\Phi^{-1}(1-\frac{1}{n})\big)\right] < \infty$ and so $\frac{X_{n:n}}{\sqrt{2\log n}} \stackrel{\text{a.s.}}{\rightarrow} .1$, an important result in asymptotics of extremes.

Example 8.14 Let $X_1, X_2, \ldots$ be iid standard Cauchy. Then $1 - F(x) \sim \frac{1}{\pi x}$. It follows that $1 - F\big(kF^{-1}(1 - \frac{1}{n})\big) \asymp \frac{1}{kn}$ for any $k > 0$. But $\sum_{n=1}^{\infty} \frac{1}{n} = \infty$; hence $\{X_{n:n}\}$ does not admit a multiplicative strong law in the Cauchy case.

Example 8.15 Let $X_1, X_2, \ldots$ be iid standard exponential. Then, for $x > 0$, $1-F(x) = e^{-x}$ and therefore $1-F\big(kF^{-1}(1-\frac{1}{n})\big) = n^{-k}$. Since $\sum_{n=1}^{\infty} n^{-k} < \infty$ for all $k > 1$, in the exponential case one has the multiplicative strong law $\frac{X_{n:n}}{\log n} \stackrel{\text{a.s.}}{\Rightarrow} 1$, another important result in the asymptotic theory of extremes.

8.5 Additive Strong Law

Strong laws in the sense of a difference rather than a ratio are called *additive strong laws*. For CDFs with $\xi(F) = \infty$, an additive strong law is stronger than a multiplicative strong law, and in some cases, an additive strong law is needed. Here is a formal definition.

Definition 8.9 The sample maximum $X_{n:n}$ of an iid sequence is said to admit an *additive strong law* if there exists a_n such that $X_{n:n} - a_n \stackrel{\text{a.s.}}{\rightarrow} 0$.

Let us start with a simple example.

Example 8.16 Suppose $X_1, \ldots, X_n$ is an iid sample from the discrete uniform distribution on $\{1, 2, \ldots, M\}$ for some fixed $M < \infty$. Then, $P(X_{n:n} < M) = (1 - \frac{1}{M})^n$ and so clearly $\sum_{n=1}^{\infty} P(X_{n:n} < M) < \infty$. By the Borel-Cantelli lemma, it follows that $X_{n:n} = M$ for all large n with probability 1 and hence $X_{n:n} - M \stackrel{\text{a.s.}}{\rightarrow} 0$. Thus, by direct verification, $X_{n:n}$ satisfies an additive strong law.

However, in general, the question of the existence of an additive strong law is entirely nontrivial. We next give a few results to either confirm that an additive strong law holds or to prove that it does not hold; see Galambos (1977) for details and proofs. In particular, Theorem 8.15 below gives a sufficient condition under which an additive strong law holds with $a_n = F^{-1}(1 - \frac{1}{n})$, an intuitive choice.

Theorem 8.13 Let $X_1, X_2, \ldots \stackrel{\text{iid}}{\sim} F$. Let $\{a_n\}$ be a nondecreasing sequence. Then $P(X_{n:n} \geq a_n \quad \text{infinitely often}) = 0$ or 1 according to whether $\sum(1 - F(a_n)) < \infty$ or $= \infty$.

Theorem 8.14 (a) Let $X_1, X_2, \ldots \stackrel{\text{iid}}{\sim} F$ and let $\{a_n\}$ be a nondecreasing sequence. Then $P(X_{n:n} \leq a_n \quad \text{i.o.}) = 0$ if $\sum(1 - F(a_n)) = \infty$ and $\sum(1 - F(a_n))e^{-n(1-F(a_n))} < \infty$.

(b) If $\frac{1}{2} \leq \frac{n(1-F(a_n))}{\log \log n} \leq 2$ and $\sum(1 - F(a_n))e^{-n(1-F(a_n))} = \infty$, then $P(X_{n:n} \leq a_n \text{ i.o.}) = 1$.

Theorem 8.15 Let $X_1, X_2, \ldots \stackrel{\text{iid}}{\sim} F$ with a density function $f(x)$, and suppose $F^{-1}(1) = \infty$. Suppose $\forall \epsilon > 0$, $\int_{-\infty}^{\infty} \frac{f(x)}{1-F(x-\epsilon)} dx < \infty$. Then $X_{n:n}$ satisfies an additive strong law. Furthermore, if $X_{n:n} - a_n \stackrel{\text{a.s.}}{\Rightarrow} 0$, then $\frac{a_n}{F^{-1}(1-\frac{1}{n})} \rightarrow 1$.

Remark. Note that Theorem 8.15 does not say that $F^{-1}(1-\frac{1}{n})$ is necessarily a choice for the sequence a_n.

We give an example of nonexistence of an additive strong law.

Example 8.17 If $X_1, X_2, \ldots$ are iid Exp(1), then we saw earlier that $\frac{X_{n:n}}{\log n} \stackrel{\text{a.s.}}{\Rightarrow} 1$. However, for given $\epsilon > 0$, $\sum(1 - F(\log n + \epsilon)) = \sum e^{-\log n - \epsilon} = \infty$, and therefore by Theorem 8.13, $P(X_{n:n} \geq \log n + \epsilon \text{ i.o.}) = 1$, implying that $X_{n:n} - \log n$ does not converge a.s. to zero. In fact, an additive strong law does not hold with any sequence $\{a_n\}$ in this case.

8.6 Dependent Sequences

The asymptotic theory of extremes for an iid sequence is classic. However, due to some statistical problems of current interest, there has been a renewed interest in the corresponding theory for dependent sequences. See Chapter 34 for some examples of such problems. Although there are some broad technical overlaps in how asymptotics for extremes of different types of dependent sequences are determined, the details differ depending on the exact nature of the sequence as well as the strength of dependence. We will mostly confine ourselves to stationary Gaussian sequences. The main message of the results is that if the autocorrelation function dies out quickly enough, then the weak limits for extremes remain the same as in the iid case. Otherwise, the weak limits change, and they can change in different ways, depending on the exact strength of dependence. In particular, interestingly, for very slowly decaying correlations, the weak limit would not be skewed; it would be normal again. We recommend Berman (1962a, b, 1964), Cramér (1962), Leadbetter (1974), and Leadbetter, Lindgren, and Rootzén (1983), for the material in this section. Other references specific to particular themes are mentioned later.

The proofs for the Gaussian case are based on an explicit bound on the difference between two multivariate normal CDFs in terms of how much their covariance matrices differ. It is of enough interest on its own; see Leadbetter (indgren, and Rootzén (1983) for a proof of it, as well as for a proof of the theorem that follows. Here is the bound.

Proposition 8.1 *Let $X_{n\times 1}$ be multivariate normal with zero means, unit variances, and correlations λ_{ij}. Let $Y_{n\times 1}$ be multivariate normal with zero means, unit variances, and correlations η_{ij}. Let $\rho_{ij} = \max(|\lambda_{ij}|, |\eta_{ij}|) < \delta < 1$. Then, for any $u_1, \cdots, u_n$,*

$$|P(X_i \le u_i, 1 \le i \le n) - P(Y_i \le u_i, 1 \le i \le n)| \le C(\delta) \sum_{1 \le i < j \le n} |\lambda_{ij} - \eta_{ij}| \, e^{-\frac{[\min(u_1, \cdots, u_n)]^2}{1+\rho_{ij}}}.$$

Here now is the main result on weak limits of maxima of stationary Gaussian sequences.

Theorem 8.16 Let $\{X_n\}_{n=1}^{\infty}$ be a zero-mean unit-variance stationary Gaussian process with autocorrelation function ρ_k. Let $b_n = \sqrt{2\log n}$, $a_n = b_n - \frac{\log\log n + \log 4\pi}{2b_n}$, and $Q(x) = e^{-e^{-x}}$, $-\infty < x < \infty$. Also let $X_{(n)} = \max_{1\le k\le n} X_k$.

(a) Suppose $\rho_k \log k \to 0$ as $k \to \infty$. Then $P(b_n[X_{(n)} - a_n] \le x) \to Q(x)$ for all x as $n \to \infty$.

(b) Suppose $\rho_k \log k \to \gamma$ for some $0 < \gamma < \infty$. Then $P(b_n[X_{(n)} - a_n] \le x) \to \int_{-\infty}^{\infty} e^{-e^{-x-\gamma+\sqrt{2\gamma}z}} \phi(z)dz$, where ϕ is the standard normal density function.

(c) Suppose $\rho_k \downarrow 0$ and $\rho_k \log k \uparrow \infty$. Then $P(\frac{1}{\sqrt{\rho_n}}[X_{(n)} - \sqrt{1-\rho_n}a_n] \le x) \to \Phi(x)$ for all x, where Φ is the standard normal CDF.

Remark. Part (a) says that even if ρ_k was just a little smaller than $O(\frac{1}{\log k})$, the dependence does not at all affect the limiting distribution of the series maximum, including the centering and the normalization constants. For the common Gaussian sequences in practice, the decay of ρ_k is in fact faster than $\frac{1}{\log k}$. Therefore, for all those processes, the maximum behaves in distribution, asymptotically, just like the iid case. In part (b), the maximum converges to a convolution of a random variable having the CDF Q (a Gumbel random variable) and a normal random variable that has variance 2γ. The importance of the normal component in the convolution increases as γ increases, roughly speaking, and when formally γ becomes ∞, the limit distribution becomes a normal itself, although with different normalizations. An example of extreme dependence is a constant sequence. In that case, the maximum is normally distributed for any n trivially. This illustrates why strong dependence might make the limit distribution a normal. It is shown in McCormick (1980) that if the monotonicity assumptions in part (c) are removed (that is, if ρ_k and $\rho_k \log k$ show oscillatory behavior), then other nonnormal weak limits can occur. See also Mittal and Ylvisaker (1975) for the version in which we have stated Theorem 8.16.

Part (a) of Theorem 8.16, which says that the weak convergence result for the maximum does not change if the dependence is weak enough, generalizes very nicely. See Loynes (1965) and Leadbetter, Lindgren, and Rootzén (1983). Weak dependence can be defined in many ways. Mixing is one of the most common forms of weak dependence; see Chapter 12 for more information. The following condition is weaker than what is known as *strong mixing*.

Condition M Let $\{X_n\}_{n=1}^{\infty}$ be a stationary sequence and let $F_{i_1,\cdots,i_p}(x_1,\cdots,x_p)$ denote the joint CDF of $(X_{i_1},\cdots,X_{i_p}), i_1 < \cdots < i_p$. We say that $\{X_n\}$ satisfies Condition M if for $i_1 < \cdots < i_p, j_1 < \cdots j_q, j_1 - i_p \ge n$, $|F_{i_1,\cdots,i_p,j_1,\cdots,j_q}(x,x,\cdots,x) - F_{i_1,\cdots,i_p}(x,\cdots,x)F_{j_1,\cdots,j_q}(x,\cdots,x)| \le \alpha(n)$, where $\alpha(n) \to 0$ as $n \to \infty$.

Theorem 8.17 Let $\{X_n\}_{n=1}^{\infty}$ be a stationary sequence satisfying Condition M and $\{Z_n\}_{n=1}^{\infty}$ the corresponding iid sequence. Suppose $P(b_n(Z_{n:n} - a_n) \leq x) \to G(x)$, a CDF, as $n \to \infty$, where $Z_{n:n}$ is the maximum of an iid sample of size n. Then, $P(b_n(X_{(n)} - a_n) \leq x) \to G(x)$ as $n \to \infty$; i.e., the weak limit of the maximum of a series of length n remains unchanged from the iid case, with the same centering and normalization.

Example 8.18 Here is an example of a stationary Gaussian sequence whose correlation structure is not covered by Theorem 8.16. Suppose $\{X_n\}$ is a stationary Gaussian sequence with mean 0, variance 1, and a constant autocorrelation function $\rho_k \equiv \rho, 0 < \rho < 1$. We can represent the members of this sequence as $X_i = \sqrt{1-\rho}Z_i + \sqrt{\rho}Z_0$, where $Z_i, i \geq 0$ is an iid standard normal sequence. The series maximum has the corresponding representation $X_{(n)} = \sqrt{1-\rho}Z_{n:n} + \sqrt{\rho}Z_0$. Now, we know from the iid theory that $G_n = b_n(Z_{n:n} - a_n)$ converges in law to a Gumbel random variable, say G. Therefore, G_n is an $O_p(1)$ sequence, and therefore $\frac{G_n}{b_n}$ is $o_p(1)$. It follows that on centering at $\sqrt{1-\rho}a_n$, the series maximum has a limiting normal distribution. Compare this with part (c) of Theorem 8.16.

In the iid standard normal case, we noted that the sample maximum admits a multiplicative as well as an additive strong law. The question naturally arises if the same holds in the dependent case as well. Pickands (1967) showed that indeed exactly the same multiplicative and additive strong laws hold for stationary Gaussian sequences under mild conditions on the rate of decay of the autocorrelation function. Mittal (1974) gives the exact rate of convergence in the additive strong law. We collect these important results below.

Theorem 8.18 Let $\{X_n\}_{n=1}^{\infty}$ be a stationary Gaussian sequence with mean 0 and variance 1. Let $b_n = \sqrt{2\log n}$.

(a) If $\rho_k = o(1)$, then $\frac{X_{(n)}}{b_n} \overset{\text{a.s.}}{\to} 1$.

(b) If $\rho_k \log k = o(1)$, then $X_{(n)} - b_n \overset{\text{a.s.}}{\to} 0$.

(c) If $\rho_k \log k = O(1)$, then

$$\limsup_{n\to\infty} \frac{\log n \left(\frac{X_{(n)}}{b_n} - 1\right)}{\log\log n} = \frac{1}{4}$$

and

$$\liminf_{n\to\infty} \frac{\log n \left(\frac{X_{(n)}}{b_n} - 1\right)}{\log\log n} = -\frac{1}{4}$$

almost surely.

8.7 Exercises

Exercise 8.1 * Give an example to show that sample maxima need not have a limiting distribution; i.e., no centering and normalization can result in a nondegenerate limit distribution.

Hint: Try the Poisson distribution.

Exercise 8.2 * In what domain of attraction are the distributions (a) $F = t_\alpha$, $\alpha > 0$; (b) $F = (1-\epsilon)N(0,1) + \epsilon C(0,1)$; and (c)$F = \chi_k^2$?

Exercise 8.3 * Simulate the distribution of the sample maximum for the $N(0, 1)$ case when $n = 50$ and compare it with the asymptotic distribution. Comment on the accuracy of the asymptotic answer as an approximation.

Exercise 8.4 * It is reasonable to expect that, for all n (or at least all large n), the sample range and sample standard deviation are positively correlated for iid samples from a CDF F. No general proof is known.

(a) Prove this for the normal case.

(b) Prove this for the exponential case.

(c) Prove this for the uniform case.

Exercise 8.5 Calculate the conditional expectations of the sample maximum given the sample mean and the sample mean given the maximum in the normal case. Here n is to be taken as fixed.

Hint: In the normal case, $\bar{X}$ and $X_{n:n} - \bar{X}$ are independent by Basu's theorem (Basu (1955)). On the other hand, given $X_{n:n}$, the other sample values act like an iid sample from the corresponding truncated population.

Exercise 8.6 Let $X_1, X_2, \ldots$ be iid Exp(1) samples.

(a) Derive an additive strong law for the sample minimum.

(b) Derive a multiplicative strong law for the sample minimum.

Exercise 8.7 Let $X_1, X_2, \ldots$ be iid $U[0, 1]$ samples.

(a) Derive an additive strong law for the sample minimum.

(b) Derive a multiplicative strong law for the sample minimum.

Exercise 8.8 Suppose $X_1, X_2, \ldots, X_k$ are iid $\text{Bin}(n, p)$ observations, where n and p are both fixed.

(a) Show that the sample maximum $X_{(k)} = n$, with probability 1, for all large k.

(b) Compute $E(X_{(k)} - n)$ with $k = 10, 25, 50$, and 100 when $n = 100$ and $p = 0.1$.

Exercise 8.9 Suppose $(X_1, X_2) \sim BVN(0, 0, 1, 1, \rho)$. Find the mean and variance of $X_{(2)} = \max(X_1, X_2)$.

Exercise 8.10 Suppose $X_1, X_2, \ldots$ are iid standard double exponential. Does $X_{n:n}$ admit a multiplicative strong law?

Exercise 8.11 Consider $X_1, X_2, \ldots$ as in Exercise 8.10. Find the limiting distribution of $W_n = X_{n:n} - X_{1:n}$.

Exercise 8.12 * Suppose $X_1, X_2, \ldots$ are iid $C(0, 1)$. Find the limiting distribution of $W_n = X_{n:n} - X_{1:n}$.

Exercise 8.13 For $X_1, X_2, \ldots$ iid $N(0, 1)$, find the limiting distribution of the midrange $\frac{X_{1:n}+X_{n:n}}{2}$.

Exercise 8.14 * Suppose $X_1, X_2, \ldots$ are iid standard normal. Does $X_{n:n}$ admit an additive strong law? Prove or disprove.

Exercise 8.15 Suppose $X_1, X_2, \ldots$ are iid standard double exponential. Does $X_{n:n}$ admit an additive strong law? Prove or disprove.

Exercise 8.16 * Suppose $X_1, X_2, \ldots$ are iid standard Cauchy. Does $X_{n:n} - (a + bn) \stackrel{a.s.}{\Rightarrow} 0$? for any a, b?

Exercise 8.17 (a) * Suppose X_i are iid $N(\mu, 1)$. Is $m_n = \frac{X_{1:n}+X_{n:n}}{2}$ a consistent estimate of μ?

(b) * Suppose X_i are iid with density $\frac{1}{2}e^{-|x-\mu|}$. Is m_n consistent for μ?

Hint: For (i), the answer is yes, and use the corresponding weak convergence result; for (ii), remarkably, the answer is no!

Exercise 8.18 * Suppose $X_1, X_2, \dots$ are iid $U[0, 1]$. Does $X_{n-1:n}$ have a limit distribution on suitable centering and norming? If so, find a centering, norming, and the limit distribution.

Exercise 8.19 * Suppose $X_1, X_2, \dots$ are iid Exp(1). Do the mean and the variance of $X_{n:n} - \log n$ converge to those of the Gumbel distribution G_3?

Hint: Show by a direct calculation that they do.

Exercise 8.20 * Suppose $\{X_n\}$ is a zero-mean, unit-variance stationary Gaussian sequence with constant correlation function $\rho_k \equiv \rho, 0 < \rho < 1$. Does $X_{(n)}$ admit an additive strong law? A multiplicative strong law?

Exercise 8.21 Plot the density function of the convolution distribution in Theorem 8.16 (part (b)) for $\gamma = .1, 1, 5$ and comment on the shapes.

Exercise 8.22 * Give an example of a correlation function that satisfies $\rho_k \to 0$ and $\rho_k \log k \to \infty$ but for which neither of these two sequences converges monotonically.

References

Basu, D. (1955). On statistics independent of a complete sufficient statistic, Sankhya, 15, 377–380.

Berman, S. (1962a). A law of large numbers for the maximum in a stationary Gaussian sequence, Ann. Math. Stat., 33, 93–97.

Berman, S. (1962b). Limiting distribution of the maximum term in sequences of dependent random variables, Ann. Math. Stat., 33, 894–908.

Berman, S. (1964). Limit theorem for the maximum term in stationary sequences, Ann. Math. Stat., 35, 502–516.

Cramér, H. (1962). On the maximum of a normal stationary stochastic process, Bull. Am. Math. Soc., 68, 512–516.

Galambos, J.(1977). *The Asymptotic Theory of Extreme Order Statistics*, Academic Press, New York.

Gnedenko, B. (1943). Sur la distribution limite du terme maximum d'une serie aleatoire, Ann. Math., 2(44), 423–453.

Hall, P. (1979). On the rate of convergence of normal extremes, J. Appl. Prob., 16(2), 433–439.

Leadbetter, M. (1974). On extreme values in stationary sequences, Z. Wahr. Verw. Geb., 28, 289–303.

Leadbetter, M., Lindgren, G., and Rootzén, H. (1983). *Extremes and Related Properties of Random Sequences and Processes*, Springer, New York.

Loynes, R. (1965). Extreme values in uniformly mixing stationary stochastic processes, Ann. Math. Stat., 36, 993–999.

McCormick, W. (1980). Weak convergence for the maximum of stationary Gaussian processes using random normalization, Ann. Prob., 8, 483–497.
Mittal, Y. (1974). Limiting behavior of maxima in stationary Gaussian sequences, Ann. Prob., 2, 231–242.
Mittal, Y. and Ylvisaker, D. (1975). Limit distribution for the maxima of stationary Gaussian processes, Stoch. Proc. Appl., 3, 1–18.
Pickands, J. (1967). Maxima of stationary Gaussian processes, Z. Wahr. Verw. Geb., 7, 190–223.
Reiss, R. (1989). *Approximate Distributions of Order Statistics, with Applications to Nonparametric Statistics*, Springer-Verlag, New York.
Sen, P.K. and Singer, J. (1993). *Large Sample Methods in Statistics: An Introduction with Applications*, Chapman and Hall, New York.

Chapter 9
Central Limit Theorems for Dependent Sequences

The assumption that observed data $X_1, X_2, \ldots$ form an independent sequence is often one of technical convenience. Real data frequently exhibit some dependence and at the least some correlation at small lags. Exact sampling distributions for fixed n are even more complicated for dependent data than in the independent case, and so asymptotics remain useful. In this chapter, we present CLTs for some important dependence structures. The cases of stationary m-dependence without replacement sampling and martingales are considered. General references for this chapter are Lehmann (1999), Ferguson (1996), Sen and Singer (1993), Hall and Heyde (1980), Billingsley (1995), and Hoeffding and Robbins (1948); specific references are given later.

9.1 Stationary m-dependence

We start with an example to illustrate that a CLT for sample means can hold even if the summands are not independent.

Example 9.1 Suppose $X_1, X_2, \ldots$ is a stationary Gaussian sequence with $E(X_i) = \mu, \text{Var}(X_i) = \sigma^2 < \infty$. Then, for each n, $\sqrt{n}(\bar{X} - \mu)$ is normally distributed and so $\sqrt{n}(\bar{X} - \mu) \xrightarrow{\mathcal{L}} N(0, \tau^2)$, provided $\tau^2 = \lim_{n\to\infty} \text{Var}(\sqrt{n}(\bar{X} - \mu)) < \infty$. But

$$\text{Var}(\sqrt{n}(\bar{X} - \mu)) = \sigma^2 + \frac{1}{n} \sum_{i \neq j = 1}^{n} \text{cov}(X_i, X_j) = \sigma^2 + \frac{2}{n} \sum_{i=1}^{n} (n - i)\gamma_i,$$

where $\gamma_i = \text{cov}(X_1, X_{i+1})$. Therefore, $\tau^2 < \infty$ if and only if $\frac{2}{n} \sum_{i=1}^{n} (n-i)\gamma_i$ has a finite limit, say ρ, in which case $\sqrt{n}(\bar{X} - \mu) \xrightarrow{\mathcal{L}} N(0, \sigma^2 + \rho)$.

A. DasGupta, *Asymptotic Theory of Statistics and Probability*,

What is going on qualitatively is that $\frac{1}{n}\sum_{i=1}^{n}(n-i)\gamma_i$ is summable when $|\gamma_i| \to 0$ adequately fast. Instances of this are when only a fixed finite number of the γ_i are nonzero or when γ_i is damped exponentially; i.e., $\gamma_i = O(a^i)$ for some $|a| < 1$. It turns out that there are general CLTs for sample averages under such conditions. The case of *m-dependence* is provided below.

Definition 9.1 A stationary sequence $\{X_n\}$ is called m-dependent for a given fixed m if $(X_1, \ldots, X_i)$ and $(X_j, X_{j+1}, \ldots)$ are independent whenever $j - i > m$.

Theorem 9.1 (m-dependent sequence) Let $\{X_i\}$ be a stationary m-dependent sequence. Let $E(X_i) = \mu$ and $\mathrm{Var}(X_i) = \sigma^2 < \infty$. Then $\sqrt{n}(\bar{X} - \mu) \xrightarrow{\mathcal{L}} N(0, \tau^2)$, where $\tau^2 = \sigma^2 + 2\sum_{i=2}^{m+1}\mathrm{cov}(X_1, X_i)$.

Remark. See Lehmann (1999) for a proof; m-dependent data arise either as standard time series models or as models in their own right. For example, if $\{Z_i\}$ are iid random variables and $X_i = a_1 Z_{i-1} + a_2 Z_{i-2}, i \geq 3$, then $\{X_i\}$ is 1-dependent. This is a simple *moving average* process of use in time series analysis. A more general m-dependent sequence is $X_i = h(Z_i, Z_{i+1}, \ldots, Z_{i+m})$ for some function h.

Example 9.2 Suppose Z_i are iid with a finite variance σ^2, and let $X_i = \frac{Z_i + Z_{i+1}}{2}$. Then, obviously, $\sum_{i=1}^{n} X_i = \frac{Z_1 + Z_{n+1}}{2} + \sum_{i=2}^{n} Z_i$.

$$\begin{aligned}
\therefore \sqrt{n}(\bar{X} - \mu) &= \frac{Z_1 + Z_{n+1}}{2\sqrt{n}} + \sqrt{n}\Big(\frac{Z_2 + \cdots + Z_n}{n} - \frac{n-1}{n}\mu\Big) - \frac{1}{\sqrt{n}}\mu \\
&= o_p(1) + \sqrt{n}\frac{n-1}{n}\Big(\frac{Z_2 + \cdots + Z_n}{n-1} - \mu\Big) + o_p(1) \\
&= \sqrt{n-1}\Big(\frac{Z_2 + \cdots + Z_n}{n-1} - \mu\Big)\sqrt{\frac{n-1}{n}} + o_p(1) \\
&\xrightarrow{\mathcal{L}} N(0, \sigma^2)
\end{aligned}$$

by Slutsky's theorem.

Remark. Notice the method of proof in this illustrative example. One writes $\sum_{i=1}^{n} X_i = \sum_{i=1}^{n} U_i + \sum_{i=1}^{n} V_i$, where $\{U_i\}$ are independent. The $\sum_{i=1}^{n} U_i$ part is dominant and produces the CLT, and the $\sum_{i=1}^{n} V_i$ part is asymptotically negligible. This is essentially the method of proof of the CLT for more general m-dependent sequences.

9.2 Sampling without Replacement

Dependent data also naturally arise in sampling without replacement from a finite population. Central limit theorems are available and we will present them shortly. But let us start with an illustrative example.

Example 9.3 Suppose, among N objects in a population, D are of type 1 and $N - D$ of type 2. A sample without replacement of size n is taken, and let X be the number of sampled units of type 1. We can regard these D type 1 units as having numerical values $X_1, \ldots, X_D = 1$ and the rest as having values $X_{D+1}, \ldots, X_N = 0$, $X = \sum_{i=1}^n X_{N_i}$, where $X_{N_1}, \ldots, X_{N_N}$ correspond to the sampled units.

Of course, X has the hypergeometric distribution

$$P(X = x) = \frac{\binom{D}{x}\binom{N-D}{n-x}}{\binom{N}{n}}, 0 \leq x \leq D, 0 \leq n - x \leq N - D.$$

Two configurations can be thought of:

(a) n is fixed, and $\frac{D}{N} \to p, 0 < p < 1$ with $N \to \infty$. In this case, by applying Stirling's approximation to $N!$ and $D!$, $P(X = x) \to \binom{n}{x} p^x (1 - p)^{n-x}$, and so $X \xrightarrow{\mathcal{L}} \text{Bin}(n, p)$.

(b) $n, N, N - n \to \infty$ and $\frac{D}{N} \to p, 0 < p < 1$. This is the case where convergence of X to normality holds.

Two subcases must be distinguished now. If $n = o(N)$, then

$$\frac{\sqrt{n}(\bar{X} - E(X_1))}{\sqrt{\text{Var}(X_1)}} \xrightarrow{\mathcal{L}} N(0, 1),$$

where $\bar{X} = \frac{\sum_{i=1}^n X_{N_i}}{n}$. Thus the same result as if the $\{X_i\}$ were iid holds.

On the other hand, if $\frac{n}{N} \to 0 < \lambda < 1$, then

$$\frac{\sqrt{n}(\bar{X} - E(X_1))}{\sqrt{\text{Var}(X_1)}} \xrightarrow{\mathcal{L}} N(0, 1 - \lambda).$$

Thus, if n is sufficiently large relative to N, then the dependence is adequate to change the asymptotic variance from 1 to something else; i.e., $1 - \lambda$.

Here is a general result; again, see Lehmann (1999) for a proof.

Theorem 9.2 For $N \geq 1$, let π_N be a finite population with numerical values $X_{N_1}, X_{N_2}, \ldots, X_{N_N}$. Let $X_{N_1}, X_{N_2}, \ldots, X_{N_n}$ be the values of the units of a sample without replacement of size n. Let $\bar{X} = \frac{1}{n}\sum_{i=1}^{n} X_{N_i}$ and $\bar{X_N} = \frac{1}{N}\sum_{i=1}^{N} X_{N_i}$. Suppose $n, N - n \to \infty$, and

(a) $\dfrac{\max_{1\leq i\leq N}(X_{N_i} - \bar{X}_N)^2}{\sum_{i=1}^{N}(X_{N_i} - \bar{X}_N)^2} \to 0$

and $\dfrac{n}{N} \to 0 < \tau < 1$ as $N \to \infty$

(b) $\dfrac{N\max_{1\leq i\leq N}(X_{N_i} - \bar{X}_N)^2}{\sum_{i=1}^{N}(X_{N_i} - \bar{X}_N)^2} = O(1)$

as $N \to \infty$.

Then

$$\frac{\bar{X} - E(\bar{X})}{\sqrt{\mathrm{Var}(\bar{X})}} \xrightarrow{\mathcal{L}} N(0, 1).$$

Example 9.4 Suppose $X_{N_1}, \ldots, X_{N_n}$ is a sample without replacement from the set $\{1, 2, \ldots, N\}$, and let $\bar{X} = \frac{1}{n}\sum_{i=1}^{n} X_{N_i}$. Then, by a direct calculation,

$$E(\bar{X}) = \frac{N+1}{2}$$

and

$$\mathrm{Var}(\bar{X}) = \frac{(N-n)(N+1)}{12n}.$$

Furthermore, $\dfrac{N\max_{1\leq i\leq N}(X_{N_i} - \bar{X}_N)^2}{\sum_{i=1}^{N}(X_{N_i} - \bar{X}_N)^2} = \dfrac{3(N-1)}{N+1} = O(1).$

Hence, by Theorem 9.2, $\dfrac{\sqrt{n}(\bar{X} - \frac{N+1}{2})}{\sqrt{\frac{(N-n)(N+1)}{12}}} \xRightarrow{\mathcal{L}} N(0, 1).$

9.3 Martingales and Examples

For an independent sequence of random variables $X_1, X_2, \ldots$, the conditional expectation of the present term of the sequence given the past terms is the same as its unconditional expectation with probability 1. Martingales let the conditional expectation depend on the past terms, but only through the immediate past. Martingales arise naturally in applications, particularly in the context of gambling, and also in statistical theory. Partial sums of zero-mean independent variables admit a central limit theorem (sometimes without having a variance), and such sequences form a martingale. This is a simple illustrative example. Central limit theorems are available for more general martingale sequences. Two extremely lucid expositions on martingales especially suitable for statisticians are Doob (1971) and Heyde (1972).

Definition 9.2 Let $\{X_n\}$ be a sequence of random variables on a probability space $(\Omega, \mathcal{F}, P)$ and $\mathcal{F}_n$ an increasing sequence of σ-fields in $\mathcal{F}$ with X_n being $\mathcal{F}_n$- measurable. $\{X_n\}$ is called a *martingale* if $E(X_{n+1}|\mathcal{F}_n) = X_n$, a.s., for every $n \geq 1$, and if each X_n is absolutely integrable.

We give a few illustrative examples.

Example 9.5 Suppose $X_1, X_2, \ldots, X_n$ are iid P_0 for some distribution P_0, and let P_1 be some other probability measure. With respect to some measure μ that dominates both of them (such a measure μ always exists), let f_0, f_1 denote the densities of P_0, P_1, and let $l_n = l_n(X_1, X_2, \ldots, X_n) = \prod \frac{f_1(X_i)}{f_0(X_i)}$ denote the likelihood ratio.

Then, letting E denote expectation under P_0,

$$E(l_{n+1}|l_1, \ldots, l_n)$$

$$= E\left(\frac{f_1(X_{n+1})}{f_0(X_{n+1})}|l_1, \ldots, l_n\right) l_n = l_n E\left(\frac{f_1(X_{n+1})}{f_0(X_{n+1})}\right)$$

$$= l_n \int \frac{f_1(x)}{f_0(x)} dP_0 = l_n \int \frac{f_1(x)}{f_0(x)} f_0 d\mu = l_n.$$

Thus, $\{l_n\}$ forms a martingale sequence.

Example 9.6 Let $Z_1, Z_2, \ldots$ be iid $N(0, 1)$ random variables and let $X_n = (Z_1 + \ldots + Z_n)^2 - n$. Then,

$$\begin{aligned}
&E(X_{n+1}|X_1, X_2, \ldots, X_n) \\
&= E[(Z_1 + \ldots + Z_n)^2 + 2Z_{n+1}(Z_1 + \ldots + Z_n) \\
&\quad + Z_{n+1}^2|X_1, X_2, \ldots, X_n] - (n+1) \\
&= X_n + n + 2(Z_1 + \ldots + Z_n)E(Z_{n+1}|X_1, X_2, \ldots, X_n) \\
&\quad + E(Z_{n+1}^2|X_1, X_2, \ldots, X_n) - (n+1) \\
&= X_n + n + 0 + 1 - (n+1) = X_n,
\end{aligned}$$

and so $\{X_n\}$ forms a martingale.

Actually, we did not use the normality of the Z_i at all, and the martingale property holds without the normality assumption as long as the Z_i are iid with mean 0 and variance 1.

Example 9.7 Let $Z_1, Z_2, \ldots$ be independent zero-mean random variables and let S_n denote the partial sum $\sum_{i=1}^{n} X_i$. Then, clearly, $E(S_{n+1}|S_1, \ldots, S_n) = S_n + E(Z_{n+1}|S_1, \ldots, S_n) = S_n + E(Z_{n+1}) = S_n$, and so $\{S_n\}$ forms a martingale.

Example 9.8 Suppose, given $\theta \in \Theta \subseteq \mathcal{R}$, $X^{(n)} = (X_1, \ldots, X_n)$ has some distribution $P_{\theta,n}$. Under squared error loss, $E(\theta|X^{(n)})$ is the Bayes estimate for θ. For notational convenience, denote $E(\theta|X^{(n)})$ by Z_n and $\sigma\{X_1, \ldots, X_n\}$, the σ-field generated by $X_1, \ldots, X_n$, by $\mathcal{F}_n$. Then, $E(Z_{n+1}|\mathcal{F}_n) = E(\theta|\mathcal{F}_{n+1})|\mathcal{F}_n = E(\theta|\mathcal{F}_n)$, as the $\mathcal{F}_n$ are increasing σ-fields. Therefore, with respect to the sequence of σ-fields $\mathcal{F}_n$, the Bayes estimates $\{E(\theta|X^{(n)})\}$ form a martingale.

Note that it was not necessary to assume that the components of $X^{(n)}$ are iid given θ. Some results on the almost sure convergence of the sequence of Bayes estimates to the true value of θ can be concluded from this martingale property that we just established.

Example 9.9 Suppose $X_1, X_2, \ldots$ are iid $N(0, 1)$ variables and $S_n = \sum_{i=1}^{n} X_i$. Since $S_n \sim N(0, n)$, its mgf $E(e^{tS_n}) = e^{nt^2/2}$. Now let $Z_n = e^{tS_n - nt^2/2}$, where t is a fixed real number. Then $E(Z_{n+1}|Z_1, \ldots, Z_n) = e^{-(n+1)t^2/2}E(e^{tS_n}e^{tX_{n+1}}|S_n) = e^{-(n+1)t^2/2}e^{tS_n}e^{t^2/2} = Z_n$. Therefore, for any real t, the sequence $\{e^{tS_n - nt^2/2}\}$ forms a martingale.

Example 9.10 (**Matching**) Suppose N people have gathered in a party, and at the end of the party their hats are returned at random. Those that get their own hats back then leave. The remaining hats are distributed again at random among the remaining guests, and so on. Let X_n denote the number of guests still present after the nth round.

By an elementary calculation, $E(X_n - X_{n+1}) = 1 \quad \forall n$. With a little more effort, one has $E(X_{n+1}|X_1, \ldots, X_n) = E(X_{n+1} - X_n + X_n|X_1, \ldots, X_n) = -1 + X_n$, whence the sequence $\{X_n + n\}$ is a martingale.

There is also a useful concept of a *reverse martingale*, which we define next.

Definition 9.3 Let $\{X_n\}$ be a sequence of random variables in some probability space $(\Omega, \mathcal{F}, P)$ and $\{\mathcal{F}_n\}$ a sequence of *decreasing* σ-fields in $\mathcal{F}$, X_n being $\mathcal{F}_n$-measurable. $\{X_n\}$ is called a *reverse martingale* if $E(X_n|\mathcal{F}_{n+1}) = X_{n+1}$, a.s., $\forall\, n \geq 1$.

For an iid sequence with zero mean, the partial sums form a martingale, but the means do not. The reverse martingale concept comes in handy here.

Example 9.11 Let $\{X_n\}$ be a sequence of iid random variables with a finite mean, which we assume without loss of generality to be zero. If we choose $\mathcal{F}_n$ to be the smallest σ-algebra generated by $\{\bar{X}_k, k \geq n\}$, then it can be shown that $E(\bar{X}_n|\mathcal{F}_{n+1}) = \bar{X}_{n+1}$, a.s. and so $\{\bar{X}_n\}$ forms a reverse martingale. Basically, the important thing to notice is that the conditional distribution of X_1, X_2 given $X_{1:2}, X_{2:2}$ is uniform over the two permutations of $X_{1:2}, X_{2:2}$, due to the iid nature of X_1, X_2. A refinement of this simple fact gives the reverse martingale property for $\{\bar{X}_n\}$.

Remark. The reverse martingale concept is useful because such a sequence satisfies certain convergence properties such as almost sure and $\mathcal{L}_1$ convergence and convergence in distribution. Thus, strong laws and central limit theorems can sometimes be established very easily if a suitable reverse martingale structure exists. Instances of this are the Kolmogorov SLLN, the CLT for U-statistics, and the SLLN for U-statistics (see Chapter 15) for treatment of U-statistics).

Example 9.12 This example indicates a paradoxical phenomenon through use of a *submartingale* concept. Submartingales are sequences (with associated σ-fields in the background, as always) such that $E(X_n|\mathcal{F}_m) \geq X_m$ when $n > m$. Likewise, there is a concept of reverse submartingales, defined in the obvious way. Convex functions of reverse martingales are reverse submartingales, with the $\mathcal{L}_1$ property assumed. Since $x \to |x|$ is convex, it would mean that for *any* constant a whatsoever, $|\bar{X}_n - a|$ would have the reverse submartingale property, and so, for instance, if X_1, X_2 are iid from any CDF F with a finite mean, then for any constant a, $E|\frac{X_1+X_2}{2} - a| \leq E|X_1 - a|$. It does seem paradoxical that, to estimate any arbitrary parameter a, it is better to take two observations than to take one from any CDF F, however unrelated F might seem to be with the parameter a.

9.4 The Martingale and Reverse Martingale CLTs

Martingale and reverse martingale CLTs have very useful unifying properties in the sense that many specific CLTs follow as special cases of martingale or reverse martingale CLTs. Statements of martingale and reverse martingale central limit theorems can be confusing because different versions are stated in different places. One source of possible confusion is that one can use random normings or nonrandom normings. We recommend Hall and Heyde (1980) for further details and proofs of martingale central limit theorems.

Theorem 9.3 Suppose $\{X_n\}$ is a zero-mean martingale sequence relative to $\mathcal{F}_n$, and let $s_n^2 = E(Y_n^2|\mathcal{F}_{n-1})$, where $Y_n = X_n - X_{n-1}$. Given t, let $\tau_t = \inf\{n : \sum_{i=1}^n s_i^2 \geq t\}$. Suppose

$$|Y_n| < K < \infty \quad \forall n,$$

$$\sum_{i=1}^{\infty} s_i^2 = \infty \quad \text{w.p.1},$$

$$\tau_t/t \stackrel{\mathcal{P}}{\Rightarrow} 1.$$

Then, $\frac{X_n}{\sqrt{n}} \stackrel{\mathcal{L}}{\Rightarrow} N(0, 1)$.

Convergence to nonnormal limits is possible, too. If one does not insist on a nonrandom norming, then the limit could be turned into a normal again. Here is one result that describes such a phenomenon.

Theorem 9.4 With notation as in the previous theorem, let $c_n = \sqrt{\text{Var}(X_n)}$, assumed to be finite for every n. Suppose

(a) $\frac{\max\{Y_j:1\leq j\leq n\}}{c_n} \stackrel{\mathcal{P}}{\Rightarrow} 0$.

(b) There exists a real-valued random variable η such that $\frac{\sum_{i=1}^n Y_i^2}{c_n^2} \stackrel{\mathcal{P}}{\Rightarrow} \eta$.

Then, $\frac{X_n}{c_n} \stackrel{\mathcal{L}}{\Rightarrow} G$, where G has the characteristic function $\psi(t) = E(e^{-t^2\eta/2})$; moreover, if $P(\eta = 0) = 0$, then $\frac{X_n}{\sqrt{\sum_{i=1}^n Y_i^2}} \stackrel{\mathcal{L}}{\Rightarrow} N(0, 1)$.

Theorem 9.5 Suppose $\{X_n\}$ is a zero-mean reverse martingale relative to $\mathcal{F}_n$, and let $s_n^2 = E(Y_n^2|\mathcal{F}_{n+1})$, where $Y_n = X_{n+1} - X_n$. Suppose

$$\frac{\sum_{k=n}^{\infty} s_k^2}{E[\sum_{k=n}^{\infty} s_k^2]} \overset{\text{a.s.}}{\Rightarrow} 1,$$

$$\frac{\sum_{k=n}^{\infty} Y_k^2}{\sum_{k=n}^{\infty} s_k^2} \overset{\text{a.s.}}{\Rightarrow} 1.$$

Then, $\frac{X_n}{\sqrt{E[\sum_{k=n}^{\infty} s_k^2]}} \overset{\mathcal{L}}{\Rightarrow} N(0, 1)$.

9.5 Exercises

Exercise 9.1 For an iid sequence $\{X_i\}$ with a finite variance, derive a CLT for $\sum_{i=1}^{n} X_i X_{i+1}$.

Exercise 9.2 Consider the stationary Gaussian autoregressive process of order 1, $X_{i+1} - \mu = \rho(X_i - \mu) + \sigma Z_i$, where Z_i are iid $N(0, 1)$. Find the limiting distribution of $\bar{X}$ assuming $|\rho| < 1$.

Exercise 9.3 A fair coin is tossed repeatedly. Let N_n denote the number of trials among the first n where a head is followed by (at least) two other heads. Find the limit distribution of N_n (e.g., if $n = 8$ and the outcomes are HTHHHHTT, then $N_n = 2$).

Exercise 9.4 * Suppose X_i are iid Exp(1). Does $\frac{1}{n}\sum_{i=1}^{n-1} \frac{X_i}{X_{i+1}}$ converge in probability?

Exercise 9.5 * Consider a finite population of some unknown size N from which n_1 units are sampled, tagged, and released. Let X be the number of tagged units in a second sample from the same population of size n_2. Let $p = \frac{n_1}{N}$ (fixed), and suppose $\frac{n_2}{N} \to \lambda$. Show that $\sqrt{n_2}(\frac{n_2}{X} - \frac{1}{p}) \xrightarrow{\mathcal{L}} N\left(0, \frac{(1-\lambda)(1-p)}{p^3}\right)$.

Exercise 9.6 Suppose $\{X_n\}$ is a martingale sequence. Show that $E(X_{n+m}|X_1, \ldots, X_n) = X_n$.

Exercise 9.7 * Consider a simple asymmetric random walk with iid steps distributed as $P(X_1 = 1) = p < 1 - p = P(X_1 = -1)$. Let $S_n = \sum_{i=1}^{n} X_i$.

(a) Show that $U_n = (\frac{1-p}{p})^{S_n}$ is a martingale.

(b) Show that, with probability 1, $\sup_n S_n < \infty$.

(c) Show that in fact $\sup_n S_n$ has a geometric distribution, and identify the parameter of the geometric distribution.

Exercise 9.8 * **(Polya's Urn)** An urn contains r red and b black balls. One ball is drawn at random from the urn, and c additional balls of the same color as the ball drawn are added to the urn. The scheme is iterated. Let Y_n be the proportion of black balls after the nth drawing and the corresponding replacement.

(a) Show that Y_n is a martingale.

(b) Does $E(Y_n^k)$ converge as $n \to \infty$? Here k is an arbitrary positive integer.

(c) Does Y_n converge in law?

Exercise 9.9 * **(Branching Process)** Let $\{Z_{ij}\}$ be a double array of iid random variables with mean μ and variance $\sigma^2 < \infty$. Let $X_0 = 1$ and $X_{n+1} = \sum_{j=1}^{X_n} Z_{nj}$. Show that

(a) $W_n = \frac{X_n}{\mu^n}$ is a martingale.

(b) $\sup_n E(W_n) < \infty$.

(c) Is $\{W_n\}$ uniformly integrable? Prove or disprove.

Exercise 9.10 Consider a stationary Markov chain on the state space $X = \{0, 1, \ldots, N\}$ with the transition density $p_{ij} = \binom{N}{j}(i/N)^j(1 - i/N)^{N-j}$.

(a) Give a physical example of a chain with such a transition density.

(b) Show that the chain $\{X_n\}$ is a martingale.

Remark. It may be useful to read Chapter 10 for basic Markov chain terminology.

Exercise 9.11 * Let U_n be a U-statistic with a kernel h of order $r \geq 1$ (U-statistics are defined in Chapter 15). Show that $\{U_n\}$ is a reverse martingale w.r.t. some suitable sequence of σ-fields; identify such a sequence of σ-fields.

Exercise 9.12 Let $\{X_i\}$ be iid from some CDF F and let F_n be the empirical CDF of the first n observations. Show that, for any given x, $\{F_n(x)\}$ is a reverse martingale. Hence, without appealing to the normal approximation of the binomial, prove the asymptotic normality of $F_n(x)$ on suitable centering and norming.

References

Billingsley, P. (1995). *Probability and Measure*, 3rd ed., John Wiley, New York.
Doob, J.L. (1971). What is a martingale?, Am. Math. Monthly, 78, 451–463.
Ferguson, T.S. (1996). *A Course in Large Sample Theory*, Chapman and Hall, London.
Hall, P. and Heyde, C. (1980). *Martingale Limit Theory and Its Application*, Academic Press, New York.
Heyde, C. (1972). Martingales: a case for a place in the statistician's repertoire, Aust. J. Stat., 14, 1–9.
Hoeffding, W. and Robbins, H. (1948). The central limit theorem for dependent random variables, Duke Math. J., 15, 773–780.
Lehmann, E.L. (1999). *Elements of Large Sample Theory*, Springer, New York.
Sen, P.K. and Singer, J. (1993). *Large Sample Methods in Statistics: An Introduction with Applications*, Chapman and Hall, New York.

Chapter 10
Central Limit Theorem for Markov Chains

Markov chains are widely used models that incorporate weak dependence and admit a very complete and elegant analytical treatment. Of the numerous excellent resources on Markov chains, we recommend Bremaud (1999), Isaacson (1976), and Norris (1997) for treatments of fundamental theory and applications. For the results here, we will assume stationarity of the chain. Gudynas (1991), Maxwell and Woodroofe (2000), and Jones (2004) are the main references for the CLTs in this chapter. Dobrushin (1956a,b) has derived some CLTs for the nonstationary case. Other specific references are given later.

10.1 Notation and Basic Definitions

Although we do not develop Markov chain theory itself in this chapter, we describe the basic notation followed throughout the chapter and also state a few important definitions that may not be known to the reader. Here is the notation.

X: state space of Markov chain,
ξ, η : elements of X,
X_n: nth state of the stationary chain, $n \geq 0$,
$P(\xi, A)$: stationary transition probability function,
$P^{(n)}(\xi, A)$: n-step transition probability function,
π: initial distribution; i.e., distribution of X_0,
Q: stationary distribution,
f: scalar function on X,
P_a: probabilities calculated under the assumption $X_0 = a$, a.s.

Next, a few definitions are given for completeness.

A. DasGupta, *Asymptotic Theory of Statistics and Probability*,

Definition 10.1 Given a state a, let T_a denote the first hitting time of a on or after time $n \geq 1$. a is called *recurrent* if $P_a(T_a < \infty) = 1$; if a is not recurrent, it is called *transient*.

Definition 10.2 A recurrent state a is called *positive recurrent* if $E_a(T_a) < \infty$ and is called *null recurrent* otherwise. A chain for which every state is positive recurrent is called a *positive recurrent chain.*

Definition 10.3 A chain is called *irreducible* if it is possible to go to any state from any other state; i.e., for all states a, b, $P_a(T_b < \infty) > 0$.

We will study limit theorems for $S_n = \frac{1}{B_n} \sum_{i=1}^n f(X_i) - A_n$, where A_n, B_n are suitable sequences of constants. Under conditions on the chain and the function f, asymptotic normality of S_n holds. It is possible, however, for certain types of functions f and for suitable initial distributions π, that convergence to nonnormal limits occurs.

10.2 Normal Limits

Normal limits for additive functionals of Markov chains are available under different sets of conditions suitable for different types of applications. Often the conditions can be hard to verify. Some results assume stronger conditions on the chain and less stringent conditions on the function f; e.g., weaker moment conditions. Others assume less on the chain and stronger conditions on f. The conditions on the chain typically may involve recurrence properties or the dependence structure of the chain. Conditions on f may typically involve finiteness of enough moments. We describe some basic terminology and conditions below; we follow Nagaev (1957) and Gudynas (1991) for notation and Theorem 10.1 below. Theorem 10.2 is easy to prove.

Definition 10.4 Let $\{X_n\}$ be a stationary Markov chain. Then, geometric ergodicity, uniform ergodicity, and ρ-mixing are defined as follows.

A. (Geometric Ergodicity) For some function $\gamma > 0$ and some $0 < \rho < 1$,

$$\sup_A |P^{(n)}(\xi, A) - Q(A)| \leq \gamma(\xi)\rho^n.$$

B. (Uniform Ergodicity) For some finite constant $\gamma > 0$ and some $0 < \rho < 1$,

$$\sup_{\xi, A} |P^{(n)}(\xi, A) - Q(A)| \leq \gamma \rho^n.$$

C. (ρ -mixing) For any $n, m \geq 1$, let $\mathcal{A}_n = \sigma(X_1, \ldots, X_n)$, $\mathcal{B}_m = \sigma(X_m, X_{m+1}, \ldots)$. Take $f \in L_2(\mathcal{A}_n, Q)$, $g \in L_2(\mathcal{B}_{n+m}, Q)$.

$$C_m = \sup_{n \geq 1} \{\text{corr}_Q(f, g) : f \in L_2(\mathcal{A}_n, Q), g \in L_2(\mathcal{B}_{n+m}, Q)\} \to 0$$

as $m \to \infty$.
We also say that the chain is *asymptotically maximally uncorrelated* if it is ρ-mixing.

Theorem 10.1 If condition (A) holds and if $\int_X \gamma(\xi) dQ(\xi) < \infty$ and $\int_X |f|^{(2+\epsilon)}(\xi) dQ(\xi) < \infty$ for some $\epsilon > 0$, or if one of conditions (B) or (C) holds and if $\int_X |f|^2(\xi) dQ(\xi) < \infty$, then, independent of the initial distribution π,

$$\left| P\left(\frac{1}{\sqrt{n}} \sum_{i=1}^{n} \left(f(X_i) - \int_X f(\xi) dQ(\xi) \right) \leq x \right) - P(N(0, \sigma^2) \leq x) \right|$$
$$\to 0 \text{ as } n \to \infty,$$

where we assume

$$\sigma^2 = \lim_{n \to \infty} E\left(\frac{1}{\sqrt{n}} \sum_{i=1}^{n} \left(f(X_i) - \int_X f(\xi) dQ(\xi) \right)\right)^2$$

exists and is in $(0, \infty)$.

Remark. Geometric ergodicity is usually hard to verify in abstract state spaces. Some methods can be found in Meyn and Tweedie (1993). Although rare in most statistical applications, if the state space of the chain happens to be finite, then one does not need any hard calculations and even uniform ergodicity would often follow. Rosenblatt (1971) gives a necessary and sufficient condition for ρ-mixing; sufficient conditions are developed in Roberts and Rosenthal (1997). However, these involve a priori verification of geometric ergodicity. Maxwell and Woodroofe (2000) give a sufficient condition for asymptotic normality in terms of the growth of the conditional expectation (as an operator) of $S_n(f) = \sum_{i=1}^{n} \left(f(X_i) - \int_X f(\xi) dQ(\xi) \right)$ given $X_0 = x$ and show that the sufficient condition is close to being necessary.

Remark. The problem has an interesting history. Markov himself had proved the CLT when the chain has three states, and Doeblin (1938) had expanded it to a general state space assuming f is bounded. Doob (1953) proved the CLT for a general state space under the moment condition $\int_X |f(\xi)|^{2+\delta} dQ(\xi) < \infty$ for some $\delta > 0$. The result above under uniform ergodicity or ρ-mixing was proved by Nagaev (1957).

Remark. The theorem's proof, as in the iid case, uses characteristic functions (cfs). Let $f_n(t)$ be the cf of $S_n = \frac{1}{\sqrt{n}} \sum_{i=1}^n (f(X_i) - \int_X f(\xi) dQ(\xi))$. It turns out that, in a neighborhood of $t = 0$,

$$f_n(t) = \psi^n(it)(1 + t\theta_1(t)) + \rho_*^n \theta_2(n, t),$$

where $0 < \rho_* < 1$ and θ_1, θ_2 are bounded in t. The function $\psi(z)$ in the equation above is not a cf but has properties similar to that of a cf, so $f_n(t)$ behaves locally like the cf of a convolution. That is enough to prove the CLT.

Direct verification of conditions (A) or (B) in Theorem 10.1 may be tedious. The following more restricted theorem is simpler to apply.

Theorem 10.2 Suppose $\{X_n\}$ is an irreducible positive recurrent chain with a countable state space X. Let $h_a(x, y) = \sum_{n=1}^{\infty} P_x(T_a \geq n, X_n = y)$. Then, regardless of the initial distribution π, for any bounded function f on the state space X,

$$\frac{1}{\sqrt{n}} \left(\sum_{i=1}^{n} f(X_i) - n\mu \right) \stackrel{\mathcal{L}}{\Rightarrow} N(0, \sigma^2),$$

where σ^2 is as in Exercise 10.2 and $\mu = \sum_{\xi} f(\xi) Q\{\xi\}$.

Remark. The condition that f is bounded can be relaxed at the expense of some additional conditions and notation.

Example 10.1 Consider as a simple example the stationary chain with state space $\{0, 1\}$ and transition probabilities $p(0, 0) = \alpha, p(1, 0) = \beta$. Let $f(x) = x$. By solving the simple (matrix) eigenvalue equations for the stationary distribution, we find the stationary probabilities to be $Q\{0\} = \frac{\beta}{1+\beta-\alpha}$, $Q\{1\} = \frac{1-\alpha}{1+\beta-\alpha}$. Consequently, $\mu = \frac{1-\alpha}{1+\beta-\alpha}$. Furthermore, all the conditions on the chain assumed in Theorem 10.2 hold.

On the other hand, by choosing $a = 0$, using the notation of Theorem 10.2, $h_a(1, y) = 1$ if $y = 0$ and $= \frac{1-\beta}{\beta}$ if $y = 1$. Plugging into the expression for σ^2 in Exercise 10.2, on algebra, one finds $\sigma^2 = \frac{1-\alpha}{(1+\beta-\alpha)^3}[(1-\alpha)^2 + 2\beta - \beta^2]$. Thus, from Theorem 10.2, $\frac{\sqrt{n}(\bar{X} - \frac{1-\alpha}{1+\beta-\alpha})}{\sqrt{\frac{1-\alpha}{(1+\beta-\alpha)^3}[(1-\alpha)^2 + 2\beta - \beta^2]}} \stackrel{\mathcal{L}}{\Rightarrow} N(0, 1)$.

10.3 Nonnormal Limits

If the distribution of $f(X_0)$ under Q is appropriately chosen (e.g., so as to force it not to have a mean), the partial sums $S_n = \sum_{i=1}^n f(X_i)$ can converge to a nonnormal limit distribution. However, if it converges, the limit has to be a stable law, as is the case for independent summands. Here is one such result on a nonnormal limit; see Gudynas (1991) for the precise technical conditions on f for the nonnormal limit to hold.

Theorem 10.3 Let $F(x) = P_Q(f(X_0) \leq x)$. Let $U_i,\ i \geq 1$ be iid with CDF $F(x)$. Suppose for some A_n, B_n that

$$\frac{\sum_{i=1}^n U_i}{B_n} - A_n \xrightarrow{\mathcal{L}} V_\alpha,$$

where V_α is a stable distribution of exponent α. Then, for suitable choices of the function f with the same A_n, B_n and the same stable law V_α,

$$S_n = \frac{\sum_{i=1}^n f(X_i)}{B_n} - A_n \xrightarrow{\mathcal{L}} V_\alpha.$$

Remark. It can be shown that the norming constant $B_n = n^{\frac{1}{\alpha}} h(n)$ for some function h that is slowly varying as $n \to \infty$.

10.4 Convergence to Stationarity: Diaconis-Stroock-Fill Bound

Under various conditions, the n-step transition probabilities $p_{ij}^{(n)}$ converge to the stationary probabilities $Q(j) = Q\{j\}$. However, the issue is very subtle. In general, any nonnegative row vector π satisfying $\pi P = \pi$ is called an invariant measure of the chain. It can happen that invariant measures exist none of which is a probability measure. If an invariant measure exists that is a probability measure, it is called a stationary distribution. An interesting question is the rapidity of the convergence of the transition probabilities to the stationary probabilities. The convergence is exponentially fast under suitable conditions on the chain. We describe certain results in that direction in this section. But first the issue of existence of a stationary distribution is addressed. For simplicity of presentation, only the finite state space case is mentioned; almost any text on Markov chains can be consulted for the simple first result below. A specific reference is Bremaud (1999).

Theorem 10.4 Let $\{X_n\}$ be a stationary Markov chain with a finite state space $\mathcal{S}$. Suppose $\{X_n\}$ is irreducible and aperiodic. Then a unique stationary distribution Q exists and satisfies each of

$$(a)\ QP = Q,$$
$$(b)\ Q(j) = \frac{1}{E_j(T_j)}.$$

Example 10.2 Consider the two-state stationary chain with state space $\{0, 1\}$, $p_{00} = \alpha$, $p_{10} = \beta$. Let $Q = (\frac{\beta}{1+\beta-\alpha}, \frac{1-\alpha}{1+\beta-\alpha})$. Then, by direct verification, $QP = P$ and, Q being a probability vector, it is the unique stationary distribution.

Let us directly verify that the transition probabilities $p_{ij}^{(n)}$ converge to $Q(j)\,\forall j$, regardless of the starting state i. For this, we will need a formula for the n-step transition matrix $P^{(n)} = P^n$. One can prove either by induction or by repeated use of the Chapman-Kolmogorov equations that

$$P^n = \frac{1}{1+\beta-\alpha}\left[\begin{pmatrix} \beta & 1-\alpha \\ \beta & 1-\alpha \end{pmatrix} + \frac{(\alpha-\beta)^n}{1+\beta-\alpha}\begin{pmatrix} 1-\alpha & \alpha-1 \\ -\beta & \beta \end{pmatrix}\right].$$

Since $|\alpha - \beta| < 1$, we notice that each row of P^n converges exponentially fast to the stationary vector Q derived above.

Example 10.3 Consider the three-state Markov chain with the rows of the transition probability matrix being $(0, \frac{1}{2}, \frac{1}{2})$, $(\frac{1}{3}, \frac{1}{4}, \frac{5}{12})$, $(\frac{2}{3}, \frac{1}{4}, \frac{1}{12})$. By considering the characteristic polynomial, it is easy to directly compute its eigenvalues to be $1, -\frac{1}{2}, -\frac{1}{6}$ and from the spectral decomposition of P obtain

$$P^n = \frac{1}{3}\begin{pmatrix} 1 & 1 & 1 \\ 1 & 1 & 1 \\ 1 & 1 & 1 \end{pmatrix} + \left(-\frac{1}{2}\right)^n \frac{1}{12}\begin{pmatrix} 8 & -4 & -4 \\ 2 & -1 & -1 \\ -10 & 5 & 5 \end{pmatrix} + \left(-\frac{1}{6}\right)^n \frac{1}{4}\begin{pmatrix} 0 & 0 & 0 \\ -2 & 3 & -1 \\ 2 & -3 & 1 \end{pmatrix}.$$

Since the modulus of the second-largest eigenvalue is $\frac{1}{2} < 1$, again we see that convergence of P^n to the invariant distribution is geometrically fast.

The phenomenon of the two simple examples above is in fact quite general, as is seen in the following theorem.

Theorem 10.5 (Diaconis, Stroock, Fill) Let $\{X_n\}$ be an irreducible aperiodic Markov chain on a finite state space X with stationary transition probabilities p_{ij} and a unique stationary distribution Q. Define

$$\tilde{p}_{ij} = \frac{p_{ji}Q(j)}{Q(i)}, \ M = P\tilde{P},$$

and let λ denote the second-largest eigenvalue of M (a number in (0, 1)). Then, for any n, i,

$$\rho^2(P^n(i), Q) \leq \frac{\lambda^n}{4Q(i)},$$

where $P^n(i)$ denotes the ith row in P^n and ρ denotes the total variation distance between the two distributions on the state space X.

Remark. Diaconis and Stroock (1991) and Fill (1991) give several other exponential bounds on convergence to stationarity in the ergodic case. They also give numerous illustrative examples, many involving various random walks.

10.5 Exercises

Exercise 10.1 Let $\{X_i\}$ be a two-state stationary Markov chain with initial probabilities $p, 1-p$ for 0, 1 and transition probabilities $P(X_{i+1} = 1|X_i = 0) = p_{01}$ and $P(X_{i+1} = 1|X_i = 1) = p_{11}$. Find the limit of $\text{Var}(\sqrt{n}(\bar{X} - \mu))$, where $\mu = 1 - p$.

Exercise 10.2 Suppose the Markov chain $\{X_i\}$ is irreducible and positive recurrent with a countable state space. Show that σ^2 in Theorem 10.1 equals, independent of the choice of the state a,

$$\sum_{x \in X} \frac{Q\{x\}}{Q\{a\}}(f(x) - \mu)^2 + 2 \sum_{\substack{x,y \in X \\ x \neq a}} Q\{x\}h_a(x, y)(f(x) - \mu)(f(y) - \mu),$$

where $\mu = \sum f(x)Q\{x\}$.

Remark. The state space X has been assumed to be countable.

Exercise 10.3 Do all Markov chains have a unique stationary distribution? Hint: Look at chains that are not irreducible.

Exercise 10.4 Suppose $\{X_i\}$ are independent integer-valued random variables. Show that $\{X_i\}$ is certainly a Markov chain. Is it necessarily stationary?

Exercise 10.5 Suppose a fair die is rolled repeatedly. Show that the following are Markov chains and evaluate $\sigma^2 = \lim_n \text{Var}(\sqrt{n}(\bar{X} - \mu))$:

(a) X_n = # sixes until the nth roll.

(b) X_n = at time n, time elapsed since the last six.

Exercise 10.6 Show that increasing functions of a (real-valued) Markov chain are Markov chains. Characterize the transition density.

Exercise 10.7 Suppose S_n is the simple random walk with $S_0 = 0$. Give an example of a function h such that h is not one-to-one and yet $h(S_n)$ is a Markov chain. Does Theorem 10.1 hold for your example?

Exercise 10.8 Consider the simple random walk S_n, again with $S_0 = 0$. Show that S_n has more than one invariant measure unless the walk is symmetric. Are any of the invariant measures finite?

Exercise 10.9 Consider the chain X_n = # of sixes until the nth roll of a balanced die. Are there any finite invariant measures?

Exercise 10.10 Consider the chain with state space $\mathcal{X}$ as the nonnegative integers and transition probabilities $P(x, x+1) = p_x$, $P(x, 0) = 1 - p_x$. Let $T_0 = \inf\{n \geq 1 : X_n = 0\}$.

(a) Show that X_n has a stationary distribution Q when it is positive recurrent.

(b) Show that then $Q(x) = \frac{P_0(T_0 > x)}{E_0(T_0)}$.

(c) Confirm if condition (A) of Theorem 10.1 holds.

Exercise 10.11 There are initially X_0 particles in a box with $X_0 \sim \text{Poi}(\mu)$. At time $n \geq 1$, any particle inside the box escapes into the atmosphere with probability p, $0 < p < 1$, and the particles inside act independently. An additional Poisson number I_n of particles immigrate into the box at time n; assume $I_n \stackrel{\text{iid.}}{\sim} \text{Poi}(\theta)$. Show that X_n = # of particles in the box at time n has a Poisson stationary distribution. Does Theorem 10.1 hold for $\bar{X}$?

Exercise 10.12 Ergodicity

Suppose $\{X_n\}$ is an irreducible chain with a countable state space $\mathcal{X}$. Assume that $\forall i \in \mathcal{X}$, the mean recurrence time $\mu_i < \infty$. Show that $\frac{1}{n}\sum_{i=1}^n f(X_i) \to \sum_{i \in \mathcal{X}} \frac{f(i)}{\mu_i}$ for any bounded function f. What mode of convergence can you assert?

Exercise 10.13 Consider a two-state chain with transition matrix $\begin{pmatrix} \alpha & 1-\alpha \\ 1-\beta & \beta \end{pmatrix}$, $0 < \alpha, \beta < 1$. Identify the unique stationary distribution and confirm if condition (A) of Theorem 10.1 holds.

Exercise 10.14 Give an example of a chain and an f such that Theorem 10.3 holds with $\alpha = 1$.

Exercise 10.15 * Consider the three-state stationary Markov chain on $\{0, 1, 2\}$ with $p_{00} = .25$, $p_{01} = .75$, $p_{10} = .75$, $p_{11} = .25$, and $p_{22} = 1$.

(a) Does the chain have a stationary distribution? If so, is it unique?
(b) Compute P^n and verify directly that its rows converge exponentially to the stationary distribution.
(c) Compute the second-largest eigenvalue of M in Theorem 10.5 and hence the upper bound of Theorem 10.5. Compare this with your exact result in part (b).

Exercise 10.16 * Suppose the transition matrix of a stationary Markov chain with a finite state space is doubly stochastic; i.e., the column vectors are probability vectors as well. Show that the uniform distribution on the state space is a stationary distribution for such a chain.

Exercise 10.17 * For the two-state Markov chain of Example 10.1, find $\text{cov}(X_n, X_{n+m})$ and find its limit, if it exists, as $m \to \infty$.

References

Bremaud, P. (1999). *Markov Chains: Gibbs Fields, Monte Carlo Simulation and Queue*, Springer, New York.
Diaconis, P. and Stroock, D. (1991). Geometric bounds for eigenvalues of Markov chains, Ann. Appl. Prob., 1(1), 36–61.
Dobrushin, R. (1956a). Central limit theorems for non-stationary Markov chains, I, Teor. Veroyatnost Primerien, 1, 72–89.
Dobrushin, R. (1956b). Central limit theorems for non-stationary Markov chains, II, Teor. Veroyatnost Primerien, 1, 365–425.
Doeblin, W. (1938). Sur deus problemes de M. Kolmogoroff concernant les chaines denombrables, Bull. Soc. Math. France, 52, 210–220.
Doob, J.L. (1953). *Stochastic Processes*, John Wiley, New York.
Fill, J. (1991). Eigenvalue bounds on convergence to stationarity for nonreversible Markov chains, Ann. Appl. Prob., 1(1), 62–87.

Gudynas, P. (1991). Refinements of the central limit theorem for homogeneous Markov chains, in *Limit Theorems of Probability Theory*, N.M. Ostianu (ed.), Akad. Nauk SSSR, Moscow, 200–218.

Isaacson, D. (1976). *Markov Chains:Theory and Applications*, John Wiley, New York.

Jones, G. (2004). On the Markov chain central limit theorem, Prob. Surveys, 1, 299–320.

Maxwell, M. and Woodroofe, M. (2000). Central limit theorems for additive functionals of Markov chains, Ann. Prob., 28(2), 713–724.

Meyn, S.P. and Tweedie, R.L. (1993). *Markov Chains and Stochastic Stability*, Springer, New York.

Nagaev, S. (1957). Some limit theorems for stationary Markov chains, Teor. Veroyat-nost Primerien, 2, 389–416.

Norris, J.R. (1997). *Markov Chains*, Cambridge University Press, Cambridge.

Roberts, G.O. and Rosenthal, J.S. (1997). Geometric ergodicity and hybrid Markov chains, Electron. Commun. Prob., 2, 13–25.

Rosenblatt, M. (1971). *Markov Processes, Structure, and Asymptotic Behavior*, Springer, New York.

Chapter 11
Accuracy of Central Limit Theorems

Suppose a sequence of CDFs $F_n \xrightarrow{\mathcal{L}} F$ for some F. Such a weak convergence result is usually used to approximate the true value of $F_n(x)$ at some fixed n and x by $F(x)$. However, the weak convergence result by itself says absolutely nothing about the accuracy of approximating $F_n(x)$ by $F(x)$ for that particular value of n. To approximate $F_n(x)$ by $F(x)$ for a given finite n is a leap of faith unless we have some idea of the error committed; i.e., $|F_n(x) - F(x)|$. More specifically, if for a sequence of random variables $X_1, \ldots, X_n$

$$\frac{\overline{X}_n - E(\overline{X}_n)}{\sqrt{\mathrm{Var}(\overline{X}_n)}} \xrightarrow{\mathcal{L}} Z \sim N(0, 1),$$

then we need some idea of the error

$$\left| P\left(\frac{\overline{X}_n - E(\overline{X}_n)}{\sqrt{\mathrm{Var}(\overline{X}_n)}} \leq x \right) - \Phi(x) \right|$$

in order to use the central limit theorem for a practical approximation with some degree of confidence. The first result for the iid case in this direction is the classic Berry-Esseen theorem (Berry (1941), Esseen (1945)). Typically, these accuracy measures give bounds on the error in the appropriate CLT for any fixed n, making assumptions about moments of X_i. Good general references for this chapter are Feller (1966), Petrov (1975), Serfling (1980), and Bhattacharya and Rao (1986). A good recent monograph is Senatov (1998). Additional specific references are given later.

A. DasGupta, *Asymptotic Theory of Statistics and Probability*,

11.1 Uniform Bounds: Berry-Esseen Inequality

In the canonical iid case with a finite variance, the CLT says that $\frac{\sqrt{n}(\bar{X}-\mu)}{\sigma}$ converges in law to the $N(0, 1)$ distribution. By Polya's theorem (see Chapter 1), therefore the uniform error $\Delta_n = \sup_{-\infty<x<\infty} |P(\frac{\sqrt{n}(\bar{X}-\mu)}{\sigma} \leq x) - \Phi(x)| \to 0$ as $n \to \infty$. Bounds on Δ_n for any given n are called *uniform bounds*. Here is the classic Berry-Esseen uniform bound; a proof can be seen in Petrov (1975).

Theorem 11.1 (Berry-Esseen) Let $X_1, \ldots, X_n$ be iid with $E(X_1) = \mu$, $\text{Var}(X_1) = \sigma^2$, and $\beta_3 = E|X_1 - \mu|^3 < \infty$. Then there exists a universal constant C, not depending on n or the distribution of the X_i, such that

$$\sup_x \left| P\left(\frac{\sqrt{n}(\bar{X} - \mu)}{\sigma} \leq x\right) - \Phi(x)\right| \leq \frac{C\beta_3}{\sigma^3\sqrt{n}}.$$

Remark. The universal constant C may be taken as $C = 0.8$. Fourier analytic proofs give the best available constant; direct proofs have so far not succeeded in producing good constants. Numerous refinements of Theorem 11.1 are known, but we will not mention them here. The constant C cannot be taken to be smaller than $\frac{3+\sqrt{10}}{6\sqrt{2\pi}} \doteq 0.41$. Better values of the constant C can be found for specific types of the underlying CDF; e.g., if it is known that the samples are iid from an exponential distribution. However, no systematic studies in this direction seem to have been done. See the chapter exercises for some examples. Also, for some specific underlying CDFs F, better rates of convergence in the CLT may be possible. For example, one can have

$$\sup_x \left| P\left(\frac{\sqrt{n}(\bar{X} - \mu)}{\sigma} \leq x\right) - \Phi(x)\right| \leq \frac{C(F)}{n}$$

for some finite constant $C(F)$. This issue will be clearer when we discuss asymptotic expansions for $F_n(x) = P_F\left(\frac{\sqrt{n}(\bar{X}-\mu)}{\sigma} \leq x\right)$.

The following is an extension of the Berry-Esseen inequality to the case of independent but not iid variables.

Theorem 11.2 Let $X_1, \ldots, X_n$ be independent random variables with $E(X_i) = \mu_i$, $\text{Var}(X_i) = \sigma_i^2$, and $E|X_i - \mu_i|^3 < \infty, i = 1, 2, \ldots, n$. Then

$$\sup_x \left| P\left(\frac{\overline{X}_n - E(\overline{X}_n)}{\sqrt{\text{Var}(\overline{X}_n)}} \le x \right) - \Phi(x) \right| \le C^* \frac{\sum_{i=1}^n E|X_i - \mu_i|^3}{\left(\sum_{i=1}^n \sigma_i^2 \right)^{3/2}}$$

for some universal constant $0 < C^* < \infty$.

Example 11.1 The Berry-Esseen bound is uniform in x, and it is valid for any $n \geq 1$. While these are positive features of the theorem, it may not be possible to establish that $\sup_x |F_n(x) - \Phi(x)| \leq \epsilon$ for some preassigned $\epsilon > 0$ by using the Berry-Esseen theorem unless n is very large. Let us see an illustrative example.

Suppose $X_1, X_2, \ldots, X_n \overset{iid}{\sim} \text{Ber}(p)$ and $n = 100$. Suppose we want the CLT approximation to be accurate to within an error of $\epsilon = 0.005$. In the Bernoulli case, $\beta_3 = E|X_i - \mu|^3 = pq(1 - 2pq)$, where $q = 1 - p$. Using $C = 0.8$, the uniform Berry-Esseen bound is

$$\Delta_n \leq \frac{0.8pq(1 - 2pq)}{(pq)^{3/2}\sqrt{n}}.$$

This is less than or equal to the prescribed $\epsilon = 0.005$ iff $pq > 0.4784$, which does not hold for any $0 < p < 1$. Even for $p = .5$, the bound is less than or equal to $\epsilon = 0.005$ only when $n > 25{,}000$, which is a very large sample size. Of course, this is not necessarily a flaw of the Berry-Esseen inequality itself because the desire to have a uniform error of at most $\epsilon = .005$ is a tough demand, and a fairly large value of n is probably needed to have such a small error in the CLT.

Example 11.2 As an example of independent variables that are not iid, consider $X_i \sim \text{Ber}(\frac{1}{i}), i \geq 1$, and let $S_n = \sum_{i=1}^n X_i$. Then, $E(S_n) = \sum_{i=1}^n \frac{1}{i}$, $\text{Var}(S_n) = \sum_{i=1}^n \text{Var}(X_i) = \sum_{i=1}^n \frac{i-1}{i^2}$ and $E|X_i - E(X_i)|^3 = \frac{(i-1)(i^2-2i+2)}{i^4}$. Therefore, from the Berry-Esseen bound for independent summands,

$$\sup_x \left| P\left(\frac{S_n - \sum_{i=1}^n \frac{1}{i}}{\sqrt{\sum_{i=1}^n \frac{i-1}{i^2}}} \le x \right) - \Phi(x) \right| \le C^* \frac{\sum_{i=1}^n \frac{(i-1)(i^2-2i+2)}{i^4}}{\left(\sum_{i=1}^n \frac{i-1}{i^2} \right)^{3/2}}.$$

Observe now that $\sum_{i=1}^n \frac{(i-1)(i^2-2i+2)}{i^4} = \sum_{i=1}^n \frac{i^3-3i^2+4i-2}{i^4} = \log n + O(1)$, and of course $\sum_{i=1}^n \frac{i-1}{i^2} = \log n + O(1)$. Substituting these back into the Berry-Esseen bound, one obtains with some minor algebra that

$$\sup_x \left| P\left(\frac{S_n - \sum_{i=1}^n \frac{1}{i}}{\sqrt{\sum_{i=1}^n \frac{i-1}{i^2}}} \leq x \right) - \Phi(x) \right| = O\left(\frac{1}{\sqrt{\log n}} \right).$$

In particular, S_n is asymptotically normal, although the X_i are not iid. From the above, with some additional manipulations it would follow that $|P\left(\frac{S_n - \log n}{\sqrt{\log n}} \leq x\right) - \Phi(x)| = O\left(\frac{1}{\sqrt{\log n}}\right)$; i.e., for large n, the mean and the variance are both approximately $\log n$. A loose interpretation is that S_n is approximately Poisson with mean $\log n$ for large n.

Various refinements of the two basic Berry-Esseen results given above are available. They replace the third absolute moment by an expectation of some more general functions. Petrov (1975) and van der Vaart and Wellner (1996) are good references for generalized Berry-Essen bounds. Here is a result in that direction.

Theorem 11.3 Let $X_1, \ldots, X_n$ be independent with $E(X_i) = \mu_i$, $\text{Var}(X_i) = \sigma_i^2$, and let $B_n = \sum_{i=1}^n \sigma_i^2$. Let $g : \mathcal{R} \to \mathcal{R}_+$ be such that g is even, both $g(x)$ and $\frac{x}{g(x)}$ are nondecreasing for $x > 0$, and $E(X_i^2 g(X_i)) < \infty$ for every i. Then,

$$\sup_x \left| P\left(\frac{\overline{X}_n - E(\overline{X}_n)}{\sqrt{\text{Var}(\overline{X}_n)}} \leq x \right) - \Phi(x) \right| \leq C^{**} \frac{\sum_{i=1}^n E(X_i^2 g(X_i))}{g(\sqrt{B_n}) B_n}$$

for some universal $0 < C^{**} < \infty$.

Remark. Consider the iid case in particular. If, for some $0 < \delta < 1$, $E(|X_1|^{2+\delta}) < \infty$, then $g(x) = |x|^\delta$ satisfies the hypotheses of Theorem 11.3 and one obtains that the rate of convergence in the CLT for iid summands is $O(\frac{1}{n^{\delta/2}})$ with $2 + \delta$ moments. But many other choices of g are possible as long as the moment condition on g is satisfied.

11.2 Local Bounds

There has been a parallel development on developing bounds on the error in the CLT at a particular x as opposed to bounds on the uniform error. Such bounds are called *local Berry-Esseen bounds*. Many different types of local bounds are available. We present here just one.

Theorem 11.4 Let $X_1, \ldots, X_n$ be independent random variables with $E(X_i) = \mu_i$, $\text{Var}(X_i) = \sigma_i^2$, and $E|X_i - \mu_i|^{2+\delta} < \infty$ for some $0 < \delta \leq 1$.

Then

$$\left| P\left(\frac{\overline{X}_n - E(\overline{X}_n)}{\sqrt{\text{Var}(\overline{X}_n)}} \leq x \right) - \Phi(x) \right| \leq \frac{D}{(1+|x|^{2+\delta})} \frac{\sum_{i=1}^n E|X_i - \mu_i|^{2+\delta}}{(\sum_{i=1}^n \sigma_i^2)^{1+\frac{\delta}{2}}}$$

for some universal constant $0 < D < \infty$.

Remark. Such local bounds are useful in proving convergence of global error criteria such as $\int_{-\infty}^{+\infty} |F_n(x) - \Phi(x)|^p dx$ or for establishing approximations to the moments of F_n. Uniform error bounds would be useless for these purposes. If the third absolute moments are finite, an explicit value for the universal constant D can be chosen to be 31. Good references for local bounds are Michel (1981), Petrov (1975), and Serfling (1980).

11.3 The Multidimensional Berry-Esseen Theorems

Bounds on the error of the central limit theorem for normalized sample means are also available in dimensions $d > 1$. Typically, the available bounds are of the form $\sup_{A \in \mathcal{A}} |P(S_n \in A) - \Phi(A)| \leq \frac{C(\mathcal{A}, F, d)}{\sqrt{n}}$ for appropriate families of sets $\mathcal{A}$, where S_n is the normalized partial sum of iid vectors $X_1, \ldots, X_n$ and F is the underlying CDF. The constant $C(\mathcal{A}, F, d)$ grows with d, and the current best rate seems to be $d^{\frac{1}{4}}$. Note that we cannot take the class of sets to be all Borel sets if $d > 1$; we must have appropriate restrictions to have meaningful results when $d > 1$. We will only mention the iid case here, although bounds are available when $X_1, \ldots, X_n$ are independent but not iid. Bhattacharya and Rao (1986), Göetze (1991), and Bentkus (2003) are good references for the results in this section.

First we need some notation. Let Z denote the standard normal distribution in d dimensions. Let $B(x, \epsilon)$ denote the ball of radius ϵ with center at x. With ρ denoting Euclidean distance, for a set $A \subseteq \mathcal{R}^d$ and $\epsilon > 0$, let $A^\epsilon = \{x \in \mathcal{R}^d : \rho(x, A) \leq \epsilon\}$ and $A_\epsilon = \{x \in A : B(x, \epsilon) \subseteq A\}$. Here, $\rho(x, A)$ denotes the distance from x to the nearest point in A. Finally, given a class $\mathcal{A}$ of sets in $\mathcal{R}^d$, let $a_d = a_d(\mathcal{A})$ denote the smallest constant a such that $\forall \epsilon > 0, A \in \mathcal{A}, \max\{\frac{P(Z \in A^\epsilon - A)}{\varepsilon}, \frac{P(Z \in A - A_\epsilon)}{\epsilon}\} \leq a$. Let $b_d = \max\{1, a_d\}$.

Then one has the following elegant bound of the Berry-Esseen type; see Bentkus (2003).

Theorem 11.5 Let $X_1, \ldots, X_n$ be normalized iid d-vectors with $E(X_1) = \mathbf{0}$ and $\mathbf{cov}(X_1) = I$. Let $S_n = \frac{X_1 + \ldots + X_n}{\sqrt{n}}$ and $\beta = E(||X_1||^3)$. Let $\mathcal{A}$ be either the class of all convex sets in $\mathcal{R}^d$ or the class of all closed balls in $\mathcal{R}^d$. Then,

$$\sup_{A \in \mathcal{A}} |P(S_n \in A) - P(Z \in A)| \le \frac{100 b_d \beta}{\sqrt{n}}.$$

Remark. Exact expressions for a_d do not seem to be known for either of these two classes of sets. However, it is known that $a_d \le 4d^{\frac{1}{4}}$ for the class of convex sets and a_d is uniformly bounded in d for the class of balls. This leads to the following corollary.

Corollary 11.1 (a) Let $\mathcal{A}$ be the class of all convex sets. Then,

$$\sup_{A \in \mathcal{A}} |P(S_n \in A) - P(Z \in A)| \le \frac{400 d^{\frac{1}{4}} \beta}{\sqrt{n}}.$$

(b) Let $\mathcal{A}$ be the class of all closed balls. Then,

$$\sup_{A \in \mathcal{A}} |P(S_n \in A) - P(Z \in A)| \le \frac{K\beta}{\sqrt{n}}$$

for some finite positive constant K, independent of d.

Establishing bounds on the error of the CLT when $d > 1$ is a very hard problem. It seems likely that the topic will see further development.

11.4 Other Statistics

Error bounds for normal approximations to many other types of statistics besides sample means are known. We present here one result for statistics that are smooth functions of means. The order of the error depends on the conditions one assumes on the nature of the function; see Bhattacharya and Rao (1986).

Definition 11.1 Let $\mathcal{G}$ denote the class of functions $g : \mathcal{R} \to \mathcal{R}$ such that g is differentiable and g' satisfies the *Lipschitz* condition $|g'(x) - g'(y)| = O(|x - y|^{\delta}|)$ for some $0 < \delta \le 1$.

Theorem 11.6 Let $X_1, \ldots, X_n$ be iid random variables with mean μ and variance $\sigma^2 < \infty$, and let $g : \mathcal{R} \to \mathcal{R} \in \mathcal{G}$ such that $g'(\mu) \neq 0$. Let $T_n = \frac{\sqrt{n}(g(\bar{X}) - g(\mu))}{\sigma |g'(\mu)|}$. Then, $\forall n \ge 2$,

$$\sup_x |P(T_n \le x) - \Phi(x)| \le K \frac{\log n}{\sqrt{n}},$$

for some finite positive constant K.

Remark. If the assumption on the function g is changed, other error bounds are obtained. One disadvantage of these error bounds is that it is hard to specify an explicit constant K in the bounds. Thus, the results are more valuable in understanding the rate of convergence in the CLT rather than as devices to really bound the error concretely.

Example 11.3 Suppose X_i are iid Ber(p) and $S_n = \sum_{i=1}^n X_i \sim \text{Bin}(n, p)$. Consider estimation of the odds $\frac{p}{1-p}$ so that the maximum likelihood estimate is $\frac{S_n}{n-S_n} = g(\bar{X})$, where $g(x) = \frac{x}{1-x}$, $0 < x < 1$. If the true value of p is different from 0, 1, we may assume that $0 < S_n < n$ for all large n, and so the MLE exists with probability 1 for all large n.

From Theorem 11.6, there is a finite positive constant K such that

$$\sup_x \left| P\left(\frac{\sqrt{n}(\frac{S_n}{n-S_n} - \frac{p}{1-p})}{\sqrt{\frac{p}{1-p}}} \le x \right) - \Phi(x) \right| \le K \frac{\log n}{\sqrt{n}}, \ \forall n.$$

By choosing $p = \frac{1}{2}$, let us try to numerically find the smallest K that works in this global error bound.

With $p = \frac{1}{2}$, define the quantity $\Lambda_n = \frac{\sqrt{n}}{\log n} \sup_x \left| P\left(\frac{\sqrt{n}(\frac{S_n}{n-S_n} - \frac{p}{1-p})}{\sqrt{\frac{p}{1-p}}} \le x\right) - \Phi(x) \right|$. We are interested in $\sup_{n \ge 2} \Lambda_n$. Because of the discreteness of the binomial distribution, $\left| P\left(\frac{\sqrt{n}(\frac{S_n}{n-S_n} - \frac{p}{1-p})}{\sqrt{\frac{p}{1-p}}} \le x\right) - \Phi(x) \right|$ is a very choppy function of x, and identifying its supremum over x needs careful computing. It is useful to observe that by simple manipulation this quantity equals $\left| \frac{\sum_{k=0}^{[\frac{n(x+\sqrt{n})}{x+2\sqrt{n}}]} \binom{n}{k}}{2^n} - \Phi(x) \right|$, where [.] denotes the integer part function.

Λ_n is also oscillating in n. Computation gives Λ_n for $n = 2, \ldots, 10$; e.g., as .51, .72, .45, .51, .46, .42, .45, .37, and .44. It keeps oscillating for larger values of n, but the value .72 corresponding to $n = 3$ is not exceeded up to $n = 100$. A numerical conjecture is that the global error bound in the theorem works in this example with $K = .75$. It would be interesting to have systematic theoretical studies of these best constant problems.

11.5 Exercises

Exercise 11.1 Let $X \sim \text{Bin}(n, p)$. Use the Berry-Esseen theorem to bound $P(X \le k)$ in terms of the standard normal CDF Φ uniformly in k. Can you give any nontrivial bounds that are uniform in both k and p?

Exercise 11.2 * Suppose $X_i \stackrel{\text{iid}}{\sim}$ lognormal(0, 1). Use the Berry-Esseen theorem to find bounds on the median of $\bar{X}$, and compare the exact value of the median with your bounds for selected values of n.

Exercise 11.3 * Suppose $X_i \stackrel{\text{iid}}{\sim} U[0, 1]$. Is it true that, for some finite constant c,

$$\left| P\left(\sqrt{12n}\left(\bar{X} - \frac{1}{2}\right) \leq x\right) - \Phi(x)\right| \leq \frac{c}{n}$$

uniformly in x?

Exercise 11.4 * Suppose $X_i \stackrel{\text{indep.}}{\sim}$ Poisson(i), $i \geq 1$.

(a) Does $\bar{X}$ have an asymptotic normal distribution?
(b) Obtain an explicit Berry-Esseen bound on the error of the CLT.
Hint: Bound $E|X_i - i|^3$ by using $E(X_i - i)^4$.

Exercise 11.5 *Suppose $X_1, \ldots, X_{10}$ are the scores in ten independent rolls of a fair die. Obtain an upper and lower bound on $P(X_1 + X_2 + \cdots + X_{10} = 30)$ by using the Berry-Esseen theorem. Compare your bounds with the exact value.

Hint: The exact value can be obtained from the probability generating function on some manipulation; for now, you can just simulate it.

Exercise 11.6 Suppose $X \sim \text{Bin}(50, .1)$. Compare the exact value of $P(X \leq 10)$ with the Poisson approximation and compare the normal approximation by using the Berry-Esseen bound.

Exercise 11.7 * Suppose $X_i \stackrel{\text{iid}}{\sim}$ Exp(1). Obtain a local bound on $|F_n(x) - \Phi(x)|$ of the form $\frac{c_n}{(1+|x|)^3}$. For which values of α can you get bounds on $\int |x|^\alpha dF_n(x)$ from these local bounds? Obtain these bounds.

Exercise 11.8 *Suppose X_i are iid with density proportional to $e^{-||x||}$ in d dimensions. For what values of n is the global bound of Theorem 11.5 .01 or less when the class $\mathcal{A}$ is the class of all convex sets?

Exercise 11.9 *Numerically compute the quantity a_d for the class of all closed balls (i.e., intervals) when $d = 1$.

Exercise 11.10 Do all functions g with a bounded second derivative belong to the class $\mathcal{G}$ of Section 11.4?

References

Bentkus, V. (2003). On the dependence of the Berry-Esseen bound on dimension, J. Stat. Planning Infer., 113(2), 385–402.

Berry, A.C. (1941). The accuracy of the Gaussian approximation to the sum of independent variates, Trans. Am. Math. Soc., 49, 122–136.

Bhattacharya, R.N. and Rao, R.R. (1986). *Normal Approximation and Asymptotic Expansions*, Robert E. Krieger, Melbourne, FL.

Esseen, C. (1945). Fourier analysis of distribution functions: a mathematical study, Acta Math., 77, 1–125.

Feller, W. (1966). *An Introduction to Probability Theory with Applications*, John Wiley, New York.

Göetze, F. (1991). On the rate of convergence in the multivariate CLT, Ann. Prob., 19(2), 724–739.

Michel, R. (1981). On the constant in the nonuniform version of the Berry-Esseen theorem, Z. Wahr. Verw. Geb., 55(1), 109–117.

Petrov, V. (1975). *Limit Theorems for Sums of Independent Random Variables* (translation from Russian), Springer-Verlag, New York.

Senatov, V. (1998). *Normal Approximations: New Results, Methods and Problems*, VSP, Utrecht.

Serfling, R. (1980). *Approximation Theorems of Mathematical Statistics*, John Wiley, New York.

van der Vaart, A. and Wellner, J. (1996). *Weak Convergence and Empirical Processes, with Applications to Statistics*, Springer-Verlag, New York.

Chapter 12
Invariance Principles

The previous chapters discuss the asymptotic behavior of the sequence of partial sums $S_n = \sum_{i=1}^{n} X_i, n \geq 1$, for an iid sequence $X_1, X_2, \cdots$, under suitable moment conditions. In particular, we have described limit distributions and laws of large numbers for centered and normalized versions of S_n. The sequence of partial sums is clearly a natural thing to consider, given that sample means are so natural in statistics and probability. The central limit theorem says that as long as some moment conditions are satisfied, at any particular large value of n, S_n acts like a normally distributed random variable. In other words, the population from which the X_i came does not matter. The delta theorem says that we can do even better. We can even identify the limit distributions of functions, $h(S_n)$, and this is nice because there are problems in which the right statistic is not S_n itself but some suitable function $h(S_n)$.

Now, the sequence S_n is obviously a discrete-time stochastic process. We can think of a continuous-time stochastic process suitably devised, say $S_n(t)$, on the interval $[0, 1]$ such that, for any n, members of the discrete sequence $\frac{S_1}{\sqrt{n}}, \frac{S_2}{\sqrt{n}}, \cdots, \frac{S_n}{\sqrt{n}}$, are the values of that continuous-time process at the discrete times $t = \frac{1}{n}, \frac{2}{n}, \cdots, \frac{n}{n} = 1$. One can ask, what is the asymptotic behavior of this sequence of continuous-time stochastic processes? And just as in the case of the discrete-time stochastic process S_n, one can look at functionals $h(S_n(t))$ of these continuous-time stochastic processes. There are numerous problems in which it is precisely some suitable functional $h(S_n(t))$ that is the appropriate sequence of statistics. The problems in which a functional of such a type is the appropriate statistic arise in estimation, testing, model selection, goodness of fit, regression, and many other common statistical contexts. The invariance principle says the remarkable thing that, once again, under limited moment conditions, the continuous-time process $S_n(t)$ will act like a suitable continuous-time Gaussian process, say $W(t)$, and any nice enough functional $h(S_n(t))$ will act like the same functional of the limiting Gaussian process $W(t)$. The original F from which the data X_i

A. DasGupta, *Asymptotic Theory of Statistics and Probability*,

came is once again not going to matter. The form of the result is always the same. So, if we can identify that limiting Gaussian process $W(t)$, and if we know how to deal with the distribution of $h(W(t))$, then we have obtained the asymptotic behavior of our statistics $h(S_n(t))$ all at one stroke. It is a profoundly useful fact in the asymptotic theory of probability that all of this is indeed a reality. This chapter deals with such invariance principles and their concrete applications in important problems.

We recommend Billingsley (1968), Hall and Heyde (1980), and Csörgo and Révész (1981) for detailed and technical treatments, Erdös and Kac (1946), Donsker (1951), Komlós, Major, and Tusnady (1975, 1976), Major (1978), Whitt (1980), and Csörgo (1984) for invariance principles for the partial sum process, Mandrekar and Rao (1989) for more general symmetric statistics, Csörgo (1984), Dudley (1984), Shorack and Wellner (1986), Wellner (1992), Csörgo and Horváth (1993), and Giné (1996) for comprehensive treatments of empirical processes and their invariance principles, Heyde (1981), Pyke (1984), and Csörgo (2002) for lucid reviews, Philipp (1979), Dudley and Philipp (1983), Révész (1976), Einmahl (1987), and Massart (1989) for the multidimensional case, and Billingsley (1956), Jain, Jogdeo, and Stout (1975), McLeish (1974, 1975), Philip and Stout (1975), Hall (1977), Sen (1978), and Merlevéde, Peligrad, and Utev (2006) for treatments and reviews of various dependent cases. Other references are given within the specific sections.

12.1 Motivating Examples

Although we only talked about a continuous-time process $S_n(t)$ that suitably interpolates the partial sums $S_1, S_2, \cdots$, another continuous-time process of immense practical utility is the so-called *empirical process*. The empirical process $F_n(t)$ counts the proportion of sample observations among the first n that are less than or equal to a given t. We will discuss it in a little more detail shortly. But first we give a collection of examples of functionals $h(S_n(t))$ or $h(F_n(t))$ that arise naturally as test statistics in important testing problems or in the theory of probability. Their exact finite sample distributions being clumsy or even impossible to write, it becomes necessary to consider their asymptotic behavior. And, here is where an invariance principle of some appropriate type comes into play and settles the asymptotics in an elegant and crisp way.

Here are a small number of examples of such functionals that arise in statistics and probability.

Example 1 $D_n = \sup_{-\infty<t<\infty} |F_n(t) - F_0(t)|$, where F_0 is a given CDF on $\mathcal{R}$.
Example 2 $C_n = \int_{-\infty}^{\infty}(F_n(t) - F_0(t))^2 dF_0(t)$.
Example 3 $M_n = \sup_{0\le t\le 1} S_n(t)$.
Example 4 $\Pi_n = \frac{1}{n}\#\{k : S_k > 0\} = \lambda(t : S_n(t) > 0)$, where λ denotes Lebesgue measure (restricted to $[0, 1]$).

D_n and C_n arise as common test statistics in goodness-of-fit problems; we will study them in greater detail in Chapter 26. M_n and Π_n arise in the theory of random walks as the maximum fortune of a fixed player up to time n and as the proportion of times that she has been ahead. There are numerous such examples of statistics that can be regarded as functionals of either a partial sum process or an empirical process, and typically they satisfy some appropriate continuity property from which point an invariance principle takes over and settles the asymptotics.

12.2 Two Relevant Gaussian Processes

We remarked earlier that $S_n(t)$ will asymptotically act like a suitable Gaussian process. So does the empirical process $F_n(t)$. These two limiting Gaussian processes are closely related and happen to be the Brownian motion and the Brownian bridge, also known as the Wiener process and the tied-down Wiener process. For purposes of completeness, we give the definition and mention some fundamental properties of these two processes.

Definition 12.1 A stochastic process $W(t)$ defined on a probability space $(\Omega, \mathcal{A}, P)$, $t \in [0, \infty)$ is called a Wiener process or the Brownian motion starting at zero if:

(i) $W(0) = 0$ with probability 1;
(ii) for $0 \le s < t < \infty$, $W(t) - W(s) \sim N(0, t - s)$;
(iii) given $0 \le t_0 < t_1 < \cdots < t_k < \infty$, the random variables $W(t_{j+1}) - W(t_j), 0 \le j \le k - 1$ are mutually independent; and
(iv) the sample paths of $W(.)$ are almost all continuous (i.e., except for a set of sample points of probability 0), as a function of t, $W(t, \omega)$ is a continuous function.

Definition 12.2 Let $W(t)$ be a Wiener process on $[0, 1]$. The process $B(t) = W(t) - tW(1)$ is called a Brownian bridge on $[0, 1]$.

The proofs of the invariance principle theorems exploit key properties of the Brownian motion and the Brownian bridge. The properties that are the most useful toward this are sample path properties and some distributional results for relevant functionals of these two processes. The most fundamental properties are given in the next theorem. These are also of paramount importance and major historical value in their own right. There is no single source where all of these properties are available or proved. Most of them can be seen in Durrett (1996) and Csörgo (2002).

Theorem 12.1 (Important Properties of the Brownian Motion) Let $W(t)$ *and* $B(t)$ denote, respectively, a Brownian motion on $[0, \infty)$ starting at zero and a Brownian bridge on $[0, 1]$. Then,

(a) $\mathrm{cov}(W(s), W(t)) = \min(s, t)$; $\mathrm{cov}(B(s), B(t)) = \min(s, t) - st$.

(b) **Karhunen-Loeve Expansion** If $Z_1, Z_2, \cdots$ is an infinite sequence of iid $N(0, 1)$ random variables, then $W(t)$ defined as $W(t) = \sqrt{2}\sum_{m=1}^{\infty} \frac{\sin([m-\frac{1}{2}]\pi t)}{[m-\frac{1}{2}]\pi} Z_m$ is a Brownian motion starting at zero and $B(t)$ defined as $B(t) = \sqrt{2}\sum_{m=1}^{\infty} \frac{\sin(m\pi t)}{m\pi} Z_m$ is a Brownian bridge on $[0, 1]$.

(c) **Scale and Time Change** $\frac{1}{\sqrt{c}} W(ct), c > 0, tW(\frac{1}{t})$ are also each Brownian motions on $[0, \infty)$.

(d) $W(t)$ is a Markov process and $W(t), W^2(t) - t, e^{\theta W(t) - \frac{\theta^2 t}{2}}, \theta \in \mathcal{R}$, are each martingales.

(e) **Unbounded Variations** On any nondegenerate finite interval, $W(t)$ is almost surely of unbounded total variation.

(f) **Rough Paths** Almost surely, sample paths of $W(t)$ are nowhere differentiable, but the paths are Holder continuous of order α for all $\alpha < \frac{1}{2}$.

(g) Almost surely, there does not exist any t_0 such that t_0 is a *point of increase* of $W(t)$ in the usual sense of analysis.

(h) **Behavior Near Zero** Almost surely, on any interval $(0, t_0), t_0 > 0, W(t)$ has infinitely many zeros.

(i) **Zero Set and Cantor Property** The set of all zeros of $W(t)$ is almost surely a closed, uncountable set of Lebesgue measure zero without any isolated points.

(j) **Strong Markov Property** If τ is a *stopping time* (w.r.t. the $W(t)$ process), then $W(t + \tau) - W(\tau)$ is also a Brownian motion on $[0, \infty)$.

(k) As $t \to \infty$, $\frac{W(t)}{t} \to 0$ with probability 1.

(l) **LIL** Almost surely, $\limsup_{t\downarrow 0} \frac{W(t)}{\sqrt{2t\log\log(1/t)}} = 1$, and $\liminf_{t\downarrow 0} \frac{W(t)}{\sqrt{2t\log\log(1/t)}} = -1$.

(m) Almost surely, $\limsup_{t\to\infty} \frac{W(t)}{\sqrt{2t\log\log t}} = 1$, and $\liminf_{t\to\infty} \frac{W(t)}{\sqrt{2t\log\log t}} = -1$.

(n) **Order of Increments** Almost surely, for any given $c > 0$,

$$\lim_{T\to\infty} \sup_{0\le t\le T-c\log T} \frac{|W(t+c\log T) - W(t)|}{c\log T} = \sqrt{\frac{2}{c}}$$

and

$$\lim_{T\to\infty} \sup_{t\ge 0, t+1\le T} \frac{|W(t+1) - W(t)|}{\sqrt{2\log T}} = 1.$$

(o) **Domination Near Zero** If $r(t) \in C[0,1]$ (the class of real continuous functions on $[0,1]$) and is such that $\inf r(t) > 0$, $r(t)$ is increasing and $\frac{r(t)}{\sqrt{t}}$ is decreasing in some neighborhood of $t = 0$, then $P(|W(t)| < r(t) \forall t \in [0, t_0]$ for some $t_0) = 1$ iff $\int_0^1 t^{-3/2} r(t) e^{-r^2(t)/(2t)} < \infty$.

(p) **Maxima and Reflection Principle** $P(\sup_{0<s\le t} W(s) \ge x) = 2P(W(t) \ge x)$.

(q) **First Arcsine Law** Let T be the point of maxima of $W(t)$ on $[0,1]$. Then T is almost surely unique, and $P(T \le t) = \frac{2}{\pi}\arcsin(\sqrt{t})$.

(r) **Last Zero and the Second Arcsine Law** Let $L = \sup\{t \in [0,1] : W(t) = 0\}$. Then $P(L \le t) = \frac{2}{\pi}\arcsin(\sqrt{t})$.

(s) **Reflected Brownian Motion** Let $X(t) = \sup_{0\le s\le t} |W(s)|$. Then $P(X(t) \le x) = 2\Phi(\frac{x}{\sqrt{t}}) - 1, x > 0$.

(t) **Loops and Self-Crossings** Given $d \ge 2$, let $W_1(t), \cdots, W_d(t)$ be independent Brownian motions starting at zero, and let $W^d(t) = (W_1(t), \cdots, W_d(t))$, called a *$d$-dimensional Brownian motion*. Then, for $d = 2$, for any given finite $k \ge 2$ and any nondegenerate time interval, almost surely there exist times $t_1, \cdots, t_k$ such that $W^d(t_1) = \cdots = W^d(t_k)$; for $d = 3$, given any nondegenerate time interval, almost surely there exist times t_1, t_2 such that $W^d(t_1) = W^d(t_2)$ (called double points or self-crossings); for $d > 3$, $W^d(t)$ has, almost surely, no double points.

(u) **Exit Time from Spheres** Let W^d be a d-dimensional Brownian motion, B the d-dimensional open unit sphere centered at the origin, and $\tau = \inf\{t > 0 : W^d(t) \notin B\}$. For a bounded function $f : \mathcal{R}^d \to \mathcal{R}$,

$E[f(W(\tau))] = \int_{\partial B} \frac{1}{||x||^d} f(x) dS(x)$, where S is the normalized surface measure on the boundary of B.

(v) **Recurrence and Transience** Let W^d be a d-dimensional Brownian motion. W^d is recurrent for $d = 1, 2$ and transient for $d \geq 3$.

12.3 The Erdös-Kac Invariance Principle

Although the invariance principle for partial sums of iid random variables is usually credited to Donsker (Donsker (1951)), Erdös and Kac (1946) contained the basic idea behind the invariance principle and also worked out the asymptotic distribution of a number of key and interesting functionals $h(S_n(t))$. We provide a glimpse into the Erdös-Kac results in this section. Erdös and Kac describe their method of proof as follows:

"The proofs of all these theorems follow the same pattern. It is first proved that the limiting distribution exists and is independent of the distribution of the X_i's; then the distribution of the X_i's is chosen conveniently so that the limiting distribution can be calculated explicitly. This simple principle has, to the best of our knowledge, never been used before."

Theorem 12.2 Let $X_1, X_2, \cdots$ be an infinite iid sequence of zero-mean random variables such that $\frac{S_n}{\sqrt{n}}$ admits the central limit theorem. Then,

$$\lim_{n\to\infty} P(n^{-1/2} \max_{1\leq k\leq n} S_k \leq x) = G_1(x), x \geq 0,$$

$$\lim_{n\to\infty} P(n^{-1/2} \max_{1\leq k\leq n} |S_k| \leq x) = G_2(x), x \geq 0,$$

$$\lim_{n\to\infty} P\left(n^{-2} \sum_{k=1}^{n} S_k^2 \leq x\right) = G_3(x), x \geq 0,$$

$$\lim_{n\to\infty} P\left(n^{-3/2} \sum_{k=1}^{n} |S_k| \leq x\right) = G_4(x), x \geq 0,$$

$$\lim_{n\to\infty} P\left(\frac{1}{n} \sum_{k=1}^{n} I_{S_k>0} \leq x\right) = \frac{2}{\pi} \arcsin(\sqrt{x}), 0 \leq x \leq 1,$$

where

$$G_1(x) = 2\Phi(x) - 1,$$
$$G_2(x) = \frac{4}{\pi} \sum_{m=0}^{\infty} \frac{(-1)^m}{2m+1} e^{-(2m+1)^2\pi^2/(8x^2)},$$

and formulas for G_3 and the Laplace transform of G_4 are provided in Erdös and Kac (1946) (they involve complicated integral representations).

Remark. It is quite interesting that for existence of a limiting distribution of the sum of squares of the partial sums, a fourth moment of the X_i's is not at all necessary, and in fact all that is needed is that the distribution F from which the X_i arise be in the domain of attraction of the normal. In particular, existence of a variance is already enough. We will see this phenomenon reemerge in the next section.

12.4 Invariance Principles, Donsker's Theorem, and the KMT Construction

Donsker (1951) provided the full generalization of the Erdös-Kac technique by providing explicit embeddings of the discrete sequence $\frac{S_k}{\sqrt{n}}, k = 1, 2, \cdots, n$ into a continuous-time stochastic process $S_n(t)$ and by establishing the limiting distribution of a general continuous functional $h(S_n(t))$. In order to achieve this, it is necessary to use a continuous mapping theorem for metric spaces, as consideration of Euclidean spaces is no longer enough. It is also useful to exploit a property of the Brownian motion known as the *Skorohod embedding theorem*. We first describe this necessary background material.

Define $C[0, 1]$ the class of all continuous real-valued functions on $[0, 1]$ and $D[0, 1]$ the class of all real-valued functions on $[0, 1]$ that are right continuous and have a left limit at every point in $[0, 1]$.

Given two functions f, g in either $C[0,1]$ or $D[0,1]$, let $\rho(f,g) = \sup_{0\leq t\leq 1}| f(t) - g(t)|$ denote the supremum distance between f and g. We will refer to ρ as the *uniform metric*. Both $C[0, 1]$ and $D[0, 1]$ are (complete) metric spaces with respect to the uniform metric ρ. Two common embeddings of the discrete sequence $\frac{S_k}{\sqrt{n}}, k = 1, 2, \cdots, n$ into a continuous-time process are the following:

$$S_{n,1}(t) = \frac{1}{\sqrt{n}}[S_{[nt]} + \{nt\}X_{[nt]+1}]$$

and

$$S_{n,2}(t) = \frac{1}{\sqrt{n}} S_{[nt]},$$

$0 \leq t \leq 1$. Here, [.] denotes the integer part and {.} the fractional part of a positive real.

The first one simply continuously interpolates between the values $\frac{S_k}{\sqrt{n}}$ by drawing straight lines, but the second one is only right continuous, with jumps at the points $t = \frac{k}{n}, k = 1, 2, \cdots, n$. For certain specific applications, the second embedding is more useful. It is because of these jump discontinuities that Donsker needed to consider weak convergence in $D[0, 1]$. It did lead to some additional technical complexities.

The main idea from this point on is not difficult. One can produce a version of $S_n(t)$, say $\widetilde{S}_n(t)$, such that $\widetilde{S}_n(t)$ is *close to* a sequence of Wiener processes $W_n(t)$. Since $\widetilde{S}_n(t) \approx W_n(t)$, if $h(.)$ is a continuous functional with respect to the uniform metric, then one can expect that $h(\widetilde{S}_n(t)) \approx h(W_n(t)) = h(W(t))$ in distribution. $\widetilde{S}_n(t)$ being a version of $S_n(t)$, $h(S_n(t)) = h(\widetilde{S}_n(t))$ in distribution, and so $h(S_n(t))$ should be close to the fixed Brownian functional $h(W(t))$ in distribution, which is the question we wanted to answer.

The results leading to Donsker's theorem are presented below; we recommend Csörgo (2003) for further details on Theorems 12.3–12.5 below.

Theorem 12.3 (Skorohod Embedding) Given $n \geq 1$, iid random variables $X_1, \cdots, X_n$, $E(X_1) = 0$, $\text{Var}(X_1) = 1$, there exists a common probability space on which one can define a Wiener process $W(t)$ starting at zero and a triangular array of nonnegative random variables $\{\tau_{1,n}, \cdots, \tau_{n,n}\}$, iid under each given n such that

$$\text{(a) } \tau_{n,n} \stackrel{\mathcal{L}}{=} \tau_{1,1},$$

$$\text{(b) } E(\tau_{1,1}) = 1,$$

$$\text{(c) } \left\{\frac{S_1}{\sqrt{n}}, \cdots, \frac{S_n}{\sqrt{n}}\right\} \stackrel{\mathcal{L}}{=} \left\{W\left(\frac{\tau_{1,n}}{n}\right), \cdots, W\left(\frac{\tau_{1,n} + \cdots + \tau_{n,n}}{n}\right)\right\}.$$

Remark. Much more general versions of the Skorohod embedding theorem are known. See, for example, Oblój (2004). The version above suffices for the following *weak invariance principle for partial sum processes*.

Theorem 12.4 Let $S_n(t) = S_{n,1}(t)$ or $S_{n,2}(t)$ as defined above. Then there exists a common probability space on which one can define a Wiener process $W(t)$ starting at zero and a sequence of processes $\{\widetilde{S}_n(t)\}, n \geq 1$, such that

(a) for each n, $S_n(t)$ and $\widetilde{S}_n(t)$ are identically distributed as processes;

(b) $\sup_{0 \leq t \leq 1} |\widetilde{S}_n(t) - W(t)| \stackrel{P}{\Rightarrow} 0$.

This leads to the famous Donsker theorem. We state a version that is slightly less general than the original result in order to avoid discussion of the so-called *Wiener measure*.

Theorem 12.5 (Donsker) Let h be a continuous functional with respect to the uniform metric ρ on $C[0, 1]$ or $D[0, 1]$ and let $S_n(t)$ be defined as either $S_{n,1}(t)$ or $S_{n,2}(t)$. Then $h(S_n(t)) \stackrel{\mathcal{L}}{\Rightarrow} h(W(t))$ as $n \to \infty$.

Example 12.1 The five examples worked out by Erdös and Kac now follow from Donsker's theorem by considering the following functionals, each of which is continuous (with the exception of h_5, even which is continuous at *almost all* $f \in C[0, 1]$) with respect to the uniform metric on $C[0, 1]$: $h_1(f) = \sup_{0 \leq t \leq 1} f(t); h_2(f) = \sup_{0 \leq t \leq 1} |f(t)|; h_3(f) = \int_0^1 f^2(t)dt; h_4(f) = \int_0^1 |f(t)|dt; h_5(f) = \lambda\{t \in [0, 1] : f(t) > 0\}$, where λ denotes Lebesgue measure. Note that the formulas for the CDF of the limiting distribution are always a separate calculation and do not follow from Donsker's theorem.

Example 12.2 Consider the functional $h(f) = \int_0^1 f^m(t)dt$, where $m \geq 1$ is an integer. Because $[0, 1]$ is a compact interval, it is easy to verify that h is a continuous functional on $C[0, 1]$ with respect to the uniform metric. Indeed, it follows simply from the algebraic identity $|x^m - y^m| = |x - y||x^{m-1} + x^{m-2}y + \cdots + y^{m-1}|$. On the other hand, by direct integration of the polygonal curve $S_{n,1}(t)$, it follows from Donsker's theorem that $n^{-1-m/2} \sum_{k=1}^n S_k^m \stackrel{\mathcal{L}}{\Rightarrow} \int_0^1 W^m(t)dt$. At first glance, it seems surprising that a nondegenerate limit distribution for partial sums of S_k^m can exist with *only* two moments (and even that is not necessary).

Other examples and classic theory on distributions of functionals of $W(t)$ can be seen in Cameron and Martin (1945), Kac (1951), Durrett (1996), and Fitzsimmons and Pitman (1999).

Contrary to the weak invariance principle described above, there are also *strong invariance principles*, which, roughly speaking, say that the partial sum process is close to a Brownian motion with probability 1. However, the exact statements need careful description, and the best available results need a fair amount of extra assumptions, as well as sophisticated methods of proof. Furthermore, somewhat paradoxically, the strong invariance principles may not lead to the desired weak convergence results unless enough assumptions have been made so that a good enough almost sure bound on the deviation between the two processes can be established. The first strong invariance principle for partial sums was obtained in Strassen (1964). Since then, a lot of literature has developed, including for the multidimensional case. Good sources for information are Strassen (1967), Komlós, Major, and Tusnady (1976), Major (1978), and Einmahl (1987). We present two results on strong approximations of partial sums below. The results may be seen in Csörgo (1984) and Heyde (1981).

Theorem 12.6 (Strassen) Given iid random variables $X_1, X_2, \cdots$ with $E(X_1) = 0$, $\text{Var}(X_1) = 1$, there exists a common probability space on which one can define a Wiener process $W(t)$ and iid random variables $Y_1, Y_2, \cdots$ such that

$$\text{(a)} \quad \left\{ S_n = \sum_{i=1}^{n} X_i, n \geq 1 \right\} \stackrel{\mathcal{L}}{=} \left\{ \widetilde{S}_n = \sum_{i=1}^{n} Y_i, n \geq 1 \right\},$$

$$\text{(b)} \quad \sup_{0 \leq t \leq 1} \frac{|\widetilde{S}_{[nt]} - W(nt)|}{\sqrt{n \log \log n}} \to 0,$$

almost surely, as $n \to \infty$.

Remark. The $\sqrt{n \log \log n}$ bound cannot be improved in general without further assumptions on the distribution of the X_i's. The next theorem says that if we assume finiteness of more than two moments of the X_i's, or even better the moment-generating function itself, then the error can be made sufficiently small to allow the weak convergence result to be derived directly from the strong invariance principle itself. The improved rate under the existence of the mgf is the famous *KMT construction* due to Komlós, Major, and Tusnady (1976).

Theorem 12.7 Given iid random variables $X_1, X_2, \cdots$ with $E(X_1) = 0$, $\text{Var}(X_1) = 1$, $E(|X_1|^{2+\delta}) < \infty$, for some $\delta > 0$, Theorem 12.6 holds with $n^{1/(2+\delta)}$ in place of $\sqrt{n \log \log n}$. If $E(X_1^4) < \infty$, the result holds with $(n \log \log n)^{1/4} \sqrt{\log n}$ in place of $\sqrt{n \log \log n}$. If the mgf $E(e^{tX_1}) < \infty$ in some neighborhood of zero, then the $O(\log n)$ rate holds, almost surely, in place of $o(\sqrt{n \log \log n})$.

12.5 Invariance Principle for Empirical Processes

Of our four motivating examples, the statistics D_n and C_n are not functionals of a partial sum process; they are functionals of the empirical process. If weak and strong invariance principles akin to the case of partial sum processes were available for the empirical process, clearly that would be tremendously helpful in settling the questions of useful asymptotics of D_n, C_n, and more generally *nice* functionals of the empirical process. It turns out that there are indeed such invariance principles for the empirical process. Although new and significant developments are still taking place, a large part of this literature is classic, dating back to at least Kolmogorov (1933) and Smirnov (1944). In this section, we provide a brief description of some major results on this and give applications. The topic is also discussed with applications in the context of goodness-of-fit tests in Chapter 26.

To describe the weak and strong approximations of the empirical process, we first need some notation and definitions. Given a sequence of iid $U[0, 1]$ random variables $U_1, U_2, \cdots$, we define the *uniform empirical process* as $G_n(t) = \frac{1}{n}\sum_{i=1}^{n} I_{U_i \leq t}$ and the normalized uniform empirical process $\alpha_n(t) = \sqrt{n}(G_n(t) - t), n \geq 1, 0 \leq t \leq 1$. For an iid sequence $X_1, X_2, \cdots$ distributed as a general F, the empirical process is defined as $F_n(x) = \frac{1}{n}\sum_{i=1}^{n} I_{X_i \leq x}$. The normalized empirical process is $\beta_n(x) = \sqrt{n}(F_n(x) - F(x)), -\infty < x < \infty$.

The weak invariance principle is very similar to that for partial sum processes and is given below; see Dudley (1984) for further details.

Theorem 12.8 Given a sequence of iid random variables $X_1, X_2, \cdots \sim F$ and h a continuous functional on $D(-\infty, \infty)$ with respect to the uniform metric, $h(\beta_n(.)) \stackrel{\mathcal{L}}{\Rightarrow} h(B_F(.))$, where $B_F(.)$ is a centered Gaussian process with covariance kernel

$$\text{cov}(B_F(x), B_F(y)) = F(x \wedge y) - F(x)F(y).$$

An important consequence of this result is the asymptotic distribution of D_n, an extremely common statistic in the goodness-of-fit literature, which we will revisit in Chapter 26.

Example 12.3 Since the functional $h(f) = \sup_x |f(x)|$ satisfies the continuity assumption under the uniform metric, it follows that $D_n = \sup_x |\beta_n(x)| \stackrel{\mathcal{L}}{\Rightarrow} \sup_x |B_F(x)|$, which is equal to the supremum of the absolute value of a Brownian bridge (on [0, 1]) in distribution. This distribution was calculated in closed form by Kolmogorov (1933) and is a classic result. Kolmogorov found the CDF of the supremum of the absolute value of a Brownian bridge to be $H(z) = 1 - \sum_{k=-\infty}^{\infty} (-1)^{k-1} e^{-2k^2 z^2}$, from which the quantiles of the limiting distribution of D_n can be (numerically) computed.

Similar to partial sum processes, there are strong invariance principles for empirical processes as well. Some of the first ideas and results were due to Brillinger (1969) and Kiefer (1972). Given a sequence of independent Brownian bridges $B_n(t), n \geq 1, 0 \leq t \leq 1$, a *Kiefer process* (with two time parameters, namely n and t) is defined as $K(n,t) = \sum_{i=1}^{n} B_i(t)$. The better-known strong approximations for the empirical process are not the initial ones; we present one of the modern strong approximations below. See Komlós, Major, and Tusnady (1975) for the method of proof, which is different from the original Skorohod embedding technique.

Theorem 12.9 Given iid random variables $X_1, X_2, \cdots \sim F$, let $\beta_n(x) = \sqrt{n}(F_n(x) - F(x)) \stackrel{\mathcal{L}}{=} \alpha_n(F(x))$. Then, there exists a probability space on which one can define a Kiefer process $K(n, F(x))$, Brownian bridges $B_n(t), n \geq 1$, and $\tilde{\beta}_n(x), n \geq 1$, such that

(a) $\{\tilde{\beta}_n(x), n \geq 1\} \stackrel{\mathcal{L}}{=} \{\beta_n(x), n \geq 1\}$;

(b) $\sup_{-\infty<x<\infty} |\tilde{\beta}_n(x) - n^{-1/2} K(n, F(x))| = O(n^{-1/2}(\log n)^2)$ almost surely;

(c) for suitable constants C_1, C_2, λ,

$$P(\sup_{1 \leq k \leq n, -\infty<x<\infty} |\sqrt{k}\tilde{\beta}_k(x) - K(k, F(x))| > C_1(\log n)^2 + z \log n) \leq C_2 e^{-\lambda z}$$

for any z and any n;

(d) for suitable constants C_1, C_2, λ,

$$P(\sup_{-\infty<x<\infty} |\tilde{\beta}_n(x) - B_n(F(x))| > n^{-1/2}(C_1 \log n + z)) \leq C_2 e^{-\lambda z}$$

for any z and any n;

(e) $\sup_{-\infty<x<\infty} |\tilde{\beta}_n(x) - B_n(F(x))| = O(n^{-1/2} \log n)$ almost surely.

Remark. No additional improvement in the rate in part (e) is possible. Multidimensional versions are available in many places; a recent reference is Massart (1989).

We now give an application of the result above.

Example 12.4 We have noted that the general (normalized) empirical process has the property that $\sup_{-\infty<x<\infty}|\beta_n(x)|$ converges in law to a distribution with CDF $H(z) = 1 - \sum_{k=-\infty}^{\infty}(-1)^{k-1}e^{-2k^2z^2}$ as $n \to \infty$. The key point is that this is the CDF of the absolute value of a Brownian bridge. We can combine this fact with part (c) of Theorem 12.9 to produce guaranteed coverage confidence bands for the CDF $F(.)$ at any given n. The form of this *nonparametric confidence band* is $F_n(x) \pm \frac{\delta}{\sqrt{n}}$, where δ is to be chosen appropriately. If $r(n) = (C_1 + \epsilon\lambda^{-1})n^{-1/2}\log n$, then we need $H(\delta - r(n)) - C_2 n^{-\epsilon}$ to be greater than or equal to the nominal coverage $1 - \alpha$. To execute this, we need values for the constants C_1, C_2, λ; they may be taken to be 100, 10, and $\frac{1}{50}$, respectively (see Csörgo and Hall (1984)). However, these values are not sharp enough to produce useful (or even nontrivial) confidence bands at moderate values of n. But the coverage property is exact; i.e., there is no need to say that the coverage is approximately $1 - \alpha$ for large n.

12.6 Extensions of Donsker's Principle and Vapnik-Chervonenkis Classes

As we have seen, an important consequence of the weak invariance principle is the derivation of the limiting distribution of $D_n = \sqrt{n}\sup_{-\infty<x<\infty}|F_n(x) - F(x)|$ for a continuous CDF F and the empirical CDF $F_n(x)$. If we let $\mathcal{F} = \{I_{-\infty,x} : x \in \mathcal{R}\}$, then the Kolmogorov-Smirnov result says that $\sqrt{n}\sup_{f\in\mathcal{F}}(E_{P_n}f - E_P f) \stackrel{\mathcal{L}}{\Rightarrow} \sup_{0\le t\le 1}|B(t)|$, P_n, P being the probability measures corresponding to F_n, F, and $B(t)$ being a Brownian bridge. Extensions of this involve studying the asymptotic behavior of $\sup_{f\in\mathcal{F}}(E_{P_n}f - E_P f)$ for much more general classes of functions $\mathcal{F}$ and the range space of the random variables X_i; they need not be $\mathcal{R}$, or $\mathcal{R}^d$ for some finite d. Examples of asymptotic behavior include derivation of laws of large numbers and central limit theorems.

There are numerous applications of these extensions. To give just one motivating example, suppose $X_1, X_2, \cdots, X_n$ are d-dimensional iid random vectors from some P and we want to test the null hypothesis that $P = P_0$ (specified). Then, a natural statistic to assess the truth of the hypothesis is $T_n = \sup_{C\in\mathcal{C}}|P_n(C) - P_0(C)|$ for a suitable class of sets $\mathcal{C}$. Now, if $\mathcal{C}$ is too rich (for example, if it is the class of all measurable sets), then clearly there cannot be any meaningful asymptotics if P_0 is absolutely continuous. On the other hand, if $\mathcal{C}$ is too small, then the statistic cannot be good enough for detecting departures from the null hypothesis. So these extensions study the question of what kinds of families $\mathcal{C}$ or function classes $\mathcal{F}$ allow meaningful asymptotics and also result in good and commonsense tests or estimators. In some sense, the topic is a study in the art of the possible.

The technical tools required for such generalizations are extremely sophisticated and have led to striking new discoveries and mathematical advances in the theory of empirical processes. Along with these advances have come numerous new and useful statistical and probabilistic applications. The literature is huge; we strongly recommend Wellner (1992), Giné (1996), Pollard (1989), and Giné and Zinn (1984) for comprehensive reviews and sources for major theorems and additional references; specific references to some results are given later. We limit ourselves to a short description of a few key results and tools.

12.7 Glivenko-Cantelli Theorem for VC Classes

We first discuss plausibility of strong laws more general than the well-known Glivenko-Cantelli theorem, which asserts that in the one-dimensional iid case $\sup_x |F_n(x) - F(x)| \stackrel{\text{a.s.}}{\rightarrow} 0$. We need a concept of combinatorial richness of a class of sets $\mathcal{C}$ that will allow us to make statements like $\sup_{C \in \mathcal{C}} |P_n(C) - P(C)| \stackrel{\text{a.s.}}{\rightarrow} 0$. A class of sets for which this property holds is called a *Glivenko-Cantelli class*. A useful such concept is the *Vapnik-Chervonenkis dimension* of a class of sets. Meaningful asymptotics will exist for classes of sets that have a finite Vapnik-Chervonenkis dimension. It is therefore critical to know what it means and what are good examples of classes of sets with a finite Vapnik-Chervonenkis dimension. A basic treatment of this is given next.

Definition 12.3 Let $A \subset \mathcal{S}$ be a fixed set and $\mathcal{C}$ a class of subsets of $\mathcal{S}$. A is said to be *shattered by* $\mathcal{C}$ if every subset U of A is the intersection of A with some member C of $\mathcal{C}$ (i.e., $\{A \cap C : C \in \mathcal{C}\} = P(A)$, where $P(A)$ denotes the power set of A).

Sometimes the phenomenon is colloquially described as *every subset of A is picked up by some member of* $\mathcal{C}$.

Definition 12.4 The Vapnik-Chervonenkis (VC) dimension of $\mathcal{C}$ is the size of the largest set A that can be shattered by $\mathcal{C}$.

Although this is already fine as a definition, a more formal definition is given by using the concept of *shattering coefficients*.

Definition 12.5 For $n \geq 1$, the nth shattering coefficient of $\mathcal{C}$ is defined to be

$$S(n, \mathcal{C}) = \max_{x_1, x_2, \cdots, x_n \in \mathcal{S}} \text{Card}\{\{x_1, x_2, \cdots, x_n\} \cap C : C \in \mathcal{C}\}.$$

That is, $S(n, \mathcal{C})$ is the largest possible number of subsets of some (wisely chosen) set of n points that can be formed by intersecting the set with members of $\mathcal{C}$. Clearly, for any n, $S(n, \mathcal{C}) \leq 2^n$.

Here is an algebraic definition of the VC dimension of a class of sets.

Definition 12.6 The VC dimension of $\mathcal{C}$ equals $VC(\mathcal{C}) = \min\{n : S(n, \mathcal{C}) < 2^n\} - 1 = \max\{n : S(n, \mathcal{C}) = 2^n\}$.

Definition 12.7 $\mathcal{C}$ is called a Vapnik-Chervonenkis (VC) class if $VC(\mathcal{C}) < \infty$.

The following remarkable result is known as *Sauer's lemma* (Sauer (1972)).

Proposition 12.1 *Either* $S(n, \mathcal{C}) = 2^n \forall n$ *or* $\forall n$, $S(n, \mathcal{C}) \leq \sum_{i=0}^{VC(\mathcal{C})} \binom{n}{i}$.

Remark. Sauer's lemma says that either a class of sets has infinite VC dimension or its shattering coefficients grow polynomially. A few other important and useful properties of the shattering coefficients are listed below; most of them are derived easily. These properties are useful for generating new classes of VC sets from known ones by using various Boolean operations.

Theorem 12.10 The shattering coefficients $S(n, \mathcal{C})$ of a class of sets $\mathcal{C}$ satisfy

(a) $S(m, \mathcal{C}) < 2^m$ for some $m \Rightarrow S(n, \mathcal{C}) < 2^n \; \forall n > m$;
(b) $S(n, \mathcal{C}) \leq (n + 1)^{VC(\mathcal{C})} \forall n \geq 1$;
(c) $S(n, \mathcal{C}^c) = S(n, \mathcal{C})$, where $\mathcal{C}^c$ is the class of complements of members of $\mathcal{C}$;
(d) $S(n, \mathcal{B} \cap \mathcal{C}) \leq S(n, \mathcal{B})S(n, \mathcal{C})$, where the $\cap$ notation means the class of sets formed by intersecting members of $\mathcal{B}$ with those of $\mathcal{C}$;
(e) $S(n, \mathcal{B} \otimes \mathcal{C}) \leq S(n, \mathcal{B})S(n, \mathcal{C})$, where the $\otimes$ notation means the class of sets formed by taking Cartesian products of members of $\mathcal{B}$ and those of $\mathcal{C}$;
(f) $S(m + n, \mathcal{C}) \leq S(m, \mathcal{C})S(n, \mathcal{C})$.

See Vapnik and Chervonenkis (1971) and Sauer (1972) for many of the parts in this theorem. Now we give a few quick examples.

Example 12.5 Let $\mathcal{C}$ be the class of all left unbounded closed intervals on the real line; i.e., $\mathcal{C} = \{(-\infty, x] : x \in \mathcal{R}\}$. To illustrate the general formula, suppose $n = 2$. What is $S(n, \mathcal{C})$? Clearly, if we pick up the larger one among

x_1, x_2, we will pick up the smaller one, too. Or we may pick up none of them or just the smaller one. So we can pick up three distinct subsets from the power set of $\{x_1, x_2\}$. The same argument shows that the general formula for the shattering coefficients is $S(n, \mathcal{C}) = n + 1$. Consequently, this is a VC class with VC dimension one.

Example 12.6 Although topologically there are just as many left unbounded intervals on the real line as there are arbitrary intervals, in the VC index they act differently. This is interesting. Thus, let $\mathcal{C} = \{(a, b) : a \leq b \in \mathcal{R}\}$. Then it is easy to establish the formula $S(n, \mathcal{C}) = 1 + \binom{n+1}{2}$. For $n = 2$, this is equal to 4, which is also 2^2. Consequently, this is a VC class with VC dimension two.

Example 12.7 The previous example says that, on $\mathcal{R}$, the class of all convex sets is a VC class. However, this is far from being true, even in two dimensions. Indeed, if we let $\mathcal{C}$ be just the class of convex polygons in the plane, it is clear geometrically that for any n, $\mathcal{C}$ can shatter n points. So, convex polygons in $\mathcal{R}^2$ have an infinite VC dimension.

More examples of exact values of VC dimensions are given in the chapter exercises. For actual applications of these ideas to concrete extensions of Donsker's principles, it is extremely useful to know what other natural classes of sets in various spaces are VC classes. The various parts of the following result are available in Vapnik and Chervonenkis (1971) and Dudley (1978, 1979).

Theorem 12.11 Each of the following classes of sets is a VC class:

(a) southwest quadrants of $\mathcal{R}^d$ (i.e., the class of all sets of the form $\prod_{i=1}^d (-\infty, x_i]$);
(b) closed half-spaces of $\mathcal{R}^d$;
(c) closed balls of $\mathcal{R}^d$;
(d) closed rectangles of $\mathcal{R}^d$;
(e) $\mathcal{C} = \{\{x \in \mathcal{R}^d : g(x) \geq 0\} : g \in G\}$, where G is a finite-dimensional vector space of real-valued functions defined on $\mathcal{R}^d$.

We can now state a general version of the familiar Glivenko-Cantelli theorem. However, to appreciate the probabilistic utility of the combinatorial concept of shattering coefficients, it is useful to see also a famous theorem of Vapnik and Chervonenkis (1971) on Euclidean spaces, which we also provide.

Theorem 12.12 Let $X_1, X_2, \cdots \stackrel{\text{iid}}{\sim} P$, a probability measure on $\mathcal{R}^d$ for some finite d. Given any class of (measurable) sets $\mathcal{C}$, for $n \geq 1, \epsilon > 0$, $P(\sup_{\mathcal{C}} |P_n(C) - P(C)| > \epsilon) \leq 8E[\text{Card}\{\{X_1, X_2, \cdots, X_n\} \cap C : C \in \mathcal{C}\}]e^{-n\epsilon^2/32} \leq 8S(n, \mathcal{C})e^{-n\epsilon^2/32}$.

Remark. This theorem implies that for classes of sets that are of the right complexity as measured by the VC dimension, the empirical measure converges to the true measure at an essentially exponential rate. This is a sophisticated generalization of the one-dimensional DKW inequality. The improved bound of the theorem is harder to implement because it involves computation of a hard expectation with respect to a sample of n observations from the underlying P. It would usually not be possible to find this expectation, although simulating the quantity $\text{Card}\{\{X_1, X_2, \cdots, X_n\} \cap C : C \in \mathcal{C}\}$ is an interesting exercise.

The general theorem is given next; see Giné (1996).

Theorem 12.13 Let P be a probability measure on a general measurable space $\mathcal{S}$ and let $X_1, X_2, \cdots \stackrel{\text{iid}}{\sim} P$. Let P_n denote the sequence of empirical measures and let $\mathcal{C}$ be a VC class of sets in $\mathcal{S}$. Then, under suitable measurability conditions, $\sup_{C \in \mathcal{C}} |P_n(C) - P(C)| \stackrel{\text{a.s.}}{\rightarrow} 0$ as $n \to \infty$.

12.8 CLTs for Empirical Measures and Applications

This result gives us hope for establishing CLTs for suitably normalized versions of $\sup_{C \in \mathcal{C}} |P_n(C) - P(C)|$ in general spaces and with general VC classes of sets. It is useful to think of this as an analog of the one-dimensional Kolmogorov-Smirnov statistic for real-valued random variables, namely $\sup_x |F_n(x) - F(x)|$. Invariance principles allowed us to conclude that the limiting distribution is related to a Brownian bridge with real numbers in $[0, 1]$ as the time parameter. Now, however, the setup is much more abstract. The space is not a Euclidean space, and the time parameter is a set or a function. So the formulation and description of the appropriate CLTs is more involved, and although suitable Gaussian processes will still emerge as the relevant processes that determine the asymptotics, they are not Brownian bridges, and they even depend on the underlying P from which we are sampling. Some of the most profound advances in the theory of statistics and probability in the twentieth century took place around this problem, resulting along the way in deep mathematical developments and completely new tools. A short description of this is provided next.

12.8.1 Notation and Formulation

First, we give some notation and definitions. The notation $(P_n - P)(f)$ would mean $\int f dP_n - \int f dP$. Here, f is supposed to belong to some suitable class of functions $\mathcal{F}$. For example, $\mathcal{F}$ could be the class of indicator functions of the members C of a class of sets $\mathcal{C}$. In that case, $(P_n - P)(f)$ would simply mean $P_n(C) - P(C)$; we have just talked about strong laws for their suprema as C varies over $\mathcal{C}$. That is a uniformity result. Likewise, we will now need certain uniformity assumptions on the class of functions $\mathcal{F}$. We assume that

$$\text{(a)}\ \sup_{f \in \mathcal{F}} |f(s)| := F(s) < \infty\, \forall s \in \mathcal{S}$$

(measurability of F is clearly not obvious but is being ignored here) and

$$\text{(b)}\ F \in L_2(P).$$

In the case of real-valued random variables and for the problem of convergence of the process $F_n(x) - F(x)$, the corresponding functions, as we noted before, are indicator functions of $(-\infty, x)$, which are uniformly bounded functions. Now the time parameter has become a function itself, and we will need to talk about uniformly bounded functionals of functions; we will use the notation

$$l_\infty(\mathcal{F}) = \{h : \mathcal{F} \to \mathcal{R} : \sup_{f \in \mathcal{F}} |h(f)| < \infty\}.$$

Furthermore, we will refer to $\sup_{f \in \mathcal{F}} |h(f)|$ as the uniform norm and denote it as $||h||_{\infty,\mathcal{F}}$.

The two other notions we need to define are those of convergence of the process $(P_n - P)(f)$ (on normalization) and of a limiting Gaussian process that will play the role of a Brownian bridge in these general circumstances.

The Gaussian process, which we will denote as $B_P(f)$, will continue to have continuous sample paths, as was the case for the ordinary Brownian bridge, but now with the time parameter being a function and continuity being with respect to $\rho_P(f, g) = \sqrt{E_P(f(X) - g(X))^2}$. B_P has mean zero, and the covariance kernel $\text{cov}(B_P(f), B_P(g)) = P(fg) - P(f)P(g) := E_P(f(X)g(X)) - E_P(f(X))\ E_P(g(X))$. Note that due to our assumption that $F \in L_2(P)$, the covariance kernel is well defined. Trajectories of our Gaussian process B_P are therefore members of $l_\infty(\mathcal{F})$, also (uniformly) continuous with respect to the norm ρ_P we have defined above.

Finally, as in the Portmanteau theorem in Chapter 1, convergence of the process $\sqrt{n}(P_n - P)(f)$ to $B_P(f)$ would mean that expectation of any functional H of $\sqrt{n}(P_n - P)(f)$ will converge to the expectation of $H(B_P(f))$,

H being a bounded and continuous functional defined on $l_\infty(\mathcal{F})$ and taking values in $\mathcal{R}$. We remind the reader that continuity on $l_\infty(\mathcal{F})$ is with respect to the uniform norm we have already defined there. A class of functions $\mathcal{F}$ for which this central limit property holds is called a *P-Donsker class*; if the property holds for *every* probability measure P on $\mathcal{S}$, it is called a *universal Donsker class*.

12.8.2 Entropy Bounds and Specific CLTs

We can now state what sorts of assumptions on our class of functions $\mathcal{F}$ will ensure that convergence occurs (i.e., a CLT holds) and what are some good applications of such CLTs. There are multiple sets of assumptions on the class of functions that ensure a CLT. Here we describe only two, one of which relates to the concept of VC classes and the second related to *metric entropy* and *packing numbers*. Since we are already familiar with the concept of VC classes, we first state a CLT based on a VC assumption of a suitable class of sets.

Definition 12.8 A family $\mathcal{F}$ of functions f on a (measurable) space $\mathcal{S}$ is called a *VC-subgraph* if the class of subgraphs of $f \in \mathcal{F}$ is a VC class of sets, where the subgraph of f is defined to be $C_f = \{(x, y), x \in \mathcal{S}, y \in \mathcal{R} : 0 \le y \le f(x) \text{ or } f(x) \le y \le 0\}$.

Theorem 12.14 Given $X_1, X_2, \cdots \stackrel{\text{iid}}{\sim} P$, a probability measure on a measurable space $\mathcal{S}$, and a family of functions $\mathcal{F}$ on $\mathcal{S}$ such that $F(s) := \sup_{f\in\mathcal{F}} |f(s)| \in L_2(P)$, $\sqrt{n}(P_n - P)(f) \stackrel{\mathcal{L}}{\Rightarrow} B_P(f)$ if $\mathcal{F}$ is a VC-subgraph family of functions.

An important application of this theorem is the following result.

Corollary 12.1 Under the other assumptions made in Theorem 12.14, $\sqrt{n}(P_n - P)(f) \stackrel{\mathcal{L}}{\Rightarrow} B_P(f)$ if $\mathcal{F}$ is a finite-dimensional space of functions or if $\mathcal{F} = \{I_C : C \in \mathcal{C}\}$, where $\mathcal{C}$ is any VC class of sets.

Theorem 12.14 beautifully connects the scope of a Glivenko-Cantelli theorem to that of a CLT via the same VC concept, modulo the extra qualification that $F \in L_2(P)$. One can see more about this key theorem in Alexander (1984, 1987) and Giné (1996).

A pretty and useful statistical application of the result above is the following example on extension (due to Beran and Millar (1986)) of the familiar Kolmogorov-Smirnov test for goodness of fit to general spaces.

Example 12.8 Let $X_1, X_2, \cdots$ be iid observations from P on some space $\mathcal{S}$, and consider testing the null hypothesis $H_0 : P = P_0$ (specified). The natural Kolmogorov-Smirnov type test statistic for this problem is $T_n = \sqrt{n} \sup_{C \in \mathcal{C}} |P_n(C) - P_0(C)|$ for a judiciously chosen family of (measurable) sets $\mathcal{C}$. Theorem 12.14 implies that T_n converges under the null in distribution to the supremum of the absolute value of the Gaussian process $|B_{P_0}(f)|$, the sup being taken over all $f = I_C, C \in \mathcal{C}$, a VC class of subsets of $\mathcal{S}$. In principle, therefore, the null hypothesis can be tested by using this Kolmogorov-Smirnov type statistic. Note, however, that the limiting Gaussian process depends on P_0. Evaluation of the critical points of the limiting distribution of T_n under the null needs more work; see Giné (1996) for more discussion and references on this computational issue.

The second CLT we will present requires the concepts of metric entropy and bracketing numbers, which we introduce next.

Definition 12.9 Let $\mathcal{F}^*$ be a space of real-valued functions defined on some space $\mathcal{S}$, and suppose $\mathcal{F}^*$ is equipped with a norm $||.||$. Let $\mathcal{F}$ be a specific subcollection of $\mathcal{F}^*$. The *covering number* of $\mathcal{F}$ is defined to be the smallest number of balls $B(g, \epsilon) = \{h : ||h - g|| < \epsilon\}$ needed to cover $\mathcal{F}$, where $\epsilon > 0$ is arbitrary but fixed, $g \in \mathcal{F}^*$, and $||g|| < \infty$.

The covering number of $\mathcal{F}$ is denoted as $N(\epsilon, \mathcal{F}, ||.||)$. $\log N(\epsilon, \mathcal{F}, ||.||)$ is called the *entropy without bracketing* of $\mathcal{F}$.

Definition 12.10 In the same setup as in the previous definition, a *bracket* is the set of functions sandwiched between two given functions l, u (i.e., a bracket is the set $\{f : l(s) \leq f(s) \leq u(s) \forall s \in \mathcal{S}\}$). It is denoted as $[l, u]$.

Definition 12.11 The *bracketing number* of $\mathcal{F}$ is defined to be the smallest number of brackets $[l, u]$ needed to cover $\mathcal{F}$ under the restriction $||l - u|| < \epsilon \, with \, \epsilon > 0$ an arbitrary but fixed number.

The bracketing nnumber of $\mathcal{F}$ is denoted as $N_{[]}(\epsilon, \mathcal{F}, ||.||)$. $\log N_{[]}(\epsilon, \mathcal{F}, ||.||)$ is called the *entropy with bracketing* of $\mathcal{F}$.

Clearly, the smaller the radius of the balls or the width of the brackets, the greater the number of balls or brackets necessary to cover the function class $\mathcal{F}$. The important thing is to pin down qualitatively the rate at which the entropy (with or without bracketing) is going to ∞ for a given $\mathcal{F}$. It turns out, as we shall see, that for many interesting and useful classes of functions $\mathcal{F}$, this rate would be of the order of $(-\log \epsilon)$, and this will, by virtue of some theorems to be given below, ensure that the class $\mathcal{F}$ is P-Donsker.

Theorem 12.15 Assume that $F \in L_2(P)$. Then, $\mathcal{F}$ is P-Donsker if either

$$\int_0^\infty \sqrt{\log N_{[]}(\epsilon, \mathcal{F}, ||.|| = L_2(P))} d\epsilon < \infty$$

or

$$\int_0^\infty \sup_Q (\sqrt{\log N(\epsilon ||F||_{2,Q}, \mathcal{F}, ||.|| = L_2(Q))}) d\epsilon < \infty,$$

where Q denotes a general probability measure on $\mathcal{S}$.

We have previously seen that if $\mathcal{F}$ is a VC-subgraph, then it is P-Donsker. It turns out that this result follows from Theorem 12.15 on the integrability of $\sup_Q \sqrt{\log N}$. What one needs is the following upper bound on the entropy without bracketing of a VC-subgraph class. See van der Vaart and Wellner (1996) for its proof.

Proposition 12.2 *Given a VC-subgraph class $\mathcal{F}$, for any probability measure Q and any $r \geq 1$, for all $0 < \epsilon < 1$, $N(\epsilon ||F||_{r,Q}, \mathcal{F}, ||.|| = L_r(Q)) \leq C(\frac{1}{\epsilon})^{r\text{VC}(\mathcal{C})}$, where the constant C depends only on* VC($\mathcal{C}$), *$\mathcal{C}$ being the subgraph class of $\mathcal{F}$.*

Here are some additional good applications of the entropy results.

Example 12.9 As mentioned above, the key to the applicability of the entropy theorems is a good upper bound on the rate of growth of the entropy numbers of the class. Such bounds have been worked out for many intuitively interesting classes. The bounds are sometimes sharp in the sense that lower bounds can also be obtained that grow at the same rate as the upper bounds. In nearly every case mentioned in this example, the derivation of the upper bound is completely nontrivial. A very good reference is van der Vaart and Wellner (1996), particularly Chapter 2.7 there.

Uniformly Bounded Monotone Functions on $\mathcal{R}$
For this function class $\mathcal{F}$, $\log N_{[]}(\epsilon, \mathcal{F}, ||.|| = L_2(P)) \leq \frac{K}{\epsilon}$, where K is a universal constant independent of P, and so this class is in fact universal P-Donsker.

Uniformly Bounded Lipschitz Functions on Bounded Intervals in $\mathcal{R}$
Let $\mathcal{F}$ be the class of real-valued functions on a bounded interval $\mathcal{I}$ in $\mathcal{R}$ that are uniformly bounded by a universal constant and are uniformly Lipschitz of some order $\alpha > \frac{1}{2}$ (i.e., $|f(x) - f(y)| \leq M|x - y|^\alpha$) uniformly in x, y and for some finite universal constant M. For this class,

$\log N_{[]}(\epsilon, \mathcal{F}, ||.|| = L_2(P)) \leq K(\frac{1}{\epsilon})^{1/\alpha}$, where K depends only on the length of $\mathcal{I}$, M, and α, and so this class is also universal P-Donsker.

Compact Convex Subsets of a Fixed Compact Set in $\mathcal{R}^{\mathbf{d}}$
Suppose S is a compact set in $\mathcal{R}^d$ for some finite d, and let $\mathcal{C}$ be the class of all compact convex subsets of S. For any absolutely continuous P, this class satisfies $\log N_{[]}(\epsilon, \mathcal{C}, ||.|| = L_2(P)) \leq K(\frac{1}{\epsilon})^{d-1}$, where K depends on S, P, and d. Here it is meant that the function class is the set of indicators of the members of $\mathcal{C}$. Thus, for $d = 2$, $\mathcal{F}$ is P-Donsker for any absolutely continuous P.

A common implication of all of these applications of the entropy theorems is that, in the corresponding setups, asymptotic goodness-of-fit tests can be constructed by using these function classes.

12.9 Dependent Sequences: Martingales, Mixing, and Short-Range Dependence

Central limit theorems for certain types of dependent sequences were described in Chapters 9 and 10. The progression to a weak and then strong invariance principles for most of those processes was achieved by the mid-1970s; some key references are Billingsley (1956), Philip and Stout (1975), Hall (1977), and Andrews and Pollard (1994). McLeish (1975) unified much of the work by introducing what he called *mixingales*. Depending on the nature of the dependence, the norming constant for an invariance principle for the partial sum process can be $\sqrt{n}$, or something more complicated involving suitable moments of $|S_n|$. Merlevéde, Peligrad, and Utev (2006) have provided a modern review, including the latest technical innovations. We provide a brief treatment of a few classic results in this section.

Theorem 12.16 Let $\{X_k\}_{k\geq 0}$ be a stationary process with mean zero and finite variance. Assume the condition

$$\sum_{n=1}^{\infty} \frac{[E(S_n^2|X_0)]^{1/2}}{n^{3/2}} < \infty.$$

Let $S_n(t) = S_{n,1}(t), 0 \leq t \leq 1$. Then there exists a common probability space on which one can define a Wiener process $W(t)$ starting at zero and a sequence of processes $\{\hat{S}_n(t)\}, n \geq 1$, such that

(a) $\{\hat{S}_n(t), n \geq 1\} \stackrel{\mathcal{L}}{=} \{S_n(t), n \geq 1\}$;

(b) $\hat{S}_n(t) \overset{\mathcal{L}}{\Rightarrow} \eta W(t)$, where η is a nonnegative random variable (on the same probability space) with $E(\eta^2) = \lim_{n\to\infty} \frac{E(S_n^2)}{n}$.

Remark. The random variable η can in fact be explicitly characterized; see Merlevéde, Peligrad, and Utev (2006). If $\{X_k\}_{k\geq 0}$ is also *ergodic*, then η is a trivial random variable and $\eta^2 = E(X_0^2) + 2\sum_{k>0} E(X_k X_0)$.

Theorem 12.16 already applies to a broad variety of time series used in practice, although generalizations with norming different from $\sqrt{n}$ in the definition of $S_n(t)$ are available. To give specific applications of this theorem, we need some definitions. In particular, we need the definition of the concept of a *mixing sequence*, introduced in Rosenblatt (1956).

Definition 12.12 Given a process $\{X_k\}_{k\geq 0}$ (not necessarily stationary), let $\alpha(j,n) = \sup\{|P(A \cap B) - P(A)P(B)|, A \in \sigma(X_k, k \leq j), B \in \sigma(X_k, k \geq j+n)\}$. $\{X_k\}_{k\geq 0}$ is called strongly mixing if $\alpha(n) := \sup_j \alpha(j,n) \to 0$ as $n \to \infty$.

Definition 12.13 Given a process $\{X_k\}_{k\geq 0}$ (not necessarily stationary), let $\rho(j,n) = \sup\{\rho(f,g) : f \in L_2(\sigma(X_k, k \leq j)), g \in L_2(\sigma(X_k, k \geq j+n))\}$, where $\rho(f,g)$ denotes the correlation between f and g. $\{X_k\}_{k\geq 0}$ is called ρ-mixing if $\rho(n) := \sup_j \rho(j,n) \to 0$ as $n \to \infty$.

Definition 12.14 $\{X_k\}_{k\geq 0}$ is called a short-range dependent one-sided linear process if $X_k = \sum_{i\geq 0} a_i Z_{k-i}$, where $\{Z_k\}_{-\infty<k<\infty}$ is a martingale difference sequence and $\sum_{i=0}^{\infty} |a_i| < \infty$.

The general inequality $4\alpha(j,n) \leq \rho(j,n)$ is true, so that ρ-mixing implies strong mixing. For stationary Gaussian processes, the two concepts are equivalent. Many other concepts of mixing-type asymptotic independence are known. Among them is the concept of ϕ-mixing, which says that $P(B|A)$ and $P(B)$ should be uniformly close together in the sense of a small difference when A and B are events separated by a large lag. We will not discuss the other types of mixing here. A good general reference is Doukhan (1994). More specifically, for invariance principles for mixing empirical processes and the rates of convergence, two good references are Yu (1994) and Arcones and Yu (1994). Here is a result that gives good applications of Theorem 12.16 above; the assumptions of zero mean and finite variance are implicit below. We have not stated parts of Theorem 12.17 below under the best currently known conditions in order to avoid possibly hard-to-verify assumptions. Merlevéde, Peligrad, and Utev (2006) can be consulted for the best conditions or references to the best conditions.

Theorem 12.17 (a) If $\{X_k\}$ is stationary and strongly mixing satisfying

$$E(X_0^4) < \infty; \sum_{n=1}^{\infty}[\alpha(2^n)]^{1/4} < \infty; \mathrm{Var}(S_n) \to \infty; E(S_n)^4 = O(\mathrm{Var}(S_n))^2,$$

then Theorem 12.16 holds.

(b) If $\{X_k\}$ is stationary and ρ-mixing satisfying $\sum_{n=1}^{\infty} \rho(2^n) < \infty$, then Theorem 12.16 holds.

(c) If $\{X_k\}$ is a stationary and short-range dependent one-sided linear process, then Theorem 12.16 holds.

Example 12.10 Numerous examples of common processes with various mixing properties are known; a very recent reference is Bradley (2005). For example, obviously, a stationary m-dependent process for any finite m is strongly mixing and also ρ-mixing.

The mixing properties of Markov chains are intriguing. For example, the remarkable fact that if $\alpha(n) < \frac{1}{4}$ for some $n \geq 1$ then a strictly stationary ergodic aperiodic chain with state space $\mathcal{R}$ is strongly mixing is attributed in Bradley (2005) to be implicitly proved in Rosenblatt (1972). Likewise, if $\rho(n) < 1$ for some $n \geq 1$, then even without stationarity, the chain is ρ-mixing, and $\rho(n) \to 0$ at an exponential rate. If the state space is countable, then the conditions on the chain can be relaxed; see, again, Bradley (2005). A stationary autoregressive process of some finite order p is strongly mixing under quite mild conditions. If the roots of the characteristic polynomial of the process are all inside the unit circle and the errors are iid with a finite mean and have a density, then the process is strongly mixing. See Chanda (1974), Whithers (1981), and Athreya and Pantula (1986); useful information is also given in Koul (1977) and Gastwirth and Rubin (1975).

Central limit theorems for martingales were described in Chapter 9. We commented there that martingale central limit theorems can be obtained with random or nonrandom norming; which one holds depends on exactly what assumptions one makes. Likewise, invariance principles for martingales (or, more generally, martingale arrays) have been obtained under various sets of conditions. One possibility is to assume a kind of Lindeberg condition; alternatively, one can assume suitable growth conditions in terms of L_p norms; see Brown (1971), McLeish (1974), and Hall (1977). Here we report presumably the first invariance principle obtained for martingale sequences by assuming a Lindeberg type condition; see Brown (1971). These are not the weakest conditions under which an invariance principle obtains; the other references above give more general theorems under somewhat weaker conditions.

We follow the notation in Brown (1971). Let $\{S_n, \mathcal{F}_n\}_{n\geq 1}$ be a martingale and let $X_n = S_n - S_{n-1}, n \geq 1$, with $S_0 = 0$. Let $E_{j-1}(.)$ denote conditional expectation given $\mathcal{F}_{j-1}$. Define

$$\sigma_n^2 = E_{n-1}(X_n^2);\ V_n^2 = \sum_{j=1}^{n} \sigma_j^2;\ s_n^2 = E(S_n^2)(= E(V_n^2)).$$

Consider the piecewise linear function $\xi_n(t)$ on $[0, 1]$ that joins the discrete set of points $(s_k^2/s_n^2, S_k/s_n), 0 \leq k \leq n$. Thus,

$$\xi_n(t) = s_n^{-1}[S_k + X_{k+1}(ts_n^2 - s_k^2)/(s_{k+1}^2 - s_k^2)],\ s_k^2 \leq ts_n^2 \leq s_{k+1}^2,$$

$0 \leq k \leq n$. The invariance principle addresses the question of weak convergence of the process $\xi_n(t)$ as an element of $C[0, 1]$.

Assume the following:

$$(1) \qquad V_n^2/s_n^2 \stackrel{P}{\Rightarrow} 1,$$

$$(2) \quad s_n^{-2} \sum_{j=1}^{n} E[X_j^2 I_{|X_j| \geq \epsilon s_n}] = o_p(1).$$

Condition (2) is the *Lindeberg condition for Martingales*. The following result is proved in Brown (1971).

Theorem 12.18 Under conditions (1) and (2) stated above, the assertion of Donsker's theorem holds.

12.10 Weighted Empirical Processes and Approximations

Recently, there has been a growth in interest in studying weighted empirical processes due to their use in modern multiple testing problems. We will see the actual uses in Chapter 34. It turns out that asymptotic behavior of weighted empirical processes is surprisingly subtle. The subtlety comes from mutual interaction of the tail of a Wiener process and that of the weighting function. For example, since the normalized uniform empirical process $\alpha_n(t) = \sqrt{n}(G_n(t) - t)$ behaves like a Brownian bridge asymptotically, one might hope that for a strictly positive function $\delta(t)$, $\frac{\alpha_n(t)}{\delta(t)}$ may behave asymptotically like the weighted Brownian bridge $\frac{B(t)}{\delta(t)}$ on $[0, 1]$. This is not true for some natural choices of the weighting function $\delta(t)$; even more, it fails because of the tail behaviors. If we truncated the interval [0, 1] at suitable rates, the intuition would in fact work. The weighting functions $\delta(t)$ that do admit an invariance principle have no simple descriptions. Characterizing them requires using deep tail properties of the Wiener sample paths. Because of the extremely high current interest in weighted empirical processes in the multiple testing literature, we give a short description of the asymptotics and invariance principles for them. We consider the case of the uniform empirical process, as the general case can be reduced to it by a time change. We recommend Csörgo and Horváth (1993) for the topic of this section.

First we need some notation. The operators that we define below are related to what are called *lower and upper functions* of Wiener processes; see Section 12.2 in this chapter for some information on what the link is.

Let

$$\mathcal{F}_{0,1} = \{\delta : \inf_{\epsilon < t < 1-\epsilon} \delta(t) > 0 \quad \forall \quad 0 < \epsilon < \frac{1}{2}, \delta \uparrow \text{near } 0, \delta \downarrow \text{near } 1\}$$

$$I(c, \delta) = \int_0^1 \frac{1}{t(1-t)} e^{-c\delta^2(t)/[t(1-t)]} dt$$

$$E(c, \delta) = \int_0^1 \frac{\delta(t)}{[t(1-t)]^{3/2}} e^{-c\delta^2(t)/[t(1-t)]} dt.$$

Whether $\frac{\alpha_n(t)}{\delta(t)}$ is uniformly asymptotically close to a correspondingly weighted Brownian bridge is determined by the two operators I, E. Here is a complete characterization; see Csörgo (2002) for Theorems 12.19–12.21.

Theorem 12.19 Let $\delta \in \mathcal{F}_{0,1}$ and let $\alpha_n(t)$ be the normalized uniform empirical process $\alpha_n(t) = \sqrt{n}(G_n(t) - t), t \in [0, 1]$. Then, there exists a sequence of Brownian bridges $B_n(t)$ (on the same space) such that $\sup_{t \in (0,1)} \frac{|\alpha_n(t) - B_n(t)|}{\delta(t)} = o_p(1)$ if and only if

$$I(c, \delta) < \infty \, \forall \, c > 0$$

or

$$E(c, \delta) < \infty \, \forall \, c > 0, \lim_{t \to 0} \delta(t)/\sqrt{t} = \lim_{t \to 1} \delta(t)/\sqrt{1-t} = \infty.$$

Example 12.11 A natural choice for the weighting function is $\delta(t) = \sqrt{t(1-t)}$; after all, the pointwise variance of $\alpha_n(t)$ is $t(1-t)$. Clearly $\sqrt{t(1-t)} \in \mathcal{F}_{0,1}$. But, as is obvious, for no $c > 0$, $I(c, \delta)$ is finite. Thus, $\frac{\alpha_n(t)}{\sqrt{t(1-t)}}$ cannot be uniformly asymptotically approximated in probability by any sequence of the correspondingly weighted Brownian bridges over the whole unit interval. The tails create the problem. In fact, for any sequence of Brownian bridges whatsoever,

$$P\left(\sup_{t\in(0,1)} \frac{|\alpha_n(t) - B_n(t)|}{\sqrt{t(1-t)}} = \infty\right) = 1.$$

If we truncate the tails, we can control the maximum weighted deviation. The next result makes it precise.

Theorem 12.20 For any $u > 0$ and any $0 < \nu \leq \frac{1}{2}$, there exist Brownian bridges $B_n(t)$ such that

$$n^{1/2-\nu} \sup_{t\in[\frac{u}{n},1-\frac{u}{n}]} \frac{|\alpha_n(t) - B_n(t)|}{[t(1-t)]^\nu} = O_p(1),$$

while, for $0 < \nu < \frac{1}{2}$,

$$n^{1/2-\nu} \sup_{t\in(0,1)} \frac{|\alpha_n(t) - B_n(t)|}{[t(1-t)]^\nu} = O_p(1).$$

Furthermore, the bounds cannot be made $o_p(1)$.

Note that we should not expect convergence in law of $\frac{\alpha_n(t)}{\sqrt{t(1-t)}}$ to $\frac{B(t)}{\sqrt{t(1-t)}}$ because of the failure to have a uniform $o_p(1)$ bound on the maximum deviation, as we just saw. The question arises as to when we can ensure a weak convergence result. The answer is essentially already contained in Theorem 12.19.

Theorem 12.21 Under the assumptions of Theorem 12.19,

$$\sup_{t\in(0,1)} \frac{|\alpha_n(t)|}{\delta(t)} \stackrel{\mathcal{L}}{\Rightarrow} \sup_{t\in(0,1)} \frac{|B(t)|}{\delta(t)},$$

where $B(t)$ is a Brownian bridge.

Remark. An important relaxation in Theorem 12.21 is that the "for all c" requirement can be relaxed to "for some c"; sometimes this can be useful.

Although Theorem 12.21 specifies the limiting distribution of $\sup_{t\in(0,1)} \frac{|\alpha_n(t)|}{\delta(t)}$ for appropriate $\delta(t)$, the following inequality of Birnbaum and Marshall (1961) is useful because it is explicit and holds for all n.

Proposition 12.3 *Suppose $\delta(t)$ is right continuous and belongs to $\mathcal{F}_{0,1}$, and $M = \int_0^1 \delta^{-2}(t)dt < \infty$. Then, $\forall\, n, x$, $P(\sup_{t\in(0,1)} \frac{|\alpha_n(t)|}{\delta(t)} > x) \leq \frac{M}{x^2}$.*

Example 12.12 The results imply that a symmetric function $\delta(t)$ is not amenable to an invariance principle for the weighted uniform empirical process if $\delta(t) \sim \sqrt{t}$ near zero. But, for example, $\delta(t) = [t(1-t)]^{1/3}$works, and $\sup_{t\in(0,1)} \frac{\alpha_n(t)}{[t(1-t)]^{1/3}}$ will converge weakly to $\sup_{t\in(0,1)} \frac{B(t)}{[t(1-t)]^{1/3}}$. Explicit evaluation of the distribution of this latter functional (or similar ones corresponding to other $\delta(t)$) usually would not be possible. However, bounds on their CDFs often would be possible.

12.11 Exercises

Exercise 12.1 For $n = 25$ and 50, approximate the probability $P(\max_{1\leq k\leq n} |S_k| > 2\sqrt{n})$ when the sample observations are iid $U[-1, 1]$. Does the $U[-1, 1]$ assumption have any role in your approximation?

Exercise 12.2 * Plot the density function of G_2, the limiting CDF of the normalized maximum of absolute partial sums.

Exercise 12.3 For $n = 25$ and 50, approximate the probability that at least 60% of the time, a simple symmetric random walk remains over the axis.

Exercise 12.4 Give examples of three functions that are members of $D[0, 1]$ but not of $C[0, 1]$.

Exercise 12.5 * Prove that each of the functionals $h_i, i = 1, 2, 3, 4$ are continuous on $C[0, 1]$ with respect to the uniform metric.

Exercise 12.6 * Why is the functional h_5 not everywhere continuous with respect to the uniform metric?

Exercise 12.7 * Approximately simulate 20 paths of a Brownian motion by using its Karhunen-Loeve expansion (suitably truncated).

Exercise 12.8 Formally prove that the CLT for partial sums follows from the strong invariance principle available if four moments are assumed to be finite.

Exercise 12.9 * Compute and plot a 95% nonparametric confidence band for the CDF based on the KMT theorem for $n = 100, 500$ when the data are simulated from $U[0, 1]$, $N[0, 1]$.

Exercise 12.10 * Find the VC dimension of the following classes of sets :

(a) southwest quadrants of $\mathcal{R}^d$;
(b) closed half-spaces of $\mathcal{R}^d$;
(c) closed balls of $\mathcal{R}^d$;
(d) closed rectangles of $\mathcal{R}^d$.

Exercise 12.11 Give examples of three nontrivial classes of sets in $\mathcal{R}^d$ that are not VC classes.

Exercise 12.12 * Design a test for testing that sample observations in $\mathcal{R}^2$ are iid from a uniform distribution in the unit square by using suitable VC classes and applying the CLT for empirical measures.

Exercise 12.13 * Find the VC dimension of all polygons in the plane with four vertices.

Exercise 12.14 * Is the VC dimension of the class of all ellipsoids of $\mathcal{R}^d$ the same as that of the class of all closed balls of $\mathcal{R}^d$?

Exercise 12.15 * Consider the class of *Box-Cox transformations* $\mathcal{F} = \{\frac{x^\lambda - 1}{\lambda}, x > 0, \lambda \neq 0\}$. Show that $\mathcal{F}$ is a VC-subgraph class (see p. 153 in van der Vaart and Wellner (1996) for hints).

Exercise 12.16 Give an example of a stationary Gaussian process for which the condition $\sum \rho(2^n) < \infty$ holds and a few examples where the condition does not hold.

Exercise 12.17 * Prove the general inequality $4\alpha(j, n) \leq \rho(j, n)$.

Exercise 12.18 * Define $\psi(j, n) = \sup\{|\frac{P(A \cap B)}{P(A)P(B)} - 1| : A \in \sigma(X_k, 0 \leq k \leq j), B \in \sigma(X_k, k \geq j + n)\}$, and call $\{X_k\}_{k \geq 0}$ ψ-mixing if $\sup_j \psi(j, n) \to 0$ as $n \to \infty$. Show that

(a) stationary m-dependence implies ψ-mixing;

(b) ψ-mixing implies ρ-mixing.

Note: Part (b) is hard.

Exercise 12.19 Suppose $X_k = \sum_{i=0}^{\infty} \frac{1}{(i+1)^2} \epsilon_{k-i}$, where ϵ_j are iid $U[-1, 1]$. Is $\{X_k\}_{k \geq 0}$ a short-range dependent process? Is X_k summable with probability 1?

Exercise 12.20 * Derive a martingale central limit theorem with nonrandom norming by using Theorem 12.18.

Exercise 12.21 * Prove that $P\left(\sup_{t \in (0,1)} \frac{|\alpha_n(t)|}{\sqrt{t(1-t)}} = \infty\right) = 1 \, \forall n \geq 1$. Hint: Look at t near zero and consider the law of the iterated logarithm.

Exercise 12.22 Give examples of functions $\delta(t)$ that satisfy the assumptions of the Birnbaum-Marshall inequality in Section 12.10.

Exercise 12.23 * Approximate the probability $P\left(\sup_{t \in (0,1)} \frac{|\alpha_n(t)|}{[t(1-t)]^{1/3}} > x\right)$ by using the weak convergence result of Theorem 12.21.

References

Alexander, K. (1984). Probability inequalities for empirical processes and a law of the iterated logarithm, Ann. Prob., 12, 1041–1067.

Alexander, K. (1987). The central limit theorem for empirical processes on Vapnik-Chervonenkis classes, Ann. Prob., 15, 178–203.

Andrews, D. and Pollard, D. (1994). An introduction to functional central limit theorems for dependent stochastic processes, Int. Stat. Rev., 62, 119–132.

Arcones, M. and Yu, B. (1994). Central limit theorems for empirical and U-processes of stationary mixing sequences, J. Theor. Prob., 1, 47–71.

Athreya, K. and Pantula, S. (1986). Mixing properties of Harris chains and autoregressive processes, J. Appl. Prob., 23, 880–892.

Beran, R. and Millar, P. (1986). Confidence sets for a multinomial distribution, Ann. Stat., 14, 431–443.

Billingsley, P. (1956). The invariance principle for dependent random variables, Trans. Am. Math. Soc., 83(1), 250–268.

Billingsley, P. (1968). *Convergence of Probability Measures*, John Wiley, New York.

Birnbaum, Z. and Marshall, A. (1961). Some multivariate Chebyshev inequalities with extensions to continuous parameter processes, Ann. Math. Stat., 32, 687–703.

Bradley, R. (2005). Basic properties of strong mixing conditions: a survey and some open problems, Prob. Surv., 2, 107–144.

Brillinger, D. (1969). An asymptotic representation of the sample df, Bull. Am. Math. Soc., 75, 545–547.

Brown, B. (1971). Martingale central limit theorems, Ann. Math. Stat., 42, 59–66.

Cameron, R. and Martin, W. (1945). Evaluation of various Wiener integrals by use of certain Sturm-Liouville differential equations, Bull. Am. Math. Soc., 51, 73–90.

Chanda, K. (1974). Strong mixing properties of linear stochastic processes, J. Appl. Prob., 11, 401–408.

Csörgo, M. and Révész, P. (1981). *Strong Approximations in Probability and Statistics*, Academic Press, New York.

Csörgo, M. and Horváth, L. (1993). *Weighted Approximations in Probability and Statistics*, John Wiley, New York.

Csörgo, M. (1984). Invariance principle for empirical processes, in *Handbook of Statistics*, P. K. Sen and P. R. Krishraiah (eds.), Vol. 4, North-Holland, Amsterdam, 431–462.

Csörgo, M. (2002). A glimpse of the impact of Paul Erdös on probability and statistics, Can. J. Stat., 30(4), 493–556.

Csörgo, S. and Hall, P. (1984). The KMT approximations and their applications, Aust. J. Stat., 26(2), 189–218.

Donsker, M. (1951). An invariance principle for certain probability limit theorems, Mem. Am. Math. Soc., 6.

Doukhan, P. (1994). *Mixing: Properties and Examples*, Lecture Notes in Statistics, Vol. 85, Springer, New York.

Dudley, R. (1978). Central limit theorems for empirical measures, Ann. Prob., 6, 899–929.

Dudley, R. (1979). Central limit theorems for empirical measures, Ann. Prob., 7(5), 909–911.

Dudley, R. and Philipp, W. (1983). Invariance principles for sums of Banach space valued random elements and empirical processes, Z. Wahr. Verw. Geb., 62, 509–552.

Dudley, R. (1984). *A Course on Empirical Processes*, Lecture Notes in Mathematics, Springer, Berlin.

Durrett, R. (1996). *Probability: Theory and Examples*, 2nd ed., Duxbury Press, Belmont, CA.

Einmahl, U. (1987). Strong invariance principles for partial sums of independent random vectors, Ann. Prob., 15(4), 1419–1440.

Erdös, P. and Kac, M. (1946). On certain limit theorems of the theory of Probability, Bull. Am. Math. Soc., 52, 292–302.

Fitzsimmons, P. and Pitman, J. (1999). Kac's moment formula and the Feynman-Kac formula for additive functionals of a Markov process, Stoch. Proc. Appl., 79(1), 117–134.

Gastwirth, J. and Rubin, H. (1975). The asymptotic distribution theory of the empiric cdf for mixing stochastic processes, Ann. Stat., 3, 809–824.

Giné, E. (1996). Empirical processes and applications: an overview, Bernoulli, 2(1), 1–28.

Giné, E. and Zinn, J. (1984). Some limit theorems for empirical processes, with discussion, Ann. Prob., 12(4), 929–998.

Hall, P. (1977). Martingale invariance principles, Ann. Prob., 5(6), 875–887.

Hall, P. and Heyde, C. (1980). *Martingale Limit Theory and Its Applications*, Academic Press, New York.

Heyde, C. (1981). Invariance principles in statistics, Int. Stat. Rev., 49(2), 143–152.

Jain, N., Jogdeo, K., and Stout, W. (1975). Upper and lower functions for martingales and mixing processes, Ann. Prob., 3, 119–145.

Kac, M. (1951). On some connections between probability theory and differential and integral equations, in Proceedings of the Second Berkeley Symposium, J. Neyman (ed.), University of California Press, Berkeley, 189–215.
Kiefer, J. (1972). Skorohod embedding of multivariate rvs and the sample df, Z. Wahr. Verw. Geb., 24, 1–35.
Kolmogorov, A. (1933). Izv. Akad. Nauk SSSR, 7, 363–372 (in German).
Komlós, J., Major, P., and Tusnady, G. (1975). An approximation of partial sums of independent rvs and the sample df: I, Z. Wahr. Verw. Geb., 32, 111–131.
Komlós, J., Major, P., and Tusnady, G. (1976). An approximation of partial sums of independent rvs and the sample df: II, Z. Wahr. Verw. Geb., 34, 33–58.
Koul, H. (1977). Behavior of robust estimators in the regression model with dependent errors, Ann. Stat., 5, 681–699.
Major, P. (1978). On the invariance principle for sums of iid random variables, J. Multivar. Anal., 8, 487–517.
Mandrekar, V. and Rao, B.V. (1989). On a limit theorem and invariance principle for symmetric statistics, Prob. Math. Stat., 10, 271–276.
Massart, P. (1989). Strong approximation for multivariate empirical and related processes, via KMT construction, Ann. Prob., 17(1), 266–291.
McLeish, D. (1974). Dependent central limit theorems and invariance principles, Ann. Prob., 2, 620–628.
McLeish, D. (1975). Invariance principles for dependent variables, Z. Wahr. Verw. Geb., 3, 165–178.
Merlevéde, F., Peligrad, M., and Utev, S. (2006). Recent advances in invariance principles for stationary sequences, Prob. Surv., 3, 1–36.
Oblój, J. (2004). The Skorohod embedding problem and its offspring, Prob. Surv., 1, 321–390.
Philipp, W. (1979). Almost sure invariance principles for sums of B-valued random variables, in *Problems in Banach Spaces*, A. Beck (ed.), Vol. II, Lecture Notes in Mathematics, Vol. 709, Springer, Berlin, 171–193.
Philipp, W. and Stout, W. (1975). Almost sure invariance principles for partial sums of weakly dependent random variables, Mem. Am. Math. Soc., 2, 161.
Pollard, D. (1989). Asymptotics via empirical processes, Stat. Sci., 4, 341–366.
Pyke, R. (1984). Asymptotic results for empirical and partial sum processes: a review, Can. J. Stat., 12, 241–264.
Révész, P. (1976). On strong approximation of the multidimensional empirical process, Ann. Prob., 4, 729–743.
Rosenblatt, M. (1956). A central limit theorem and a strong mixing condition, Proc. Natl. Acad. Sci. USA, 42, 43–47.
Rosenblatt, M. (1972). Uniform ergodicity and strong mixing, Z. Wahr. Verw. Geb., 24, 79–84.
Sauer, N. (1972). On the density of families of sets, J. Comb. Theory Ser. A, 13, 145–147.
Sen, P.K. (1978). An invariance principle for linear combinations of order statistics, Z. Wahr. Verw. Geb., 42(4), 327–340.
Shorack, G. and Wellner, J. (1986). *Empirical Processes with Applications to Statistics*, John Wiley, New York.
Smirnov, N. (1944). Approximate laws of distribution of random variables from empirical data, Usp. Mat. Nauk., 10, 179–206.

Strassen, V. (1964). An invariance principle for the law of the iterated logarithm, Z. Wahr. Verw. Geb., 3, 211–226.

Strassen, V. (1967). Almost sure behavior of sums of independent random variables and martingales, in *Proceedings of the Fifth Berkeley Symposium*, L. Le Cam and J. Neyman (eds.), Vol. 1 University of California Press, Berkeley, 315–343.

van der Vaart, A. and Wellner, J. (1996). *Weak Convergence and Empirical Processes*, Springer-Verlag, New York.

Vapnik, V. and Chervonenkis, A. (1971). On the uniform convergence of relative frequencies of events to their probabilities, Theory Prob. Appl., 16, 264–280.

Wellner, J. (1992). Empirical processes in action: a review, Int. Stat. Rev., 60(3), 247–269.

Whithers, C. (1981). Conditions for linear processes to be strong mixing, Z. Wahr. Verw. Geb., 57, 477–480.

Whitt, W. (1980). Some useful functions for functional limit theorems, Math. Oper. Res., 5, 67–85.

Yu, B. (1994). Rates of convergence for empirical processes of stationary mixing sequences, Ann. Prob., 22, 94–116.

Chapter 13
Edgeworth Expansions and Cumulants

We now consider the important topic of writing asymptotic expansions for the CDFs of centered and normalized statistics. When the statistic is a sample mean, let $Z_n = \frac{\sqrt{n}(\bar{X}-\mu)}{\sigma}$ and $F_n(x) = P(Z_n \leq x)$, where $X_1, \ldots, X_n$ are iid with a CDF F having mean μ and variance $\sigma^2 < \infty$.

The CLT says that $F_n(x) \to \Phi(x)$ for every x, and the Berry-Esseen theorem says $|F_n(x) - \Phi(x)| = O(\frac{1}{\sqrt{n}})$ uniformly in x if X has three moments. If we change the approximation $\Phi(x)$ to $\Phi(x) + \frac{C_1(F)p_1(x)\phi(x)}{\sqrt{n}}$ for some suitable constant $C_1(F)$ and a suitable polynomial $p_1(x)$, we can assert that

$$|F_n(x) - \Phi(x) - \frac{C_1(F)p_1(x)\phi(x)}{\sqrt{n}}| = O\left(\frac{1}{n}\right)$$

uniformly in x. Expansions of the form

$$F_n(x) = \Phi(x) + \sum_{s=1}^{k} \frac{q_s(x)}{\sqrt{n}^s} + o(n^{-k/2}) \text{ uniformly in } x$$

are known as Edgeworth expansions for Z_n. One needs some conditions on F and enough moments of X to carry the expansion to k terms for a given k. Excellent references for the main results on Edgeworth expansions in this chapter are Barndorff-Nielsen and Cox (1991), Hall (1992), and Lahiri (2003). Edgeworth expansions are available for very general types of statistics. Albers (2001), Bentkus et al. (1997), Bickel (1974), Bhattacharya and Ghosh (1978), and Hall (1992) are good general references for such general Edgeworth expansions. The coefficients in the Edgeworth expansion for means depend on the cumulants of F, which share a functional relationship with the sequence of moments of F. Cumulants are also useful in many other contexts; see, for example, Chapter 14. We also present some basic theory

A. DasGupta, *Asymptotic Theory of Statistics and Probability*,

of cumulants in this chapter. Stuart and Ord (1994) is an excellent reference for further details about cumulants. Other specific references are given in the following sections.

13.1 Expansion for Means

The CLT for means fails to capture possible skewness in the distribution of the mean for a given finite n because all normal distributions are symmetric. By expanding the CDF to the next term, the skewness can be captured. Expansion to another term also adjusts for the kurtosis. Although expansions to any number of terms are available under existence of enough moments, usually an expansion to two terms after the leading term is of the most practical importance. Indeed, expansions to three terms or more can be unstable due to the presence of the polynomials in the expansions. We present the two-term expansion next; see Hall (1992) or Lahiri (2003) for proofs. We first state a major assumption that distinguishes the case of lattice distributions from nonlattice ones.

Cramér's Condition A CDF F on the real line is said to satisfy Cramér's condition if $\limsup_{t\to\infty} |E_F(e^{itX})| < 1$.

Remark. All absolutely continuous distributions satisfy Cramér's condition. On the other hand, purely singular distributions do not. All lattice distributions are purely singular. The expansion to be stated below does not hold for lattice distributions. An intuitive reason for this is that, for lattice distributions, the true CDF of Z_n has jump discontinuities for every n. On the other hand, the Edgeworth expansion is a smooth function. So, it cannot approximate the true CDF of Z_n with the level of accuracy claimed in Theorem 13.1 below for lattice cases. To summarize, lattice distributions do not meet Cramér's condition and neither do they admit the expansion stated in Theorem 13.1. There are Edgeworth expansions for Z_n in the lattice case, but they include extra terms in order to account for the jumps in the true CDF of Z_n. These extra terms need additional messy notation; see Esseen (1945) for details.

Theorem 13.1 (Two-Term Edgeworth Expansion; i.e., $k = 2$) Suppose F satisfies the Cramér condition and $E_F(X^4) < \infty$. Then

$$F_n(x) = \Phi(x) + \frac{C_1(F)p_1(x)\phi(x)}{\sqrt{n}} + \frac{C_2(F)p_2(x) + C_3(F)p_3(x)}{n}\phi(x) + O(n^{-\frac{3}{2}})$$

uniformly in x, where

$$\begin{aligned}
C_1(F) &= \frac{E(X-\mu)^3}{6\sigma^3} \text{ (skewness correction)},\\
C_2(F) &= \frac{\frac{E(X-\mu)^4}{\sigma^4}-3}{24} \text{ (kurtosis correction)},\\
C_3(F) &= \frac{C_1^2(F)}{72},\\
p_1(x) &= (1-x^2),\\
p_2(x) &= 3x-x^3, \text{ and}\\
p_3(x) &= 10x^3-15x-x^5.
\end{aligned}$$

Remark. The Edgeworth expansion is derived from an expansion for the logarithm of the characteristic function of Z_n. The Cramér condition ensures that the characteristic function can be expanded in the required manner. An expansion for F_n, the CDF, is obtained by a Fourier inversion of the expansion for the characteristic function. That is the basic idea, but the details are nontrivial. To carry the Edgeworth expansion to k terms, we need to evaluate a sequence of polynomials, as in Theorem 13.1. The highest-order one is of degree $3k-1$. Theoretical descriptions of these polynomials can be seen in Petrov (1975), Bhattacharya and Denker (1990), and Hall (1992), among other places. The coefficients in the Edgeworth expansion can be rewritten in terms of the *cumulants* of F; we discuss cumulants in Section 13.6 in this chapter. It is useful to mention here that the corresponding formal two-term expansion for the density of Z_n is given by

$$\phi(z)+\frac{1}{\sqrt{n}}\tilde{p_1}(z)\phi(z)+\frac{1}{n}\tilde{p_2}(z)\phi(z),$$

where $\tilde{p_1}(z)=C_1(F)(z^3-3z)$, $\tilde{p_2}(z)=C_3(F)(z^6-15z^4+45z^2-15)+C_2(F)(z^4-6z^2+3)$.

Example 13.1 How accurate is an Edgeworth expansion? We cannot answer this question precisely in general. One would suspect that if F is skewed or n is small, then Edgeworth expansions with one or even two terms may not be very accurate. Let us see an illustrative example. Suppose $X_1, X_2, \ldots, X_n \overset{\text{iid}}{\sim}$ Exp(1). Table 13.1 explains the accuracy of the one-term and the two-term Edgeworth expansions and also the CLT. We choose $n=5$ to test the Edgeworth expansions and the CLT at a small sample size.

Table 13.1 Accuracy of Edgeworth expansions and CLT

x	Exact $F_n(x)$	$\Phi(x)$	1-term Edgeworth	2-term Edgeworth
0	0.560	0.50	0.559	0.559
0.4	0.701	0.655	0.702	0.700
1	0.847	0.841	0.841	0.849
2	0.959	0.977	0.953	0.959

So the two-term expansion is occasionally slightly better, but the one-term or the two-term expansion is always better than the CLT approximation. In particular, note the inaccuracy of the CLT when $x = 0$, caused by the CLT's inability to capture the skewness in F_n. When the skewness is taken into account by the Edgeworth expansions, the approximation becomes almost exact.

13.2 Using the Edgeworth Expansion

One of the uses of an Edgeworth expansion in statistics is approximation of the power of a test. In the one-parameter regular exponential family, the natural sufficient statistic is a sample mean, and standard tests are based on this statistic. So the Edgeworth expansion for sample means of iid random variables can be used to approximate the power of such tests. Here is an example.

Example 13.2 Suppose $X_1, X_2, \ldots, X_n \overset{\text{iid}}{\sim} \text{Exp}(\sigma)$ and we wish to test $H_0 : \sigma = 1$ vs. $H_1 : \sigma > 1$. The UMP test rejects H_0 for large values of $\sum_{i=1}^n X_i$. If the cutoff value is found by using the CLT, then the test rejects H_0 for $\bar{X} > 1 + \frac{k}{\sqrt{n}}$, where $k = z_\alpha$. The power at an alternative σ equals

$$\begin{aligned}
\text{power} &= P_\sigma\left(\bar{X} > 1 + \frac{k}{\sqrt{n}}\right) = P_\sigma\left(\frac{\bar{X} - \sigma}{\frac{\sigma}{\sqrt{n}}} > \frac{1 + \frac{k}{\sqrt{n}} - \sigma}{\frac{\sigma}{\sqrt{n}}}\right) \\
&= 1 - P_\sigma\left(\frac{\bar{X} - \sigma}{\frac{\sigma}{\sqrt{n}}} \leq \frac{\sqrt{n}(1 - \sigma)}{\sigma} + \frac{k}{\sigma}\right) \\
&= 1 - F_\sigma\left(\frac{\sqrt{n}(1 - \sigma)}{\sigma} + \frac{k}{\sigma}\right) \to 1 \text{ as } n \to \infty,
\end{aligned}$$

where $F_\sigma = F_{n,\sigma}$ is the true CDF under σ and involves an incomplete Gamma function. For a more useful approximation, the Edgeworth expansion is used. For example, the general one-term Edgeworth expansion for sample means

$$F_n(x) = \Phi(x) + \frac{C_1(F)(1-x^2)\phi(x)}{\sqrt{n}} + O\left(\frac{1}{n}\right)$$

can be used to approximate the power expression above. Algebra reduces the one-term Edgeworth expression to the formal approximation

$$\text{power} \approx \Phi\left(\frac{\sqrt{n}(\sigma-1)-1}{\sigma}\right) + \frac{1}{3\sqrt{n}}\left[\frac{(\sqrt{n}(\sigma-1)-k)^2}{\sigma^2} - 1\right]$$
$$\phi\left(\frac{(\sqrt{n}(\sigma-1)-k)}{\sigma}\right).$$

This is a much more useful approximation than simply saying that for large n the power is close to 1.

13.3 Edgeworth Expansion for Sample Percentiles

It was seen in Chapter 7 that, under mild conditions, sample percentiles corresponding to iid data are asymptotically normal. Thus we have a central limit theorem. It is natural to seek higher-order expansions to the CDF of a normalized percentile akin to Edgeworth expansions for sample means. The expansion is similar in form to the one for sample means; however, the associated polynomials are more involved. But now there are no moments involved, as one would anticipate. The main requirement is that, at the corresponding population percentile, the CDF F should be differentiable enough times. We give the two-term expansion below; see Reiss (1989) for a proof.

Theorem 13.2 Suppose $X_i \overset{\text{iid}}{\sim} F$, and let $0 < \alpha < 1$ be fixed. Let $\xi_\alpha = F^{-1}(\alpha)$, and suppose F is three times differentiable at ξ_α.

Let

$$\sigma_\alpha = \sqrt{\alpha(1-\alpha)},\ p_\alpha = F'(\xi_\alpha),\ a_1 = \frac{1-2\alpha}{3\sigma_\alpha},\ a_2 = -\frac{\alpha^3+(1-\alpha)^3}{4\sigma_\alpha^2},$$

$$b_1 = \frac{\alpha - \{n\alpha\}}{\sigma_\alpha},\ b_2 = \frac{(1-\alpha)^2\{n\alpha\} + \alpha^2(1-\{n\alpha\})}{2\sigma_\alpha^2},$$

$$d_1 = \frac{\sigma_\alpha F''(\xi_\alpha)}{p_\alpha^2},\ d_2 = \frac{\sigma_\alpha^2 F^{(3)}(\xi_\alpha)}{p_\alpha^3}.$$

In the equations above, $\{.\}$ denotes the fractional part function.

Let $F_n(x) = P_F(\sqrt{n}(X_{[n\alpha]:n} - \xi_\alpha)\frac{p_\alpha}{\sigma_\alpha} \leq x)$. Then,

$$F_n(x) = \Phi(x) + \frac{R_1(\dot{x})}{\sqrt{n}}\phi(x) + \frac{R_2(x)}{n}\phi(x) + O(n^{-\frac{3}{2}})$$

uniformly in x, where

$$R_1(x) = \left(\frac{d_1}{2} - a_1\right)x^2 - 2a_1 - b_1,$$

$$R_2(x) = -\left(\frac{d_1}{2} - a_1\right)^2 \frac{x^5}{2} - \left(\frac{5}{2}a_1^2 + a_2 + a_1 b_1 - \frac{1}{2}b_1 d_1 - \frac{1}{6}d_2\right)x^3$$
$$- \left(\frac{15}{2}a_1^2 + 3a_1 b_1 + 3a_2 + \frac{1}{2}b_1^2 + b_2\right)x.$$

Remark. The expansion would be somewhat different if the sample percentile were defined as $X_{[n\alpha]+1:n}$. Both definitions are used in the literature. Note that p_α has to be > 0 for a valid expansion, indeed even for a central limit theorem. However, the higher-order derivatives of F at ξ_α can be zero. More general results can be seen in Reiss (1976).

Example 13.3 Suppose $X_1, \ldots, X_n$ are iid $C(0, 1)$ with density $f(x) = \frac{1}{\pi(1+x^2)}$. Then, following the notation of Theorem 13.2, if $\alpha = \frac{1}{2}$, on calculation, $\sigma_\alpha = \frac{1}{2}$, $a_1 = 0$, $a_2 = -\frac{1}{4}$, $p_\alpha = \frac{1}{\pi}$, $b_1 = 0$, $b_2 = \frac{1}{2}$, $d_1 = 0$, $d_2 = -\frac{\pi^2}{2}$. Substitution of these values leads to $R_1(x) \equiv 0$, $R_2(x) = \frac{x}{4} + \frac{3-\pi^2}{12}x^3$. Thus,

$$P_{C(0,1)}\left(\sqrt{n}X_{[\frac{n}{2}]:n} \leq \frac{\pi}{2}x\right) = \Phi(x) + \frac{1}{n}\left(\frac{x}{4} + \frac{3-\pi^2}{12}x^3\right)\phi(x) + O(n^{-\frac{3}{2}})$$

uniformly in x.

13.4 Edgeworth Expansion for the *t*-statistic

The t-test and the t confidence interval are among the most used tools of statistical methodology. As such, an Edgeworth expansion for the CDF of the t-statistic for general populations is interesting and useful, and we present it next. The development here is from Hall (1987).

Theorem 13.3 Let $X_1, \ldots, X_n$ be iid with CDF F having mean μ, variance σ^2, and $E(X_1-\mu)^4 < \infty$. Assume that F satisfies Cramér's condition. Let $\beta_1(F) = \frac{E(X_1-\mu)^3}{\sigma^3}$, $\gamma = \frac{E(X_1-\mu)^4}{\sigma^4} - 3$, and

$$P_1(x) = \frac{\beta_1(F)(2x^2+1)}{6},$$

$$P_2(x) = -x\left[\frac{\beta_1^2(F)}{18}(x^4+2x^2-3) - \frac{\gamma}{12}(x^2-3) + \frac{1}{4}(x^2+3)\right].$$

Then the CDF of the t-statistic $T_n = \frac{\sqrt{n}(\bar{X}-\mu)}{s}$ admits the expansion $H_n(x) = P(T_n \le x) = \Phi(x) + \frac{P_1(x)\phi(x)}{\sqrt{n}} + \frac{P_2(x)\phi(x)}{n} + o\left(\frac{1}{n}\right)$ uniformly in x.

Remark. The $o\left(\frac{1}{n}\right)$ error can be improved if one makes more stringent moment assumptions; see Bhattacharya and Ghosh (1978).

Example 13.4 The Edgeworth expansion for the t-statistic can be used to approximate the coverage probability of the t confidence interval for nonnormal data. It is interesting how accurately the Edgeworth expansion predicts the effect of nonnormality on the t-interval's coverage. Some specific nonnormal populations are considered below.

Consider the coverage probability of the t confidence interval $\bar{X} \pm t_{\frac{\alpha}{2},n-1}\frac{s}{\sqrt{n}}$ with iid data from some CDF F. The coverage equals $P(-t_{\frac{\alpha}{2}} \le T_n \le t_{\frac{\alpha}{2}})$, which cannot be found analytically for nonnormal F. A two-term Edgeworth expansion gives the approximation:

$$\begin{aligned} \text{coverage} &\approx \Phi\left(t_{\frac{\alpha}{2}}\right) - \Phi\left(-t_{\frac{\alpha}{2}}\right) \\ &\quad + \frac{[P_1\left(t_{\frac{\alpha}{2}}\right) - P_1\left(-t_{\frac{\alpha}{2}}\right)]\phi\left(t_{\frac{\alpha}{2}}\right)}{\sqrt{n}} + \frac{[P_2\left(t_{\frac{\alpha}{2}}\right) - P_2\left(-t_{\frac{\alpha}{2}}\right)]\phi\left(t_{\frac{\alpha}{2}}\right)}{n} \\ &= 2\Phi\left(t_{\frac{\alpha}{2}}\right) - 1 - \frac{2t_{\frac{\alpha}{2}}\phi\left(t_{\frac{\alpha}{2}}\right)}{n}\left\{\frac{\beta_1^2}{18}\left(t_{\frac{\alpha}{2}}^4 + 2t_{\frac{\alpha}{2}}^2 + 3\right)\right. \\ &\quad \left. -\frac{\gamma}{12}\left(t_{\frac{\alpha}{2}}^2 - 3\right) + \frac{1}{4}\left(t_{\frac{\alpha}{2}}^2 + 3\right)\right\}. \end{aligned}$$

The $2\Phi\left(t_{\frac{\alpha}{2}}\right) - 1$ term is $\approx 1-\alpha$, the nominal value, so we may be able to predict whether the coverage is $< 1-\alpha$ by looking at the algebraic sign of the expression in {}. The algebraic sign depends on α and the true F. Here are two illustrative examples.

Example 13.5 Let $F = U[-1, 1]$. Then $\beta_1 = 0$ and $\gamma = -\frac{6}{5}$. The expression in {} reduces to $\frac{1}{10}(t^2 - 3) + \frac{1}{4}(t^2 + 3) = \frac{7}{20}t^2 + \frac{9}{20} \geq 0 \ \forall t \geq 0$, so we would suspect that the t-interval when applied to uniform data gives a coverage $< 1 - \alpha$. Basu and DasGupta (1993) give details about the coverage of the t-interval for uniform data. The table below gives the simulated coverage of the t-interval when $\alpha = .05$.

n	simulated coverage
10	.914
20	.932
30	.941
60	.948

Thus the actual coverages are $< 1-\alpha$, as the Edgeworth expansion suggests.

Example 13.6 F = lognormal(0, 1). In this case, $\beta_1 = \frac{e^{\frac{9}{2}} - 3e^{\frac{5}{2}} + 2e^{\frac{3}{2}}}{(e^2 - e)^{\frac{3}{2}}} = 38.25$ and $\gamma = \frac{e^8 - 4e^5 + 6e^3 - 3e^2}{(e^2 - e)^2} = 110.94$. The expression in {} reduces to $2.12(t^4 + 2t^2 + 3) - 9.24(t^2 - 3) + .25(t^2 + 3)$. This can be seen to be ≥ 0 for all $t \geq 0$. So again we may suspect that the coverage is $< 1 - \alpha$. The table below gives the simulated coverage of the t-interval when $\alpha = .05$.

n	simulated coverage
20	.87
50	.91
200	.93

Again, the coverage is indeed $< 1 - \alpha$, as the Edgeworth expansion would suggest.

13.5 Cornish-Fisher Expansions

For constructing asymptotically correct confidence intervals for a parameter on the basis of an asymptotically normal statistic, the first-order approximation to the quantiles of the statistic (suitably centered and normalized) comes from using the central limit theorem. Just as Edgeworth expansions produce more accurate expansions for the CDF of the statistic than does just the central limit theorem, higher-order expansions for the quantiles produce more accurate approximations than does just the normal quantile. These higher-order expansions for quantiles are essentially obtained from recursively inverted Edgeworth expansions, starting with the normal quantile as the initial approximation. They are called Cornish-Fisher expansions. We briefly present the case of sample means and the t-statistic. First, we need some notation.

Given a CDF F with a finite mgf $\psi(t)$ in some open neighborhood of zero, let $K(t) = \log \psi(t)$ denote the *cumulant generating function*. Thus, $K(t) = \sum_{r=1}^{\infty} \kappa_r \frac{t^r}{r!}$, where κ_r is the rth cumulant of F. If σ^2 denotes the variance of F, then the standardized cumulants are the quantities $\rho_r = \frac{\kappa_r}{\sigma^r}$. See Barndorff-Nielsen and Cox (1991) for further details on the following theorem.

Theorem 13.4 Let $X_1, \ldots, X_n$ be iid with CDF F having a finite mgf in some open neighborhood of zero. Assume that F satisfies Cramér's condition. Let $Z_n = \frac{\sqrt{n}(\bar{X}-\mu)}{\sigma}$ and $H_n(x) = P_F(Z_n \leq x)$. Then,

$$H_n^{-1}(\alpha) = z_\alpha + \frac{(z_\alpha^2 - 1)\rho_3}{6\sqrt{n}} + \frac{(z_\alpha^3 - 3z_\alpha)\rho_4}{24n} - \frac{(2z_\alpha^3 - 5z_\alpha)\rho_3^2}{36n} + O(n^{-\frac{3}{2}}),$$

where $P(Z > z_\alpha) = \alpha$, $Z \sim N(0, 1)$.

Example 13.7 Let $W_n \sim \chi_n^2$ and $Z_n = \frac{W_n - n}{\sqrt{2n}} \stackrel{\mathcal{L}}{\Rightarrow} N(0, 1)$ as $n \to \infty$, so a first-order approximation to the upper αth quantile of W_n is just $n + z_\alpha\sqrt{2n}$. The Cornish-Fisher expansion should produce a more accurate approximation. To verify this, we will need the standardized cumulants, which are $\rho_3 = 2\sqrt{2}$, $\rho_4 = 12$. Now substituting into Theorem 13.4, we get the two-term Cornish-Fisher expansion $\chi_{n,\alpha}^2 = n + z_\alpha\sqrt{2n} + \frac{2(z_\alpha^2-1)}{3} + \frac{z_\alpha^3 - 7z_\alpha}{9\sqrt{2n}}$.

The accuracy of the two-term Cornish-Fisher expansion is very impressive. At the very small sample size $n = 5$, the exact 99th percentile of W_n is 15.09; the central limit theorem approximation is 12.36, while the two-term Cornish-Fisher expansion above gives the approximation 15.07. Thus, the Cornish-Fisher approximation is almost exact, while the normal approximation is poor. See Barndorff-Nielsen and Cox (1991) for further numerics.

By inverting the Edgeworth expansion for the t-statistic, similar Cornish-Fisher expansions for the quantiles of the t-statistic can be obtained. Here is the formal result.

Theorem 13.5 Consider the setup of Theorem 13.3. Then,

$$
\begin{aligned}
H_n^{-1}(\alpha) =& z_\alpha - \frac{1}{\sqrt{n}} \frac{\beta_1}{6}(2z_\alpha^2 + 1) + \frac{1}{n}\left[z_\alpha \left(\frac{\gamma+3}{4} - \frac{5\beta_1^2}{72} \right) \right. \\
& \left. + z_\alpha^3 \left(\frac{1}{2} + \frac{11\beta_1^2}{36} - \frac{\gamma+3}{12} \right) + z_\alpha^5 \frac{5\beta_1^2}{18} \right] + o\left(\frac{1}{n}\right)
\end{aligned}
$$

Remark. Under the assumption of additional moments, the error term can be improved to $O(n^{-\frac{3}{2}})$. If the underlying CDF F is symmetric around μ, then $\beta_1 = 0$, and the two-term Cornish-Fisher expansion gives a quantile approximation *larger* than z_α for the quantile of the t-statistic if, for example, the kurtosis $\gamma \leq 3$, as one can see from Theorem 13.5 above.

13.6 Cumulants and Fisher's k-statistics

We saw in the previous sections that the coefficients in the Edgeworth expansion and also the Cornish-Fisher expansion for distributions and quantiles of the mean (and also some other statistics) can be conveniently represented in terms of the cumulants of the underlying CDF. For two-term expansions, only the first four cumulants are needed; more would be needed for still higher-order expansions. However, that is not the only use of cumulants in statistics. Cumulants of common parametric distribution families often have neat properties. These can be used for model fitting or to model departures from a fixed parametric model. For example, cumulants of order higher than 2 are zero for normal distributions. It is common in applied sciences (e.g., signal processing and physical chemistry) to model distributions by assuming structures on their cumulants. The term cumulant was coined by Fisher (1931), although they were introduced by Thiele 40 years before Fisher coined the cumulant term. Fisher's letter to the editors of the American Math Monthly (Fisher (1931)) in response to C. C. Grove's article (Grove (1930)) in the same journal is an interesting account of Fisher's defense against the suggestion that the credit for developing the algebra and the properties of cumulants should be Thiele's rather than Fisher's. We recommend Stuart and Ord (1994) for a classic discussion of cumulants. Harvey (1972) and Doss (1973) give some different algebraic characterizations. We start with the definition and recursive representations of the sequence of cumulants of a distribution.

Definition 13.1 Let $X \sim F$ have a finite mgf $\psi(t)$ in some neighborhood of zero, and let $K(t) = \log \psi(t)$ when it exists. The rth cumulant of X

(or of F) is defined as $\kappa_r = \frac{d^r}{dt^r} K(t)|_{t=0}$. Equivalently, the cumulants of X are the coefficients in the power series expansion $K(t) = \sum_{n=1}^{\infty} \kappa_n \frac{t^n}{n!}$ within the radius of convergence of $K(t)$.

By equating coefficients in $e^{K(t)}$ with those in $\psi(t)$, it is easy to express the first few moments (and therefore the first few central moments) in terms of the cumulants. Indeed, letting $c_i = E(X^i), \mu = E(X) = c_1, \mu_i = E(X - \mu)^i, \sigma^2 = \mu_2$, one obtains the expressions

$$\begin{aligned}
&c_1 = \kappa_1; c_2 = \kappa_2 + \kappa_1^2; c_3 = \kappa_3 + 3\kappa_1\kappa_2 + \kappa_1^3;\\
&c_4 = \kappa_4 + 4\kappa_1\kappa_3 + 3\kappa_2^2 + 6\kappa_1^2\kappa_2 + \kappa_1^4;\\
&c_5 = \kappa_5 + 5\kappa_1\kappa_4 + 10\kappa_2\kappa_3 + 10\kappa_1^2\kappa_3 + 15\kappa_1\kappa_2^2 + 10\kappa_1^3\kappa_2 + \kappa_1^5;\\
&c_6 = \kappa_6 + 6\kappa_1\kappa_5 + 15\kappa_2\kappa_4 + 15\kappa_1^2\kappa_4 + 10\kappa_3^2 + 60\kappa_1\kappa_2\kappa_3 + 20\kappa_1^3\kappa_3\\
&\qquad + 15\kappa_2^3 + 45\kappa_1^2\kappa_2^2 + 15\kappa_1^4\kappa_2 + \kappa_1^6.
\end{aligned}$$

The corresponding expressions for the central moments are much simpler:

$$\begin{aligned}
\sigma^2 = \kappa_2;\ \mu_3 = \kappa_3; \mu_4 = \kappa_4 + 3\kappa_2^2; \mu_5 = \kappa_5 + 10\kappa_2\kappa_3;\\
\mu_6 = \kappa_6 + 15\kappa_2\kappa_4 + 10\kappa_3^2 + 15\kappa_2^3.
\end{aligned}$$

In general, the cumulants satisfy the recursion relations

$$\kappa_n = c_n - \sum_{k=1}^{n-1} \binom{n-1}{k-1} c_{n-k}\kappa_k.$$

These result in the specific expressions

$$\begin{aligned}
\kappa_2 = \mu_2; \kappa_3 = \mu_3; \kappa_4 = \mu_4 - 3\mu_2^2; \kappa_5 = \mu_5 - 10\mu_2\mu_3;\\
\kappa_6 = \mu_6 - 15\mu_2\mu_4 - 10\mu_3^2 + 30\mu_2^3.
\end{aligned}$$

The higher-order ones are quite complex, and also the corresponding expressions in terms of the c_j are more complex.

Example 13.8 Suppose $X \sim N(\mu, \sigma^2)$. Of course, $\kappa_1 = \mu, \kappa_2 = \sigma^2$. Since $K(t) = t\mu + \frac{t^2\sigma^2}{2}$, a quadratic, all derivatives of $K(t)$ of order higher than 2 vanish. Consequently, $\kappa_r = 0 \,\forall\, r > 2$.

If $X \sim \text{Poisson}(\lambda)$, then $K(t) = \lambda(e^t - 1)$, and therefore all derivatives of $K(t)$ are equal to λe^t. It follows that $\kappa_r = \lambda \,\forall\, r \geq 1$. These are two interesting special cases with neat structure and have served as the basis for stochastic modeling. For example, departure from normality can be modeled as all cumulants vanishing for r larger than some r_0.

Fisher (1929) made extremely important contributions to the calculation of moments of sample moments by using the notation of cumulants and also provided formal calculations that form the basis of the extremely complex formulas. In the process, he defined statistics called *k-statistics*, which are functions of sample moments and unbiasedly estimate the sequence of cumulants. We denote the k-statistics as $k_1, k_2, \ldots$, etc. Fisher showed that representations of these k-statistics in terms of power sums are amenable to some clever algebraic manipulations. By using the k-statistics and the mutual relations between cumulants and central moments, one can write down unbiased estimates of central moments of a population. Fisher also gave formal algebraic rules for writing expressions for product moments of k-statistics, (i.e., expectations of the form $E(k_1-\kappa_1)^m(k_2-\kappa_2)^n(k_3-\kappa_3)^p \cdots$) in terms of cumulants. These are in turn useful for deriving formal expressions for moments of functions of sample moments. See an illustrative example below.

Definition 13.2 Suppose $X_1, X_2, \cdots, X_n$ are iid observations from a CDF F. The rth k-statistic is a permutation-invariant function k_r of $X_1, X_2, \cdots, X_n$ such that $E(k_r) = \kappa_r$, assuming the rth moment of F exists.

The k-statistics are unique (as functions). Using the notation $s_r = \sum_{i=1}^n X_i^r$, expressions for the first four k-statistics are

$$\begin{aligned}
k_1 &= \frac{1}{n}s_1; k_2 = \frac{1}{n(n-1)}[ns_2 - s_1^2]; \\
k_3 &= \frac{1}{n(n-1)(n-2)}[n^2 s_3 - 3ns_1s_2 + 2s_1^3]; \\
k_4 &= \frac{1}{n(n-1)(n-2)(n-3)}[(n^2+n^3)s_4 - 4(n+n^2)s_1s_3 - 3(n^2-n)s_2^2 \\
&\quad + 12ns_1^2s_2 - 6s_1^4].
\end{aligned}$$

More elaborate tables can be seen in Stuart and Ord (1994). Alternative general representations can be seen in Good (1977).

The algebra of cumulants and k-statistics can be used to produce formal asymptotic expansions for moments of statistics that are smooth functions of sample moments. This can be viewed as an alternative to approximations based on delta theorems; see Chapter 3. Just as the delta-theorem-based moment approximations require growth and boundedness conditions on suitable derivatives, similarly the formal expansions based on cumulants and k-statistics cannot be given rigorous error estimates without additional assumptions. Nevertheless, as a practical matter, the expansions can be useful. Here is an example to illustrate the kind of formal expansion one does.

Example 13.9 Suppose $X_1, X_2, \cdots, X_n$ are iid samples from a CDF F and consider the sample coefficient of variation $T_n = \frac{s}{\bar{X}}$, s being the sample standard deviation. We can write T_n in terms of k-statistics as $T_n = \frac{\sqrt{k_2}}{k_1} = \frac{\sqrt{\kappa_2}}{\kappa_1}[1 + \frac{k_2-\kappa_2}{\kappa_2}]^{1/2}[1 + \frac{k_1-\kappa_1}{\kappa_1}]^{-1}$. Squaring both sides and taking formal expansions,

$$T_n^2 = \frac{\kappa_2}{\kappa_1^2}\left[1 + \frac{k_2-\kappa_2}{\kappa_2}\right]\left[1 + \frac{k_1-\kappa_1}{\kappa_1}\right]^{-2} = \frac{\kappa_2}{\kappa_1^2}\left[1 + \frac{k_2-\kappa_2}{\kappa_2}\right]$$

$$\left[1 - \frac{2(k_1-\kappa_1)}{\kappa_1} + \frac{3(k_1-\kappa_1)^2}{\kappa_1^2} - \frac{4(k_1-\kappa_1)^3}{\kappa_1^3} + \frac{5(k_1-\kappa_1)^4}{\kappa_1^4} - \cdots\right].$$

Taking formal expectations on both sides,

$$E\left(\frac{s^2}{\bar{X}^2}\right) = \frac{\sigma^2}{\mu^2}\left[1 + \frac{3}{\kappa_1^2}E(k_1-\kappa_1)^2 - \frac{4}{\kappa_1^3}E(k_1-\kappa_1)^3 + \frac{5}{\kappa_1^4}E(k_1-\kappa_1)^4 \right.$$
$$\left. - \frac{2}{\kappa_1\kappa_2}E(k_1-\kappa_1)(k_2-\kappa_2) + \cdots\right].$$

The various central and product central moments of low-order k-statistics can be derived using Fisher's method; in fact, they are already available in Fisher (1929) and in Stuart and Ord (1994), among other places. Plugging those in and performing some more algebra, one can write formal expansions of the following type:

$$E\left[\frac{s}{\bar{X}}\right] = \frac{\sigma}{\mu}\left[1 - \frac{1}{4(n-1)} + \frac{1}{n}\{\kappa_2/\kappa_1^2 - \kappa_3/(2\kappa_1\kappa_2)\right.$$
$$\left. -\kappa_4/(8\kappa_2^2)\} + \cdots\right],$$

$$E\left[\frac{s}{\bar{X}} - E\left(\frac{s}{\bar{X}}\right)\right]^2 = \frac{\sigma^2}{\mu^2}\left[\frac{1}{2(n-1)} + \frac{1}{n}\{\kappa_2/\kappa_1^2 + \kappa_4/(4\kappa_2^2)\right.$$
$$\left. -\kappa_3/(\kappa_1\kappa_2)\} + \cdots\right].$$

Note that the expansions are formal and do not even make sense if $\mu = 0$.

13.7 Exercises

Exercise 13.1 * Suppose $X_i \stackrel{\text{iid}}{\sim} U[0, 1]$.

(a) Find the exact density of $\overline{X}$ by Fourier inversion of its characteristic function.

(b) Find a one-term and a two-term Edgeworth expansion for the CDF of $\overline{X}$. Is there anything special about them?

(c) Compare the expansion to the exact CDF for $n = 3, 6, 10, 20$.

Exercise 13.2 Suppose $X_i \stackrel{\text{iid}}{\sim} \text{Exp}(1)$.

(a) Find the error $\Delta_n(x)$ of the two-term Edgeworth expansion for the CDF of $\overline{X}$.

(b) Compute for $n = 10, 20, 50, 100$, as accurately as possible, the constant $c_n = n^{3/2} \sup_x |\Delta_n(x)|$.

(c) According to theory, $c_n = O(1)$. Do your numerical values for c_n seem to stabilize?

Exercise 13.3 Suppose X_i are iid discrete uniform on $\{1, 2, \ldots, N\}$. Does the Cramér condition hold?

Exercise 13.4 * Suppose $X_i \stackrel{\text{iid}}{\sim} .99\delta_0 + .01N(0, 1)$, where δ_0 denotes point mass at 0. Does the Cramér condition hold?

Exercise 13.5 Suppose X_i are iid geometric with pmf $P(X_1 = x) = p(1-p)^x, x = 0, 1, 2, \ldots$. Does the Cramér condition hold?

Exercise 13.6 Suppose $X_i \stackrel{\text{iid}}{\sim} N(\mu, 1)$.

(a) Find a two-term Edgeworth expansion for the CDF of the sample median, say M. Adopt the definition $M = X_{\frac{n}{2}:n}$.

(b) Hence find a two-term expansion for $P(\sqrt{n}|M - \mu| \le x)$.

(c) Compare it with $P(\sqrt{n}|\overline{X} - \mu| \le x)$. Is $P(\sqrt{n}|M - \mu| \le x)$ always smaller?

Exercise 13.7 * Suppose $X_i \stackrel{\text{iid}}{\sim} U[0, 1]$.

(a) Show that the exact density of the sample median $M = X_{\frac{n}{2}:n}$ is a Beta density.

(b) Find a two-term Edgeworth expansion for the CDF of M. Does the $\frac{1}{\sqrt{n}}$ term vanish? Why or why not?

(c) Compare the expansion to the exact CDF of M for $n = 3, 6, 10, 20$.

Exercise 13.8 Suppose $X_i \stackrel{\text{iid}}{\sim} f(x) = .5\phi(x - k) + .5\phi(x + k)$.

(a) Is f necessarily unimodal? If not, characterize those k for which f is not unimodal.

(b) Find a two-term Edgeworth expansion for the CDF of the t-statistic. Is there any effect of the lack of unimodality of f on how the distribution of the t-statistic behaves?

Exercise 13.9 Suppose $X_i \stackrel{\text{iid}}{\sim} \text{Beta}(\frac{1}{2}, \frac{1}{2})$, a U-shaped density.

(a) Approximate the coverage probability $P(\sqrt{n}|\overline{X} - \frac{1}{2}| \leq 1.4)$ by a two-term Edgeworth expansion for $n = 10, 20, 30$.

(b) How accurate is the approximation? Simulate the true coverage and compare the two.

Exercise 13.10 Suppose $X_i \stackrel{\text{iid}}{\sim} U[0, 1]$. Approximate $P(\prod_{i=1}^n X_i > .1)$ up to an error of $O(n^{-3/2})$ and compare it with the exact value of that probability for some selected n. Hint: $-\log(X_i) \sim \text{Exp}(1)$.

Exercise 13.11 Suppose $X_i \stackrel{\text{iid}}{\sim} U[0, 1]$. Approximate the 95th percentile of $\sum_{i=1}^n X_i$ by the two-term Cornish-Fisher expansion for $n = 3, 6, 12$, and compare the approximations with the correct value obtained by a careful simulation.

Exercise 13.12 *Approximate the 95th percentile of the t-statistic for iid Exp(1) observations for $n = 5, 15, 25$ by the two-term Cornish-Fisher expansion, and compare the approximations with the correct value obtained by a careful simulation.

Exercise 13.13 Let f be a density symmetric about μ. Show that $\kappa_{2r+1} = 0$ for all $r > 0$.

Exercise 13.14 Find the sequence of cumulants of a general Gamma density.

Exercise 13.15 * Find the sequence of cumulants of the $U[-a, a]$ distribution.

Exercise 13.16 Characterize the sequence of cumulants of a Bernoulli distribution.

Exercise 13.17 Characterize the sequence of cumulants of a geometric distribution.

Exercise 13.18 Are cumulants defined for a t-density?

Exercise 13.19 * Let F be a CDF with a compact support. Is F uniquely determined by its sequence of cumulants?

Exercise 13.20 Is a normal distribution the only possible distribution with vanishing cumulants for all order > 2?

Exercise 13.21 Give a direct proof that the first four k-statistics are indeed unbiased for the corresponding cumulants.

References

Albers, W. (2001). *From A to Z: Asymptotic Expansions by van Zwet*, IMS Lecture Notes Monograph Series, Vol. 36, Institute of Mathematical Statistics, Beachwood, OH, 2–20.

Barndorff-Nielsen, O.E. and Cox, D.R. (1991). *Asymptotic Techniques for Use in Statistics*, Chapman and Hall, London.

Basu, S. and DasGupta, A. (1993). Robustness of standard confidence intervals for location parameters under departure from normality, Ann. Stat., 23(4), 1433–1442.

Bentkus, V., Goetze, F. and van Zwet, W.R. (1997). An Edgeworth expansion for symmetric statistics, Ann. Stat., 25(2), 851–896.

Bhattacharya, R.N. and Denker, M. (1990). *Asymptotic Statistics*, Birkhauser, Basel.

Bhattacharya, R.N. and Ghosh, J.K. (1978). On the validity of the formal Edgeworth expansion, Ann. Stat., 2, 434–451.

Bickel, P.J. (1974). Edgeworth expansions in nonparametric statistics, Ann. Stat., 2, 1–20.

Doss, D. (1973). Moments in terms of cumulants and vice versa, Am. Stat., 27, 239–240.

Esseen, C. (1945). Fourier analysis of distribution functions, Acta Math., 77, 1–125.

Fisher, R.A. (1929). Moments and product moments of sampling distributions, Proc. London Math. Soc., 2, 199–238.

Fisher, R.A. (1931). Letter to the Editor, Am. Math Mon., 38, 335–338.

Good, I.J. (1977). A new formula for k-statistics, Ann. Stat., 5, 224–228.

Grove, C.C. (1930). Review of Statistical Methods for Research Workers by R. A. Fisher, 3rd Ed., Am. Math Mon., 37, 547–550.

Hall, P. (1987). Edgeworth expansions for Student's t-statistic under minimal moment conditions, Ann. Prob., 15(3), 920–931.

Hall, P. (1992). *The Bootstrap and Edgeworth Expansion*, Springer-Verlag, New York.

Harvey, D. (1972). On expressing moments in terms of cumulants and vice versa, Am. Stat., 26, 38–39.

Lahiri, S.N. (2003). *Resampling Methods for Dependent Data*, Springer-Verlag, New York.

Petrov, V.V. (1975). *Limit Theorems for Sums of Independent Random Variables* (Translation from Russian), Springer-Verlag, New York.

Reiss, R.D. (1976). Asymptotic expansions for sample quantiles, Ann. Prob., 4(2), 249–258.

Reiss, R. (1989). *Approximate Distributions of Order Statistics*, Springer-Verlag, New York.

Stuart, A. and Ord, J. (1994). *Kendall's Advanced Theory of Statistics*, 6th Ed., Halsted Press, New York.

Chapter 14
Saddlepoint Approximations

We saw in the previous chapter that the one-term Edgeworth expansion improves on the normal approximation to the CDF of a mean by performing a skewness correction, and the two-term expansion improves it further by also adjusting for kurtosis. Precisely, the $O(\frac{1}{\sqrt{n}})$ rate of the Berry-Esseen theorem is improved to uniform errors of the orders of $\frac{1}{n}$ and $n^{-\frac{3}{2}}$ by the one- and two-term Edgeworth expansions. However, the actual approximations themselves, at whichever term they are terminated, are not bona fide CDFs. In particular, they can and do take values outside the [0, 1] range. Empirical experience has shown that approximations of tail areas by Edgeworth expansions can be quite inaccurate. Edgeworth expansions are particularly vulnerable to inaccuracies for extreme tail-area approximations and for *small* n. Alternatives to Edgeworth expansions that address and sometimes even circumvent these Edgeworth difficulties have been suggested. One approach that has proved particularly successful in many examples is the so-called saddlepoint approximation. Saddlepoint approximations can be found for both the density and the tail probability of statistics of various kinds, including nonlinear statistics, and sometimes for dependent data. They have illuminating connections to Edgeworth expansions, embedding in exponential families, and also to classical complex analysis. The approach was pioneered by Henry Daniels; see Daniels (1954). Jensen (1995), Barndorff-Nielsen and Cox (1979), Daniels (1987), Reid (1988), and Wood (2000) provide excellent exposition with history, various variants and technical approaches, examples, and comparisons. For other technical material related to the results in this chapter, we recommend Lugannani and Rice (1980), Barndorff-Nielsen (1983), McCullagh (1987), and Kolassa (2003). Goutis and Casella (1999) explain the derivation of the saddlepoint approximation in lucid and elementary terms. More specific references are given in the following sections.

A. DasGupta, *Asymptotic Theory of Statistics and Probability*,

14.1 Approximate Evaluation of Integrals

The saddlepoint approximations we discuss can be derived in a number of ways. Proofs using complex analytic arguments use an idea central to *Laplace's method* for approximate evaluation of integrals with large parameters. Another proof uses known results from the theory of Edgeworth expansions, and this proof has a greater statistical flavor. Approximate evaluation of integrals is a recurring theme in statistics. Also, the main idea of Laplace's theorem is the main idea of complex analytic derivations of saddlepoint approximations, so we will first present a brief description of some key results on approximate evaluation of integrals, including in particular Laplace's method and *Watson's lemma*. These will be followed by a number of illustrative examples. Olver (1997) is a modern reference for methods of approximate evaluation of integrals.

The basic problem we address is to approximately evaluate the integral $I(x) = \int_a^b e^{xg(t)} f(t)dt$, for large x (i.e., as $x \to \infty$). We are aiming for results like $I(x) \sim \hat{I}(x)$ for some explicit function $\hat{I}(x)$, with the notation $\sim$ meaning that the ratio converges to 1 when $x \to \infty$. The form of the approximating function $\hat{I}(x)$ is going to depend on the behavior of the two functions f, g, for example whether g is monotone or not, and what the boundary behaviors of f, g are. So, let us explore a few special cases before we present a theorem.

Suppose g was strictly monotone and had a nonzero derivative in the interval (a, b). Then, an integration by parts would give

$$I(x) = \frac{1}{x}\frac{f(t)}{g'(t)}e^{xg(t)}|_a^b - \frac{1}{x}\int_a^b \frac{d}{dt}\frac{f(t)}{g'(t)}e^{xg(t)}dt.$$

The first term would be the dominant term if one of $f(a), f(b)$ is nonzero, and so, in that case,

$$I(x) \sim \frac{1}{x}\frac{f(b)}{g'(b)}e^{xg(b)} - \frac{1}{x}\frac{f(a)}{g'(a)}e^{xg(a)}.$$

This is the simplest situation. In practice, g may not be strictly monotone and may have zero derivative at one or more points in the interval (a, b). In such a case, we cannot integrate by parts as above. The idea behind Laplace's method is that if g has a maximum at some (unique) c in (a, b), and if $f(c), g''(c) \neq 0$, then due to the large magnitude of the parameter x, the dominant part of the integral will come from a neighborhood of c. By expanding g around c up to the quadratic term in the usual Taylor expansion, one would obtain a normal density. The integral is not going to change much if the range of the integration is changed to the whole real line. On normalizing the normal density, a $\sqrt{2\pi}$ factor would come in and the approximation is going to be

$$I(x) \sim \frac{\sqrt{2\pi} f(c)e^{xg(c)}}{\sqrt{-xg''(c)}}, x \to \infty.$$

The intermediate steps of this derivation require that a, b are not stationary points of g, that $g'(c) = 0$, and $g''(c) < 0$. If either of the two boundary points a, b is a stationary point of g, then the approximating function will change to accommodate contributions from the boundary stationary points. The approximation is more complicated when the interior local maximum of g is not unique. In that case, $I(x)$ has to be partitioned into subintervals separating the various maxima and the terms added to obtain the final approximation to $I(x)$.

We can see that the exact form of the approximation will depend on the details of the behaviors of f, g. Theorem 14.1 below puts together a number of special cases.

Theorem 14.1 Let $I(x) = \int_a^b e^{xg(t)} f(t)dt$.

(a) Suppose f is bounded and continuous on (a, b), and $f(a)f(b) \neq 0$. Suppose also that g is strictly monotone and differentiable and that $\frac{f(a)}{g'(a)}, \frac{f(b)}{g'(b)}$ both exist as finite reals, defined as limits if either endpoint is infinite. Assume also that $\int_a^b e^{xg(t)}dt$ exists for all $x > 0$. Then,

$$I(x) \sim \frac{1}{x}\frac{f(b)}{g'(b)}e^{xg(b)} - \frac{1}{x}\frac{f(a)}{g'(a)}e^{xg(a)}.$$

(b) Suppose f is bounded and continuous on (a, b), that g has a unique maximum at some c in the open interval (a, b), g is differentiable in some neighborhood of c, $g''(c)$ exists and is < 0, and that $f(c) \neq 0$. Then,

$$I(x) \sim \frac{\sqrt{2\pi} f(c)e^{xg(c)}}{\sqrt{-xg''(c)}}.$$

(c) Suppose f is bounded and continuous on (a, b), and g is twice differentiable. Let $a < x_1 < x_2 < \cdots < x_k < b$ be all local maxima of g in the open interval (a, b). Assume that $f(x_i) \neq 0, g'(x_i) = 0, g''(x_i) < 0, i = 1, \cdots, k$. Then,

$$I(x) \sim \sum_{i=1}^{k} \left[\frac{\sqrt{2\pi} f(x_i) e^{xg(x_i)}}{\sqrt{-xg''(x_i)}} \right] + \frac{1}{2} \frac{\sqrt{2\pi} f(a) e^{xg(a)}}{\sqrt{-xg''(a)}} I_{g'(a)=0,\ g''(a)<0}$$
$$+ \frac{1}{2} \frac{\sqrt{2\pi} f(b) e^{xg(b)}}{\sqrt{-xg''(b)}} I_{g'(b)=0,\ g''(b)<0},$$

the derivatives at the endpoints a, b being defined as limits if either endpoint is infinite.

Expansions to more terms beyond the dominant term are available, but computing the coefficients of the higher terms is a laborious exercise; see Olver (1997) for Theorem 14.1 and also higher-order versions and the proofs.

A related result that is also useful in asymptotic evaluation of integrals and often plays a useful complementary role to Laplace's theorem is the following theorem, popularly known as *Watson's lemma.*

Theorem 14.2 Let $\lambda > -1$ and let g be a continuous function such that $g(0) \neq 0$ and such that it admits the Maclaurin expansion $g(t) = \sum_{n=0}^{\infty} \frac{g^{(n)}(0)}{n!} t^n$ in some open interval around zero. Assume also that there exist constants $-\infty < c < \infty$, $M > 0$, such that $|t^\lambda g(t)| \leq Me^{ct}\ \forall t \geq 0$. Then, as $x \to \infty$,

$$I(x) = \int_0^\infty e^{-xt} t^\lambda g(t) dt \sim \sum_{n=0}^{\infty} \frac{g^{(n)}(0) \Gamma(\lambda + n + 1)}{n! x^{\lambda+n+1}}.$$

In the equation above, the notation $\sim$ means that the infinite series can be truncated at any finite term, with the remainder being of a smaller order than the last term retained.

Let us see a number of illustrative examples of various kinds.

Example 14.1 Consider the function $I(x) = \int_0^1 t^x (\cos \pi t)^n dt$, where n is a fixed positive integer. We rewrite this in the form $\int_0^1 e^{xg(t)} f(t) dt$, with $g(t) = \log t$, $f(t) = (\cos \pi t)^n$. Note that $g(t)$ is strictly monotone, and so, by the approximation formula in the monotone case (see above), $I(x) \sim \frac{1}{xg'(1)} (-1)^n e^{xg(1)} - \frac{1}{x} t^{x+1}|_{t=0} = \frac{(-1)^n}{x}$.

Example 14.2 Consider the function $I(x) = \int_{-\infty}^{\infty} e^{x(t-e^t)} dt$, which actually equals $\frac{\Gamma(x)}{x^x}$ on a change of variable in $I(x)$. We find an asymptotic approximation for it as $x \to \infty$ by using Laplace's method with $f(t) = 1$ and

$g(t) = t - e^t$. The only saddlepoint of g is $t = 0$, and $g''(t) = -e^t$. Therefore, by an elementary application of Laplace's theorem, $I(x) \sim \frac{\sqrt{2\pi}e^{-x}}{x}$.

Example 14.3 Example 14.2 above is essentially Stirling's approximation, but let us rederive Stirling's approximation for $n!$ directly by writing it as a Gamma function and then using Laplace's approximation again. Toward this, note that

$$n! = \int_0^\infty e^{-z} z^n dz = \int_0^\infty e^{n \log z} e^{-z} dz = n^{n+1} \int_0^\infty e^{n(\log t - t)} dt.$$

We can now use $f(t) = 1$, $g(t) = \log t - t$ in Laplace's method, and plugging the second derivative of g into the approximation produces the familiar Stirling approximation

$$n! \sim e^{-n} n^{n+\frac{1}{2}} \sqrt{2\pi}.$$

Example 14.4 Consider the integral $I(x) = \int_0^\pi e^{-x\cos(t)} dt$, the integrand being an oscillatory function. We identify $f(t) = 1$, $g(t) = -\cos(t)$. There is no stationary point of g in the interior of the interval $[0, \pi]$. The boundary points are both stationary points, of which the point $a = 0$ does not count since $g''(0) = 1 > 0$. But $g''(1) = -1 < 0$. Therefore, from Theorem 14.2, $I(x) \sim \frac{1}{2} \frac{\sqrt{2\pi}e^{-x\cos(\pi)}}{\sqrt{-x\cos(\pi)}} = \sqrt{\frac{\pi}{2}} \frac{e^x}{\sqrt{x}}$. For example, when $x = 10$, the exact value of $I(x)$ is 8845.83 and the approximation from the Laplace method is 8729.81, the ratio being 1.013.

Example 14.5 Let $I(x) = \int_0^\infty e^{-xt} \sin t dt$. The function $\sin t$ being oscillatory for large t, we approximate $I(x)$ by using Watson's lemma rather than using the general formula for the case of monotone g. By an integration by parts, $I(x) = 1 - x \int_0^\infty e^{-xt} \cos t dt$, and we apply Watson's lemma on $\int_0^\infty e^{-xt} \cos t dt$ using $\lambda = 0$, $g(t) = \cos t$. Indeed, since the odd derivatives of g at zero are all zero and the even derivatives are $g^{(2n)}(0) = (-1)^n$, Watson's lemma gives the approximation

$$I(x) = \int_0^\infty e^{-xt} \sin t dt \sim 1 - x \sum_{n=0}^\infty \frac{(-1)^n}{x^{2n+1}} = 1 - \sum_{n=0}^\infty \frac{(-1)^n}{x^{2n}}$$

$$= 1 - \frac{1}{1 + \frac{1}{x^2}} = \frac{1}{1 + x^2}.$$

Thus, the full infinite series exactly equals the integral $I(x)$, and if the series is terminated at any finite term, then the remainder is of a smaller order than the last term that was retained (smaller order in the sense of being $o(.)$).

Example 14.6 Here is an application of Watson's lemma to a problem of interest in statistics and probability. Let $R(x) = \frac{1-\Phi(x)}{\phi(x)}$ denote the so-called *Mill's ratio*. The problem is to determine an asymptotic expansion for $R(x)$. Toward this, note that

$$1 - \Phi(x) = \int_x^\infty \phi(t)dt = \int_0^\infty \phi(t+x)dt = \phi(x)\int_0^\infty e^{-xt-t^2/2}dt.$$

Applying Watson's lemma with $\lambda = 0$, $g(t) = e^{-t^2/2}$, and on noting that the successive derivatives $g^{(n)}(0)$ equal $H_n(0)$, where H_n is the nth Hermite polynomial, one gets the asymptotic expansion $R(x) \sim \frac{1}{x} - \frac{1}{x^3} + \frac{3}{x^5} - \ldots$, where the $\sim$ notation means that the series can be truncated at any finite term, with a remainder of a smaller order than the last term kept. In particular, $xR(x) \to 1$ as $x \to \infty$.

14.2 Density of Means and Exponential Tilting

Saddlepoint approximations were brought into statistical consciousness in the pioneering article by Daniels (1954), where the case of the mean $\bar{X}$ of an iid sample was treated. The technical ideas come through most transparently in this case, and besides, sample means of iid random variables are special statistics anyway. This is the case we present first.

We first need some notation. Let $X_1, X_2, \cdots, X_n$ be iid observations from a density $f(x)$ in $\mathcal{R}$. The density $f(x)$ could be with respect to the counting measure; thus, the results are applicable to integer-valued random variables also, such as Poisson variables. Suppose the moment generating function (mgf) of f, $\psi(t) = E(e^{tX}) = \int e^{tx} f(x)d\mu(x)$ exists for t in some neighborhood of the origin, and let $K(t) = \log \psi(t)$ denote the *cumulant generating function* (cgf). For given $\bar{x}$, let $\hat{\phi} = \hat{\phi}(\bar{x})$ denote a saddlepoint of K; i.e., a root of the equation $K'(\hat{\phi}) = \bar{x}$. We will need it to be unique. For given n, let $f_n(u)$ denote the density of $\bar{X}$. Then the classic saddlepoint approximation to f_n says the following.

Theorem 14.3 $f_n(\bar{x}) = \sqrt{\frac{n}{2\pi K''(\hat{\phi})}} e^{n[K(\hat{\phi})-\hat{\phi}\bar{x}]}(1 + O(n^{-1}))$ for any given value $\bar{x}$.

Remark. Two specific references for this theorem are Reid (1988) and Daniels (1954). The error can be made $O(n^{-1})$ uniformly in the argument with some conditions on the parent density f; see Daniels (1954). The approximation on the right-hand side cannot be negative, which is a pleasant property. However, usually it does not integrate to one. One may renormalize the approximation by multiplying by a constant $c = c(n)$; however, c needs to be found by numerical integration. The expansion can be carried to higher-order terms, in which case the error of the approximation goes down in powers of $\frac{1}{n}$ rather than powers of $\frac{1}{\sqrt{n}}$. This gives a hint that the saddlepoint approximation may be more accurate at small sample sizes. The coefficient of the $\frac{1}{n}$ term is explicitly given in Daniels (1954); the coefficient depends on the *standardized cumulants* defined through

$$\rho_3(t) = K^{(3)}(t)/(K''(t))^{\frac{3}{2}},\ \rho_4(t) = K^{(4)}(t)/(K''(t))^2.$$

The coefficient of the $O(n^{-1})$ term equals $\frac{\rho_4(\hat{\phi})}{8} - \frac{5}{24}[\rho_3(\hat{\phi})]^2$. Notice that the coefficient in general depends on the particular argument at which we want to approximate the density. For some parent densities, it does not. And, even if it does, one may be able to make the $O(n^{-1})$ relative error rate uniform, as we mentioned earlier. Later, Good (1957) obtained the coefficient of the $\frac{1}{n^2}$ term.

The result in Theorem 14.3 can be derived in a number of ways. Perhaps a derivation via Laplace's method gives the best explanation for *why* the result should hold. The idea is to write the density $f_n(\bar{x})$ in the form of the inversion formula of the Laplace transform of $\bar{X}$ and then argue that the exponential term in the integrand in the inversion formula falls off sharply away from the saddlepoint. An application of complex analytic techniques similar to Laplace's method formally produces the approximation given in the theorem. Here is a sketch of the technical argument.

Suppose then that the underlying density f has a finite mgf in some bilateral neighborhood $-c < t < c (c > 0)$ of zero. The mgf of the sum $\sum_{i=1}^n X_i$ is $e^{nK(t)}$ for each t in this neighborhood. By the Laplace transform inversion formula, the density for the mean $\bar{X} = \frac{\sum_{i=1}^n X_i}{n}$ is

$$f_n(\bar{x}) = \frac{1}{2\pi i}\int_{\tau-i\infty}^{\tau+i\infty} e^{nK(\frac{t}{n})-t\bar{x}}dt = \frac{n}{2\pi i}\int_{\tau-i\infty}^{\tau+i\infty} e^{n[K(t)-t\bar{x}]}dt,$$

where τ is larger in absolute value than each singularity of $K(t)$. Suppose now that the function $K(t) - t\bar{x}$ has a unique saddlepoint given by the root of the equation $K'(t) = \bar{x}$; this is the point we denoted as $\hat{\phi}$. Then, analogous to the Laplace approximation result *but requiring delicate complex analysis*, one gets the first-order approximation

$$f_n(\bar{x}) \approx \frac{n}{2\pi} \frac{\sqrt{2\pi}\, e^{n[K(\hat{\phi})-\hat{\phi}\bar{x}]}}{\sqrt{nK''(\hat{\phi})}} = \sqrt{\frac{n}{2\pi K''(\hat{\phi})}} e^{n[K(\hat{\phi})-\hat{\phi}\bar{x}]},$$

which is the expression (other than the error term) in Theorem 14.3. There should be no illusion about the need to use delicate complex analysis in a rigorous proof of these steps; Daniels (1954) has given such a rigorous proof.

14.2.1 Derivation by Edgeworth Expansion and Exponential Tilting

However, better statistical motivation is obtained by alternative derivations that use either an Edgeworth expansion itself or a technique known as *exponential tilting*, which embeds the parent density in a one-parameter exponential family and chooses the parameter value in such a way that the approximation gets one more free order of accuracy. That is, the order of the *relative* error is $O(n^{-1})$. These are very clearly explained in Reid (1988) and Goutis and Casella (1999). However, the exponential tilting method is sufficiently important that we give a brief explanation of it below.

The point is that any arbitrary density function can always be embedded in a one-parameter exponential family. Suppose $f(x)$ is a density function. We can embed it in the one-parameter exponential family by defining a family of densities $f_\theta(x) = e^{\theta x - K(\theta)} f(x)$, where θ is any real for which the mgf of the density $f(x)$ exists and the function $K(\theta)$ is such that $e^{-K(\theta)}$ normalizes $f_\theta(x)$ to integrate to 1. Note that, very conveniently, $K(\theta)$ is in fact just the cumulant generating function of f. Now, if we take an iid sample of size n from a one-parameter exponential family density, then the density of the mean $\bar{X}$ can be written in closed form as

$$f_{n,\theta}(\bar{x}) = e^{-n[K(\theta)-\theta\bar{x}]} f_n(\bar{x}),$$

where $f_n(.)$ stands for the density of the mean of n iid observations from the original density f. Note that it is this function $f_n(.)$ for which we seek an approximation. We can rewrite this latest equation as

$$f_n(\bar{x}) = f_{n,\theta}(\bar{x}) e^{n[K(\theta)-\theta\bar{x}]}.$$

Curiously, we do not have any θ on the left-hand side of this equation, but we have it on the right-hand side of the equation. This is precisely what saves us.

What it means is that θ is for us to choose in just the right manner. We choose it so that the means match; i.e., under the exponential family density f_θ, the mean of $\bar{X}$ is exactly equal to the specific point at which we wish to approximate the f_n function. It follows from the property of the strict monotonicity of the mean as a function of θ in the exponential family that there is a unique such θ at which the mean exactly matches the specific $\bar{x}$. Indeed, this unique θ is just the saddlepoint satisfying the equation $K'(\theta) = \bar{x}$. Having chosen θ to be the saddlepoint of $K(.)$, the last step is to use the Edgeworth expansion for the density of the sample mean, which we previously saw in Chapter 13. We specialize that Edgeworth expansion to approximate the function $f_{n,\theta}(\bar{x})$. Recall that to implement that Edgeworth expansion for the density of the mean, we need the coefficients $C_1(F)$, $C_2(F)$, and $C_3(F)$, which involve the moments of the underlying density. The underlying density being an exponential family density, its moments are easily written in terms of the function $K(.)$ and its derivatives. Indeed, the coefficients C_1, C_2, C_3 are given by

$$C_1 = K^{(3)}(\theta)/[6(K''(\theta))^{3/2}]; C_2 = K^{(4)}(\theta)/[24(K''(\theta))^2]; C_3 = C_1^2/72.$$

On plugging in these expressions into the Edgeworth expansion for the density of the mean (not the CDF), some more algebra then reduces the entire right-hand side of the equation $f_n(\bar{x}) = f_{n,\theta}(\bar{x})e^{n[K(\theta)-\theta\bar{x}]}$ to exactly the saddlepoint density formula in Theorem 14.3.

14.3 Some Examples

Let us see some examples of application of the saddlepoint formula.

Example 14.7 Suppose the X_i are iid $N(0, 1)$. Then, $\psi(t) = e^{\frac{t^2}{2}}$, and so $K(t) = \frac{t^2}{2}$. The unique saddlepoint $\hat{\phi}$ solves $K'(\hat{\phi}) = \hat{\phi} = \bar{x}$, giving the approximation as the expression

$$f_n(\bar{x}) = \sqrt{\frac{n}{2\pi}} e^{n[\frac{\bar{x}^2}{2} - \bar{x}^2]} = \sqrt{\frac{n}{2\pi}} e^{-n\frac{\bar{x}^2}{2}}.$$

Very interestingly, the supposed approximation $f_n(\bar{x})$ in fact coincides with the exact density of $\bar{X}$, namely the $N(0, \frac{1}{n})$ density.

Example 14.8 Consider the case where the X_i are iid Poi(λ). Then, $\psi(t) = e^{\lambda[e^t - 1]}$, giving $K(t) = \lambda[e^t - 1]$. The unique saddlepoint $\hat{\phi}$ solves $K'(\hat{\phi}) = \lambda e^{\hat{\phi}} = \bar{x}$, i.e., $\hat{\phi} = \log \frac{\bar{x}}{\lambda}$. This produces the one-term saddlepoint approximation

$$f_n(\bar{x}) = \sqrt{\frac{1}{2\pi n}} e^{-n\lambda} \frac{\lambda^{n\bar{x}}}{\bar{x}^{n\bar{x}+\frac{1}{2}}}.$$

By a scale change, the approximation to the density of the sum $\sum X_i$ is

$$h_n(k) = \sqrt{\frac{1}{2\pi}} e^{-n\lambda} \frac{e^k (n\lambda)^k}{k^{k+\frac{1}{2}}}.$$

Contrasting this with the exact distribution of the sum, which is Poi($n\lambda$), we realize that the saddlepoint approximation substitutes the familiar one-term Stirling's approximation in place of $k!$, a very interesting outcome. Note that, unlike in the previous example, the one-term approximation is no longer exact in the Poisson case.

Example 14.9 Next, consider the case where the X_i are iid Exp(λ) with density $f(x) = \frac{1}{\lambda} e^{-\frac{x}{\lambda}}$. Then, the mgf exists for $|t| < \frac{1}{\lambda}$ and equals $\psi(t) = \frac{1}{1-\lambda t}$, so $K(t) = -\log(1-\lambda t)$, and the unique saddlepoint $\hat{\phi}$ equals $\hat{\phi} = \frac{1}{\lambda} - \frac{1}{\bar{x}}$. On substituting the various quantities, this produces the saddlepoint approximation

$$f_n(\bar{x}) = \sqrt{\frac{n}{2\pi \bar{x}^2}} e^{n[\log \frac{\bar{x}}{\lambda} - \frac{\bar{x}}{\lambda} + 1]} = \sqrt{\frac{n}{2\pi}} e^{-\frac{n\bar{x}}{\lambda}} \bar{x}^{n-1} e^n / \lambda^n.$$

Recalling that $S_n = n\bar{X}$ has exactly a Gamma density with parameters n and λ, once again it turns out that the saddlepoint approximation essentially replaces $(n-1)!$ in the exact density of $\bar{X}$ by its one-term Stirling's approximation. If we renormalize, the saddlepoint approximation becomes exact.

Remark. Renormalizing the one-term saddlepoint approximation in order to make it integrate to 1 is not only an intuitively correct thing to do but in fact also improves the order of the relative error in the approximation. The renormalized version will *sometimes* have a relative error of the order $O(n^{-2})$; see Reid (1988). It was proved in Daniels (1954) that the only three absolutely continuous cases where the saddlepoint approximation to the density of $\bar{X}$, perhaps with a renormalization, is exact are the Gamma, normal, and the inverse Gaussian cases. To our knowledge, no corresponding result is known in the lattice case.

14.4 Application to Exponential Family and the Magic Formula

An immediate application of the saddlepoint density approximation result is to maximum likelihood estimation of the mean function in the natural exponential family. Thus, suppose $f(x|\theta) = e^{\theta x - d(\theta)}h(x)$ with respect to some dominating measure ν. Then, based on an iid sample $X_1, X_2, \cdots, X_n, \sum X_i$ is the minimal sufficient statistic, and if the true θ is an interior point of the natural parameter space, then for all large n, $\bar{X}$ is the unique MLE of $\mu = E_\theta(X) = d'(\theta)$. Therefore, the saddlepoint approximation to the density of the MLE can be immediately obtained by Theorem 14.3. Indeed, if $l(\theta)$ denotes the joint likelihood function and $\hat{\theta}$ the unique MLE of the natural parameter θ (which has a one-to-one relation with the unique MLE of μ), and if $I(\theta) = d''(\theta)$ denotes the Fisher information based on one observation, then the saddlepoint approximation to the density of the MLE of μ works out to the following expression.

Theorem 14.4 Let $f(x|\theta)$ be as above, and suppose an MLE of $\mu = E_\theta(X)$ exists with probability 1 under the true θ. Let $\hat{\theta}$ denote the corresponding unique MLE of θ. Then, the density of the MLE of μ satisfies

$$f_n(\hat{\mu}|\theta) = \sqrt{\frac{n}{2\pi d''(\hat{\theta})}} \frac{l(\theta)}{l(\hat{\theta})}(1 + O(n^{-1})),$$

where $l(\theta)$ is the joint likelihood function based on all n observations.

Remark. The statement of Theorem 14.4 is to be understood as saying that the ratio $\frac{l(\theta)}{l(\hat{\theta})}$, by virtue of sufficiency, is actually a function of $\hat{\mu}$ and μ, and likewise $d''(\hat{\theta})$ is a function of $\hat{\mu}$. As usual, the saddlepoint approximation will not generally integrate to 1. Actually, an analogous approximation to the density of the MLE holds in some other models that have an inherent group structure; see Barndorff-Nielsen (1983). On renormalization, these approximations often turn out to be so accurate that the colorful term *magic formula* was coined for the renormalized saddlepoint approximation to the density of the MLE.

14.5 Tail Area Approximation and the Lugannani-Rice Formula

Approximation of tail probabilities is of great interest to statisticians, for example for approximation of p-values or coverage probabilities of confidence intervals. A number of approximations to tail probabilities using the saddlepoint idea have been worked out over the years. An obvious idea is to take the saddlepoint approximation to the density given in the previous section and integrate it. This will normally require numerical integration and calculation of the saddlepoint at many values of the integrand, depending on which particular method of numerical integration is used. Another possibility is to take the Fourier inversion formula for the tail probability itself as the starting step and apply saddlepoint methods to that formula. A few tail-area approximations are available using this latter approach. Perhaps the most well known among them is the *Lugannani-Rice formula*, which has the reputation of often being extremely accurate over the entire range of the argument; that is, the formula often produces sharp approximations to tail areas in the *central limit domain*, which consists of arguments of the order $\frac{1}{\sqrt{n}}$ away from the mean value, as well as in the *large-deviation domain*, which consists of arguments that are just simply $O(1)$. We present the Lugannani-Rice formula below. Daniels (1987) gives a careful comparative analysis of the various approximations and also gives many illustrative examples.

Toward this development, we first need some notation. Let $X_1, X_2, \cdots, X_n$ be iid with density $f(x)$ and cgf $K(t)$; we continue to make the same assumptions about the existence of $K(t)$ and use $\hat{\phi}$ to again denote the unique saddlepoint of K. Let

$$\lambda_r = \frac{K^{(r)}(\hat{\phi})}{K''(\hat{\phi})^{r/2}},\ K^{(r)} = K^{(r)}(\hat{\phi}),$$

$$\hat{W} = \sqrt{2[\hat{\phi}K'(\hat{\phi}) - K(\hat{\phi})]},\ \hat{U} = \hat{\phi}\sqrt{K^{(2)}},$$

$$\hat{\xi} = \sqrt{2n[\hat{\phi}\bar{x} - K(\hat{\phi})]}\text{sgn}(\hat{\phi}),\ \hat{Z} = \hat{\phi}\sqrt{nK^{(2)}}.$$

As usual, also let ϕ and Φ denote the standard normal PDF and CDF, respectively.

Theorem 14.5 (The Lugannani-Rice Formula) Suppose a saddlepoint $\hat{\phi}$ exists and is unique for all $\bar{x}, n$. Then,

(a) for fixed $\bar{x} \neq E(X_1)$,

$$Q_n(\bar{x}) = P(\bar{X} > \bar{x}) = 1 - \Phi(\sqrt{n}\hat{W}) + \frac{1}{\sqrt{n}}\phi(\sqrt{n}\hat{W}) \times \left(\frac{1}{\hat{U}} - \frac{1}{\hat{W}}\right)(1 + O(n^{-1}));$$

(b) for $\bar{x} = E(X_1) + O(\frac{1}{\sqrt{n}})$,

$$Q_n(\bar{x}) = 1 - \Phi(\hat{\zeta}) + \phi(\hat{\zeta})\left(\frac{1}{\hat{Z}} - \frac{1}{\hat{\zeta}} + O(n^{-3/2})\right);$$

(c) for $\bar{x} = E(X_1)$,

$$Q_n(\bar{x}) = \frac{1}{2} - \frac{\lambda_3(0)}{6\sqrt{2\pi n}} + O(n^{-3/2}),$$

where $\lambda_3(0)$ means λ_3 with the value zero substituted for $\hat{\phi}$ (see the notation that was defined above).

Remark. The positive attributes of the Lugannani-Rice formula are the validity of the approximation for both $\bar{x} > E(X_1)$ and $\bar{x} < E(X_1)$ without needing a change in the formula, its relative algebraic simplicity, and most of all its frequent ability to produce accurate tail approximations, including in the far tail, for very small sample sizes. Some of the other formulas for approximating tail areas are different over the domains $\bar{x} > E(X_1)$ and $\bar{x} < E(X_1)$ and are notationally and algebraically significantly more complex. Daniels (1987) is a very good review of these methods. Numerical reports of tail areas and their various approximations are also provided in that article. All approximations turn out to be extremely accurate in the examples cited, and no practically important difference in quality is seen among the various approximations. So, in a nutshell, the special status of the Lugannani-Rice formula is mainly due to its relative simplicity and the novel method of its derivation in the original Lugannani and Rice (1980) article. Next, we present two examples.

Example 14.10 Suppose $X_1, X_2, \cdots$ are iid $N(0, 1)$. We have seen before that in this case $K(t) = \frac{t^2}{2}$ and the unique saddlepoint is $\hat{\phi} = \bar{x}$. Plugging into their respective definitions,

$$\hat{W} = \hat{U} = \bar{x}, \hat{Z} = \sqrt{n}\bar{x}, \hat{\zeta} = \sqrt{n}\bar{x}.$$

Therefore, for the large deviation range $\bar{x} = O(1)$, by applying part (a) of Theorem 14.5, the Lugannani-Rice approximation to the tail probability $P(\bar{X} > \bar{x})$ works out to $1 - \Phi(\sqrt{n}\bar{x})$ and is therefore exact. Likewise, in the central limit domain also, the approximation turns out to be exact because $\hat{Z} = \hat{\zeta}$. Finally, at $\bar{x} = 0$, the Lugannani-Rice approximation gives exactly the answer $\frac{1}{2}$ as $\lambda_3 = 0$, $K(t)$ being a quadratic function of t. So, as in the case of density approximation, we again find the approximation for the tail area to be exact in the normal case.

Example 14.11 Suppose $X_1, X_2, \cdots$ are iid Exp(λ). We have seen before that in this case $K(t) = -\log(1 - \lambda t)$ and the unique saddlepoint is $\hat{\phi} = \frac{1}{\lambda} - \frac{1}{\bar{x}}$. By plugging into the respective formulas,

$$\hat{W} = \sqrt{2\left[\frac{\bar{x}}{\lambda} - 1 - \log\frac{\lambda}{\bar{x}}\right]}\operatorname{sgn}(\bar{x} - \lambda),$$

$$\hat{U} = \frac{\bar{x}}{\lambda} - 1, \hat{Z} = \sqrt{n}\left(\frac{\bar{x}}{\lambda} - 1\right),$$

$$\hat{\zeta} = \sqrt{n}\hat{W}.$$

The Lugannani-Rice approximation resulting from these formulas is not a simple analytic expression, but it can be easily computed. Since the exact tail area is an incomplete Gamma function, the Lugannani-Rice approximation does not work out to an expression that can be analytically renormalized to be exact. Limited numerical calculation in Lugannani and Rice (1980) shows the percentage error to be less than 0.5%, which is impressive.

We close this section with the corresponding Lugannani-Rice formula in the lattice case.

Theorem 14.6 (The Lattice Case) Suppose $X_1, X_2, \cdots$ are iid, distributed on a lattice. With $\hat{\zeta}$ as before and $\bar{z} = (1 - e^{-\hat{\phi}})\sqrt{nK^{(2)}}$, in the central limit domain (i.e., $\bar{x} = E(X_1) + O(n^{-1/2})$),

$$Q_n(\bar{x}) = 1 - \Phi(\hat{\zeta}) + \phi(\hat{\zeta})\left[\frac{1}{\bar{z}} - \frac{1}{\hat{\zeta}} + O(n^{-3/2})\right].$$

Remark. A continuity-corrected version of Theorem 14.6 is derived in Daniels (1987), but its practical superiority over the uncorrected version above seems to be in question.

Example 14.12 The following table reports the exact value, the ordinary and the continuity-corrected normal approximations, and the Lugannani-Rice approximation above to a Poisson tail probability for selected values of n and $\bar{x}$, assuming that λ, the Poisson parameter, is 1.

n	$\bar{x}$	Exact	Normal	Cont. Correct	Saddlepoint
5	1.5	.1334	.1318	.1318	.1803
	1.75	.0681	.0468	.0588	.0812
	2.50	.0020	.0004	.0004	.0033
25	1.1	.2998	.3085	.3085	.3344
	1.5	.0092	.0062	.0062	.0116
	1.75	.0004	.0001	.0001	.0004

We see from this table that when the tail probability is small or very small, the Lugannani-Rice formula gives a substantially better approximation. But when the tail probability is moderate or large, actually the normal approximations are better. A reason for this is that the jump terms that come associated with an Edgeworth expansion for the tail area in the lattice case have not been captured in the saddlepoint approximation. The lattice case is a bit different as regards the accuracy of the Lugannani-Rice formula.

14.6 Edgeworth vs. Saddlepoint vs. Chi-square Approximation

It is generally believed that the Lugannani-Rice formula provides a better approximation than an Edgeworth expansion up to the $\frac{1}{\sqrt{n}}$ term, and especially so in the extreme tails. There is credible empirical experience to support this for sample means. No systematic studies have been done for nonlinear statistics. Having said that, the Edgeworth expansions have a neat structural form, which the saddlepoint approximation does not. We saw in several of our examples that the saddlepoint approximation can basically only be computed but not written down as a concrete formula. The Edgeworth expansions are aesthetically more nice. Moreover, in the lattice case, the Edgeworth expansion includes extra terms to account for the jump discontinuities, and this can make it more accurate.

One drawback of Edgeworth expansions is that they are not themselves CDFs. They are not even monotone because of the involvement of the polynomials in the expansions. Hall (1983b) suggests approximations of the CDF of a normalized sum by a chi-square distribution. This is likely to pick up any positive skewness component in the true distribution, as chi-square distributions are positively skewed. Furthermore, by definition, they are CDFs. Hall (1983b) shows that an appropriate chi-square approximation is second-order accurate; that is, provided four moments are finite and the underlying CDF satisfies Cramér's condition (see Chapter 13), the $\frac{1}{\sqrt{n}}$ term vanishes in the expansion. See Hall (1983b) for a precise description of this result.

There do not seem to be any firm reasons to recommend any of the three methods universally over the others. It is potentially risky to use the Edgeworth expansion to approximate very small tail probabilities. However, it has succeeded in predicting with astounding accuracy the performance of various types of procedures. Instances of these success stories can be seen in the bootstrap literature (see Chapter 29). A loose recommendation is that, for small tail areas, the saddlepoint approximation may be used, especially if the sample size is small and the distribution is nonlattice. Otherwise, the Edgeworth expansion or a chi-square approximation may be used.

14.7 Tail Areas for Sample Percentiles

Having addressed the question of the sample mean, a smooth linear statistic, the next natural question is whether similar developments are possible for other statistics, possibly nonsmooth. In the previous chapter, we saw Edgeworth expansions for the CDFs of sample percentiles in the iid case. It turns out that saddlepoint approximations are also available; the derivation, obtained by using the usual connection to the CDF of a binomial distribution, can be seen in Ma and Robinson (1998).

Once again, we first need some notation. Let $X_1, X_2, \cdots$ be iid with CDF $F(.)$ and let x be a fixed real and let $0 < p < 1$. Assume that np is an integer. Let

$$\alpha = F(x);\ \Omega_\alpha = \Omega_\alpha(p) = (1-p)\log\frac{1-p}{1-\alpha} + p\log\frac{p}{\alpha};\ \omega_\alpha = \sqrt{2\Omega_\alpha};$$

$$u_\alpha = u_\alpha(p) = \sqrt{\frac{1-p}{p}\frac{p-\alpha}{1-\alpha}};\ \tilde{\omega}_\alpha = \omega_\alpha + \frac{1}{n\omega_\alpha}\log\frac{u_\alpha}{\omega_\alpha}.$$

Theorem 14.7 Let X_{np} denote the pth quantile of $X_1, X_2, \cdots, X_n$. Then, for $x < F^{-1}(p)$ and bounded away from zero,

$$P(X_{np} \leq x) = [1 - \Phi(\sqrt{n}\tilde{\omega}_\alpha)](1 + O(n^{-1}))$$

uniformly in x.

We consider an example of this result.

Example 14.13 Suppose $X_1, X_2, \cdots$ are iid $C(0,1)$, and suppose $p = \frac{1}{2}$. Assume that n is even, so that the formula in Theorem 14.7 can be used (for odd n, the formula changes).

By a simple calculation, $\omega_\alpha = \sqrt{-\log(4\alpha(1-\alpha))}$ and $u_\alpha = \frac{\frac{1}{2}-\alpha}{1-\alpha}$. On plugging these into the approximation formula above, we can get approximations to the probabilities that a Cauchy sample median is $\leq x$ for a given x. On the other hand, the exact value of such a probability can be found by using the usual connection to the CDF of a binomial distribution; i.e., $P(X_{n/2} \leq x) = P(\text{Bin}(n, F(x)) \geq \frac{n}{2})$. A comparison of the approximation to the exact probability is given in the table below.

n	x	Exact	Approx.
10	-1	.0781	.0172
	$-.5$	.2538	.0742
	$-.2$	.4622	.2945
25	-1	.0107	.0002
	$-.5$	.1309	.0051
	$-.1$	.5312	.2186

The approximation is of poor practical quality in this example unless n is quite large and the area being approximated is a small-tail area.

14.8 Quantile Approximation and Inverting the Lugannani-Rice Formula

The classic asymptotic expansions to the quantiles of an asymptotically normal statistic go by the name of *Cornish-Fisher expansions*. We discussed the case of the mean in the chapter on Edgeworth expansions. Hall (1983a) considered inversion of the Edgeworth expansion itself for quantile approximations; see also Barndorff-Nielsen and Cox (1989). It is interesting that the classical Cornish-Fisher expansions are retrieved from the Edgeworth expansion inversion. Since saddlepoint approximations are competitors to the Edgeworth expansion, it seems very natural to want to invert the saddlepoint approximation to a tail area and thereby obtain a quantile approximation. It is interesting that this approach is frequently numerically successful in accurately approximating the quantiles of the mean. The development below is from Arevalillo (2003).

First, we need a substantial amount of notation. Let $X_1, X_2, \cdots$ be iid with CDF F and density $f(x)$. Let $T_n(X_1, X_2, \cdots, X_n, F)$ be asymptotically standard normal. Let F_n be the CDF of T_n, M_n its mgf, assumed to exist in a neighborhood of zero, and $K_n(t) = \frac{1}{n}\log M_n(t\sqrt{n})$, the scaled cgf. Let

$$z_n(\tau) = \tau\sqrt{nK_n''(\tau)},\ \xi_n(\tau) = \sqrt{2n[\tau s/\sqrt{n} - K_n(\tau)]}\text{sgn}(\tau),$$

where s, τ are related by the saddlepoint equation $s/\sqrt{n} = K_n'(\tau)$, and let

$$r_n(\tau) = \xi_n(\tau) + \frac{1}{\xi_n(\tau)} \log \frac{z_n(\tau)}{\xi_n(\tau)}.$$

Theorem 14.8 Assume that $T_n(X_1, X_2, \cdots, X_n, F)$ admits a density function. Fix $0 < \alpha < 1$. Let

$$\tau^* = z_\alpha + \frac{z_\alpha - \sqrt{n}K_n'(z_\alpha/\sqrt{n})}{K_n''(z_\alpha/\sqrt{n})}, \; z_\alpha^* = \sqrt{n}K_n'(\tau^*/\sqrt{n})$$

and

$$s^* = z_\alpha^* + \frac{\xi_n(\tau^*/\sqrt{n})(z_\alpha - r_n(\tau^*/\sqrt{n}))}{\tau^*}.$$

Then, $F_n^{-1}(1-\alpha) = s^* + O(n^{-1})$.

Remark. Notice the extremely inconvenient aspect of the theorem that the mgf of T_n has to be known in an analytical form. This limits the practical scope of saddlepoint approximations of quantiles to very special statistics, such as sample means. The other unappealing feature of the approximation is its untidy formula in comparison with the structured and neat formulas in the Cornish-Fisher expansions. However, the availability of the saddlepoint quantile approximation under only mild conditions provides us with a competitor to Cornish-Fisher expansions in a practical problem. Example 14.15 shows that the saddlepoint quantile approximation can be very accurate.

We present two examples of Theorem 14.8.

Example 14.14 Suppose the observations are iid from a normal distribution, and assume without any loss of generality that it is the standard normal. Let T_n be $\sqrt{n}\bar{X}$. In this case, $K_n(t) = \frac{t^2}{2}$, leading to

$$\xi_n(\tau) = \sqrt{n}\tau = z_n(\tau) = r_n(\tau),$$
$$\tau^* = z_\alpha = z_{\alpha^*},$$

and hence $s^* = z_\alpha$; i.e., in the normal case, the saddlepoint approximation of the quantiles of the sample mean is exact, an expected result.

Example 14.15 Suppose the observations are iid from a standard double exponential density, and let $T_n = \frac{\sqrt{n}\bar{X}}{\sqrt{2}}$, so that T_n is asymptotically standard normal. Arevalillo (2003) reports the following numerically computed exact values, the normal approximation, and the saddlepoint approximation to the 99th percentile of T_n. The exact values are found by using a known formula for the CDF of the sum of iid double exponentials; see Arevalillo (2003).

Notice that the saddlepoint approximation is extremely accurate even for $n = 10$ and that the normal approximation is not so even for $n = 25$. However, the extra accuracy of the saddlepoint approximation comes at a computational price.

n	Exact	Normal approx.	Saddlepoint
10	2.393	2.326	2.394
15	2.372	2.326	2.373
25	2.354	2.326	2.355

14.9 The Multidimensional Case

Saddlepoint approximations for the density $f_n(\bar{x})$ of the mean in the general d-dimensional iid case are available. The approximation itself is not formally more complex than the one-dimensional case. However, explicitly writing the coefficients of the error terms, and even the $O(n^{-1})$ term, requires substantial notation in terms of the joint cumulants of the coordinates of the d-dimensional underlying random vector. It is too notationally complex beyond $d = 2$. See McCullagh (1987), Reid (1988), and Barndorff-Nielsen and Cox (1979) for Theorem 14.9 below and also for more detailed treatment of the multidimensional case.

The approximation itself is described without much notational difficulty. Suppose $\mathbf{X_i}$ are iid from a d-dimensional density $f(x)$. Let $\psi(t) = E(e^{t'X})$ be its mgf, assumed to exist in some open ball around the origin. Let $K(t) = \log \psi(t)$ be the cgf; here the word *cumulant* means multivariate cumulants. Given a point $\bar{x}$, let $\hat{\phi}$ denote the saddlepoint, defined as a root of the saddlepoint equation $\nabla K(\hat{\phi}) = \bar{x}$. We assume the saddlepoint exists and is unique. Let $\nabla^2 K$ denote the Hessian matrix of the cgf K and $|\nabla^2 K|$ its determinant.

Theorem 14.9 $f_n(\bar{x}) = \frac{1}{(2\pi)^{d/2}} \sqrt{\frac{n}{|\nabla^2 K(\hat{\phi})|}} e^{n[K(\hat{\phi}) - \hat{\phi}'\bar{x}]}(1 + O(n^{-1}))$.

Remark. It cannot be guaranteed in general that the $O(n^{-1})$ relative error is uniform in the argument $\bar{x}$. In other words, the n^{-1} coefficient in general depends on the specific $\bar{x}$.

Feasibility of an analytic expression of the approximation formula rests on the ability to write the d-dimensional mgf and the saddlepoint in closed form. This can be difficult or impossible if the components of the underlying random vector have a dependent structure.

14.10 Exercises

Exercise 14.1 Use Laplace's method to approximate $\int_0^\infty \frac{e^{-t}}{\alpha+t}dt$ as $\alpha \to \infty$.

Exercise 14.2 Use Laplace's method to approximate $\int_0^{\pi/2} te^{-x\sin t}dt$ as $x \to \infty$.

Exercise 14.3 Use Laplace's method to approximate the standard normal tail probability $P(Z > z)$, $Z \sim N(0, 1)$.

Exercise 14.4 For each of the following cases, plot the exact density and the saddlepoint approximation of the density of $\bar{X}$ for $n = 5, 10, 20$: $F = \exp(1)$; $F = U[0, 1]$; $F = \text{Poi}(1)$; $F = \text{Geo}(.5)$. Discuss the visual accuracy of the approximation.

Exercise 14.5 * The *standard inverse normal density* is given by $f(x) = \frac{1}{\sqrt{2\pi}}x^{-3/2}e^{-\frac{(x-\mu)^2}{2\mu^2 x}}$, $x > 0$. Find the saddlepoint approximation to the density of $\bar{X}$. Plot it for $n = 5, 20$, $\mu = 1, 5$. Obtain the exact density by a careful simulation and plot it also for $n = 5, 20$, $\mu = 1, 5$. Comment on your plots.

Exercise 14.6 Show that the Beta(α, α) density belongs to the one-parameter exponential family, and obtain the saddlepoint approximation to the density of the MLE of α.

Exercise 14.7 * Suppose f is the density of $N(\theta, 1)$ conditioned on it being positive. Obtain the saddlepoint approximation to the density of the MLE of θ.

Exercise 14.8 * Suppose f is the pmf of a Poi(λ) distribution conditioned on it being strictly positive. Obtain the saddlepoint approximation to the density of the MLE of λ and plot it for $\lambda = 1$. Obtain the exact density by a simulation and plot it. Comment on your plots. Note: The exact density can in fact be found by intricate combinatorics.

Exercise 14.9 * Suppose $X_1, X_2, \cdots$ are iid from the $U[-1, 1]$ density.

(a) Obtain the cumulant generating function.

(b) Write the equations for deriving the saddlepoint.

(c) Compute the saddlepoint by a numerical root-solving procedure for some selected values of n and with $\bar{x} = 1$.

(d) Compute the Lugannani-Rice approximation to the tail probability $P(\bar{X} > \bar{x})$.

Exercise 14.10 Is the Lugannani-Rice approximation to the CDF of the mean in the large deviation domain itself necessarily a valid CDF?

Exercise 14.11 What is the most fundamental difference between the Edgeworth approximation and the Lugannani-Rice approximation to the density of $\bar{X}$ in the lattice case?

Exercise 14.12 * Compute the exact expression, normal approximation, continuity-corrected normal approximation, and Lugannani-Rice approximation to the density of $\bar{X}$ in the Geo(.5) case with $\bar{x} = 2.25, 2.5, 2.75$ and $n = 5, 25$.

Exercise 14.13 Suppose observations are iid from a standard normal density. Compute the saddlepoint approximation to the CDF of the sample median at $x = -1, -.5, -.2$ for $n = 3, 8, 15$. Compare it with the exact values and the values obtained from the usual normal approximation.

Exercise 14.14 How does the saddlepoint approximation to the CDF of a percentile change when $x > F^{-1}(p)$?

Exercise 14.15 * Compute the saddlepoint approximation and the exact values of the 95th and the 99th percentiles of the mean of an iid sample from the Exp (1) density for $n = 3, 10, 25$.

Exercise 14.16 * Is the saddlepoint approximation to the quantiles of an asymptotically standard normal statistic always monotone in α?

References

Arevalillo, J. (2003). Inverting a saddlepoint approximation, Stat. Prob. Lett., 61, 421–428.

Barndorff-Nielsen, O.E. (1983). On a formula for the distribution of the maximum likelihood estimator, Biometrika, 70, 343–365.

Barndorff-Nielsen, O.E. and Cox, D.R. (1979). Edgeworth and saddlepoint approximations with statistical applications, J.R. Stat. Soc. Ser. B, 41, 279–312.

Barndorff-Nielsen, O.E. and Cox, D.R. (1989). *Asymptotic Techniques for Use in Statistics*, Chapman and Hall, London.

Daniels, H.E. (1954). Saddlepoint approximations in statistics, Ann. Math. Stat., 25, 631–650.

Daniels, H.E. (1987). Tail probability approximations, Int. Stat. Rev., 55, 37–48.
Good, I.J. (1957). Saddlepoint methods for the multinomial distribution, Ann. Math. Stat., 28, 861–881.
Goutis, C. and Casella, G. (1999). Explaining the saddlepoint approximation, Am. Stat., 53(3), 216–224.
Hall, P. (1983a). Inverting an Edgeworth expansion, Ann. Stat., 11(2), 569–576.
Hall, P. (1983b). Chi-square approximations to the distribution of a sum of independent random variables, Ann. Prob., 11, 1028–1036.
Jensen, J.L. (1995). *Saddlepoint Approximations*, Oxford University Press, Oxford.
Kolassa, J. (2003). Multivariate saddlepoint tail probability approximations, Ann. Stat., 31(1), 274–286.
Lugannani, R. and Rice, S. (1980). Saddlepoint approximation for the distribution of the sum of independent random variables, Adv. Appl. Prob., 12, 475–490.
Ma, C. and Robinson, J. (1998). Saddlepoint approximation for sample and bootstrap quantiles, Aust. N. Z. J. Stat., 40(4), 479–486.
McCullagh, P. (1987). *Tensor Methods in Statistics*, Chapman and Hall, London.
Olver, F.J. (1997). *Asymptotics and Special Functions*, A.K. Peters Ltd., Wellesley, MA.
Reid, N. (1988). Saddlepoint methods and statistical inference, Stat. Sci., 3(2), 213–227.
Wood, A.T.A. (2000). Laplace/saddlepoint approximations, in *Symposium in Honour of Ole E. Barndorff-Nielsen*, Memoirs, Vol. 16, University of Aarhus, 110–115.

Chapter 15
U-statistics

The limiting distributions of linear statistics $\frac{1}{n}\sum_{i=1}^{n} h(X_i)$ are handled by the canonical CLT under suitable moment conditions. In applications, frequently one encounters statistics that are not linear. There is no master theorem of asymptotic theory that can give the limiting distribution of an arbitrary statistic. Different techniques are needed to handle different types of statistics. However, a class of statistics called U-statistics admit a common CLT and settle the question of limiting distributions of a variety of useful nonlinear statistics. U-statistics are also often regarded as the next interesting class of statistics after the linear statistics. We present the basic results about U-statistics in this chapter.

Suppose that $h(x_1, x_2, ..., x_r)$ is some real-valued function of r arguments $x_1, x_2, ..., x_r$. The arguments $x_1, x_2, ..., x_r$ can be real or vector valued. Now suppose $X_1, \ldots, X_n$ are iid observations from some CDF F, and for a given $r \geq 1$ we want to estimate or make inferences about the parameter $\theta = \theta(F) = E_F h(X_1, X_2, ..., X_r)$. We assume $n \geq r$. Of course, one unbiased estimate is $h(X_1, X_2, ..., X_r)$ itself. But one should be able to find a better unbiased estimate if $n > r$ because $h(X_1, X_2, ..., X_r)$ does not use all of the sample data. For example, if the X_i are real valued, then the set of order statistics $X_{(1)}, ..., X_{(n)}$ is always sufficient and the Rao-Blackwellization $E\left[h(X_1, X_2, ..., X_r)|X_{(1)}, ..., X_{(n)}\right]$ is a better unbiased estimate than $h(X_1, X_2, ..., X_r)$. Indeed, in this case

$$E\left[h(X_1, X_2, ..., X_r)|X_{(1)}, ..., X_{(n)}\right] = \frac{1}{\binom{n}{r}} \sum_{1\leq i_1<i_2<\cdots<i_r\leq n} h(X_{i_1}, X_{i_2}, ..., X_{i_r}).$$

Statistics of this form are called U-statistics (U for unbiased), and h is called the kernel and r its order. They were introduced in Hoeffding (1948).

A. DasGupta, *Asymptotic Theory of Statistics and Probability*,

15.1 Examples

Definition 15.1 For an iid sample $X_1, \ldots, X_n$, a U-statistic of order r with kernel h is defined as

$$U = U_n = \frac{1}{\binom{n}{r}} \sum_{1 \le i_1 < i_2 < \cdots < i_r \le n} h(X_{i_1}, X_{i_2}, \ldots, X_{i_r}).$$

Remark. We will assume that h is permutation symmetric in order that U has that property as well.

Example 15.1 Suppose, $r = 1$. Then the linear statistic $\frac{1}{n}\sum_{i=1}^n h(X_i)$ is clearly a U-statistic. In particular, $\frac{1}{n}\sum_{i=1}^n X_i^k$ is a U-statistic for any k.

Example 15.2 Let $r = 2$ and $h(x_1, x_2) = \frac{1}{2}(x_1 - x_2)^2$. Then, on calculation,

$$\frac{1}{2\binom{n}{2}} \sum\sum_{i<j} (X_i - X_j)^2 = \frac{1}{n-1} \sum_{i=1}^{n} (X_i - \bar{X})^2.$$

Thus the sample variance $s^2 = \frac{1}{n-1}\sum_{i=1}^{n-1}(X_i - \bar{X})^2$ is a U-statistic.

Example 15.3 Let x_0 be a fixed real, $r = 1$, and $h(x) = I_{x \le x_0}$. Then $U = \frac{1}{n}\sum_{i=1}^n I_{X_i \le x_0} = F_n(x_0)$, the empirical CDF at x_0. Thus $F_n(x_0)$ for any specified x_0 is a U-statistic.

Example 15.4 Let $r = 2$ and $h(x_1, x_2) = |x_1 - x_2|$. Then the corresponding $U = \frac{1}{\binom{n}{2}} \sum\sum_{i<j} |x_i - x_j|$. This is the well-known Gini mean difference very widely used in the economics and inequality literature. It is an unbiased estimate of $E|X_1 - X_2|$.

Example 15.5 Let $r = 2$ and $h(X_1, X_2) = I_{X_1 + X_2 > 0}$. The corresponding $U = \frac{1}{\binom{n}{2}} \sum\sum_{i<j} I_{X_i + X_j > 0}$. If $X_1, \ldots, X_n$ are observations from a CDF F that is continuous and symmetric about 0, then U has a distribution symmetric about $\frac{1}{2}$. If F is a location shift by $\mu > 0$ of some F_0 symmetric about 0, then U will have a center at $\tau > \frac{1}{2}$. Thus U can be used as a test statistic for testing that the location parameter of a symmetric density is zero. In fact, U is related to the one-sample Wilcoxon statistic.

Example 15.6 Let $r = 2$, $z_i = (x_i, y_i)$, and $h(z_1, z_2) = I_{(x_1-x_2)(y_1-y_2)>0}$. The corresponding $U = \frac{1}{\binom{n}{2}} \sum\sum_{i<j} I_{(X_i-X_j)(Y_i-Y_j)>0}$, where $(X_i, Y_i), i = 1, ..., n$ are a bivariate iid sample. This statistic U measures the fraction of concordant pairs (i, j) of individuals, where concordance means $X_i < X_j \Rightarrow Y_i < Y_j$ and $X_i > X_j \Rightarrow Y_i > Y_j$. In fact, $2U - 1$ is the well-known τ-statistic of Kendall.

15.2 Asymptotic Distribution of U-statistics

The summands in the definition of a U-statistic are not independent. Hence, neither the exact distribution theory nor the asymptotics are straightforward. Hajek had the brilliant idea of projecting U onto the class of linear statistics of the form $\frac{1}{n}\sum_{i=1}^{n} h(X_i)$. It turns out that the projection is the dominant part and determines the limiting distribution of U. The main theorems can be seen in Serfling (1980), Lehmann (1999), and Lee (1990). We have the following definition of the Hajek projection.

Definition 15.2 The Hajek projection of $U - \theta$ onto the class of statistics $\sum_{i=1}^{n} h(X_i)$ is

$$\hat{U} = \hat{U}_n = \sum_{i=1}^{n} E(U - \theta | X_i).$$

Remark. It can be shown that $\hat{U} = \frac{r}{n}\sum_{i=1}^{n} h_1(X_i)$, where $h_1(x) = E(h(X_1, ..., X_r)|X_1 = x) - \theta$.

Theorem 15.1 Suppose that the kernel h is twice integrable; i.e., $Eh^2(X_1, ..., X_r) < \infty$. Let $X_1, X_2, ..., X_r, Y_2, ..., Y_r$ be iid observations from F, and let $\zeta_1 = \text{cov}(h(X_1, ..., X_r), h(X_1, Y_2, ..., Y_r))$. Assume that $0 < \zeta_1 < \infty$. Then $\sqrt{n}(U - \theta - \hat{U}) \xrightarrow{\mathcal{P}} 0$ and $\sqrt{n}(U - \theta) \xrightarrow{\mathcal{L}} N(0, r^2\zeta_1)$.

Remark. A proof can be seen in Serfling (1980). The fact that $\sqrt{n}(U - \theta - \hat{U}) \xrightarrow{P} 0$ can be verified either on a case-by-case basis or by using a general theorem that essentially asks that $\frac{\text{Var}(U)}{\text{Var}(\hat{U})} \to 1$ as $n \to \infty$. $\text{Var}(\hat{U})$ is typically the easier to compute. There are exact expressions for $\text{Var}(U)$, but an approximation might be enough. For example, it is known that if $\zeta_1 > 0$, then $\text{Var}(U) = \frac{r^2\zeta_1}{n} + O(n^{-2})$.

Example 15.7 Here is an example where we compute the projection $\hat{U}$ and verify directly that $\sqrt{n}(U - \theta\hat{U}) \xrightarrow{P} 0$. Suppose $h(x_1, x_2) = \frac{1}{2}(x_1 - x_2)^2$ so that $U = \frac{1}{n-1}\sum_{i=1}^{n}(X_i - \bar{X})^2$. We calculate below $\hat{U} = \sum_j E(U|X_j) - (n-1)\sigma^2$. Assume without loss of generality that $E_F(X_1) = 0$. Then

$$
\begin{aligned}
E(U|X_j) &= \frac{1}{n-1} E\left(\sum X_i^2 - n\bar{X}^2 | X_j\right) \\
&= \frac{1}{n-1} E\left(X_j^2 + \sum_{i \neq j} X_i^2 - \frac{1}{n}\sum_i \sum_k X_i X_k | X_j\right) \\
&= \frac{1}{n-1} E\left(X_j^2 + \sum_{i \neq j} X_i^2 - \frac{1}{n}\left\{\sum_{i \neq j}\sum_{k \neq j} X_i X_k + 2\sum_{k \neq j} X_j X_k + X_j^2\right\} | X_j\right) \\
&= \frac{1}{n-1}\left[X_j^2 + (n-1)E(X_1^2) - \frac{n-1}{n}E(X_1^2) - \frac{2(n-1)}{n}X_j E(X_1) - \frac{1}{n}X_j^2\right] \\
&= \frac{1}{n}X_j^2 + \frac{n-1}{n}\sigma^2
\end{aligned}
$$

using that, for $i \neq j$, $E(X_i|X_j) = E(X_i) = 0$. Therefore,

$$
\begin{aligned}
\sqrt{n}(U_n - \hat{U}_n) &= \sqrt{n}\left(\frac{\sum X_j^2}{n-1} - \frac{n}{n-1}\bar{X}^2 - \frac{1}{n}\sum X_j^2\right) \\
&= \sqrt{n}\left(\frac{\sum X_j^2}{n(n-1)} - \frac{(\sqrt{n}\bar{X})^2}{n-1}\right) \\
&= o_p(1) + o_p(1) \\
&= o_p(1)
\end{aligned}
$$

since we assume $E(X_1) = 0$ and $E(X_1^2) < \infty$.

Example 15.8 Let $h(x_1, x_2) = I_{x_1 + x_2 > 0}$. Recall that this corresponds to the one-sample Wilcoxon test statistic. It is used to test the hypothesis that the location parameter of a symmetric density is zero. We need θ and ζ_1. Under the null hypothesis, obviously $\theta = \frac{1}{2}$. Also,

$$\begin{aligned}\zeta_1 &= \text{cov}_{H_0}(h(X_1, X_2), h(X_1, Y_2)) \\ &= P_{H_0}(X_1 + X_2 > 0, X_1 + Y_2 > 0) - \frac{1}{4} \\ &= P_{H_0}(X_1 - X_2 > 0, X_1 - Y_2 > 0) - \frac{1}{4} \\ &= \frac{1}{3} - \frac{1}{4} = \frac{1}{12},\end{aligned}$$

where X_1, X_2, Y_2 are iid from F. Therefore, if F is continuous and symmetric, then $\sqrt{n}\left(U_n - \frac{1}{2}\right) \xrightarrow{\mathcal{L}} N\left(0, \frac{1}{12}\right)$. Thus a completely distribution-free test of the null hypothesis is obtained.

Example 15.9 The Gini mean difference $G = \frac{2}{n(n-1)} \sum\sum |X_i - X_j|$ is of wide use in economics. This is a U-statistic with kernel $h(x_1, x_2) = |x_1 - x_2|$. From the CLT for U-statistics, it follows that $\sqrt{n}(G - E|X_1 - X_2|) \xrightarrow{\mathcal{L}} N(0, 4\zeta_1)$, where ζ_1 may be calculated to be $\zeta_1 = \text{Var}_F\left(2XF(X) - X - 2\int_{(-\infty,X]} y\,dF(y)\right)$. Notice that ζ_1 is not distribution-free and that in general ζ_1 can only be computed numerically.

15.3 Moments of U-statistics and the Martingale Structure

A relatively simple calculation shows that the variance of a U-statistic can be written as a finite combinatorial sum, but higher-order moments are too complicated except in convenient examples. However, asymptotic expansions for moments or moments of the difference of the U-statistic and its Hájek projection can be obtained. We report a few of these expressions and results below; again, see Serfling (1980) for more details and proofs.

Theorem 15.2 (a) For a U-statistic with a kernel h of order 2, $\text{Var}(U_n) = \frac{4(n-2)}{n(n-1)}\zeta_1 + \frac{2}{n(n-1)}\text{Var}(h)$.

(b) $\text{Var}(\sqrt{n}(U_n - \theta))$ is nonincreasing in n.

(c) $\text{Var}(\sqrt{n}(U_n - \theta)) = r^2\zeta_1 + O(n^{-1})$.

(d) If h has k absolute moments, then $E|U_n - \theta|^k = O(n^{-\frac{k}{2}})$.

(e) If h has k absolute moments, then $E(U_n - \theta - \hat{U}_n)^k = O(n^{-k})$.

Example 15.10 Let $h(x, y) = \frac{(x-y)^2}{2}$, so that $U_n = \frac{1}{n-1}\sum_{i=1}^{n}(X_i - \bar{X})^2$, namely the sample variance. Then, by elementary algebra, $h_1(x) = \frac{x^2-2x\mu+\mu^2+\sigma^2}{2}$, so that $\zeta_1 = \frac{\mu_4-\sigma^4}{4}$. Also, $\text{Var}(h) = \frac{\mu_4+\sigma^4}{2}$. From part (a) of Theorem 15.2, on substitution, one gets $\text{Var}(s^2) = \frac{\mu_4-\sigma^4}{n} + \frac{2\sigma^4}{n(n-1)}$.

If h has two moments, then part (e) of Theorem 15.2 suggests that the difference between $U_n - \theta$ and $\hat{U}_n$ should be asymptotically negligible. In fact, the difference converges in that case a.s. to zero, and U_n would converge a.s. to θ. But, in fact, the almost sure convergence of U_n to θ holds under weaker conditions due to the martingale structure present in the U-statistics. This was noticed by Hoeffding himself (Hoeffding (1948)), and later a more useful *reverse martingale* property was noticed and exploited in Berk (1966). Berk's result follows.

Theorem 15.3 (a) Suppose h has one absolute moment. Then $\{U_n\}$ is a reverse martingale.

(b) $U_n \stackrel{\text{a.s.}}{\Rightarrow} \theta$, if h has one absolute moment.

15.4 Edgeworth Expansions

Both Berry-Esseen theorems and Edgeworth expansions for U-statistics of order 2 are available. Grams and Serfling (1973) established the $O(n^{-1/2+\epsilon})$ rate for any $\epsilon > 0$, for the error of the CLT for nondegenerate U-statistics, under the assumption that the kernel h has finite moments of all orders. The anticipated $O(n^{-\frac{1}{2}})$ rate was obtained in Bickel (1974) for bounded kernels. The result below is due to Callaert and Janssen (1978).

Theorem 15.4 Suppose h is of order 2, and $|h|$ has three moments. Then, there exists a universal constant $C, 0 < C < \infty$ such that $\sup_x |P(\sqrt{n}(U_n - \theta)/(2\zeta_1) \le x) - \Phi(x)| \le C\frac{E|h-\theta|^3}{(2\zeta_1)^3}/\sqrt{n}, n \ge 2$.

With additional moments, the Berry-Esseen theorem can be improved to an Edgeworth expansion with an error of the order $o(\frac{1}{n})$. The Edgeworth expansion requires an additional technical condition, which needs to be verified on a case-by-case basis. See Callaert et al. (1980) for the condition that ensures expansion of the characteristic function that leads to the Edgeworth expansion below.

We need the following notation. Let

$$h_1(x) = E_F(h(X, Y)|X = x), \psi(x, y) = h(x, y) - h_1(x) - h_1(y),$$

$$\kappa_3 = \frac{1}{(2\zeta_1)^3}[Eh_1(X)^3 + 3Eh_1(X)h_1(Y)\psi(X,Y)],$$

$$\kappa_4 = \frac{1}{(2\zeta_1)^4}[Eh_1(X)^4 - 3(2\zeta_1)^4 + 12Eh_1(X)^2h_1(Y)\psi(X,Y)$$
$$+ 12Eh_1(Y)h_1(Z)\psi(X,Y)\psi(X,Z)],$$

where $X, Y, Z \overset{\text{iid}}{\sim} F$.

Theorem 15.5 Suppose h is of order 2 and $K_n(x) = \Phi(x) - \phi(x)[\frac{\kappa_3}{6\sqrt{n}}(x^2 - 1) + \frac{\kappa_4}{24n}(x^3 - 3x) + \frac{\kappa_3^2}{72n}(x^5 - 10x^3 - 15x)]$. Then, $P(\sqrt{n}(U_n - \theta)/(2\zeta_1) \leq x) = K_n(x) + o(\frac{1}{n})$ uniformly in x.

Remark. Loh (1996) obtains Edgeworth expansions with higher order accuracy for U-statistics in certain dependent data situations.

Example 15.11 Suppose $X_1, X_2, \ldots$ are iid observations with mean 0 and variance 1. Let U_n be the sample variance $\frac{1}{n-1}\sum_{i=1}^n (X_i - \bar{X})^2$ with the kernel $h(x,y) = \frac{(x-y)^2}{2} - 1$, so as to make it have mean 0. On calculation, $h_1(x) = \frac{x^2-1}{2}$, $\psi(x,y) = \frac{(x-y)^2}{2} - \frac{x^2-1}{2} - \frac{y^2-1}{2} - 1$, $4\zeta_1^2 = \frac{\mu_4-1}{4}$, $Eh_1(X)^3 = \frac{\mu_6-3\mu_4+2}{8}$, $Eh_1(X)^4 = \frac{\mu_8-4\mu_6+6\mu_4-3}{16}$,
$Eh_1(X)h_1(Y)\psi(X,Y) = -\frac{\mu_3^2}{4}$, $Eh_1(X)^2h_1(Y)\psi(X,Y) = -\frac{(\mu_5-2\mu_3)\mu_3}{8}$,
$Eh_1(Y)h_1(Z)\psi(X,Y)\psi(X,Z) = \frac{\mu_3^2}{4}$.
Substituting these into the definitions,

$$\kappa_3 = \frac{\mu_6 - 3\mu_4 - 6\mu_3^2 + 2}{(\mu_4 - 1)^{\frac{3}{2}}},$$

$$\kappa_4 = \frac{\mu_8 - 4\mu_6 - 3\mu_4^2 + 12\mu_4 + 96\mu_3^3 - 24\mu_3\mu_5 - 6}{(\mu_4 - 1)^2}.$$

This leads to the two-term Edgeworth expansion $K_n(x)$ as given in Theorem 15.5.

15.5 Nonnormal Limits

In some examples, the limiting distribution of U_n is not normal because $\zeta_1 = 0$. The limits in such cases are generally complicated. We state one main result.

Theorem 15.6 Let $h_2(x_1, x_2) = E_F(h(X_1, X_2, \cdots, X_r)|X_1 = x_1, X_2 = x_2)$. Let $\zeta_2 = \text{Var}_F(h_2(X_1, X_2))$. If $\zeta_1 = 0, \zeta_2 > 0$ and $E_F(h^2(X_1, X_2, \cdots, X_r)) < \infty$, then for appropriate constants $\lambda_1, \lambda_2, \cdots$ and iid variables $W_1, W_2, \cdots \sim \chi^2(1)$,

$$n(U_n - \theta) \xrightarrow{\mathcal{L}} T,$$

where $T \sim \sum_{i=1}^{\infty} \lambda_i (W_i - 1)$.

Remark. See Serfling (1980) for additional details on this theorem. The constants λ_i have to be found on a case-by-case basis.

Example 15.12 Let $h(x_1, x_2) = \frac{1}{2}(x_1 - x_2)^2$ so that $U_n = s^2 = \frac{1}{n-1}\sum(X_i - \bar{X})^2$. By an easy calculation, $\zeta_1 = \frac{\mu_4 - \sigma^4}{4}$, so that $\sqrt{n}(s^2 - \sigma^2) \to N(0, \mu_4 - \sigma^4)$ provided $\mu_4 - \sigma^4 > 0$. Sometimes $\mu_4 - \sigma^4 = 0$. In such a case, s^2 does not have a normal limit. Also, by a direct calculation, $\zeta_2 = \frac{\mu_4 + \sigma^4}{2} = \sigma^4 > 0$. Thus, Theorem 15.6 will apply. For example, if $F = \text{Bin}(1, \frac{1}{2})$, then indeed $\mu_4 = \sigma^4 = \frac{1}{16}$ and $\zeta_2 = \frac{1}{16}$. It can be verified that $\lambda_1 = -\frac{1}{4}$, $\lambda_i = 0 \,\forall i > 1$ in this case, and so $n(s^2 - \frac{1}{4}) \xrightarrow{\mathcal{L}} \frac{1}{4}(1 - W)$, where $W \sim \chi^2(1)$. (See Serfling (1980) for the derivation of $\lambda_1, \lambda_2, \cdots$) Thus, if $F = \text{Bin}(1, p)$, then for $p = \frac{1}{2}$, $n(s^2 - \frac{1}{4}) \xrightarrow{\mathcal{L}} \frac{1}{4}(1 - W)$, and for $p \neq \frac{1}{2}$, $\sqrt{n}(s^2 - pq) \xrightarrow{\mathcal{L}} N(0, pq^4 + qp^4 - p^2q^2)$, where $q = 1 - p$.

15.6 Exercises

Exercise 15.1 L_2 Projections

Let U be a fixed random variable and $\mathcal{L}$ the class of linear statistics, all with finite second moments. Rigorously characterize the L_2 projection of U onto $\mathcal{L}$.

Exercise 15.2 Hajek's Lemma

Let $X_1, X_2, \cdots$ be a sequence of independent random variables. Let $U_n = U(X_1, X_2, \cdots, X_n)$ have finite variance for every n. Let $\mathcal{L}$ be the class of all random variables of the form $\sum_{i=1}^{n} h(X_i)$ with $h(X_i)$ having finite variance for all i. Using Exercise 15.1, derive the L_2 projection of U onto $\mathcal{L}$.

Exercise 15.3 (a) Let U_n be a sequence with finite variance and $\hat{U}_n$ its L_2 projection onto a linear class $\mathcal{L}$ containing constant random variables as particular members. Show that

$$\frac{U_n - E(U_n)}{\sqrt{\text{Var}(U_n)}} - \frac{\hat{U}_n - E\hat{U}_n}{\sqrt{\text{Var}(\hat{U}_n)}} \xrightarrow{\mathcal{P}} 0$$

if $\frac{\text{Var}(U_n)}{\text{Var}(\hat{U}_n)} \to 1$.

(b) Hence, U_n and $\hat{U}_n$ have the "same asymptotic distributions."

Exercise 15.4 Show that sample moments, sample variances, the Gini mean difference, and the empirical CDF all have asymptotically normal distributions.

Exercise 15.5 The Asymptotic Normality Theorem Understates the Variance

(a) Compute the exact variance of the Gini mean difference for iid $N(0, 1)$ data with $n = 20, 40$.

(b) Compare the result in (a) with the variance given by the limiting distribution you obtained in Exercise 15.4.

Exercise 15.6 * Suppose X_i are iid $N(\mu, 1)$. Give an example of a prior G such that the posterior mean

(a) is a U-statistic;

(b) is not a U-statistic.

Exercise 15.7 * Suppose X_i are iid $f(x - \mu)$, where $f(x) = f(-x)$. Suppose T_n is the proportion of pairs $i < j$ for which $X_i + X_j \geq 0$. Show that $\sqrt{n}(T_n - \frac{1}{2}) \xrightarrow{\mathcal{L}} N(0, \frac{1}{3})$, independent of f.

Exercise 15.8 * Suppose X_i are iid $N(\mu, \sigma^2)$. Give a direct proof that the Gini mean difference converges a.s. to $\frac{2\sigma}{\sqrt{\pi}}$.

Exercise 15.9 Let U_n be the sample variance of n iid observations from a CDF F with four moments. Prove directly that $\sqrt{n}(U_n - \sigma^2)$ has a variance decreasing in n.

Exercise 15.10 * Let U_n be the sample variance again as in the previous exercise. Show directly that $\{U_n\}$ is a reverse martingale, and identify a sequence of σ-fields w.r.t. which the reverse martingale property holds.

Exercise 15.11 Use the two-term Edgeworth expansion for s^2 to approximate the probability $P(s^2 \leq t)$ when $X_1, \ldots, X_n$ are iid $N(0, 1)$ and compare it with the exact value of that probability from the χ^2 distribution. Plot the exact and the approximated probabilities as a function of t.

Exercise 15.12 * Derive the two-term Edgeworth expansion for the Gini mean difference when $X_1, \ldots, X_n$ are iid $U[0, 1]$ and plot it against the CLT approximation.

References

Berk, R. (1966). Limiting behavior of posterior distributions when the model is incorrect, Ann. Math. Stat., 37, 51–58.

Bickel, P.J. (1974). Edgeworth expansions in nonparametric statistics, Ann. Stat., 2, 1–20.

Callaert, H. and Janssen, P.(1978). The Berry-Esseen theorem for U-statistics, Ann. Stat., 6(2), 417–421.

Callaert, H., Janssen, P., and Veraverbeke, N. (1980). An Edgeworth expansion for U-statistics, Ann. Stat., 8(2), 299–312.

Grams, W. and Serfling, R. (1973). Convergence rates for U-statistics and related statistics, Ann. Stat., 1, 153–160.

Hoeffding, W. (1948). A class of statistics with asymptotically normal distribution,Ann. Math. Stat., 19, 293–325.

Lee, A.J. (1990). *U-statistics: Theory and Practice*, Marcel Dekker, New York.

Lehmann, E.L. (1999). *Elements of Large Sample Theory*, Springer, New York.

Loh, W.L. (1996). An Edgeworth expansion for U-statistics with weakly dependent observations, Stat. Sinica, 6(1), 171–186.

Serfling, R. (1980). *Approximation Theorems of Mathematical Statistics*, John Wiley, New York.

Chapter 16
Maximum Likelihood Estimates

Many think that maximum likelihood is the greatest conceptual invention in the history of statistics. Although in some high-or infinite-dimensional problems, computation and performance of maximum likelihood estimates (MLEs) are problematic, in a vast majority of models in practical use, MLEs are about the best that one can do. They have many asymptotic optimality properties that translate into fine performance in finite samples. We treat MLEs and their asymptotic properties in this chapter. We start with a sequence of examples, each illustrating an interesting phenomenon.

16.1 Some Examples

Example 16.1 In smooth regular problems, MLEs are asymptotically normal with a $\sqrt{n}$-norming rate. For example, if $X_1, \ldots, X_n$ are iid $N(\mu, 1)$, $-\infty < \mu < \infty$, then the MLE of μ is $\hat{\mu} = \bar{X}$ and $\sqrt{n}(\hat{\mu} - \mu) \xrightarrow{\mathcal{L}} N(0, 1), \forall \mu$.

Example 16.2 Let us change the problem somewhat to $X_1, X_2, \ldots, X_n \overset{\text{iid}}{\sim} N(\mu, 1)$ with $\mu \geq 0$. Then the MLE of μ is

$$\hat{\mu} = \begin{cases} \bar{X} \text{ if } \bar{X} \geq 0 \\ 0 \text{ if } \bar{X} < 0 \end{cases};$$

i.e., $\hat{\mu} = \bar{X} I_{\bar{X} \geq 0}$. If the true $\mu > 0$, then $\hat{\mu} = \bar{X}$ a.s. for all large n and $\sqrt{n}(\hat{\mu} - \mu) \xrightarrow{\mathcal{L}} N(0, 1)$. If the true $\mu = 0$, then we still have consistency; in fact, still $\hat{\mu} \xrightarrow{\text{a.s.}} \mu = 0$. Let us now look at the question of the limiting distribution of $\hat{\mu}$. Denote $Z_n = \sqrt{n}\bar{X}$ so that $\hat{\mu} = \frac{Z_n I_{Z_n \geq 0}}{\sqrt{n}}$.

A. DasGupta, *Asymptotic Theory of Statistics and Probability*,

Let $x < 0$. Then $P_0(\sqrt{n}\hat{\mu} \le x) = 0$. Let $x = 0$. Then $P_0(\sqrt{n}\hat{\mu} \le x) = \frac{1}{2}$. Let $x > 0$. Then

$$\begin{aligned} P_0(\sqrt{n}\hat{\mu} \le x) &= P(Z_n I_{Z_n \ge 0} \le x) \\ &= \frac{1}{2} + P(0 < Z_n \le x) \\ &\to \Phi(x), \end{aligned}$$

so

$$P(\sqrt{n}\hat{\mu} \le x) \to \begin{cases} 0 & \text{for } x < 0 \\ \frac{1}{2} & \text{for } x = 0 \\ \Phi(x) & \text{for } x > 0 \end{cases}.$$

The limit distribution of $\sqrt{n}\hat{\mu}$ is thus not normal; it is a mixed distribution.

Example 16.3 Consider the case where $X_1, X_2, \ldots, X_n \stackrel{\text{iid}}{\sim} N(\mu, \sigma^2)$ with μ known to be an integer. For the argument below, existence of an MLE of μ is implicitly assumed, but this can be directly proved by considering tail behavior of the likelihood function $l(\mu, \sigma^2)$.

Let $\hat{\mu}$ = MLE of μ. Then, by standard calculus, $\hat{\sigma}^2 = \frac{1}{n}\sum(X_i - \hat{\mu})^2$ is the MLE of σ^2.

Consider, for integer μ, the ratio

$$\begin{aligned} \frac{l(\mu, \sigma^2)}{l(\mu - 1, \sigma^2)} &= e^{\frac{1}{2\sigma^2}\left\{\sum(x_i - \mu + 1)^2 - \sum(x_i - \mu)^2\right\}} \\ &= e^{\frac{1}{2\sigma^2}\{n + 2\sum(x_i - \mu)\}} \\ &= e^{\frac{n}{2\sigma^2} + \frac{2n(\bar{X} - \mu)}{2\sigma^2}} \\ &= e^{\frac{n}{2\sigma^2}\{2(\bar{X} - \mu) + 1\}} \\ &\ge 1 \end{aligned}$$

iff $2(\bar{X} - \mu) + 1 \ge 0$ iff $\mu \le \bar{X} + \frac{1}{2}$. In the interval $(\bar{X} - \frac{1}{2}, \bar{X} + \frac{1}{2}]$, there is a unique integer. It is the integer closest to $\bar{X}$. This is the MLE of μ. Now let us look at the asymptotic behavior of the MLE $\hat{\mu}$.

$$
\begin{aligned}
P(\hat{\mu} \neq \mu) &= P(\text{integer closest to } \bar{X} \text{ is} \neq \mu) \\
&= P\left(\bar{X} > \mu + \frac{1}{2}\right) + P\left(\bar{X} < \mu - \frac{1}{2}\right) \\
&= 2P\left(\bar{X} > \mu + \frac{1}{2}\right) \\
&= 2P\left(\frac{\sqrt{n}(\bar{X} - \mu)}{\sigma} > \frac{\sqrt{n}}{2\sigma}\right) \\
&= 2\left(1 - \Phi(\frac{\sqrt{n}}{2\sigma})\right) \\
&\sim 2\phi\left(\frac{\sqrt{n}}{2\sigma}\right)\frac{2\sigma}{\sqrt{n}} \\
&= \frac{4\sigma}{\sqrt{2\pi n}} e^{-\frac{n}{8\sigma^2}}.
\end{aligned}
$$

For any $c > 0$, $\sum_n \frac{e^{-cn}}{\sqrt{n}} < \infty$. Therefore, $\sum_n P(\hat{\mu} \neq \mu) < \infty$ and so, by the Borel-Cantelli lemma, $\hat{\mu} = \mu$ a.s. for all large n. Thus there is no asymptotic distribution of $\hat{\mu}$ in the usual sense.

Example 16.4 We do not need a closed-form formula for figuring out the asymptotic behavior of MLEs. In smooth regular problems, MLEs will be jointly asymptotically normal with a $\sqrt{n}$-norming. Suppose $X_1, X_2, \ldots, X_n \stackrel{\text{iid}}{\sim} \Gamma(\alpha, \lambda)$ with density $\frac{e^{-\lambda x} x^{\alpha-1} \lambda^{\alpha}}{\Gamma(\alpha)}$. Then the likelihood function is

$$
l(\alpha, \lambda) = \frac{e^{-\lambda \sum x_i} (\prod x_i)^{\alpha} \lambda^{n\alpha}}{(\Gamma(\alpha))^n}, \quad \alpha, \lambda > 0,
$$

so

$$
L = \log l(\mu, \sigma) = \alpha \log P - \lambda \sum x_i + n\alpha \log \lambda - n \log \Gamma(\alpha),
$$

where $P = \prod x_i$.

The likelihood equations are

$$
0 = \frac{\partial L}{\partial \alpha} = \log P + n \log \lambda - n\Psi(\alpha),
$$

$$
0 = \frac{\partial L}{\partial \lambda} = -\sum x_i + \frac{n\alpha}{\lambda},
$$

where $\Psi(\alpha) = \frac{\Gamma'(\alpha)}{\Gamma(\alpha)}$ is the di-Gamma function. From solving $\frac{\partial L}{\partial \lambda} = 0$, one gets $\hat{\lambda} = \frac{\hat{\alpha}}{\bar{X}}$, where $\hat{\alpha}$ is the MLE of α. Existence of MLEs of $\hat{\alpha}, \hat{\lambda}$ can be directly concluded from the behavior of $l(\alpha, \lambda)$. Using $\hat{\lambda} = \frac{\hat{\alpha}}{\bar{X}}$, $\hat{\alpha}$ satisfies

$$\log P + n \log \hat{\alpha} - n \log \bar{X} - n\Psi(\hat{\alpha}) = \log P - n \log \bar{X} - n(\Psi(\hat{\alpha}) - \log \hat{\alpha}) = 0.$$

The function $\Psi(\alpha) - \log \alpha$ is strictly monotone and continuous with range $\supset (-\infty, 0)$, so there is a unique $\hat{\alpha} > 0$ at which $n(\Psi(\hat{\alpha}) - \log \hat{\alpha}) = \log P - n \log \bar{X}$. This is the MLE of α. It can be found only numerically, and yet, from general theory, one can assert that $\sqrt{n}(\hat{\alpha} - \alpha, \hat{\lambda} - \lambda) \xrightarrow{\mathcal{L}} N(\underset{\sim}{0}, \Sigma)$ for some covariance matrix Σ.

Example 16.5 In nonregular problems, the MLE is not asymptotically normal and the norming constant is usually not $\sqrt{n}$. For example, if $X_1, X_2, \ldots, X_n \overset{\text{iid}}{\sim} U[0, \theta]$, then the MLE $\hat{\theta} = X_{(n)}$ satisfies $n(\theta - \hat{\theta}) \xrightarrow{\mathcal{L}} \text{Exp}(\theta)$.

Example 16.6 This example shows that MLEs need not be functions of a minimal sufficient statistic. Suppose $X_1, X_2, \ldots, X_n \overset{\text{iid}}{\sim} U[\mu - \frac{1}{2}, \mu + \frac{1}{2}]$. Then the likelihood function is

$$l(\mu) = I_{\mu - \frac{1}{2} \leq X_{(1)} \leq X_{(n)} \leq \mu + \frac{1}{2}} = I_{X_{(n)} - \frac{1}{2} \leq \mu \leq X_{(1)} + \frac{1}{2}},$$

so any function of $X_1, \ldots, X_n$ that is in the interval $[X_{(n)} - \frac{1}{2}, X_{(1)} + \frac{1}{2}]$ is an MLE; e.g., $e^{-\bar{X}^2}(X_{(n)} - \frac{1}{2}) + (1 - e^{-\bar{X}^2})(X_{(1)} + \frac{1}{2})$ is an MLE, but it is not a function of $(X_{(1)}, X_{(n)})$, the minimal sufficient statistic.

Example 16.7 This example shows that MLEs of different parameters can have limit distributions with different norming rates.

Suppose $X_1, X_2, \ldots, X_n \overset{\text{iid}}{\sim} \text{Exp}(\mu, \sigma)$ with density $\frac{1}{\sigma} e^{-\frac{x-\mu}{\sigma}}$, $x \geq \mu$. By simple calculus, the MLEs $\hat{\mu}, \hat{\sigma}$ are

$$\hat{\mu} = X_{(1)},$$

$$\hat{\sigma} = \frac{1}{n} \sum (X_i - X_{(1)}) = \bar{X} - X_{(1)}.$$

We can assume $\mu = 0$ and $\sigma = 1$ for the following calculations, from which the case of general μ, σ follows.

If $\mu = 0, \sigma = 1$, then $nX_{(1)} \sim \text{Exp}(1)$. Thus, for the general case, $n(\hat{\mu}-\mu) \xrightarrow{\mathcal{L}} \text{Exp}(\sigma)$. On the other hand, if $\mu = 0, \sigma = 1$, then $\sqrt{n}(\hat{\sigma}-1) = \sqrt{n}(\bar{X}-X_{(1)}-1) = \sqrt{n}(\bar{X}-1)-\sqrt{n}X_{(1)} = \sqrt{n}(\bar{X}-1)-\frac{nX_{(1)}}{\sqrt{n}} \xrightarrow{\mathcal{L}} N(0,1)$ by the CLT and Slutsky's theorem. Thus, for the general case, $\sqrt{n}(\hat{\sigma}-\sigma) \xrightarrow{\mathcal{L}} N(0,\sigma^2)$. Note the different norming constants for $\hat{\mu}, \hat{\sigma}$.

16.2 Inconsistent MLEs

The first example of an MLE being inconsistent was provided by Neyman and Scott (1948). It is by now a classic example and is known as the Neyman-Scott example. That first example shocked everyone at the time and sparked a flurry of new examples of inconsistent MLEs, including those offered by LeCam (1953) and Basu (1955). Note that what makes the Neyman-Scott example work is that, compared with the number of parameters, there aren't enough data to kill the bias of the MLE. It is possible to find adjustments to the MLE or suitable Bayesian estimates in many of these problems that do have the consistency property; see Ghosh (1994) for examples and also some general techniques.

Example 16.8 Let X_{ij} For $i = 1, 2, \ldots, n$ and $j = 1, 2, \ldots, k$ be independent with $X_{ij} \sim N(\mu_i, \sigma^2)$. Note that this is basically a balanced one-way ANOVA design where we assume k is fixed and $n \to \infty$, so the sample sizes of the groups are (probably) big, but the number of groups is bigger. We want to estimate the common variance of the groups. By routine calculus, the MLEs are

$$\hat{\mu}_i = \bar{X}_i \text{ and } \hat{\sigma}^2 = \frac{1}{nk}\sum_{i=1}^{n}\sum_{j=1}^{k}(X_{ij} - \bar{X}_i)^2.$$

It is the MLE of σ^2 that is inconsistent. Indeed,

$$\hat{\sigma}^2 = \frac{1}{nk}\sum_{i=1}^{n}\sum_{j=1}^{k}(X_{ij} - \bar{X}_i)^2 = \frac{1}{n}\frac{1}{k}\sum_{i=1}^{n}\left(\sum_{j=1}^{k}(X_{ij} - \bar{X}_i)^2\right) = \frac{1}{n}\frac{1}{k}\sum_{i=1}^{n}\sigma^2 W_i,$$

where the W_i are independent χ^2_{k-1}. By the WLLN,

$$\frac{\sigma^2}{k}\frac{1}{n}\sum_{i=1}^{n} W_i \xrightarrow{\mathcal{P}} \frac{\sigma^2}{k}(k-1).$$

Hence, the MLE for σ^2 does not converge to σ^2! It is the bias that is making the estimate inconsistent. If we kill the bias by multiplying by $\frac{k}{k-1}$, the new estimator is consistent; i.e., if we "adjust" the MLE and use

$$\frac{1}{n(k-1)}\sum_{i=1}^{n}\sum_{j=1}^{k}(X_{ij}-\bar{X}_i)^2,$$

then we return to consistency. In these sorts of problems, where the number of observations and the number of free parameters grow at the same rate, maximum likelihood often runs into problems. However, these problems are hard for any school of thought.

16.3 MLEs in the Exponential Family

It is part of the statistical folklore that MLEs cannot be beaten asymptotically. One needs to be careful in making such a statement. Under various conditions, MLEs are indeed asymptotically optimal and asymptotically normal. But it is important to remember that the conditions needed are *not* just on the probability model; there must also be conditions imposed on the competing estimates for optimality.

There are other potential problems. Maximum likelihood estimates may not exist for all samples. In such cases, one can only talk about asymptotic behavior and optimality of estimates that are quasi-MLEs. Careful exposition of the technical issues and proofs may be seen in Perlman (1983), Bickel and Doksum (2001), Lehmann and Casella (1998), and Brown (1986). Computing the MLE can also be a difficult numerical exercise in general; the EM algorithm is a popular tool for this; see McLachlan and Krishnan (1997).

We start with a familiar model, namely exponential families; things are relatively uncomplicated in this case. For the sake of completeness, we state the definition and a few basic facts about the exponential family.

Definition 16.1 Let $f(x|\theta) = e^{\theta T(x)-\psi(\theta)}h(x)d\mu(x)$, where μ is a positive σ-finite measure on the real line and $\theta \in \Theta = \{\theta : \int e^{\theta T(x)}h(x)d\mu(x) < \infty\}$. Then, f is said to belong to the one-parameter exponential family with natural parameter space Θ. The parameter θ is called the natural parameter of f.

The following are some standard facts about a density in the one-parameter exponential family.

Proposition 16.1 (a) *For $\theta \in \Theta^0$, the interior of Θ, all moments of $T(X)$ exist, and $\psi(\theta)$ is infinitely differentiable at any such θ. Furthermore, $E_\theta(T) = \psi'(\theta)$ and $\text{Var}_\theta(T) = \psi''(\theta)$.*

(b) *Given an iid sample of size n from f, $\sum_{i=1}^n T(X_i)$ is minimally sufficient.*

(c) *The Fisher information function exists, is finite at all $\theta \in \Theta^0$, and equals $I(\theta) = \psi''(\theta)$.*

(d) *The following families of distributions belong to the one-parameter exponential family:*

$$N(\mu, 1), N(0, \sigma^2), \text{Ber}(p), \text{Bin}(n, p), n \text{ fixed},$$

$$\text{Poi }(\mu), \text{Geo }(p), \text{Exp }(\lambda), \text{Gamma }(\alpha, \lambda), \alpha \text{ fixed}, \text{Gamma }(\alpha, \lambda), \ \lambda \text{ fixed}.$$

Theorem 16.1 Let $X_1, X_2, \ldots, X_n \overset{\text{iid}}{\sim} f(x|\theta) = e^{\theta T(x)-\psi(\theta)}h(x)d\mu(x)$. Let the true $\theta = \theta_0 \in \Theta^o$ (i.e., the interior of the natural parameter space). Assume $\psi''(\theta) > 0 \ \forall \theta \in \Theta^o$. Then, for all large n, w.p. 1, a unique MLE of θ exists, is consistent, and is asymptotically normal.

Proof. The likelihood function is

$$l(\theta) = e^{\theta \Sigma T(x_i) - n\psi(\theta)},$$

$$\Rightarrow L(\theta) = \log l(\theta) = n[\theta \bar{T} - \psi(\theta)].$$

Therefore, the likelihood equation is

$$L'(\theta) = n[\overline{T} - \psi'(\theta)] = 0 \Leftrightarrow \overline{T} - E_\theta(T(X_1)) = 0.$$

Now $T(X_1)$ has a finite mean and hence, by the SLLN,

$$\overline{T} \xrightarrow{\text{a.s.}} E_{\theta_0}T(X_1) = \psi'(\theta_0).$$

Hence, for all large n, w.p. 1, $\overline{T}$ is in the interior of the range of the function $\theta \to \psi(\theta)$. On the other hand, $E_\theta(T(X)) = \psi'(\theta)$ is a strictly monotone increasing function of θ because $\psi''(\theta) = \text{Var}_\theta(T(X)) > 0$. Therefore, for all large n, w.p. 1, there exists a unique θ such that $E_\theta T(X) = \overline{T}$. This is the MLE of θ and is characterized as the unique root of $\psi'(\theta) = \overline{T} \Leftrightarrow \hat{\theta} = (\psi')^{-1}(\overline{T})$.

By the continuous mapping theorem, $\hat{\theta} \xrightarrow[\mathcal{P}_{\theta_0}]{\text{a.s.}} \theta_0$. By the central limit theorem, $\overline{T}$ is asymptotically normal. By the delta theorem, a smooth function of an asymptotically normal sequence is also asymptotically normal. Indeed, since $\sqrt{n}(\bar{T} - \psi'(\theta_0)) \stackrel{\mathcal{L}}{\Rightarrow} N(0, \psi''(\theta_0))$, and since $\hat{\theta} = (\psi')^{-1}(\overline{T})$, a direct application of the delta theorem implies that

$$\sqrt{n}(\hat{\theta} - \theta_0) \xrightarrow[\mathcal{P}_{\theta_0}]{\mathcal{L}} N(0, I^{-1}(\theta_0)),$$

where $I(\theta_0) = \psi''(\theta_0)$.

Remark. So this is a success story for the MLE: strong consistency and asymptotic normality hold. Nevertheless, even in this successful case, it is not true that this estimate gives the uniformly best limit distribution. It is possible to find a competing estimate, $\hat{\hat{\theta}}$, that converges to some other limit distribution that has a smaller variance for some particular θ. We will discuss these important subtleties in Section 16.7.

16.4 More General Cases and Asymptotic Normality

In general, we may have problems with the existence of the MLE, even for large samples. What can we get in such cases? We will need a laundry list of assumptions. If we are satisfied with something that is merely consistent, the list is shorter. If we want something that is also asymptotically normal, then the list of assumptions gets longer. This list has come to be known as the Cramér-Rao conditions; see Lehmann and Casella (1998) and Lehmann (1999) for a proof of the next theorem.

Theorem 16.2 (Cramér-Rao conditions) Assume $X_1, X_2, \ldots, X_n \stackrel{\text{iid}}{\sim} P_\theta$ and $\frac{dP_\theta}{d\mu} = f(x|\theta)$ for some σ-finite μ (e.g., Lebesgue measure in the continuous case or counting measure in the discrete case). Assume the following conditions:

(A1) Identifiability; i.e., $P_{\theta_1} = P_{\theta_2} \Leftrightarrow \theta_1 = \theta_2$.

(A2) $\theta \in \Theta =$ an open interval in the real line.

(A3) $S = \{x : f(x|\theta) > 0\}$ is free of θ.

(A4) $\forall x \in S$, $\frac{d}{d\theta} f(x|\theta)$ exists; i.e., the likelihood function is smooth as a function of the parameter.

Let $\theta_0 \in \Theta^0$ be the true value of θ. Then there exists a sequence of functions $\hat{\theta_n} = \hat{\theta_n}(X_1, \ldots, X_n)$ such that:

(i) $\hat{\theta}_n$ is a root of the likelihood equation $L'(\theta) = 0$ for all large n, where $L(\theta)$ is $\log l(\theta) = \Sigma \log f(x_i|\theta)$.

(ii) P_{θ_0}(the root $\hat{\theta}_n$ is a local maximum of $l(\theta)$) $\longrightarrow 1$ as $n \longrightarrow \infty$.

(iii) $\hat{\theta_n} \xrightarrow{P_{\theta_0}} \theta_0$.

Remark. This theorem does not say which sequence of roots of $L'(\theta) = 0$ should be chosen to ensure consistency in the case of multiple roots. It does not even guarentee that for any given n, however large, the likelihood function $l(\theta)$ has any local maxima at all. This specific theorem is useful in *only* those cases where $L'(\theta) = 0$ has a *unique* root for all n.

Since consistency is regarded as a weak positive property, and since in statistics one usually wants to make actual inferences such as confidence interval construction, it is important to have weak convergence results in addition to consistency. As we remarked earlier, establishing weak convergence results requires more conditions.

The issues and the results in the multiparameter case are analogous to those in the one-parameter case. As in the one-parameter case, in general one can only assert consistency and asymptotic normality of suitable sequences of roots of the likelihood equation. We state here the asymptotic normality result directly for the multiparameter case, from which the one-parameter case follows as a special case. For a complete list of the regularity conditions needed to prove the following theorem and also for a proof, see Lehmann and Casella (1998). We refer to the list of all assumptions as *multiparameter Cramér-Rao conditions for asymptotic normality*. A problem in which these conditions are all satisfied is usually called a *regular parametric problem*.

Theorem 16.3 Under the multiparameter Cramér-Rao conditions for asymptotic normality, there exists a sequence of roots of the likelihood equation that is consistent and satisfies $\sqrt{n}(\hat{\theta}_n - \theta_0) \xrightarrow{\mathcal{L}} N(0, I^{-1}(\theta_0))$, where $I(\theta) = ((I_{ij}(\theta)))$ with $I_{ij}(\theta) = -E_\theta \left[\frac{\partial^2}{\partial\theta_i \partial\theta_j} \log f(X|\theta) \right]$ is the Fisher information matrix.

Remark. As an example of what the conditions are, one of the conditions for Theorem 16.3 is that $\frac{\partial^3}{\partial\theta_i\partial\theta_j\partial\theta_k}\log f(x|\theta)$ exists for all x in $S = \{x : f(x|\theta) > 0\}$ and

$$\frac{\partial^3}{\partial\theta_i\partial\theta_j\partial\theta_k}\int_S \log f(x|\theta)dx = \int_S \left\{\frac{\partial^3}{\partial\theta_i\partial\theta_j\partial\theta_k}\log f(x|\theta)\right\} dx.$$

Remark. Theorem 16.3 applies to any distribution in the exponential family for which $\psi''(\theta)$ is positive and finite for every θ in the interior of the natural parameter space. There are also *multiparameter exponential families*, very similar to the one-parameter version, for which the theorem holds, but we will not treat them here.

16.5 Observed and Expected Fisher Information

Consider the regular one-parameter problem with iid observations from a density $f(x|\theta)$ w.r.t. some dominating measure μ. According to Theorem 16.3, the "MLE" $\hat{\theta}_n$ is asymptotically normal with mean θ and variance $\frac{1}{nI(\theta)}$, where $I(\theta) = -E_\theta(\frac{\partial^2}{\partial\theta^2}\log f(X|\theta))$. Since the observations X_i are iid, by Kolmogorov's SLLN, the average $\frac{1}{n}\sum_{i=1}^n -\frac{\partial^2}{\partial\theta^2}\log f(X_i|\theta) \stackrel{\text{a.s.}}{\Rightarrow} I(\theta)$. Thus, as a matter of providing an estimate of the variance of the MLE, it is very reasonable to provide the estimate $\frac{1}{\sum_{i=1}^n -\frac{\partial^2}{\partial\theta^2}\log f(X_i|\theta)|_{\theta=\hat{\theta}}}$, where $\hat{\theta}$ is the MLE of θ. The quantity $\frac{1}{n}\sum_{i=1}^n -\frac{\partial^2}{\partial\theta^2}\log f(X_i|\theta)$ is called the *observed Fisher information.* Its expectation, which is just the Fisher information function $I(\theta)$, is called the *expected Fisher information.* It is natural to ask whether $\frac{1}{nI(\hat{\theta})}$ or $\frac{1}{\sum_{i=1}^n -\frac{\partial^2}{\partial\theta^2}\log f(X_i|\theta)|_{\theta=\hat{\theta}}}$ gives a better estimate of the true variance of the MLE.

We present two examples to help understand this question.

Example 16.9 Suppose $X_1, \ldots, X_n$ are iid from a distribution in the one-parameter exponential family with density $f(x|\theta) = e^{\theta T(x)-\psi(\theta)}h(x)(d\mu)$. Then, $\frac{\partial^2}{\partial\theta^2}\log f(x|\theta) = -\psi''(\theta)$. Thus, $I(\theta)$ and $\frac{1}{n}\sum_{i=1}^n -\frac{\partial^2}{\partial\theta^2}\log f(X_i|\theta)$ are both equal to $\psi''(\theta)$, and so use of the observed or the expected Fisher information leads to the same estimate for the variance of $\hat{\theta}_n$.

Example 16.10 Suppose $X_1, \ldots, X_n$ are iid from the Cauchy distribution $C(\theta, 1)$. Then, $f(x|\theta) = \frac{1}{\pi(1+(x-\theta)^2)}$ and $\frac{\partial^2}{\partial\theta^2}\log f(x|\theta) = \frac{2((x-\theta)^2-1)}{(1+(x-\theta)^2)^2}$. On doing the necessary integration, $I(\theta) = \frac{1}{2}$. Thus the estimate of the variance of the MLE based on the expected information is $\frac{2}{n}$. However, it is clear that the observed information method would produce an estimated variance that depends on the actual observed data. Over repeated sampling, it will have typically an asymptotically normal distribution itself, but its performance as an estimate of the true variance relative to the constant estimate $\frac{2}{n}$ can be accurately understood only by careful simulation.

Some interesting facts are revealed by a simulation. For $n = 20$ and the true θ value equal to 0, a simulation of size 500 was conducted to inquire into the performance of the variance estimates discussed above. The estimate based on the expected Fisher information is $\frac{2}{n} = .1$. The true variance of the MLE when $n = 20$ is .1225 according to the simulation. Thus the expected Fisher information method produces an underestimate of the variance by about 16%. The variance estimate produced by the observed information method gives an average estimate of .1071 over the 500 simulations. Thus, the bias is significantly lower. However, the variability in the variance estimate over the simulations is high. While the smallest variance estimate produced by the observed information method is .0443, the largest one is .9014. The heaviness of the Cauchy tail impacts the variance of the variance estimate as well. The estimate produced by the observed information method has a smaller bias than the one based on expected information but can go wild from time to time and is perhaps risky. The expected information estimate, on the other hand, is just a constant estimate $\frac{2}{n}$ and is not prone to fluctuations caused by a whimsical Cauchy sample. This example illustrates the care needed in assessing the accuracy of maximum likelihood estimates; the problem is harder than it is commonly believed to be.

16.6 Edgeworth Expansions for MLEs

The central limit theorem gives a first-order approximation to the distribution of the MLE under regularity conditions. More accurate approximations can be obtained by Edgeworth expansions of higher order. In the exponential family, where the MLE is a linear statistic, the expansion is a bit easier to state. For more general regular densities, the assumptions are complex and many, and the expansion itself is notationally more messy. References for these expansions are Pfanzagl (1973), Bhattacharya and Ghosh (1978), and Bai and Rao (1991). We present the expansions in two cases below, namely the case of the exponential family and a more general regular case. Of these, in the exponential family, the MLE of the mean function is the sample mean itself, and so an Edgeworth expansion follows from general expansions for sample means as in Chapter 13; a specific reference for the next theorem is Pfanzagl (1973).

Theorem 16.4 Let $X_1, X_2, \ldots, X_n \stackrel{\text{iid}}{\sim} f_\theta(x) = e^{\theta x - \psi(\theta)} h(x)dx$. Consider estimation of $E_\theta(X) = \psi(\theta)$, and let $F_n(x) = P_\theta\left(\frac{\sqrt{n}(\bar{X} - \psi'(\theta))}{\sqrt{\psi''(\theta)}} \leq x\right)$. Then,

$$F_n(x) = \Phi(x) + \frac{p_1(x,\theta)\phi(x)}{\sqrt{n}} + \frac{p_2(x,\theta)\phi(x)}{n} + O(n^{-3/2})$$

uniformly in x, where

$$p_1(x,\theta) = c_1(1 - x^2),\ p_2(x,\theta) = c_2(3x - x^3) + \frac{c_1^2}{72}(10x^3 - 15x - x^5)$$

with $c_1 = \frac{\psi^{(3)}(\theta)}{6(\psi''(\theta))^{3/2}}$, $c_2 = \frac{\psi^{(4)}(\theta)}{24(\psi''(\theta))^2}$.

Example 16.11 Suppose $X_1, X_2, \ldots, X_n \stackrel{\text{iid}}{\sim} f_\theta(x) = \theta e^{\theta x}, x < 0, \theta > 0$. Then, $\psi(\theta) = -\log\theta$, $c_1 = -\frac{1}{3}$, $c_2 = \frac{1}{4}$. Thus, the MLE $\bar{X}$ of $E_\theta(X) = -\frac{1}{\theta}$ satisfies the expansion

$$P_\theta\left(\sqrt{n}\left(\bar{X} + \frac{1}{\theta}\right) \leq \frac{x}{\theta}\right) = \Phi(x) - \frac{(1-x^2)\phi(x)}{3\sqrt{n}} + \frac{\left(\frac{3x - x^3}{4} + \frac{10x^3 - 15x - x^5}{648}\right)\phi(x)}{n} + O(n^{-3/2})$$

uniformly in x. For ease of reference, we will denote the two-term expansion by $H(n,x)$. As a test of the expansion's numerical accuracy, suppose the true $\theta = 1$ and we want to approximate $P(\sqrt{n}(\bar{X} + \frac{1}{\theta}) \leq \frac{x}{\theta})$. Since $-\sum_{i=1}^n X_i$ is Gamma with shape parameter n and scale parameter 1, on computation one finds the following exact values and approximations obtained from the two-term expansion above using $n = 30$:

$$x = .5; \text{exact} = .675; H(n,x) = .679;$$
$$x = 2.0; \text{exact} = .988; H(n,x) = .986;$$
$$x = 3.0; \text{exact} = 1.000; H(n,x) = 1.0001.$$

Thus, the expansion is quite accurate at the sample size of $n = 30$. This example brings out an undesirable feature of Edgeworth expansions, that they are not CDFs and can take values < 0 or > 1 as it does here when $x = 3.0$.

A general Edgeworth expansion for the MLE under a variety of regularity conditions is given in Pfanzagl (1973). The conditions are too many to state here. However, the expansion is explicit. We give the expansion below. Pfanzagl (1973) gives examples of families of densities that satisfy the conditions required for the validity of his theorem. We first need some more notation.

For a given density $f_\theta(x)$, let

$$l_\theta = \log f_\theta(x), \dot{l}_\theta = \frac{\partial}{\partial\theta} l_\theta, \ddot{l}_\theta = \frac{\partial^2}{\partial\theta^2} l_\theta, l_\theta^{(3)} = \frac{\partial^3}{\partial\theta^3} l_\theta,$$

$$\rho_{20} = E_\theta[\dot{l}_\theta]^2, \rho_{11} = -E_\theta[\ddot{l}_\theta], \rho_{30} = -E_\theta[\dot{l}_\theta]^3,$$
$$\rho_{12} = -E_\theta[l_\theta^{(3)}], \rho_{21} = 2E_\theta[\dot{l}_\theta \ddot{l}_\theta].$$

With this notation, we have the following expansion for the CDF of the MLE; for notational simplicity, we present only the one-term expansion in the general case.

Theorem 16.5 Let $X_1, X_2, \ldots, X_n \overset{\text{iid}}{\sim} f_\theta(x)$. Under the regularity conditions on f_θ as in Pfanzagl (1973), the MLE $\hat{\theta}_n$ of θ satisfies $F_n(x, \theta) = P_\theta(\frac{\sqrt{n}(\hat{\theta}_n - \theta)}{\beta} \leq x) = \Phi(x) + \frac{q_1(x,\theta)\phi(x)}{\sqrt{n}} + o(\frac{1}{\sqrt{n}})$ uniformly in x and uniformly in compact neighborhoods of the given θ, where $q_1(x, \theta) = a_{10} + a_{11}x^2$, with $a_{10} = -\frac{\rho_{30}}{6\rho_{20}^{3/2}}, a_{11} = -a_{10} + \frac{\rho_{12}\sqrt{\rho_{20}}}{2\rho_{11}^2} - \frac{\rho_{21}}{2\sqrt{\rho_{20}}\rho_{11}}, \beta = \frac{\sqrt{\rho_{20}}}{\rho_{11}}$.

Example 16.12 Consider maximum likelihood estimation of the location parameter of a Cauchy distribution. The regularity conditions needed for an application of Theorem 16.5 are met; see Pfanzagl (1973). We have $f_\theta(x) = \frac{1}{\pi(1+(x-\theta)^2)}$. Because of the fact that the density is symmetric, the coefficients $\rho_{12}, \rho_{21}, \rho_{30}$ (i.e., those whose subscripts add to an odd integer) are all zero. Therefore, $a_{10} = a_{11} = 0$, and so it follows that $F_n(x, \theta) = \Phi(x) + o(\frac{1}{\sqrt{n}})$ uniformly in x; i.e., the CLT approximation is *second-order accurate*. This is interesting and is a consequence of the symmetry.

16.7 Asymptotic Optimality of the MLE and Superefficiency

It was first believed as folklore that the MLE under regularity conditions on the underlying distribution is asymptotically the best for every value of θ; i.e., if an MLE $\hat{\theta}_n$ exists and $\sqrt{n}(\hat{\theta}_n - \theta) \xrightarrow[P_\theta]{\mathcal{L}} N(0, I^{-1}(\theta))$, and if another competing sequence T_n satisfies $\sqrt{n}(T_n - \theta) \xrightarrow[P_\theta]{\mathcal{L}} N(0, V(\theta))$, then for every

θ, $V(\theta) \geq \frac{1}{I(\theta)}$. It was a major shock when in 1952 Hodges gave an example that destroyed this belief and proved it to be false even in the normal case. Hodges, in a private communication to LeCam, produced an estimate T_n that beats the MLE $\overline{X}$ locally at some θ, say $\theta = 0$. The example can be easily refined to produce estimates T_n that beat $\overline{X}$ at any given finite set of values of θ. Later, in a very insightful result, LeCam (1953) showed that this can happen only on Lebesgue-null sets of θ. If, in addition, we insist on using only such estimates T_n that have a certain smoothness property (to be made precise later), then the inequality $V(\theta) < \frac{1}{I(\theta)}$ cannot materialize at all. So, to justify the folklore that MLEs are asymptotically the best, one not only needs regularity conditions on $f(x|\theta)$ but must also restrict attention to only those estimates that are adequately nice (and Hodges' estimate is not). An excellent reference for this topic is van der Vaart (1998).

Example 16.13 Let $X_1, X_2, \ldots, X_n \overset{\text{iid}}{\sim} N(\theta, 1)$. Define an estimate T_n as

$$T_n = \begin{cases} \bar{X} & \text{if } \left|\bar{X}\right| > n^{-\frac{1}{4}} \\ a\bar{X} & \text{if } \left|\bar{X}\right| \leq n^{-\frac{1}{4}} \end{cases},$$

where $0 \leq a < 1$.
To derive the limiting distribution of T_n, notice that

$$P_\theta(\sqrt{n}\,|T_n - \theta| \leq c)$$

$$= P_\theta(\sqrt{n}\,|T_n - \theta| \leq c, \left|\overline{X}\right| \leq n^{-\frac{1}{4}}) + P_\theta(\sqrt{n}\,|T_n - \theta| \leq c, \left|\overline{X}\right| > n^{-\frac{1}{4}}).$$

If $\theta = 0$, then the second term goes to zero and so the limit distribution is determined from the first term. For $\theta \neq 0$, the situation reverses. It follows that $\sqrt{n}(T_n - \theta) \xrightarrow[\mathcal{P}_\theta]{\mathcal{L}} N(0, 1)$ if $\theta \neq 0$ and $\xrightarrow[\mathcal{P}_\theta]{\mathcal{L}} N(0, a^2)$ if $\theta = 0$. Thus, if we denote by $V(\theta)$ the asymptotic variance of T_n, then $V(\theta) = \frac{1}{I(\theta)}$ for $\theta \neq 0$, and $V(\theta) = \frac{a^2}{I(\theta)}$ for $\theta = 0$. Therefore, $V(\theta) \leq \frac{1}{I(\theta)}$ for every θ, and $V(\theta) < \frac{1}{I(\theta)}$ at $\theta = 0$.

Remark. The Hodges estimate T_n is what we call a *shrinkage estimate* these days. Because $V(\theta) \leq \frac{1}{I(\theta)}\ \forall\theta$ and $V(\theta) < \frac{1}{I(\theta)}$ at $\theta = 0$, the asymptotic relative efficiency (ARE) of T_n with respect to $\overline{X}$ is $\geq 1, \forall\theta$ and > 1 when $\theta = 0$. Such estimates, which have a smaller asymptotic variance than the MLE locally at some θ and never a larger asymptotic variance at any θ, are called *superefficient*.

It is clear, however, that T_n has certain undesirable features. First, as a function of $X_1, ..., X_n$, T_n is not smooth. Second, $V(\theta)$ is not continuous in θ. However, what transpires is that something is very seriously wrong with T_n. The mean squared error of T_n behaves erratically. For a given n as a function of θ, $nE_\theta(T_n - \theta)^2$ sharply leaps over $nE(\overline{X} - \theta)^2 = 1$ for values of $\theta \approx 0$. The values of θ at which this occurs change with n. At any given θ, the leaps vanish for large n. But, for any n, the leaps reoccur at other values of θ close to 0. Thus, the superefficiency of T_n is being purchased at the cost of a sharp spike in the mean squared error at values of θ very close to 0. DasGupta (2004) shows that

$$\liminf_{n\to\infty} \sup_{|\theta| \le n^{-\frac{1}{4}}} \sqrt{n} E_\theta (T_n - \theta)^2 \ge \frac{1}{2}.$$

Notice the $\sqrt{n}$-norming in the result as opposed to the norming by n for the equalizer minimax estimate $\bar{X}$.

16.8 Hajek-LeCam Convolution Theorem

The superefficiency phenomenon, it turns out, can only happen on Lebesgue-null subsets of Θ. It cannot happen at all if furthermore attention is restricted to estimators that are distributionally smooth in the following sense.

Definition 16.2 Let T_n be an estimate sequence for a vector function $\psi(\theta)$ such that $\sqrt{n}(T_n - \psi(\theta)) \xrightarrow[\mathcal{P}_\theta]{\mathcal{L}} \mu_\theta$. T_n is called a regular estimating sequence if, for all finite h,

$$\sqrt{n}\left(T_n - \psi\left(\theta + \frac{h}{\sqrt{n}}\right)\right) \xrightarrow[\mathcal{P}_{\theta+\frac{h}{\sqrt{n}}}]{\mathcal{L}} \mu_\theta.$$

Remark. Thus, for a regular estimating sequence T_n, changing the parameter ever so slightly would not change the limit distribution at all. Among such estimates, we cannot find one that is superefficient.

Theorem 16.6 Suppose $X_1, X_2, \ldots, X_n \overset{\text{iid}}{\sim} P_\theta \ll \mu, \theta \in \Theta$. Suppose $f(x|\theta) = \frac{dP_\theta}{d\mu}(x) > 0$ for every x, θ, and $\nabla_\theta f(x|\theta)$ exists for every x, θ. Suppose also that

$$0 < I_{ij}(\theta) = E_\theta \left[-\frac{\partial^2}{\partial\theta_i \partial\theta_j} \log f(X|\theta) \right] < \infty$$

for every θ, is continuous at every θ, and $I^{-1}(\theta)$ exists for every θ. Let $\psi(\theta)$ be any differentiable function of θ with gradient vector $\nabla\psi(\theta)$. Let T_n be any regular estimate sequence of $\psi(\theta)$ with $\sqrt{n}(T_n - \psi(\theta)) \xrightarrow[\mathcal{P}_\theta]{\mathcal{L}} \mu_\theta$. Then there exists a (unique) probability distribution ν_θ such that μ_θ admits the convolution representation $\mu_\theta = N(0, (\nabla\psi)I^{-1}(\theta)(\nabla\psi)') * \nu_\theta$. In particular, if μ_θ has a covariance matrix, say $\sum_\theta$, then $\sum_\theta \geq (\nabla\psi)I^{-1}(\theta)(\nabla\psi)'$ in the sense $\sum_\theta - (\nabla\psi)I^{-1}(\theta)(\nabla\psi)'$ is non negative definite (nnd) at every θ.

In the absence of the regularity of the estimate sequence T_n, we can assert something a bit weaker.

Theorem 16.7 Assume the conditions in the previous theorem on P_θ, $I(\theta)$, and $\psi(\theta)$. Suppose $\sqrt{n}(T_n - \psi(\theta)) \xrightarrow[\mathcal{P}_\theta]{\mathcal{L}} \mu_\theta$. Then, for almost all θ (Lebesgue), μ_θ admits the convolution representation $\mu_\theta = N(0, (\nabla\psi)I^{-1}(\theta)(\nabla\psi)') * \nu_\theta$.

Remark. These theorems are collectively known as the Hajek-LeCam convolution theorem. See van der Vaart (1998) for greater details and proofs. The second theorem says that even without regularity of the competing estimates T_n, superefficiency can occur only on sets of θ of Lebesgue measure 0. This result of Lucien LeCam is regarded as one of the most insightful results in theoretical statistics.

We give an example of a nonregular estimate for illustration.

Example 16.14 Let $X_1, X_2, \ldots, X_n \overset{\text{iid}}{\sim} N_p(\theta, I)$. The MLE of θ is $\overline{X}$. In 1961, James and Stein showed that $T_n = \left(1 - \frac{p-2}{n\|\overline{X}\|^2}\right)\overline{X}$ has a smaller mean squared error than $\overline{X}$ at every θ, provided $p \geq 3$; i.e., $E_\theta\|T_n - \theta\|^2 < E_\theta\|\overline{X} - \theta\|^2 = \frac{p}{n}, \forall\theta$. The James-Stein estimate (James and Stein (1961)) T_n has the property that $E_{\frac{h}{\sqrt{n}}}\|T_n - \frac{h}{\sqrt{n}}\|^2 < E_{\frac{h}{\sqrt{n}}}\|\overline{X} - \frac{h}{\sqrt{n}}\|^2, \forall h$. It follows that the limit distribution of $\sqrt{n}(\overline{X} - \frac{h}{\sqrt{n}})$ does not have a smaller covariance matrix than that of $\sqrt{n}(T_n - \frac{h}{\sqrt{n}})$. The James-Stein estimate does not have the property of regularity. And it is exactly at $\theta = 0$ that the estimate T_n is nonregular; i.e., $\sqrt{n}(T_n - \frac{h}{\sqrt{n}}) \xrightarrow[\mathcal{P}_{\frac{h}{\sqrt{n}}}]{\mathcal{L}} \mu_h$ for some distribution μ_h that really does depend on h. In fact, one can describe μ_h. It is the same as the distribution of $(1 - \frac{p-2}{\|Z\|^2})Z - h$, where $Z \sim N_p(0, I)$.

16.9 Loss of Information and Efron's Curvature

In exponential families, the maximum likelihood estimate based on n iid observations is itself a sufficient statistic. Since we think of sufficient statistics as capturing all the information about the parameter present in the sample, it would mean that the loss of information caused by summarizing the full data into the MLE is zero in exponential families. How does one formalize this question for nonexponential families and give a quantification of the loss of information suffered by the MLE and relate it to something of actual statistical relevance? Efforts to show that the maximum likelihood estimate leads to the least amount of information lost by a one-dimensional summary started with the seminal second-order efficiency theory of Rao (1961, 1962, 1963). More recently, Efron (1975) gave a theory of curvature of parametric families that attempts to connect the information loss question with how nonexponential a family is. The idea is that the more *curved* a parametric family is, the greater is the information loss suffered by the MLE. We present a few results in this direction below.

Definition 16.3 Let $P_\theta << \mu$ be a family of dominated measures with corresponding densities $f_\theta(x)$ in a Euclidean space. Assuming that all the required derivatives and the expectations exist, let

$$l_\theta(x) = \log f_\theta(x),\ \nu_{11}(\theta) = E_\theta\left[\frac{\partial}{\partial\theta}l_\theta\frac{\partial^2}{\partial\theta^2}l_\theta\right],\ \nu_{02}(\theta) = E_\theta\left[\frac{\partial^2}{\partial\theta^2}l_\theta\right]^2 - I^2(\theta),$$

where $I(\theta)$ denotes the Fisher information at θ. The curvature of $\{P_\theta\}$ at θ is defined as

$$\gamma_\theta = \sqrt{\frac{\nu_{02}(\theta)}{I^2(\theta)} - \frac{\nu_{11}^2(\theta)}{I^3(\theta)}}.$$

Remark. A detailed geometric justification for the name *curvature* is given in Efron (1975). The *curvature* γ_θ defined above works out to zero in the regular exponential family, which acts like a straight line in the space of all probability distributions on the given Euclidean space. Nonexponential families have nonzero (at some values of θ) γ_θ and act like curves in the space of all probability distributions. Hence the name *curvature*. Before explaining a theoretical significance of γ_θ in terms of information loss suffered by the MLE, let us see a few examples.

Example 16.15 Suppose $f_\theta(x)$ is a member of the exponential family with $f_\theta(x) = e^{\theta T(x)-\psi(\theta)}h(x)(d\mu)$. Then, $l_\theta(x) = \theta T(x) - \psi(\theta) + \log h(x)$ and hence $\frac{\partial}{\partial\theta}l_\theta = T(x) - \psi'(\theta)$, $\frac{\partial^2}{\partial\theta^2}l_\theta = -\psi''(\theta)$. Therefore, the Fisher information function $I(\theta) = \psi''(\theta)$. On the other hand, $\nu_{02}(\theta) = E_\theta[\frac{\partial^2}{\partial\theta^2}l_\theta]^2 - I^2(\theta) = 0$, and also $\nu_{11}(\theta) = E_\theta[\frac{\partial}{\partial\theta}l_\theta \frac{\partial^2}{\partial\theta^2}l_\theta] = -\psi''(\theta)E_\theta[T(X) - \psi'(\theta)] = 0$, as $E_\theta[T(X)] = \psi'(\theta)$. It follows from the definition of the curvature that $\gamma_\theta = 0$.

Example 16.16 Consider a general location-parameter density $f_\theta(x) = g(x-\theta)$ with support of g as the entire real line. Then, writing $\log g(x) = h(x)$, $l_\theta = h(x-\theta)$ and by direct algebra $I(\theta) = \int \frac{g'^2}{g}$, $\nu_{02}(\theta) = \int gh''^2 - (\int \frac{g'^2}{g})^2$, $\nu_{11}(\theta) = -\int h'h''g$. All these integrals are on $(-\infty, \infty)$, and the expressions are independent of θ. Consequently, the curvature γ_θ is also independent of θ.

For instance, if $f_\theta(x)$ is the density of the central t-distribution with location parameter θ and m degrees of freedom, then, on the requisite integrations, the different quantities are

$$I(\theta) = \frac{m+1}{m+3}, \nu_{02}(\theta) = \frac{m+1}{m+3}\left[\frac{(m+2)(m^2+8m+19)}{m(m+5)(m+7)} - \frac{m+1}{m+3}\right],$$
$$\nu_{11}(\theta) = 0.$$

On plugging into the definition of γ_θ, one finds that $\gamma_\theta^2 = \frac{6(3m^2+18m+19)}{m(m+1)(m+5)(m+7)}$; see Efron (1975). As $m \to \infty$, $\gamma_\theta \to 0$, which one would expect since the t-distribution converges to the normal when $m \to \infty$, and the normal has zero curvature by the previous example. For the Cauchy case corresponding to $m = 1$, γ_θ^2 works out to 2.5. The curvature across the whole family as m varies between 1 and ∞ is a bounded decreasing function of m. The curvature becomes unbounded when $m \to 0$.

We now present an elegant result connecting curvature to the loss of information suffered by the MLE when f_θ satisfies certain structural and regularity assumptions. The density f_θ is assumed to belong to the *curved exponential family*, as defined below.

Definition 16.4 Suppose for $\theta \in \Theta \subseteq \mathcal{R}$, $f_\theta(x) = e^{\eta' T(x)-\psi(\eta)}h(x)(d\mu)$, where $\eta = \eta(\theta)$ for some specified function from Θ to a Euclidean space $\mathcal{R}^k$. Then f_θ is said to belong to the curved exponential family with carrier μ.

Remark. If η varies in the entire set $\{\eta : \int e^{\eta' T(x)}h(x)d\mu < \infty\}$, then the family would be a member of the exponential family. By making the natural parameter η a function of a common underlying parameter θ, the exponential family density has been restricted to a subset of lower dimension. In the curved exponential family, the different components of the natural parameter vector of an exponential family density are tied together by a common underlying parameter θ.

Example 16.17 Consider the $N(\theta, \theta^2)$ density with $\theta \neq 0$. These form a subset of the two-parameter $N(\mu, \sigma^2)$ densities, with $\mu(\theta) = \theta$ and $\sigma^2(\theta) = \theta^2$. Writing out the $N(\theta, \theta^2)$ density, it is seen to be a member of the curved exponential family with $T(x) = (x^2, x)$ and $\eta(\theta) = (-\frac{1}{2\theta^2}, \frac{1}{\theta})$.

Example 16.18 Consider Gamma densities for which the mean is known to be 1. They have densities of the form $f_\theta(x) = \frac{e^{-x/\theta} x^{1/\theta - 1}}{\theta^{1/\theta}\Gamma(\frac{1}{\theta})}$. This is a member of the curved exponential family with $T(x) = (x, \log x)$ and $\eta(\theta) = (-\frac{1}{\theta}, \frac{1}{\theta})$. Here is the principal theorem on information loss by the MLE in curved exponential families.

Theorem 16.8 Suppose $f_\theta(x)$ is a member of the curved exponential family and that the characteristic function $\psi_\theta(t)$ of f_θ is in $\mathcal{L}_p$ for some $p \geq 1$. Let $\hat{\theta}_n$ denote the MLE of θ based on n iid observations from f_θ, $I(\theta)$ the Fisher information based on f_θ, and $I_{n,0}(\theta)$ the Fisher information obtained from the exact sampling distribution of $\hat{\theta}_n$ under θ. Then, $\lim_{n\to\infty}(nI(\theta) - I_{n,0}(\theta)) = I(\theta)\gamma_\theta^2$. In particular, the limiting loss of information suffered by the MLE is finite at any θ at which the curvature γ_θ is finite.

Remark. This is the principal theorem in Efron (1975). Efron's interpretation of this result is that the information obtained from n samples if one uses the MLE would equal the information obtained from $n - \gamma_\theta^2$ samples if the full sample is used. The interpretation hinges on using Fisher information as the criterion. However, γ_θ has other statistical significance; e.g., in testing hypothesis problems. In spite of the controversy about whether γ_θ has genuine inferential relevance, it seems to give qualitative insight into the wisdom of using methods based on the maximum likelihood estimate when the minimal sufficient statistic is multidimensional.

16.10 Exercises

Exercise 16.1 * For each of the following cases, write or characterize the MLE and describe its asymptotic distribution and consistency properties.

(a) $X_1, \ldots, X_n$ are iid with density

$$f(x|\sigma_1, \sigma_2) = \begin{cases} ce^{-\frac{x}{\sigma_1}}, & x > 0 \\ ce^{\frac{x}{\sigma_2}}, & x < 0 \end{cases},$$

each of σ_1, σ_2 being unknown parameters.

Remark. This is a standard way to produce a skewed density on the whole real line.

(b) $X_i, 1 \leq i \leq n$ are independent Poi(λx_i), the x_i being fixed covariates.

(c) $X_1, X_2, \cdots, X_m$ are iid $N(\mu, \sigma_1^2)$ and $Y_1, Y_2, \cdots, Y_n$ are iid $N(\mu, \sigma_2^2)$, and all $m + n$ observations are independent.

(d) m classes are represented in a sample of n individuals from a multinomial distribution with an unknown number of cells θ and equal cell probabilities $\frac{1}{\theta}$.

Exercise 16.2 Suppose $X_1, \ldots, X_n$ are p-vectors uniformly distributed in the ball $B_r = \{x : ||x||_2 \leq r\}; r > 0$ is an unknown parameter. Find the MLE of r and its asymptotic distribution.

Exercise 16.3 * Two independent proofreaders A and B are asked to read a manuscript containing N errors; $N \geq 0$ is unknown. n_1 errors are found by A alone, n_2 by B alone, and n_{12} by both. What is the MLE of N? What kind of asymptotics are meaningful here?

Exercise 16.4 * **(Due to C. R. Rao)** In an archaeological expedition, investigators are digging up human skulls in a particular region. They want to ascertain the sex of the individual from the skull and confirm that there is no demographic imbalance. However, determination of sex from an examination of the skull is inherently not an error-free process.

Suppose they have data on n skulls, and for each one they have classified the individual as being a male or female. Model the problem, and write the likelihood function for the following types of modeling.

(a) The error percentages in identifying the sex from the skull are assumed known.

(b) The error percentages in identifying the sex from the skull are considered unknown but are assumed to be parameters independent of the basic parameter p, namely the proportion of males in the presumed population.

(c) The error percentages in identifying the sex from the skull are considered unknown, and they are thought to be functions of the basic parameter p. The choice of the functions is also a part of the model.

Investigate, under each type of modeling, existence of the MLE of p, and write a formula if possible under the particular model.

Exercise 16.5 * **Missing Data** The number of fires reported in a week to a city fire station is Poisson with some mean λ. The city station is supposed to report the number each week to the state central office. But they do not bother to report it if their number of reports is less than three.

Suppose you are employed at the state central office and want to estimate λ. Model the problem, and write the likelihood function for the following types of modeling.

(a) You ignore the weeks on which you did not get a report from the city office.

(b) You do not ignore the weeks on which you did not get a report from the city office, and you know that the city office does not send its report only when the number of incidents is less than three.

(c) You do not ignore the weeks on which you did not get a report from the city office, and you do not know that the city office does not send its report only when the number of incidents is less than three.

Investigate, under each type of modeling, existence of the MLE of λ, and write a formula if possible under the particular model.

Exercise 16.6 * Find a location-scale parameter density $\frac{1}{\sigma} f(\frac{x-\mu}{\sigma})$ for which the MLE of σ is $\frac{1}{n} \sum |X_i - M|$, where M is the median of the sample values $X_1, \ldots, X_n$. Find the asymptotic distribution of the MLE under this f (challenging!).

Exercise 16.7 * Consider the polynomial regression model $y_i = \beta_0 + \sum_{j=1}^{m} \beta_j x_i^j + \sigma e_i$, where e_i are iid $N(0, 1)$. What is the MLE of m?

Exercise 16.8 * Suppose $X_1, \ldots, X_{m+n}$ are independent, with $X_1, \ldots, X_m \sim N(\mu_1, \sigma^2)$, $X_{m+1}, \ldots, X_{m+n} \sim N(\mu_2, \sigma^2)$, where $\mu_1 \le \mu_2$ and σ^2 are unknown. Find the MLE of (μ_1, μ_2) and derive its asymptotic distribution when $\mu_1 < \mu_2$, $\mu_1 = \mu_2$.

Exercise 16.9 If $X_1, \ldots, X_n$ are iid Poi(λ), show that $\frac{1}{\bar{X}+\frac{1}{n}}$ is second-order unbiased for $\frac{1}{\lambda}$.

Exercise 16.10 * Find the limiting distribution of the MLE of (μ, σ, α) for the three-parameter Gamma density

$$\frac{e^{-\frac{(x-\mu)}{\sigma}}(x-\mu)^{\alpha-1}}{\sigma^{\alpha}\Gamma(\alpha)}, \quad x \geq \mu, \quad \alpha, \sigma > 0, \quad -\infty < \mu < \infty.$$

Exercise 16.11 Suppose $X_1, \ldots, X_n$ are iid Exp(λ). Find the MLE of the expected residual life $E(X_1 - t|X_1 > t)$ and its asymptotic distribution.

Exercise 16.12 * Suppose $X_1, \ldots, X_n$ are BV$N(\mu_1, \mu_2, \sigma_1, \sigma_2, \rho)$, all five parameters being unknown. Find the MLE of $P(X_{11} > \mu_1, X_{12} > \mu_2)$, where $\binom{X_{11}}{X_{12}} = X_1$, and find its asymptotic distribution.

Exercise 16.13 * Derive a closed-form expression for the mean squared error $R(\theta, T_n)$ of the Hodges superefficient estimate, and show that $\limsup_{|\theta| \leq n^{-\frac{1}{4}}} nR(\theta, T_n) = \infty$.

Exercise 16.14 * Suppose $X_1, \ldots, X_n$ are iid with density

$$p\frac{1}{\sqrt{2\pi}}e^{-\frac{(x-\mu)^2}{2}} + (1-p)\frac{1}{\sqrt{2\pi}\sigma}e^{-\frac{1}{2\sigma^2}(x-\mu)^2},$$

where $0 < p < 1$ is known. Show that MLEs for μ, σ do not exist. How would you estimate μ, σ? What is the asymptotic distribution of your estimates?

Exercise 16.15 * Suppose X_i are iid $N(\mu, 1)$, where μ is known to be a positive integer. Let $g : \mathcal{R} \to \mathcal{R}$ be the function

$$g(x) = \begin{cases} x & \text{if } x \text{ is a prime} \\ -x & \text{if } x \text{ is not a prime} \end{cases}.$$

(a) Is $\bar{X}$ consistent for μ?

(b) Is $g(\bar{X})$ consistent for $g(\mu)$?

Exercise 16.16 Suppose $X_i \overset{\text{indep.}}{\sim}$ Poi(λ^i) (thus X_i are not iid), $1 \leq i \leq n$.

(a) What is the MLE of λ?

(b) What is the asymptotic distribution of the MLE of λ?

Exercise 16.17 *

(a) Suppose X_i are iid $N(\mu, 1)$, but the collector rounds the X_i to Y_i, the nearest integer. Is $\bar{Y}$ consistent for μ?

(b) Find a consistent estimate for μ based on the Y_i.

Exercise 16.18 * Suppose X_i are iid nonnegative random variables and X_i are recorded as the integer closest to the X_i, say Y_i. Give a necessary and sufficient condition for $\bar{Y} \stackrel{\mathcal{P}}{\longrightarrow} E(X_1)$.

Exercise 16.19 Suppose $X_i \stackrel{\text{iid}}{\sim} \text{Poi}(\lambda)$, where $\lambda > 0$ is known to be an integer.

(a) Find the MLE $\hat{\lambda}$ of λ.
(b) * What is $\lim_{n\to\infty} \text{Var}(\hat{\lambda})$?

Exercise 16.20 * Show that, for iid $C(\theta, 1)$ data, the statistic $\frac{1}{\sum_{i=1}^n \frac{\partial^2}{\partial\theta^2} \log f_\theta(X_i)}$ is asymptotically normal. Find the appropriate centering, norming, and the variance of the asymptotic normal distribution.

Exercise 16.21 Compute the curvature of the $N(\theta, \theta^4)$ family.

Exercise 16.22 Compute the curvature of the family of Gamma densities with a known mean c.

Exercise 16.23 * For estimation of a Poisson mean λ, find the limiting information lost by s^2, the sample variance, compared with the information in the full sample. Is it finite? Is it bounded?

Exercise 16.24 Simulate the exact variance of the MLE of a double exponential mean based on 20 iid samples, and compare the estimates based on expected and observed Fisher information with this exact value. Comment on the bias and the variability of these two estimates.

Exercise 16.25 * Is the central limit theorem for the MLE of a logistic mean second-order accurate?

Exercise 16.26 * Derive a two-term Edgeworth expansion for the MLE of the shape parameter of a Gamma distribution assuming the scale parameter is 1.

Exercise 16.27 * Derive a one-term Edgeworth expansion for the MLE of θ in the $N(\theta, \theta^2)$ distribution.

References

Bai, Z.D. and Rao, C.R. (1991). Edgeworth expansion of a function of sample means, Ann. Stat., 19(3), 1295–1315.

Basu, D. (1955). An inconsistency of the method of maximum likelihood, Ann. Math. Stat., 26, 144–145.

Bhattacharya, R.N. and Ghosh, J.K. (1978). On the validity of the formal Edgeworth expansion, Ann. Stat., 2, 434–451.

Bickel, P.J. and Doksum, K. (2001). *Mathematical Statistics: Basic Ideas and Selected Topics*, Vol. I, Prentice-Hall, Upper Saddle River, NJ.

Brown, L.D. (1986). *Fundamentals of Statistical Exponential Families*, IMS Lecture Notes Monograph Series, Vol. 9, Institute of Mathematical Statistics, Hayward, CA.

DasGupta, A. (2004). On the risk function of superefficient estimates, preprint.

Efron, B. (1975). Defining the curvature of a statistical problem, with applications to second order efficiency, Ann. Stat., 3(6), 1189–1242.

Ghosh, M. (1994). On some Bayesian solutions of the Neyman-Scott problem, Statistical Decision Theory and Related Topics, Vol. V, J. Berger and S.S. Gupta (eds.), Springer-Verlag, New York, 267–276.

James, W. and Stein, C. (1961). Estimation with quadratic loss, Proceedings of the Fourth Berkeley Symposium on Mathematical Statistics and Probability, I, J. Neyman (ed.), University of California, Berkeley, 361–379.

LeCam, L. (1953). On some asymptotic properties of maximum likelihood estimates and related Bayes estimates, Univ. Calif. Publ., 1, 277–330.

Lehmann, E.L. and Casella, G. (1998). *Theory of Point Estimation*, 2nd ed., Springer, New York.

McLachlan, G. and Krishnan, T. (1997). *The EM Algorithm and Extensions*, John Wiley, New York.

Neyman, J. and Scott, E. (1948). Consistent estimates based on partially consistent observations, Econometrica, 16, 1–32.

Perlman, M.D. (1983). The limiting behavior of multiple roots of the likelihood equation, in *Recent Advances in Statistics*, M. Rizui, J.S. Rustagis and D. Siegmund (eds.), Academic Press, New York, 339–370.

Pfanzagl, J. (1973). The accuracy of the normal approximation for estimates of vector parameters, Z. Wahr. Verw. Geb., 25, 171–198.

Rao, C.R. (1961). Asymptotic efficiency and limiting information, *Proceeding of the Fourth Berkeley Symposium on Mathematical Statistics and Probability*, J. Neyman (ed.), Vol. I, University of California, Berkeley, 531–545.

Rao, C.R. (1962). Efficient estimates and optimum inference procedures in large samples, J. R. Stat. Soc. Ser. B, 24, 46–72.

Rao, C.R. (1963). Criteria of estimation in large samples, Sankhya Ser. A, 25, 189–206.

van der vaart, A. (1998). *Superefficiency: Festschrift for Lucien LeCam*, Springer, New York, 397–410.

Chapter 17
M Estimates

The material in the previous chapter shows that MLEs (provided they exist) are asymptotically optimal under (many) conditions on P_θ and the class of estimates under consideration. But by definition an MLE requires a specific model. In the 1960s, systematic efforts started toward suggesting and studying estimates that do not need one specific model but behave reasonably well in a whole neighborhood of a specific model. For example, consider the location-parameter problem $X_1, X_2, \ldots, X_n \stackrel{\text{iid}}{\sim} F(x - \theta)$, where $F(x) = (1 - \epsilon)\Phi(x) + \epsilon H(x)$ for H belonging to some suitable class of CDFs. Omnibus estimates, known as *M estimates*, were the outcome of this work. M estimates provide a crisp mathematical formulation as well as theory for robustness in statistics, and they also unified asymptotic theory in an unprecedented way. Although we discuss only the iid case for one dimension in this chapter, M estimates have been introduced and studied for many other situations (e.g., the multivariate case and regression models). Some selected references for M estimates in non-iid situations are Huber (1973), Maronna (1976), and Carroll (1978).

The name M estimate is apparently due to the fact that they are defined in a way formally analogous to MLEs, although they are omnibus estimates. Excellent references for the main results in this chapter are Serfling (1980), Sen and Singer (1993), Huber (1981), and van der Vaart (1998). Other specific references are given in the following sections.

M estimates have been defined in the literature in two different ways. Sometimes the two definitions are equivalent but not always. We present both definitions below.

Definition 17.1 Let $X_1, X_2, \ldots, X_n \stackrel{\text{iid}}{\sim} F$. An M estimate T_n is a solution of the equation $\sum_{i=1}^n \psi(x_i, t) = \int \psi(x, t)\, dF_n(x) = 0$ for a specific function ψ, where F_n is the empirical CDF.

A. DasGupta, *Asymptotic Theory of Statistics and Probability*,

Remark. By appropriate choices of ψ, different kinds of estimates are obtained as M estimates. Thus the range of application of the M estimate theory is very broad.

Definition 17.2 Let $X_1, X_2, \ldots, X_n \overset{\text{iid}}{\sim} F$. An M estimate T_n is a minimizer of $\sum_{i=1}^{n} \rho(x_i, t) = \int \rho(x, t) dF_n(x)$ for a specific function ρ.

Remark. If the function $\rho(x, t)$ is partially differentiable with respect to t, then a minimizer of $\sum_{i=1}^{n} \rho(x_i, t)$ would be a root of the equation $\sum_{i=1}^{n} \psi(x_i, t) = 0$ with $\psi(x, t) = \frac{\partial}{\partial t}\rho(x, t)$. The equation can have multiple roots.

17.1 Examples

Example 17.1 (a) If $\psi(x, t) = x - t$, then $T_n = \bar{X}$.

(b) If $\psi(x, \theta) = -\frac{d}{d\theta} \log f(x|\theta)$ for some specific family of densities $f(x|\theta)$, then T_n is a root of the likelihood equation

$$\frac{d}{d\theta} \log \left(\prod_{i=1}^{n} f(x_i|\theta) \right) = 0.$$

(c) If $\psi(x, t) = \psi_0(x - t)$, where

$$\psi_0(z) = \begin{cases} -1 & \text{for } z < 0 \\ 0 & \text{for } z = 0 \\ \frac{p}{1-p} & \text{for } z > 0 \end{cases}$$

for a fixed $0 < p < 1$, then usually the equation $\sum_{i=1}^{n} \psi(x_i, t) = 0$ will not have any roots. But one can use as T_n approximate roots; i.e., T_n such that $\sum_{i=1}^{n} \psi(x_i, T_n) = o_p(1)$. One example is the sample pth percentile, assuming that the population pth percentile, say ξ_p, is unique.

(d) A special case is $\psi(x, t) = \psi_0(x - t)$ with $\psi_0(z) = k\text{sgn}(z)$, $(k > 0)$. This lets one handle sample medians as special cases of the M estimate theory.

(e) By combining examples (a) and (d), Huber (1964) suggested the ψ function $\psi(x, t) = \psi_o(x - t)$, where

$$\psi_0(z) = \begin{cases} z & \text{if } |z| \leq k \\ k & \text{if } z > k \\ -k & \text{if } z < -k. \end{cases}$$

Huber's motivation was that an unbounded ψ function results in estimates that have undesirable properties (e.g., undue influence by a small number of outliers). By using the Huber ψ function, one has a bounded ψ, and also the corresponding M estimate T_n is a root of the likelihood function when $X_1, X_2, \ldots, X_n \overset{\text{iid}}{\sim} f(x-\theta)$, where

$$f(z) = \begin{cases} ce^{-\frac{z^2}{2}} & \text{for } \mid z \mid \leq k \\ ce^{\frac{k^2}{2}-k|z|} & \text{for } \mid z \mid > k. \end{cases}$$

Thus one is looking effectively at MLEs when the density has a normal body and a double exponential tail. It turns out that in fact this particular M estimate due to Huber has an interesting minimaxity property. Specifically,

$$\sqrt{n}(T_n - \theta) \xrightarrow{\mathcal{L}} N(0, \sigma^2_{0,F}), \quad \text{and if } \sqrt{n}(\hat{\theta}_{n,\psi} - \theta) \xrightarrow{\mathcal{L}} N(0, \sigma^2_{\psi,F})$$

for another M estimate $\hat{\theta}_{n,\psi}$, then

$$\sup_{F \in \mathcal{F}} \sigma^2_{0,F} \leq \sup_{F \in \mathcal{F}} \sigma^2_{\psi,F}$$

for suitable families of CDFs $\mathcal{F}$.

(f) M estimates, or equivalently ψ functions that amount to entirely ignoring observations at the tails, have also been suggested. For these estimates, the ψ function is actually equal to zero outside of a compact interval. Consequently, the estimates are called *redescending M estimates*. Many such ψ functions have been suggested and their performance investigated numerically. A few of them are

i. $\psi(x,t) = \psi_o(x-t)$, where, $\psi_0(z) = \begin{cases} z & \text{if } |z| \leq a \\ a\,\text{sgn}(z) & \text{if } a \leq |z| \leq b \\ a\frac{r-|z|}{r-b}\text{sgn}(z) & \text{if } b \leq |z| \leq r. \\ 0 & \text{if } |z| > r \end{cases}$

This ψ-function is the 45° straight line near zero, a flat line thereafter, and then a negatively sloped line merging with zero and staying zero from that point on. Thus it is not differentiable at the juncture points where the straight lines change.

ii. A specific redescending M estimate with a differentiable ψ function corresponds to $\psi(x,t) = \psi_o(x-t)$, where

$$\psi_0(z) = z(r^2 - z^2)^2 I_{|z| \le r}.$$

This estimate is known as *Tukey's bi-weight estimate* and is a special estimate in the robustness literature.

Redescending M estimates can pose computational complexities such as multiple roots of the defining equation or sometimes singularity troubles. It is a pretty common practice to use the root closest to the sample median in the case of multiple roots or use a one-step Newton-Raphson iteration starting with the sample median as an initial value.

17.2 Consistency and Asymptotic Normality

M estimates can be defined as solutions to estimating equations $\sum_{i=1}^n \psi(x_i, t) = 0$ or as the maximizer of $-\sum_{i=1}^n \rho(x_i, t)$, where ρ is the primitive of ψ with respect to t. Under the two different formulations, different sets of conditions on $\psi(\rho)$ ensure consistency, strong consistency, and asymptotic normality of the M estimate. The variety of conditions can be confusing, so we will deal essentially with one clean set of conditions. The various sets of conditions can be seen in Huber (1964), Serfling (1980), and van der Vaart (1998). The difficulties with the ψ approach are that:

(a) The asymptotic behavior of a root of $\int \psi(t, x) dF_n(x)$ depends on the global behavior of $\lambda_F(t) = \int \psi(t, x) dF(x)$, in particular whether $\int \psi(t, x) dF(x) = 0$ has a unique root.

(b) The equation $\int \psi(t, x) dF_n(x) = 0$ may not have any exact roots.

(c) There can be multiple roots of $\int \psi(t, x) dF_n(x) = 0$, in which case a rule is required to select one. If $\psi(t, x)$ is continuous and strictly monotone in t, and if $\lambda_F(t) = 0$ has a unique root t_0, then $\lambda_{F_n}(t) = 0$ will have a unique root and the M estimate is uniquely defined and consistent. Monotonicity and continuity of ψ are frequently assumed.

Example 17.2 (a) Let $\psi(x, t) = \psi_0(x - t)$, where $\psi_0(z) = z$. This is continuous and strictly monotone.

(b) Let $\psi(x, t) = \psi_0(x - t)$, where $\psi_0(z) = \text{sgn}(z)$. This is not continuous and monotone but not strictly monotone.

(c) Let $\rho(x, t) = -\log f(x_i - t)$ and $\psi(x, t) = -\frac{\frac{d}{dt} f(x_i - t)}{f(x_i - t)}$. Thus ψ need not be continuous and is strictly monotone iff $f(z)$ is log concave or log convex. For instance, the Cauchy density or any t-density is not log concave, but the normal is.

Theorem 17.1 Assume that:

(i) $\lambda_F(t) = \int \psi(t, x)dF(x) = 0$ has a unique root t_0.

(ii) ψ is continuous and either bounded or monotone. Then $\lambda_{F_n(t)} = 0$ admits a sequence of roots $\hat{\theta}_n$ such that $\hat{\theta}_n \xrightarrow{\text{a.s.}} t_0$.

Remark. Serfling (1980) gives a proof of this result. Alternatively, one can entirely give up the idea of identifying consistent roots of $\lambda_{F_n}(t) = 0$ and look for approximate roots. For example, starting with an initial $\sqrt{n}$-consistent estimate $\hat{\theta}_n$, one can look at the one-step Newton-Raphson estimate $\delta_n = \hat{\theta}_n - \frac{\psi_{F_n}(\hat{\theta}_n)}{\psi'_{F_n}(\hat{\theta}_n)}$, where $\psi'_{F_n}(t) = \frac{\partial}{\partial t} \sum \psi(x_i - t)$. Consistency (and usually even asymptotic normality) of δ_n is automatic, but δ_n is not a root of $\lambda_{F_n}(t) = 0$. Further iteration of δ_n is not going to change any first-order asymptotic properties. What happens for a given n can only be found on a case-by-case basis via simulation.

Assume that t_0 is a unique root of $\lambda_F(t) = 0$, and $\hat{\theta}_n$ is a consistent sequence of roots of $\lambda_{F_n}(t) = 0$. Then, by expanding $\lambda_{F_n}(\hat{\theta}_n)$ around t_0,

$$0 = \lambda_{F_n}(\hat{\theta}_n) = \lambda_{F_n}(t_0) + (\hat{\theta}_n - t_0)\lambda'_{F_n}(t_0) + \frac{(\hat{\theta}_n - t_0)^2}{2}\lambda''_{F_n}(\theta_n^*)$$

$$\Longrightarrow (\hat{\theta}_n - t_0) = -\frac{\lambda_{F_n}(t_0)}{\lambda'_{F_n}(t_0) + \frac{\hat{\theta}_n - t_0}{2}\lambda''_{F_n}(\theta_n^*)}$$

$$\Longrightarrow \sqrt{n}(\hat{\theta}_n - t_0) = -\frac{\sqrt{n}\lambda_{F_n}(t_0)}{\lambda'_{F_n}(t_0) + \frac{\hat{\theta}_n - t_0}{2}\lambda''_{F_n}(\theta_n^*)}.$$

Now, $\lambda_{F_n}(t_0)$ is a sample mean and admits (hopefully) a CLT, $\lambda'_{F_n}(t_0)$ is also a sample mean and admits (hopefully) the WLLN, and $(\hat{\theta}_n - t_0)$ is $o_p(1)$. So if we can control $\sup_{|t-t_0|<\epsilon} \lambda''_{F_n}(t)$, then by Slutsky's theorem

$$\sqrt{n}(\hat{\theta}_n - t_0) \xrightarrow{\mathcal{L}} N\left(0, \frac{\int \psi^2(x, t_0)dF(x)}{(\int \psi'(x, t)dF(x) \mid_{t=t_0})^2}\right).$$

Here is a formal result.

Theorem 17.2 Assume

(i) $\lambda_F(t) = 0$ has a unique root t_0,
(ii) $\psi(x, t)$ is monotone in t,
(iii) $\lambda'_F(t_0)$ exists and is $\neq 0$, and
(iv) $\int \psi^2(x, t)dF(x) < \infty$ in some neighborhood of t_0 and is continuous at t_0.

Then, any sequence of roots $\hat{\theta}_n$ of $\lambda_{F_n}(t) = 0$ satisfies

$$\sqrt{n}(\hat{\theta}_n - t_0) \xrightarrow{\mathcal{L}} N\left(0, \frac{\int \psi^2(x, t_0)dF(x)}{(\int \psi'(x, t)dF(x))^2 \mid_{t=t_0}}\right).$$

We have essentially proved this theorem above, but a formal proof can be seen in Serfling (1980).

Example 17.3 The Huber M estimate for the location-parameter case is characterized by solving $\sum_{i=1}^n \psi_0(x_i - t) = 0$, where

$$\psi_0(z) = \begin{cases} z & \text{if } |z| \leq k \\ k & \text{if } z > k \\ -k & \text{if } z < -k. \end{cases}$$

If we make the (somewhat more than needed) assumption that $f(x) > 0$ for all x, then the Huber M estimate $\hat{\theta}_{n,H}$, defined as either the argmax of $-\sum_{i=1}^n \rho(x_i - t)$ or any root of $\sum_{i=1}^n \psi_0(x_i - t) = 0$, satisfies, on calculation,

$$\sqrt{n}(\hat{\theta}_{n,H} - \theta) \overset{\mathcal{L}}{\Rightarrow} N(0, \sigma_H^2),$$

where

$$\sigma_H^2 = \frac{\int_{-k}^k x^2 f(x)dx + k^2 F(-k) + k^2(1 - F(k))}{(F(k) - F(-k))^2}.$$

Remark. Notice that $\hat{\theta}_{n,H}$ is asymptotically normal regardless of whether F has a variance or not. This is an attractive property of $\hat{\theta}_{n,H}$. Of course, an important practical question is what value of k should be used. The Huber estimate has a minimax property in the location-parameter case, as mentioned before, when the class of null distributions is a suitable ϵ contamination of the standard normal distribution; see Huber (1964). Given a specific $\epsilon > 0$, the value of k that leads to the minimax solution is uniquely determined and satisfies the equation $\frac{\epsilon}{1-\epsilon} = \frac{2\phi(k)}{k} - 2(1 - \Phi(k))$. Thus, for example, if $\epsilon = .05$, then k is (approximately) 1.4. However, it is also true that by changing k not too drastically from this minimax choice, one can keep the increase in maximum asymptotic variance over the contamination class in excellent control. Changing the value of k from the exact minimax choice can give further protection against tails that do not look like the least favorable exponential tail; for example, thicker tails. Thus, the choice of k should perhaps be left somewhat flexible. One has to play with it to some extent and use external information about tails one is concerned about in the specific problem.

Example 17.4 As a further illustration of the previous example, suppose the observations are iid from the density $f(x - \theta)$, where $f(x) = \frac{e^{-x}}{(1+e^{-x})^2}$, the standard logistic density. The integral $\int_{-k}^{k} x^2 f(x)dx$ can in fact be calculated in closed form by using the *polylogarithm* function. Hence, the asymptotic variance of the Huber estimate can be found in closed form for any value of k, although the final expression is considerably messy. On numerical calculation using the closed-form expression, one can check that $k = .77$ results in the minimum variance under this logistic model. But any k from about .5 to 1 gives an asymptotic variance not greatly more than the minimum value 1.65 attained when $k = .77$. Thus, we see that indeed there is some room to be flexible with k.

17.3 Bahadur Expansion of M Estimates

The Taylor heuristics enunciated earlier also indicate what the Bahadur expansion for M estimates should be. See He and Shao (1996) and Bose (1998) for versions of the following theorem.

Theorem 17.3 Assume that $\lambda_F(t) = 0$ has a unique root t_0 and that, for $|t - t_0| \leq \epsilon$, $\psi(t, x)$ is Lipschitz of order 1 in t. Assume also that $\lambda_F'(t_0)$ exists and is not equal to 0. Then any consistent sequence of roots of $\lambda_{F_n}(t) = 0$, say $\hat{\theta}_n$, admits the expansion

$$\hat{\theta}_n = t_0 - \frac{1}{n}\sum_{i=1}^{n} \frac{\psi(X_i, t_0)}{\lambda_F'(t_0)} + o_p\left(\frac{1}{\sqrt{n}}\right).$$

Remark. The Bahadur expansion produces the joint asymptotic multivariate normal distribution for any finite number of M estimates, each corresponding ψ function satisfying the theorem's assumption.

Remark. Although not of any great interest to statisticians, the $o_p(\frac{1}{\sqrt{n}})$ rate for the remainder can be improved to a suitable a.s. rate under more stringent assumptions; see Carroll (1978). The case of a multidimensional parameter is covered in Carroll (1978).

Example 17.5 As an illustration, consider the Huber estimate that has the ψ function $\psi(x,t) = \psi_0(x-t)$, where $\psi_0(z)$ is as in Example 17.3. Assume that the underlying CDF F is a location-parameter CDF with density $f(x-\theta)$, and assume further that $f(.)$ itself is symmetric (i.e., an even function). Then, $\lambda_F(t) = E_\theta \psi_0(x-\theta)$ has the root $t_0 = \theta$. Furthermore,

$$\lambda_F(t) = \int_{|x-t|\leq k} (x-t)f(x-\theta)dx + kP_\theta(X > k+t) - kP_\theta(X < t-k)$$

and hence, by an application of routine calculus, $\lambda'_F(t) = -\int_{t-k}^{t+k} f(x-\theta)dx$, which gives $\lambda'_F(t_0) = 1 - 2F(k)$. Substituting into the Bahadur representation result above, the Huber estimate is seen to have the asymptotic linear representation

$$\hat{\theta}_{n,H} = \theta + \frac{1}{n(2F(k)-1)} \sum_{i=1}^{n} \psi_0(X_i - \theta) + o_p\left(\frac{1}{\sqrt{n}}\right).$$

This can be used to derive the joint asymptotic distribution of $\hat{\theta}_{n,H}$ and another M estimate; e.g., $\bar{X}$.

Example 17.6 As a second illustration, consider the sample pth percentile for a general $0 < p < 1$. We have remarked before that it is a special M estimate. To make notation slightly simpler than it would be otherwise, assume that $X_1, X_2, \ldots, X_n$ are iid F with a density function f and that f is strictly positive at the population pth percentile ξ_p. The ψ function corresponding to the sample pth percentile is $\psi(x,t) = \psi_0(x-t)$, where

$$\psi_0(z) = \begin{cases} -1 & \text{for } z < 0 \\ 0 & \text{for } z = 0 \\ \frac{p}{1-p} & \text{for } z > 0. \end{cases}$$

By a direct integration, therefore,

$$\lambda_F(t) = -\int_{x<t} f(x)dx + \frac{p}{1-p}\int_{x>t} f(x)dx = \frac{p-F(t)}{1-p}.$$

Therefore, $\lambda_F(t) = 0$ at $t_0 = \xi_p$, and $\lambda_F'(t) = -\frac{f(t)}{1-p}$. Now, substituting into the general Bahadur expansion result above, the sample pth percentile has the representation

$$\begin{aligned}
F_n^{-1}(p) =& \xi_p + \frac{1}{\frac{nf(\xi_p)}{1-p}} \sum_{i=1}^{n} \psi_0(X_i - \xi_p) + o_p\left(\frac{1}{\sqrt{n}}\right) \\
=& \xi_p + \frac{1-p}{nf(\xi_p)} \left[\frac{p}{1-p} \sum I_{X_i > \xi_p} - \sum I_{X_i \le \xi_p} \right] + o_p\left(\frac{1}{\sqrt{n}}\right) \\
=& \xi_p + \frac{1-p}{nf(\xi_p)} \frac{1}{1-p} \sum_{i=1}^{n} (p - I_{X_i \le \xi_p}) + o_p\left(\frac{1}{\sqrt{n}}\right) \\
=& \xi_p + \frac{1}{n} \sum_{i=1}^{n} Y_i + o_p\left(\frac{1}{\sqrt{n}}\right),
\end{aligned}$$

where $Y_i = \frac{p - I_{X_i \le \xi_p}}{f(\xi_p)}$. This is the representation we previously saw in Chapter 6.12.

17.4 Exercises

Exercise 17.1 * Find the asymptotic variance of the Huber estimate for the normal and the double exponential location parameter cases, and in each case find the value of k that minimizes the asymptotic variance if one such k exists.

Exercise 17.2 * By using the Bahadur expansion, derive the value of $\lim_{n\to\infty} P_\theta(|\hat{\theta}_{n,H} - \theta| \le |\bar{X} - \theta|)$ when the observations are iid from $N(\theta, 1)$ and from $\frac{1}{2}e^{-|x-\theta|}$. Hint: Consider $P(|X| \le |Y|)$ in a bivariate normal distribution.

Exercise 17.3 Find the joint asymptotic distribution of $\bar{X}$ and the Huber estimate when samples are iid standard double exponential with location parameter μ. What is the value of the correlation parameter in the asymptotic covariance matrix?

Exercise 17.4 * Find an asymptotically correct $100(1-\alpha)\%$ confidence interval by using the Huber estimate when samples are iid $N(\mu, 1)$, and simulate its coverage for $n = 20, 30, 50$.

Exercise 17.5 * (a) Consider the MLE of a Cauchy location parameter as an M estimate, and derive a Bahadur expansion under a general F.

(b) For which of these models would the estimate of part (a) be consistent for estimating the mean: $N(\theta, 1)$; $\frac{1}{2}e^{-|x-\theta|}$; or Gamma$(\alpha, \theta)$?

(c) By using the Bahadur expansion of part (a), derive the limiting distribution of this estimate under the first two of the three models in part (b).

Exercise 17.6 (a) Derive a Bahadur expansion for the Tukey biweight redescending M estimate.

(b) Hence find its asymptotic efficiency with respect to the MLE for the following models: $N(\theta, 1)$; $\frac{1}{2}e^{-|x-\theta|}$; $C(\theta, 1)$.

Exercise 17.7 * (a) Is a convex combination of the mean and the median an M estimate?

(b) Are posterior means M estimates?

Exercise 17.8 Give an example of a ψ function for which $\int \psi(x, t)dF(x) = 0$ has more than one root.

Exercise 17.9 * Prove or disprove that M estimates are scale equivariant; i.e., $\hat{\theta}_n(cX_1, \ldots, cX_n) = c\hat{\theta}_n(X_1, \ldots, X_n)$ for all $c > 0$.

Exercise 17.10 * By using respectively the Bahadur representations for the median and the Huber estimate, find the asymptotic distribution of a convex combination of the mean, the median, and the Huber estimate when samples are from a location-parameter logistic distribution, and find the best estimate of this form by optimizing over the coefficients and the Huber tuning parameter k. What is this estimate's asymptotic efficiency with respect to the logistic MLE?

References

Bose, A. (1998). Bahadur representation of *M* estimates, Ann. Stat., 26(2), 771–777.

Carroll, R.J. (1978). On the asymptotic distribution of multivariate M estimates, J. Multivar Anal., 8(3), 361–371.

He, X. and Shao, Q-M. (1996). A general Bahadur representation of M estimators and its application to linear regression with nonstochastic designs, Ann. Stat., 24(6), 2608–2630.

Huber, P.J. (1964). Robust estimation of a location parameter, Ann. Math. Stat., 35, 73–101.

Huber, P.J. (1973). Robust regression; asymptotics, conjectures and Monte Carlo, Ann. Stat., 1, 799–821.

Huber, P.J. (1981). *Robust Statistics*, John Wiley, New York.

Maronna, R. (1976). Robust M estimation of multivariate location and scatter, Ann. Stat., 4(1), 51–67.

Sen, P.K. and Singer, J. (1993). *Large Sample Methods in Statistics: An Introduction with Applications*, Chapman and Hall, New York.
Serfling, R. (1980). *Approximation Theorems of Mathematical Statistics*, John Wiley, New York.
van der Vaart, A. (1998). *Asymptotic Statistics*, Cambridge University Press, Cambridge.

Chapter 18
The Trimmed Mean

The trimmed mean, as the name suggests, is a mean of a sample when a certain proportion of the extreme order statistics are trimmed from each tail. For $0 < \alpha_1, \alpha_2 < \frac{1}{2}$, the trimmed mean $\bar{X}_{n,\alpha_1,\alpha_2}$ is defined as

$$\bar{X}_{n,\alpha_1,\alpha_2} = \frac{1}{n - [n\alpha_1] - [n\alpha_2]} \sum_{i=[n\alpha_1]+1}^{n-[n\alpha_2]} X_{i:n},$$

where $[y]$ denotes the integer part of y. Usually one trims symmetrically, so that $\alpha_1 = \alpha_2$. In that case, we will use the notation $\bar{X}_{n,\alpha}$. Note that symmetric trimming is not natural as an estimate of the mean or more generally the location parameter of an asymmetric population.

Remark. $\bar{X}$ and the sample median are both examples of trimmed means. Some remarkable asymptotic properties of $\bar{X}_{n,\alpha}$ were proved in the 1960s and 1970s; see Bickel and Lehmann (1975) and Lehmann (1983). These results showed that the trimmed mean with $\alpha \approx 0.1$ is nearly as good as $\bar{X}$ for approximately normal data and far safer than $\bar{X}$ for long-tailed data.

18.1 Asymptotic Distribution and the Bahadur Representation

The following fundamental theorems on trimmed means are proved in Lehmann (1983).

Theorem 18.1 Let $X_1, X_2, \ldots, X_n \stackrel{\text{iid}}{\sim} f(x - \theta)$, where $f(z)$ is continuous and strictly positive $\forall z \in \mathcal{R}$, and $f(z) = f(-z)\ \forall z \in \mathcal{R}$. Then

$$\sqrt{n}(\bar{X}_{n,\alpha} - \theta) \stackrel{\mathcal{L}}{\Rightarrow} N(0, \sigma_\alpha^2),$$

where $\sigma_\alpha^2 = \frac{2}{(1-2\alpha)^2}\left(\int_0^{F^{-1}(1-\alpha)} z^2 f(z)dz + \alpha\{F^{-1}(1-\alpha)\}^2\right)$.

Example 18.1 Using the expression for σ_α^2, one can compute the asymptotic relative efficiency of $\bar{X}_{n,\alpha}$ with respect to $\bar{X}$ for a given F. The table below provides numerical values of the ARE for some selected F.

Table (ARE)

	$F = t(m)$			$F = (1-\epsilon)N(0,1) + \epsilon N(0,9)$ *	
	α			α	
m	.05	.125	ϵ	0.05	.125
3	1.7	1.91	0.25	1.4	1.66
5	1.2	1.24	0.05	1.2	1.19
∞	0.99	0.94	0.01	1.04	0.98
			0	0.99	.94

* F is called a Tukey distribution.

Remark. It seems that even a moderate amount of trimming provides much better efficiency than $\bar{X}$ for long-tailed data without much loss at all for normal data. For n not too small, some trimming (about 0.10 for each tail) is a good idea.

Analogous to the median, the trimmed mean is also asymptotically linear in the sense that it admits a Bahadur type linear representation. We will state it only for the symmetrically trimmed mean, although the representation is available for general trimmed means.

Theorem 18.2 Let $X_1, X_2, \ldots, X_n \overset{\text{iid}}{\sim} F(x-\theta)$. Let $0 < \alpha < 1$ be fixed. Then the trimmed mean $\bar{X}_{n,\alpha}$ admits the representation

$$\bar{X}_{n,\alpha} = \theta + \frac{1}{n}\sum_{i=1}^{n} Z_i + o_p\left(\frac{1}{\sqrt{n}}\right),$$

where $Z_i = \frac{1}{1-2\alpha}(F^{-1}(\alpha)I_{X_i-\theta<F^{-1}(\alpha)} + (X_i-\theta)I_{F^{-1}(\alpha)\le X_i-\theta\le F^{-1}(1-\alpha)} + F^{-1}(1-\alpha)I_{X_i-\theta>F^{-1}(1-\alpha)})$.

As always, the Bahadur representation is useful for writing the joint asymptotic distribution of the trimmed mean and some other suitable statistics. See Section 18.4 for a concrete application of the Bahadur representation.

18.2 Lower Bounds on Efficiencies

Some elegant lower bounds on the ARE of $\bar{X}_{n,\alpha}$ with respect to $\bar{X}$ are available in the nonparametrics literature. We state two bounds below; both are proved in Lehmann (1983).

Theorem 18.3 Let $X_1, X_2, \ldots, X_n \overset{\text{iid}}{\sim} f(x-\theta)$. Assume that $f(\cdot)$ is unimodal in addition to being symmetric, continuous, and positive on its support. Then

$$\text{ARE}(\bar{X}_{n,\alpha}, \bar{X}) \geq \frac{1}{1+4\alpha}.$$

Remark. The bound cannot be improved and is attained in the limit for a uniform distribution; see Lehmann (1983).

One can remove the assumption of unimodality and prove the following bound.

Theorem 18.4 Let $X_1, X_2, \ldots, X_n \overset{\text{iid}}{\sim} F(x-\theta)$, where, under F, X and $-X$ have the same distribution. Then

$$\text{ARE}(\bar{X}_{n,\alpha}, \bar{X}) \geq (1-2\alpha)^2.$$

Remark. This bound is attained at a suitable mixed distribution F_0 with an atom at $F_0^{-1}(1-\alpha)$; see Lehmann (1983).

18.3 Multivariate Trimmed Mean

In the one-dimensional case, trimmed means are means of observations left after having deleted a certain number of extreme order statistics from each tail. Since there is no single natural ordering of multivariate data, trimmed means can mean a lot of different things in more than one dimension. Nevertheless, the basic idea of deleting observations that seem discordant can still be formalized and the asymptotic theory worked out for means of observations after having deleted such discordant observations.

We follow the development in Arcones (1995). We take as a multivariate trimmed mean the mean of those observations that are inside a ball of some suitable radius centered at the sample mean. The center can be chosen to be much more general statistics (M estimates), but the description is more complicated in those cases.

Definition 18.1 Let $0 < \alpha < 1$ be fixed. Given sample vectors $X_1, \ldots, X_n$ taking values in $\mathcal{R}^d$, let $K_n(x) = \frac{1}{n}\sum_{i=1}^n I_{||X_i - \bar{X}|| \le x}$ and $\lambda_n = \inf\{x : K_n(x) \ge 1 - \alpha\}$. Then a *multivariate trimmed mean* is defined as $\hat{\theta}_{n,\alpha} = \frac{1}{n - n\alpha}\sum_{i=1}^n I_{||X_i - \bar{X}|| \le \lambda_n}$.

Under mild conditions on the underlying distribution, such a multivariate trimmed mean is asymptotically d-variate normal. To describe the asymptotic distribution, we need quite a bit of notation, which we introduce below.

Let

$\theta_0 = E(X), \phi(x) = x, l_{\theta,\lambda}(x) = xI_{||x-\theta|| \le \lambda}, L_{\theta,\lambda}(x) = E(l_{\theta,\lambda})$, $h_{\theta,\lambda}(x) = I_{||x-\theta|| \le \lambda}, H_{\theta,\lambda} = E(h_{\theta,\lambda}), K(x) = P(||X - \theta_0|| \le x), \lambda_0 = K^{-1}(1 - \alpha)$ (assumed to exist and such that $K'(\lambda_0) > 0, K(\lambda_0) = 1 - \alpha$), $D_{d \times d} = \left(\left(\frac{\partial L_i}{\partial \theta_j}\right)\right)|_{\theta_0,\lambda_0}, e_i = \frac{\partial L_i}{\partial \lambda}|_{\theta_0,\lambda_0}, e = (e_1, \ldots, e_d), b = \nabla_\theta H|_{\theta_0,\lambda_0}, c = \frac{\partial H}{\partial \lambda}|_{\theta_0,\lambda_0}, T_0 = \frac{1}{1-\alpha}E(l(\theta_0, \lambda_0))$.

With this notation, the following asymptotic representation of the multivariate trimmed mean T_n holds (Arcones (1995)).

Theorem 18.5 Let P_n denote the expectation operator w.r.t. the empirical measure F_n and let P denote expectation with respect to F. Then,

$$\sqrt{n}(T_n - T_0) = \sqrt{n}(P_n - P)\left[l_{\theta_0,\lambda_0} + D\phi - \left(\frac{1}{c}h_{\theta_0,\lambda_0} + \frac{1}{c}b'\phi\right)e\right] + o_p(1),$$

and hence $\sqrt{n}(T_n - T_0)$ is asymptotically normally distributed.

Example 18.2 As an illustration of Theorem 18.5, we work out in this example the asymptotic distribution of the multivariate trimmed mean when the samples are iid from a d-dimensional $N(\theta, I)$ distribution. Without loss of generality, we let the true $\theta_0 = \mathbf{0}$. To find the asymptotic distribution, we need to evaluate the vectors e and b, the matrix D, and the constant c in Theorem 18.5. From symmetry considerations, it is easy to see that $e = 0$. Thus, we only need to find the matrix D. Again, because of symmetry, D is diagonal, and, in fact, it is even true that it is a multiple of the identity matrix. Denoting the χ_d^2 CDF by $Q_d(.)$, the common diagonal element is

$$c_d = \frac{1}{(2\pi)^{\frac{d}{2}}}\left[\int_{||x|| \le \lambda_0} e^{-\frac{x'x}{2}}dx - \frac{1}{d}\int_{||x|| \le \lambda_0} ||x||^2 e^{-\frac{x'x}{2}}dx\right]$$

$$= Q_d(\lambda_0^2) - \left[Q_d(\lambda_0^2) - \frac{2\lambda_0^d e^{-\frac{\lambda_0^2}{2}}}{d 2^{\frac{d}{2}} \Gamma(\frac{d}{2})} \right] = \frac{2\lambda_0^d e^{-\frac{\lambda_0^2}{2}}}{d 2^{\frac{d}{2}} \Gamma(\frac{d}{2})}$$

on integrating each term in the preceding line separately. The integrations are done by using the fact that, under $\theta_0 = 0$, $||X||^2 \sim \chi_d^2$, and by integrating the second term by parts and then some algebraic simplification.

Thus, the Bahadur representation of $\sqrt{n}(T_n - T_0)$ works out to

$$\sqrt{n}T_n = \sqrt{n}\left[\frac{1}{n}\sum_{i=1}^{n} g(X_i) - E(g(X))\right] + o_p(1),$$

where $g(x) = xI_{||x|| \leq \lambda_0} + c_d x$. It follows that, under $\theta_0 = 0$, $\sqrt{n}T_n \stackrel{\mathcal{L}}{\Rightarrow} N_d(\mathbf{0}, R)$, where R is the covariance matrix of $g(X)$.

We now only need to evaluate the matrix R. From symmetry considerations, again R is a multiple of the identity matrix with the common diagonal element being the second moment of $X_1 I_{||X|| \leq \lambda_0} + c_d X_1$. Now,

$$\begin{aligned}
E(X_1 I_{||X|| \leq \lambda_0} + c_d X_1)^2 &= E(X_1^2 I_{||X|| \leq \lambda_0} + c_d^2 X_1^2 + 2c_d X_1^2 I_{||X|| \leq \lambda_0}) \\
&= c_d^2 + (1 + 2c_d) E(X_1^2 I_{||X|| \leq \lambda_0}) \\
&= c_d^2 + (1 + 2c_d)\left(Q_d(\lambda_0^2) - \frac{2\lambda_0^d e^{-\frac{\lambda_0^2}{2}}}{d 2^{\frac{d}{2}} \Gamma(\frac{d}{2})} \right) \\
&= (1 + 2c_d)(1 - \alpha) - c_d - c_d^2 = \gamma_d,
\end{aligned}$$

where the last line follows by using the definition of c_d. Thus, $\sqrt{n}T_n \stackrel{\mathcal{L}}{\Rightarrow} N_d(\mathbf{0}, \gamma_d I)$.

The value of γ_d for a few values of the dimension d are given below, where $\alpha = .05$.

d	2	3	4	5	6	7
γ_d	1.062	1.041	1.029	1.020	1.013	1.008

Note that γ_d appears to converge to 1 as the dimension $d \to \infty$.

18.4 The 10-20-30-40 Rule

Trimmed means have excellent asymptotic efficiency in normal and t models but not very good efficiency in the double exponential model. The mean does not have good efficiencies in either the t or the double exponential model; similarly, the median fails to have good efficiencies in t models. It is interesting that by taking a convex combination of the mean, the median, a trimmed mean, and the Huber estimate, one can attain good asymptotic efficiencies across a broad spectrum of models. It is shown in Banerjee and DasGupta (2005) that an estimate that combines the mean, a 5% trimmed mean, the Huber estimate with $k = 1$, and the median in the ratio 10-20-30-40 has such an attractive efficiency property. Note that the efficiencies are *not* 1, only very good. That is, the estimate is not *adaptive* but only semiadaptive. But it is far less complex than what full adaptivity would demand; see, e.g., Hogg (1974), Stone (1975), and Koul and Susarla (1983).

The key to finding a good convex combination is to use the Bahadur representation for each estimate and thereby obtain the joint asymptotic distribution of the four estimates. There is nothing special about taking a convex combination of four estimates; any number of estimates can be used as long as we can find the joint asymptotic distribution. Once we know the joint distribution, the asymptotic distribution of any linear combination follows. The inverse efficiency of a linear combination in a specific model satisfying the Cramér-Rao conditions is obtained by dividing the asymptotic variance of the linear combination by the reciprocal of the Fisher information in that model. The 10-20-30-40 rule gives good efficiencies across a broad spectrum of location models.

Of the four estimates under consideration, the mean is exactly linear. The Bahadur representation for the median was presented in Chapter 7, that for the Huber estimate in Chapter 17, and for the trimmed mean it is stated in this chapter. Putting all of these together, we have the following theorem.

Theorem 18.6 Let $X_1, \ldots, X_n \stackrel{\text{iid}}{\sim} f(x - \theta)$, and suppose $f(.)$ is such that each of $M_n = F_n^{-1}(\frac{1}{2})$, $\bar{X}_{n,\alpha}$, and $\hat{\theta}_{n,H}$ admits a Bahadur expansion. Suppose also that f has a finite variance. Then,

$$\sqrt{n}(\bar{X} - \theta, M_n - \theta, \bar{X}_{n,\alpha} - \theta, \hat{\theta}_{n,H} - \theta) \stackrel{\mathcal{L}}{\Rightarrow} N_4(\mathbf{0}, \Sigma)$$

for an appropriate Σ. Furthermore, $\sqrt{n}(c_1\bar{X} + c_2 M_n + c_3\bar{X}_{n,\alpha} + c_4\hat{\theta}_{n,H} - \theta(c_1 + c_2 + c_3 + c_4)) \stackrel{\mathcal{L}}{\Rightarrow} N(0, c'\Sigma c)$.

Remark. The exact Σ depends on the choice of α and on k (of the Huber estimate). Using Theorem 18.6, the efficiency of the 10-20-30-40 rule in a few specific models is listed below.

Example 18.3 Consider *generalized exponential models* with $f_\gamma(z) = ce^{-|z|^\gamma}$, $1 \leq \gamma \leq 2$, and t-distributions with m degrees of freedom as the trial models. Together, they allow a very wide selection of tails. The normal, double exponential, and heavy-tailed t-densities are all covered. The table below gives the asymptotic relative efficiency of the 10–20–30–40 rule with respect to the MLEs in some of these models.

Model	f_1	$f_{1.1}$	$f_{1.3}$	$f_{1.5}$	$f_{1.75}$	f_2	t_3	t_5	t_7	t_{10}
Efficiency	.870	.944	.996	.987	.938	.871	.955	.964	.951	.934

It is seen that the estimate has a particularly impressive efficiency at heavy and intermediate tails. As regards the lower efficiency of the 10-20-30-40 rule for the normal and the double exponential models, *it is actually not easy to get very high efficiency simultaneously in those two models without using complicated adaptation schemes.*

18.5 Exercises

Exercise 18.1 * Show that, for every $0 < \alpha < \frac{1}{2}$, there exists a Huber estimate with a suitable k such that it has the same asymptotic variance as $\bar{X}_{n,\alpha}$ under a given f and vice versa.

Exercise 18.2 * Find the trimmed mean that maximizes efficiency for double exponential data, subject to 90% efficiency for normal data. Are you happy with the value of the resulting maximum efficiency?

Remark. Refer to Example 18.3.

Exercise 18.3 Calculate the efficiency of the 12.5% trimmed mean for t-distributions and plot it as a function of the degrees of freedom.

Exercise 18.4 Suggest another reasonable definition of a trimmed mean for multivariate data. Would you be able to carry out the asymptotics with your definition?

Exercise 18.5 * Identify an f for which the efficiency lower bound of Theorem 18.2. is attained.

Exercise 18.6 * Find the efficiency of the 10-20-30-40 rule for f in the Tukey, logistic, and t_4 models.

References

Arcones, M. (1995). Asymptotic normality of multivariate trimmed means, Stat. Prob. Lett., 25(1), 43–53.

Banerjee, M. and DasGupta, A. (2005). The 10-20-30-40 rule and its impressive efficiency properties, preprint.

Bickel, P.J. and Lehmann, E.L. (1975). Descriptive statistics for nonparametric models, II: Location, Ann. Stat., 3(5), 1045–1069.

Hogg, R. (1974). Adaptive robust procedures: a partial review and some suggestions for future applications and theory, J.Am. Stat. Assoc., 69, 909–927.

Koul, H.L. and Susarla, V. (1983). Adaptive estimation in linear regression, Stat. Decisions, 1(4–5), 379–400.

Lehmann, E.L. (1983). *Theory of Point Estimation*, John Wiley, New York.

Stone, C.J. (1975). Adaptive maximum likelihood estimators of a location parameter, Ann. Stat., 3, 267–284.

Chapter 19
Multivariate Location Parameter and Multivariate Medians

There are a variety of interesting extensions of the concept of the univariate median to the multivariate case. Some of these do not require the structure of a location parameter for developing the distribution theory, whereas others do. Often some form of symmetry in the population is needed to have a meaningful definition or a distribution theory. We start with a list of some definitions of multidimensional medians that have been offered and their properties under some suitable types of symmetry. Of the numerous types of multidimensional medians that are available in the literature, we mention only the L_1 median, the vector of coordinatewise medians, the Oja median, and peeled medians. The main theorems in this chapter are derived in Hettmansperger and McKean (1998). Various statistical uses and generalizations and history are available in Brown (1983), Chaudhuri (1996), and Small (1990).

19.1 Notions of Symmetry of Multivariate Data

Definition 19.1 A distribution F on $\mathcal{R}^p$ is said to be coordinatewise symmetric around a point $\mu \in \mathcal{R}^p$ if $X - \mu$ and $\mu - X$ have the same distribution under F.

Definition 19.2 Let $X \sim F$ and suppose that, for any $(p \times p)$ orthogonal matrix Γ, $Y = \Gamma X$ also has the same distribution F. Then F is called spherically symmetric. If F has a density with respect to a p-dimensional Lebesgue measure, then the density f is of the form $f(x) = f(x'x)$.

The $N_p(0, I)$ is an example of a spherically symmetric law. The $N_p(\mu, I)$ is spherically symmetric around μ; i.e., if $X \sim N_p(\mu, I)$, then $X - \mu$ and $\Gamma(X - \mu)$ are identically distributed for all orthogonal Γ. If $X - \mu$ is spherically symmetric, then $X_1 - \mu_1, \cdots, X_p - \mu_p$ cannot be independent unless they are $N(0, \sigma^2)$ for some σ^2.

A. DasGupta, *Asymptotic Theory of Statistics and Probability*,

Definition 19.3 Let $X \sim F$, where F is spherically symmetric around some μ. Let A be a nonsingular matrix. Then the distribution of $A(X-\mu)+b$ is elliptically symmetric around b.

Examples of elliptically symmetric distributions are uniform distributions on or inside ellipsoids and the multivariate normal and t-distributions. Elliptically symmetric distributions that have a density have densities of the form $f(x) = |\Sigma|^{-1/2} f((x-b)'\Sigma^{-1}(x-b))$.

19.2 Multivariate Medians

Many of the multivariate medians are defined through minimization of an appropriate criterion function $\sum \rho(X_i, \mu)$ and correspondingly an estimating equation $\sum \psi(X_i, \mu) = 0$. Thus they can be regarded as multidimensional M estimates.

We list a few criterion functions and the corresponding medians. We consider bivariate data for simplicity.

1. Minimize $\sqrt{\sum \left\{(x_{i_1}-\mu_1)^2 + (x_{i_2}-\mu_2)^2\right\}} = D_1(\mu)$.

   This is just traditional least squares and results in $\hat{\mu} = \begin{pmatrix}\hat{\mu}_1\\ \hat{\mu}_2\end{pmatrix} = \begin{pmatrix}\bar{X}_1\\ \bar{X}_2\end{pmatrix} = \bar{X}$.

2. Pluck the $\sqrt{}$ sign and put it inside. That is, minimize

$$\sum_i \sqrt{\left\{(x_{i_1}-\mu_1)^2 + (x_{i_2}-\mu_2)^2\right\}} = \sum_i ||X_i - \mu||_2 = D_2(\mu).$$

   The L_1 median is defined as $\operatorname{argmin}_\mu D_2(\mu)$.

3. Put the $\sqrt{}$ sign further inside. That is, minimize

$$\sum_i \left\{\sqrt{(x_{i_1}-\mu_1)^2} + \sqrt{(x_{i_2}-\mu_2)^2}\right\} = \sum_i (|x_{i_1}-\mu_1| + |x_{i_2}-\mu_2|) = D_3(\mu).$$

   The problem is now one-dimensional, and $D_3(\mu)$ is minimized at the vector of coordinate medians.

The minimization of a criterion $D(\mu)$ will sometimes correspond to $S(\mu) = \nabla_\mu D(\mu) = 0 \Leftrightarrow \sum \psi(X_i - \mu) = 0$ for some ψ. This is the similarity to M estimates.

Of these different estimates, the L_1 median is special. We state a few key properties of the L_1 median; they are not hard to prove.

Proposition 19.1 (a) The L_1 median of a dataset is unique with probability 1 if X_i are samples from an absolutely continuous distribution.

(b) If we denote the L_1 median of $X_1, \cdots, X_n$ as $\hat{\mu}(X_1, \cdots, X_n)$, then, for any orthogonal Γ,

$$\sum_i ||\Gamma X_i - \Gamma\hat{\mu}||_2 = \sum_i ||X_i - \hat{\mu}||_2 \leq \sum_i ||X_i - \mu||_2$$
$$= \sum_i ||\Gamma X_i - \Gamma\mu||_2.$$

Thus $\hat{\mu}(\Gamma X_1, \cdots, \Gamma X_n) = \Gamma(\hat{\mu}(X_1, \cdots, X_n))$.

(c) The L_1 median is, however, not affine or even scale invariant; i.e., $\hat{\mu}(AX_1, \cdots, AX_n) \neq A(\hat{\mu}(X_1, \cdots, X_n))$ for a nonsingular matrix A in general.

Example 19.1 Consider bivariate data, and for simplicity suppose $n = 3$.

Consider now the criterion function $D_4(\mu) = \sum_{1\leq i<j\leq n(=3)}$ Area $(\Delta(X_i, X_j, \mu))$, where $\Delta(X_i, X_j, \mu)$ stands for the triangle in the plane with X_i, X_j, μ as the vertices. Then $\text{argmin}_\mu D_4(\mu)$ is called the Oja simplicial median. In higher dimensions, the areas of triangles are replaced by volumes

$$D_4(\mu) = \sum_{1\leq i_1<i_2\cdots<i_p\leq n} \text{Vol}(\Delta(X_{i_1}, \cdots, X_{i_p}, \mu))$$

and $\hat{\mu}_{Oja} = \text{argmin}_\mu D_4(\mu)$.

Remark. $\hat{\mu}_{Oja}$ is affine invariant but not necessarily unique. Multivariate medians are also sometimes defined through a process of repeated peeling (also known as stripping) of the convex hull of a set of sample vectors $X_1, \ldots, X_n$. The peeled median is obtained according to some well-defined selection rule when no further peeling can be done.

An unfortunate feature of peeled medians is that the peeled median is not a continuous function of $X_1, \ldots, X_n$. Therefore, the distribution theory is very difficult. In addition, there is no obvious corresponding population analog for a peeled median. However, it follows from the definition of a convex hull that the peeled median is affine invariant.

19.3 Asymptotic Theory for Multivariate Medians

The multidimensional medians we discussed are point estimates for multidimensional location parameters under appropriate symmetry of the distribution. Usually, one wants to go beyond point estimation and conduct hypothesis tests and construct confidence sets. The asymptotic theory of the multidimensional medians would be needed for those purposes. We address the question of deriving limiting distributions of multidimensional medians in this section. Although we often focus on the bivariate case for illustration, the theorems will generally cover the case of a general dimension k. We look at estimates that satisfy an *estimating equation*.

Let $S(\theta)$ be a function such that the multivariate median $\hat{\theta}_n$ solves $S(\theta) = 0$. For example, in the bivariate case, $\hat{\theta}_n$ solves

$$\begin{pmatrix} S_1(\theta) \\ S_2(\theta) \end{pmatrix} = \begin{pmatrix} 0 \\ 0 \end{pmatrix}.$$

We take $X_1, \ldots, X_n \sim F(x - \theta)$ and make the following assumptions:

(a) $S_i(\theta)$ are coordinatewise monotone for all i.

(b) $E_0 S(0) = 0$.

(c) For some p.d. matrix A, $\frac{S(0)}{\sqrt{n}} \xrightarrow{\mathcal{L}} N_k(0, A)$.

Define

$$B = \nabla_\theta E_{F_\theta} \frac{1}{n} S(0) \mid_{\theta=0} .$$

Note that B is a $k \times k$ matrix if $\theta \in R^k$. We have the following theorem on the limiting distribution of the root of $S(\theta) = 0$; for a proof of it, and also the derivation of A and B in the next subsection, see Hettmansperger and McKean (1998).

Theorem 19.1 If the true value $\theta_0 = 0$ and A, B are nonsingular, then $\sqrt{n}\hat{\theta}_n \xrightarrow{\mathcal{L}} N_k(0, B^{-1}AB^{-1})$.

Remark. In applications, the matrices A and B have to be evaluated on a case-by-case basis, and this task may not be easy. Sometimes a general description is possible, as we show below. Note that both A and B are functionals of F, and for certain types of F, the evaluation of A, B is harder.

19.4 The Asymptotic Covariance Matrix

For simplicity, consider the bivariate case and suppose $S(\theta)$ has the form

$$S(\theta) = \begin{pmatrix} S_1(\theta) \\ S_2(\theta) \end{pmatrix} = \begin{pmatrix} \Sigma\psi(x_{i1} - \theta_1) \\ \Sigma\psi(x_{i2} - \theta_2) \end{pmatrix},$$

where $x_i = \begin{pmatrix} x_{i_1} \\ x_{i_2} \end{pmatrix}$, $1 \le i \le n$, are the sample observations. Then,

$$A = \begin{pmatrix} E\psi^2(X_{11}) & E\psi(X_{11})\psi(X_{12}) \\ E\psi(X_{11})\psi(X_{12}) & E\psi^2(X_{12}) \end{pmatrix}$$

and

$$B = \begin{pmatrix} E\psi^{'}(X_{11}) & 0 \\ 0 & E\psi^{'}(X_{12}) \end{pmatrix}.$$

Here, $E(\cdot)$ denote $E_{F_{\theta_0}}$, where we have assumed that θ_0 is equal to 0.

Example 19.2 Consider $\hat{\theta}_n$ equal to the vector of coordinate medians. In this case, the corresponding ψ function is $\psi(t) = \text{sgn}(t)$. Then, one easily finds $B^{-1}AB^{-1}$ to be equal to

$$B^{-1}AB^{-1} = \begin{pmatrix} \frac{1}{4f_1^2(0)} & \frac{E\ \text{sgn}(X_1)\text{sgn}(X_2)}{4f_1(0)f_2(0)} \\ \frac{E\ \text{sgn}(X_1)\text{sgn}(X_2)}{4f_1(0)f_2(0)} & \frac{1}{4f_2^2(0)} \end{pmatrix},$$

where f_1 and f_2 are the marginal densities of the coordinates of $X_1 = \begin{pmatrix} X_{11} \\ X_{12} \end{pmatrix}$.
If F is continuous, then the off-diagonal element only requires calculation of $P_{\theta=0}(X_1 > 0, X_2 > 0) - P_{\theta=0}(X_1 < 0, X_2 < 0)$. For example, if $X_1 \sim BVN(0, 0, \sigma_1^2, \sigma_2^2, \rho)$, then $P(X_1 > 0, X_2 > 0) = P(X_1 < 0, X_2 < 0) = \frac{1}{4} + \frac{1}{2\pi}\arcsin(\rho)$ by a well-known formula.

Definition 19.4 If $\sqrt{n}(\hat{\theta}_{in} - \theta_0) \xrightarrow{\mathcal{L}} N_2(0, \Sigma_i)$, $i = 1, 2$, then we define the ARE of $\hat{\theta}_{2n}$ with respect to $\hat{\theta}_{1n}$ for bivariate data as

$$\left(\frac{|\Sigma_1|}{|\Sigma_2|}\right)^{1/2}.$$

Table 19.1 Effect of a correlation on the asymptotic efficiency of the vector of medians with respect to the mean

ρ	0	0.1	0.2	0.4	0.5	0.7	0.9	0.999
$\text{ARE}(\hat{\theta}_n, \bar{X})$	0.64	0.63	0.63	0.60	0.58	0.52	0.4	0.2

For k dimensions, the definition is

$$\left(\frac{|\Sigma_1|}{|\Sigma_2|}\right)^{1/k}.$$

This definition corresponds to ratios of volumes of asymptotic confidence sets of equal nominal coverages constructed from the respective asymptotic distributions.

Example 19.2 (continued) Consider the $BVN(0, 0, \sigma_1^2, \sigma_2^2, \rho)$ case. Then, we immediately have

$$\text{ARE}(\hat{\theta}_n, \bar{X}) = \frac{2}{\pi}\sqrt{\frac{1-\rho^2}{1-(\frac{2}{\pi}\arcsin(\rho))^2}}.$$

Table 19.1 illustrates the damaging effect of a correlation on the asymptotic efficiency of the vector of coordinatewise medians with respect to the mean.

19.5 Asymptotic Covariance Matrix of the L_1 Median

Suppose $X_1, \ldots, X_n \stackrel{\text{iid}}{\sim} F(x-\theta)$, and assume that the true $\theta_0 = 0$. Then A and B admit the following formulas:

$$A = E_{\theta=0}\frac{XX^{\mathrm{T}}}{||X||^2} \text{ and } B = E_{\theta=0}\frac{I - \frac{XX^{\mathrm{T}}}{||X||^2}}{||X||};$$

see Hettmansperger and McKean (1998). These formulas simplify in the spherically symmetric case.

Corollary 19.1 Let $X_1, \ldots, X_n \stackrel{\text{iid}}{\sim} F(x-\theta)$, and suppose that F is spherically symmetric around the zero vector. Assume that the true θ_0 equals 0. For the case $k = 2$, let (r, θ) denote the polar coordinates of X. Then $A = \frac{1}{2}I$ and $B = \frac{1}{2}E(1/r)I$. Consequently, we have

$$\text{ARE}(L_1 \text{ median}, \bar{X}) = \frac{1}{4} E(r^2)(E(1/r))^2.$$

Remark. If the density of X is $f(x) = h(r)$, then it is easy to see that $r \sim 2\pi r h(r)$, $r > 0$. This fact helps us calculate the moments and the inverse moments of r, which are required in the efficiency calculation.

The corresponding ARE for k dimensions works out to

$$\begin{aligned}
\text{ARE}(L_1 \text{ median}, \bar{X}) &= \left(\frac{k-1}{k}\right)^2 E(r^2)(E(r^{-1}))^2 \\
&\geq \left(\frac{k-1}{k}\right)^2 (E(r))^2(E(r^{-1}))^2 \\
&= \left(\frac{k-1}{k}\right)^2 (E(r)E(r^{-1}))^2 \\
&\geq \left(\frac{k-1}{k}\right)^2 \longrightarrow 1 \text{ as } k \to \infty.
\end{aligned}$$

Furthermore, in the special $N_k(0, I)$ case, the ARE works out to the exact formula

$$\text{ARE}(L_1 \text{ median}, \bar{X}) = \left(\frac{k-1}{k}\right)^2 k \left(\frac{\Gamma(\frac{k-1}{2})}{\sqrt{2}\Gamma(\frac{k}{2})}\right)^2.$$

Table 19.2 gives some numerical values when $F = N_k(0, I)$.

It is also possible to explicitly calculate the matrices A, B in the asymptotic distribution of the L_1 median when the population is elliptically symmetric. By an orthogonal transformation, the scale matrix can be diagonalized; i.e., if the population density is of the form $f((x-\theta)'\Sigma^{-1}(x-\theta))$, then there is an orthogonal matrix P such that $P\Sigma P'$ is a diagonal matrix. The ARE of the L_1 median with respect to $\bar{X}$ remains unchanged by this orthogonal transformation, and so we will assume that, in the special bivariate case, $\Sigma = \begin{pmatrix} 1 & 0 \\ 0 & \sigma^2 \end{pmatrix}$. By transforming to the polar coordinates (r, θ) again, the expressions for A, B in general are

Table 19.2 ARE when $F = N_k(0, I)$

k	2	4	6	8	10
ARE(L_1 median, $\bar{X}$)	0.785	0.884	0.920	0.940	0.951

$$A = \begin{pmatrix} E\cos^2\theta & E\sin\theta\cos\theta \\ & E\sin^2\theta \end{pmatrix}, \quad B = \begin{pmatrix} Er^{-1}\sin^2\theta & -Er^{-1}\sin\theta\cos\theta \\ & Er^{-1}\cos^2\theta \end{pmatrix}.$$

For example, in the $BV\ N(0, 0, 1, \sigma^2, 0)$ case,

$$E\cos^2\theta = \frac{1}{1+\sigma}, \quad E\sin^2\theta = \frac{\sigma}{1+\sigma}, \quad E\sin\theta\cos\theta = 0,$$

$$Er^{-1}\sin\theta\cos\theta = 0,$$

$$Er^{-1}\cos^2\theta = \frac{\sqrt{\pi}}{2\sqrt{2}} \sum_{j=0}^{\infty} \left[\frac{(2j+2)!(2j)!}{2^{4j+1}(j!)^2[(j+1)!]^2} \right]^2 (1-\sigma^2)^j,$$

$$Er^{-1} = \frac{\sqrt{\pi}}{\sqrt{2}} \sum_{j=0}^{\infty} \left[\frac{(2j)!}{2^{2j}(j!)^2} \right]^2 (1-\sigma^2)^j,$$

$$Er^{-1}\sin^2\theta = Er^{-1} - Er^{-1}\cos^2\theta,$$

where $Er^{-1}\cos^2\theta$ is as above. These expressions lead to $B^{-1}AB^{-1}$ and hence the asymptotic covariance matrix of the L_1 median. See Hettmansperger and McKean (1998) for all of the expressions in this section.

Table 19.3 illustrates the ARE of the L_1 median with respect to the sample mean vector in the $BV\ N(0, 0, 1, \sigma^2, 0)$ case.

Remark. The numerical values show that the L_1 median is not competitive when the variances of the coordinate variables are very different. It is more competitive for spherically symmetric data rather than elliptically symmetric data, and particularly so in high dimensions. L_1 medians are very competitive with $\bar{X}$ even for normal data and are more robust than $\bar{X}$ for nonnormal data.

Table 19.3 ARE(L_1 median, $\bar{X}$)

σ	1	0.8	0.6	0.4	0.2	0.01
ARE	0.785	0.783	0.773	0.747	0.678	0.321

Table 19.4 ARE($\hat{\theta}_{Oja}, \bar{X}$)

k	1	2	3	5	10
a_k	$\frac{\pi}{2}$	$\frac{4}{\pi}$	$\frac{3\pi}{8}$	$\frac{45\pi}{128}$	$\frac{65536}{19845\pi}$
ARE($\hat{\theta}_{Oja}, \bar{X}$)	0.637	0.617	0.612	0.608	0.607

Remark. The Oja simplicial median's asymptotic distribution is also multivariate normal. The asymptotic distribution can be worked out using the theory of U-statistics; see Nadar, Hettmansperger, and Oja (2003). We give an example to illustrate its performance.

Example 19.3 Suppose $X_i \stackrel{\text{iid}}{\sim} N_k(\theta_0, \Sigma)$. Then the Oja median $\hat{\theta}_{Oja}$ has a limiting distribution given by

$$\sqrt{n}(\hat{\theta}_{Oja} - \theta_0) \xrightarrow{\mathcal{L}} N_k(0, a_k \Sigma),$$

where $a_k = \Gamma(\frac{k}{2})\Gamma(\frac{k}{2}+1)/(\Gamma(\frac{k+1}{2}))^2 > 1$.

Some values of a_k are given in Table 19.4.

19.6 Exercises

Exercise 19.1 * Prove that with probability 1 the L_1 median is unique for $k > 1$ if the samples are from an absolutely continuous distribution.

Exercise 19.2 *

(a) Simulate 20 samples from a standard bivariate normal distribution.
(b) Compute the mean, the L_1 median, the Oja median, and the peeled median for your simulated dataset (plotting the data would be useful).

Exercise 19.3 * Find an expression for the ARE of the L_1 median with respect to the mean when samples are from the bivariate t-density

$$\frac{c}{(1 + \frac{(x-\mu)'\Sigma^{-1}(x-\mu)}{m})^{\frac{m}{2}+1}}, \quad m > 2, \quad \text{where } \Sigma = \begin{pmatrix} 1 & \rho \\ \rho & 1 \end{pmatrix}.$$

Exercise 19.4 * Suppose samples are from a $BVN(\mu_1, \mu_2, 1, 1, \rho)$-density. Characterize the values of ρ for which the Oja median is more efficient than the L_1 median.

Exercise 19.5 * Suggest a confidence region for μ based on the L_1 median when samples are from the $N_k(\mu, I)$-density, and suggest another one based on the Oja median. Is one of them always smaller (in the sense of volume) for the same nominal coverage level?

Exercise 19.6 Suppose $X = \binom{X_1}{X_2}$ has a spherical t-distribution. Find the density of $r = \sqrt{X_1^2 + X_2^2}$. How hard is this calculation if it is an elliptical t?

References

Brown, B.M. (1983). Statistical uses of the spatial median, J.R. Stat. Soc. Ser. B, 45(1), 25–30.

Chaudhuri, P. (1996). On a geometric notion of quantiles for multivariate data, J. Am. Stat. Soc., 91(434), 862–872.

Hettmansperger, T.P. and McKean, J.W. (1998). *Robust Nonparametric Statistical Methods*, John Wiley, New York.

Nadar, M., Hettmansperger, T.P., and Oja, H. (2003). The asymptotic covariance matrix of the Oja median, Stat. Prob. Lett., 64(4), 431–442.

Small, C.G. (1990). A survey of multidimensional medians, Int. Stat. Rev., 58, 263–277.

Chapter 20
Bayes Procedures and Posterior Distributions

Although maximum likelihood forms the principal conceptual framework of estimation, Bayes procedures are clearly practically useful in many problems. Besides, there are well-known theorems in decision theory that say that *reasonable* procedures should be Bayes or generalized Bayes with respect to some prior distribution. To the philosophically committed Bayesian, the posterior distribution is much more relevant than the sampling distribution of the Bayes rule. On the other hand, to those who consider Bayes procedures as useful tools, sampling properties are relevant, too. We treat asymptotics of the posterior distribution of the parameter and the sampling distributions of the Bayes procedure in this chapter.

The asymptotic behavior of Bayes procedures and posterior distributions can be casually summarized easily. Asymptotically, Bayes procedures are equivalent to an optimal frequentist procedure with the prior having no effect, and posteriors are asymptotically normal, again without the prior having an effect. But the exact meanings of these statements are quite subtle, and it is not really true that the prior has no effect on the asymptotic behavior of Bayes procedures. Certain properties of a Bayes procedure *are* affected, even asymptotically, by the choice of the prior, whereas others are not. It depends on how deep one wants to look.

Careful treatment of aspects of Bayesian asymptotics is available in Bickel and Doksum (2001), Hartigan (1983), and Schervish (1995). LeCam (1986) also gives a rigorous treatment of certain aspects. Lehmann and Casella (1998) and van der Vaart (1998) are also useful references. Lucid expositions can be found in Berger (1985) and Wasserman (2004). Infinite-dimensional problems are discussed in Diaconis and Freedman (1986), Freedman (1991), Wasserman (1998), and Walker (2004). A novel method of higher-order Bayesian asymptotics is presented in Woodroofe (1992). Other specific references are given in the following sections.

A. DasGupta, *Asymptotic Theory of Statistics and Probability*,

20.1 Motivating Examples

We start with two motivating examples.

Example 20.1 Let $X_1, X_2, \ldots, X_n \overset{\text{iid}}{\sim} N(\theta, \sigma^2)$, $\sigma^2 > 0$ known, and $\theta \sim N(\eta, \tau^2)$. The posterior distribution of θ given $X^{(n)} = (X_1, \cdots, X_n)$ is

$$N\left(\frac{\tau^2}{\sigma^2/n+\tau^2}\bar{X} + \frac{\sigma^2/n}{\sigma^2/n+\tau^2}\eta, \frac{\tau^2\sigma^2/n}{\sigma^2/n+\tau^2}\right)$$
$$= N\left(\bar{X} + \frac{\sigma^2/n}{\sigma^2/n+\tau^2}(\eta - \bar{X}), \frac{\tau^2\sigma^2/n}{\sigma^2/n+\tau^2}\right).$$

Also, the posterior distribution of $\sqrt{n}(\theta - \bar{X})$ is

$$N\left(\frac{\sigma^2/\sqrt{n}}{\sigma^2/n+\tau^2}(\eta - \bar{X}), \frac{\tau^2\sigma^2}{\sigma^2/n+\tau^2}\right).$$

Suppose the true value of θ is θ_0; i.e., $X_1, X_2, \ldots, X_n \overset{\text{iid}}{\sim} N(\theta_0, \sigma^2)$. Then $\bar{X} \overset{\text{a.s.}}{\to} \theta_0$ and $\frac{\tau^2\sigma^2}{\sigma^2/n+\tau^2} \to \sigma^2$. If we denote the sequence of normal distributions above by Π_n and the $N(0, \sigma^2)$ distribution by Q, then it follows from Scheffe's theorem (see Chapter 2) that the total variation distance is

$$d_{\text{TV}}(\Pi_n, Q) \to 0 \text{ a.s.}, \forall \theta_0.$$

In particular,

$$\sup_{-\infty<t<\infty} \left| P(\sqrt{n}(\theta - \bar{X})/\sigma \le t | X^{(n)}) - \Phi(t) \right| \to 0 \quad \text{a.s.}, \forall \theta_0;$$

i.e., the CDF of the posterior of $\sqrt{n}(\theta - \bar{X})/\sigma$ is asymptotically *uniformly* close to the $N(0, 1)$ CDF, regardless of the choice of the prior parameters η, τ^2. In this sense, in this example, the prior did not matter.

Example 20.2 Suppose $X_1, X_2, \ldots, X_n$ are iid Ber$(1, p)$ observations and $p \sim$ Beta(a, b). Then the posterior density of p is $\frac{p^{a+n\bar{X}-1}(1-p)^{b+n(1-\bar{X})-1}\Gamma(a+b+n)}{\Gamma(a+n\bar{X})\Gamma(b+n(1-\bar{X}))}$, and therefore the posterior density of $\psi = \sqrt{n}(p - \bar{X})$ is

$$\pi_n(\psi) = \frac{\left(\bar{X} + \psi/\sqrt{n}\right)^{a+n\bar{X}-1}\left(1 - \bar{X} - \psi/\sqrt{n}\right)^{b+n(1-\bar{X})-1}\Gamma(a+b+n)}{\Gamma(a+n\bar{X})\Gamma(b+n(1-\bar{X}))}.$$

On using Stirling's approximation to $\Gamma(a+n\bar{X})$, $\Gamma(b+n(1-\bar{X}))$, $\Gamma(a+b+n)$, and on using the expansions

$$\log\left(1+\psi/(\sqrt{n}\bar{X})\right) = \psi/(\sqrt{n}\bar{X}) - \psi^2/(2n\bar{X}^2) + o(n^{-3/2})$$

and

$$\log\left(1+\psi/(\sqrt{n}(1-\bar{X}))\right) = \psi/\left(\sqrt{n}(1-\bar{X})\right) - \psi^2/(2n(1-\bar{X})^2) + o(n^{-3/2})$$

a.s., one gets after some algebra that $\log(\pi_n(\psi)) \stackrel{\text{a.s.}}{\Rightarrow} \frac{1}{\sqrt{2\pi p_0(1-p_0)}} e^{-\psi^2/(2p_0(1-p_0))}$ if the true value is $p = p_0$. Thus, we again see that the density of $\sqrt{n}\frac{p-p_0}{\sqrt{p_0(1-p_0)}}$ converges a.s. to the $N(0,1)$ density under $p = p_0$, regardless of the choice of the prior parameters.

A general version of the two previous examples is called the Bernstein-von Mises theorem and is the strongest version known for asymptotic normality of posterior distributions. However, many conditions are needed for proving it in general. For a proof of the Bernstein-von Mises theorem, see Bickel and Doksum (2001).

20.2 Bernstein-von Mises Theorem

Under the iid assumption and further regularity conditions of the Cramér-Rao type, asymptotic normality of the posterior can be asserted in a very strong sense. However, a rigorous statement needs care, and particular attention is necessary when the iid assumption is dropped, as the result loses some of its strength without the iid assumption.

Theorem 20.1 Let $X_1, X_2, \ldots, X_n \stackrel{\text{iid}}{\sim} f(x|\theta) = \frac{dP_\theta}{d\mu}$, and let $\theta \sim \pi(\theta)$, a density. Assume that:

(a) Θ is open.
(b) $\int \frac{\partial^2}{\partial\theta^2} f(x|\theta)d\mu(x) = 0 \quad \forall\theta \in \Theta$.
(c) $0 < I(\theta) < \infty \quad \forall\theta$, where $I(\theta)$ is the Fisher information function.
(d) If θ_0 denotes the true value, then $\pi(\theta_0) > 0$, and $\pi(\theta)$ is continuous in $\theta_0 \pm \epsilon$ for some $\epsilon > 0$.
(e) $\exists k(x)$ such that $|\frac{\partial^2}{\partial\theta^2} f(x|\theta)| \leq k(x) \quad \forall\theta$ in $\theta_0 \pm \epsilon$ for some $\epsilon > 0$ and such that $E_{\theta_0}(k(X)) < \infty$.

Let $\hat{\theta}$ be any strongly consistent sequence of roots of the likelihood equation and denote $\psi = \sqrt{n}(\theta - \hat{\theta})$. Then the posterior of ψ, $\pi(\psi|X_1, \ldots, X_n)$ satisfies:

(A) $|\pi(\psi|X_1, \ldots, X_n) - \sqrt{I(\theta_0)}\phi(\psi\sqrt{I(\theta_0)})| \overset{\text{a.s.}}{\Rightarrow} 0$ (under P_{θ_0}).
(B) Provided $I(\theta)$ is continuous,

$$|\pi(\psi|X_1, \ldots, X_n) - \sqrt{I(\hat{\theta}_n)}\phi\left(\psi\sqrt{I(\hat{\theta}_n)}\right)| \overset{\text{a.s.}}{\Rightarrow} 0.$$

Corollary 20.1 Let $A \subseteq \Theta$ be any Borel set. Then,

$$\sup_A |P(\psi \in A|X_1, \ldots, X_n) - P(\psi \in A|\psi \sim N(0, I^{-1}(\theta_0)))| \overset{\text{a.s.}}{\Rightarrow} 0.$$

In particular, if θ is one-dimensional, then

$$\sup_{-\infty<x<\infty} |P(\psi \leq x|X_1, \ldots, X_n) - P(\psi \leq x|\psi \sim N(0, I^{-1}(\theta_0)))| \overset{\text{a.s.}}{\Rightarrow} 0.$$

Remark. This is what is casually referred to as asymptotic normality of the posterior. Again, one can use instead the $N(0, I^{-1}(\hat{\theta}_n))$ in the corollary, provided the Fisher information function is continuous.

What happens if, conditional on θ, X_i are not iid? There is still asymptotic normality of the posterior under regularity conditions, but in a weaker sense. See Heyde and Johnstone (1979) for results on asymptotic Gaussianity of the posterior under fairly general time series models. Here is one specific result without the iid assumption. Note that the theorem is weaker than what we had in the iid situation because we can now only assert convergence in probability under the true θ_0, and we must restrict ourselves to compact subsets.

Theorem 20.2 With the same notation as above and $L_n(\theta) = \log l_n(\theta) = \log f(X_1, \ldots, X_n|\theta)$, assume:

(a) Θ is open.
(b) At the true value θ_0, $\pi(\theta_0) > 0$ and $\pi(\theta)$ is continuous in $\theta_0 \pm \epsilon$ for some $\epsilon > 0$.
(c) For all n, $\frac{\partial^2}{\partial\theta^2}L_n(\theta)$ exists and is continuous in $\theta_0 \pm \epsilon$ for some $\epsilon > 0$.
(d) There is a strongly consistent MLE $\hat{\theta}_n$ that satisfies the likelihood equation.

(e) For every $\eta > 0$, $\exists \epsilon > 0$ such that $P_{\theta_0}\left(\sup_{|\theta - \theta_0| \le \epsilon} \left| 1 - \frac{L_n''(\theta)}{L_n''(\hat{\theta}_n)} \right| > \eta \right) \to 0$.

(f) $\frac{1}{L_n''(\hat{\theta}_n)} \xrightarrow{P_{\theta_0}} 0$.

Let $\psi = \sqrt{n}(\theta - \hat{\theta}_n)$ and $\mathcal{A}$ the class of all compact Borel sets in Θ. Then

$$\sup_{A \in \mathcal{A}} |P(\psi \in A | X_1, \ldots, X_n) - P(\psi \in A | \psi \sim N(0, I^{-1}(\theta_0)))| \xrightarrow{P_{\theta_0}} 0.$$

Remark. This says that unless we have bad luck, we will still be able to uniformly approximate posterior probabilities of compact sets by the corresponding normal probability when n is large. In particular, if θ is one-dimensional and $-\infty < \psi_1 \le \psi_2 < \infty$, then $P(\psi_1 \le \psi \le \psi_2 | X_1, \ldots, X_n)$ can be approximated by the corresponding normal probability uniformly in ψ_1, ψ_2, with P_{θ_0}-probability tending to 1. For statistical purposes, this may often be enough.

An alternative version that gives convergence in total variation on the average is given in the next theorem; a proof can be seen in LeCam (1986).

To state LeCam's theorem, we need his notion of locally quadratic likelihoods, which is defined below.

Definition 20.1 Let μ be a σ-finite measure on a measurable space $\mathcal{X}$, and let $\mathcal{X}^n$ and μ^n respectively denote the n-fold product space and the product measure corresponding to μ. Let $\{P_{\theta,n} : \theta \in \mathcal{R}^k\}$ be a family of probability measures on $\mathcal{X}^n$, and let $f_{\theta,n} = \frac{dP_{\theta,n}}{d\mu^n}$ denote the Radon-Nikodym derivative, assumed to exist. The family $\{P_{\theta,n}\}$ is called LAQ (*locally asymptotically quadratic*) at θ_0 with rate δ_n if there exist random vectors m_n and random matrices K_n such that

(a) under $\theta = \theta_0$, K_n is a.s. positive definite, and

(b) $\log \frac{f_{\theta_0 + \delta_n t_n, n}}{f_{\theta_0, n}} - [t_n' m_n - \frac{1}{2} t_n' K_n t_n] \overset{\text{a.s.}}{\Rightarrow} 0$ under $\theta = \theta_0$ for any bounded sequence of vectors t_n in $\mathcal{R}^k$.

A version of the Bernstein-von Mises theorem that is often called *LeCam's version of the Bernstein-von Mises theorem* is stated next. We have not stated it under the weakest possible conditions. Furthermore, there are certain other versions of this theorem that are also called LeCam's version of the Bernstein-von Mises theorem.

Theorem 20.3 Let $\{P_{\theta,n} : \theta \in \mathcal{R}^k\}$ be LAQ and π a prior density for θ. Let $\Pi_{n,x}$ denote the posterior measure given the data $x \in \mathcal{X}^n$ and $G_{n,x}$ the $N_k(m_n, K_n^{-1})$ distribution (measure). Assume:

(a) $\pi(\theta_0) > 0$ and π is continuous in θ.

(b) For some compact neighborhood K, there exist uniformly consistent tests for testing $\theta = \theta_0$ against $\theta \notin K$.

(c) For some $a > 0$, $\lim_{\epsilon \to 0} \frac{1}{\lambda(B(\theta_0,\epsilon))} \int_{B(\theta_0,\epsilon)} |\pi(\theta) - a| d\theta = 0$, where $\lambda(.)$ denotes the Lebesgue measure on $\mathcal{R}^k$.

(d) There is a sequence of strongly consistent and asymptotically efficient estimates $\hat{\theta}_n$ such that, for any $\epsilon > 0$, there exists a finite positive b with the property that the marginal probability of the set $\{(x, \theta) \in \mathcal{X}^n \bigotimes \mathcal{R}^k : ||\hat{\theta}_n - \theta|| > b\}$ is less than ϵ for all large n.

Then, the average total variation distance between $\Pi_{n,x}$ and the $N_k(m_n, K_n^{-1})$ converges to zero under the sequence of $P_{\theta_0,n}$ distributions; i.e.,

$$\int d_{\mathrm{TV}}(\Pi_{n,x}, G_{n,x}) dP_{\theta_0,n}(x) \to 0$$

as $n \to \infty$.

Remark. As stated before, this is LeCam's version of the Bernstein-von Mises theorem (LeCam (1986)). The scope of application of Theorem 20.3 is to data that are independent but not iid or to dependent data such as stationary time series or stationary Markov chains; see LeCam (1986).

20.3 Posterior Expansions

Analogous to Edgeworth expansions for CDFs of statistics that admit a central limit theorem (e.g., sample means), expansions for the posterior CDF of parameters are available under various conditions on the data-generating mechanism and on the density of the prior. Just like Edgeworth expansions, the higher-order terms involve the standard normal density and polynomials in the argument of the CDF. The coefficients of the polynomials depend on *empirical cumulants* of the MLE and various derivatives of the prior density. These formulas typically tend to be complicated. There are a variety of methods to derive these posterior expansions. One way or the other, they involve a Taylor expansion. A specifically novel method of deriving such posterior expansions in exponential family setups is given in Woodroofe (1992). An important earlier reference is Johnson (1967). We only present two-term expansions for the CDF of the posterior below, but expansions to any given order can be written *in principle* provided the prior density is adequately many times continuously differentiable near the true value. In particular, analyticity in a neighborhood will be enough.

We start with the case of the natural parameter in the one-parameter exponential family. Consider iid observations from a density in the one-parameter exponential family $f(x|\theta) = e^{\theta T(x)-q(\theta)} d\mu$. Let t denote the observed value of the minimal sufficient statistic $\frac{1}{n}\sum_{i=1}^{n} T(X_i)$ after n observations, and let $\hat{\theta}$ denote the MLE of θ, which exists with probability 1 under the true $\theta = \theta_0$ for all large n if θ_0 is an interior point of the natural parameter space. Suppose θ has a density $\rho(\theta)$ (with respect to the Lebesgue measure), which we assume to be thrice continuously differentiable in a neighborhood of θ_0. Let $\sigma^2(\theta) = q''(\theta)$. Define

$$
\begin{aligned}
c_{00} =& \rho(\hat{\theta}); c_{01} = \rho'(\hat{\theta})/\sigma(\hat{\theta}); c_{10} = -q^{(3)}(\hat{\theta})/(6[\sigma(\hat{\theta})]^3);\\
c_{02} =& \rho''(\hat{\theta})/(\sigma(\hat{\theta})^2); c_{20} = \rho(\hat{\theta})[q^{(3)}(\hat{\theta})]^2/(72[\sigma(\hat{\theta})]^6);\\
c_{11} =& -\frac{1}{6}\rho'(\hat{\theta})q^3(\hat{\theta})/[\sigma(\hat{\theta})]^4 - \rho(\hat{\theta})q^{(4)}(\hat{\theta})/(24[\sigma(\hat{\theta})]^4).
\end{aligned}
$$

Then, one has the following one-and two-term expansions for the CDF of the posterior of $\psi = \sqrt{n}\sigma(\hat{\theta})[\theta - \hat{\theta}]$ (Johnson (1967)).

Theorem 20.4 Assume that $\rho(\theta_0) > 0$. Let $G_n(x) = P(\psi \le x|t)$. Then, for almost all infinite sequences $\{x_1, x_2, \cdots\}$ under the true $\theta = \theta_0$,

$$G_n(x) = \Phi(x) + \frac{\gamma_1(x)\phi(x)}{\sqrt{n}} + O(n^{-1})$$

uniformly in x and

$$G_n(x) = \Phi(x) + \frac{\gamma_1(x)\phi(x)}{\sqrt{n}} + \frac{\gamma_2(x)\phi(x)}{n} + O(n^{-3/2})$$

uniformly in x, where

$$\gamma_1(x) = -c_{00}^{-1}[c_{10}(x^2+2) + c_{01}]$$

and

$$\gamma_2(x) = -c_{00}^{-1}[(15c_{20} + 3c_{11} + c_{02})x + (5c_{20} + c_{11})x^3 + c_{20}x^5].$$

Remark. Note that θ is centered at the MLE of the natural parameter in the definition of ψ. One can ask if the result holds by centering it on another estimate (e.g., the mean, median, or the mode of the posterior) or by centering it at the true value θ_0. Without further conditions on the prior density, recentering at the mean of the posterior may not work. Likewise, recentering at the true value may also be invalid. The point is that the posterior expansion is derived by an expansion of the loglikelihood, and the MLE is the right choice for the center of that expansion. If an alternative estimate differs from the MLE by an amount $o(n^{-1/2})$, then the recentering can work. See the next section for more on this. Another important point is that the methods in Johnson (1967) actually apply to any Borel set and not just sets of the form $(-\infty, x)$, so that in the exponential family case, the Bernstein-von Mises theorem follows by employing Johnson's methods.

Next, we give an example of the expansion above.

Example 20.3 Suppose $X_1, X_2, \cdots$ are iid $N(\theta, 1)$ and the density of the prior satisfies the conditions made above. Then, obviously, $q(\theta) = \frac{\theta^2}{2}$, which leads to $c_{00} = \rho(\bar{x}), c_{01} = \rho'(\bar{x}), c_{02} = \rho''(\bar{x}), c_{10} = c_{11} = c_{20} = 0$. This leads to

$$P(\sqrt{n}(\theta - \bar{x}) \leq x | x_1, x_2, \cdots, x_n) = \Phi(x) - \frac{\rho'(\bar{x})}{\rho(\bar{x})} \frac{\phi(x)}{\sqrt{n}} - \frac{\rho''(\bar{x})}{\rho(\bar{x})} \frac{x\phi(x)}{n} + O(n^{-3/2})$$

uniformly in x.

Notice that if one formally differentiates the right-hand side with respect to x and then integrates the derivative on multiplying it by x, one reproduces the Brown identity for the posterior mean (Brown (1971)), namely $E(\theta | x_1, x_2, \cdots, x_n) = \bar{x} + \frac{1}{n} \frac{m'(\bar{x})}{m(\bar{x})}$, with the prior density replacing the marginal.

A drawback of the previous theorem is that it only treats the case of the natural parameter in the exponential family. But usually the mean of the minimal sufficient statistic is a more important function to estimate than the natural parameter. For example, in the Bernoulli case, the success probability p is considered to be a more basic parameter than the natural parameter, which is $\log \frac{p}{1-p}$. An extension of the previous theorem to general parameters and to more general smooth densities is given in the next theorem; see Johnson (1970). In particular, it handles the mean function in the one-parameter exponential family.

Toward this, let $a_{kn}(\theta) = \frac{1}{k!}\sum_{i=1}^{n}\frac{\partial^k}{\partial\theta^k}\log f(x_i|\theta)$, and let $\sigma^2(\hat{\theta}) = -2a_{2n}(\hat{\theta})$.

Define

$$\begin{aligned} c_{00} =& \rho(\hat{\theta}); c_{01} = \rho(\hat{\theta})/\sigma(\hat{\theta}); c_{10} = \rho(\hat{\theta})a_{3n}(\hat{\theta})/[\sigma(\hat{\theta})]^3; \\ c_{02} =& \rho''(\hat{\theta})/\sigma^2(\hat{\theta}); c_{20} = \rho(\hat{\theta})a_{3n}^2(\hat{\theta})/(2[\sigma(\hat{\theta})]^6); \\ c_{11} =& \rho(\hat{\theta})a_{4n}(\hat{\theta})/[\sigma(\hat{\theta})]^4 + \rho'(\hat{\theta})a_{3n}(\hat{\theta})/[\sigma(\hat{\theta})]^4. \end{aligned}$$

Let ψ be as before; i.e., $\psi = \sqrt{n}\sigma(\hat{\theta})[\theta - \hat{\theta}]$. Then, under conditions on the underlying $f(x|\theta)$ and the same conditions as before on the prior density $\rho(\theta)$, an expansion for the posterior CDF of ψ holds with the same order of the error and uniformly in the argument. The conditions on $f(x|\theta)$ can be manipulated. Perhaps the most natural set of conditions are those that ensure the strong consistency of the MLE under the true $\theta = \theta_0$.

Theorem 20.5 Assume that $\rho(\theta_0) > 0$ and that ρ is thrice continuously differentiable in a neighborhood of θ_0. Under conditions 1 through 9 in Johnson (1970) (which are almost the same as the Cramér-Rao regularity conditions for the asymptotic normality of the MLE), the previous theorem holds.

Example 20.4 Suppose we have n independent observations from the Bernoulli pmf $f(x|p) = p^x(1-p)^{(1-x)}$, $x = 0, 1$. The two-term expansion for the posterior CDF can be written, but the polynomials in the expansion are too messy for the expansion to have an aesthetic appeal. We work out the one-term expansion below. By direct calculation using their definitions, with $\hat{p} = \frac{\sum X_i}{n}$,

$$\sigma^2(\hat{p}) = \frac{1}{\hat{p}(1-\hat{p})}, \quad a_{3n}(\hat{p}) = \frac{1-2\hat{p}}{3[\hat{p}(1-\hat{p})]^2}.$$

Substituting,

$$c_{01} = \frac{\rho'(\hat{p})}{\sqrt{\hat{p}(1-\hat{p})}}, \quad c_{10} = \frac{\rho(\hat{p})(1-2\hat{p})}{3\sqrt{\hat{p}(1-\hat{p})}},$$

and thus $\gamma_1(x) = -1/\rho(\hat{p}) \times \left[\rho'(\hat{p})\sqrt{\hat{p}(1-\hat{p})} + \frac{\rho(\hat{p})(1-2\hat{p})(x^2+2)}{3\sqrt{\hat{p}(1-\hat{p})}}\right]$.

The one-term expansion is

$$P\left(\frac{\sqrt{n}}{\sqrt{\hat{p}(1-\hat{p})}}(p-\hat{p}) \le x|X_1, \cdots, X_n\right) = \Phi(x) + \frac{\gamma_1(x)\phi(x)}{\sqrt{n}} + O(n^{-1})$$

uniformly in x. Note that if the prior density has a zero derivative at $\frac{1}{2}$, and if the MLE is exactly equal to $\frac{1}{2}$, then (and only then) the $\frac{1}{\sqrt{n}}$ term vanishes and the ordinary normal approximation to the posterior is second-order accurate.

Example 20.5 Even the one-term expansion for the CDF of the posterior is complicated and could be time-consuming to compute when the underlying density is such that there is no one-dimensional sufficient statistic. In fact, one-dimensional sufficient statistics exist with probability 1 at all sample sizes only in the one-parameter exponential family (Brown (1964)). In this example, we work out the different quantities necessary for the one-term expansion of the posterior CDF of a Cauchy location parameter. Since $f(x|\theta) = \frac{1}{\pi(1+(\theta-x)^2)}$, by direct differentiations and some algebra,

$$a_{2n}(\theta) = \sum \frac{(\theta - x_i)^2 - 1}{(1 + (\theta - x_i)^2)^2},$$

$$a_{3n}(\theta) = \frac{2}{3} \sum \frac{3(\theta - x_i) - (\theta - x_i)^3}{(1 + (\theta - x_i)^2)^3}.$$

The MLE $\hat{\theta}$ exists with probability 1 at all sample sizes. The components of $\gamma_1(x)$ are found from $\sigma^2(\hat{\theta}) = -2a_{2n}(\hat{\theta})$ and $a_{3n}(\hat{\theta})$ with the expressions given above for the functions a_{2n}, a_{3n}. Clearly, even the one-term expansion is complicated.

20.4 Expansions for Posterior Mean, Variance, and Percentiles

Accurate approximations to posterior summaries, such as the mean, variance, and percentiles, are useful because, except for special priors, they cannot be written in closed-form. Also, closed form approximations with error bounds lead to an understanding of how Bayesian measures differ from their frequentist counterparts. In this section, we present expansions for the mean, variance, and percentiles of the posterior. The error bounds for these expansions are absolute; approximations with relative error bounds due to Tierney and Kadane (1986) are presented in the next section. The following expansions for the regular case are derived in Johnson (1970); see Ghosal and Samanta (1997) for appropriate versions of the Bernstein-von Mises theorem and higher-order expansions for posterior moments and percentiles in nonregular cases.

Theorem 20.6 Assume conditions 1 through 9 in Johnson (1970). Let θ_0 denote the true value of θ.

(a) If $\int |\theta|\rho(\theta)d\theta < \infty$, $\rho(\theta_0) > 0$, and ρ is thrice continuously differentiable in a neighborhood of θ_0, then

$$E(\theta|X_1, \cdots, X_n) = \hat{\theta} + \frac{1}{n\sigma(\hat{\theta})}[\rho'(\hat{\theta})/\rho(\hat{\theta}) + 6a_{3n}(\hat{\theta})] + O(n^{-2})$$

for almost all sequences $\{X_1, X_2, \cdots\}$ under the true $\theta = \theta_0$.

(b) If $\int \theta^2\rho(\theta)d\theta < \infty$, $\rho(\theta_0) > 0$, and ρ is four times continuously differentiable in a neighborhood of θ_0, then

$$E[(\theta - \hat{\theta})^2|X_1, \cdots, X_n] = \frac{1}{n\sigma^2(\hat{\theta})} + O(n^{-2})$$

for almost all sequences $\{X_1, X_2, \cdots\}$ under the true $\theta = \theta_0$.

(c) Suppose $\rho(\theta_0) > 0$ and ρ is thrice continuously differentiable in a neighborhood of θ_0. Let $\theta_{n,\alpha}$ denote the αth percentile of the posterior of θ. Then, for almost all sequences $\{X_1, X_2, \cdots\}$ under the true $\theta = \theta_0$,

$$\theta_{n,\alpha} = \hat{\theta} + \frac{z_\alpha}{\sqrt{n}\sigma(\hat{\theta})} + \frac{\tau(\alpha)}{n\sigma(\hat{\theta})} + O(n^{-3/2})$$

uniformly in α, $0 < \alpha_1 \le \alpha \le \alpha_2 < 1$, where $z_\alpha = \Phi^{-1}(\alpha)$ and

$$\tau(\alpha) = c_{00}^{-1}[c_{01} + 2c_{10} + c_{10}z_\alpha^2].$$

Example 20.6 The general asymptotic expansions given in Theorem 20.6 cannot be proved without all those smoothness and moment assumptions on the prior density $\rho(\theta)$. But, in a given case, the corresponding expansions could be obtained by direct arguments. An interesting instance is the estimation of a normal mean with a uniform prior. Thus, suppose that the observations are iid $N(\theta, 1)$ and that $\theta \sim U[-1, 1]$. Then, the marginal density of $\bar{X}$ is

$$m(\bar{x}) = \Phi\left(\sqrt{n}(\bar{x} + 1)\right) - \Phi(\sqrt{n}(\bar{x} - 1)),$$

and so, by the Brown identity,

$$E(\theta|X_1, \cdots, X_n) = \bar{X} + \frac{1}{\sqrt{n}} \frac{\phi\left(\sqrt{n}(\bar{x} + 1)\right) - \phi\left(\sqrt{n}(\bar{x} - 1)\right)}{\Phi\left(\sqrt{n}(\bar{x} + 1)\right) - \Phi\left(\sqrt{n}(\bar{x} - 1)\right)}.$$

Suppose now that θ_0, the true value of θ, is in the open interval $(-1,1)$. Then, with P_{θ_0} probability equal to 1, so is $\bar{X}$ for all large n. Therefore, it follows from tail properties of the standard normal CDF that, for all $k > 0$, $E(\theta|X_1, \cdots, X_n) = \bar{X} + o(n^{-k})$ almost surely. Note that if θ_0 *is* strictly between -1 and 1, then the smoothness conditions of the general theorem do apply. But the general theorem only gives the much weaker conclusion that $E(\theta|X_1, \cdots, X_n) = \bar{X} + O(n^{-2})$ almost surely. The compact support of the prior and that the true value is not a boundary point of the support of the prior permit a stronger conclusion, namely that in fact exponential convergence is occurring. But the general theorem is unable to pick it up.

Example 20.7 Consider again estimation of the binomial p parameter, and suppose p has the Jeffreys prior, which is the Beta(1/2, 1/2) density $\rho(p) = \frac{1}{\pi\sqrt{p(1-p)}}$. We have previously worked out c_{00}, c_{01}, and c_{10} for a general smooth prior. Plugging into them, for the special Jeffreys prior, an expansion for the αth posterior percentile is

$$\begin{aligned} p_{n,\alpha} &= \hat{p} + \frac{z_\alpha\sqrt{\hat{p}(1-\hat{p})}}{\sqrt{n}} + \frac{1}{n}\left[\rho'(\hat{p})/\rho(\hat{p}) + \frac{2}{3}(1-2\hat{p}) + (1-2\hat{p})z_\alpha^2\right] \\ &\quad + O(n^{-3/2}) \\ &= \hat{p} + \frac{z_\alpha\sqrt{\hat{p}(1-\hat{p})}}{\sqrt{n}} + \frac{1}{n}\,[(2\hat{p}-1)/(2\hat{p}(1-\hat{p})) \\ &\quad + (1-2\hat{p})\left(\frac{2}{3} + z_\alpha^2\right)\Bigg] + O(n^{-3/2}), \end{aligned}$$

where $\hat{p} = \frac{\sum X_i}{n}$. Notice that if $\hat{p}$ is exactly equal to $\frac{1}{2}$, then the $\frac{1}{n}$ term completely vanishes and the normal approximation to the posterior percentile is third-order accurate. The exact posterior percentile can be computed numerically, the posterior being a Beta density again, namely the Beta$(\sum X_i + 1/2, n - \sum X_i + 1/2)$ density. If actual sample data are available, then a numerical check of the accuracy of the asymptotic expansion can be performed.

20.5 The Tierney-Kadane Approximations

The Tierney-Kadane approximations are to the expansions in the previous sections what saddlepoint approximations are to Edgeworth expansions in classical statistics. Suppose $l(\theta) = \sum \log f(x_i|\theta)$ is the loglikelihood function based on n iid observations from the density $f(x|\theta)$; f need only be adequately smooth in θ, and there is no implication that it is a density with respect to the Lebesgue measure. Let $g(\theta)$ be a given function of θ. If θ has a prior density $\rho(\theta)$, then the posterior mean of $g(\theta)$ is the ratio

$$E(g(\theta)|x_1, \cdots, x_n) = \frac{\int g(\theta)e^{l(\theta)}\rho(\theta)d\theta}{\int e^{l(\theta)}\rho(\theta)d\theta}$$

$$= \frac{\int e^{nL^*(\theta)}d\theta}{\int e^{nL(\theta)}d\theta},$$

where $nL(\theta) = l(\theta) + \log\rho(\theta)$ and $nL^*(\theta) = nL(\theta) + \log g(\theta)$.

Suppose $\hat{\theta}$ is a unique maximum of $L(\theta)$ and $\hat{\theta}^*$ is the unique maximum of $L^*(\theta)$. Then, by applying the Laplace approximation separately to the numerator and the denominator in the expression for $E(g(\theta)|x_1, \cdots, x_n)$ above, one obtains the approximation

$$E(g(\theta)|x_1, \cdots, x_n) = \frac{\sigma^*}{\sigma}e^{n[L^*(\hat{\theta}^*)-L(\hat{\theta})]}(1 + O(n^{-2})),$$

where $\sigma^2 = -1/L''(\hat{\theta})$ and $\sigma^{*2} = -1/L^{*''}(\hat{\theta}^*)$. Normally, the numerator and the denominator in the expression for the posterior mean would both have $O(n^{-1})$ terms. On combining the separate expansions for the numerator and the denominator, the $\frac{1}{n}$ term cancels because the coefficients of the $\frac{1}{n}$ term in the numerator and the denominator are themselves equal up to the order $O(n^{-1})$ with probability 1 under the true θ_0; see Tierney and Kadane (1986).

Note that *unlike* the expansion for the posterior mean in the previous section, the approximation above comes with a *relative error* of the order n^{-2}. This is the Tierney-Kadane (1986) approximation to the posterior mean of a smooth function. It has been found to be impressively accurate in sufficiently many test cases. It is important to keep in mind, however, that a lot of assumptions go into the validity of the final claim and also that the coefficient of the n^{-2} term is not known to be an absolute constant. The coefficient depends on the actual data, and the absoluteness of the constant does not seem to have been established. The conditions required on ρ and g are that they should be six times continuously differentiable in a neighborhood of the true $\theta = \theta_0$ and that so is the underlying $f(x|\theta)$. Thus, the Tierney-Kadane approximation requires stronger assumptions on all components of the problem than the expansions due to Johnson (1970).

20.6 Frequentist Approximation of Posterior Summaries

As in frequentist statistics, summarizing the posterior by means of a few quantities, such as the mean, median, or variance, is useful to Bayesians. A natural question is whether a posterior summary is asymptotically close to some appropriate frequentist summary. One can talk about numerical closeness for specific data or closeness as statistics in the sense of their frequentist distributions. We discussed approximations for specific data by means of asymptotic expansions in the previous section. Now we turn to approximations in terms of distributions, considering them as statistics.

Theorem 20.7 Suppose, given θ, $X_1, \ldots, X_n$ are iid P_θ and the conditions of the Bernstein-von Mises theorem hold. Let $\hat{\theta}_M$ denote the median of the posterior distribution of θ. Then

$$\sqrt{n}(\hat{\theta}_M - \hat{\theta}_n) \stackrel{\text{a.s.}}{\rightarrow} 0 \text{ for all } \theta_0,$$

where $\hat{\theta}_n$ is as in the Bernstein-von Mises theorem.

Remark. See Bickel and Doksum (2001) for a proof. Here is an explanation for why the result is true. Since the median is a linear functional on the set of probability measures, the median of the posterior distribution of $\sqrt{n}(\theta - \hat{\theta})$ is $\sqrt{n}(\hat{\theta}_M - \hat{\theta})$. But the Bernstein-von Mises theorem says, as a corollary,

$$\sup_{-\infty<t<\infty} \left| P(\sqrt{n}(\theta - \hat{\theta}) \leq t | X_1, \ldots, X_n) - P(N(0, I^{-1}(\theta_0)) \leq t) \right| \stackrel{\text{a.s.}}{\rightarrow} 0, \ \forall \theta_0.$$

It follows that each fixed quantile of the sequence of random distributions $\pi_{n,X^{(n)}}$, the posterior of θ given $X^{(n)} = (X_1, \ldots, X_n)$, converges to the corresponding quantile of the $N(0, I^{-1}(\theta_0))$ distribution, and this convergence is a.s. P_{θ_0}, $\forall \theta_0$, because the convergence in the Bernstein-von Mises theorem is so.

Historically, posterior means are more popular than posterior medians. Because the posterior is approximately normal for large n, the asymptotic behavior of the mean and the median of the posterior ought to be the same. But an extra condition is needed (again, see Bickel and Doksum (2001)).

Theorem 20.8 Suppose that the conditions of the Bernstein-von Mises theorem hold. If in addition $\int \theta^2 \pi(\theta)\, d\theta < \infty$, then

$$\sqrt{n}(E(\theta | X^{(n)}) - \hat{\theta}_n) \stackrel{\text{a.s.}}{\rightarrow} 0, \ \forall \theta_0.$$

From here, a version of the limiting frequentist distribution of $\hat{\theta}_M$ and $E(\theta|X^{(n)})$ follows. Here is the result.

Theorem 20.9 Suppose, given θ, $X_1, \ldots, X_n$ are iid P_θ, that the conditions of the Bernstein-von Mises theorem hold, and $\int \theta^2 \pi(\theta)\, d\theta < \infty$. Then, under P_θ,

$$\text{(a)} \qquad \sqrt{n}(\hat{\theta}_M - \theta) \xrightarrow{\mathcal{L}} N(0, I^{-1}(\theta)),$$

$$\text{(b)} \qquad \sqrt{n}(E(\theta|X^{(n)}) - \theta) \xrightarrow{\mathcal{L}} N(0, I^{-1}(\theta)).$$

Remark. There is a cute way to interpret the asymptotic normality of the posterior and its link to the asymptotic normality of a frequentist optimal estimate, say the MLE. Denoting the true value of θ as θ itself, we know that $\sqrt{n}(\theta - \hat{\theta}) \xrightarrow{\mathcal{L}} N(0, I^{-1}(\theta))$ (for almost all sequences $X_1, X_2, \ldots$); this is the sequence of posteriors. On the other hand, from standard asymptotics for MLEs, $\sqrt{n}(\hat{\theta} - \theta) \xrightarrow{\mathcal{L}} N(0, I^{-1}(\theta))$. The two statements $\sqrt{n}(\theta - \hat{\theta}) \xrightarrow{\mathcal{L}} N(0, I^{-1}(\theta))$ and $\sqrt{n}(\hat{\theta} - \theta) \xrightarrow{\mathcal{L}} N(0, I^{-1}(\theta))$ can each be obtained from the other by switching the location of θ and $\hat{\theta}$ in $\sqrt{n}(\theta - \hat{\theta})$, although they are sequences of distributions on different spaces! The asymptotics of the MLE and the asymptotics of the posterior come together in this beautiful way.

We next give a discrete example (see Freedman (1963) for nontraditional Bayesian asymptotics in discrete cases) to show that things can fall apart if the true value of θ is not supported by the prior $\pi(\theta)$.

Example 20.8 Let $X_1, \ldots, X_n|\theta$ be iid $N(\theta, 1)$, and suppose G is a two-point prior with $G\{1\} = G\{-1\} = \frac{1}{2}$. Then

$$E(\theta|X_1, \ldots, X_n) = \frac{e^{-\frac{n}{2}(1-\bar{X})^2}\frac{1}{2} - e^{-\frac{n}{2}(1+\bar{X})^2}\frac{1}{2}}{e^{-\frac{n}{2}(1-\bar{X})^2}\frac{1}{2} + e^{-\frac{n}{2}(1+\bar{X})^2}\frac{1}{2}}$$

$$= \frac{e^{n\bar{X}} - e^{-n\bar{X}}}{e^{n\bar{X}} + e^{-n\bar{X}}} = \frac{e^{S_n} - e^{-S_n}}{e^{S_n} + e^{-S_n}} = \frac{e^{2S_n} - 1}{e^{2S_n} + 1},$$

where $S_n = \sum_{i=1}^n X_i$.

Suppose now that the true value θ_0 is actually 0; then θ_0 is not in the support of G. Basically, because of this, $E(\theta|X_1, \ldots, X_n)$ will get confused about what the true θ is and one will see wild swings in its values from one

n to another n. To understand why, we need the following lemma, which can be seen in Chung (2001).

Lemma 20.1 $P(\text{the set of limit points of } \frac{S_n}{\sqrt{2n\log\log n}} = [-1, 1]) = 1$. So the sequence $\frac{S_n}{\sqrt{2n\log\log n}}$ will come arbitrarily close to any $c \in [-1, 1]$ infinitely often, with probability 1. Consequently, $E(\theta|X_1, \ldots, X_n)$ will come arbitrarily close to any real v in $[-1, 1]$ infinitely often, with probability 1. Thus, if one simulates $N(0, 1)$ data and continuously updates the posterior mean after every new sample value, then the sequence of posterior means swings between ± 1. In other words, more data do not help. This can also be seen by using the theory of recurrent random walks.

20.7 Consistency of Posteriors

The previous results show that under sufficiently many conditions, the posterior distribution of $\sqrt{n}(\theta - \hat{\theta})$ converges to a nondegenerate normal distribution. What would happen if we do not normalize; i.e., what happens to the sequence of posteriors of θ by itself? Then the sequence of posteriors of θ becomes asymptotically very spikey and almost a point mass at the true value of θ. We first give an example.

Example 20.9 Suppose $X_1, \ldots, X_n|\theta \stackrel{\text{iid}}{\sim} N(\theta, 1)$ and $\theta \sim N(0, 1)$. Then $\pi_{n,X^{(n)}} =$ the posterior of $\theta = N(\frac{n\bar{X}}{n+1}, \frac{1}{n+1})$. This is a sequence of random distributions. But if the true value of θ is θ_0, then

$$P\left(N\left(\frac{n\bar{X}}{n+1}, \frac{1}{n+1}\right) \leq t\right) - P(N(\theta_0, 0) \leq t) \stackrel{\text{a.s.}}{\Rightarrow} 0,$$

where the notation $P(N(\frac{n\bar{X}}{n+1}, \frac{1}{n+1}) \leq t)$ means $P(W_n \leq t)$ with $W_n \sim N(\frac{n\bar{X}}{n+1}, \frac{1}{n+1})$ and $(N(\theta_0, 0) \leq t)$ means a random variable equal to θ_0 w.p. 1 is less than or equal to t. So, for almost all sequences $(X_1, X_2, \ldots)$,

$$\pi_{n,X^{(n)}} \xrightarrow{\mathcal{L}} \delta_{\{\theta_0\}},$$

where $\delta_{\{\theta_0\}}$ is the point mass at θ_0.

Here is another example of this phenomenon.

Example 20.10 Suppose $X \sim \text{Bin}(n, p)$ and p has a Beta(a, b) prior. Then the posterior density of p given $X = x$ is the Beta$(a + x, b + n - x)$ density.

Consider any fixed $\epsilon > 0$. Then, by a familiar connection between Beta and binomial CDFs,

$$\begin{aligned} P(p \le p_0 - \epsilon | X = x) &= P(\text{Bin}(a + b + n - 1, p_0 - \epsilon) > a + x - 1) \\ &= P\left(\tfrac{1}{n}\text{Bin}(a + b + n - 1, p_0 - \epsilon) > \tfrac{a+x-1}{n}\right). \end{aligned}$$

Since $\frac{1}{n}\text{Bin}(a + b + n - 1, p_0 - \epsilon) \overset{P}{\Rightarrow} p_0 - \epsilon$ and $\frac{X}{n} \Rightarrow p_0$ almost surely under $p = p_0$, it follows that the probability above converges almost surely to zero. Analogously, $P(p \ge p_0 + \epsilon | X = x) \overset{\text{a.s.}}{\Rightarrow} 0$. That is, almost surely, the posterior probability of any fixed neighborhood of the true value p_0 converges to 1.

The phenomenon evidenced in the previous two examples is in fact quite general. Here is a well-known result.

Theorem 20.10 (Doob (1949)) Let the sample space $\mathcal{X}$ for each X_i be a subset of $\mathbb{R}^p$ for some $p < \infty$. Let $\theta \in \Theta$ be real valued and have a prior distribution G. Then the sequence of posteriors $\pi_{n,X^{(n)}}$ satisfies $\pi_{n,X^{(n)}} \overset{\mathcal{L}}{\longrightarrow} \delta_{\{\theta_0\}}$ almost everywhere G; i.e., $\exists A \subseteq \Theta$ such that $P_G(\theta \in A) = 1$ and $\forall\, \theta_0 \in A$,

$$P_{\theta_0}^{\infty}(P_{\pi_{n,X^{(n)}}}(\theta \le t) \to 1) = 1 \text{ for all } t > \theta_0,$$

and

$$P_{\theta_0}^{\infty}(P_{\pi_{n,X^{(n)}}}(\theta \le t) \to 0) = 1 \text{ for all } t < \theta_0.$$

A potential criticism of this result is that we can only guarantee consistency for almost all values of θ_0 under the prior G. Additional conditions can be imposed in order to remove the restriction.

Remark. Theorem 20.10 is stated in some other forms as well in the literature.

20.8 The Difference between Bayes Estimates and the MLE

We saw above that, under appropriate conditions, $E(\theta|X^{(n)}) - \hat{\theta}_n$ and even $\sqrt{n}(E(\theta|X^{(n)}) - \hat{\theta}_n)$ converges in P_{θ_0}-probability to zero. So $E(\theta|X^{(n)}) - \hat{\theta}_n = o_p(\frac{1}{\sqrt{n}})$ under P_{θ_0}, and we will have to use norming rates faster than $\sqrt{n}$ for $E(\theta|X^{(n)}) - \hat{\theta}_n$ to have a nondegenerate limit distribution. There is an effect of the prior in these results, as we shall see; the prior will not drop out. But first let us see an example for illustration

Example 20.11 Suppose $X_1, \ldots, X_n|\theta \overset{\text{iid}}{\sim} N(\theta, 1)$ and $\theta \sim N(0, 1)$. Then $E(\theta|X^{(n)}) = \frac{n\bar{X}}{n+1}$. Using the MLE $\bar{X}$ for $\hat{\theta} = \hat{\theta}_n$, we have

$$E(\theta|X^{(n)}) - \hat{\theta} = \frac{n\bar{X}}{n+1} - \bar{X} = -\frac{\bar{X}}{n+1}$$

$$\begin{aligned}\Rightarrow n\left(E(\theta|X^{(n)}) - \hat{\theta}\right) &= -\tfrac{n}{n+1}\bar{X} = -\tfrac{n}{n+1}(\bar{X} - \theta_0) - \tfrac{n}{n+1}\theta_0 \\ &= -(\bar{X} - \theta_0) + \tfrac{\bar{X}-\theta_0}{n+1} - \theta_0 + \tfrac{\theta_0}{n+1},\end{aligned}$$

where θ_0 is the true value.

$$\begin{aligned}\Rightarrow \sqrt{n}\left(n\left(E(\theta|X^{(n)}) - \hat{\theta}\right) + \theta_0\right) &= \sqrt{n}(\theta_0 - \bar{X}) + \tfrac{\sqrt{n}(\bar{X}-\theta_0)}{n+1} + \tfrac{\sqrt{n}\theta_0}{n+1} \\ &= \sqrt{n}(\theta_0 - \bar{X}) + O_p(\tfrac{1}{n}) + O_p(\tfrac{1}{\sqrt{n}}) \\ &= \sqrt{n}(\theta_0 - \bar{X}) + o_p(1) \\ &\xrightarrow[\mathcal{P}_{\theta_0}]{\mathcal{L}} N(0, 1);\end{aligned}$$

i.e., under P_{θ_0}, the limit distribution is $N(0, 1)$.

So we have in this example that the difference between the posterior mean and the MLE has a nondegenerate limit distribution given by

$$\sqrt{n}\left(n\left(E(\theta|X^{(n)}) - \hat{\theta}\right) + \theta_0\right) \xrightarrow{\mathcal{L}} N(0, 1) \quad \forall\, \theta_0.$$

In particular, if $\theta_0 = 0$, then $n^{\frac{3}{2}}\left(E(\theta|X^{(n)}) - \hat{\theta}\right) \xrightarrow{\mathcal{L}} N(0, 1)$. The argument generalizes, as we will see in the next section.

20.9 Using the Brown Identity to Obtain Bayesian Asymptotics

We will discuss the issue of a nondegenerate limit distribution for the difference between Bayes estimates and a frequentist estimate only for the case of normal data, although without too much additional effort, similar results can be established for the exponential family. The main tool we use is the representation of posterior means known as the Brown identity (Brown (1971)). The limiting distribution results presented in this section are not available elsewhere in the literature.

Proposition 20.1 Let $X_1, \ldots, X_n|\theta \overset{\text{iid}}{\sim} N(\theta, 1)$ and let θ have a prior distribution G. Let $m_n(t) = m_{n,G}(t) = \int \frac{\sqrt{n}}{\sqrt{2\pi}} e^{-\frac{n}{2}(t-\theta)^2} dG(\theta)$ denote the marginal density of $\bar{X}$. Then

$$E(\theta|X^{(n)}) = \bar{X} + \frac{1}{n}\frac{m_n'(\bar{X})}{m_n(\bar{X})}.$$

A proof can be obtained by noting that

$$\begin{aligned}
E(\theta|X^{(n)}) &= \frac{\int \theta \frac{\sqrt{n}}{\sqrt{2\pi}} e^{-\frac{n}{2}(\bar{X}-\theta)^2} dG(\theta)}{\int \frac{\sqrt{n}}{\sqrt{2\pi}} e^{-\frac{n}{2}(\bar{X}-\theta)^2} dG(\theta)} \\
&= \frac{\int (\bar{X} + \theta - \bar{X}) \frac{\sqrt{n}}{\sqrt{2\pi}} e^{-\frac{n}{2}(\bar{X}-\theta)^2} dG(\theta)}{\int \frac{\sqrt{n}}{\sqrt{2\pi}} e^{-\frac{n}{2}(\bar{X}-\theta)^2} dG(\theta)} \\
&= \bar{X} + \frac{\int (\theta - \bar{X}) \frac{\sqrt{n}}{\sqrt{2\pi}} e^{-\frac{n}{2}(\bar{X}-\theta)^2} dG(\theta)}{\int \frac{\sqrt{n}}{\sqrt{2\pi}} e^{-\frac{n}{2}(\bar{X}-\theta)^2} dG(\theta)} \\
&= \bar{X} + \frac{1}{n}\frac{\int n(\theta - \bar{X}) \frac{\sqrt{n}}{\sqrt{2\pi}} e^{-\frac{n}{2}(\theta-\bar{X})^2} dG(\theta)}{\int \frac{\sqrt{n}}{\sqrt{2\pi}} e^{-\frac{n}{2}(\theta-\bar{X})^2} dG(\theta)} \\
&= \bar{X} + \frac{1}{n}\frac{\frac{d}{d\bar{X}} \int \frac{\sqrt{n}}{\sqrt{2\pi}} e^{-\frac{n}{2}(\theta-\bar{X})^2} dG(\theta)}{\int \frac{\sqrt{n}}{\sqrt{2\pi}} e^{-\frac{n}{2}(\theta-\bar{X})^2} dG(\theta)} \\
&= \bar{X} + \frac{1}{n}\frac{m_n'(\bar{X})}{m_n(\bar{X})}.
\end{aligned}$$

Now, let us specialize to the case where the prior G has a density, say $\pi(\theta)$. Then, on making a change of variable, we have

$$m_n(\bar{X}) = \int \frac{\sqrt{n}}{\sqrt{2\pi}} e^{-\frac{n}{2}(\bar{X}-\theta)^2} \pi(\theta) d\theta$$
$$= \frac{1}{\sqrt{2\pi}} \int e^{-\frac{z^2}{2}} \pi(\bar{X} + \frac{z}{\sqrt{n}}) dz$$

and likewise

$$m_n'(\bar{X}) = \int \frac{1}{\sqrt{2\pi}} e^{-\frac{z^2}{2}} \pi' \left(\bar{X} + \frac{z}{\sqrt{n}} \right) dz.$$

Under adequate conditions on the prior density $\pi(\theta)$, it follows from these two expressions that

$$m_n(\bar{X}) \xrightarrow{P_{\theta_0}} \pi(\theta_0)$$

and

$$m_n'(\bar{X}) \xrightarrow{P_{\theta_0}} \pi'(\theta_0).$$

From the Brown identity,

$$n\left(E(\theta|X^{(n)}) - \bar{X}\right) = \frac{m_n'(\bar{X})}{m_n(\bar{X})} = \psi_n(\bar{X}) \quad \text{(say)}.$$

Under suitable conditions, $\psi_n(\bar{X})$ is approximately equal to the function $\nu(\bar{X}) = \dfrac{\pi'(\bar{X})}{\pi(\bar{X})}$ in a sense that we shall see below.

Hence,

$$n\left(E(\theta|X^{(n)}) - \bar{X}\right) = \psi_n(\bar{X})$$
$$= \nu(\bar{X}) + \psi_n(\bar{X}) - \nu(\bar{X})$$
$$= \left(\nu(\bar{X}) - \nu(\theta_0)\right) + \nu(\theta_0) + \left(\psi_n(\bar{X}) - \nu(\bar{X})\right)$$

$$\Rightarrow \sqrt{n}\left(n\left(E(\theta|X^{(n)}) - \bar{X}\right) - \nu(\theta_0)\right) = \sqrt{n}\left(\nu(\bar{X}) - \nu(\theta_0)\right)$$
$$+\sqrt{n}\left(\psi_n(\bar{X}) - \nu(\bar{X})\right)$$

$\Rightarrow \sqrt{n}\left(n\left(E(\theta|X^{(n)}) - \bar{X}\right) - \nu(\theta_0)\right) \xrightarrow{\mathcal{L}} N(0, (\nu'(\theta_0))^2)$, provided the prior density $\pi(\theta)$ is such that $\sqrt{n}(\psi_n(\bar{X}) - \nu(\bar{X})) \xrightarrow{P_{\theta_0}} 0$ and $\nu'(\theta_0) \neq 0$.

If $\nu'(\theta_0) = 0$, then the normal limit distribution obviously does not hold and a little more work is needed.

We state the following general result.

Theorem 20.11 Suppose $X_1, \ldots, X_n|\theta \overset{\text{iid}}{\sim} N(\theta, 1)$ and $\theta \sim \pi(\theta)$, a density. Suppose π is three times differentiable and $\pi'', \pi^{(3)}$ are uniformly bounded. Let $\nu(\theta) = \frac{\pi'(\theta)}{\pi(\theta)}$. Let θ_0 be a fixed value, and suppose $\pi(\theta_0) > 0$ and $\nu'(\theta_0) \neq 0$. Then

$$\sqrt{n}\left[n\left(E(\theta|X^{(n)}) - \bar{X}\right) - \nu(\theta_0)\right] \xrightarrow[\mathcal{P}_{\theta_0}]{\mathcal{L}} N\left(0, (\nu'(\theta_0))^2\right).$$

In particular, if $\pi'(\theta_0) = 0$ but $\pi(\theta_0)$ and $\pi''(\theta_0)$ are not equal to zero, then

$$n^{\frac{3}{2}}\left(E(\theta|X^{(n)}) - \bar{X}\right) \xrightarrow[\mathcal{P}_{\theta_0}]{\mathcal{L}} N\left(0, \left(\frac{\pi''(\theta_0)}{\pi(\theta_0)}\right)^2\right).$$

Proof. The proof uses the argument sketched above. First,

$$\begin{aligned} m_n(\bar{X}) &= \int \phi(z)\pi\left(\bar{X} + \frac{z}{\sqrt{n}}\right) dz \\ &= \int \phi(z)\left[\pi(\bar{X}) + \frac{z}{\sqrt{n}}\pi'(\bar{X}) + \frac{z^2}{2n}\pi''(x^*)\right] dz, \end{aligned}$$

where x^* is between $\bar{X}$ and $\bar{X} + \frac{z}{\sqrt{n}}$. So

$$m_n(\bar{X}) = \pi(\bar{X}) + O_p\left(\frac{1}{n}\right)$$

since $\int z\phi(z)dz = 0$ and π'' is uniformly bounded. Similarly,

$$\begin{aligned} m_n'(\bar{X}) &= \frac{d}{d\bar{X}}\int \phi(z)\pi\left(\bar{X} + \frac{z}{\sqrt{n}}\right) dz \\ &= \int \phi(z)\pi'\left(\bar{X} + \frac{z}{\sqrt{n}}\right) dz \\ &= \int \phi(z)\left[\pi'(\bar{X}) + \frac{z}{\sqrt{n}}\pi''(\bar{X}) + \frac{z^2}{2n}\pi^{(3)}(x^{**})\right] dz, \end{aligned}$$

where x^{**} is between $\bar{X}$ and $\bar{X} + \frac{z}{\sqrt{n}}$. Therefore,

$$m_n'(\bar{X}) = \pi'(\bar{X}) + O_p\left(\frac{1}{n}\right)$$

since $\int z\phi(z)dz = 0$ and $\pi^{(3)}$ is uniformly bounded. So

$$\begin{aligned} n\left[E(\theta|X^{(n)}) - \bar{X}\right] &= \frac{m_n'(\bar{X})}{m_n(\bar{X})} \\ &= \frac{\pi'(\bar{X}) + O_p(\frac{1}{n})}{\pi(\bar{X}) + O_p(\frac{1}{n})} \end{aligned}$$

by the Brown identity. Thus,

$$\begin{aligned} &\sqrt{n}\left(n\left[E(\theta|X^{(n)}) - \bar{X}\right] - \frac{\pi'(\theta_0)}{\pi(\theta_0)}\right) \\ &= \sqrt{n}\left(\frac{\pi'(\bar{X}) + O_p(\frac{1}{n})}{\pi(\bar{X}) + O_p(\frac{1}{n})} - \frac{\pi'(\theta_0)}{\pi(\theta_0)}\right) \\ &= \sqrt{n}\left(\frac{\pi'(\bar{X}) + O_p(\frac{1}{n})}{\pi(\bar{X}) + O_p(\frac{1}{n})} - \frac{\pi'(\bar{X})}{\pi(\bar{X})} + \frac{\pi'(\bar{X})}{\pi(\bar{X})} - \frac{\pi'(\theta_0)}{\pi(\theta_0)}\right) \\ &= \sqrt{n}\left(\frac{\pi'(\bar{X})}{\pi(\bar{X})} - \frac{\pi'(\theta_0)}{\pi(\theta_0)}\right) + \sqrt{n}\, O_p\left(\frac{1}{n}\right) \\ &\xrightarrow{\mathcal{L}} N(0, (\nu'(\theta_0))^2) \end{aligned}$$

by the delta theorem and Slutsky's theorem.

Remark. Theorem 20.11 shows that the frequentist limit distribution of the difference between $E(\theta|X^{(n)})$ and $\hat{\theta}$ indeed does depend on the prior, although $E(\theta|X^{(n)})$ and $\hat{\theta}$ (typically) have identical limit distributions free of the particular prior.

Thus, Bayesian asymptotics are subtle in that the answers to some questions do depend on the prior. The prior does not wash out in every limit distribution question.

20.10 Testing

In point estimation, typically a frequentist procedure such as the MLE and a Bayes procedure will give similar answers asymptotically. But testing problems can give rise to a discrepancy. This has been well-known for a long time, the most common example being due to Dennis Lindley (Lindley (1957)). The discrepancies generally arise for point null hypothesis testing. The discrepancies can disappear if the Bayesian's formulation is suitably changed. We only present an example, which nevertheless illustrates the nature of the discrepancy that arises.

Example 20.12 Suppose $X_1, \ldots, X_n|\theta \overset{\text{iid}}{\sim} N(\theta, 1)$ and we have the problem of testing $H_0 : \theta = 0$ vs. $H_1 : \theta \neq 0$. As a specific prior, suppose $P(\theta = 0) = \lambda > 0$ and $\theta|\theta \neq 0$ has density $\frac{e^{-\frac{\theta^2}{2\tau^2}}}{\tau\sqrt{2\pi}}$. The posterior distribution is therefore a mixed distribution. By straightforward calculation,

$$P(\theta = 0|X^{(n)}) = \frac{\lambda\phi(\sqrt{n}\bar{X})}{\lambda\phi(\sqrt{n}\bar{X}) + (1-\lambda)m_n(\sqrt{n}\bar{X})},$$

where as before $m_n(\cdot)$ denotes the marginal density of $\bar{X}$; i.e.,

$$m_n(t) = \frac{\sqrt{n}}{2\pi\tau} \int e^{-\frac{n}{2}(t-\theta)^2} e^{-\frac{\theta^2}{2\tau^2}} \, d\theta.$$

$m_n(\sqrt{n}\bar{X})$ can be written in closed form, and it follows after some algebra that

$$\frac{m_n(\sqrt{n}\bar{X})}{\phi(\sqrt{n}\bar{X})} = O_p\left(\frac{1}{\sqrt{n}}\right) \text{ under } H_0.$$

So

$$\begin{aligned}
\tilde{p}_n &= P(H_0|X^{(n)}) \\
&= \frac{1}{1 + \frac{1-\lambda}{\lambda}\frac{m_n(\sqrt{n}\bar{X})}{\phi(\sqrt{n}\bar{X})}} \\
&= \frac{1}{1 + \frac{1-\lambda}{\lambda}O_p(\frac{1}{\sqrt{n}})} \\
&\xrightarrow{P_{H_0}} 1.
\end{aligned}$$

So the posterior measure of evidence for the null converges in probability to 1 under H_0. On the other hand, the traditional frequentist measure of evidence for the null is the p-value computed from the UMPU test that rejects H_0 for large values of $|\sqrt{n}\bar{X}|$. Indeed, the p-value is

$$p_n = 2\left(1 - \Phi(|\sqrt{n}\bar{X}|)\right).$$

Under H_0, for every n, $p_n \sim U[0, 1]$ and so $p_n \xrightarrow[H_0]{\mathcal{L}} U[0, 1]$. But obviously $\tilde{p}_n \xrightarrow[H_0]{\mathcal{L}} \delta_{\{1\}}$, the point mass at 1. So, from the same data, the extent of evidentiary support for H_0 would look very different to a Bayesian and a frequentist, in one case the limiting distribution being a point mass and in the other case it being a uniform distribution.

A discrepancy arises under the alternative also, but in a slightly different way. Thus, suppose the true value $\theta_0 \neq 0$. Then it can be shown by a straightforward calculation that $\frac{\tilde{p}_n}{p_n} \xrightarrow{P_{\theta_0}} \infty$. So even under the alternative, the Bayesian's evidence for the null is factors of magnitude larger than the frequentist's chosen measure of evidence, namely the p-value.

20.11 Interval and Set Estimation

Bayesian and frequentist inferences come together once again asymptotically in the confidence set problem. In an elegant and strong sense, in regular problems, a Bayesian's confidence set and a frequentist's confidence set of the same nominal coverage will look visually indistinguishable as $n \to \infty$.

We need some definitions.

Definition 20.2 Let $(X_1, \ldots, X_n)|\theta \sim P_{\theta,n}$ and let $\theta \sim \pi(\theta)$, a density with respect to Lebesgue measure. Let $\pi_{n,X^{(n)}}$ denote the posterior density of θ given $X^{(n)}$. For a given α, $0 < \alpha < 1$, the $100(1-\alpha)\%$ HPD (highest posterior density) credible set is defined as

$$C_{\alpha,n} = \{\theta : \pi_{n,X^{(n)}}(\theta) \geq R_\alpha(n, X^{(n)})\},$$

where $R_\alpha(n, X^{(n)})$ is such that $P(\theta \in C_{\alpha,n}|X^{(n)}) = 1 - \alpha$.

Remark. This is the standard set a Bayesian uses when a set is needed to estimate θ. Note that $C_{\alpha,n}$ need not be convex or even connected.

Next, consider a set estimate constructed from frequentist considerations.

Definition 20.3 Let $\hat{\theta}_n$ be a point estimate such that

$$\sqrt{n}(\hat{\theta}_n - \theta) \xrightarrow[P_{\theta,n}]{\mathcal{L}} N_k(0, I^{-1}(\theta)),$$

where θ is k-dimensional. Consider a fixed α, $0 < \alpha < 1$. Define

$$S_{n,\alpha} = \{\theta : n(\hat{\theta}_n - \theta)' I(\hat{\theta}_n)(\hat{\theta}_n - \theta) \leq \chi^2_{\alpha,k}\},$$

where $\chi^2_{\alpha,k}$ is the $100(1-\alpha)$th percentile of the χ^2_k-distribution.

Remark. $S_{n,\alpha}$ is known as the Wald confidence set. What happens is that $C_{\alpha,n} \approx S_{n,\alpha}$ as sets in a well-defined sense.

Definition 20.4 Let $A, B \subseteq \mathbb{R}^k$. The Hausdorff distance between A and B is defined as

$$\rho(A, B) = \max\{\sup_{x\in A} \inf_{y\in B} \|x - y\|, \sup_{y\in B} \inf_{x\in A} \|x - y\|\}.$$

Remark. $\rho(A, B) \leq \epsilon$ only if every point in A is within ϵ distance of some point in B and vice versa.

The following theorem is implicit in Bickel and Doksum (2001).

Theorem 20.12 Let $X_1, \ldots, X_n$ be iid given θ, $\theta \in \Theta \subseteq \mathbb{R}^k$. Suppose an estimate $\hat{\theta}_n$ as in the definition of $S_{n,\alpha}$ exists. Suppose also that the conditions of the Bernstein-Von Mises theorem hold. Let θ_0 be the true value of θ. Then $\rho(C_{n,\alpha}, S_{n,\alpha}) \stackrel{\text{a.s.}}{\Rightarrow} 0$ under P_{θ_0}.

Remark. Thus, in every "regular" problem, the Bayesian HPD credible set and the Wald confidence set would look visually almost identical for large n.

20.12 Infinite-Dimensional Problems and the Diaconis-Freedman Results

Although much of everyday statistical methodology is parametric, with a finite-dimensional parameter, there are also a lot of problems of wide interest that come with infinite-dimensional parameters. Historically, we have referred to them as *nonparametric problems*. Examples include estimation of an underlying density in a Euclidean space based on a sequence of iid observations, estimation of a regression function assuming only shape or smoothness properties, or testing for goodness of fit of an infinite parametric hypothesis such as unimodality, etc. Just as in the parametric case, one certainly can attempt Bayesian approaches to such infinite-dimensional problems also. However, one now has to put prior distributions on infinite-dimensional spaces; for example, prior distributions on densities, function spaces, distributions, stochastic processes, etc. The posteriors would also correspondingly be distributions on infinite-dimensional spaces. One can ask the same questions as in the finite-dimensional problems. Is there an analog of a Bernstein- von Mises theorem? Does the posterior usually accumulate near the truth in large samples? Do data swamp the influence of the prior as the sample size increases? Computation, although not addressed in this book, is also an issue.

For the Bayesian, the infinite-dimensional problems are fundamentally different. All the questions are much harder to answer, and in some cases the known answers differ from the finite-dimensional counterparts. Something that appears to be reasonable at first glance may end up being undesirable. For example, one could lose consistency because something very subtle did not work out exactly right. For the theoretical Bayesian, there are still a lot of questions that are unresolved, or not resolved in a way that contributes to intuition. In this section, we give a brief introduction to some of the most fundamental results on infinite-dimensional Bayesian asymptotics. The material is taken from Diaconis and Freedman (1986) and Walker (2004). We recommend Wasserman (1998) for a well-written review. Barron, Schervish, and Wasserman (1999) contains substantive additional technical information. Freedman (1991) demonstrates that there cannot be a Bernstein-von Mises theorem in the finite-dimensional sense even for the normal case when there are infinitely many normal means to estimate.

The first result, due to Diaconis and Freedman (1986), takes about the simplest nonparametric inference problem and shows that, with seemingly reasonable priors, the Bayes estimate can fail to be even consistent. To describe the result, we first need to define Ferguson's *Dirichlet priors* on a CDF F (Ferguson (1973)). Dirichlet priors can be defined on general spaces, but we will only define them for a CDF on the real line.

Definition 20.5 Let P be a probability distribution with CDF F on $\mathcal{R}$ and α a nonzero and finite measure on $\mathcal{R}$. We say that P (or synonymously F) has a *Dirichlet process prior with parameter* α if for any (measurable) partition $A_1, A_2, \cdots, A_k$ of $\mathcal{R}$, $(P(A_1), P(A_2), \cdots, P(A_k)) \sim \text{Dir}(\alpha(A_1), \alpha(A_2), \cdots, \alpha(A_k))$, where Dir $(c_1, c_2, \cdots, c_k)$ denotes the Dirichlet distribution on the simplex with the parameter vector $(c_1, c_2, \cdots, c_k)$. If $P(F)$ has a Dirichlet process prior, we write $P(F) \sim DP(\alpha)$.

Remark. The main advantage of Dirichlet process priors is that they allow closed-form calculations of the posterior and many other Bayesian measures. Dirichlet priors have been criticized on a number of grounds, the main one being the property that realizations of Dirichlet process priors are almost surely discrete. Thus, they do not represent honest beliefs if we think that our data are coming from a PDF. The discrete realization property also leads to certain technical disadvantages in some inference problems. Despite these drawbacks, Dirichlet process priors have been quite popular and have given rise to other proven useful priors on distributions such as mixtures of Dirichlet processes. Other priors, including those that lead to absolutely continuous realizations with probability 1, have been suggested. see Freedman (1963), Dubins and Freedman (1966), and Mauldin, Sudderth, and Williams (1992). Also see Doksum (1974) for alternatives to Dirichlet process priors and Ferguson, Phadia, and Tiwari (1992) for a well-written exposition on priors in infinite-dimensional problems.

We now present a result that illustrates the complexities that can arise in infinite-dimensional problems that at first appear to be simple and straightforward.

Theorem 20.13 Suppose, given θ and a CDF F, that $X_i = \theta + \epsilon_i, i \geq 1$, with ϵ_i being iid F. Suppose θ and F are independent with respective marginals $\theta \sim \pi(\theta)$, a density on $\mathcal{R}$, and $F \sim DP(C(0, 1))$.

(a) If $\pi(\theta)$ is infinitely differentiable and greater than 0 at all θ, then there exists a density function h and a positive constant $\gamma = \gamma(h)$ such that almost surely, when the true θ equals 0 and the true distribution of ϵ_i is 1 with density h, the posterior distribution of θ accumulates near $\pm\gamma$; i.e., for any $c > 0$,

$$P_{\theta=0,h}[P_{\pi_n}(|\theta - \gamma| < c) + P_{\pi_n}(|\theta + \gamma| < c) \to 1] = 1,$$

where π_n denotes the posterior of θ given $X_1, \cdots, X_n$.

(b) For large n, about 50% of the time the posterior accumulates near γ and about 50% of the time the posterior accumulates near $-\gamma$; i.e., given $c, \eta, \delta > 0$, for $n \geq n_0(h, c, \eta, \delta)$,

$$P_{\theta=0,h}[P_{\pi_n}(|\theta - \gamma| < c) > 1 - \eta] > \frac{1}{2} - \delta$$

and

$$P_{\theta=0,h}[P_{\pi_n}(|\theta + \gamma| < c) > 1 - \eta] > \frac{1}{2} - \delta.$$

Furthermore, for both part (a) and part (b), h can be chosen to be symmetric, infinitely differentiable, and with a unique finite global maximum at zero.

Remark. This is one of the principal results in Diaconis and Freedman (1986). The particular density h offered in Diaconis and Freedman (1986) has three local maxima, with a global maximum at zero. The inconsistency phenomenon of Theorem 20.13 cannot hold if the true density is symmetric and *strictly unimodal*; i.e., if it strictly increases and then strictly decreases. The parameter measure of the Dirichlet process prior can be somewhat generalized; e.g., it may be taken as any t-density without affecting the results. Now, peculiarly, the posterior correctly recognizes the true sampling distribution, which in the setup of the theorem woul be the distribution with density h. Thus, about half of the time, the posterior for θ accumulates near the wrong value γ, the posterior for F accumulates near h shifted by γ, although the true F is 1 with density h itself, and yet the two errors made by the posteriors cancel each other out and the overall sampling distribution is recognized correctly! The posterior is confused about each of θ and ϵ but not about the convolution $\theta + \epsilon$.

Perhaps the most telling story of the result is that Bayesians must be extremely careful about their priors in infinite-dimensional problems. One may compute an answer and have it lead to a wrong inference in seemingly nice setups. On the other hand, as regards the Diaconis-Freedman result itself, a skeptic could ask whether an h such as the one offered is *likely* to be a true model for the errors. It may or may not be. The fact remains that frequentists have very simple consistent estimates for θ in this setup (e.g., the median). Yet, a Bayesian using quite conventional priors could fall into a trap. The need to be careful is the main message of the Diaconis-Freedman result, although it may take a lot of mathematics to ensure that nothing bad is going to happen with a particular prior.

Remark. The Diaconis-Freedman result is a cautionary result for someone wanting to be a Bayesian in an infinite-dimensional problem. The prior used by Diaconis and Freedman does not satisfy a certain property, now understood to be a basic requirement for positive results on posterior consistency.

If $X_1, X_2, \cdots$ are iid from a density f_0, then a prior on the true density is said to satisfy the *Kullback-Leibler property* if the prior probability of every Kullback-Leibler neighborhood of the true density must be strictly positive. If one takes the view that no knowledge of the true density is available, then the Kullback-Leibler property means that every neighborhood of any density under the Kullback-Leibler divergence is strictly positive. Although the Diaconis-Freedman prior does not have this property, some other priors do; see, e.g., Barron, Schervish, and Wasserman (1999) and Walker (2004). It turns out that the Kullback-Leibler property *alone* is not enough to ensure strong consistency of the posterior. Here, strong consistency means that with probability 1, under the true f_0, the posterior probability of any neighborhood of f_0 converges to 1 as $n \to \infty$, with neighborhoods being defined in terms of total variation distance, i.e., neighborhoods are sets of the form $U = \{f : d_{\text{TV}}(f, f_0) < \delta\}$. Various different sets of sufficient conditions (in addition to the Kullback-Leibler property) that ensure strong posterior consistency are available. We refer the reader to Walker (2004) for a review.

20.13 Exercises

Exercise 20.1 Suppose samples are iid from $N(\theta, \sigma^2)$, σ^2 is known, and $\theta \sim N(\eta, \tau^2)$. Using the notation of Example 20.1, give a direct proof that $d_{\text{TV}}(\Pi_n, Q) \xrightarrow[\mathcal{P}_{\theta_0}]{a.s.} 0$ for all θ_0 using only the Kolmogorov strong law.

Exercise 20.2 * Suppose that in Exercise 20.1, σ^2 was unknown. Under what conditions will the Bernstein-von Mises theorem hold for $\psi = \frac{\sqrt{n}(\theta - \bar{X})}{\sigma}$?

Exercise 20.3 * Consider the setup of Exercise 20.1. Let $\psi^* = \sqrt{n}(\theta - M)$, where M is the sample median. Is there a Bernstein-von Mises theorem for ψ^* as well?

Exercise 20.4 * Consider the setup of Exercise 20.1, and let $\sigma = 1$. Define $N = \inf\{n : d_{\text{TV}}(\Pi_n, Q) \leq .05\}$. Is $E(N) < \infty$?

Exercise 20.5 * Consider the setup of Exercise 20.1. Since $d_{\text{TV}}(\Pi_n, Q) \xrightarrow[\mathcal{P}_{\theta_0}]{a.s.} 0$, it also converges to zero in P_{θ_0}-probability. Prove this directly by calculating $P_{\theta_0}(d_{\text{TV}}(\Pi_n, Q) > \epsilon)$.

Exercise 20.6 * Suppose a Brownian motion with linear drift is observed at the times $\frac{i}{n}, i = 1, 2, \dots, n$. Prove the Bernstein-von Mises theorem directly assuming an $N(0, 1)$ prior for the drift coefficient.

Exercise 20.7 * Can there be a Bernstein-von Mises theorem in the iid case when priors are improper? What fundamental difficulties arise?

Exercise 20.8 * Suppose $X \sim \text{Bin}(n, p)$ and p has the Jeffrey prior, which is $\text{Beta}(\frac{1}{2}, \frac{1}{2})$. Let $\hat{p} = \frac{X}{n}$ and $\psi = \frac{\sqrt{n}(p-\hat{p})}{\sqrt{p(1-p)}}$. Prove the Bernstein-von Mises theorem directly.

Exercise 20.9 Suppose iid observations are obtained from a $\text{Poi}(\lambda)$ distribution.

(a) Derive a one-term expansion for the posterior CDF of λ.
(b) Simulate $n = 10$ observations from Poi(1). Calculate $P(\lambda \leq 1.1|X_1, ..., X_n)$ exactly for an Exp(1) prior and from the one-term asymptotic expansion. Comment on the expansion's accuracy.

Exercise 20.10 Suppose iid observations are obtained from an exponential density with mean θ.

(a) Derive one- and two-term expansions for the posterior CDF of θ.
(b) Simulate $n = 15$ observations from Exp(1). Calculate $P(\theta \leq 1.1|X_1, ..., X_n)$ exactly for an *inverse* Exp(1) prior and from the one- and two-term expansions. Comment on the expansions' accuracy.

Exercise 20.11 Suppose iid observations are obtained from $\text{Ber}(p)$.

(a) Derive a one-term expansion for the posterior mean of p.
(b) Suppose $p \sim \text{Beta}(\alpha, \alpha)$. Calculate the exact values and the approximation from the one-term expansion of the posterior mean of p for $n = 10; \alpha = .5, 1, 2, 5; X = \sum X_i = 1, ..., 9$.

Exercise 20.12 * Suppose that the observations are iid from $N(\theta, 1)$ and $\theta \sim \text{Beta}(\frac{1}{2}, \frac{1}{2})$. Is $E(\theta|X_1, ..., X_n) = \bar{X} + o(n^{-k})$ for all $k > 0$?

Exercise 20.13 * Suppose that the observations are iid from a double exponential with location parameter μ and $\mu \sim U[-1, 1]$.

(a) What is an MLE of μ?

(b) Is $E(\mu|X_1, ..., X_n) = \hat{\mu} + o(n^{-k})$ for all $k > 0$, where $\hat{\mu}$ denotes an MLE of μ. If such a result holds, will it hold for all MLEs?

Exercise 20.14 * Suppose iid observations are obtained from $N(\theta, 1)$ and $\theta \sim N(\mu, \tau^2)$. Derive the asymptotic (frequentist) distribution of the αth percentile of the posterior of θ (of course, on suitable centering and norming, etc.).

Exercise 20.15 * Generalize the previous problem to a general model and a general prior under enough regularity conditions.

Exercise 20.16 Suppose iid observations are obtained from $N(\theta, 1)$ and $\theta \sim C(0, 1)$. Compute the Tierney-Kadane approximation to the posterior mean of θ and compare it with the exact value by using $n = 10$ simulated samples from $N(0, 1)$.

Exercise 20.17 * Suppose iid observations are obtained from Ber(p), and $p \sim$ Beta(α, α). Compute the Tierney-Kadane approximation to the posterior mean of $\frac{p}{1-p}$, and compare it with the exact value by using $n = 15$ samples from Ber(.4) and with $\alpha = .5, 1, 2, 5$.

Remark. Note that the prior thinks that p should be about .5 if $\alpha = 2$ or 5.

Exercise 20.18 * Suppose a CDF F on the real line has the $DP(\alpha)$ prior and that n iid observations from F are available. Find the posterior of F.

Exercise 20.19 * The Diaconis-Freedman inconsistency result cannot hold if the parameter measure α of the Dirichlet prior is absolutely continuous with an everywhere logconvex density. Give half a dozen examples of such a measure α.

Exercise 20.20 * Derive a formula for the Hausdorff distance between two general spheres in the general dimension.

Exercise 20.21 * Consider the setup of Exercise 20.1, and let ρ_H denote the Hausdorff distance between the HPD interval and the usual z-interval. Derive the limit distribution of ρ_H (be careful about norming).

Exercise 20.22 * Consider iid samples from $N(\theta, 1)$, and suppose θ has a standard double exponential prior. Where does $\hat{\theta} - \bar{X}$ converge in law when the true $\theta_0 = 0$ if $\hat{\theta}$ = posterior mean, posterior mode?

Exercise 20.23 * Consider iid samples from $N(\theta, 1)$, and suppose θ has a $U[0, 1]$ prior. Where does $\hat{\theta} - \bar{X}$ converge in law when the true $\theta_0 = 0, 1$ if $\hat{\theta} =$ posterior mean? What fundamental differences arise when $0 < \theta_0 < 1$?

Exercise 20.24 Does a Brown identity hold for improper priors? For the multivariate case?

Exercise 20.25 * Use the Brown identity to prove Brown's second identity

$$\int R(\theta, \hat{\theta}) dG(\theta) = \frac{\sigma^2}{n} - \frac{\sigma^4}{n^2} \int \frac{(m'(x))^2}{m(x)} dx$$

when samples are iid $N(\theta, \sigma^2)$, $\theta \sim G$, $m(x)$ is the marginal density of $\bar{X}$, $\hat{\theta}$ is the posterior mean, and $R(\theta, \hat{\theta})$ denotes the mean squared error.

Exercise 20.26 * Consider the setup of Exercise 20.25 and suppose G belongs to $\mathcal{G} = \{G : G \text{isabsolutelycontinuous, unimodal, andsymmetric around0}\}$. Does $\sup_{G \in \mathcal{G}} \int R(\theta, \hat{\theta}) dG(\theta) \to 0$ as $n \to \infty$?

Exercise 20.27 Let $X|\theta \sim P_\theta$ and let $\theta \sim G$. Suppose $\delta(X)$ is the posterior mean of some parametric function $\psi(\theta)$ and $b_G(\theta) = E_\theta(\delta(X) - \psi(\theta))$ its bias. Show that the Bayes risk of $\delta(X)$ equals $-\int b_G(\theta) dG(\theta)$.

References

Barron, A., Schervish, M., and Wasserman, L. (1999). The consistency of posterior distributions in nonparametric problems, Ann. Stat., 27, 536–561.

Berger, J. (1985). *Statistical Decision Theory and Bayesian Analysis*, 2nd ed. Springer-Verlag, New York.

Bickel, P.J. and Doksum, K. (2001). *Mathematical Statistics: Basic Concepts and Selected Ideas*, Vol. I, Prentice-Hall, Upper Saddle River, NJ.

Brown, L. (1964). Sufficient statistics in the case of independent random variables, Ann. Math. Stat., 35, 1456–1474.

Brown, L. (1971). Admissible estimators, recurrent diffusions, and insoluble boundary value problems, Ann. Math. Stat., 42, 855–903.

Chung, K.L. (2001). *A Course in Probability Theory*, 3rd. Academic Press, San Diego, CA.

Diaconis, P. and Freedman, D. (1986). On inconsistent Bayes estimates of location, Ann. Stat., 14, 68–87.

Doksum, K. (1974). Tailfree and neutral random probabilities and their posterior distributions, Ann. Prob., 2, 183–201.

Doob, J.L. (1949). Application of the theory of martingales, Colloq. Int. C. N. R. S., 13.

Dubins, L. and Freedman, D. (1966). Random distribution functions, in *Proceedings of the Fifth Berkeley Symposium*, L. LeCam and J. Neyman (eds.), Vol. 2, University of California Press, Berkeley, 183–214.

Ferguson, T. (1973). A Bayesian analysis of some nonparametric problems, Ann. Stat., 1, 209–230.

Ferguson, T., Phadia, E., and Tiwari, R. (1992). *Bayesian Nonparametric Inference*, Lecture Notes in Mathematical Statistics, Vol. 17, Institute of Mathematical Statistics, Hayward, CA, 127–150.

Freedman, D. (1963). On the asymptotic behavior of Bayes estimates in the discrete case, Ann. Math. Stat., 34, 1386–1403.

Freedman, D. (1991). On the Bernstein-von Mises theorem with infinite- dimensional parameters, Ann. Stat., 27, 1119–1140.

Ghosal, S. and Samanta, T. (1997). Asymptotic expansions of posterior distributions in nonregular cases, Ann. Inst. Stat. Math., 49, 181–197.

Hartigan, J. (1983). *Bayes Theory*, Springer-Verlag, New York.

Heyde, C.C. and Johnstone, I.M. (1979). On asymptotic posterior normality for stochastic processes, J. R. Stat. Soc. B, 41(2), 184–189.

Johnson, R.A. (1967). An asymptotic expansion for posterior distributions, Ann. Math. Stat., 38, 1899–1906.

Johnson, R.A. (1970). Asymptotic expansions associated with posterior distributions, Ann. Math. Stat., 41, 851–864.

LeCam, L. (1986). *Asymptotic Methods in Statistical Decision Theory*, Springer-Verlag, New York.

Lehmann, E.L. and Casella, G. (1998). *Theory of Point Estimation*, 2nd ed., Springer, New York.

Lindley, D. (1957). A statistical paradox, Biometrika, 44, 187–192.

Mauldin, R., Sudderth, W., and Williams, S. (1992). Pólya trees and random distributions, Ann. Stat., 20, 1203–1221.

Schervish, M. (1995). *Theory of Statistics*, Springer-Verlag, New York.

Tierney, L. and Kadane, J.B. (1986). Accurate approximations for posterior moments and marginal densities, J. Am. Stat. Assoc., 81, 82–86.

Walker, S.G. (2004). Modern Bayesian asymptotics, Stat. Sci., 19, 111–117.

Wasserman, L. (1998). *Asymptotic Properties of Nonparametric Bayesian Procedures*, Lecture Notes in Statistics, Vol. 133, Springer, New York, 293–304.

Wasserman, L. (2004). *All of Statistics: A Concise Course in Statistical Inference*, Springer-Verlag, New York.

Woodroofe, M. (1992). Integrable expansions for posterior distributions for one parameter Exponential families, Stat. Sinica, 2(1), 91–111.

van der Vaart, A. (1998). *Asymptotic Statistics*, Cambridge University Press, Cambridge.

Chapter 21
Testing Problems

Testing is certainly another fundamental problem of inference. It is interestingly different from estimation. The accuracy measures are fundamentally different and the appropriate asymptotic theory is also different. Much of the theory of testing has revolved somehow or other around the Neyman-Pearson lemma, which has led to a lot of ancillary developments in other problems in mathematical statistics. Testing has led to the useful idea of local alternatives, and like maximum likelihood in estimation, likelihood ratio, Wald, and Rao score tests have earned the status of default methods, with a neat and quite unified asymptotic theory. For all these reasons, a treatment of testing is essential. We discuss the asymptotic theory of likelihood ratio, Wald, and Rao score tests in this chapter. Principal references for this chapter are Bickel and Doksum (2001), Ferguson (1996), and Sen and Singer (1993). Many other specific references are given in the following sections.

21.1 Likelihood Ratio Tests

The likelihood ratio test is a general omnibus test applicable, in principle, in most finite-dimensional parametric problems. Thus, let $X^{(n)} = (X_1, \ldots, X_n)$ be the observed data with joint distribution $P_\theta^n \ll \mu_n, \theta \in \Theta$, and density $f_\theta(x^{(n)}) = dP_\theta^n/d\mu_n$. Here, μ_n is some appropriate σ-finite measure on $\mathcal{X}_n$, which we assume to be a subset of a Euclidean space. For testing $H_0 : \theta \in \Theta_0$ vs. $H_1 : \theta \in \Theta - \Theta_0$, the likelihood ratio test (LRT) rejects H_0 for small values of

$$\Lambda_n = \frac{\sup_{\theta \in \Theta_0} f_\theta(x^{(n)})}{\sup_{\theta \in \Theta} f_\theta(x^{(n)})}.$$

The motivation for Λ_n comes from two sources:

A. DasGupta, *Asymptotic Theory of Statistics and Probability*,

(a) The case where H_0, and H_1 are each simple, for which a most powerful (MP) test is found from Λ_n by the Neyman-Pearson lemma.

(b) The intuitive explanation that, for small values of Λ_n, we can better match the observed data with some value of θ outside of Θ_0.

Likelihood ratio tests are useful because they are omnibus tests and because otherwise optimal tests for a given sample size n are generally hard to find outside of the exponential family. However, the LRT is not a universal test. There are important examples where the LRT simply cannot be used because the null distribution of the LRT test statistic depends on nuisance parameters. Also, the exact distribution of the LRT statistic is very difficult or impossible to find in many problems. Thus, asymptotics become really important. But the asymptotics of the LRT may be nonstandard under nonstandard conditions.

Although we only discuss the case of a parametric model with a fixed finite-dimensional parameter space, LRTs and their useful modifications have been studied where the number of parameters grows with the sample size n and in nonparametric problems. See, e.g., Portnoy (1988), Fan, Hung, and Wong (2000), and Fan and Zhang (2001). We start with a series of examples that illustrate various important aspects of the likelihood ratio method.

21.2 Examples

Example 21.1 Let $X_1, X_2, \ldots, X_n \stackrel{\text{iid}}{\sim} N(\mu, \sigma^2)$, and consider testing

$$H_0 : \mu = 0 \text{ vs. } H_1 : \mu \neq 0.$$

Let $\theta = (\mu, \sigma^2)$. Then,

$$\Lambda_n = \frac{\sup_{\theta \in \Theta_0} (1/\sigma^n) \exp\left(-\frac{1}{2\sigma^2} \sum_i (X_i - \mu)^2\right)}{\sup_{\theta \in \Theta} (1/\sigma^n) \exp\left(-\frac{1}{2\sigma^2} \sum_i (X_i - \mu)^2\right)} = \left(\frac{\sum_i (X_i - \overline{X}_n)^2}{\sum_i X_i^2}\right)^{n/2}$$

by an elementary calculation of MLEs of θ under H_0 and in the general parameter space. By another elementary calculation, $\Lambda_n < c$ is seen to be equivalent to $t_n^2 > k$, where

$$t_n = \frac{\sqrt{n}\overline{X}_n}{\sqrt{\frac{1}{n-1} \sum_i (X_i - \overline{X}_n)^2}}$$

is the t-statistic. In other words, the t-test is the LRT. Also, observe that

$$
\begin{aligned}
t_n^2 &= \frac{n\overline{X}_n^2}{\frac{1}{n-1}\sum_i (X_i - \overline{X}_n)^2} \\
&= \frac{\sum_i X_i^2 - \sum_i (X_i - \overline{X}_n)^2}{\frac{1}{n-1}\sum_i (X_i - \overline{X}_n)^2} \\
&= \frac{(n-1)\sum_i X_i^2}{\sum_i (X_i - \overline{X}_n)^2} - (n-1) \\
&= (n-1)\Lambda_n^{-2/n} - (n-1).
\end{aligned}
$$

This implies

$$
\begin{aligned}
& \Lambda_n = \left(\frac{n-1}{t_n^2 + n - 1}\right)^{n/2} \\
\Rightarrow \quad & \log \Lambda_n = \frac{n}{2} \log \frac{n-1}{t_n^2 + n - 1} \\
\Rightarrow & -2\log \Lambda_n = n \log \left(1 + \frac{t_n^2}{n-1}\right) \\
& = n\left(\frac{t_n^2}{n-1} + o_p\left(\frac{t_n^2}{n-1}\right)\right) \\
& \xrightarrow{\mathcal{L}} \chi_1^2
\end{aligned}
$$

under H_0 since $t_n \xrightarrow{\mathcal{L}} N(0, 1)$ under H_0.

Example 21.2 Consider a multinomial distribution $MN(n, p_1, \cdots, p_k)$. Consider testing

$$
H_0 : p_1 = p_2 = \cdots = p_k \text{ vs. } H_1 : H_0 \; is \; not \; true.
$$

Let $n_1, \cdots, n_k$ denote the observed cell frequencies. Then, by an elementary calculation,

$$\Lambda_n = \Pi_{i=1}^{k} \left(\frac{n}{kn_i} \right)^{n_i}$$

$$\Rightarrow -\log \Lambda_n = n \left(\log \frac{k}{n} \right) + \sum_{i=1}^{k} n_i \log n_i.$$

The exact distribution of this is a messy discrete object, and so asymptotics will be useful. We illustrate the asymptotic distribution for $k = 2$. In this case,

$$-\log \Lambda_n = n \log 2 - n \log n + n_1 \log n_1 + (n - n_1) \log(n - n_1).$$

Let $Z_n = (n_1 - n/2)/\sqrt{n/4}$. Then,

$$-\log \Lambda_n = n \log 2 - n \log n$$

$$+ \frac{n + \sqrt{n} Z_n}{2} \log \frac{n + \sqrt{n} Z_n}{2} + \frac{n - \sqrt{n} Z_n}{2} \log \frac{n - \sqrt{n} Z_n}{2}$$

$$= -n \log n + \frac{n + \sqrt{n} Z_n}{2} \log(n + \sqrt{n} Z_n) + \frac{n - \sqrt{n} Z_n}{2} \log(n - \sqrt{n} Z_n)$$

$$= \frac{n + \sqrt{n} Z_n}{2} \log \left(1 + \frac{Z_n}{\sqrt{n}} \right) + \frac{n - \sqrt{n} Z_n}{2} \log \left(1 - \frac{Z_n}{\sqrt{n}} \right)$$

$$= \frac{n + \sqrt{n} Z_n}{2} \left(\frac{Z_n}{\sqrt{n}} - \frac{Z_n^2}{2n} + o_p \left(\frac{Z_n^2}{n} \right) \right)$$

$$+ \frac{n - \sqrt{n} Z_n}{2} \left(-\frac{Z_n}{\sqrt{n}} - \frac{Z_n^2}{2n} + o_p \left(\frac{Z_n^2}{n} \right) \right)$$

$$= \frac{Z_n^2}{2} + o_p(1).$$

Hence $-2 \log \Lambda_n \xrightarrow{\mathcal{L}} \chi_1^2$ under H_0 as $Z_n \xrightarrow{\mathcal{L}} N(0, 1)$ under H_0.

Remark. The popular test in this problem is the Pearson chi-square test that rejects H_0 for large values of

$$\frac{\sum_{i=1}^{k}(n_i - n/k)^2}{n/k}.$$

Interestingly, this test statistic, too, has an asymptotic χ_1^2 distribution, as does the LRT statistic.

Example 21.3 This example shows that the chi-square asymptotics of the LRT statistic fail when parameters have constraints or somehow there is a boundary phenomenon.

Let $X_1, X_2, \ldots, X_n \stackrel{\text{iid}}{\sim} N_2(\theta, I)$, where the parameter space is restricted to $\Theta = \{\theta = (\theta_1, \theta_2) : \theta_1 \geq 0, \theta_2 \geq 0\}$. Consider testing

$$H_0 : \theta = 0 \text{ vs. } H_1 : \; H_0 \; is \; not \; true.$$

We would write the n realizations as $x_i = (x_{1i}, x_{2i})^T$, and the sample average would be denoted by $\overline{x} = (\overline{x}_1, \overline{x}_2)$. The MLE of θ is given by

$$\hat{\theta} = \begin{pmatrix} \overline{x}_1 \vee 0 \\ \overline{x}_2 \vee 0 \end{pmatrix}.$$

Therefore, by an elementary calculation,

$$\Lambda_n = \frac{\exp(-\sum_i x_{1i}^2/2 - \sum_i x_{2i}^2/2)}{\exp(-\sum_i (x_{1i} - \overline{x}_1 \vee 0)^2/2 - \sum_i (x_{2i} - \overline{x}_2 \vee 0)^2/2)}.$$

Case 1: $\overline{x}_1 \leq 0, \overline{x}_2 \leq 0$. In this case, $\Lambda_n = 1$ and $-2\log\Lambda_n = 0$.

Case 2: $\overline{x}_1 > 0, \overline{x}_2 \leq 0$. In this case, $-2\log\Lambda_n = n\overline{x}_1^2$.

Case 3: $\overline{x}_1 \leq 0, \overline{x}_2 > 0$. In this case, $-2\log\Lambda_n = n\overline{x}_2^2$.

Case 4: $\overline{x}_1 > 0, \overline{x}_2 > 0$. In this case, $-2\log\Lambda_n = n\overline{x}_1^2 + n\overline{x}_2^2$.

Now, under H_0, each of the four cases above has a 1/4 probability of occurrence. Therefore, under H_0,

$$-2\log\Lambda_n \xrightarrow{\mathcal{L}} \frac{1}{4}\delta_{\{0\}} + \frac{1}{2}\chi_1^2 + \frac{1}{4}\chi_2^2$$

in the sense of the mixture distribution being its weak limit.

Example 21.4 In bioequivalence trials, a brand name drug and a generic are compared with regard to some important clinical variable, such as average drug concentration in blood over a 24 hour time period. By testing for a difference, the problem is reduced to a single variable, often assumed to be normal. Formally, one has $X_1, X_2, \ldots, X_n \stackrel{\text{iid}}{\sim} N(\mu, \sigma^2)$, and the bioequivalence hypothesis is

$$H_0 : |\mu| \leq \epsilon \text{ for some specified } \epsilon > 0.$$

Inference is *always* significantly harder if known constraints on the parameters are enforced. Casella and Strawderman (1980) is a standard introduction to the normal mean problem with restrictions on the mean; Robertson, Wright, and Dykstra (1988) is an almost encyclopedic exposition.

Coming back to our example, to derive the LRT, we need the restricted MLE and the nonrestricted MLE. We will assume here that σ^2 is known. Then the MLE under H_0 is

$$\hat{\mu}_{H_0} = \bar{X} \text{ if } |\bar{X}| \leq \epsilon$$

$$= \epsilon \operatorname{sgn}(\bar{X}) \text{ if } |\bar{X}| > \epsilon.$$

Consequently, the LRT statistic Λ_n satisfies

$$-2 \log \Lambda_n = 0 \text{ if } |\bar{X}| \leq \epsilon$$

$$= \frac{n}{\sigma^2}(\bar{X} - \epsilon)^2 \text{ if } \bar{X} > \epsilon$$

$$= \frac{n}{\sigma^2}(\bar{X} + \epsilon)^2 \text{ if } \bar{X} < -\epsilon.$$

Take a fixed μ. Then,

$$P_\mu(|\bar{X}| \leq \epsilon) = \Phi\left(\frac{\sqrt{n}(\epsilon - \mu)}{\sigma}\right) + \Phi\left(\frac{\sqrt{n}(\epsilon + \mu)}{\sigma}\right) - 1,$$

$$P_\mu(\bar{X} < -\epsilon) = \Phi\left(-\frac{\sqrt{n}(\epsilon + \mu)}{\sigma}\right),$$

and

$$P_\mu(\bar{X} > \epsilon) = 1 - \Phi\left(\frac{\sqrt{n}(\epsilon - \mu)}{\sigma}\right).$$

From these expressions, we get:
Case 1: If $-\epsilon < \mu < \epsilon$, then $P_\mu(|\bar{X}| \leq \epsilon) \to 1$ and so $-2\log\Lambda_n \xrightarrow{\mathcal{L}} \delta_{\{0\}}$.

Case 2(a): If $\mu = -\epsilon$, then both $P_\mu(|\bar{X}| \leq \epsilon)$ and $P_\mu(\bar{X} < -\epsilon)$ converge to 1/2. In this case, $-2\log\Lambda_n \xrightarrow{\mathcal{L}} \frac{1}{2}\delta_{\{0\}} + \frac{1}{2}\chi_1^2$; i.e., the mixture of a point mass and a chi-square distribution.

Case 2(b): Similarly, if $\mu = \epsilon$, then again $-2\log\Lambda_n \xrightarrow{\mathcal{L}} \frac{1}{2}\delta_{\{0\}} + \frac{1}{2}\chi_1^2$.

Case 3: If $\mu > \epsilon$, then $P_\mu(|\bar{X}| \leq \epsilon) \to 0$ and $P_\mu(\bar{X} > \epsilon) \to 1$. In this case, $-2\log\Lambda_n \xrightarrow{\mathcal{L}} NC\chi^2(1, (\mu-\epsilon)^2/\sigma^2)$, a noncentral chi-square distribution.

Case 4: Likewise, if $\mu < -\epsilon$, then $-2\log\Lambda_n \xrightarrow{\mathcal{L}} NC\chi^2(1, (\mu+\epsilon)^2/\sigma^2)$.

So, unfortunately, even the null asymptotic distribution of $-\log\Lambda_n$ depends on the exact value of μ.

Example 21.5 Consider the Behrens-Fisher problem with

$$X_1, X_2, \ldots, X_m \overset{\text{iid}}{\sim} N(\mu_1, \sigma_1^2),$$
$$Y_1, Y_2, \ldots, Y_n \overset{\text{iid}}{\sim} N(\mu_2, \sigma_2^2),$$

and all $m+n$ observations are independent. We want to test

$$H_0 : \mu_1 = \mu_2 \text{ vs. } H_1 : \mu_1 \neq \mu_2.$$

Let $\hat{\mu}, \hat{\sigma}_1^2, \hat{\sigma}_2^2$ denote the restricted MLE of μ, σ_1^2, and σ_2^2, respectively, under H_0. They satisfy the equations

$$\hat{\sigma}_1^2 = (\bar{X}-\hat{\mu})^2 + \frac{1}{m}\sum_i (X_i - \bar{X})^2,$$

$$\hat{\sigma}_2^2 = (\bar{Y}-\hat{\mu})^2 + \frac{1}{n}\sum_j (Y_j - \bar{Y})^2,$$

$$0 = \frac{m(\bar{X}-\hat{\mu})}{\hat{\sigma}_1^2} + \frac{n(\bar{Y}-\hat{\mu})}{\hat{\sigma}_2^2}.$$

This gives

$$-2\log\Lambda_n = \left(\frac{\frac{1}{m}\sum(X_i-\bar{X})^2}{\frac{1}{m}\sum(X_i-\bar{X})^2+(\bar{X}-\hat{\mu})^2}\right)^{\frac{m}{2}} \left(\frac{\frac{1}{n}\sum(Y_j-\bar{Y})^2}{\frac{1}{n}\sum(Y_j-\bar{Y})^2+(\bar{Y}-\hat{\mu})^2}\right)^{\frac{n}{2}}.$$

However, the LRT fails as a matter of practical use because its distribution depends on the nuisance parameter σ_1^2/σ_2^2, which is unknown.

Example 21.6 In point estimation, the MLEs in nonregular problems behave fundamentally differently with respect to their asymptotic distributions. For example, limit distributions of MLEs are not normal (see Chapter 15.12). It is interesting, therefore, that limit distributions of $-2\log\Lambda_n$ in nonregular cases still follow the chi-square recipe, but with different degrees of freedom.

As an example, suppose $X_1, X_2, \ldots, X_n \overset{\text{iid}}{\sim} U[\mu-\sigma, \mu+\sigma]$, and suppose we want to test $H_0 : \mu = 0$. Let $W = X_{(n)} - X_{(1)}$ and $U = \max(X_{(n)}, -X_{(1)})$. Under H_0, U is complete and sufficient and W/U is an ancillary statistic. Therefore, by Basu's theorem (Basu (1955)), under H_0, U and W/U are independent. This will be useful to us shortly. Now, by a straightforward calculation,

$$\Lambda_n = \left(\frac{W}{2U}\right)^n,$$

$$\Rightarrow -2\log\Lambda_n = 2n(\log 2U - \log W).$$

Therefore, under H_0,

$$Ee^{it(-2\log\Lambda_n)} = Ee^{2nit(\log 2U - \log W)} = \frac{Ee^{2nit(-\log W)}}{Ee^{2nit(-\log 2U)}}$$

by the independence of U and W/U. On doing the calculation, this works out to

$$Ee^{it(-2\log\Lambda_n)} = \frac{n-1}{n(1-2it)-1} \to \frac{1}{1-2it}.$$

Since $(1-2it)^{-1}$ is the characteristic function of the χ_2^2 distribution, it follows that $-2\log\Lambda_n \xrightarrow{\mathcal{L}} \chi_2^2$.

Thus, in spite of the nonregularity, $-2\log\Lambda_n$ is asymptotically chi-square, but we gain a degree of freedom! This phenomenon holds more generally in nonregular cases.

Example 21.7 Consider the problem of testing for equality of two Poisson means. Thus, let $X_1, X_2, \ldots, X_m \overset{\text{iid}}{\sim} \text{Poi}(\mu_1)$ and $Y_1, Y_2, \ldots, Y_n \overset{\text{iid}}{\sim} \text{Poi}(\mu_2)$, where X_i's and Y_j's are independent. Suppose we wish to test $H_0 : \mu_1 = \mu_2$. We assume $m, n \to \infty$ in such a way that

$$\frac{m}{m+n} \to \lambda \text{ with } 0 < \lambda < 1.$$

The restricted MLE for $\mu = \mu_1 = \mu_2$ is

$$\frac{\sum_i X_i + \sum_j Y_j}{m+n} = \frac{m\bar{X} + n\bar{Y}}{m+n}.$$

Therefore, by an easy calculation,

$$\Lambda_{m,n} = \frac{\left((m\bar{X}+n\bar{Y})/(m+n)\right)^{m\bar{X}+n\bar{Y}}}{\bar{X}^{m\bar{X}}\bar{Y}^{n\bar{Y}}}$$

$$\Rightarrow -\log\Lambda_{m,n} = m\bar{X}\log\bar{X} + n\bar{Y}\log\bar{Y}$$

$$-(m\bar{X}+n\bar{Y})\log\left(\frac{m}{m+n}\bar{X} + \frac{n}{m+n}\bar{Y}\right).$$

The asymptotic distribution of $-\log\Lambda_{m,n}$ can be found by a two-term Taylor expansion by using the multivariate delta theorem. This would be a more direct derivation, but we will instead present a derivation based on an asymptotic technique that is useful in many problems. Define

$$Z_1 = Z_{1,m} = \frac{\sqrt{m}(\bar{X}-\mu)}{\sqrt{\mu}}, \quad Z_2 = Z_{2,m} = \frac{\sqrt{n}(\bar{Y}-\mu)}{\sqrt{\mu}},$$

where $\mu = \mu_1 = \mu_2$ is the common value of μ_1 and μ_2 under H_0. Substituting in the expression for $\log\Lambda_{m,n}$, we have

$$-\log \Lambda_{m,n} = (m\mu + \sqrt{m\mu}Z_1)\left[\log\left(1+\frac{Z_1}{\sqrt{m\mu}}\right) + \log\mu\right]$$

$$+(n\mu + \sqrt{n\mu}Z_2)\left[\log\left(1+\frac{Z_2}{\sqrt{n\mu}}\right) + \log\mu\right]$$

$$-((m+n)\mu + \sqrt{m\mu}Z_1 + \sqrt{n\mu}Z_2)$$

$$\left[\log\left(1+\frac{m}{m+n}\frac{Z_1}{\sqrt{m\mu}} + \frac{n}{m+n}\frac{Z_2}{\sqrt{n\mu}}\right) + \log\mu\right]$$

$$= (m\mu + \sqrt{m\mu}Z_1)\left(\frac{Z_1}{\sqrt{m\mu}} - \frac{Z_1^2}{2m\mu} + O_p(m^{-3/2}) + \log\mu\right)$$

$$+(n\mu + \sqrt{n\mu}Z_2)\left(\frac{Z_2}{\sqrt{n\mu}} - \frac{Z_2^2}{2n\mu} + O_p(n^{-3/2}) + \log\mu\right)$$

$$-((m+n)\mu + \sqrt{m\mu}Z_1 + \sqrt{n\mu}Z_2)\left(\frac{m}{m+n}\frac{Z_1}{\sqrt{m\mu}} + \frac{n}{m+n}\frac{Z_2}{\sqrt{n\mu}}\right.$$

$$-\frac{m}{(m+n)^2}\frac{Z_1^2}{2\mu} - \frac{n}{(m+n)^2}\frac{Z_2^2}{2\mu}$$

$$\left.-\frac{\sqrt{mn}}{(m+n)^2}\frac{Z_1Z_2}{\mu} + O_p(\min(m,n)^{-3/2}) + \log\mu\right).$$

On further algebra from above, by breaking down the terms, and on cancellation,

$$-\log \Lambda_{m,n} = Z_1^2\left(\frac{1}{2} - \frac{m}{2(m+n)}\right) + Z_2^2\left(\frac{1}{2} - \frac{n}{2(m+n)}\right)$$

$$-\frac{\sqrt{mn}}{m+n}Z_1Z_2 + o_p(1).$$

Assuming that $m/(m+n) \to \lambda$, it follows that

$$-\log \Lambda_{m,n} = \frac{1}{2}\left(\sqrt{\frac{n}{m+n}}Z_1 - \sqrt{\frac{m}{m+n}}Z_2\right)^2 + o_p(1)$$

$$\xrightarrow{\mathcal{L}} \frac{1}{2}(\sqrt{1-\lambda}N_1 - \sqrt{\lambda}N_2)^2,$$

where N_1, N_2 are independent $N(0, 1)$. Therefore, $-2 \log \Lambda_{m,n} \xrightarrow{\mathcal{L}} \chi_1^2$ since $\sqrt{1-\lambda}N_1 - \sqrt{\lambda}N_2 \sim N(0, 1)$.

Example 21.8 This example gives a general formula for the LRT statistic for testing about the natural parameter vector in the q-dimensional multiparameter exponential family. What makes the example special is a link of the LRT to the Kullback-Leibler divergence measure in the exponential family.

Suppose $X_1, X_2, \ldots, X_n \overset{\text{iid}}{\sim} p(x|\theta) = e^{\theta' T(x)} c(\theta) h(x)(d\mu)$. Suppose we want to test $\theta \in \Theta_0$ vs. $H_1 : \theta \in \Theta - \Theta_0$, where Θ is the full natural parameter space. Then, by definition,

$$-2 \log \Lambda_n = -2 \log \frac{\sup_{\theta \in \Theta_0} c(\theta)^n \exp(\theta' \sum_i T(x_i))}{\sup_{\theta \in \Theta} c(\theta)^n \exp(\theta' \sum_i T(x_i))}$$

$$= -2 \sup_{\theta_0 \in \Theta_0} [n \log c(\theta_0) + n\theta_0' \overline{T}] + 2 \sup_{\theta \in \Theta} [n \log c(\theta) + n\theta' \overline{T}]$$

$$= 2n \sup_{\theta \in \Theta} \inf_{\theta_0 \in \Theta_0} [(\theta - \theta_0)' \overline{T} + \log c(\theta) - \log c(\theta_0)]$$

on a simple rearrangement of the terms. Recall now (see Chapter 2) that the Kullback-Leibler divergence, $K(f, g)$, is defined to be

$$K(f, g) = E_f \log \frac{f}{g} = \int \log \frac{f(x)}{g(x)} f(x) \mu(dx).$$

It follows that

$$K(p_\theta, p_{\theta_0}) = (\theta - \theta_0)' E_\theta T(X) + \log c(\theta) - \log c(\theta_0)$$

for any $\theta, \theta_0 \in \Theta$. We will write $K(\theta, \theta_0)$ for $K(p_\theta, p_{\theta_0})$. Recall also that, in the general exponential family, the unrestricted MLE $\hat{\theta}$ of θ exists for all large n, and furthermore, $\hat{\theta}$ is a moment estimate given by $E_{\theta=\hat{\theta}}(T) = \overline{T}$. Consequently,

$$-2 \log \Lambda_n = 2n \sup_{\theta \in \Theta} \inf_{\theta_0 \in \Theta_0} [(\hat{\theta} - \theta_0)' E_{\hat{\theta}}(T) + \log c(\hat{\theta}) - \log c(\theta_0) +$$

$$(\theta - \hat{\theta})' E_{\hat{\theta}}(T) + \log c(\theta) - \log c(\hat{\theta})]$$

$$= 2n \sup_{\theta \in \Theta} \inf_{\theta_0 \in \Theta_0} [K(\hat{\theta}, \theta_0) + (\theta - \hat{\theta})' E_{\hat{\theta}}(T) + \log c(\theta) - \log c(\hat{\theta})]$$

$$= 2n \inf_{\theta_0 \in \Theta_0} K(\hat{\theta}, \theta_0) + 2n \sup_{\theta \in \Theta} [(\theta - \hat{\theta})' \overline{T}_n + \log c(\theta) - \log c(\hat{\theta})].$$

The second term above vanishes as the supremum is attained at $\theta = \hat{\theta}$. Thus, we ultimately have the identity

$$-2 \log \Lambda_n = 2n \inf_{\theta_0 \in \Theta_0} K(\hat{\theta}, \theta_0).$$

The connection is beautiful. Kullback-Leibler divergence being a measure of distance, $\inf_{\theta_0 \in \Theta_0} K(\hat{\theta}, \theta_0)$ quantifies the disagreement between the null and the estimated value of the true parameter θ when estimated by the MLE. The formula says that when the disagreement is small, one should accept H_0.

The link to the K-L divergence is special for the exponential family; for example, if

$$p_\theta(x) = \frac{1}{\sqrt{2\pi}} e^{-\frac{1}{2}(x-\theta)^2},$$

then, for any θ, θ_0, $K(\theta, \theta_0) = (\theta - \theta_0)^2/2$ by a direct calculation. Therefore,

$$-2 \log \Lambda_n = 2n \inf_{\theta_0 \in \Theta_0} \frac{1}{2} (\hat{\theta} - \theta)^2 = n \inf_{\theta_0 \in \Theta_0} (\bar{X} - \theta_0)^2.$$

If H_0 is simple (i.e., $H_0 : \theta = \theta_0$), then the expression above simplifies to $-2 \log \Lambda_n = n(\bar{X} - \theta_0)^2$, just the familiar chi-square statistic.

21.3 Asymptotic Theory of Likelihood Ratio Test Statistics

Suppose $X_1, X_2, \ldots, X_n \overset{\text{iid}}{\sim} f(x|\theta)$, densities with respect to some dominating measure μ. We present the asymptotic theory of the LRT statistic for two types of null hypotheses.

Suppose $\theta \in \Theta \subseteq \mathbb{R}^q$ for some $0 < q < \infty$, and suppose as an affine subspace of $\mathbb{R}^q$, $\dim \Theta = q$. One type of null hypothesiswe consider is

$H_0 : \theta = \theta_0$ (specified). A second type of null hypothesis is that for some $0 < r < q$ and functions $h_1(\theta), \ldots h_r(\theta)$, linearly independent,

$$h_1(\theta) = h_2(\theta) = \cdots = h_r(\theta) = 0.$$

For each case, under regularity conditions to be stated below, the LRT statistic is asymptotically a central chi-square under H_0 with a fixed degree of freedom, including the case where H_0 is composite.

Example 21.9 Suppose $X_{11}, \ldots, X_{1n_1} \stackrel{\text{iid}}{\sim}$ Poisson (θ_1), $X_{21}, \ldots, X_{2n_2} \stackrel{\text{iid}}{\sim}$ Poisson (θ_2) and $X_{31}, \ldots, X_{3n_3} \stackrel{\text{iid}}{\sim}$ Poisson (θ_3), where $0 < \theta_1, \theta_2, \theta_3 < \infty$, and all observations are independent. An example of H_0 of the first type is $H_0 : \theta_1 = 1, \theta_2 = 2, \theta_3 = 1$. An example of H_0 of the second type is $H_0 : \theta_1 = \theta_2 = \theta_3$. The functions h_1, h_2 may be chosen as $h_1(\theta) = \theta_1 - \theta_2, h_2(\theta) = \theta_2 - \theta_3$.

In the first case, $-2 \log \Lambda_n$ is asymptotically χ^2_3 under H_0, and in the second case it is asymptotically χ^2_2 under each $\theta \in H_0$. This is a special example of a theorem originally stated by Wilks (1938); also see Lawley (1956). The degree of freedom of the asymptotic chi-square distribution under H_0 is the number of independent constraints specified by H_0; it is useful to remember this as a general rule.

Theorem 21.1 Suppose $X_1, X_2, \ldots, X_n \stackrel{\text{iid}}{\sim} f_\theta(x) = f(x|\theta) = \frac{dP_\theta}{d\mu}$ for some dominating measure μ. Suppose $\theta \in \Theta$, an affine subspace of $\mathcal{R}^q$ of dimension q. Suppose that the conditions required for asymptotic normality of any strongly consistent sequence of roots $\hat{\theta}_n$ of the likelihood equation hold (see Chapter 15.12). Consider the problem $H_0 : \theta = \theta_0$ against $H_1 : \theta \neq \theta_0$. Let

$$\Lambda_n = \frac{\prod_{i=1}^{n} f(x_i|\theta_0)}{l(\hat{\theta}_n)},$$

where $l(\cdot)$ denotes the likelihood function. Then $-2 \log \Lambda_n \xrightarrow{\mathcal{L}} \chi^2_q$ under H_0.

The proof can be seen in Ferguson (1996), Bickel and Doksum (2001), or Sen and Singer (1993).

Remark. As we saw in our illustrative examples earlier in this chapter, the theorem can fail if the regularity conditions fail to hold. For instance, if the null value is a boundary point in a constrained parameter space, then the result will fail.

Remark. This theorem is used in the following way. To use the LRT with an exact level α, we need to find $c_{n,\alpha}$ such that $P_{H_0}(\Lambda_n < c_{n,\alpha}) = \alpha$. Generally, Λ_n is so complicated as a statistic that one cannot find $c_{n,\alpha}$ exactly. Instead, we use the test that rejects H_0 when

$$-2\log\Lambda_n > \chi^2_{\alpha,q}.$$

Theorem 21.1 implies that $P(-2\log\Lambda_n > \chi^2_{\alpha,q}) \to \alpha$ and so $P(-2\log\Lambda_n > \chi^2_{\alpha,q}) - P(\Lambda_n < c_{n,\alpha}) \to 0$ as $n \to \infty$.

Under the second type of null hypothesis, the same result holds with the degree of freedom being r. Precisely, the following theorem holds.

Theorem 21.2 Assume all the regularity conditions in the previous theorem. Define $H_{q\times r} = H(\theta) = \left(\frac{\partial}{\partial\theta_i}h_j(\theta)\right)_{\substack{1\le i\le q\\1\le j\le r}}$. It is assumed that the required partial derivatives exist. Suppose for each $\theta \in \Theta_0$ that H has full column rank. Define

$$\Lambda_n = \frac{\sup_{\theta:h_j(\theta)=0,1\le j\le r} l(\theta)}{l(\hat{\theta}_n)}.$$

Then, $-2\log\Lambda_n \xrightarrow{\mathcal{L}} \chi^2_r$ under each θ in H_0.

A proof can be seen in Sen and Singer (1993).

21.4 Distribution under Alternatives

To find the power of the test that rejects H_0 when $-2\log\Lambda_n > \chi^2_{\alpha,d}$ for some d, one would need to know the distribution of $-2\log\Lambda_n$ at the particular $\theta = \theta_1$ value where we want to know the power. But the distribution under θ_1 of $-2\log\Lambda_n$ for a fixed n is also generally impossible to find, so we may appeal to asymptotics.

However, there cannot be a nondegenerate limit distribution on $[0,\infty)$ for $-2\log\Lambda_n$ under a fixed θ_1 in the alternative. The following simple example illustrates this difficulty.

Example 21.10 Suppose $X_1, X_2, \ldots, X_n \overset{\text{iid}}{\sim} N(\mu, \sigma^2)$, μ, σ^2 both unknown, and we wish to test $H_0: \mu = 0$ vs. $H_1: \mu \neq 0$. We saw earlier that

$$\Lambda_n = \left(\frac{\sum(X_i - \bar{X})^2}{\sum X_i^2}\right)^{n/2} = \left(\frac{\sum(X_i - \bar{X})^2}{\sum(X_i - \bar{X})^2 + n\bar{X}^2}\right)^{n/2}$$

$$\Rightarrow -2\log\Lambda_n = n\log\left(1 + \frac{n\bar{X}^2}{\sum(X_i - \bar{X})^2}\right) = n\log\left(1 + \frac{\bar{X}^2}{\frac{1}{n}\sum(X_i - \bar{x})^2}\right).$$

Consider now a value $\mu \neq 0$. Then $\bar{X}^2 \stackrel{\text{a.s.}}{\to} \mu^2(> 0)$ and $\frac{1}{n}\sum(X_i - \bar{X})^2 \stackrel{\text{a.s.}}{\to} \sigma^2$. Therefore, clearly $-2\log\Lambda_n \stackrel{\text{a.s.}}{\to} \infty$ under each fixed $\mu \neq 0$. There cannot be a bona fide limit distribution for $-2\log\Lambda_n$ under a fixed alternative μ.

However, if we let μ depend on n and take $\mu_n = \frac{\Delta}{\sqrt{n}}$ for some fixed but arbitrary Δ, $0 < \Delta < \infty$, then $-2\log\Lambda_n$ still has a bona fide limit distribution under the sequence of alternatives μ_n.

Sometimes alternatives of the form $\mu_n = \frac{\Delta}{\sqrt{n}}$ are motivated by arguing that one wants the test to be powerful at values of μ close to the null value. Such a property would correspond to a sensitive test. These alternatives are called *Pitman alternatives*.

The following result holds. We present the case $h_i(\theta) = \theta_i$ for notational simplicity.

Theorem 21.3 Assume the same regularity conditions as in Theorems 21.1 and 21.2. Let $\theta_{q\times 1} = (\theta_1, \eta)'$, where θ_1 is $r \times 1$ and η is $(q - r) \times 1$, a vector of nuisance parameters. Let $H_0 : \theta_1 = \theta_{10}$ (specified) and let $\theta_n = (\theta_{10} + \frac{\Delta}{\sqrt{n}}, \eta)$. Let $I(\theta)$ be the Fisher information matrix, with $I_{ij}(\theta) = -E\left(\frac{\partial^2}{\partial\theta_i\partial\theta_j}\log f(X|\theta)\right)$. Denote $V(\theta) = I^{-1}(\theta)$ and $V_r(\theta)$ the upper $(r \times r)$ principal submatrix in $V(\theta)$. Then $-2\log\Lambda_n \xrightarrow[P_{\theta_n}]{\mathcal{L}} NC\chi^2(r, \Delta' V_r^{-1}(\theta_0, \eta)\Delta)$, where $NC\chi^2(r, \delta)$ denotes the noncentral χ^2 distribution with r degrees of freedom and noncentrality parameter δ.

Remark. Theorem 21.3 is essentially proved in Sen and Singer (1993). It can be restated in an obvious way for the testing problem $H_0 : g(\theta) = (g_i(\theta), \ldots, g_r(\theta)) = 0$.

Example 21.11 Let $X_1, \ldots, X_n$ be iid $N(\mu, \sigma^2)$. Let $\theta = (\mu, \sigma^2) = (\theta_1, \eta)$, where $\theta_1 = \mu$ and $\eta = \sigma^2$. Thus $q = 2$ and $r = 1$. Suppose we want to test $H_0 : \mu = \mu_0$ vs. $H_1 : \mu \neq \mu_0$. Consider alternatives $\theta_n = (\mu_0 + \frac{\Delta}{\sqrt{n}}, \eta)$. By a familiar calculation,

$$I(\theta) = \begin{pmatrix} \frac{1}{\eta} & 0 \\ 0 & \frac{1}{2\eta^2} \end{pmatrix} \quad \Rightarrow \quad V(\theta) = \begin{pmatrix} \eta & 0 \\ 0 & 2\eta^2 \end{pmatrix} \text{ and } V_r(\theta) = \eta.$$

Therefore, by Theorem 21.3, $-2\log\Lambda_n \xrightarrow[\mathcal{P}_{\theta_n}]{\mathcal{L}} NC\chi^2(1, \frac{\Delta^2}{\eta}) = NC\chi^2(1, \frac{\Delta^2}{\sigma^2})$.

Remark. One practical use of Theorem 21.3 is in approximating the power of a test that rejects H_0 if $-2\log\Lambda_n > c$.

Suppose we are interested in approximating the power of the test when $\theta_1 = \theta_{10} + \epsilon$. Formally, set $\epsilon = \frac{\Delta}{\sqrt{n}} \Rightarrow \Delta = \epsilon\sqrt{n}$. Use $NC\chi^2(1, \frac{n\epsilon^2}{V_r(\theta_{10},\eta)})$ as an approximation to the distribution of $-2\log\Lambda_n$ at $\theta = (\theta_{10}+\epsilon, \eta)$. Thus the power will be approximated as $P(NC\chi^2(1, \frac{n\epsilon^2}{V_r(\theta_{10},\eta)}) > c)$.

21.5 Bartlett Correction

We saw that, under regularity conditions, $-2\log\Lambda_n$ has asymptotically a $\chi^2(r)$ distribution under H_0 for some r. Certainly, $-2\log\Lambda_n$ is not exactly distributed as $\chi^2(r)$. In fact, even the expectation is not r. It turns out that $E(-2\log\Lambda_n)$ under H_0 admits an expansion of the form $r(1 + \frac{a}{n} + o(n^{-1}))$. Consequently, $E\left(\frac{-2\log\Lambda_n}{1+\frac{a}{n}+R_n}\right) = r$, where R_n is the remainder term in $E(-2\log\Lambda_n)$. Moreover, $E\left(\frac{-2\log\Lambda_n}{1+\frac{a}{n}}\right) = r + O(n^{-2})$. Thus, we gain higher-order accuracy in the mean by rescaling $-2\log\Lambda_n$ to $\frac{-2\log\Lambda_n}{1+\frac{a}{n}}$. For the absolutely continuous case, the higher-order accuracy even carries over to the χ^2 approximation itself,

$$P\left(\frac{-2\log\Lambda_n}{1+\frac{a}{n}} \le c\right) = P(\chi^2(r) \le c) + O(n^{-2}).$$

Due to the rescaling, the $\frac{1}{n}$ term that would otherwise be present has vanished. This type of rescaling is known as the *Bartlett correction*. See Bartlett (1937), Wilks (1938), McCullagh and Cox (1986), and Barndorff-Nielsen and Hall (1988) for detailed treatments of Bartlett corrections in general circumstances.

21.6 The Wald and Rao Score Tests

Competitors to the LRT are available in the literature. They are general and can be applied to a wide selection of problems. Typically, the three procedures are asymptotically first-order equivalent. See Wald (1943) and Rao (1948) for the first introduction of these procedures.

We define the Wald and the score statistics first. We have not stated these definitions under the weakest possible conditions. Also note that establishing the asymptotics of these statistics will require more conditions than are needed for just defining them.

Definition 21.1 Let $X_1, X_2, \ldots, X_n$ be iid $f(x|\theta) = \frac{dP_\theta}{d\mu}, \theta \in \Theta \subseteq \mathcal{R}^q$, μ a σ-finite measure. Suppose the Fisher information matrix $I(\theta)$ exists at all θ. Consider $H_0 : \theta = \theta_0$. Let $\hat{\theta}_n$ be the MLE of θ. Define $Q_W = n(\hat{\theta}_n - \theta_0)' I(\hat{\theta}_n)(\hat{\theta}_n - \theta_0)$. This is the Wald test statistic for $H_0 : \theta = \theta_0$.

Definition 21.2 Let $X_1, X_2, \ldots, X_n$ be iid $f(x|\theta) = \frac{dP_\theta}{d\mu}, \theta \in \Theta \subseteq \mathcal{R}^q$, μ a σ-finite measure. Consider testing $H_0 : \theta = \theta_0$. Assume that $f(x|\theta)$ is partially differentiable with respect to each coordinate of θ for every x, θ, and the Fisher information matrix exists and is invertible at $\theta = \theta_0$. Let $U_n(\theta) = \sum_{i=1}^n \frac{\partial}{\partial\theta} \log f(x_i|\theta)$. Define $Q_S = \frac{1}{n} U_n'(\theta_0) I^{-1}(\theta_0) U_n(\theta_0)$; Q_S is the Rao score test statistic for $H_0 : \theta = \theta_0$.

Remark. As we saw in Chapter 15.12, the MLE $\hat{\theta}_n$ is asymptotically multivariate normal under the Cramér-Rao regularity conditions. Therefore, $n(\hat{\theta}_n - \theta_0)' I(\hat{\theta}_n)(\hat{\theta}_n - \theta_0)$ is asymptotically a central chi-square, provided $I(\theta)$ is smooth at θ_0. The Wald test rejects H_0 when $Q_W = n(\hat{\theta}_n - \theta_0)' I(\hat{\theta}_n)(\hat{\theta}_n - \theta_0)$ is larger than a chi-square percentile.

On the other hand, under simple moment conditions on the *score function* $\frac{\partial}{\partial\theta} \log f(x_i|\theta)$, by the CLT, $\frac{U_n(\theta)}{\sqrt{n}}$ is asymptotically multivariate normal, and therefore the Rao score statistic $Q_S = \frac{1}{n} U_n'(\theta_0) I^{-1}(\theta_0) U_n(\theta_0)$ is asymptotically a chi-square. The score test rejects H_0 when Q_S is larger than a chi-square percentile. Notice that in this case of a simple H_0, the χ^2 approximation of Q_S holds under fewer assumptions than would be required for Q_W or $-2 \log \Lambda_n$. Note also that, to evaluate Q_S, computation of $\hat{\theta}_n$ is not necessary, which can be a major advantage in some applications. Here are the asymptotic chi-square results for these two statistics. A version of the two parts of Theorem 21.4 below is proved in Serfling (1980) and Sen and Singer (1993). Also see van der Vaart (1998).

Theorem 21.4 (a) If $f(x|\theta)$ can be differentiated twice under the integral sign, $I(\theta_0)$ exists and is invertible, and $\{x : f(x|\theta) > 0\}$ is independent of θ, then under $H_0 : \theta = \theta_0$, $Q_S \xrightarrow{\mathcal{L}} \chi^2_q$, where q is the dimension of Θ.

(b) Assume the Cramér-Rao regularity conditions for the asymptotic multivariate normality of the MLE, and assume that the Fisher information matrix is continuous in a neighborhood of θ_0. Then, under $H_0 : \theta = \theta_0$, $Q_W \xrightarrow{\mathcal{L}} \chi^2_q$ with q as above.

Remark. Q_W and Q_S can be defined for composite nulls of the form $H_0 : g(\theta) = (g_1(\theta), \ldots, g_r(\theta)) = 0$ and, once again, under appropriate conditions, Q_W and Q_S are asymptotically χ^2_r for any $\theta \in H_0$. However, now the definition of Q_S requires calculation of the restricted MLE $\hat{\theta}_{n,H_0}$ under H_0. Again, consult Sen and Singer (1993) for a proof. Comparisons between the LRT, the Wald test, and Rao's score test have been made using higher-order asymptotics. See Mukerjee and Reid (2001) in particular.

21.7 Likelihood Ratio Confidence Intervals

The usual duality between testing and confidence intervals says that the acceptance region of a test with size α can be inverted to give a confidence set of coverage probability at least $(1-\alpha)$. In other words, suppose $A(\theta_0)$ is the acceptance region of a size α test for $H_0 : \theta = \theta_0$, and define $S(x) = \{\theta_0 : x \in A(\theta_0)\}$. Then $P_{\theta_0}(S(x) \ni \theta_0) \geq 1 - \alpha$ and hence $S(x)$ is a $100(1-\alpha)\%$ confidence set for θ.

This method is called the inversion of a test. In particular, the LRT, the Wald test, and the Rao score test can all be inverted to construct confidence sets that have asymptotically a $100(1 - \alpha)\%$ coverage probability.

The confidence sets constructed from the LRT, the Wald test, and the score test are respectively called the likelihood ratio, Wald, and score confidence sets. Of these, the Wald and the score confidence sets are ellipsoids because of how the corresponding test statistics are defined. The likelihood ratio confidence set is typically more complicated. Here is an example.

Example 21.12 Suppose $X_i \stackrel{\text{iid}}{\sim} \text{Bin}(1, p)$, $1 \leq i \leq n$. For testing $H_0 : p = p_0$ vs. $H_1 : p \neq p_0$, the LRT statistic is

$$\Lambda_n = \frac{p_0^x(1-p_0)^{n-x}}{\sup_p p^x(1-p)^{n-x}}, \qquad \text{where } x = \sum_{i=1}^n X_i$$
$$= \frac{p_0^x(1-p_0)^{n-x}}{(\frac{x}{n})^x(1-(\frac{x}{n}))^{n-x}} = \frac{p_0^x(1-p_0)^{n-x}n^n}{x^x(n-x)^{n-x}}.$$

Thus, the LR confidence set is of the form $S_{\mathrm{LR}}(x) := \{p_0 : \Lambda_n \geq k\} = \{p_0 : p_0^x(1-p_0)^{n-x} \geq k^*\} = \{p_0 : x \log p_0 + (n-x)\log(1-p_0) \geq \log k^*\}$.

The function $x \log p_0 + (n-x)\log(1-p_0)$ is concave in p_0, and therefore $S_{\mathrm{LR}}(x)$ is an interval. The interval is of the form $[0, u]$ or $[l, 1]$ or $[l, u]$ for $0 < l, u < 1$. However, l, u cannot be written in closed form; see Brown, Cai, and DasGupta (2001) for asymptotic expansions for them.

Next, $Q_W = n(\hat{p} - p_0)^2 I(\hat{p})$, where $\hat{p} = \frac{x}{n}$ is the MLE of p and $I(p) = \frac{1}{p(1-p)}$ is the Fisher information function. Therefore

$$Q_W = n\frac{(\frac{x}{n} - p_0)^2}{\hat{p}(1-\hat{p})}.$$

The Wald confidence interval is $S_W = \{p_0 : \frac{n(\hat{p}-p_0)^2}{\hat{p}(1-\hat{p})} \leq \chi_\alpha^2\} = \{p_0 : (\hat{p} - p_0)^2 \leq \frac{\hat{p}(1-\hat{p})}{n}\chi_\alpha^2\} = \{p_0 : |\hat{p} - p_0| \leq \frac{\chi_\alpha}{\sqrt{n}}\sqrt{\hat{p}(1-\hat{p})}\} = [\hat{p} - \frac{\chi_\alpha}{\sqrt{n}}\sqrt{\hat{p}(1-\hat{p})}, \hat{p} + \frac{\chi_\alpha}{\sqrt{n}}\sqrt{\hat{p}(1-\hat{p})}]$.
This is the textbook confidence interval for p.

For the score test statistic, we need $U_n(p) = \sum_{i=1}^n \frac{\partial}{\partial p} \log f(x_i|p) = \sum_{i=1}^n \frac{\partial}{\partial p}[x_i \log p + (1-x_i)\log(1-p)] = \frac{x-np}{p(1-p)}$. Therefore, $Q_S = \frac{1}{n}\frac{(x-np_0)^2}{p_0(1-p_0)}$ and the score confidence interval is $S_S = \{p_0 : \frac{(x-np_0)^2}{np_0(1-p_0)} \leq \chi_\alpha^2\} = \{p_0 : (x - np_0)^2 \leq n\chi_\alpha^2 p_0(1-p_0)\} = \{p_0 : p_0^2(n^2 + n\chi_\alpha^2) - p_0(2nx + n\chi_\alpha^2) + x^2 \leq 0\} = [l_S, u_S]$, where l_S, u_S are the roots of the quadratic equation $p_0^2(n^2 + n\chi_\alpha^2) - p_0(2nx + n\chi_\alpha^2) + x^2 = 0$.

These intervals all have the property

$$\lim_{n\to\infty} P_{p_0}(S \ni p_0) = 1 - \alpha.$$

See Brown, Cai, and DasGupta (2001) for a comparison of their exact coverage properties; they show that the performance of the Wald confidence interval is extremely poor and that the score and the likelihood ratio intervals perform much better.

21.8 Exercises

For each of the following Exercises 21.1 to 21.18, derive the LRT statistic and find its limiting distribution under the null if there is a meaningful one.

Exercise 21.1 $X_1, X_2, \ldots, X_n \stackrel{\text{iid}}{\sim} \text{Ber}(p)$, $H_0 : p = \frac{1}{2}$ vs. $H_1 : p \neq \frac{1}{2}$. Is the LRT statistic a monotone function of $|X - \frac{n}{2}|$?

Exercise 21.2 $X_1, X_2, \ldots, X_n \stackrel{\text{iid}}{\sim} N(\mu, 1)$, $H_0 : a \leq \mu \leq b$, H_1 : not H_0.

Exercise 21.3 $*X_1, X_2, \ldots, X_n \stackrel{\text{iid}}{\sim} N(\mu, 1)$, H_0 : μ is an integer, H_1 : μ is not an integer.

Exercise 21.4 * $X_1, X_2, \ldots, X_n \stackrel{\text{iid}}{\sim} N(\mu, 1)$, H_0 : μ is rational, H_1 : μ is irrational.

Exercise 21.5 $X_1, X_2, \ldots, X_n \stackrel{\text{iid}}{\sim} \text{Ber}(p_1)$, $Y_1, Y_2, \ldots, Y_m \stackrel{\text{iid}}{\sim} \text{Ber}(p_2)$, $H_0 : p_1 = p_2$, $H_1 : p_1 \neq p_2$. Assume all $m + n$ observations are independent.

Exercise 21.6 $X_1, X_2, \ldots, X_n \stackrel{\text{iid}}{\sim} \text{Ber}(p_1)$, $Y_1, Y_2, \ldots, Y_m \stackrel{\text{iid}}{\sim} \text{Ber}(p_2)$; $H_0 : p_1 = p_2 = \frac{1}{2}$, H_1 : not H_0. All observations are independent.

Exercise 21.7 $X_1, X_2, \ldots, X_n \stackrel{\text{iid}}{\sim} \text{Poi}(\mu)$, $Y_1, Y_2, \ldots, Y_m \stackrel{\text{iid}}{\sim} \text{Poi}(\lambda)$, $H_0 : \mu = \lambda = 1$, H_1 : not H_0. All observations are independent.

Exercise 21.8 * $X_1, X_2, \ldots, X_n \stackrel{\text{iid}}{\sim} N(\mu_1, \sigma_1^2)$, $Y_1, Y_2, \ldots, Y_m \stackrel{\text{iid}}{\sim} N(\mu_2, \sigma_2^2)$, $H_0 : \mu_1 = \mu_2, \sigma_1 = \sigma_2$, H_1 : not H_0 (i.e., test that two normal distributions are identical). Again, all observations are independent.

Exercise 21.9 $X_1, X_2, \ldots, X_n \stackrel{\text{iid}}{\sim} N(\mu, \sigma^2)$, $H_0 : \mu = 0, \sigma = 1$, H_1 : not H_0.

Exercise 21.10 $X_1, X_2, \ldots, X_n \stackrel{\text{iid}}{\sim} BVN(\mu_1, \mu_2, \sigma_1, \sigma_2, \rho)$, $H_0 : \rho = 0$, $H_1 : \rho \neq 0$.

Exercise 21.11 $X_1, X_2, \ldots, X_n \stackrel{\text{iid}}{\sim} MVN(\mu, I)$, $H_0 : \mu = 0$, $H_1 : \mu \neq 0$.

Exercise 21.12 * $X_1, X_2, \ldots, X_n \stackrel{\text{iid}}{\sim} MVN(\mu, \Sigma)$, $H_0 : \mu = 0$, $H_1 : \mu \neq 0$.

Exercise 21.13 $X_1, X_2, \ldots, X_n \overset{\text{iid}}{\sim} MVN(\mu, \Sigma)$, $H_0 : \Sigma = I$, $H_1 : \Sigma \neq I$.

Exercise 21.14 * $X_{ij} \overset{\text{indep.}}{\sim} U[0, \theta_i]$, $j = 1, 2, \ldots, n, i = 1, 2, \ldots, k$, H_0 : θ_i are equal, H_1 : not H_0.

Remark. Notice the exact chi-square distribution.

Exercise 21.15 $X_{ij} \overset{\text{indep.}}{\sim} N(\mu_i, \sigma^2)$, $H_0 : \mu_i$ are equal, H_1 : not H_0.

Remark. Notice that you get the usual F-test for ANOVA.

Exercise 21.16 * $X_1, X_2, \ldots, X_n \overset{\text{iid}}{\sim} c(\alpha)\text{Exp}[-|x|^\alpha]$, where $c(\alpha)$ is the normalizing constant, $H_0 : \alpha = 2$, $H_1 : \alpha \neq 2$.

Exercise 21.17 $(X_i, Y_i), i = 1, 2, \ldots, n \overset{\text{iid}}{\sim} BVN(\mu_1, \mu_2, \sigma_1, \sigma_2, \rho)$, H_0 : $\mu_1 = \mu_2$, $H_1 : \mu_1 \neq \mu_2$.

Remark. This is the *paired t-test.*

Exercise 21.18 $X_1, X_2, \ldots, X_n \overset{\text{iid}}{\sim} N(\mu_1, \sigma_1^2)$, $Y_1, Y_2, \ldots, Y_m \overset{\text{iid}}{\sim} N(\mu_2, \sigma_2^2)$, $H_0 : \sigma_1 = \sigma_2$, $H_1 : \sigma_1 \neq \sigma_2$. Assume all observations are independent.

Exercise 21.19 * Suppose $X_1, X_2, \ldots, X_n$ are iid from a two-component mixture $pN(0, 1) + (1 - p)N(\mu, 1)$, where p, μ are unknown.

Consider testing $H_0 : \mu = 0$ vs. $H_1 : \mu \neq 0$. What is the LRT statistic? Simulate the value of the LRT by gradually increasing n, and observe its slight decreasing trend as n increases.

Remark. It has been shown by Jon Hartigan that in this case $-2 \log \Lambda_n \to \infty$.

Exercise 21.20 * Suppose $X_1, X_2, \ldots, X_n$ are iid from a two-component mixture $pN(\mu_1, 1) + (1 - p)N(\mu_2, 1)$, and suppose we know that $|\mu_i| \leq 1, i = 1, 2$. Consider testing $H_0 : \mu_1 = \mu_2$ vs. $H_1 : \mu_1 \neq \mu_2$ when $p, 0 < p < 1$ is unknown.

What is the LRT statistic? Simulate the distribution of the LRT under the null by gradually increasing n.

Remark. The asymptotic null distribution is nonstandard but known.

Exercise 21.21 *For each of Exercises 21.1, 21.6, 21.7, and 21.9–21.11, simulate the exact distribution of $-2 \log \Lambda_n$ for $n = 10, 20, 40, 100$ under the null and compare its visual match to the limiting null distribution.

Exercise 21.22 *For each of Exercises 21.1, 21.5–21.7, and 21.9, simulate the exact power of the LRT at selected alternatives and compare it with the approximation to the power obtained from the limiting nonnull distribution as indicated in the text.

Exercise 21.23 * For each of Exercises 21.1, 21.6, 21.7, and 21.9, compute the constant a of the Bartlett correction of the likelihood ratio as indicated in the text.

Exercise 21.24 * Let $X_1, X_2, \ldots, X_n \stackrel{\text{iid}}{\sim} N(\mu, \sigma^2)$. Plot the Wald and Rao score confidence sets for $\theta = (\mu, \sigma)$ for a simulated dataset. Observe the visual similarity of the two sets by gradually increasing n.

Exercise 21.25 Let $X_1, X_2, \ldots, X_n \stackrel{\text{iid}}{\sim} \text{Poi}(\mu)$. Compute the likelihood ratio, Wald, and Rao score confidence intervals for μ for a simulated dataset. Observe the similarity of the limits of the intervals for large n.

Exercise 21.26 * Let $X_1, X_2, \ldots, X_n \stackrel{\text{iid}}{\sim} \text{Poi}(\mu)$. Derive a two-term expansion for the expected lengths of the Wald and Rao score intervals for μ. Plot them for selected n and check if either interval is usually shorter than the other.

Exercise 21.27 Consider the linear regression model $E(Y_i) = \beta_0 + \beta_1 x_i, i = 1, 2, \ldots, n$, with Y_i being mutually independent, distributed as $N(\beta_0 + \beta_1 x_i, \sigma^2)$, σ^2 unknown. Find the likelihood ratio confidence set for (β_0, β_1).

References

Barndorff-Nielsen, O. and Hall, P. (1988). On the level error after Bartlett adjustment of the likelihood ratio statistic, Biometrika, 75, 374–378.

Bartlett, M. (1937). Properties of sufficiency and statistical tests, Proc. R. Soc. London Ser. A, 160, 268–282.

Basu, D. (1955). On statistics independent of a complete sufficient statistic, Sankhya, Ser. A, 15, 377–380.

Bickel, P.J. and Doksum, K. (2001). *Mathematical Statistics: Basic Ideas and Selected Topics*, Prentice-Hall, Upper Saddle River, NJ.

Brown, L., Cai, T., and DasGupta, A. (2001). Interval estimation for a binomial proportion, Stat. Sci., 16(2), 101–133.

Casella, G. and Strawderman, W.E. (1980). Estimating a bounded normal mean, Ann. Stat., 9(4), 870–878.

Fan, J., Hung, H., and Wong, W. (2000). Geometric understanding of likelihood ratio statistics, J. Am. Stat. Assoc., 95(451), 836–841.

Fan, J. and Zhang, C. (2001). Generalized likelihood ratio statistics and Wilks' phenomenon, Ann. Stat., 29(1), 153–193.

Ferguson, T.S. (1996). *A Course in Large Sample Theory*, Chapman and Hall, London.

Lawley, D.N. (1956). A general method for approximating the distribution of likelihood ratio criteria, Biometrika, 43, 295–303.

McCullagh, P. and Cox, D. (1986). Invariants and likelihood ratio statistics, Ann. Stat., 14(4), 1419–1430.

Mukerjee, R. and Reid, N. (2001). Comparison of test statistics via expected lengths of associated confidence intervals, J. Stat. Planning Infer., 97(1), 141–151.

Portnoy, S. (1988). Asymptotic behavior of likelihood methods for Exponential families when the number of parameters tends to infinity, Ann. Stat., 16, 356–366.

Rao, C.R. (1948). Large sample tests of statistical hypotheses concerning several parameters with applications to problems of estimation, Proc. Cambridge Philos. Soc., 44, 50–57.

Robertson, T., Wright, F.T. and Dykstra, R.L. (1988). *Order Restricted Statistical Inference*, John Wiley, New York.

Sen, P.K. and Singer, J. (1993). *Large Sample Methods in Statistics: An Introduction with Applications*, Chapman and Hall, New York.

Serfling, R. (1980). *Approximation Theorems of Mathematical Statistics*, Wiley, New York.

van der Vaart, A. (1998). *Asymptotic Statistics*, Cambridge University Press, Cambridge.

Wald, A. (1943). Tests of statistical hypotheses concerning several parameters when the number of observations is large, Trans. Am. Math. Soc., 5, 426–482.

Wilks, S. (1938). The large sample distribution of the likelihood ratio for testing composite hypotheses, Ann. Math. Stat., 9, 60–62.

Chapter 22
Asymptotic Efficiency in Testing

In estimation, an agreed-on basis for comparing two sequences of estimates whose mean squared error each converges to zero as $n \to \infty$ is to compare the variances in their limit distributions. Thus, if $\sqrt{n}(\hat{\theta}_{1,n} - \theta) \xrightarrow{\mathcal{L}} N(0, \sigma_1^2(\theta))$ and $\sqrt{n}(\hat{\theta}_{2,n} - \theta) \xrightarrow{\mathcal{L}} N(0, \sigma_2^2(\theta))$, then the asymptotic relative efficiency (ARE) of $\hat{\theta}_{2,n}$ with respect to $\hat{\theta}_{1,n}$ is defined as $\frac{\sigma_1^2(\theta)}{\sigma_2^2(\theta)}$.

One can similarly ask what should be a basis for comparison of two sequences of tests based on statistics $T_{1,n}$ and $T_{2,n}$ of a hypothesis $H_0 : \theta \in \Theta_0$. Suppose we use statistics such that large values of them correspond to rejection of H_0; i.e., H_0 is rejected if $T_n > c_n$. Let α, β denote the type 1 error probability and the power of the test, and let θ denote a specific alternative. Suppose $n(\alpha, \beta, \theta, T)$ is the smallest sample size such that

$$P_\theta(T_n \geq c_n) \geq \beta \quad \text{and} \quad P_{H_0}(T_n \geq c_n) \leq \alpha.$$

Two tests based on $T_{1,n}$ and $T_{2,n}$ can be compared through the ratio $\frac{n(\alpha,\beta,\theta,T_1)}{n(\alpha,\beta,\theta,T_2)}$, and $T_{1,n}$ is preferred if this ratio is ≤ 1. The threshold sample size $n(\alpha, \beta, \theta, T)$ is difficult or impossible to calculate even in the simplest examples. Furthermore, the ratio can depend on particular choices of α, β, and θ.

Fortunately, if $\alpha \to 0$ $\beta \to 1$, or $\theta \to \theta_0$ (an element of the boundary $\partial\Theta_0$), then the ratio (generally) converges to something that depends on θ alone or is just a constant.

The three respective measures of efficiency correspond to approaches by Bahadur, Hodges and Lehmann, and Pitman; see Pitman (1948), Hodges and Lehmann (1956), and Bahadur (1960). Other efficiency measures, due to Chernoff, Kallenberg, and others, are hybrids of these three approaches. Rubin and Sethuraman (1965) offer measures of asymptotic relative efficiency in testing through the introduction of loss functions and priors in a formal decision-theory setting. Chernoff (1952), Kallenberg (1983), Rubin and Sethuraman (1965), Serfling (1980), and van der Vaart (1998) are excellent

A. DasGupta, *Asymptotic Theory of Statistics and Probability*,

references for the technical material in this chapter. For overall expositions of asymptotic efficiency in testing, see DasGupta (1998), Basu (1956), and Singh (1984).

Definition 22.1 Let $X_1, \ldots, X_n$ be iid observations from a distribution P_θ, $\theta \in \Theta$. Suppose we want to test $H_0 : \theta \in \Theta_0$ vs. $H_1 : \theta \in \Theta - \Theta_0$. Let $T_n = T_n(X_1, \ldots, X_n)$ be a sequence of statistics such that we reject H_0 for large values of T_n. Precisely, fix $0 < \alpha < 1$, $0 < \beta < 1$, $\theta \in \Theta - \Theta_0$. Let $c_n = c_n(\theta, \beta)$ be defined as $P_\theta(T_n > c_n) \leq \beta \leq P_\theta(T_n \geq c_n)$. The size of the test is defined as $\alpha_n(\theta, \beta) = \sup_{\theta_0 \in \Theta_0} P_{\theta_0}(T_n \geq c_n)$. Let $N_T(\alpha, \beta, \theta) = \inf\{n : \alpha_m(\theta, \beta) \leq \alpha \quad \forall m \geq n\}$.

Thus $N_T(\alpha, \beta, \theta)$ is the smallest sample size beyond which the test based on the sequence T_n has power β at the specified alternative θ and size $\leq \alpha$. The quantity $N_T(\alpha, \beta, \theta)$ is difficult (and mostly impossible) to calculate for given α, β, and θ. To calculate $N_T(\alpha, \beta, \theta)$, the exact distribution of T_n under any fixed θ and for all given n has to be known. There are very few problems where this is the case.

For two given sequences of test statistics T_{1n} and T_{2n}, we define $e_{T_2,T_1}(\alpha, \beta, \theta) = \frac{N_{T_1}(\alpha,\beta,\theta)}{N_{T_2}(\alpha,\beta,\theta)}$. Let

$$
\begin{aligned}
e_B(\beta, \theta) &= \lim_{\alpha \to 0} e_{T_2,T_1}(\alpha, \beta, \theta), \\
e_{HL}(\alpha, \theta) &= \lim_{\beta \to 1} e_{T_2,T_1}(\alpha, \beta, \theta), \\
e_P(\alpha, \beta, \theta_0) &= \lim_{\theta \to \theta_0} e_{T_2,T_1}(\alpha, \beta, \theta), \text{ where } \theta_0 \in \partial\Theta_0,
\end{aligned}
$$

assuming the limits exist.

e_B, e_{HL}, and e_P respectively are called the Bahadur, Hodges-Lehmann, and Pitman efficiencies of the test based on T_2 relative to the test based on T_1.

Typically, $e_{\mathrm{B}}(\beta, \theta)$ depends just on θ, $e_{\mathrm{HL}}(\alpha, \theta)$ also depends just on θ, and $e_{\mathrm{P}}(\alpha, \beta, \theta_0)$ depends on neither α nor β. Of these, e_{P} is the easiest to calculate in most applications, and e_{B} can be very hard to find. It is interesting that comparisons based on e_{B}, e_{HL}, and e_{P} can lead to different conclusions.

22.1 Pitman Efficiencies

The Pitman efficiency is easily calculated by a fixed recipe under frequently satisfied conditions that we present below. It is also important to note that the Pitman efficiency works out to just the asymptotic efficiency in the point estimation problem, with T_{1n} and T_{2n} being considered as the respective estimates. Testing and estimation come together in the Pitman approach. We state two theorems describing the calculation of the Pitman efficiency. The second of these is simpler in form and suffices for many applications, but the first one is worth knowing. It addresses more general situations. See Serfling (1980) for further details on both theorems.

Conditions A

(1) For some sequence of functions $\mu_n(\theta)$, $\sigma_n^2(\theta)$, and some $\delta > 0$,

$$\sup_{|\theta-\theta_0|\leq\delta} \sup_z |P_\theta\left(\frac{T_n - \mu_n(\theta)}{\sigma_n(\theta)} \leq z\right) - \Phi(z)| \to 0$$

as $n \to \infty$. This is a locally uniform asymptotic normality condition. Usually, μ_n and σ_n can be taken to be the exact mean and standard deviation of T_n or the counterparts in the CLT for T_n.

(2) $\mu_n'(\theta_0) > 0$.

(3) $\frac{\sqrt{n}\sigma_n(\theta_0)}{\mu_n'(\theta_0)} = O(1)$.

(4) If $|\theta_n - \theta_0| = O(\frac{1}{\sqrt{n}})$, then $\frac{\mu_n'(\theta_n)}{\mu_n'(\theta_0)} \to 1$.

(5) If $|\theta_n - \theta_0| = O(\frac{1}{\sqrt{n}})$, then $\frac{\sigma_n(\theta_n)}{\sigma_n(\theta_0)} \to 1$.

Theorem 22.1 Suppose T_{1n} and T_{2n} each satisfy conditions A. Then

$$e_P(T_2, T_1) = \left(\frac{\lim_{n\to\infty} \frac{\sqrt{n}\sigma_{1n}(\theta_0)}{\mu_{1n}'(\theta_0)}}{\lim_{n\to\infty} \frac{\sqrt{n}\sigma_{2n}(\theta_0)}{\mu_{2n}'(\theta_0)}}\right)^2.$$

Remark. In many applications, σ_{1n}, σ_{2n} are fixed functions σ_1, σ_2 and μ_{1n}, μ_{2n} are each the same fixed function μ. In such a case, $e_P(T_2, T_1)$ works out to the ratio $\frac{\sigma_1^2(\theta_0)}{\sigma_2^2(\theta_0)}$. If $\sigma_1(\theta)$, $\sigma_2(\theta)$ have the interpretation of being the asymptotic variance of T_{1n}, T_{2n}, then this will result in $e_P(T_2, T_1)$ being the same as the asymptotic efficiency in the estimation problem.

Conditions B

Let $\theta_0 \in \partial\Theta_0$. Let $-\infty < h < \infty$ and $\theta_n = \theta_0 + \frac{h}{\sqrt{n}}$.

(1) There exist functions $\mu(\theta)$, $\sigma(\theta)$, such that, for all h,

$$\frac{\sqrt{n}(T_n - \mu(\theta_n))}{\sigma(\theta_n)} \xrightarrow[P_{\theta_n}]{\mathcal{L}} N(0, 1).$$

(2) $\mu'(\theta_0) > 0$.

(3) $\sigma(\theta_0) > 0$ and $\sigma(\theta)$ is continuous at θ_0.

Remark. Condition (1) does not follow from pointwise asymptotic normality of T_n. Neither is it true that if (1) holds, then with the same choice of $\mu(\theta)$ and $\sigma(\theta)$, $\frac{\sqrt{n}(T_n-\mu(\theta))}{\sigma(\theta)} \xrightarrow{\mathcal{L}} N(0, 1)$.

Theorem 22.2 Suppose T_{1n} and T_{2n} each satisfy conditions B. Then

$$e_P(T_2, T_1) = \frac{\sigma_1^2(\theta_0)}{\sigma_2^2(\theta_0)} \left[\frac{\mu_2'(\theta_0)}{\mu_1'(\theta_0)}\right]^2.$$

Example 22.1 Suppose $X_1, \ldots, X_n$ are iid $N(\theta, \sigma^2)$, where $\sigma^2 > 0$ is known. We want to test that the mean θ is zero. Choose the test statistic $T_n = \frac{\bar{X}}{s}$. Let $\mu(\theta) = \frac{\theta}{\sigma}$ and $\sigma(\theta) = 1$. Then

$$\begin{aligned}\frac{\sqrt{n}(T_n - \mu(\theta_n))}{\sigma(\theta_n)} &= \sqrt{n}\left(\frac{\bar{X}}{s} - \frac{\theta_n}{\sigma}\right) \\ &= \sqrt{n}\left(\frac{\bar{X} - \theta_n + \theta_n}{s} - \frac{\theta_n}{\sigma}\right) \\ &= \sqrt{n}\left(\frac{\bar{X} - \theta_n}{s}\right) + \sqrt{n}\theta_n\left(\frac{1}{s} - \frac{1}{\sigma}\right).\end{aligned}$$

Of these, the second term goes in probability to zero and the first term is asymptotically $N(0, 1)$ under P_{θ_n}, so (1) is satisfied. But it is actually not true that $\frac{\sqrt{n}(T_n-\mu(\theta))}{\sigma(\theta)} = \sqrt{n}(\frac{\bar{X}}{S} - \frac{\theta}{\sigma})$ is asymptotically $N(0, 1)$.

We give a few examples illustrating the application of Theorems 22.1 and 22.2.

Example 22.2 Suppose $X_1, X_2, \ldots, X_n \stackrel{\text{iid}}{\sim} F(x - \theta)$, where F is absolutely continuous with density $f(x)$. Suppose we want to test $H_0 : \theta = 0$ against $H_1 : \theta > 0$. We assume $F(-x) = 1 - F(x)$ for any x, $f(0) > 0$, and f is continuous at 0. For a technical reason pertaining to an application of the Berry-Esseen theorem, we also make the assumption $E_F|X|^3 < \infty$. This is stronger than what we need to assume.

A well-known test for H_0 is the so-called sign test, which uses the test statistic T_n = proportion of sample values > 0 and rejects H_0 for large values of T_n. We will denote the sign test statistic by $S = S_n$. Thus, if $Z_i = I_{X_i>0}$, then $S = \frac{1}{n}\sum Z_i = \bar{Z}$. We wish to calculate the Pitman efficiency of the sign test with respect to the test that uses $\bar{X}$ as the test statistic and rejects H_0 for large values of $\bar{X}$. For this, because Pitman efficiencies are dependent on central limit theorems for the test statistics, we will need to assume that $\sigma_F^2 = \text{Var}_F(X) < \infty$. We will denote the mean statistic simply as $\bar{X}$. To calculate $e_P(S, \bar{X})$, we verify conditions A in Theorem 22.1.

For $T_n = S$, first notice that $E_\theta(Z_1) = P_\theta(X_1 > 0) = F(\theta)$. Also $\text{Var}_\theta(Z_1) = F(\theta)(1 - F(\theta))$. We choose $\mu_n(\theta) = F(\theta)$ and $\sigma_n^2(\theta) = \frac{F(\theta)(1-F(\theta))}{n}$. Therefore, $\mu_n'(\theta) = f(\theta)$ and $\mu_n'(\theta_0) = \mu_n'(0) = f(0) > 0$. Next, $\frac{\sqrt{n}\sigma_n(\theta)}{\mu_n'(\theta)} = \frac{\sqrt{F(\theta)(1-F(\theta))}}{f(\theta)}$ implies that $\frac{\sqrt{n}\sigma_n(\theta_0)}{\mu_n'(\theta_0)} = \frac{1}{2f(0)}$ and so obviously $\frac{\sqrt{n}\sigma_n(\theta_0)}{\mu_n'(\theta_0)} = O(1)$. If $\theta_n = \theta_0 + \frac{h_n}{\sqrt{n}} = \frac{h_n}{\sqrt{n}}$, where $h_n = O(1)$, then $\frac{\mu_n'(\theta_n)}{\mu_n'(\theta_0)} = \frac{f(\theta_n)}{f(\theta_0)} = \frac{f(\frac{h_n}{\sqrt{n}})}{f(0)} \to 1$ as f is continuous. It only remains to verify that, for some $\delta > 0$,

$$\sup_{|\theta-\theta_0|\leq\delta} \sup_{z\in\mathcal{R}} |P_\theta\left(\frac{T_n - \mu_n(\theta)}{\sigma_n(\theta)} \leq z\right) - \Phi(z)| \to 0.$$

Notice now that $\frac{S-\mu_n(\theta)}{\sigma_n(\theta)} = \frac{\sqrt{n}(\bar{Z}-E_\theta Z_1)}{\sqrt{\text{Var}_\theta(Z_1)}}$ and so, by the Berry-Esseen theorem,

$$\sup_{z\in\mathcal{R}} \left|P\left(\frac{S - \mu_n(\theta)}{\sigma_n(\theta)} \leq z\right) - \Phi(z)\right| \leq \frac{c}{\sqrt{n}} \frac{E_\theta|Z_1 - E_\theta Z_1|^3}{(\text{Var}_\theta(Z_1))^{3/2}}$$

for some absolute constant $0 < c < \infty$.

Trivially, $E_\theta|Z_1 - E_\theta Z_1|^3 = F(\theta)(1 - F(\theta))[1 - 2F(\theta)(1 - F(\theta))]$. Thus

$$\sup_{z\in\mathcal{R}} \left|P\left(\frac{S - \mu_n(\theta)}{\sigma_n(\theta)} \leq z\right) - \Phi(z)\right| \leq \frac{c}{\sqrt{n}} \frac{1 - 2F(\theta)(1 - F(\theta))}{\sqrt{F(\theta)(1 - F(\theta))}}.$$

Clearly, $\frac{1-2F(\theta_0)(1-F(\theta_0))}{\sqrt{F(\theta_0)(1-F(\theta_0))}} = 1$ and F is continuous. Thus, for sufficiently small $\delta > 0$, $\frac{1-2F(\theta)(1-F(\theta))}{\sqrt{F(\theta)(1-F(\theta))}} < 2$ if $|\theta - \theta_0| \leq \delta$. This proves that, for $T_n = S$, conditions A are satisfied.

For $T_n = \bar{X}$, choose $\mu_n(\theta) = \theta$ and $\sigma_n^2(\theta) = \frac{\sigma_F^2}{n}$. Conditions A are easily verified here, too, with these choices of $\mu_n(\theta)$ and $\sigma_n(\theta)$.

Therefore, by Theorem 22.1,

$$e_P(S, \bar{X}) = \left[\frac{\lim_n \sqrt{n}\frac{\sigma_F}{\sqrt{n}}}{\lim_n \frac{\sqrt{n}\sqrt{\frac{F(\theta_0)(1-F(\theta_0))}{n}}}{f(\theta_0)}}\right]^2 = 4\sigma_F^2 f^2(0).$$

Notice that $e_P(S, \bar{X})$ equals the asymptotic relative efficiency of the sample median with respect to $\bar{X}$ in the estimation problem (see Chapter 7).

Example 22.3 Again let $X_1, X_2, \ldots, X_n \overset{\text{iid}}{\sim} F(x - \theta)$, where $F(x) = 1 - F(-x)$ for all x, has a density f, and f is positive and continuous at 0. We want to test $H_0 : \theta = \theta_0 = 0$ against $H_1 : \theta > 0$. We assume $\sigma_F^2 = \text{Var}_F(X) < \infty$. We will compare the t-test with the test based on $\bar{X}$ in this example by calculating $e_P(t, \bar{X})$. Recall that the t-test rejects H_0 for large values of $\frac{\sqrt{n}\bar{X}}{s}$.

We verify in this example conditions B for both $\frac{\bar{X}}{s}$ and $\bar{X}$. Then let $T_{2n} = \frac{\bar{X}}{s}$. Choose $\mu(\theta) = \frac{\theta}{\sigma_F}$ and $\sigma(\theta) = 1$. Therefore $\mu'(\theta_0) = \frac{1}{\sigma_F} > 0$, and $\sigma(\theta)$ is obviously continuous at $\theta = \theta_0$. By the CLT and Slutsky's theorem, one can verify that

$$\frac{T_n - \mu(\theta_n)}{\sigma(\theta_n)} \xrightarrow[P_{\theta_n}]{\mathcal{L}} N(0, 1).$$

For $T_{1n} = \bar{X}$, it is easily proved that conditions B hold with $\mu(\theta) = \theta$ and $\sigma(\theta) = \sigma_F$. Therefore, by Theorem 22.2,

$$e_P(t, \bar{X}) = \frac{\sigma_1^2(\theta_0)}{\sigma_2^2(\theta_0)} \left[\frac{\mu_2'(\theta_0)}{\mu_1'(\theta_0)}\right]^2 = \frac{\sigma_F^2}{1} \left[\frac{\frac{1}{\sigma_F}}{1}\right]^2 = 1.$$

This says that in the Pitman approach there is asymptotically no loss in estimating σ_F by s even though σ_F is considered to be known, and the t-test has efficiency 1 with respect to the test based on the mean and this is true for all F as defined in this example. We shall later see that this is not true in the Bahadur approach.

Another reputable test statistic in the symmetric location-parameter problem is $W = \frac{1}{\binom{n}{2}} \sum\sum_{i \neq j} I_{X_i + X_j > 0}$. The test that rejects H_0 for large values of W is called the Wilcoxon test. By verifying conditions B, we can show that $e_P(W, \bar{X}) = 12\sigma_F^2\left[\int f^2(x)dx\right]^2$. It turns out that W has remarkably good Pitman efficiencies with respect to $\bar{X}$ and is generally preferred to the sign test (see Chapter 24).

The following bounds are worth mentioning.

Proposition 22.1 *(1)* $\inf_F e_P(S, \bar{X}) = \frac{1}{3}$, *where the infimum is over all F that are symmetric, absolutely continuous, and unimodal, is symmetric, absolutely continuous, and unimodal.*

(2) $\inf_{\{F: F \text{ is symmetric}\}} e_P(W, \bar{X}) = \frac{108}{125}$.

Remark. Of course, as long as F is such that $e_P(t, \bar{X}) = 1$, the results above can be stated in terms of $e_P(S, t)$ and $e_P(W, t)$ as well.

Example 22.4 We provide a table of Pitman efficiencies $e_P(S, \bar{X})$ and $e_P(W, \bar{X})$ for some specific choices of F. The values are found from direct applications of the formulas given above.

f	$e_P(S, \bar{X})$	$e_P(W, \bar{X})$
$\frac{1}{\sqrt{2\pi}} e^{-\frac{x^2}{2}}$	$\frac{2}{\pi}$	$\frac{3}{\pi}$
$\frac{1}{2} e^{-\lvert x \rvert}$	2	$\frac{3}{2}$
$\frac{1}{2} I_{-1 \le x \le 1}$	$\frac{1}{3}$	1

Remark. The table reveals that as F gets thicker tailed, the test based on $\bar{X}$ becomes less desirable.

22.2 Bahadur Slopes and Bahadur Efficiency

The results in the previous section give recipes for explicit calculation of the Pitman efficiency $e_P(T_2, T_1)$ for two sequences of tests T_{1n} and T_{2n}. We now describe a general method for calculation of the Bahadur efficiency $e_B(T_2, T_1)$. The recipe will take us into the probabilities of large deviation under the null hypothesis. Large-deviation probabilities under a distribution P_0 are probabilities of the form $P_{H_0}(T_n \ge t)$ when T_n converges in probability to, say, zero. For fixed $t > 0$, $P_{H_0}(T_n \ge t)$ typically converges to zero at an exponential rate. Determining this rate exactly is at the heart of calculating Bahadur efficiencies, and except for specified types of statistics T_n, calculation of the large-deviation rate is a very difficult mathematical problem. We will discuss more general large-deviation problems in the next chapter. For now, we discuss large-deviation rates for very special types of statistics T_n in just the real-valued case. First we describe some notation.

Consider first the case of a simple null hypothesis $H_0 : \theta = \theta_0$. Let $\{T_n\}$ be a specified sequence of test statistics such that H_0 is rejected for large values of T_n.

Define

$$\begin{aligned} I_n(t) &= -\frac{2}{n}\log P_{\theta_0}(T_n > t); \\ L_n &= P_{\theta_0}(T_n > t_n), \text{ where } t_n \text{ is the observed value of } T_n; \\ K_n &= -\frac{2}{n}\log L_n = -\frac{2}{n}\log P_{\theta_0}(T_n > t_n) = I_n(t_n). \end{aligned}$$

Note that L_n is simply the p-value corresponding to the sequence $\{T_n\}$.

Definition 22.2 Suppose $I(t)$ is a fixed continuous function such that $I_n(t) \to I(t)$ pointwise and that, for fixed θ, $T_n \xrightarrow{\text{a.s.}} \psi(\theta)$ for some function $\psi(\theta)$. The Bahadur slope of $\{T_n\}$ at θ is defined to be $I(\psi(\theta))$. $I(t)$ is called the *rate function* of $\{T_n\}$.

Remark. We will work out the rate function $I(t)$ in many examples. The link of Bahadur efficiencies to the p-value is described in the following elegant theorem.

Theorem 22.3 Let $\{T_{1n}\}$, $\{T_{2n}\}$ be two sequences of test statistics for $H_0 : \theta = \theta_0$, and suppose H_0 is rejected for large values of $T_{in}, i = 1, 2$. Suppose T_{in} has Bahadur slope $m_i(\theta)$ at the alternative θ. Then the Bahadur efficiency is

$$e_{\text{B}}(T_2, T_1, \beta, \theta) = \frac{m_2(\theta)}{m_1(\theta)}.$$

Remark. See Serfling (1980) for this theorem. Notice that this theorem says that, provided each sequence $\{T_{in}\}$ admits a limit $I(t)$ for the associated sequence of functions $I_n(t)$ and admits a law of large numbers under θ, the Bahadur efficiency depends only on θ, although according to its definition it could depend on β.

The next questions concern what is special about the quantity $I(\psi(\theta))$ and why it is called a slope. Note that as $I_n(t) \to I(t)$ pointwise, $I(t)$ is continuous, $T_n \xrightarrow{\text{a.s.}} \psi(\theta)$, and I_n is a sequence of monotone functions, $K_n = I_n(t_n) \xrightarrow{\text{a.s.}} I(\psi(\theta))$, under θ. But $K_n = -\frac{2}{n}\log L_n$, where L_n is the p-value, if we use $\{T_n\}$ as the sequence of test statistics. If, for a range of successive values of n, the points (n, K_n) are plotted for a simulated sample from P_θ, then the plot would look approximately like a straight line with slope $I(\psi(\theta))$. This is why $I(\psi(\theta))$ is known as a slope, and this is also why $I(\psi(\theta))$ is a special quantity of importance in the Bahadur efficiency theory.

To summarize, in order to calculate $e_B(T_2, T_2)$, we need to establish an SLLN for $\{T_{in}\}, i = 1, 2$, and we need to identify the function $I(t)$ for each of $\{T_{in}\}, i = 1, 2$. Thus, what are involved are laws of large numbers and analysis of large-deviation probabilities under the null. The first task is usually simple. The second task is in general very difficult, although, for specialized types of statistics, methods for calculating $I(t)$ have been obtained. We will come to this issue later.

Example 22.5 Suppose that $X_1, X_2, \ldots, X_n \stackrel{\text{iid}}{\sim} N(\theta, 1)$ and the null hypothesis is $H_0 : \theta = 0$. Suppose the test statistic is $T_n = \bar{X}$. Then the rate function is $I(t) = -2 \lim_n \frac{1}{n} \log P_\theta(T_n > t)$. Under $\theta_0 = 0$, $\sqrt{n}\bar{X} \sim N(0, 1)$, and so $P_0(T_n > t) = P_0(\sqrt{n}\bar{X} > t\sqrt{n}) = 1 - \Phi(t\sqrt{n})$. For fixed $t > 0$,

$$1 - \Phi(t\sqrt{n}) \sim \frac{\phi(t\sqrt{n})}{t\sqrt{n}} = \frac{\frac{1}{\sqrt{2\pi}} e^{-\frac{nt^2}{2}}}{t\sqrt{n}}$$

$\therefore \frac{1}{n} \log P_0(T_n > t) = -\frac{t^2}{2} + o(1) \Longrightarrow I(t) = t^2$. Also, under a general θ, $T_n \xrightarrow{\text{a.s.}} \psi(\theta) = \theta$, and so the slope of $\bar{X}$ at an alternative θ is $I(\psi(\theta)) = \theta^2$. In this case, therefore, we can compute the Bahadur slope directly.

The following theorem describes how to find the rate function $I(t)$ in general when T_n is a sample mean.

Theorem 22.4 (Cramér-Chernoff) Suppose $Y_1, \ldots, Y_n$ are iid zero-mean random variables with an mgf (moment generating function) $M(z) = E(e^{zY_1})$ assumed to exist for all z. Let $k(z) = \log M(z)$ be the cumulant generating function of Y_1. Then, for fixed $t > 0$,

$$\lim_n -\frac{2}{n} \log P(\bar{Y} > t) = I(t) = -2 \inf_{z>0}(k(z) - tz) = 2 \sup_{z>0}(tz - k(z)).$$

Remark. See Serfling (1980) or van der Vaart (1998) for a proof. In a specific application, one has to find the cgf (cumulant generating function) of Y_1 to carry out this agenda. It can be shown by simple analysis from Theorem 22.4 that $I(t)$ is increasing and convex for $t \geq E(Y_1)$. This is an important mathematical property of the rate function and is useful in proving various results on large deviations (see Chapter 23).

Where is the function $tz - k(z)$ coming from? Toward this end, note that

$$\begin{aligned} P(\bar{Y} > t) &= P\left(\frac{1}{n}\sum_{i=1}^{n}(Y_i - t) > 0\right) \\ &= P\left(\sum_{i=1}^{n}(Y_i - t) > 0\right) = P\left(e^{z\sum_{i=1}^{n}(Y_i - t)} > 1\right) \quad \text{for positive } z \\ &\leq E\left(e^{z\sum_{i=1}^{n}(Y_i - t)}\right) = \left(e^{-tz}M(z)\right)^n. \end{aligned}$$

This gives

$$\begin{aligned} &\frac{1}{n}\log P(\bar{Y} > t) \leq \log M(z) - tz = k(z) - tz \\ &\Longrightarrow \limsup_n \frac{1}{n}\log P(\bar{Y} > t) \leq \inf_{z>0}(k(z) - tz). \end{aligned}$$

It takes some work to show that $\liminf_n \frac{1}{n}\log P(\bar{Y} > t) \geq q \inf_{z>0}(k(z) - tz)$, which gives the Cramér-Chernoff theorem; see, e.g., Serfling (1980).

Let us now use this theorem to compute the Bahadur slope of some test statistics in some selected hypothesis-testing problems.

Example 22.6 Again let $X_1, X_2, \ldots, X_n \overset{\text{iid}}{\sim} N(\theta, 1)$, and suppose we test $H_0 : \theta = 0$ using $T_n = \bar{X}$. To find $\lim_n -\frac{2}{n}\log P_{H_0}(T_n > t)$, we use the Cramér-Chernoff theorem by identifying $Y_i \overset{\text{iid}}{\sim} N(0, 1)$ (i.e., the distribution of Y_i is that of X_i under H_0), so $M(z) = E_{H_0}(e^{zX_1}) = e^{\frac{z^2}{2}} \Longrightarrow k(z) = \frac{z^2}{2}$, which gives $tz - k(z) = tz - \frac{z^2}{2} \Longrightarrow \frac{d}{dz}\left(tz - \frac{z^2}{2}\right) = t - z = 0$ at $z = t$. Therefore, for $t > 0$, $\sup_{t>0}(tz - k(z)) = t^2 - \frac{t^2}{2} = \frac{t^2}{2}$. This gives $I(t) = \lim -\frac{2}{n}\log P_{H_0}(T_n > t) = t^2$ by the Cramér-Chernoff theorem.

Example 22.7 Let $X_1, X_2, \ldots, X_n \overset{\text{iid}}{\sim} N(\theta, \sigma^2)$, where σ^2 is known and assumed to be 1. For testing $H_0 : \theta = 0$, we have seen that the Bahadur slope of $\bar{X}$ is θ^2. Let $T = T_n$ be the t-statistics $\frac{\bar{X}}{s} = \frac{\bar{X}}{\sqrt{\frac{1}{n-1}\sum(X_i - \bar{X})^2}}$. Previously we saw that $e_P(t, \bar{X}) = 1$. The basic reason that T and $\bar{X}$ are equally asymptotically efficient in the Pitman approach is that $\sqrt{n}\bar{X}$ and $\frac{\sqrt{n}\bar{X}}{s}$ have the same limiting $N(0, 1)$ distribution under H_0. More precisely, for any fixed t,

$P_{H_0}(\sqrt{n}\bar{X} > t) = 1 - \Phi(t)$; i.e., $\lim_n P_{H_0}(\bar{X} > \frac{t}{\sqrt{n}}) = \lim_n P_{H_0}(\frac{\bar{X}}{s} > \frac{t}{\sqrt{n}}) = 1 - \Phi(t)$. But Bahadur slopes are determined by the rate of exponential convergence of $P_{H_0}(\bar{X} > t)$ and $P_{H_0}(\frac{\bar{X}}{s} > t)$. The rates of exponential convergence are different. Thus the t-statistic has a different Bahadur slope from $\bar{X}$. In fact, the Bahadur slope of the t-statistic in this problem is $\log(1 + \theta^2)$, so that $e_B(T, \bar{X}) = \frac{\log(1+\theta^2)}{\theta^2}$. In the rest of this example, we outline the derivation of the slope $\log(1 + \theta^2)$ for the t-statistic.

For simplicity of the requisite algebra, we will use the statistic $\tilde{T} = \frac{\bar{X}}{\sqrt{\frac{1}{n}\sum(X_i-\bar{X})^2}}$; this change does not affect the Bahadur slope. Now,

$$\begin{aligned} P_{H_0}(\tilde{T} > t) &= \frac{1}{2} P_{H_0}\left(\tilde{T}^2 > t^2\right) \\ &= \frac{1}{2} P\left(\frac{z_1^2}{\sum_{i=2}^{n} z_i^2} > t^2\right), \end{aligned}$$

where $z_1, \ldots, z_n$ are iid $N(0, 1)$; such a representation is possible because $n\bar{X}^2$ is χ_1^2, $\sum(X_i - \bar{X})^2$ is $\chi^2_{(n-1)}$, and the two are independent. Therefore,

$$\begin{aligned} P_{H_0}(\tilde{T} > t) &= \frac{1}{2} P\left(z_1^2 - t^2 \sum_{i=2}^{n} z_i^2 > 0\right) = \frac{1}{2} P(e^{z(z_1^2 - t^2 \sum_{i=2}^{n} z_i^2)} > 1)(z > 0) \\ &\leq \frac{1}{2} E\left(e^{z(z_1^2 - t^2 \sum_{i=2}^{n} z_i^2)}\right) \\ &\Longrightarrow \log P_{H_0}(\tilde{T} > t) \leq \inf_{z>0} \left\{ -\log 2 + \log E\left(e^{z[z_1^2 - t^2 \sum_{i=2}^{n} z_i^2]}\right)\right\}. \end{aligned}$$

By direct calculation, $\log E(e^{z[z_1^2 - t^2 \sum_{i=2}^{n} z_i^2]}) = -\frac{1}{2}\log(1 - 2z) - \frac{n-1}{2}\log(1 + 2t^2 z)$, and by elementary calculus, the minimum value of this over $z > 0$ is $-\frac{1}{2}\log\left(\frac{1+t^2}{1-t^2}\right) - \frac{n-1}{2}\log\left(\frac{n-1}{n}(1 + t^2)\right)$, which implies $\limsup_n \frac{1}{n}\log P_{H_0}(\tilde{T} > t) \leq -\frac{1}{2}\log(1 + t^2)$.

In fact, it is also true that $\liminf_n \frac{1}{n}\log P_{H_0}(\tilde{T} > t) \geq -\frac{1}{2}\log(1 + t^2)$. Together, these give $I(t) = \lim_n -\frac{2}{n}\log P_{H_0}(\tilde{T} > t) = \log(1 + t^2)$. At a fixed θ, $\tilde{T} \stackrel{a.s.}{\Rightarrow} \theta = \psi(\theta)$ (as σ was assumed to be 1), implying that the Bahadur slope of $\tilde{T}$ is $I(\psi(\theta)) = \log(1 + \theta^2)$.

Example 22.8 This example is intriguing because it brings out unexpected phenomena as regards the relative comparison of tests by using the approach of Bahadur slopes.

Suppose $X_1, X_2, \ldots, X_n \stackrel{\text{iid}}{\sim} F(X - \theta)$, where $F(\cdot)$ is continuous and $F(-x) = 1 - F(x)$ for all x. Suppose we want to test $H_0 : \theta = 0$. A test statistic we have previously discussed is $S = S_n = \frac{1}{n}\sum_{i=1}^{n} I_{X_i>0}$. This is equivalent to the test statistic $T = T_n = \frac{1}{n}\sum_{i=1}^{n} \text{sgn}(X_i)$, where $\text{sgn}(X) = \pm 1$ according to whether $X > 0$ or $X < 0$. To apply the Cramér-Chernoff theorem, we need the cumulant generating function (cdf) of $Y_1 = \text{sgn}(X_1)$ under H_0. This is $k(z) = \log E_{H_0}[e^{zY_1}] = \log \frac{e^z + e^{-z}}{2} = \log\cosh(z)$. Therefore, $k(z) - tz = \log\cosh(z) - tz$ and $\frac{d}{dz}(k(z) - tz) = \tanh(z) - t = 0$ when $z = \text{arctanh}(t)$. Therefore, the rate function $I(t)$ by the Cramér-Chernoff theorem is $I(t) = -2\log(\cosh(\text{arctanh}(t))) + 2t\text{arctanh}(t)$. Furthermore,

$$\begin{aligned}\bar{Y} = \frac{1}{n}\sum_{i=1}^{n} \text{sgn}(X_i) &\xrightarrow{\text{a.s.}} E_\theta(\text{sgn}(X_1)) \\ &= P_\theta(X_1 > 0) - P_\theta(X_1 < 0) \\ &= 1 - 2F(-\theta) \\ &= 2F(\theta) - 1 = \psi(\theta).\end{aligned}$$

Thus, the slope of the sign test under a given F and a fixed alternative θ is

$$\begin{aligned}I(\psi(\theta)) = &-2\log(\cosh(\text{arctanh}\,(2F(\theta) - 1))) \\ &+2(2F(\theta) - 1)\,\text{arctanh}\,(2F(\theta) - 1).\end{aligned}$$

This general formula can be applied to any specific F. For the CDF of $f(x) = \frac{1}{2}e^{-|x|}$, for $\theta > 0$,

$$\begin{aligned}F(\theta) &= \int_{-\infty}^{\theta} \frac{1}{2}e^{-|x|}dx \\ &= \frac{1}{2} + \int_{0}^{\theta} \frac{1}{2}e^{-|x|}dx \\ &= 1 - \frac{1}{2}e^{-\theta}.\end{aligned}$$

Plugging this into the general formula above, the slope of the sign test for the double exponential case is $-2\log(\cosh(\text{arctanh}\,(1 - e^{-\theta}))) + 2(1 - e^{-\theta})\,\text{arctanh}\,(1 - e^{-\theta})$.

We can compare this slope with the slope of competing test statistics. As competitors, we choose $\bar{X}$ and the sample median M_n. We derive here the slope of $\bar{X}$ for the double exponential case. To calculate the slope of $\bar{X}$, note that

$$E_{H_0}(e^{zX_1}) = \int_{-\infty}^{\infty} e^{zx} \frac{1}{2} e^{-|x|} dx = \frac{1}{1-z^2}, \quad |z| < 1.$$

So, $k(z) - tz = \log \frac{1}{1-z^2} - tz$, which is minimized at $z = \frac{\sqrt{t^2+1}-1}{t}$ (for $t > 0$). Therefore, by the Cramér-Chernoff theorem, the rate function $I(t) = 2\sqrt{1+t^2} + 2\log \frac{2\sqrt{1+t^2}-2}{t^2} - 2$. Since $\bar{X} \xrightarrow{\text{a.s.}} \theta$, the slope of $\bar{X}$ is $2\sqrt{1+\theta^2} + 2\log \frac{2\sqrt{1+\theta^2}-2}{\theta^2} - 2$. As regards the slope of M_n, it cannot be calculated from the Cramér-Chernoff theorem. However, the rate function $\lim_n -\frac{2}{n} \log P_{H_0}(M_n > t)$ can be calculated directly by analyzing binomial CDFs. It turns out that the slope of M_n is $-\log(4pq)$, where $p = 1 - F(\theta)$ and $q = 1 - p = F(\theta)$.

Taking the respective ratios, one can compute $e_{\text{B}}(S, \bar{X}, \theta)$, $e_{\text{B}}(M_n, \bar{X}, \theta)$, and $e_{\text{B}}(S, M_n, \theta)$. For θ close to $\theta_0 = 0$, S and M_n are more efficient than $\bar{X}$; however (and some think it is counterintuitive), $\bar{X}$ becomes more efficient than S and M_n as θ drifts away from θ_0. For example, when $\theta = 1.5, e_{\text{B}}(M, \bar{X}, \theta) < 1$. This is surprising because for the double exponential case, M_n is the MLE and $\bar{X}$ is not even asymptotically efficient as a point estimate of θ. Actually, what is even more surprising is that the test based on $\bar{X}$ is asymptotically optimal in the Bahadur sense as $\theta \to \infty$.

The next result states what exactly asymptotic optimality means. See Bahadur (1967), Brown (1971), and Kallenberg (1978) for various asymptotic optimality properties of the LRT. Apparently, Charles Stein never actually published his results on optimality of the LRT, although the results are widely known. Cohen, Kemperman, and Sackrowitz (2000) argue that LRTs have unintuitive properties in certain families of problems. A counter view is presented in Perlman and Wu (1999).

Theorem 22.5 (Stein-Bahadur-Brown) Suppose $X_1, \ldots, X_n \overset{\text{iid}}{\sim} P_\theta, \theta \in \Theta$. Assume Θ is finite, and consider testing $H_0 : \theta \in \Theta_0$ vs. $H_1 : \theta \in \Theta - \Theta_0$.

(a) For any sequence of test statistics T_n, the Bahadur slope $m_T(\theta)$ satisfies $m_T(\theta) \leq 2\inf_{\theta_0 \in \Theta_0} K(\theta, \theta_0)$, where $K(\theta, \theta_0)$ is the Kullback-Leibler distance between P_θ and P_{θ_0}.

(b) The LRT (likelihood ratio test) statistic Λ_n satisfies $m_\Lambda(\theta) = 2\inf_{\theta_0 \in \Theta_0} K(\theta, \theta_0)$.

Remark. This says that if Θ is finite, then the LRT is Bahadur optimal at every fixed alternative θ.

Example 22.9 The Bahadur efficiency approach treats the type 1 and type 2 errors unevenly. In some problems, one may wish to treat the two errors evenly. Such an approach was taken by Chernoff. We illustrate the Chernoff approach by an example. The calculations are harder than in the Bahadur approach, the reason being that one now needs large deviation rates under both H_0 and H_1. Suppose $X_1, X_2, \ldots, X_n \stackrel{\text{iid}}{\sim} \text{Bin}(1, p)$ and we wish to test $H_0 : p = p_0$ vs. $H_1 : p = p_1$, where $p_1 > p_0$. Suppose we reject H_0 for large values of $\bar{X}$. The mgf of X_i is $q + pe^z$. Therefore, the minimum of $k(z) - tz = \log(q + pe^z) - tz$ is attained at z satisfying $pe^z(1-t) = qt$. Plugging into the Cramér-Chernoff theorem, we get, for $t > p_0$, $P_{H_0}(\bar{X} > t) \approx e^{-nK(t,p_0)}$, where $K(a, b) = a \log \frac{a}{b} + (1-a) \log \frac{1-a}{1-b}$. Analogously, $P_{H_1}(\bar{X} < t) \approx e^{-nK(t,p_1)}$ for $t < p_1$. Thus, for $p_0 < t < p_1$, $\alpha_n(t) = P_{H_0}(\bar{X} > t)$ and $\gamma_n(t) = P_{H_1}(\bar{X} < t)$ satisfy $\alpha_n(t) \approx e^{-nK(t,p_0)}$ and $\gamma_n(t) \approx e^{-nK(t,p_1)}$. If we set $K(t, p_1) = K(t, p_0)$, then $\alpha_n(t) \approx \gamma_n(t)$. This is an attempt to treat the two errors evenly. The unique t satisfying $K(t, p_0) = K(t, p_1)$ is

$$t = t(p_0, p_1) = \frac{\log \frac{q_0}{q_1}}{\log \frac{q_0}{q_1} + \log \frac{p_1}{p_0}},$$

where we write q_i for $1 - p_i$. Plugging back, $\alpha_n(t(p_0, p_1))$ and $\gamma_n(t(p_0, p_1))$ are each $\approx e^{-nK(t(p_0,p_1),p_0)}$. This quantity $K(t(p_0, p_1), p_0) = K(t(p_0, p_1), p_1)$ is called the Chernoff index of the test based on $\bar{X}$. The function $K(t(p_0, p_1), p_0)$ is a complicated function of p_0 and p_1. It can be shown easily that it is approximately equal to $\frac{(p_1 - p_0)^2}{8p_0q_0}$ when $p_1 \approx p_0$.

We close with a general theorem on the rate of exponential decrease of a convex combination of the error probabilities.

Theorem 22.6 (Chernoff) Consider testing a null hypothesis against an alternative H_1. Suppose we reject H_0 for large values of some statistics $T_n = \sum_{i=1}^n Y_i$, where $Y_i \stackrel{\text{iid}}{\sim} P_0$ under H_0 and $Y_i \stackrel{\text{iid}}{\sim} P_1$ under H_1. Let $\mu_0 = E_{H_0}(Y_1)$ and $\mu_1 = E_{H_1}(Y_1)$. Let $0 < \lambda < 1$ and $t > 0$. Define $\alpha_n(t) = P_{H_0}(T_n > t)$ and $\gamma_n(t) = P_{H_1}(T_n \le t)$, $\theta_n = \inf_{t>0}(\lambda\alpha_n(t) + (1 - \lambda)\gamma_n(t))$, and $\log \rho = \inf_{\mu_0 \le t \le \mu_1}[\max\{\inf_z(k_0(z) - tz), \inf_z(k_1(z) - tz)\}]$, where $k_i(z) = \log E_{H_i} e^{zY_1}$. Then $\frac{\log \theta_n}{n} \to \log \rho$.

Remark. See Serfling (1980) for this theorem; $\log \rho$ is called the Chernoff index of the statistic T_n.

22.3 Bahadur Slopes of U-statistics

There is some general theory about Bahadur slopes of U-statistics. They are based on large-deviation results for U-statistics. See Arcones (1992) for this entire section. This theory is, in principle, useful because a variety of statistics in everyday use are U-statistics. The general theory is in general hard to implement except approximately, and it may even be more efficient to try to work out the slopes by direct means in specific cases instead of appealing to this general theory.

Here is a type of result that is known.

Theorem 22.7 Let $U_n = U_n(X_1, \ldots, X_n)$ be a U-statistic with kernel h and order r. Let $\psi(x) = E[h(X_1, \ldots, X_r)|X_1 = x]$, and assume U_n is nondegenerate; i.e., $\tau^2 = E[\psi^2(X_1)] > 0$. If $|h(X_1, \ldots, X_r)| \leq M < \infty$, then, for any $\gamma_n = o(1)$ and $t > E[h(X_1, \ldots, X_r)]$, $\lim_n \frac{1}{n} \log P(U_n > t + \gamma_n)$ is an analytic function for $|t - E(h)| \leq B$ for some $B > 0$ and admits the expansion $\lim_n \frac{1}{n} \log P(U_n > t + \gamma_n) = \sum_{j=2}^{\infty} c_j (t - Eh)^j$, where c_j are appropriate constants, of which $c_2 = -\frac{1}{2r^2\tau^2}$.

Remark. The assumption of boundedness of the kernel h is somewhat restrictive. The difficulty with full implementation of the theorem is that there is a simple formula for only c_2. The coefficient c_3 has a known complicated expression, but none exist for c_j for $j \geq 4$. The practical use of the theorem is in approximating $\frac{1}{n} \log P(U_n > t)$ by $-\frac{1}{2r^2\tau^2}(t - E(h))^2$ for $t \approx E(h)$.

Example 22.10 Recall that the sample variance $\frac{1}{n-1} \sum_{i=1}^{n} (X_i - \bar{X})^2$ is a U-statistic with the kernel $h(X_1, X_2) = \frac{1}{2}(X_1 - X_2)^2$. Suppose $X_1, X_2, \ldots, X_n \stackrel{\text{iid}}{\sim} U[0, 1]$. This enables us to meet the assumption that the kernel h is uniformly bounded. Then, by trivial calculation, $E(h) = \frac{1}{12}$, $\psi(X) = \frac{X^2}{2} - \frac{X}{2} + \frac{1}{12}$, and $\tau^2 = \frac{1}{720}$. Therefore, by a straight application of Theorem 22.7, $\lim_n \frac{1}{n} \log P[\frac{1}{n-1} \sum (X_i - \bar{X})^2 - \frac{1}{12} > c] = -90c^2(1 + o(1))$ as $c \to 0$.

Remark. An example of a common U-statistic for which Theorem 22.7 is hard to apply (even when the X_i are uniformly bounded) is the Wilcoxon statistic $T_n = \frac{1}{n(n-1)} \sum\sum_{i \neq j} I_{X_i + X_j > 0}$. Note that T_n is a U-statistic and the kernel is always bounded. Still, it is difficult to calculate the large-deviation rate for T_n from Theorem 22.7. It turns out that for certain types of F, the large-deviation rate for T_n can be worked out directly. This reinforces the remark we made before that sometimes it is more efficient to attack the problem directly than to use the general theorem.

Remark. The case where $\tau^2 = 0$ also has some general results on the large-deviation rate of U_n. The results say that the rate can be represented in the form $\sum_{j=2}^{\infty} c_j (t - E(h))^{\frac{j}{2}}$.

22.4 Exercises

Exercise 22.1 Consider iid observations $X_1, X_2, \ldots$ from $N(\mu, 1)$ and consider the test that rejects $H_0 : \mu \leq 0$ for large values of $\sqrt{n}\bar{X}$. Find an expression for $N_T(\alpha, \beta, \mu)$ as defined in the text.

Exercise 22.2 Consider again iid observations $X_1, X_2, \ldots$ from $N(\mu, 1)$ and consider the test that rejects $H_0 : \mu \leq 0$ for large values of the sample median. For $\alpha = .05, \beta = .95, \mu = 1$, give a numerical idea of the value of $N_T(\alpha, \beta, \mu)$ as defined in the text.

Exercise 22.3 In the previous exercise, take successively smaller values of $\mu \to 0$ and numerically approximate the limit of the ratio $e_{median,\bar{X}}(\alpha, \beta, \mu)$, still using $\alpha = .05, \beta = .95$. Do you get a limit that is related to the theoretical Pitman efficiency?

Exercise 22.4 Suppose $X_1, X_2, \ldots \stackrel{\text{iid}}{\sim} U[0, \theta]$, and consider the statistic $T_n = \bar{X}$. Do conditions A hold? With what choice of μ_n, σ_n?

Exercise 22.5 * For the symmetric location-parameter problem, derive a formula for $e_P(S, \bar{X})$ and $e_P(W, \bar{X})$ when $f(x) = c(\alpha)\exp[-|x|^\alpha], \alpha > 0$, and $c(\alpha)$ is the normalizing constant. Then plot $e_P(W, \bar{X})$ vs. $e_P(S, \bar{X})$ and identify those densities in the family for which (i) W is more efficient than S and (ii) W is more efficient than both S and $\bar{X}$.

Exercise 22.6 * Consider the family of densities as in the previous exercise, but take $1 \leq \alpha \leq 2$. Find the average value of $e_P(W, \bar{X})$ and $e_P(S, \bar{X})$ in this family by averaging over α.

Exercise 22.7 * Find distributions F for which the bounds of Proposition 22.1 are attained. Are these distributions unique?

Exercise 22.8 * Find the Bahadur slope of the sample median for iid observations from the location-parameter double exponential density.

Remark. Note that this has to be done directly, as the Cramér-Chernoff result cannot be used for the median.

Exercise 22.9 * For each of the following cases, find the Bahadur slope of the sample mean. Then, simulate a sample, increase n by steps of 10, draw a scatterplot of (n, K_n) values, and eyeball the slope to see if it roughly matches the Bahadur slope:

(a) Exp(θ), $H_0 : \theta = 1$.
(b) Gamma(α, θ), $H_0 : \theta = 1, \alpha$ known.
(c) $U[0, \theta]$, $H_0 : \theta = 1$.
(d) $N(0, \sigma^2)$, $H_0 : \sigma = 1$.
(e) $N(\theta, \theta)$; $H_0 : \theta = 1$.
(f) Poisson(λ); $H_0 : \lambda = 1$.

(a) Double Exponential$(\mu, 1)$, $H_0 : \mu = 0$.
(b) Logistic$(\mu, 1)$, $H_0 : \mu = 0$.

Exercise 22.10 * Show directly that in the general continuous one-parameter exponential family, the natural sufficient statistic attains the lower bound of the *Bahadur-Brown-Stein theorem* stated in the text.

Exercise 22.11 * Compute the *exact* Chernoff index numerically for the binomial case when the two errors are treated evenly, and compare it with the approximation $\frac{(p_1-p_0)^2}{8p_0q_0}$.

Exercise 22.12 * By using the general theorem on Bahadur slopes of U-statistics, derive a one-term expansion for $\lim_n \frac{1}{n} \log P_F(U_n > \sigma^2(F) + c)$ when U_n is the sample variance and F is a general Beta(m, m) distribution with an integer-valued m.

References

Arcones, M. (1992). Large deviations for U statistics, J.Multivar. Anal., 42(2), 299–301.
Bahadur, R.R. (1960). Stochastic comparison of tests, Ann. Math. Stat., 31, 276–295.
Bahadur, R.R. (1967). Rates of convergence of estimates and test statistics, Ann. Math. Stat., 38, 303–324.
Basu, D. (1956). On the concept of asymptotic efficiency, Sankhya, 17, 193–196.
Brown, L. (1971). Non-local asymptotic optimality of appropriate likelihood ratio tests, Ann. Math. Stat., 42, 1206–1240.
Chernoff, H. (1952). A measure of asymptotic efficiency for tests of a hypothesis based on the sum of observations, Ann. Math. Stat., 23, 493–507.
Cohen, A., Kemperman, J.H.B., and Sackrowitz, H. (2000). Properties of likelihood inference for order restricted models, J. Multivar. Anal., 72(1), 50–77.

DasGupta, A. (1998). Asymptotic Relative Efficiency, in *Encyclopedia of Biostatistics*, P. Armitage and T. Colton (eds.), Vol. I, John Wiley, New York.
Hodges, J.L. and Lehmann, E.L. (1956). The efficiency of some nonparametric competitors of the t test, Ann. Math. Stat., 27, 324–335.
Kallenberg, W.C.M. (1983). Intermediate efficiency: theory and examples, Ann. Stat., 11(1), 170–182.
Kallenberg, W.C.M. (1978). *Asymptotic Optimality of Likelihood Ratio Tests*, Mathematical Centre Tracts, Vol. 77,Mathematisch Centrum, Amsterdam.
Perlman, M.D. and Wu, L. (1999). The emperor's new tests, Stat. Sci., 14(4), 355–381.
Pitman, E.J.G. (1948). *Lecture Notes on Nonparametric Statistical Inference*, Columbia University, New York.
Rubin, H. and Sethuraman, J. (1965). Bayes risk efficiency, Sankhya Ser.A, 27, 347–356.
Serfling, R. (1980). *Approximation Theorems of Mathematical Statistics*, John Wiley, New York.
Singh, K. (1984). Asymptotic comparison of tests—a review, in *Handbook of Statistics*, P.K. Sen and P.R. Krishnaiah (eds.), Vol. 4, North-Holland, Amsterdam, 173–184.
van der Vaart, A. (1998). *Asymptotic Statistics*, Cambridge University Press, Cambridge.

Chapter 23
Some General Large-Deviation Results

The theory and the examples in the preceding chapter emphasized large-deviation rates for statistics T_n that have a specific structure, such as a sample mean or a U-statistic, in the one-dimensional case with iid data. Furthermore, the large-deviation rates discussed there only addressed the problem of tail probabilities. Sometimes, one wishes to find sharp approximations for densities or probabilities of a fixed value. These are called large-deviation rates for local limit theorems. In this chapter, we present some results on large-deviation rates for more general problems, including the vector-valued case, the case of certain types of dependent data, and the large-deviation problem for local limit theorems. It should be emphasized that the large-deviation problem for the vector-valued case is significantly harder than for the case of scalar statistics.

There are numerous good books and references on large deviations, starting from the elementary concepts to very abstract treatments. For our purposes, excellent references are Groeneboom (1980), Bucklew (2004), Varadhan (2003), den Hollander (2000), and Dembo and Zeitouni (1998). Other specific references are given later.

23.1 Generalization of the Cramér-Chernoff Theorem

For sample means of iid scalar random variables, in the previous chapter we described the rate of exponential convergence of probabilities of the form $P(\bar{X} > \mu + t)$, where μ is the mean and t a fixed positive number. It should be mentioned that one may also consider $P(\bar{X} > \mu + t)$ itself rather than its logarithm, which is what appears in the statement of the Cramér-Chernoff theorem. The quantity $P(\bar{X} > \mu + t)$ itself is a sequence of the type $c_n e^{-nI(t)}$ for some suitable sequence c_n; c_n does not converge to zero at an exponential rate. However, pinning down its exact asymptotics is a much harder problem. Since it converges at a less than exponential rate, on taking a logarithm, the

A. DasGupta, *Asymptotic Theory of Statistics and Probability*,

function $I(t)$ becomes the leading term in an expansion for the logarithm of the tail probability, and the asymptotics of the sequence c_n can be avoided. We will continue to look at the logarithm of large-deviation probabilities in this chapter in order to have this advantage of not having to worry about the difficult sequence c_n. However, we look at events *more general* than tail events, such as $\{\bar{X} > \mu + t\}$. The disadvantage in this generalization is that we will no longer be able to assert the existence of an actual limit for the normalized (by n) logarithmic large-deviation probabilities. Here is a generalization of the basic Cramér-Chernoff theorem, although still for sample means of iid real-valued random variables.

Theorem 23.1 Let $X_1, X_2, \cdots$ be an iid sequence with mean zero and mgf $M(z)$, assumed to exist for all z. Let $k(z) = \log M(z)$, $I(t) = \sup_z[tz - k(z)]$, and let F, G denote general (measurable) closed and open subsets of $\mathcal{R}$, respectively. Then,

$$\text{(a)} \quad \limsup_{n\to\infty}\left[\frac{1}{n}\log P(\bar{X} \in F)\right] \leq -I(F),$$

$$\text{(b)} \quad \liminf_{n\to\infty}\left[\frac{1}{n}\log P(\bar{X} \in G)\right] \geq -I(G),$$

where $I(C) = \inf_{t\in C} I(t)$.

Remark. This fundamental result on large-deviation theory is proved in many places; see, e.g., Bucklew (2004). The function $I(t)$ is a bowl-shaped function with its minimum value at the mean (namely zero). This says something interesting about the asymptotics of the logarithmic large-deviation probabilities. Suppose we take a set F away from the mean value, such as $F = [1, 2]$. The asymptotics of the logarithmic large-deviation probability $\log P(\bar{X} \in F)$ are going to be the same as the asymptotics of, for example, $\log P(1 \leq \bar{X} \leq 10)$. Because of the bowl shape of the $I(t)$ function, the logarithmic asymptotics are going to be determined by just a single point, namely the point in the set F that is the closest to the mean zero.

Example 23.1 Suppose $X_1, X_2, \cdots$ are iid $C(0, 1)$. In this case, $\bar{X}$ is also distributed as $C(0, 1)$ for all n. So, $\log P(\bar{X} \in F)$ is just a fixed number and does not even converge to zero. And indeed, in this case, $M(z) \equiv \infty \,\forall\, z \neq 0$, while for $z = 0$, $tz - k(z) = 0$. This implies that formally the function $I(t)$ is always zero.

Example 23.2 Here is an example where the function $I(t)$ is related to the concept of *entropy* of the underlying distribution. Suppose $X_1, X_2, \cdots$ are iid Ber(p). The entropy of an integer-valued random variable X is defined as $H(X) = -\sum_j p(j)\log p(j)$. So, for a Bernoulli variable, the entropy is $p\log(\frac{1}{p}) + (1-p)\log(\frac{1}{1-p})$. One can use entropy to define a distance between two integer-valued distributions. The entropy divergence between two integer-valued random variables X, Y is defined as $H(X||Y) = -\sum_j p(j)\log(\frac{q(j)}{p(j)})$, $p(.), q(.)$ being the two mass functions (consult Chapter 2 for details). If $X \sim$ Ber(p) and $Y \sim$ Ber(t), then $H(X||Y) = p\log(\frac{p}{t}) + (1-p)\log(\frac{1-p}{1-t})$. Now, on calculation, the $I(t)$ function in our large-deviation context turns out to be $H(Y||X)$. Of course, if t is not in [0,1], $I(t)$ would be infinite.

Example 23.3 Suppose $X_1, X_2, \cdots$ are iid $N(0,1)$. We had commented before that the $I(t)$ function is bowl shaped. The simplest bowl-shaped function one can think of is a quadratic. And indeed, in the $N(0,1)$ case, $M(z) = e^{z^2/2}$, and so $I(t) = \sup_z[tz - \frac{z^2}{2}] = \frac{t^2}{2}$, a convex quadratic.
Remark. What is the *practical* value of a large-deviation result? The practical value is that in the absence of a large-deviation result, we will approximate a probability of the type $P(\bar{X} > \mu + t)$ by a CLT approximation, as that would be almost the only approximation we would be able to think of. Indeed, it is common practice to do so in applied work. For *fixed* t, the CLT approximation is not going to give an accurate approximation to the true value of $P(\bar{X} > \mu + t)$. In comparison, an application of the Cramér-Chernoff theorem is going to produce a more accurate approximation, although whether it really is more accurate depends on the value of t. See Groeneboom and Oosterhoff (1977, 1980, 1981) for extensive finite sample numerics on the accuracy of large-deviation approximations.

23.2 The Gärtner-Ellis Theorem

For sample means of iid random variables, the mgf is determined from the mgf of the underlying distribution itself, and that underlying mgf determines the large-deviation rate function $I(t)$. When we give up the iid assumption, there is no longer one underlying mgf that determines the final rate function, if there is one. Rather, one has a *sequence* of mgfs, one for each n, corresponding to whatever statistics T_n we consider. It turns out that, despite this complexity, a large-deviation rate can be established without the restrictions of independence or the sample mean structure. This greatly expands the scope of application, but at the expense of considerably more subtlety in exactly what assumptions are needed for which result. The Gärtner-Ellis theorem, a special case of which we present below, is regarded as a major advance in large-deviation theory due to its huge range of applications.

Theorem 23.2 Let $\{T_n\}$ be a sequence of random variables taking values in a Euclidean space $\mathcal{R}^d$. Let $M_n(z) = E[e^{z'T_n}]$, $k_n(z) = \log M_n(z)$, and $\phi(z) = \lim_{n\to\infty} \frac{1}{n} k_n(z)$, assuming the limit exists as an extended real-valued function. Let $I(t) = \sup_z [t'z - \phi(z)]$. Let F and G denote general (measurable) sets that are closed and open, respectively. Assume the following about the function $\phi(z)$:

(a) $D_\phi = \{z : \phi(z) < \infty\}$ has a nonempty interior.
(b) ϕ is differentiable in the interior of D_ϕ.
(c) If ξ_n is a sequence in the interior of D_ϕ approaching a point ξ on the boundary of D_ϕ, then the length of the gradient vector of ϕ at ξ_n converges to ∞.
(d) The origin belongs to the interior of D_ϕ.

Then,

$$\text{(a)} \quad \limsup_{n\to\infty} \frac{1}{n} \log P(T_n \in nF) \le -\inf_{t\in F} I(t).$$

$$\text{(b)} \quad \liminf_{n\to\infty} \frac{1}{n} \log P(T_n \in nG) \ge -\inf_{t\in G} I(t).$$

Without any of the assumptions (i)–(iv) above, for (measurable) compact sets C,

$$\text{(c)} \quad \limsup_{n\to\infty} \frac{1}{n} \log P(T_n \in nC) \le -\inf_{t\in C} I(t).$$

Remark. See Bucklew (2004) for a proof of this theorem. It is important to understand that the assumptions (some or even all) (a)–(d) may fail in simple-looking applications. In such cases, only the bound for compact sets in part (c) can be used. However, the Gärtner-Ellis theorem has been successfully used in many probabilistic setups where independence is lacking; for example, to obtain large-deviation rates for functionals of Markov chains.

Example 23.4 This is still an example corresponding to sample means of an iid sequence, but the variables are multidimensional. Let $X_1, X_2, \cdots$ be iid $N_d(\mathbf{0}, \mathbf{S})$, $\mathbf{S}$ p.d. The mgf of the $N_d(\mathbf{0}, \mathbf{S})$ distribution is $e^{\frac{z'\Sigma z}{2}}$. Let T_n be the nth partial sum $\sum_{i=1}^n X_i$. Then the $\phi(z)$ function is just $\frac{z'\Sigma z}{2}$.

Correspondingly, the $I(t)$ function is $\frac{t'\Sigma t}{2}$, so the large-deviation rate is going to be determined by the single point in the set, say C, under consideration at which the convex function $t'\Sigma t$ is the smallest. Imagine then a set C sitting separated from the origin in d dimensions. With Σ as the orientation, keep drawing ellipsoids centered at the origin until for the first time the ellipsoid is just large enough to touch the set C. The point where it touches determines the large-deviation rate. This is a beautiful geometric connection.

The next result gives the assumptions in a formally different form. We describe a large-deviation result for tail probabilities of general real-valued statistics T_n. The results mimic the Gärtner-Ellis theorem. We need some notation. Let $m_n(t) = E(e^{tT_n})$; $k_n(t) = \log M_n(t)$; $c_0(t) = \lim_n n^{-1}k_n(t)$; $c_1(t) = \lim_n n^{-1}k_n'(t)$; $c_2(t) = \lim_n n^{-1}k_n''(t)$.

We make the following assumptions:

(a) For some $\delta > 0$, $-\delta < t < \delta$, $m_n(t), c_0(t), c_1(t), c_2(t)$ exist and are finite.
(b) For $-\delta < t < \delta$, $|n^{-1}k_n'(t) - c_1(t)| = O(n^{-1})$.
(c) For $-\delta < t < \delta$, $c_2(t) > 0$.
(d) $k_n^{(3)}(t)$ exists for $-\delta < t < \delta$, and for $t_n \to t$, $n^{-1}k_n^{(3)}(t_n) = O(1)$.

Theorem 23.3 Under conditions (a)–(d), for $-\delta < t < \delta$, $\lim_n -\frac{1}{n}\log P(T_n > nt) = th - c_0(h)$, where $h = h(t)$ is the unique solution of the equation $t = c_1(h)$.

Remark. The statement of this theorem is basically the same as in the Gärtner-Ellis theorem. Notice the similarity to saddlepoint approximations of tail probabilities, discussed in Chapter 14. Indeed, here the point h is the unique saddlepoint of the function $tz - c_0(z)$. The theorem can be extended to cover $P(T_n > nt_n)$, where t_n converges to a fixed t. The hard part is the actual implementation of the theorem because usually it will be hard to write an analytical formula for $m_n(t)$ when T_n is a complicated statistic, making it hard in turn to find formulas for $c_0(t)$ and $c_1(t)$. In such cases, the large-deviation rate given in the theorem has to be computed numerically. Recently, there has been a lot of innovative work on self-normalized large-deviation probabilities. A key reference is Shao (1997).

We now give two examples of Theorem 23.3.

Example 23.5 Suppose $X_i \stackrel{\text{iid}}{\sim} f(x|\theta) = e^{\theta T(x) - \Psi(\theta)}h(x)(d\mu)$. This is the general one-parameter exponential family. Consider the statistic $T_n = T(X_1) + \ldots + T(X_n)$, where $T(.)$ is as above. By direct calculation, $c_0(t) = \Psi(\theta + t) - \Psi(\theta)$ and $c_1(t) = \Psi'(\theta + t)$. Thus, the h in the statement of the theorem solves $t = \Psi'(\theta + h)$; i.e., $h = (\Psi')^{-1}(t) - \theta$. It follows that $\lim_n -\frac{1}{n} \log P(\frac{T_n}{n} > t) = t[(\Psi')^{-1}(t) - \theta] + \Psi(\theta) - \Psi((\Psi')^{-1}(t))$.

Example 23.6 Here is an example applying Theorem 23.3 when T_n is not a sum of IID random variables. Suppose $X_i \stackrel{\text{iid}}{\sim} N(\mu, \theta)$, where $\theta = \text{Var}(X) > 0$. Suppose we want to test $H_0 : \theta = \theta_0$ vs. $H_1 : \theta > \theta_0$ using the test statistic $T_n = \frac{n}{n-1} \sum_{i=1}^{n} (x_i - \bar{x})^2$. We are going to get the Bahadur slope of T_n by using Theorem 23.3 and the results in the previous chapter. The Bahadur slope equals, from Theorem 23.3, $2m(\theta)$, where $m(\theta) = h(\theta)\Psi(\theta) - c_0(h(\theta))$, where $c_0(.)$ is as in the theorem, $h = h(\theta)$ is also as in the theorem, and Ψ is the a.s. limit of $\frac{T_n}{n}$ (i.e., $\Psi(\theta) = \theta$). Because $\sum_{i=1}^{n} (x_i - \bar{x})^2 \sim \theta \chi_{n-1}^2$ under P_θ, we can calculate its mgf and hence we can calculate explicitly $c_0(t)$, $c_1(t)$. Indeed, $c_0(t) = -\frac{1}{n} \log(1 - 2\theta_0 t)$, $c_1(t) = \frac{\theta_0}{1 - 2\theta_0 t}$, for $t < \frac{1}{2\theta_0}$. Plugging into the statement of the general theorem and noting that $\Psi(\theta) = \theta$, some algebra gives $m(\theta) = \frac{\theta}{\theta_0} + \log \frac{\theta}{\theta_0} - 1$ for $\theta > \theta_0$. The Bahadur slope is twice this expression.

23.3 Large Deviation for Local Limit Theorems

Let T_n be a given sequence of statistics. Sometimes it is necessary to find a sharp approximation for $P(T_n = k)$ for some fixed k or the density of T_n at some fixed k. We give two theorems covering these two cases, followed by examples.

Again, we need some notation first. For a given sequence of statistics T_n, let $\Psi_n(z) = E(e^{zT_n})$ denote the Laplace transform of T_n, where it is assumed that $\Psi_n(z)$ exists and is finite in some $\Omega \subseteq \mathbb{C}$, where $\mathbb{C}$ is the complex plane such that, for some $\delta > 0$, $\Omega \supseteq \{z \in \mathbb{C} : |\text{Re}(z)| < \delta\}$. Let $k_n(z) = \frac{1}{n} \log \Psi_n(z)$ and $\gamma_n(t) = \sup_{|s|<\delta} [ts - k_n(s)] = -\inf_{|s|<\delta} [k_n(s) - ts]$. The nonlattice case is covered by Theorem 23.4 below and the lattice case in Theorem 23.5; see Chaganty and Sethuraman (1993) for the list of exact conditions and proofs.

Theorem 23.4 Let m_n be a given sequence of the form $m_n = k_n'(\tau_n)$, where $\limsup_n |\tau_n| < \delta$. Then, under some additional conditions, the density of T_n at m_n satisfies $f_n(m_n) = \sqrt{\frac{n}{2\pi k_n''(\tau_n)}} e^{-n\gamma_n(m_n)}(1 + O(n^{-1}))$.

Remark. A rough interpretation of this result is that $-\frac{1}{n} \log f_n(m_n) \approx \gamma_n(m_n)$; γ_n is called the *entropy function* of T_n.

The corresponding lattice-valued result is the following.

Theorem 23.5 Suppose T_n is a sequence of statistics with T_n taking values in the lattice $\Omega_n = \{a_n + kh_n, k \in Z\}$ for some $h_n > 0$, $a_n \in \mathcal{R}$. Let m_n be of the form $m_n = k_n'(\tau_n)$, where τ_n is as in Theorem 23.4. Then, under some additional conditions,

$$\frac{\sqrt{n}}{h_n} P\left(\frac{T_n}{n} = m_n\right) = \frac{1}{\sqrt{2\pi k_n''(\tau_n)}} e^{-n\gamma_n(m_n)}(1 + O(n^{-1})).$$

Remark. If T_n takes nonnegative integer values, then the corresponding lattice has $a_n = 0$ and $h_n = 1$ with $P(T_n = k) = 0$ for $k < 0$. The Poisson distribution fits this situation. The additional conditions on the Laplace transform of T_n can be tedious to verify. See Chaganty and Sethuraman (1993) for verifications in some examples.

Example 23.7 Suppose $X_i \stackrel{\text{iid}}{\sim} \text{Poi}(\lambda)$ and $T_n = \sum_{i=1}^n X_i$. Then

$$\Psi_n(z) = E(e^{zT_n}) = \sum_{x=0}^{\infty} \frac{e^{-n\lambda}(n\lambda)^x}{x!} e^{zx}$$

$$= e^{n\lambda(e^z - 1)},$$

$$k_n(z) = \frac{1}{n} n\lambda(e^z - 1) = \lambda(e^z - 1),$$

$$k_n'(z) = \lambda e^z,$$

$$k_n''(z) = \lambda e^z.$$

Furthermore, $\gamma_n(t) = \sup_s (ts - \lambda(e^s - 1)) = t \log \frac{t}{\lambda} - \lambda(\frac{t}{\lambda} - 1) = \lambda + t \log \frac{t}{\lambda} - t$ by elementary calculus. Suppose now, for some τ_n, $m_n = k_n'(\tau_n) = \lambda e^{\tau_n}$; i.e., $\tau_n = \log \frac{m_n}{\lambda}$. Therefore, from Theorem 23.5,

$$\sqrt{n} P\left(\frac{T_n}{n} = m_n\right) = \sqrt{n} P(\bar{X} = m_n) = \frac{1}{\sqrt{2\pi k_n''(\tau_n)}} e^{-n\gamma_n(m_n)}(1 + O(n^{-1}))$$

$$= \frac{1}{\sqrt{2\pi m_n}} e^{-n(\lambda + m_n \log \frac{m_n}{\lambda} - m_n)}(1 + O(n^{-1})).$$

As a specific example, suppose $\lambda = 1$ and suppose $m_n = 2$. Then the approximation above reduces to

$$P(\bar{X} = 2) = \frac{1}{\sqrt{4\pi n}} e^{-n[1+2\log 2-2]}(1+O(n^{-1})) = \frac{1}{\sqrt{4\pi}} \frac{e^n}{4^n\sqrt{n}}(1+O(n^{-1})).$$

But we can calculate $P(\bar{X} = 2)$ exactly. Indeed,

$$P(\bar{X} = 2) = P(T_n = 2n) = \frac{e^{-n\lambda}(n\lambda)^{2n}}{(2n)!} = \frac{e^{-n}n^{2n}}{(2n)!}$$

$$= \frac{e^{-n}n^{2n}}{e^{-2n}(2n)^{2n+\frac{1}{2}}\sqrt{2\pi}} = \frac{1}{\sqrt{4\pi}} \frac{e^n}{4^n\sqrt{n}}(1 + O(n^{-1}))$$

by using the Stirling approximation.

It is interesting that, on using Stirling's approximation, the exact value of the probability coincides with the large- deviation local limit approximation we worked out above. Again, see Chapter 14 for similar phenomena in that context.

Example 23.8 Let $X_i \stackrel{\text{iid}}{\sim} U[1, 2, \ldots, m]$ for some $m > 1$. This is the discrete uniform distribution. An example would be the roll of a fair die. Suppose $T_n = \sum_{i=1}^{n} X_i$ and we want to find an approximation for $P(\frac{T_n}{n} = m_n)$. First, note that the distribution of T_n can be found exactly. Toward this end, note that the pgf (probability generating function) of X_1 is $E(s^{X_1}) = \frac{\sum_{i=1}^{m} s^i}{m}$. Therefore, the pgf of T_n is $E(s^{T_n}) = \frac{(\sum_{i=1}^{m} s^i)^n}{m^n}$, and for any given k, $P(T_n = k) = \frac{1}{m^n}\times$(coefficient of s^k in $(\sum_{i=1}^{m} s^i)^n$) $= \frac{1}{m^n}$(coefficient of s^k in $s^n(1 - s^m)^n(1 - s)^{-n}$. But by using binomial expansions, for $|s| < 1$,

$$s^n(1 - s^m)^n(1 - s)^{-n} = s^n \sum_{i=0}^{n}(-1)^i \binom{n}{i} s^{mi} \sum_{j=0}^{\infty} \binom{n + j - 1}{j} s^j.$$

Letting $n + m + j = k$, the double sum above reduces to $\sum_{i=0}^{[\frac{k-n}{m}]}(-1)^i \binom{n}{i}$ $\binom{k-mi-1}{n-1} s^k$, so $P(T_n = k) = \frac{1}{m^n} \sum_{i=0}^{[\frac{k-n}{m}]}(-1)^i \binom{n}{i}\binom{k-mi-1}{n-1}$.

Coming to the approximation given by the large-deviation local limit theorem, we need the quantities $\Psi_n(z)$, $k_n(z)$, $k_n^{'}(z)$, $k_n^{''}(z)$ and $\gamma_n(t)$. We have

$$\Psi_n(z) = \frac{1}{m^n}\left(e^z \frac{e^{mz}-1}{e^z-1}\right)^n,$$

$$k_n(z) = \frac{1}{n}\log \Psi_n(z) = z + \log(e^{mz}-1) - \log(e^z-1) - \log m,$$

$$k_n^{'}(z) = 1 + \frac{me^{mz}}{e^{mz}-1} - \frac{e^z}{e^z-1},$$

$$k_n^{''}(z) = m\frac{me^{mz}(e^{mz}-1) - me^{2mz}}{(e^{mz}-1)^2} + \frac{e^z}{(e^z-1)^2},$$

$$\gamma_n(t) = \sup_s[ts - k_n(s)] = \sup_s[ts - s - \log(e^{ms}-1) + \log(e^s-1) + \log m].$$

Clearly, $\gamma_n(t)$ cannot be written in a closed form. For large m, one has the approximation $\gamma_n(t) \approx (t-1-m)\log(1+\frac{1}{m-t}) - \log(m-t) + \log m$ when $t < m$. To apply the local limit theorem approximation for a given m_n, we find τ_n such that $m_n = k_n^{'}(\tau_n)$, and the approximation is obtained as

$$P\left(\frac{T_n}{n} = m_n\right) = P(T_n = nm_n) = \frac{1}{\sqrt{2n\pi k_n^{''}(\tau_n)}}e^{-n\gamma_n(m_n)}(1 + O(n^{-1})).$$

In executing this, τ_n and $\gamma_n(m_n)$ have to be found numerically or replaced by approximations, as we did for $\gamma_n(t)$ above.

Let us see an application of this general development to the fair die case.

Example 23.9 Consider $n = 10$ rolls of a fair die; note that here $m = 6$. Suppose we want an approximation for $P(T_n = 15)$. The exact value of this probability is 0.000033 using the formula for the exact distribution of T_n given above. To apply the large-deviation local limit theorem, we need τ_n such that $m_n = k_n^{'}(\tau_n)$, $k_n^{''}(\tau_n)$, and $\gamma_n(m_n)$, where $m_n = \frac{15}{10} = 1.5$. As we remarked, τ_n and $\gamma_n(m_n)$ cannot be written in closed form. It is interesting to check the accuracy of the large-deviation result by numerically computing τ_n, $k_n^{''}(\tau_n)$, and $\gamma_n(m_n)$. The numerical solution is $\tau_n = -1.08696$. This gives $k_n^{''}(\tau_n) = 0.714648$ and $\gamma_n(m_n) = 0.838409$. Plugging into the approximation formula $\frac{1}{\sqrt{2n\pi k_n^{''}(\tau_n)}}e^{-n\gamma_n(m_n)}$, one obtains the value 0.00003407. This matches the exact value 0.000033 very well.

23.4 Exercises

Exercise 23.1 * For each of the following densities in the exponential family, derive the limit of $-\frac{1}{n}\log P(T_n > nt)$, where $T_n = \sum_{i=1}^n T(X_i)$ is the natural sufficient statistic for that density:

(a) $f(x|\mu) = \frac{1}{\text{Beta}(\mu, \alpha)} e^{-\mu x}(1 - e^{-x})^{\alpha - 1}, x, \mu, \alpha > 0, \alpha$ known.

(b) $f(x|\mu) = \frac{e^{-\mu x}}{\Gamma(\mu)\zeta(\mu)(e^{e^{-x}} - 1)}, x > 0, \mu > 1.$

(c) $f(x|\mu) = \frac{e^{-\mu x}}{-x Ei(-\mu)}, x > 1, \mu > 0$, where $Ei(.)$ denotes the exponential integral function.

(d) $f(x|\theta) = \frac{e^{\theta \cos x}}{I_0(\theta)}, 0 < x < 1, -\infty < \theta < \infty$, where $I_0(.)$ denotes the corresponding Bessel function.

(e) $f(x|\theta) = \frac{e^{-\theta^2 x^2}}{(1 - \Phi(\theta))\pi e^{\theta^2}(1 + x^2)}, -\infty < x < \infty, \theta > 0.$

(f) $f(x|\theta) = \frac{x^{\alpha-1} e^{-\theta(x + \frac{1}{x})}}{2K_\alpha(2\theta)}, x, \theta, \alpha > 0, \alpha$ known, where $K_\alpha(.)$ denotes the corresponding Bessel function.

Exercise 23.2 * Characterize the rate function $I(t)$ for $T_n = ||\bar{X}||$ when X_i are iid $N_p(0, I)$.

Exercise 23.3 * Characterize the rate function $I(t)$ for $T_n = ||\bar{X}||$ when X_i are iid uniform in the unit ball $B_p(0, 1)$.

Exercise 23.4 Characterize the rate function $I(t)$ when T_n is the mean of an iid Poisson sequence.

Exercise 23.5 * Suppose X_i are iid according to a uniform distribution in a d-dimensional ellipsoid $\{x : x'AX \leq 1\}$.

(a) Do all the assumptions of the Gärtner-Ellis theorem hold?

(b) Characterize the rate function $I(t)$.

Exercise 23.6 * Suppose X_i are iid $N(\mu, 1)$ and Y is an independent exponential with mean λ. Suppose $T_n = \sum_{i=1}^n X_i + Y$. Which of the assumptions in the Gärtner-Ellis theorem hold, and which do not?

Exercise 23.7 * Let $W_n \sim \chi^2(n)$. Do a straight CLT approximation for $P(W_n > (1+c)n), c > 0$, and do an approximation using the Cramér-Chernoff theorem. Conduct a study of the pairs (n, c) for which the Cramér-Chernoff approximation is numerically more accurate.

Exercise 23.8 Evaluate the density approximation for the mean of iid exponential random variables as in Theorem 23.4.

Exercise 23.9 Evaluate the density approximation for the mean of iid $U[0, 1]$ random variables as in Theorem 23.4.

Exercise 23.10 Evaluate the density approximation for the mean of iid random variables with a triangular density on [0, 1] as in Theorem 23.4.

Exercise 23.11 * Evaluate the density approximation for the mean of iid random variables with the *Huber density* on $(-\infty, \infty)$ as in Theorem 23.4.

Exercise 23.12 Evaluate the local limit approximation for the mean of iid random variables with the simple trinomial distribution $P(X = 0) = P(X = \pm 1) = \frac{1}{3}$ as in Theorem 23.5.

Exercise 23.13 Evaluate the local limit approximation for the mean of iid geometric (p) random variables as in Theorem 23.5.

Exercise 23.14 * Evaluate the local limit approximation for the mean of iid random variables with the Poi(λ) distribution truncated below 2.

Exercise 23.15 * By using the exact formula given in the text, find the exact pmf of the sum of 20 rolls of a fair die, and investigate if the local limit approximation of Theorem 23.5 or a straight CLT approximation gives a more accurate approximation to the probabilities of the values in the range [80, 100].

References

Bucklew, J. (2004). *Introduction to Rare Event Simulation*, Springer, New York.

Chaganty, N.R. and Sethuraman, J. (1993). Strong large deviation and local limit theorems, Ann. Prob., 3, 1671–1690.

Dembo, A. and Zeitouni, O. (1998). *Large Deviations: Techniques and Applications*, Springer-Verlag, New York.

den Hollander, F. (2000). *Large Deviations*, Fields Institute Monograph, American Mathematical Society, Providence, RI.

Groeneboom, P. (1980). *Large Deviations and Asymptotic Efficiencies*, Mathematisch Centrum, Amsterdam.
Groeneboom, P. and Oosterhoff, J. (1977). Bahadur efficiency and probabilities of large deviations, Stat. Neerlandica, 31(1), 1–24.
Groeneboom, P. and Oosterhoff, J. (1980). *Bahadur Efficiency and Small Sample Efficiency: A Numerical Study*, Mathematisch Centrum, Amsterdam.
Groeneboom, P. and Oosterhoff, J. (1981). Bahadur efficiency and small sample efficiency, Int. Stat. Rev., 49(2), 127–141.
Shao, Q-M (1997). Self-normalized large deviations, Ann. Prob., 25(1), 265–328.
Varadhan, S.R.S. (2003). *Large Deviations and Entropy*, Princeton University Press, Princeton, NJ.

Chapter 24
Classical Nonparametrics

The development of nonparametrics originated from a concern about the approximate validity of parametric procedures based on a specific narrow model when the model is questionable. Procedures that are reasonably insensitive to the exact assumptions that one makes are called robust. Such assumptions may be about a variety of things. They may be about an underlying common density assuming that the data are iid; they may be about the dependence structure of the data itself; in regression problems, they may be about the form of the regression function; etc. For example, if we assume that our data are iid from a certain $N(\theta, 1)$ density, then we have a specific parametric model for our data. Statistical models are always, at best, an approximation. We do not believe that the normal model is *the correct* model. So, if we were to use a procedure that had excellent performance under the normal model but fell apart for models similar to normal but different from the normal in aspects that a statistician would find hard to pin down, then the procedure would be considered risky. For example, tails of underlying densities are usually hard to pin down. Nonparametric procedures provide a certain amount of robustness to departure from a narrow parametric model, at the cost of a suboptimal performance in the parametric model. It is important to understand, however, that what we commonly call nonparametric procedures do not provide robustness with regard to all characteristics. For instance, a nonparametric test may retain the type I error rate approximately under various kinds of models but may not retain good power properties under different kinds of models. The implicit robustness is limited, and it always comes at the cost of some loss of efficiency in fixed parametric models. There is a trade-off.

As a simple example, consider the t-test for the mean μ of a normal distribution. If normality holds, then, under the null hypothesis $H_0 : \mu = \mu_0$,

A. DasGupta, *Asymptotic Theory of Statistics and Probability*,

$$P_{\mu_0}\left(\frac{\sqrt{n}(\overline{X}_n - \mu_0)}{s} > t_{\alpha,n-1}\right) = \alpha$$

for all n, μ_0, and σ. However, if the population is not normal, neither the size nor the power of the t-test remains the same as under the normal case. If these change substantially, we have a robustness problem. However, as we will see later, by making a minimal number of assumptions (specifically, no parametric assumptions), we can develop procedures with some sort of a safety net. Such methods would qualify for being called nonparametric methods.

There are a number of texts that discuss classical nonparametric estimators and tests in various problems. We recommend Hajek and Sidak (1967), Hettmansperger (1984), Randles and Wolfe (1979), and Lehmann and Romano (2005), in particular. A recent review of asymptotic theory of common nonparametric tests is Jurevckova (1995). Other specific references are given in the following sections.

24.1 Some Early Illustrative Examples

We start with three examples to explain the ideas of failure of narrowly focused parametric procedures in broader nonparametric models and the possibility of other procedures that have some limited validity independent of specific parametric models.

Example 24.1 Let F be a CDF on $\mathcal{R}$. For $0 < p < 1$, let ξ_p denote the pth percentile of the distribution F. That is,

$$\xi_p = \inf\{x : F(x) \geq p\}.$$

Let F_n be the empirical CDF given by

$$F_n(x) = \frac{1}{n}\sum_{i=1}^{n} I_{\{X_i \leq x\}}.$$

Suppose we estimate the percentile ξ_p by inverting the empirical CDF. That is,

$$\hat{\xi}_p = F_n^{-1}(p) = \inf\{x : F_n(x) \geq p\}.$$

Then, it can be shown that, under minimal assumptions, the estimator $\hat{\xi}_p$, a distribution-free estimate of the corresponding population quantile, is strongly consistent for ξ_p; see Hettmansperger (1984), for example Thus, $F_n^{-1}(p)$ at least gives us consistency without requiring any rigid parametric assumptions. It would qualify for being called a nonparametric procedure.

Example 24.2 Consider the t confidence interval $\overline{X}_n \pm t_{\alpha/2,n-1}\frac{s}{\sqrt{n}}$ denoted by C_n. If $X_1, \ldots, X_n$ are iid observations from $N(\mu, \sigma^2)$, then $P_{\mu,\sigma}(C_n \ni \mu) \equiv 1 - \alpha$ for all n, μ, σ. That is, the coverage probability is exact. But what happens if $X_1, \ldots, X_n$ are iid observations from a general distribution F? More precisely, what can be asserted about the coverage probability, $P_F(C_n \ni \mu(F))$? If we can assume that $\mathbb{E}_F X^2 < \infty$, then it can be shown that

$$\lim_{n\to\infty} P_F(C_n \ni \mu(F)) = 1 - \alpha.$$

That is, for fixed F and $\epsilon > 0$, there exists a number $N = N(F, \alpha, \epsilon) > 0$ such that

$$n > N \Longrightarrow |P_F(C_n \ni \mu(F)) - (1 - \alpha)| < \epsilon.$$

However, the asymptotic validity is not uniform in F. That is, if $\mathfrak{F}$ denotes the set of all CDFs with finite second moment, then

$$\lim_{n\to\infty} \inf_{F\in\mathfrak{F}} P_F(C_n \ni \mu) = 0.$$

This is an example of the failure of a parametric procedure under completely nonparametric models.

Example 24.3 This is a classical example of a nonparametric confidence interval for a quantile. Let θ denote the median of a distribution F. Suppose $X_{(1)} < \ldots < X_{(n)}$ are the order statistics of an iid sample from a continuous CDF F. For a fixed k, $0 \le k \le \frac{n-1}{2}$, using the notation $\{U_i\}$ to denote an iid $U[0, 1]$ sample,

$$\begin{aligned} P_F(X_{(k+1)} \le \theta \le X_{(n-k)}) &= P_F\left(F\left(X_{(k+1)}\right) \le F(\theta) \le F\left(X_{(n-k)}\right)\right) \\ &= P(U_{(k+1)} \le 0.5 \le U_{(n-k)}) \\ &= P(k + 1 \le S_n < n - k) \\ &= P(k + 1 \le \text{Bin}(n, 0.5) < n - k), \end{aligned}$$

where $S_n = \#\{i : U_i \leq 0.5\}$. We can choose k such that this probability is $\geq 1-\alpha$. This translates to a nonparametric confidence interval for θ. Notice that the only assumption we have used here is that F is continuous (this assumption is needed to perform the quantile transformation).

But, the nonparametric interval does not perform as well as a t-interval if F is a normal CDF.

24.2 Sign Test

This is perhaps the earliest example of a nonparametric testing procedure. In fact, the test was apparently discussed by Laplace in the 1700s. The sign test is a test for the median of any continuous distribution without requiring any other assumptions.

Let $X_1, \ldots, X_n$ be iid samples from an (absolutely) continuous distribution F. Let $\theta = \theta(F)$ be the median of the distribution. Consider testing $H_0 : \theta = \theta_0$ versus the one-sided alternative $H_1 : \theta > \theta_0$. Define the statistic

$$S_n = \sum_{i=1}^{n} I_{\{X_i > \theta_0\}}.$$

Then large values of S_n would indicate that H_1 is true, and so the sign test rejects the null when $S_n > k = k(n, \alpha)$, where this k is chosen so that $P_{H_0}(S_n > k) \leq \alpha$. Under H_0, S_n has a Bin(n, $1/2$) distribution and so $k = k(n, \alpha)$ is just a quantile from the appropriate binomial distribution. Thus, the sign test is a size α test for the median θ for any sample size n and any continuous CDF.

The next question concerns how the sign test performs relative to a competitor; e.g., the t-test. Of course, to make a comparison with the t-test, we must have F such that the mean exists and equals the median. A good basis for comparison is when $F = N(\theta, \sigma^2)$.

Suppose that $X_1, \ldots, X_n$ are iid observations from $N(\theta, \sigma^2)$ for some unknown θ and σ. We wish to test $H_0 : \theta = \theta_0$ against a one- or two-sided alternative. Each of the tests reject H_0 if $T_n \geq c_n$, where T_n is the appropriate test statistic. The two power functions for the case of one-sided alternatives are, respectively,

$$P_{\theta,\sigma}\left(\frac{\sqrt{n}(\overline{X}_n - \theta_0)}{s} > t\right) \quad \text{and} \quad P_{\theta,\sigma}\left(S_n > k(n, \alpha, \theta_0)\right).$$

The former probability is a noncentral t-probability, and the latter is a binomial probability. We wish to compare the two power functions.

The point is that, at a fixed alternative θ, if α remains fixed, then, for large n, the power of both tests is approximately 1 and there would be no way to practically compare the two tests. Perhaps we can see how the powers compare for $\theta \approx \theta_0$. The idea is to take $\theta = \theta_n \to \theta_0$ at such a rate that the limiting power of the tests is strictly between α and 1. If the two powers converge to different values, then we can take the ratio of the limits as a measure of efficiency. The idea is due to E.J.G. Pitman (Pitman (1948)). We have discussed this concept of efficiency, namely the Pitman efficiency, in detail in Chapter 22.

Example 24.4 Let $X_1, \ldots, X_n$ be iid observations from $N(\theta, \sigma^2)$. Suppose we wish to test $H_0 : \theta = \theta_0$. Let T denote the t-test and S denote the sign test. Then $e_P(S, T) = \frac{2}{\pi} \approx 0.637 < 1$. That is, the precision that the t-test achieves with 637 observations is achieved by the sign test with 1000 observations. This reinforces our earlier comment that while nonparametric procedures enjoy a certain amount of validity in broad models, they cannot compete with parametric optimal procedures in specified parametric models.

The sign test, however, cannot get arbitrarily bad with respect to the t-test under some restrictions on the CDF F, as is shown by the following result, although the t-test can be arbitrarily bad with respect to the sign test.

Theorem 24.1 (Hodges and Lehmann (1956)) Let $X_1, \ldots, X_n$ be iid observations from any distribution with density $f(x - \theta)$, where $f(0) > 0$, f is continuous at 0, and $\int z^2 f(z)\, dz < \infty$. Then $e_P(S, T) \geq \frac{1}{3}$ and $e_P(S, T) = \frac{1}{3}$ when f is any symmetric uniform density.

Remark. We learn from this result that the sign test has an asymptotic efficiency with respect to the t-test that is bounded away from zero for a fairly large class of location-parameter CDFs but that the minimum efficiency is only $\frac{1}{3}$, which is not very good. We will later discuss alternative nonparametric tests for the location-parameter problem that have much better asymptotic efficiencies.

24.3 Consistency of the Sign Test

Definition 24.1 *Let $\{\varphi_n\}$ be a sequence of tests for $H_0 : F \in \Omega_0$ v. $H_1 : F \in \Omega_1$. Then $\{\varphi_n\}$ is consistent against the alternatives Ω_1 if*

(i) $\mathbb{E}_F(\varphi_n) \to \alpha \in (0, 1) \ \forall \ F \in \Omega_0$,
(ii) $\mathbb{E}_F(\varphi_n) \to 1 \ \forall \ F \in \Omega_1$.

As in estimation, consistency is a rather weak property of a sequence of tests. However, something must be fundamentally wrong with the test for it not to be consistent. If a test is inconsistent against a large class of alternatives, then it is considered an undesirable test.

Example 24.5 For a parametric example, let $X_1, \ldots, X_n$ be an iid sample from the Cauchy distribution, $C(\theta, 1)$. For all $n \geq 1$, we know that $\overline{X}_n$ also has the $C(\theta, 1)$ distribution. Consider testing the hypothesis $H_0 : \theta = 0$ versus $H_1 : \theta > 0$ by using a test that rejects for large $\overline{X}_n$. The cutoff point, k, is found by making $P_{\theta=0}(\overline{X}_n > k) = \alpha$. But k is simply the αth quantile of the $C(0, 1)$ distribution. Then the power of this test is given by

$$P_\theta(\overline{X}_n > k) = P(C(\theta, 1) > k) = P(\theta + C(0, 1) > k) = P(C(0, 1) > k - \theta).$$

This is a fixed number not dependent on n. Therefore, the power $\not\to 1$ as $n \to \infty$, and so the test is not consistent even against parametric alternatives.

Remark. A test based on the median would be consistent in the $C(\theta, 1)$ case.

The following theorem gives a sufficient condition for a sequence of tests to be consistent.

Theorem 24.2 Consider a testing problem $H_0 : F \in \Omega_0$ vs. $H_1 : F \in \Omega_1$. Let $\{V_n\}$ be a sequence of test statistics and $\{k_n\}$ a sequence of numbers such that

$$P_F(V_n \geq k_n) \to \alpha < 1 \ \forall F \in \Omega_0.$$

For a test that rejects H_0 when $V_n \geq k_n$, suppose:

- Under any $F \in \Omega_0 \cup \Omega_1$, $V_n \stackrel{P}{\Rightarrow} \mu(F)$, some suitable functional of F.
- For all $F \in \Omega_0$, $\mu(F) = \mu_0$, and for all $F \in \Omega_1$, $\mu(F) > \mu_0$.
- Under H_0, $\frac{\sqrt{n}(V_n - \mu_0)}{\sigma_0} \stackrel{\mathcal{L}}{\Rightarrow} N(0, 1)$ for some $0 < \sigma_0 < \infty$.

Then the sequence of tests is consistent against $H_1 : F \in \Omega_1$.

Proof. We can take $k_n = \frac{\sigma_0 z_\alpha}{\sqrt{n}} + \mu_0$, where z_α is a standard normal quantile. With this choice of $\{k_n\}$,

$$P_F(V_n \geq k_n) \to \alpha \ \ \forall F \in \Omega_0.$$

The power of the test is

$$Q_n = P_F(V_n \geq k_n) = P_F(V_n - \mu(F) \geq k_n - \mu(F)).$$

Since we assume $\mu(F) > \mu_0$, it follows that $k_n - \mu(F) < 0$ for all large n and for all $F \in \Omega_1$. Also, $V_n - \mu(F)$ converges in probability to 0 under any F, and so $Q_n \to 1$. Since the power goes to 1, the test is consistent against any alternative F in Ω_1. □

Corollary 24.1 If F is an absolutely continuous CDF with unique median $\theta = \theta(F)$, then the sign test is consistent for tests on θ.

Proof. Recall that the sign test rejects $H_0 : \theta(F) = \theta_0$ in favor of $H_1 : \theta(F) > \theta_0$ if $S_n = \sum I_{\{X_i > \theta_0\}} \geq k_n$. If we choose $k_n = \frac{n}{2} + z_\alpha \sqrt{\frac{n}{4}}$, then, by the ordinary central limit theorem, we have

$$P_{H_0}(S_n \geq k_n) \to \alpha.$$

Then the consistency of the sign test follows from Theorem 24.2 by letting

(i) $k_n = \frac{1}{2} + z_\alpha \sqrt{\frac{1}{4n}}$,
(ii) $V_n = \frac{S_n}{n}$,
(iii) $\mu_0 = \frac{1}{2}$, $\sigma_0 = \frac{1}{2}$,
(iv) $\mu(F) = 1 - F(\theta_0) > \frac{1}{2}$ for all F in the alternative. □

24.4 Wilcoxon Signed-Rank Test

Recall that Hodges and Lehmann proved that the sign test has a small positive lower bound of $\frac{1}{3}$ on the Pitman efficiency with respect to the t-test in the class of densities with a finite variance, which is not satisfactory (see Theorem 24.1). The problem with the sign test is that it only considers whether an observation is $> \theta_0$ or $\leq \theta_0$, not the magnitude. A nonparametric test that incorporates the magnitudes as well as the signs is called the Wilcoxon signed-rank test; see Wilcoxon (1945).

Definition 24.2 *Given a generic set of n numbers $z_1, \ldots, z_n$, the rank of a particular z_i is defined as*

$$R_i = \#\{k : z_k \leq z_i\}.$$

Suppose that $X_1, \ldots, X_n$ are the observed data from some location parameter distribution $F(x - \theta)$, and assume that F is symmetric. Let $\theta = \text{med}(F)$. We want to test $H_0 : \theta = 0$ against $H_1 : \theta > 0$. We start by ranking $|X_i|$ from the smallest to the largest, giving the units ranks $R_1, \ldots, R_n$. Then the Wilcoxon signed-rank statistic is defined to be the sum of these ranks that correspond to originally positive observations. That is,

$$T = \sum_{i=1}^{n} R_i I_{\{X_i > 0\}}.$$

If we define $W_i = I_{\{|X|_{(i)} \quad \text{corresponds to some positive } X_j\}}$, then we have an alternative expression for T, namely

$$T = \sum_{i=1}^{n} i W_i.$$

To do a test, we need the null distribution of T. It turns out that, under H_0, the $\{W_i\}$ have a relatively simple joint distribution.

Theorem 24.3 Under H_0, $W_1, \ldots, W_n$ are iid Bernoulli $\frac{1}{2}$ variables.

This, together with the representation of T above and Lyapunov's CLT (which we recall below), leads to the asymptotic null distribution of T. See Hettmansperger (1984) for the formal details in the proofs.

Theorem 24.4 (Lyapunov's CLT) For $n \geq 1$, let $X_{n1}, \ldots, X_{nn}$ be a sequence of independent random variables such that $\mathbb{E}X_{ni} = 0 \ \forall i$, $\text{Var}\left(\sum_i X_{ni}\right) = 1$, and $\mathbb{E}|X_{ni}|^3 \to 0$. Then

$$\sum_{i=1}^{n} X_{ni} \overset{\mathcal{L}}{\Rightarrow} N(0, 1).$$

Thus, under H_0, the statistic T is a sum of independent, but not iid, random variables. It follows from Lyapunov's theorem, stated above, that T is asymptotically normal. Clearly

$$E_{H_0} T = \frac{n(n+1)}{4} \quad \text{and} \quad \text{Var}_{H_0} T = \frac{n(n+1)(2n+1)}{24}.$$

The results above imply the following theorem.

Theorem 24.5 Let $X_1, \ldots, X_n$ be iid observations from $F(x - \theta)$, where F is continuous, and $F(x) = 1 - F(-x)$ for all x. Under $H_0 : \theta = 0$,

$$\frac{T - \frac{n(n+1)}{4}}{\sqrt{\frac{n(n+1)(2n+1)}{24}}} \stackrel{\mathcal{L}}{\Rightarrow} N(0, 1).$$

Therefore, the signed-rank test can be implemented by rejecting the null hypothesis, $H_0 : \theta = 0$ if

$$T > \frac{n(n+1)}{4} + z_\alpha \sqrt{\frac{n(n+1)(2n+1)}{24}}.$$

The other option would be to find the *exact* finite sample distribution of T under the null. This can be done in principle, but the CLT approximation works pretty well.

We work out the exact distribution of T_n under the null due to its classic nature. Recall that $T_n = \sum_{i=1}^n i W_i$, where W_i are iid Bernoulli $\frac{1}{2}$ random variables. Let $M = \frac{n(n+1)}{2}$. The probability generating function of T_n is

$$\psi_n(t) = E_{H_0} t^{T_n} = \sum_{k=0}^{M} t^k P(T_n = k).$$

If we can find $\psi_n(t)$ and its power series representation, then we can find $P(T_n = k)$ by equating the coefficients of t^k from each side. But

$$\sum_{k=0}^{M} t^k P(T_n = k) = E_{H_0} t^{T_n} = E_{H_0} t^{\sum k W_k} = \prod_k E_{H_0} t^{k W_k}$$
$$= \prod_{k=1}^{n} \left(\frac{1}{2} + \frac{1}{2} t^k\right) = \frac{1}{2^n} \prod_{k=1}^{n} (1 + t^k) = \frac{1}{2^n} \sum_{k=0}^{M} c_{k,n} t^k,$$

where the sequence $\{c_{k,n}\}$ is determined from the coefficients of t^k in the expansion of the product. From here, we can get the distribution of T_n by setting $P(T_n = k) = c_{k,n}/2^n$. This cannot be done by hand but is easily done in any software package unless n is large, such as $n > 30$. Tables of $P_{H_0}(T_n = k)$ are also widely available.

Remark. If $X_i X_1, \ldots, X_n F(x-\theta)$, where $F(\cdot)$ is symmetric but $\theta \neq 0$ (i.e., under the alternative), then T_n no longer has the representation of the form $T_n = \sum_{j=1}^n Z_j$ for independent $\{Z_j\}$. In this case, deriving the asymptotic distribution of T_n is more complicated. We will do this later by using the theory of U-statistics.

Meanwhile, for sampling from a completely arbitrary continuous distribution, say $H(x)$, there are formulas for the mean and variance of T_n; see Hettmansperger (1984) for proofs. These formulas are extremely useful, and we provide them next.

Theorem 24.6 Let H be a continuous CDF on the real line. Suppose X_1, X_2, X_3 $X_1, \dots, X_n H$. Define the four quantities

$$
\begin{aligned}
p_1 &= P_H(X_1 > 0) = 1 - H(0), \\
p_2 &= P_H(X_1 + X_2 > 0) = \int_{-\infty}^{\infty} [1 - H(-x_2)]\, dH(x_2), \\
p_3 &= P_H(X_1 + X_2 > 0, X_1 > 0) = \int_{0}^{\infty} [1 - H(-x_1)]\, dH(x_1), \\
p_4 &= P_H(X_1 + X_2 > 0, X_1 + X_3 > 0) = \int_{-\infty}^{\infty} [1 - H(-x_1)]^2\, dH(x_1).
\end{aligned}
$$

Then, for the Wilcoxon signed-rank statistic T_n,

$$
\begin{aligned}
E_H(T_n) &= np_1 + \frac{n(n-1)}{2} p_2, \\
\mathrm{Var}_H(T_n) &= np_1(1-p_1) + \frac{n(n-1)}{2} p_2(1-p_2) + 2n(n-1)(p_3 - p_1 p_2) \\
&\quad + n(n-1)(n-2)(p_4 - p_2^2).
\end{aligned}
$$

Example 24.6 Suppose H is symmetric; i.e., $H(-x) = 1 - H(x)$. In this case, $H(0) = 1/2$ and so $p_1 = 1/2$. Also, $p_2 = 1/2$ as $X_1 + X_2$ is symmetric if X_1 and X_2 are independent and symmetric. Therefore,

$$
E_H(T_n) = \frac{n}{2} + \frac{n(n-1)}{2} \times \frac{1}{2} = \frac{n(n+1)}{4}.
$$

Notice that this matches the expression given earlier. Likewise, $p_3 = 3/8$ and $p_4 = 1/3$. Plugging into the variance formula above, we get

$$
\mathrm{Var}_H(T_n) = \frac{n(n+1)(2n+1)}{24}.
$$

Again, this matches the variance expression we derived earlier.

Remark. It can be shown that, for any continuous H, $p_3 = \frac{p_1^2 + p_2}{2}$.

Since T_n takes into account the magnitude as well as the sign of the sample observations, we expect that overall it may have better efficiency properties than the sign test. The following striking result was proved by Hodges and Lehmann in 1956.

Theorem 24.7 (Hodges and Lehmann (1956)) Define the family of CDFs $\mathfrak{F}$ as

$$\mathfrak{F} = \left\{ F : F \text{ is continuous}, \, f(z) = f(-z), \, \sigma_F^2 = \int z^2 f(z)\, dz < \infty \right\}.$$

Suppose $X_1, \ldots, X_n X_1, \ \ldots, \ X_n F(x - \theta)$. Then the Pitman efficiency of the Wilcoxon signed-rank test, T, with respect to the t-test, t, is

$$e_{\mathrm{P}}(T, t) = 12\sigma_F^2 \left(\int f^2(z)\, dz \right)^2.$$

Furthermore,

$$\inf_{F \in \mathfrak{F}} e_{\mathrm{P}}(T, t) = \frac{108}{125} = .864,$$

attained at F such that $f(x) = b(a^2 - x^2)$, $|x| < a$, where $a = \sqrt{5}$ and $b = 3\sqrt{5}/20$.

Remark. Notice that the worst-case density f is not one of heavy tails but one with no tails at all (i.e., it has a compact support). Also note that the minimum Pitman efficiency is .864 in the class of symmetric densities with a finite variance, a very respectable lower bound.

Example 24.7 The following table shows the value of the Pitman efficiency for several distributions that belong to the family of CDFs $\mathfrak{F}$ defined in Theorem 24.7. They are obtained by direct calculation using the formula given above. It is interesting that, even in the normal case, the Wilcoxon test is 95% efficient with respect to the t-test.

F	$e_{\mathrm{P}}(T, t)$
$N(0, 1)$	0.95
$U(-1, 1)$	1.00
$f(x) = \frac{x^2}{4} e^{-\lvert x \rvert}$	1.26

24.5 Robustness of the t Confidence Interval

If $X_1, \ldots, X_n X_1, \ldots, X_n N(\theta, \sigma^2)$, then an exact $100(1 - \alpha)\%$ confidence interval for θ is the famous t confidence interval, C_n, with limits given by

$$\overline{X}_n \pm t_{\frac{\alpha}{2}, n-1} \frac{s}{\sqrt{n}}$$

and with the property $P_{\theta,\sigma}(C_n \ni \theta) = 1 - \alpha \, \forall n, \theta, \sigma$. However, if the population is nonnormal, then the exact distribution of the statistic $t_n = \frac{\sqrt{n}(\overline{X}_n - \theta)}{s}$ is *not* t. Consequently, the coverage probability may not be $1 - \alpha$, even approximately, for finite n. Asymptotically, the $1 - \alpha$ coverage property holds for any population with a finite variance.

Precisely, if $X_1, \ldots, X_n X_1, \ldots, X_n F$ with $\mu = E_F X_1$ and $\sigma^2 = \text{Var}_F X_1 < \infty$, then

$$t_n = \frac{\sqrt{n}(\overline{X}_n - \mu)}{s} = \frac{\sqrt{n}(\overline{X}_n - \mu)/\sigma}{s/\sigma} \stackrel{\mathcal{L}}{\Rightarrow} N(0, 1)$$

since the numerator converges in law to $N(0, 1)$ and the denominator converges in probability to 1. Furthermore, for any given α, $t_{\frac{\alpha}{2}, n-1} \to z_{\frac{\alpha}{2}}$ as $n \to \infty$. Hence,

$$P_F(C_n \ni \mu) = P_F\left(|t_n| \le t_{\frac{\alpha}{2}, n-1}\right) \longrightarrow P(|Z| \le z_{\alpha/2}) = 1 - \alpha.$$

That is, given a specific F and fixed α and ϵ,

$$1 - \alpha - \epsilon \le P_F(C_n \ni \mu) \le 1 - \alpha + \epsilon$$

for all $n \ge N = N(F, \alpha, \epsilon)$.

However, if we know only that F belongs to some large class of distributions, $\mathfrak{F}$, then there are no guarantees about the uniform validity of the coverage. In fact, it is possible to have

$$\lim_{n \to \infty} \inf_{F \in \mathfrak{F}} P_F(C_n \ni \mu) = 0.$$

The t confidence interval is a very popular procedure, routinely used for all types of data in practical statistics. An obviously important question is whether this is a safe practice, or, more precisely, when it is safe. The literature has some surprises. The t-interval is not unsafe, as far as coverage is concerned, for heavy-tailed data, at least when symmetry is present. The paper by Logan et al. (1973) has some major surprises as regards the asymptotic behavior of the t-statistic for heavy-tailed data. However, the t-interval can have poor coverage properties when the underlying distribution is skewed.

Example 24.8 A Mathematica® simulation was done to check the coverage probabilities of the nominal 95% t-interval for various distributions. The table below summarizes the simulation.

n	$N(0,1)$	$U(0,1)$	$C(0,1)$	Exp(1)	$\log N(0,1)$
10	0.95	0.949	0.988	0.915	0.839
25	0.95	0.949	0.976	0.916	0.896

The table indicates that, for heavy-tailed or light-tailed symmetric distributions, the t-interval does not have a significant coverage bias for large samples. In fact, for heavy tails, the coverage of the t-interval could be $> 1-\alpha$ (of course, at the expense of interval width). However, for skewed data, the t-interval has a deficiency in coverage, even for rather large samples.

The previous example helps motivate our discussion. Now we get to the current state of the theory on this issue.

Definition 24.3 *The family of scale mixtures of normal distributions is defined as*

$$\mathfrak{F} = \left\{ f(x) = \int_0^\infty \frac{1}{\tau\sqrt{2\pi}} e^{-\frac{x^2}{2\tau^2}}\, dG(\tau),\ \text{where } G \text{ is a probability measure} \right\}.$$

The family contains many symmetric distributions on the real line that are heavier tailed than the normal. In particular, t, double exponential, logistic, and hyperbolic cosine, for example, all belong to the family $\mathfrak{F}$. Even all symmetric stable distributions belong to $\mathfrak{F}$.

Here is a rather surprising result that says that the coverage of the t confidence interval in the normal scale mixture class is often better than the nominal level claimed.

Theorem 24.8 (Benjamini (1983)) Let $X_i = \mu + \sigma Z_i$, where Z_i $X_1, \ldots, X_n f \in \mathfrak{F}$ for $i = 1, \ldots, n$. Let C_n be the t-interval given by the formula $\overline{X}_n \pm t\frac{s}{\sqrt{n}}$, where t stands for the t percentile $t_{\alpha/2,n-1}$. Then

$$\inf_{f\in\mathfrak{F}} P_{\mu,\sigma,f}(C_n \ni \mu) = P_{\mu,\sigma,\varphi}(C_n \ni \mu),$$

provided $t \geq 1.8$, where φ is the standard normal density.

Remark. The cutoff point 1.8 is not a mathematical certainty, so the theorem is partially numerical. If $\alpha = 0.05$, then $t \geq 1.8$ for all $n \geq 2$. If $\alpha = 0.10$, the result holds for all $n \leq 11$.

This theorem states that the t-interval is safe when the tails are heavy. The natural question that now arises is: What can happen when the tails are light?

The following theorem (see Basu and DasGupta (1995)) uses the family of symmetric unimodal densities given by

$$\mathfrak{F}_{su} = \{f : f(z) = f(-z),\ f \text{ is unimodal}\}$$

Theorem 24.9 Let $X_i = \mu + \sigma Z_i$, where $Z_i X_1, \ldots, X_n f \in \mathfrak{F}_{su}$.

(a) If $t < 1$, then

$$\inf_{f \in \mathfrak{F}_{su}} P_{\mu,\sigma,f}(C_n \ni \mu) = 0 \ \forall n \geq 2.$$

(b) For all $n \geq 2$, there exists a number τ_n such that, for $t \geq \tau_n$,

$$\inf_{f \in \mathfrak{F}_{su}} P_{\mu,\sigma,f}(C_n \ni \mu) = P_{\mu,\sigma,U[-1,1]}(C_n \ni \mu).$$

Remark. The problem of determining the value of the infimum above for $1 \leq t \leq \tau_n$ remains unsolved. The theorem also shows that t-intervals with small nominal coverage are arbitrarily bad over the family $\mathfrak{F}_{su}$, while those with a high nominal coverage are quite safe because the t-interval performs quite well for uniform data.

Example 24.9 The values of τ_n cannot be written down by a formula, but difficult calculations can be done to get them for a given value of n, as shown in the following table. In the table, $1 - \alpha$ is the nominal coverage when the coefficient t equals τ_n.

n	2	5	7	10
τ_n	1.00	1.92	2.00	2.25
$1 - \alpha$	0.50	0.85	0.90	0.95

For example, for all $n \geq 10$, the infimum in Theorem 24.9 is attained at symmetric uniform densities if $\alpha = 0.05$. The next table shows the actual values of the infimum coverage in the symmetric unimodal class for various sample sizes and significance levels.

$\alpha \downarrow, n \rightarrow$	2	3	5	7	10
0.2	0.75	0.77	–	–	–
0.1	0.86	0.87	0.89	–	–
0.05	0.92	0.92	0.93	0.94	0.945
0.01	0.98	0.98	0.98	0.983	0.983

Remark. The numerics and the theorems presented indicate that the coverage of the t-interval can have a significant negative bias if the underlying population F is skewed, although for any F with finite variance, we know it to be asymptotically correct. That is,

$$\lim_{n\to\infty.} P_F(C_n \ni \mu) = 1 - \alpha.$$

They also indicate that for data from symmetric densities, regardless of tail, the t-interval is quite safe.

We can give a theoretical explanation for why the t-interval is likely to have a negative bias in coverage for skewed F. This explanation is provided by looking at a higher-order expansion of the CDF of the t-statistic under a general F with some moment conditions. This is the previously described *Edgeworth expansion* in Chapter 13. We recall it below.

Theorem 24.10 Let $X_1, \ldots, X_n X_1, \ldots, X_n F$ with $\mathbb{E}_F X_1^4 < \infty$. Assume that F satisfies the Cramér condition. Define

$$t_n = \frac{\sqrt{n}(\overline{X}_n - \mu(F))}{s},$$
$$\gamma = \frac{E_F(X-\mu)^3}{\sigma^3(F)},$$
$$\kappa = \frac{E(X-\mu)^4}{\sigma^4(F)} - 3.$$

Then

$$P_F(t_n \le t) = \Phi(t) + \frac{p_1(t,F)\varphi(t)}{\sqrt{n}} + \frac{p_2(t,F)\varphi(t)}{n} + o(n^{-1}),$$

where

$$p_1(t,F) = \frac{\gamma(1+2t^2)}{6},$$
$$p_2(t,F) = t\left[\frac{\kappa(t^2-3)}{12} - \frac{\gamma^2}{18}(t^4+2t^2-3) - \frac{1}{4}(t^2+3)\right].$$

Corollary 24.2 If C_n denotes the t-interval, then, by Theorem 24.10,

$$\begin{aligned} P_F(C_n \ni \mu(F)) &= P_F(|t_n| \leq t) \\ &= P_F(t_n \leq t) - P_F(t_n \leq -t) \\ &= 2\Phi(t) - 1 + \frac{2t}{n}\varphi(t) \\ &\quad \left\{ \frac{\kappa}{12}(t^2 - 3) - \frac{\gamma^2}{18}(t^4 + 2t^2 - 3) - \frac{1}{4}(t^2 + 1) \right\}. \end{aligned}$$

The corollary shows that, when $|\gamma|$ is large, the coverage is likely to have a negative bias and fall below $1 - \alpha \approx 2\Phi(t) - 1$.

Going back to the asymptotic correctness of the t-interval, for any F with a finite variance, we now show that the validity is not uniform in F.

Theorem 24.11 Let $\mathfrak{F} = \{F : \text{Var}_F(X) < \infty\}$. Then

$$\inf_{F \in \mathfrak{F}} P_F(C_n \ni \mu(F)) = 0, \ \ \forall n \geq 2.$$

Proof. Fix n and take a number c such that $e^{-n} < c < 1$. Let $p_n = p_n(c) = \frac{-\log(c)}{n}$. Take the two-point distribution $F = F_{n,c}$ with

$$P_F(X = p_n) = 1 - p_n \ \text{ and } \ P_F(X = p_n - 1) = p_n.$$

Then $\mu(F) = \mathbb{E}_F(X) = 0$ and $\text{Var}_F(X) < \infty$. Now, if all the sample observations are equal to p_n, then the t-interval is just the single point p_n and hence

$$\begin{aligned} P_F(C_n \not\ni \mu(F)) &\geq P_F(X_i = p_n, \ \forall i \leq n) \\ &= (1 - p_n)^n = \left(1 + \frac{\log(c)}{n}\right)^n. \end{aligned}$$

But this implies that, for any fixed $n \geq 2$,

$$\sup_{F \in \mathfrak{F}} P_F(C_n \not\ni \mu(F)) \geq \left(1 + \frac{\log(c)}{n}\right)^n \Rightarrow \inf_{F \in \mathfrak{F}} P_F(C_n \ni \mu(F)) = 0$$

by now letting $c \to 1$ □.

Remark. The problem here is that we have no control over the skewness in the class $\mathfrak{F}$. In fact, the skewness of the two-point distribution F used in the proof is

$$\gamma_F = \frac{2p_n - 1}{\sqrt{p_n(1-p_n)}} \to -\infty \text{ as } c \to 1.$$

It turns out that with minimal assumptions like a finite variance, no intervals can be produced that are *uniformly* (in F) asymptotically correct and yet nontrivial.

To state this precisely, recall the duality between testing and confidence set construction. If $\{\varphi_n\}$ is any (nonrandomized) sequence of test functions, then inversion of the test produces a confidence set C_n for the parameter. The coverage of C_n is related to the testing problem by

$$P_{\theta_0}(C_n \ni \theta_0) = 1 - E_{\theta_0}(\varphi_n).$$

Bahadur and Savage (1956) proved that for sufficiently rich convex families of distribution functions F, there cannot be any tests for the mean that have a uniformly small type I error probability and nontrivial (nonzero) power at the same time. This result is considered to be one of the most important results in testing and interval estimation theory.

24.6 The Bahadur-Savage Theorem

Theorem 24.12 (Bahadur and Savage (1956)) Let $\mathfrak{F}$ be any family of CDFs such that

(a) $E_F|X| < \infty$ for all $F \in \mathfrak{F}$,
(b) for any real number r, there is an $F \in \mathfrak{F}$ such that $\mu(F) = E_F(X) = r$, and
(c) if $F_1, F_2 \in \mathfrak{F}$, then for any $0 < \lambda < 1$, $\lambda F_1 + (1-\lambda)F_2 \in \mathfrak{F}$.

Suppose $X_1, X_2, \ldots, X_n$ are iid from some $F \in \mathfrak{F}$ and $C_n = C_n(X_1, X_2, \ldots, X_n)$ a (measurable) set. If there exists an $F_0 \in \mathfrak{F}$ such that $P_{F_0}(C_n$ is bounded from below$) = 1$, then $\inf_{F \in \mathfrak{F}} P_F(C(n) \ni \mu(F)) = 0$.

Remark. Examples of families of distributions that satisfy the conditions of the Bahadur-Savage theorem are

- the family of all distributions with a finite variance,
- the family of all distributions with all moments finite, and
- the family of all distributions with an (unknown) compact support.

It is a consequence of the Bahadur-Savage theorem that in general we cannot achieve uniform asymptotic validity of the t-interval over rich convex classes. It is natural to ask what additional assumptions will ensure that the t-interval is uniformly asymptotically valid or, more generally, what assumptions are needed for any uniformly asymptotically valid interval to exist at all. Here is a positive result; notice how the skewness is controlled in this next result. See Lehmann and Romano (2005) for a proof.

Theorem 24.13 Fix a number $b \in (0, \infty)$. Define the family of CDFs

$$\mathfrak{F}_b = \left\{ F : \frac{E_F|X - \mu(F)|^3}{\sigma^3(F)} \leq b \right\}.$$

Then the t-interval is uniformly asymptotically correct over $\mathfrak{F}_b$.

24.7 Kolmogorov-Smirnov and Anderson Confidence Intervals

A second theorem on existence of uniformly asymptotically valid intervals for a mean is due to T.W. Anderson (see Lehmann and Romano (2005)). This construction makes the assumption of a known compact support. The construction depends on the classical goodness-of-fit test due to Kolmogorov and Smirnov, summarized below; see Chapter 26 for more details.

Suppose $X_1, \ldots, X_n X_1, \ldots, X_n F$ and we wish to test $H_0 : F = F_0$. The commonsense estimate of the unknown CDF is the empirical CDF F_n. From the Glivenko-Cantelli theorem, we know that

$$\|F_n - F\|_\infty = \sup_x |F_n(x) - F(x)| \longrightarrow 0, \text{ a.s.}$$

However, the statistic

$$D_n = \sqrt{n}\|F_n - F\|_\infty$$

has a nondegenerate limit distribution, and, for every n, if the true CDF F is continuous, then D_n has the remarkable property that its distribution is completely independent of F.

The quickest way to see this property is to notice the identity:

$$D_n \stackrel{\mathcal{L}}{=} \sqrt{n} \max_{1 \leq i \leq n} \max \left\{ \frac{i}{n} - U_{(i)}, U_{(i)} - \frac{i-1}{n} \right\},$$

where $U_{(1)} \leq \ldots \leq U_{(n)}$ are order statistics of an independent sample from $U(0, 1)$ and the relation $=_{\mathcal{L}}$ denotes "equality in law."

Therefore, given $\alpha \in (0, 1)$, there is a well-defined $d = d_{\alpha,n}$ such that, for any continuous CDF F, $P_F(D_n > d) = \alpha$. Thus,

$$\begin{aligned}
1 - \alpha &= P_F(D_n \leq d) \\
&= P_F\left(\sqrt{n}\|F_n - F\|_\infty \leq d\right) \\
&= P_F\left(|F_n(x) - F(x)| \leq \frac{d}{\sqrt{n}} \,\forall x\right) \\
&= P_F\left(-\frac{d}{\sqrt{n}} \leq F_n(x) - F(x) \leq \frac{d}{\sqrt{n}} \,\forall x\right) \\
&= P_F\left(F_n(x) - \frac{d}{\sqrt{n}} \leq F(x) \leq F_n(x) + \frac{d}{\sqrt{n}} \,\forall x\right).
\end{aligned}$$

This gives us a "confidence band" for the true CDF F. More precisely, the $100(1 - \alpha)\%$ Kolmogorov-Smirnov confidence band for the CDF F is

$$KS_{n,\alpha} : \max\left\{0, F_n(x) - \frac{d}{\sqrt{n}}\right\} \leq F(x) \leq \min\left\{1, F_n(x) + \frac{d}{\sqrt{n}}\right\}.$$

Remark. The computation of $d = d_{\alpha,n}$ is quite nontrivial, but tables are available. See Chapter 26.

Anderson constructed a confidence interval for $\mu(F)$ using the Kolmogorov-Smirnov band for F. The interval is constructed as follows:

$$C_A = \left\{\mu : \mu = \mu(H) \text{ for some } H \in KS_{n,\alpha}\right\}.$$

That is, this *interval* contains all μ that are the mean of a *KS-plausible distribution*. With the compactness assumption, the following theorem holds; see Lehmann and Romano (2005).

Theorem 24.14 Let $X_1, \ldots, X_n X_1, \ldots, X_n F$. Suppose F is continuous and supported on a known compact interval $[a, b]$. Then, for any $\alpha \in (0, 1)$ and for any n,

$$P_F(C_A \ni \mu(F)) \geq 1 - \alpha.$$

This interval can be computed by finding the associated means for the upper and lower bounds of the KS confidence band.

Remark. So again, with suitable assumptions, in addition to finiteness of variance, uniformly asymptotically valid intervals for the mean exist.

24.8 Hodges-Lehmann Confidence Interval

The Wilcoxon signed-rank statistic T_n can be used to construct a point estimate for the point of symmetry of a symmetric density, and from it one can construct a confidence interval.

Suppose $X_1, \ldots, X_n X_1, \ldots, X_n F$, where F has a symmetric density centered at θ. For any pair i, j with $i \leq j$, define the Walsh average $W_{ij} = \frac{1}{2}(X_i + X_j)$ (see Walsh (1959)). Then the Hodges-Lehmann estimate $\hat{\theta}$ is defined as

$$\hat{\theta} = \text{med}\left\{W_{ij} : 1 \leq i \leq j \leq n\right\}.$$

A confidence interval for θ can be constructed using the distribution of $\hat{\theta}$. The interval is found from the following connection with the null distribution of T_n.

Let a be a number such that $P_{\theta=0}(T_n \geq N - a) \leq \frac{\alpha}{2}$, where $N = \frac{n(n+1)}{2}$ is the number of Walsh averages. Let $W_{(1)} \leq \ldots \leq W_{(N)}$ be the ordered Walsh averages. Then, for all continuous symmetric F,

$$P_F\left(W_{(a+1)} \leq \theta(F) \leq W_{(N-a)}\right) \geq 1 - \alpha.$$

This is the Hodges-Lehmann interval for θ.

Remark. We cannot avoid calculation of the N Walsh averages for this method. Furthermore, we must use a table to find a. However, we can approximate a by using the asymptotic normality of T_n:

$$\tilde{a} = \frac{n(n+1)}{4} - \frac{1}{2} - z_{\alpha/2}\sqrt{\frac{n(n+1)(2n+1)}{24}}.$$

Alternatively, we can construct a confidence interval for θ based on the Hodges-Lehmann estimate using its asymptotic distribution; see Hettmansperger (1984).

Theorem 24.15 Let $X_1, \ldots, X_n X_1, \ldots, X_n F(x - \theta)$, where f, the density of F, is symmetric around zero. Let $\hat{\theta}$ be the Hodges-Lehmann estimator of θ. Then, if $f \in \mathcal{L}^2$,

$$\sqrt{n}(\hat{\theta} - \theta) \stackrel{\mathcal{L}}{\Rightarrow} N(0, \tau_F^2),$$

where

$$\tau_F^2 = \frac{1}{12\|f\|_2^4}.$$

Clearly, this asymptotic result can be used to construct a confidence interval for θ in the usual way. That is,

$$P_{\theta,F}\left(\hat{\theta} - \frac{z_{\alpha/2}}{\sqrt{12n\|f\|_2^4}} \leq \theta \leq \hat{\theta} + \frac{z_{\alpha/2}}{\sqrt{12n\|f\|_2^4}}\right) \to 1 - \alpha.$$

Of course, the point of nonparametrics is to make minimal assumptions about the distribution F. Therefore, in general, we do not know f and hence we cannot know $\|f\|_2$. However, if we can estimate $\|f\|_2$, then we can simply plug it into the asymptotic variance formula.

24.9 Power of the Wilcoxon Test

Unlike the null case, the Wilcoxon signed-rank statistic T does not have a representation as a sum of independent random variables under the alternative. So the asymptotic nonnull distribution of T, which is very useful for approximating the power, does not follow from the CLT for independent summands. However, T still belongs to the class of U-statistics, and hence our previously described CLTs for U-statistics can be used to derive the asymptotic nonnull distribution of T and thereby get an approximation to the power of the Wilcoxon signed-rank test.

Example 24.10 We have previously seen exact formulas for $E_H T_n$ and $\text{Var}_H T_n$ under an arbitrary distribution H. These are now going to be useful for approximation of the power. Suppose $X_1, \ldots, X_n X_1, \ldots, X_n F(x - \theta)$ and we want to test $H_0 : \theta = 0$. Take an alternative $\theta > 0$. The power of T_n at θ is

$$\begin{aligned}
\beta(\theta) &= P_\theta(T_n > k_{n,\alpha}) \\
&= P_\theta\left(\frac{T_n - E_\theta(T_n)}{\sqrt{\text{Var}_\theta(T_n)}} > \frac{k_{n,\alpha} - E_\theta(T_n)}{\sqrt{\text{Var}_\theta(T_n)}}\right) \\
&\approx 1 - \Phi\left(\frac{k_{n,\alpha} - E_\theta(T_n)}{\sqrt{\text{Var}_\theta(T_n)}}\right),
\end{aligned}$$

where the normal approximation is made from the CLT for U-statistics. Whether the approximation is numerically accurate is a separate issue.

24.10 Exercises

Exercise 24.1 Prove that the quantile $F_n^{-1}(p)$ is a strongly consistent estimate of $F^{-1}(p)$ under very minimal assumptions.

Exercise 24.2 For each of the following cases, explicitly determine inclusion of how many order statistics of the sample are needed for an exact nonparametric confidence interval for the median of a density: $n = 20, \alpha = .05, n = 50, \alpha = .05, n = 50, \alpha = .01$.

Exercise 24.3 Find the Pitman efficiency of the sign test w.r.t. the t-test for a triangular, a double exponential, and a logistic density.

Exercise 24.4 * Find the Pitman efficiency of the sign test w.r.t. the t-test for a t-density with a general degree of freedom ≥ 3 and plot it.

Exercise 24.5 * Is the sign test consistent for *any* continuous CDF F? Prove if true or give a concrete counterexample.

Exercise 24.6 * Find the third and fourth moments of the Wilcoxon signed-rank statistic, and hence derive an expression for its skewness and kurtosis. Do they converge to the limiting normal case values?

Exercise 24.7 Tabulate the exact distribution of the Wilcoxon signed-rank statistic when $n = 3, 5, 10$.

Exercise 24.8 * Analytically evaluate the coefficients p_1, p_2, p_3, p_4 when H is a double exponential density centered at a general θ.

Exercise 24.9 Simulate the coverage probability for the nominal 95% t confidence interval when the underlying true density is the mixture $.9N(0, 1) + .1N(0, 9)$, the double exponential, and the t-density with 5 degrees of freedom. Use $n = 10, 25, 50$.

Exercise 24.10 * Suppose $U \sim U[0, 1]$ and that the underlying true density is U^β. How does the coverage probability of the t-interval behave when β is a large positive number?

Exercise 24.11 * Analytically approximate the coverage probability of the nominal 95% t confidence interval when the underlying true density is an extreme value density $e^{-e^x}e^x$ by using the Edgeworth expansion of the t-statistic.

Exercise 24.12 * Rigorously establish a method of explicitly computing the Anderson confidence interval.

Exercise 24.13 Find the limiting distribution of the Hodges-Lehmann estimate when the underlying true density is uniform, triangular, normal, and double exponential. Do you see any relation to the tail?

Exercise 24.14 * By using the asymptotic nonnull distribution, compute an approximate value of the power of the Wilcoxon signed-rank test in the $N(\theta, 1)$ model and plot it. Superimpose it on the power of the t-test. Compare them.

References

Bahadur, R.R. and Savage, L.J. (1956). The nonexistence of certain statistical procedures in nonparametric problems, Ann. Math. Stat., 27, 1115–1122.

Basu, S. and DasGupta, A. (1995). Robustness of standard confidence intervals for location parameters under departure from normality, Ann. Stat., 23(4), 1433–1442.

Benjamini, Y. (1983). Is the t test really conservative when the parent distribution is long tailed?, J. Am. Stat. Assoc., 78(383), 645–654.

Hajek, J. and Sidak, Z. (1967). *Theory of Rank Tests*, Academic Press, New York.

Hettmansperger, T. (1984). *Statistical Inference Based on Ranks*, John Wiley, New York.

Hodges, J.L. and Lehmann, E.L. (1956). The efficiency of some nonparametric competitors of the t test, Ann. Math. Stat., 27, 324–335.

Jurevckova, J. (1995). Jaroslav Hajek and asymptotic theory of rank tests: Mini Symposium in Honor of Jaroslav Hajek, Kybernetika (Prague), 31(3), 239–250.

Lehmann, E.L. and Romano, J. (2005). *Testing Statistical Hypotheses*, 3rd ed., Springer, New York.

Logan, B.F., Mallows, C.L., Rice, S.O., and Shepp, L. (1973). Limit distributions of self-normalized sums, Ann. Prob., 1, 788–809.

Pitman, E.J.G. (1948). *Lecture Notes on Nonparametric Statistical Inference*, Columbia University, New York.

Randles, R.H. and Wolfe, D.A. (1979). *Introduction to the Theory of Nonparametric Statistics*, John Wiley, New York.

Walsh, J.E. (1959). Comments on "The simplest signed-rank tests," J. Am. Stat. Assoc., 54, 213–224.

Wilcoxon, F. (1945). Individual comparisons by ranking methods, Biometrics Bull., 1(6), 80–83.

Chapter 25
Two-Sample Problems

Often in applications we wish to compare two distinct populations with respect to some property. For example, we may want to compare the average salaries of men and women at an equivalent position. Or, we may want to compare the average effect of one treatment with that of another. We may want to compare their variances instead of the mean, or we may even want to compare the distributions themselves. Problems such as these are called two-sample problems. In some sense, the two-sample problem is more important than the one-sample problem. We recommend Hajek and Sidak (1967), Hettmansperger (1984), Randles and Wolfe (1979), and Romano (2005) for further details on the material in this chapter. Additional specific references are given in the following sections.

We start with the example of a common two-sample parametric procedure in order to introduce a well-known hard problem called the Behrens-Fisher problem.

Example 25.1 (Two-Sample t-test) Let $X_1, \ldots, X_m \overset{\text{iid}}{\sim} N(\mu_1, \sigma^2)$ and $Y_1, \ldots, Y_n \overset{\text{iid}}{\sim} N(\mu_2, \sigma^2)$, where all $m + n$ observations are independent. Then the two-sample t-statistic is

$$T_{m,n} = \frac{\overline{X}_n - \bar{Y}}{s\sqrt{\frac{1}{m} + \frac{1}{n}}}, \quad \text{where } s^2 = \frac{(m-1)s_1^2 + (n-1)s_2^2}{m+n-2}.$$

Under $H_0 : \mu_1 = \mu_2$, $T_{m,n} \sim t_{m+n-2}$. If $m, n \to \infty$, then $T_{m,n} \overset{\mathcal{L}}{\Rightarrow} N(0, 1)$.

More generally, if $X_1, \ldots, X_m \overset{\text{iid}}{\sim} F$ and $Y_1, \ldots, Y_n \overset{\text{iid}}{\sim} G$ and F, G have equal mean and variance, then by the CLT and Slutsky's theorem, we still have $T_{m,n} \overset{\mathcal{L}}{\Rightarrow} N(0, 1)$ as $m, n \to \infty$. The asymptotic level and the power of the two-sample t-test are the same for any F, G with equal variance, as they would be when F, G are both normal.

A. DasGupta, *Asymptotic Theory of Statistics and Probability*,

Of course, the assumption of equal variance is not a practical one. However, the corresponding problem with unequal variances, known as the Behrens-Fisher problem, has many difficulties. We discuss it in detail next.

25.1 Behrens-Fisher Problem

Suppose $X_1, \ldots, X_m \stackrel{\text{iid}}{\sim} N(\mu_1, \sigma_1^2)$ and $Y_1, \ldots, Y_n \stackrel{\text{iid}}{\sim} N(\mu_2, \sigma_2^2)$, where all $m + n$ observations are independent. We wish to test $H_0 : \mu_1 = \mu_2$ in the presence of possibly unequal variances. We analyze four proposed solutions to this problem. There is by now a huge body of literature on the Behrens-Fisher problem. We recommend Lehmann (1986), Scheffe (1970), and Linnik (1963) for overall exposition of the Behrens-Fisher problem. Here are the four ideas we want to explore.

I. Let $\Delta = \mu_1 - \mu_2$. Then

$$\bar{Y} - \overline{X}_n \sim N\left(\Delta, \frac{\sigma_1^2}{m} + \frac{\sigma_2^2}{n}\right).$$

Also,

$$\frac{1}{\sigma_1^2}\sum_{i=1}^{m}(X_i - \overline{X}_n)^2 + \frac{1}{\sigma_2^2}\sum_{j=1}^{n}(Y_j - \bar{Y})^2 \sim \chi^2_{m+n-2}.$$

Now, define

$$t = t_{m,n} = \frac{\sqrt{m+n-2}(\bar{Y} - \overline{X}_n - \Delta)}{\sqrt{\left(\frac{\sigma_1^2}{m} + \frac{\sigma_2^2}{n}\right)\left[\sigma_1^{-2}\sum_i (X_i - \overline{X}_n)^2 + \sigma_2^{-2}\sum_j (Y_j - \bar{Y})^2\right]}}.$$

Letting $\theta = \sigma_2^2/\sigma_1^2$, we can simplify the expression above to get

$$t = \frac{\sqrt{m+n-2}(\bar{Y} - \overline{X}_n - \Delta)}{\sqrt{\left(1 + \frac{m}{n}\theta\right)\left[\frac{m-1}{m}s_1^2 + \frac{n-1}{m\theta}s_2^2\right]}}.$$

However, t is not a "statistic" because it depends on the unknown θ. This is unfortunate because if θ were known (i.e., if we knew the ratio of the two variances), then the statistic t could be used to test the hypothesis $\Delta = \Delta_0$.

II. Consider the two-sample t-statistic suitable for the equal-variance case. That is, consider

$$T = T_{m,n} = \frac{\bar{Y} - \overline{X}_n - \Delta_0}{\sqrt{s_u^2 \left(\frac{1}{m} + \frac{1}{n}\right)}},$$

where $s_u^2 = \frac{(m-1)s_1^2+(n-1)s_2^2}{m+n-2}$. We know that, under $H_0 : \Delta = \Delta_0$, the distribution of T is exactly t_{m+n-2} only if $\sigma_1 = \sigma_2$.

But what happens for large samples in the case $\sigma_1 \neq \sigma_2$? By a simple application of Slutsky's theorem, it is seen that, when $\sigma_1 \neq \sigma_2$, if $m, n \to \infty$ in such a way that $\frac{m}{m+n} \to \rho$, then

$$T_{m,n} \stackrel{\mathcal{L}}{\Rightarrow} N\left(0, \frac{(1-\rho)+\rho\theta}{\rho+(1-\rho)\theta}\right) \quad \text{under } H_0.$$

Notice that, if $\rho = \frac{1}{2}$, then $T_{m,n} \stackrel{\mathcal{L}}{\Rightarrow} N(0, 1)$. That is, if m and n are large and $m \approx n$, then $T_{m,n}$ can be used to construct a test. However, if m and n are very different, one must also estimate θ.

III. We next consider the likelihood ratio test for the Behrens-Fisher problem $H_0 : \mu_1 = \mu_2$, $H_1 : \mu_1 \neq \mu_2$. The LRT statistic is $\lambda = -2 \log \Lambda$, where

$$\Lambda = \frac{\sup_{H_0} l(\mu_1, \mu_2, \sigma_1, \sigma_2)}{\sup_{H_0 \cup H_1} l(\mu_1, \mu_2, \sigma_1, \sigma_2)},$$

where $l(.)$ denotes the likelihood function. Then H_0 is rejected for large values of λ. The statistic Λ itself is a complicated function of the data. Of course, the denominator is found by plugging in the unconstrained MLEs

$$\hat{\mu}_1 = \overline{X}_n, \quad \hat{\mu}_2 = \bar{Y}, \quad \hat{\sigma}_1^2 = \frac{1}{m}\sum_i (X_i - \overline{X}_n)^2, \ \hat{\sigma}_2^2 = \frac{1}{n}\sum_j (Y_j - \bar{Y})^2.$$

Let $\hat{\mu}$ be the MLE of the common mean $\mu\, (= \mu_1 = \mu_2)$, under H_0. Then, the MLEs of the two variances under H_0 are

$$\hat{\sigma}_1^2 = \frac{1}{m}\sum_i (X_i - \hat{\mu})^2 \quad \text{and} \quad \hat{\sigma}_2^2 = \frac{1}{n}\sum_j (Y_j - \hat{\mu})^2.$$

It can be shown that the MLE $\hat{\mu}$ is *one* of the roots of the cubic equation

$$A\mu^3 + B\mu^2 + C\mu + D = 0,$$

where

$$\begin{aligned}
A &= -(m+n), \\
B &= (m+2n)\overline{X}_n + (n+2m)\bar{Y}, \\
C &= \frac{m(n-1)}{n}s_2^2 + \frac{n(m-1)}{m}s_1^2, \\
D &= m\overline{X}_n\left(\frac{n-1}{n}s_2^2 + \bar{Y}^2\right) + n\bar{Y}\left(\frac{m-1}{m}s_1^2 + \overline{X}_n^2\right).
\end{aligned}$$

In the event that the equation above has three real roots, the actual MLE has to be picked by examination of the likelihood function. The MLE is the unique root if the equation above has only one real root.

Therefore, the numerator of Λ is not analytically expressible, but at least asymptotically it can be used because we have a CLT for λ under the null (see Chapter 21).

IV. The final proposed solution of the Behrens-Fisher problem is due to Welch (Welch (1949)). We know that, under the null, $\bar{Y} - \overline{X}_n \sim N(0, \frac{\sigma_1^2}{m} + \frac{\sigma_2^2}{n})$. Welch considered the statistic

$$W = W_{m,n} = \frac{\overline{X}_n - \bar{Y}}{\sqrt{\frac{s_1^2}{m} + \frac{s_2^2}{n}}} = \frac{\overline{X}_n - \bar{Y} \div \sqrt{\frac{\sigma_1^2}{m} + \frac{\sigma_2^2}{n}}}{\sqrt{\frac{s_1^2}{m} + \frac{s_2^2}{n}} \div \sqrt{\frac{\sigma_1^2}{m} + \frac{\sigma_2^2}{n}}}. \quad (\star)$$

It is clear that W is *not* of the form $N(0,1)/\sqrt{\chi_d^2/d}$, with the two variables being independently distributed. Let D^2 be the square of the denominator in the right-hand side of $(\star)$. Welch wanted to write $D^2 \approx \chi_f^2/f$ by choosing an appropriate f. Since the means already match (i.e., $E_{H_0}(D^2)$ is already 1), Welch decided to match the second moments. The following formula for f then results:

$$f = \frac{(\lambda_1\sigma_1^2 + \lambda_2\sigma_2^2)^2}{\frac{\lambda_1^2\sigma_1^4}{m-1} + \frac{\lambda_2^2\sigma_2^4}{n-1}}, \qquad \text{where } \lambda_1 = \frac{1}{m}, \lambda_2 = \frac{1}{n}.$$

If we plug in s_1^2 and s_2^2 for σ_1^2 and σ_2^2, respectively, we get Welch's random degree of freedom t-test. That is, we perform a test based on Welch's procedure by comparing W to a critical value from the $t_{\hat{f}}$ distribution, where $\hat{f} = f(s_1^2, s_2^2)$.

Example 25.2 The behavior of Welch's test has been studied numerically and theoretically. As regards the size of the Welch test for normal data, the news is good. The deviation of the actual size from the nominal α is small, even for small or moderate m, n. The following table was taken from Wang (1971). Let M_α denote the maximum deviation of the size of the Welch test from α. In this example, $\alpha = 0.01$ and $\theta = \frac{\sigma_2^2}{\sigma_1^2}$.

m	n	$1/\theta$	$M_{0.01}$
5	21	2	0.0035
7	7	1	0.0010
7	13	4	0.0013
7	19	2	0.0015
13	13	1	0.0003

Regarding the power of the Welch test, it is comparable to the likelihood ratio test. See, for example, Table 2 in Best and Rayner (1987).

Remark. Pfanzagl (1974) has proved that the Welch test has some local asymptotic power optimality property. It is also very easy to implement. Its size is very close to the nominal level α. Due to these properties, the Welch test has become quite widely accepted as the standard solution to the Behrens-Fisher problem. It is not clear, however, that the Welch test is even size-robust when the individual groups are not normal.

In the case where no reliable information is known about the distributional shape, we may want to use a nonparametric procedure. That is our next topic.

25.2 Wilcoxon Rank Sum and Mann-Whitney Test

Suppose $X_1, \ldots, X_m \overset{\text{iid}}{\sim} F(x - \mu)$ and $Y_1, \ldots, Y_n \overset{\text{iid}}{\sim} F(y - \nu)$, where $F(\cdot)$ is symmetric about zero. In practical applications, we often want to know if $\Delta = \nu - \mu > 0$. This is called the problem of testing for treatment effect.

Let $R_i = \text{rank}(Y_i)$ among all $m+n$ observations. The Wilcoxon rank-sum statistic is $U = \sum_{i=1}^n R_i$. Large values of U indicate that there is indeed a treatment effect; i.e., that $\Delta > 0$ (Mann and Whitney (1947)). To execute the test, we need the smallest value ξ such that $P_{H_0}(U > \xi) \leq \alpha$.

In principle, the null distribution of U can be found from the joint distribution of $(R_1, \ldots, R_n)$. In particular, if $N = m + n$, then the marginal and two-dimensional distributions are, respectively,

$$P(R_i = a) = \frac{1}{N}, \ \forall\, a = 1, \ldots, N,$$
$$P(R_i = a, R_j = b) = \frac{1}{N(N-1)}, \ \forall\, a \neq b.$$

It follows immediately that

$$E_{H_0}(U) = \frac{n(N+1)}{2} \quad \text{and} \quad \mathrm{Var}_{H_0}(U) = \frac{mn(N+1)}{12}.$$

This will be useful for the asymptotic distribution, which we discuss a little later. As for the exact null distribution of U, we have the following proposition; see Hettmansperger (1984).

Proposition 25.1 Let $p_{m,n}(k) = P_{H_0}(U = k + n(n+1)/2)$. Then the following recursive relation holds:

$$p_{m,n}(k) = \frac{n}{m+n} p_{m,n-1}(k-m) + \frac{m}{m+n} p_{m-1,n}(k).$$

From here, the exact finite sample distributions can be numerically computed for moderate values of m, n.

There is an interesting way to rewrite U. From its definition,

$$R_i = \#\{k : Y_k \leq Y_i\} + \#\{j : X_j < Y_i\},$$

which implies

$$U \equiv \sum_{i=1}^{n} R_i = \frac{n(n+1)}{2} + \#\{(i, j) : X_j < Y_i\}.$$

Then the statistic

$$W = U - \frac{n(n+1)}{2} = \sum_{i=1}^{n} R_i - \frac{n(n+1)}{2} = \#\{(i, j) : X_j < Y_i\}$$

is called the Mann-Whitney statistic. Note the obvious relationship between the Wilcoxon rank-sum statistic U and the Mann-Whitney statistic W, in particular

$$U = W + \frac{n(n+1)}{2}.$$

Therefore, $E_{H_0}(W) = \frac{mn}{2}$ and $\text{Var}_{H_0}(W) = \frac{mn(N+1)}{12}$.

The Mann-Whitney test rejects H_0 for large values of W and is obviously equivalent to the Wilcoxon rank-sum test.

It follows from one-sample U-statistics theory that, under the null hypothesis, W is asymptotically normal; see Hettmansperger (1984) for the formal details of the proof.

Theorem 25.1 Let $X_1, \ldots, X_m \overset{\text{iid}}{\sim} F(x - \mu)$, $Y_1, \ldots, Y_n \overset{\text{iid}}{\sim} F(y - \nu)$, all $m+n$ observations independent. Suppose that $F(\cdot)$ is symmetric about zero. Under the null hypothesis $H_0 : \Delta = \nu - \mu = 0$,

$$\frac{W - E_{H_0}(W)}{\sqrt{\text{Var}_{H_0}(W)}} \overset{\mathcal{L}}{\Rightarrow} N(0, 1).$$

Therefore, a cutoff value for the α-level test can be found via the CLT:

$$k_\alpha = \frac{mn}{2} + \frac{1}{2} + z_\alpha \sqrt{\frac{mn(N+1)}{12}},$$

where the additional $\frac{1}{2}$ that is added is a continuity correction.

Recall that a point estimate due to Hodges and Lehmann is available in the one-sample case based on the signed-rank test. In particular, it was the median of the Walsh averages (see Chapter 24). We can do something similar in the two-sample case.

$$W = \#\{(i, j) : X_j < Y_i\} = \#\{(i, j) : D_{ij} \equiv Y_i - X_j > 0\}.$$

This motivates the estimator $\hat{\Delta} = \#\{D_{ij}\}$. It turns out that an interval containing adequately many order statistics of the D_{ij} has a guaranteed coverage for all m, n; see Hettmansperger (1984) for a proof. Here is the theorem.

Theorem 25.2 Let $k = k(m, n, \alpha)$ be the largest number such that $P_{H_0}(W \leq k) \leq \alpha/2$. Then $(D_{(k+1)}, D_{(mn-k)})$ is a $100(1-\alpha)\%$ confidence interval for Δ under the shift model above.

Remark. Tables can be used to find k for a given case; see, for example, Milton (1964). This is a widely accepted nonparametric confidence interval for the two-sample location-parameter problem.

A natural question would now be how this test compares to, say, the t-test. It can be shown that the Mann-Whitney test is consistent for any two distributions such that $P(Y > X) > \frac{1}{2}$. That is, if $X_1, \ldots, X_m \overset{\text{iid}}{\sim} G$ and $Y_1, \ldots, Y_n \overset{\text{iid}}{\sim} H$ and if $X \sim G$, $Y \sim H$, and X, Y are independent, then consistency holds if

$$P_{G,H}(Y > X) > \frac{1}{2}.$$

Pitman efficiencies, under the special shift model, can be found, and they turn out to be the same as in the one-sample case. See Lehmann (1986) for details on the consistency and efficiency of the Mann-Whitney test.

25.3 Two-Sample U-statistics and Power Approximations

The asymptotic distribution of W under the alternative is useful for approximating the power of the Mann-Whitney test. Under the null, we used one-sample U-statistics theory to approximate the distribution. However, under the alternative, there are two underlying distributions, namely $F(x-\mu)$ and $F(y-\nu)$. So the asymptotic theory of one-sample U-statistics cannot be used under the alternative. We will need the theory of two-sample U-statistics. See Serfling (1980) for derivations of the basic formulas needed in this section.

Below we will consider two-sample U-statistics in a much more general scenario than the location-shift model.

Definition 25.1 Fix $0 < r_1, r_2 < \infty$, and let $h(x_1, \ldots, x_{r_1}, y_1, \ldots, y_{r_2})$ be a real-valued function on $\mathbb{R}^{r_1+r_2}$. Furthermore, assume that h is permutation invariant among the x's and among the y's separately. Let $X = X_1, \ldots, X_m \overset{\text{iid}}{\sim} F$ and $Y = Y_1, \ldots, Y_n \overset{\text{iid}}{\sim} G$, and suppose all observations are independent. Define

$$U_{m,n} = U_{m,n}(X, Y) = \frac{1}{\binom{m}{r_1}\binom{n}{r_2}} \sum h\left(X_{i_1}, \ldots, X_{i_{r_1}}, Y_{j_1}, \ldots, Y_{j_{r_2}}\right),$$

where the sum is over all $1 \leq i_1 \leq \ldots \leq i_{r_1} \leq m$, $1 \leq j_1 \leq \ldots \leq j_{r_2} \leq n$. Then $U_{m,n}$ is called a two-sample U-statistic with kernel h and indices r_1 and r_2.

Example 25.3 Let $r_1 = r_2 = 1$ and $h(x, y) = I_{y>x}$. Then it is easy to verify that $U_{m,n}$, based on this kernel, is the Mann-Whitney test statistic W.

As in the one-sample case, $U_{m,n}$ is asymptotically normal under suitable conditions. We use the following notation:

$$\begin{aligned}
\theta &= \theta(F, G) = E_{F,G} h(X_1, \ldots, X_{r_1}, Y_1, \ldots, Y_{r_2}), \\
h_{10}(x) &= E_{F,G}[h(X_1, \ldots, X_{r_1}, Y_1, \ldots, Y_{r_2}) | X_1 = x], \\
h_{01}(y) &= E_{F,G}[h(X_1, \ldots, X_{r_1}, Y_1, \ldots, Y_{r_2}) | Y_1 = y], \\
\zeta_{10} &= \mathrm{Var}_F h_{10}(X), \\
\zeta_{01} &= \mathrm{Var}_G h_{01}(Y).
\end{aligned}$$

Theorem 25.3 Assume that $E_{F,G} h^2 < \infty$ and $\zeta_{10}, \zeta_{01} > 0$. Also, assume that $m, n \to \infty$ in such a way that $\frac{m}{m+n} \to \lambda \in (0, 1)$. Then

$$\sqrt{n}(U_{m,n} - \theta) \stackrel{\mathcal{L}}{\Rightarrow} N(0, \sigma_{F,G}^2),$$

where

$$\sigma_{F,G}^2 = \frac{r_1^2 \zeta_{10}}{\lambda} + \frac{r_2^2 \zeta_{01}}{1 - \lambda}.$$

Remark. Sometimes it is more convenient to use the true variance of $U_{m,n}$ and the version that states

$$\frac{U_{m,n} - \theta}{\sqrt{\mathrm{Var}_{F,G}(U_{m,n})}} \stackrel{\mathcal{L}}{\Rightarrow} N(0, 1).$$

Recall that if $r_1 = r_2 = 1$ and $h(x, y) = I_{y>x}$, then $U_{m,n}$ is the Mann-Whitney test statistic W. In this case, we have exact formulas for θ and $\mathrm{Var}_{F,G}(U_{m,n})$.

Let $X_1, X_2 \stackrel{\text{iid}}{\sim} F$ and $Y_1, Y_2 \stackrel{\text{iid}}{\sim} G$, and define

$$\begin{aligned}
p_1 &= P_{F,G}(Y_1 > X_1) = \int (1 - G(x)) f(x)\, dx, \\
p_2 &= P_{F,G}(Y_1 > X_1, Y_2 > X_1) = \int (1 - G(x))^2 f(x)\, dx, \\
p_3 &= P_{F,G}(Y_1 > X_1, Y_1 > X_2) = \int F^2(y) g(y)\, dy.
\end{aligned}$$

Proposition 25.2 Let W be the Mann-Whitney statistic. Then

$$\begin{aligned}
E_{F,G}(W) &= mnp_1, \\
\mathrm{Var}_{F,G}(W) &= mn(p_1 - p_1^2) + mn(n-1)(p_2 - p_1^2) + mn(m-1)(p_3 - p_1^2).
\end{aligned}$$

See Hettmansperger (1984) for the above formulas.

Remark. We can use these formulas to compute the approximate quantiles of W by approximating $\frac{W-E(W)}{\sqrt{\text{Var}(W)}}$ by a $N(0,1)$. Another option is to use $\sigma^2_{F,G}$ in place of the exact variance of W. Here, $\sigma^2_{F,G}$ is the asymptotic variance, as defined before. To use the alternative expression $\sigma^2_{F,G}$, we need to compute ζ_{10} and ζ_{01}. For this computation, it is useful to note that

$$\zeta_{10} = \text{Var}_F G(X) \quad \text{and} \quad \zeta_{01} = \text{Var}_G F(Y).$$

Remark. Note that the exact variance of W, as well as ζ_{10} and ζ_{01}, are functionals of F and G, which we cannot realistically assume to be known. In practice, all of these functionals must be estimated (usually by a plug-in estimator; e.g., in the formula for p_1, replace F by some suitable $\hat{F}_m$ and G by some suitable $\hat{G}_n$, so that, for example, $\hat{p}_1 = \int(1 - \hat{G}_n(x))\, d\hat{F}_m(x))$.

25.4 Hettmansperger's Generalization

So far, we have considered the two cases $N(\theta_1, \sigma_1^2)$ vs. $N(\theta_2, \sigma_2^2)$ and $F(x-\mu)$ vs. $F(y-\nu)$. For the first case, we have settled on Welch's solution. For the second, we like the Mann-Whitney test. However, even this second model does not allow the scale parameters to differ. This is our next generalization.

Let $X_i \overset{\text{iid}}{\sim} F\left(\frac{x-\mu}{\rho}\right)$, $1 \le i \le m$, and $Y_j \overset{\text{iid}}{\sim} F\left(\frac{y-\nu}{\tau}\right)$, $1 \le j \le n$, where ρ and τ are unknown and, as usual, we assume that all observations are independent. We wish to test $H_0 : \mu = \nu$.

A test due to Hettmansperger (1973) with reasonable properties is the following. Let $S = \#\{j : Y_j > \#(X_i)\}$ and $S^* = \#\{i : X_i < \#(Y_j)\}$. Also, let $\sigma = \tau/\rho$. Then define

$$T_\sigma = \sqrt{n}\left(\frac{S}{n} - \frac{1}{2}\right) \div \sqrt{\frac{1 + n/(m\sigma^2)}{4}},$$

$$T_\sigma^* = \sqrt{m}\left(\frac{S^*}{m} - \frac{1}{2}\right) \div \sqrt{\frac{1 + m\sigma^2/n}{4}}.$$

Then the test statistic is $T = \min\{T_1, T_1^*\}$. This test rejects $H_0 : \mu = \nu$ if $T > z_\alpha$, a normal quantile.

Theorem 25.4 Let $X_1, \ldots, X_m \overset{\text{iid}}{\sim} F\left(\frac{x-\mu}{\rho}\right)$ and $Y_1, \ldots, Y_n \overset{\text{iid}}{\sim} F\left(\frac{y-\nu}{\tau}\right)$, with all parameters unknown and all $(m+n)$ observations independent.

Assume that F is absolutely continuous, 0 is the unique median of F, and that there exists $\lambda \in (0, 1)$ such that $\frac{m}{m+n} \to \lambda$. Consider testing $H_0 : \mu = \nu$ against $H_1 : \nu > \mu$. Then

$$\lim_{m,n\to\infty} P_{H_0}(T > z_\alpha) \leq \alpha.$$

Proof. (Sketch) By a standard argument involving Taylor expansions, for all $\sigma > 0$, each of T_σ and T_σ^* are asymptotically $N(0, 1)$. From the monotonicity in σ, it follows that $T_\sigma > T_1$ and $T_\sigma^* < T_1^*$ when $\sigma > 1$ and the inequalities are reversed for $\sigma \leq 1$. Therefore, when $\sigma > 1$,

$$P(T_1 > c) \leq P(T_\sigma > c),$$

while

$$P(T_1^* > c) \geq P(T_\sigma^* > c).$$

If we set $c \equiv z_\alpha$, then, when $\sigma > 1$, we get

$$P(T_1^* > z_\alpha) \geq P(T_\sigma^* > z_\alpha) \approx \alpha \approx P(T_\sigma > z_\alpha) \geq P(T_1 > z_\alpha),$$

where $\approx$ holds due to the asymptotic normality of T_σ and T_σ^*. Similarly, when $\sigma \leq 1$, we get the opposite string of inequalities. Consequently,

$$\begin{aligned} P(T > z_\alpha) &= P(\min\{T_1, T_1^*\} > z_\alpha) \\ &= P(T_1 > z_\alpha, T_1^* > z_\alpha) \\ &\leq \min\left\{P(T_1 > z_\alpha), P(T_1^* > z_\alpha)\right\} \\ &\approx \alpha. \end{aligned}$$

This explains why the limiting size of the test is $\leq \alpha$.

Remark. Theorem 25.4 says that the Hettmansperger test is asymptotically distribution-free under H_0 and asymptotically conservative with regard to size. An approximation for the true size of this test is

$$\alpha_{\text{true}} \approx 1 - \Phi\left(z_\alpha \sqrt{\frac{1+c}{\sigma^2 + c}} \max\{1, \sigma\}\right),$$

where $c = \lim \frac{n}{m}$. Note that the right-hand side is equal to α when $\sigma = 1$. See Hettmansperger (1973) for a derivation of this approximation.

Simulations show that, if $\sigma \approx 1$, then the true size is approximately α. However, if σ is of the order 2 or 1/2, then the test is severely conservative.

In summary, the test proposed by Hettmansperger is reasonable to use if

- the two populations have the same shape,
- the two populations have approximately the same scale parameters, and
- $F(0) = \frac{1}{2}$, although F need not be symmetric.

25.5 The Nonparametric Behrens-Fisher Problem

This is the most general version of the two-sample location problem, with as few assumptions as possible. We want to construct a test that is, at least asymptotically, distribution-free under H_0 and consistent against a broad class of alternatives. The model is as follows. Suppose $X_1, \ldots, X_m \overset{\text{iid}}{\sim} F$ and $Y_1, \ldots, Y_n \overset{\text{iid}}{\sim} G$, where F, G are arbitrary distribution functions. To avoid the difficulties of ties, we assume that F and G are (absolutely) continuous. Let $\mu = \#(F)$ and $\nu = \#(G)$. We want to test $H_0 : \mu = \nu$ vs. $H_1 : \nu > \mu$. See Fligner and Policello (1981) and Brunner and Munzel (2000) for the development in this section. Johnson and Weerahandi (1988) and Ghosh and Kim (2001) give Bayesian solutions to the Behrens-Fisher problem, while Babu and Padmanabhan (2002) use resampling ideas for the nonparametric Behrens-Fisher problem.

It turns out that a Welch-type statistic, which was suitable for the ordinary Behrens-Fisher problem, is used here. But unlike the Welch statistic used in the case of two normal populations, this test will use the ranks of the X_i's and the Y_j's.

Let Q_i and R_j denote the ranks of $X_{(i)}$ and $Y_{(j)}$ among all $m+n$ observations, respectively. Here, as usual, $X_{(i)}$ and $Y_{(j)}$ denote the respective order statistics. Also, let $P_i = Q_i - i$ and $O_j = R_j - j$. That is,

$$P_i = \#\{j : Y_j \le X_{(i)}\},$$

$$O_j = \#\{i : X_i \le Y_{(j)}\}.$$

Our test function comes from a suitable function $T = T_{m,n}$ of the vector of ranks of the Y sample in the combined sample. We would like to choose T in such a way that whatever F and G are,

$$\frac{T - \theta_{F,G}}{\tau_{F,G}} \overset{\mathcal{L}}{\Rightarrow} N(0,1), \quad \text{where} \begin{cases} \{ \begin{matrix} \theta_{F,G} = E_{F,G}(T) \\ \tau_{F,G}^2 = \mathrm{Var}_{F,G}(T) \end{matrix} \end{cases}.$$

There is some additional complexity here because $\mu = \nu$ does not force the two distributions to be the same. Thus, we need to estimate the variance $\tau_{F,G}^2$

under the null for general F, G. If we can do this and, moreover, if $\theta_{F,G} = \theta_0$, a fixed number, then we can use a standardized test statistic such as

$$\frac{T - \theta_0}{\hat{\tau}_{F,G}}.$$

So, the choice of T will be governed by the ease of finding θ_0 and $\hat{\tau}_{F,G}$ under H_0. The statistic that is used is

$$T = \frac{1}{mn} \sum_{i=1}^{m} P_i.$$

It turns out that mnT is actually the same as the Mann-Whitney statistic. For such a choice of T, under H_0, $\theta_{F,G} = \theta_0 = 1/2$. Also, $\tau_{F,G}^2$ has the exact formula given earlier in this chapter. Of course, this formula involves the unknown F, G, so we need to plug in the empirical CDFs $\hat{F}_m$ and $\hat{G}_n$. Doing so, we get the test statistic

$$\hat{W} = \frac{\frac{W}{mn} - \frac{1}{2}}{\hat{\tau}_{F,G}} = \frac{\sum_i P_i - \sum_j O_j}{2\sqrt{\sum_i (P_i - \overline{P})^2 + \sum_j (O_j - \overline{O})^2 + \overline{P} \times \overline{O}}}.$$

Then the test rejects H_0 for large values of $\hat{W}$. If $k = k_{m,n}(\alpha)$ is the cutoff for the Mann-Whitney statistic itself, then, specifically, the test that rejects H_0 for $\hat{W} > k$ has the property that if $F = G$ under H_0 (which means that equality of medians forces equality of distributions), then the size is still α. If $F \neq G$ under H_0, then some more assumptions are needed to maintain the size and for reasonable consistency properties.

Observe that

$$\begin{aligned}\lim_{m,n\to\infty} P_{F,G}\left(\hat{W} > k\right) &= \lim_{m,n\to\infty} P_{F,G}\left(\frac{W/mn - 1/2}{\hat{\tau}_{F,G}} > k\right) \\ &= \lim_{m,n\to\infty} P_{F,G}\left(\frac{W/mn - \theta_{F,G} + \theta_{F,G} - 1/2}{\hat{\tau}_{F,G}} > k\right) \end{aligned} \qquad (\star\star)$$

But, from the general two-sample U-statistics theory, we know that

$$\frac{W/mn - \theta_{F,G}}{\hat{\tau}_{F,G}} \stackrel{\mathcal{L}}{\Rightarrow} N(0, 1),$$

and $k = k_{m,n}(\alpha) \to z_\alpha$. Then, clearly, the limit above is equal to α if and only if $\theta_{F,G} = \frac{1}{2}$ under H_0. Recall now that

$$\theta_{F,G} = \int F\,dG = P_{F,G}(Y > X) = P_{F,G}(Y - X > 0).$$

This is $\frac{1}{2}$ under H_0 if and only if $Y - X$ has median zero. If $X \sim F$ and $Y \sim G$ and F, G are symmetric about some μ and ν, then $Y - X$ is symmetric about 0 when $\mu = \nu$. In that case, $P_{F,G}(Y - X > 0) = \frac{1}{2}$ holds automatically and the size of $\hat{W}$ is asymptotically maintained.

Also, from ($\star\star$) above, we see that the power under any F, G converges to 1 if and only if $\theta_{F,G} > \frac{1}{2}$. That is, the test $\hat{W}$ is consistent against those alternatives (F, G) for which

$$\int F\,dG = P_{F,G}(Y - X > 0) > \frac{1}{2}.$$

Remark. If we interpret our hypothesis of interest as

$$H_0 : P_{F,G}(Y - X > 0) = \frac{1}{2} \quad \text{vs.} \; H_1 : P_{F,G}(Y - X > 0) > \frac{1}{2},$$

then, without any assumptions, the test based on $\hat{W}$ maintains its size asymptotically and is consistent.

Example 25.4 Below are some tables giving the size and power of the $\hat{W}$-test for selected F, G, m, n. For the size values, it is assumed that $G(y) = F(y/\sigma)$, and for the power values, it is assumed that $G(y) = F(y - \Delta)$.

Size($\times$1000) table: $\alpha = 0.05$, $m = 11$, and $n = 10$

F	σ	W	$\hat{W}$	Welch
	0.1	81	48	18
	0.25	69	54	52
$N(0,1)$	1	50	48	47
	4	71	54	47
	10	82	62	52
	0.1	75	51	45
	0.25	65	54	49
Double Exp	1	50	48	46
	4	67	54	45
	10	84	62	49
	0.1	69	51	26
	0.25	62	54	26
$C(0, 1)$	1	49	48	25
	4	64	52	25
	10	80	62	30

Power (×1000) table: $\alpha = 0.05$, $m = 25$, and $n = 20$

F	Δ	W	$\hat{W}$	Welch
	0.1	195	209	207
$N(0, 1)$	0.2	506	520	523
	0.3	851	860	870
	0.1	150	159	128
Double Exp	0.2	403	411	337
	0.3	791	797	699
	0.1	140	145	49
$C(0, 1)$	0.2	348	352	97
	0.3	726	723	88

Notice the nonrobustness of Welch's test, which we had commented on earlier.

25.6 Robustness of the Mann-Whitney Test

A natural question is: What happens to the performance of the Mann-Whitney test itself under general F and G? Here we consider only the asymptotic size of the test. Similar calculations can be done to explore the robustness of the asymptotic power.

The true size of W is

$$\begin{aligned} P_{F,G}(W > k_{m,n}(\alpha)) &\approx P_{F,G}\left(\frac{W - mn/2}{\sqrt{mn(m+n+1)/12}} > z_\alpha\right) \\ &= P_{F,G}\left(\frac{W - mn\theta_{F,G} + mn(\theta_{F,G} - 1/2)}{\sqrt{mn(m+n+1)/12}} > z_\alpha\right). \end{aligned}$$

Suppose now that $\theta_{F,G} = 1/2$. As mentioned above, symmetry of each of F and G will imply $\theta_{F,G} = 1/2$ under H_0. In such a case,

$$\begin{aligned} \alpha_{\text{true}} &\equiv P_{F,G}(W > k_{m,n}(\alpha)) \\ &\approx P_{F,G}\left(\frac{W - mn\theta_{F,G}}{\sqrt{mn(m+n+1)/12}} > z_\alpha\right) \\ &= P_{F,G}\left(\frac{W - mn\theta_{F,G}}{\sqrt{v(p_1, p_2, p_3)}} \times \frac{\sqrt{v(p_1, p_2, p_3)}}{\sqrt{mn(m+n+1)/12}} > z_\alpha\right) \end{aligned}$$

where

$$v(p_1, p_2, p_3) = mn(p_1 - p_2^2) + mn(n-1)(p_2 - p_1^2) + mn(m-1)(p_3 - p_1^2),$$

where $v(p_1, p_2, p_3)$ denotes the exact variance of W and the p_i are as defined earlier in Section 25.3. Then the right-hand side has the limit

$$\longrightarrow 1 - \Phi\left(\frac{z_\alpha}{\sqrt{12[\lambda(p_3 - p_1^2) + (1-\lambda)(p_2 - p_1^2)]}}\right) \text{ as } m, n \to \infty,$$

and $\lambda = \lim_{m,n\to\infty} \frac{m}{m+n}$. Notice that, in general, this is not equal to α.

Example 25.5 In the case where $G(y) = F(y/\sigma)$, where F is symmetric and absolutely continuous, it follows from the known formulas for p_i that

$$\lim_\sigma (p_2 - p_1^2) = \begin{cases} \frac{1}{4}, & \sigma \to 0 \\ 0, & \sigma \to \infty, \end{cases}$$

$$\lim_\sigma (p_3 - p_1^2) = \begin{cases} 0, & \sigma \to 0 \\ \frac{1}{4}, & \sigma \to \infty \end{cases}$$

(see below for a proof). Plugging into the formula above for α_{true}, if $G(y) = F(y/\sigma)$, then

$$\lim_\sigma \alpha_{\text{true}} = \begin{cases} 1 - \Phi\left(\frac{z_\alpha}{\sqrt{3(1-\lambda)}}\right), & \sigma \to 0 \\ 1 - \Phi\left(\frac{z_\alpha}{\sqrt{3\lambda}}\right), & \sigma \to \infty \end{cases},$$

and the limit is between these values for $0 < \sigma < \infty$. If, for example, $\alpha = 0.05$ and $\lambda = 1/2$, then $\lim_\sigma \alpha_{\text{true}}$ varies between 0.05 and 0.087, which is reasonably robust.

For illustration, we prove the limiting behavior of α_{true} as $\sigma \to 0, \infty$ more carefully. Since we assume $G(y) = F(y/\sigma)$ and F, G are symmetric, it follows that $\theta_{F,G} = \int F\,dG = 1/2$. To evaluate the limits of α_{true}, we need only evaluate the limits of $p_2 - p_1^2$ and $p_3 - p_1^2$ as $\sigma \to 0, \infty$. First,

$$\begin{aligned} p_2 = P_{F,G}(Y_1 > X_1, Y_2 > X_1) &= \int (1 - G(x))^2\,dF(x) \\ &= \int (1 - F(x/\sigma))^2\,dF(x) \\ &= \int_{x>0} (1 - F(x/\sigma))^2\,dF(x) + \int_{x<0} (1 - F(x/\sigma))^2\,dF(x). \end{aligned}$$

Now, $F(x/\sigma) \to 1$ as $\sigma \to 0$ for all $x > 0$, and $F(x/\sigma) \to 0$ as $\sigma \to \infty$ for all $x < 0$. So, by the Lebesgue dominated convergence theorem (DCT), as $\sigma \to 0$, we get $p_2 \to 0 + 1/2 = 1/2$. If $\sigma \to \infty$, then for all x, $F(x/\sigma) \to 1/2$, so, again by the DCT, $p_2 \to 1/4$. Next, for p_3, when $\sigma \to 0$, we get

$$p_3 = \int F^2(x)\,dG(x) = \int F^2(\sigma x)\,dF(x) \to \frac{1}{4}.$$

When $\sigma \to \infty$,

$$\begin{aligned}\int F^2(\sigma x)\,dF(x) &= \int_{x>0} F^2(\sigma x)\,dF(x) + \int_{x<0} F^2(\sigma x)\,dF(x) \\ &\longrightarrow \frac{1}{2} + 0 = \frac{1}{2}.\end{aligned}$$

Since $p_1 = 1/2$, plugging into the formula for $\lim_\sigma \alpha_{\text{true}}$, we get the desired result,

$$\lim_\sigma \alpha_{\text{true}} = \begin{cases} 1 - \Phi\left(\frac{z_\alpha}{\sqrt{3(1-\lambda)}}\right), & \sigma \to 0 \\ 1 - \Phi\left(\frac{z_\alpha}{\sqrt{3\lambda}}\right), & \sigma \to \infty. \end{cases}$$

25.7 Exercises

Exercise 25.1 Give a rigorous proof that the two-sample t-statistic converges to $N(0, 1)$ in distribution if the variances are equal and finite.

Exercise 25.2 Simulate a sample of size $m = n = 25$ from the $N(0, 1)$ and the $N(0, 10)$ distributions, and compute the MLEs of the common mean and the two variances.

Exercise 25.3 * Give a necessary and sufficient condition that the cubic equation for finding the MLE of the common mean of two normal populations has one real root.

Exercise 25.4 * For each of the following cases, simulate the random degree of freedom of Welch's test: $m = n = 20, \sigma_1 = \sigma_2 = 1; m = 10, n = 50, \sigma_1 = \sigma_2 = 1; m = n = 20, \sigma_1 = 3, \sigma_2 = 1$.

Exercise 25.5 Compare by a simulation the power function of Welch's test with the two-sample t-test when the populations are normal with variances $1, 4$ and the sample sizes are $m = n = 10, 20$. Use a small grid for the values of the means.

Exercise 25.6 * Derive an expression for what Welch's degree of freedom would have been if he had tried to match a percentile instead of the second moment.

Exercise 25.7 Use the recursion relation given in the text to analytically write the distribution of the Mann-Whitney statistic when $m = n = 3$, assuming that the null is true.

Exercise 25.8 * Give a rigorous proof that the Mann-Whitney test is consistent under the condition stated in the text.

Exercise 25.9 * Analytically find the mean and the variance of the Mann-Whitney statistic under a normal shift model.

Exercise 25.10 For the normal location-scale model, approximate the true type I error rate of Hettmansperger's conservative test and investigate when it starts to diverge from the nominal value .05 with $m = n = 20, 30, 50$.

Exercise 25.11 * What is the limiting type I error of the Mann-Whitney test when the null density is a normal and the alternative is a uniform? Use the general expression given in the text.

Exercise 25.12 * What is the limiting type I error of the Mann-Whitney test when the null density is a normal and the alternative is an exponential?

References

Babu, G.J. and Padmanabhan, A.R. (2002). Resampling methods for the nonparametric Behrens-Fisher problem, Sankhya, special issue in memory of D. Basu, 64(3), Pt. I, 678–692.
Best, D.J. and Rayner, J.C. (1987). Welch's approximate solution to the Behrens- Fisher problem, Technometrics, 29(2), 205–210.
Brunner, E. and Munzel, U. (2000). The nonparametric Behrens-Fisher problem: asymptotic theory and a small-sample approximation, Biometrical J., 1, 17–25.
Fligner, M.A. and Policello, G.E. (1981). Robust rank procedures for the Behrens-Fisher problem, J. Am. Stat. Assoc., 76, 162–168.
Ghosh, M. and Kim, Y.H. (2001). The Behrens-Fisher problem revisited: a Bayes-Frequentist synthesis, Can. J. Stat., 29(1), 5–17.
Hajek, J. and Sidak, Z. (1967). *Theory of Rank Tests*, Academic Press, New York.
Hettmansperger, T. (1973). A large sample conservative test for location with unknown scale parameters, J. Am. Stat. Assoc., 68, 466–468.
Hettmansperger, T. (1984). *Statistical Inference Based on Ranks*, John Wiley, New York.
Johnson, R.A. and Weerahandi, S. (1988). A Bayesian solution to the multivariate Behrens-Fisher problem, J. Am. Stat. Assoc., 83, 145–149.
Lehmann, E.L. (1986). *Testing Statistical Hypotheses*, 2nd ed., John Wiley, New York.
Lehmann, E.L. and Romano, J. (2005). *Testing Statistical Hypotheses*, 3rd ed., Springer, New York.
Linnik, J.V. (1963). On the Behrens-Fisher problem, Bull. Inst. Int. Stat., 40, 833–841.

Mann, H.B. and Whitney, D.R. (1947). On a test whether one of two random variables is stochastically larger than the other, Ann. Math. Stat., 18, 50–60.
Milton, R.C. (1964). An extended table of critical values of the Mann-Whitney (Wilcoxon) two-sample statistic, J. Am. Stat. Assoc., 59, 925–934.
Pfanzagl, J. (1974). On the Behrens-Fisher problem, Biometrika, 61, 39–47.
Randles, R.H. and Wolfe, D.A. (1979). *Introduction to the Theory of Nonparametric Statistics*, John Wiley, New York.
Scheffe, H. (1970). Practical solutions of the Behrens-Fisher problem, J. Am. Stat. Assoc., 65, 1501–1508.
Serfling, R. (1980). *Approximation Theorems of Mathematical Statistics*, John Wiley, New York.
Wang, Y.Y. (1971). Probabilities of the type I errors of the Welch test for the Behrens-Fisher problem, J. Am. Stat. Assoc., 66, 605–608.
Welch, B. (1949). Further notes on Mrs. Aspin's tables, Biometrika, 36, 243–246.

Chapter 26
Goodness of Fit

Suppose $X_1, \ldots, X_n$ are iid observations from a distribution F on a Euclidean space, say $\mathbb{R}$. We discuss two types of goodness-of-fit problems: (i) test $H_0 : F = F_0$, a completely specified distribution and (ii) $F \in \mathcal{F}$, where $\mathcal{F}$ is a suitable family of distributions, possibly indexed by some finite-dimensional parameter. Problem (i) would be called the *simple goodness-of-fit problem* and problem (ii) the *composite goodness-of-fit problem* or, synonymously, *goodness of fit with estimated parameters*. It is the composite problem that is of greater interest in practice, although the simple problem can potentially arise in some situations. For example, one may have a hunch that F is uniform on some interval $[a, b]$ or that F is Bernoulli with parameter $\frac{1}{2}$. The simple goodness-of-fit problem has generated a vast amount of literature that has had a positive impact on the composite problem. So the methodologies and the theory for the simple case are worth looking at. We start with the simple goodness-of-fit problem and test statistics that use the empirical CDF $F_n(x)$ (EDF). Of the enormous amount of literature on goodness of fit, we recommend D'Agostino and Stephens (1986) for treatment of a variety of problems, including the simple null case, and Stephens (1993) as a very useful review, although primarily for the composite case, which is discussed in chapter 28. Stuart and Ord (1991) and Lehmann (1999) provide lucid presentations of some of the principal goodness-of-fit techniques.

Theory and methodology of goodness of fit were revolutionized with the advances in empirical process theory and the theory of central limit theorems on Banach spaces. The influence of these developments in what seems, at first glance, to be abstract probability theory on the goodness-of-fit literature was twofold. First, scattered results with case-specific proofs could be unified, with a very transparent understanding of what is really going on. Second, these developments led to development of new tests because tools were now available to work out the asymptotic theory of the new test procedures. We recommend del Barrio, Deheuvels, and van de Geer (2007)

A. DasGupta, *Asymptotic Theory of Statistics and Probability*,

for a comprehensive overview of these modern aspects of goodness of fit. In fact, we will discuss some of it in this chapter.

26.1 Kolmogorov-Smirnov and Other Tests Based on F_n

We know that, for large n, F_n is "close" to the true F. For example, by the Gilvenko-Cantelli theorem, $\sup |F_n(x) - F(x)| \xrightarrow{\text{a.s.}} 0$. So if $H_0 : F = F_0$ holds, then we should be able to test H_0 by studying the deviation between F_n and F_0. Any choice of a discrepancy measure between F_n and F_0 would result in a test. The utility of the test would depend on whether one can work out the distribution theory of the test statistic. A collection of discrepancy measures that have been proposed are the following:

$$\begin{aligned}
&D_n^+ = \sup_{-\infty<t<\infty}(F_n(t) - F_0(t)),\\
&D_n^- = \sup_{-\infty<t<\infty}(F_0(t) - F_n(t)) = -\inf_{-\infty<t<\infty}(F_n(t) - F_0(t)),\\
&D_n = \sup_{-\infty<t<\infty} |F_n(t) - F_0(t)| = \max(D_n^+, D_n^-),\\
&V_n = D_n^+ + D_n^-,\\
&C_n = \int (F_n(t) - F_0(t))^2 dF_0(t),\\
&A_n = \int \tfrac{(F_n(t)-F_0(t))^2}{F_0(t)(1-F_0(t))} dF_0(t),\\
&w_n = w_{n,k,g} = \int (F_n(t) - F_0(t))^k g(F_0(t)) dF_0(t),\\
&D_n(g) = \sup_{-\infty<t<\infty} \tfrac{|F_n(t)-F_0(t)|}{g(F_0(t))},
\end{aligned}$$

where $g : [0, 1] \to \mathbb{R}^+$ is some fixed function and $k \geq 1$ is a fixed positive integer. The tests corresponding to D_n, V_n, C_n, and A_n are respectively known as the Kolmogorov-Smirnov, the Kuiper, the Cramér-von Mises, and the Anderson-Darling tests. The tests corresponding to w_n and $D_n(g)$ are usually referred to as weighted Cramér-von Mises and weighted Kolmogorov-Smirnov tests.

26.2 Computational Formulas

D_n, C_n, and A_n are the most common among the test statistics listed above. It can be shown that D_n, C_n, and A_n are equal to the following simple expressions. Let $X_{(1)} < X_{(2)} < \cdots < X_{(n)}$ be the order statistics of the sample and let $U_i = F_0(X_{(i)})$. Then, assuming F_0 is continuous,

$$D_n = \max_{1\leq i\leq n} \max \left\{ \frac{i}{n} - U_{(i)}, U_{(i)} - \frac{i-1}{n} \right\},$$

$$C_n = \frac{1}{12n} + \sum_{i=1}^{n} \left(U_{(i)} - \frac{2i-1}{n} \right)^2,$$

$$A_n = -n - \frac{1}{n} \left[\sum_{i=1}^{n} (2i-1)(\log U_{(i)} + \log(1 - U_{(n-i+1)})) \right].$$

Remark. It is clear from these computational formulas that, for every fixed n, the sampling distributions of D_n, C_n, and A_n under F_0 do not depend on F_0, provided F_0 is continuous. Indeed, one can prove directly by making the quantile transformation $U = F_0(X)$ that all the test statistics listed above have sampling distributions (under H_0) independent of F_0, provided F_0 is continuous. For small n, the true sampling distributions can be worked out exactly by discrete enumeration. The quantiles for some particular levels have been numerically worked out for a range of values of n. Accurate approximations to the 95th and 99th percentiles of D_n for $n \geq 80$ are $\frac{1.358}{\sqrt{n}}$ and $\frac{1.628}{\sqrt{n}}$.

26.3 Some Heuristics

For an iid $U[0, 1]$ sample $Z_1, \ldots, Z_n$, let $U_n(t) = \frac{1}{n} \sum_{i=1}^{n} I_{Z_i \leq t}$. Recall that we call $U_n(t)$ a *uniform empirical process*. Suppose F_0 is a fixed CDF on $\mathcal{R}$, and $X_1, \ldots, X_n$ are iid samples from F_0. Then, defining $Z_i = F_0(X_i)$, $Z_1, \ldots, Z_n$ are iid $U[0, 1]$. Therefore,

$$\begin{aligned}
&\sup_t |F_n(t) - F_0(t)| = \sup_t \left| \tfrac{1}{n} \sum I_{X_i \leq t} - F_0(t) \right| \\
&= \sup_t \left| \tfrac{1}{n} \sum I_{F_0(X_i) \leq F_0(t)} - F_0(t) \right| = \sup_t \left| \tfrac{1}{n} \sum I_{Z_i \leq F_0(t)} - F_0(t) \right| \\
&\stackrel{\mathcal{L}}{=} \sup_t |U_n(F_0(t)) - F_0(t)| = \sup_{0 \leq t \leq 1} |U_n(t) - t|,
\end{aligned}$$

and therefore, for every n, $\sqrt{n} \sup_t |F_n(t) - F_0(t)| \stackrel{\mathcal{L}}{=} \sqrt{n} \sup_{0 \leq t \leq 1} |U_n(t) - t|$ under F_0. So, for every n, D_n has the same distribution as $\sqrt{n} \sup_{0 \leq t \leq 1} |U_n(t) - t|$. Define $X_n(t) = \sqrt{n}(U_n(t) - t), 0 \leq t \leq 1$. Recall from Chapter 12 that $X_n(0) = X_n(1) = 0$, and $X_n(t)$ converges to a Gaussian process, $B(t)$, with $E(B(t)) = 0, \forall t$ and $\text{cov}(B(s), B(t)) = s \wedge t - st, 0 \leq s, t \leq 1$, the *Brownian bridge* on $[0, 1]$. By the invariance principle, the distribution of $D_n = \sqrt{n} \sup_{0 \leq t \leq 1} |U_n(t) - t|$ converges to the distribution of $\sup_{0 \leq t \leq 1} |B(t)|$. We have seen in Chapter 12 that a rigorous development requires the use of weak convergence theory on metric spaces.

26.4 Asymptotic Null Distributions of D_n, C_n, A_n, and V_n

Asymptotic theory of EDF-based tests is now commonly handled by using empirical process techniques and weak convergence theory on metric spaces. We recommend Shorack and Wellner (1986), Pollard (1989), Martynov (1992), Billingsley (1999), and del Barrio, Deheuvels, and van de Geer (2007), apart from Chapter 12 in this text, for details on techniques, statistical aspects, and concrete applications. The following fundamental results follow as consequences of the invariance principle for empirical processes, which we treated in Chapter 12.

Theorem 26.1 Let $X_1, X_2, \ldots \stackrel{\text{iid}}{\sim} F_0$, and let $D_n = \sup_t |F_n(t) - F_0(t)|$ and $C_n = \int_{-\infty}^{\infty}(F_n(t) - F_0(t))^2 dF_0(t)$. Then, assuming that F_0 is continuous,

$$\sqrt{n}D_n \stackrel{\mathcal{L}}{\Rightarrow} \sup_{0\le t\le 1} |B(t)|,$$

$$nC_n \stackrel{\mathcal{L}}{\Rightarrow} \int_0^1 B^2(t)dt, \; nA_n \stackrel{\mathcal{L}}{\Rightarrow} \int_0^1 \frac{B^2(t)}{t(1-t)}dt,$$

$$P_{F_0}(\sqrt{n}D_n^+ \le \lambda_2, \sqrt{n}D_n^- \le \lambda_1) \to P(-\lambda_1 \le \inf_{0\le t\le 1} B(t) \le \sup_{0\le t\le 1} B(t) \le \lambda_2),$$

$$\sqrt{n}V_n \stackrel{\mathcal{L}}{\Rightarrow} \sup_{0\le t\le 1} B(t) - \inf_{0\le t\le 1} B(t).$$

Now, the question reduces to whether one can find the distributions of these four functionals of $B(.)$. Fortunately, the answer is affirmative. The distributions of the first and fourth functionals (i.e., the distributions of the supremum of the absolute value and of the range) can be found by applying the *reflection principle* (see Chapter 12). The two other statistics are quadratic functionals of $B(.)$, and their distributions can be found by using the *Karhunen-Loéve expansion* (see Chapter 12) of $B(t)$ and then writing the integrals in these two functionals as a linear combination of independent chi-square random variables. Since the characteristic function of a chi-square distribution is known, one can also write the characteristic function of the Brownian quadratic functional itself. Rather remarkably, a Fourier inversion can be done, and one can arrive at closed-form expressions for the CDFs that are the CDFs of the limiting distributions we want. See del Barrio, Deheuvels, and van de Geer (2007) for the technical details. We record below some of these closed-form expressions for the limiting CDFs.

Corollary 26.1

$$\lim_{n} P_{F_0}(\sqrt{n} D_n \le \lambda) = 1 - 2\sum_{j=1}^{\infty}(-1)^{j-1}e^{-2j^2\lambda^2},$$

$$\lim_{n\to\infty} P_{F_0}(nC_n > x) = \frac{1}{\pi}\sum_{j=1}^{\infty}(-1)^{j+1}\int_{(2j-1)^2\pi^2}^{4j^2\pi^2}\sqrt{\frac{-\sqrt{y}}{\sin(\sqrt{y})}}\,\frac{e^{-\frac{xy}{2}}}{y}dy,$$

$$\lim_{n\to\infty} P_{F_0}(\sqrt{n} D_n^+ \le \lambda_2, \sqrt{n} D_n^- \le \lambda_1)$$
$$= 1 - \sum_{k=1}^{\infty}\left\{e^{-2[k\lambda_2+(k-1)\lambda_1]^2} + e^{-2[(k-1)\lambda_2+k\lambda_1]^2} - 2e^{-2k^2(\lambda_1+\lambda_2)^2}\right\}.$$

Remark. The CDF of the limiting distribution of $\sqrt{n}V_n$ is the CDF of the sum in the joint CDF provided in the last part above. An expression for the CDF of the limiting distribution of nA_n can also be found on using the fact that it is the CDF of the infinite linear combination $\sum_{j=1}^{\infty}\frac{Y_j}{j(j+1)}$, Y_j being iid chi-squares with one degree of freedom.

26.5 Consistency and Distributions under Alternatives

The tests introduced above based on the empirical CDF F_n all have the pleasant property that they are consistent against any alternative $F \neq F_0$. For example, the Kolmogorov-Smirnov statistic D_n has the property that $P_F(\sqrt{n}D_n > G_n^{-1}(1-\alpha)) \to 1, \forall F \neq F_0$, where $G_n^{-1}(1-\alpha)$ is the $(1-\alpha)$th quantile of the distribution of $\sqrt{n}D_n$ under F_0. To explain heuristically why this should be the case, consider a CDF $F_1 \neq F_0$, so that there exists a such that $F_1(a) \neq F_0(a)$. Let us suppose that $F_1(a) > F_0(a)$. First note that $G_n^{-1}(1-\alpha) \to \lambda$, where λ satisfies $P(\sup_{0\le t\le 1}|B(t)| \le \lambda) = 1-\alpha$. So

$$\begin{aligned}
&P_{F_1}\left(\sqrt{n}D_n > G_n^{-1}(1-\alpha)\right)\\
&= P_{F_1}\left(\sup_t|\sqrt{n}(F_n(t) - F_0(t))| > G_n^{-1}(1-\alpha)\right)\\
&= P_{F_1}\left(\sup_t|\sqrt{n}(F_n(t) - F_1(t)) + \sqrt{n}(F_1(t) - F_0(t))| > G_n^{-1}(1-\alpha)\right)\\
&\ge P_{F_1}\left(|\sqrt{n}(F_n(a) - F_1(a)) + \sqrt{n}(F_1(a) - F_0(a))| > G_n^{-1}(1-\alpha)\right)\\
&\to 1
\end{aligned}$$

as $n \to \infty$ since $\sqrt{n}(F_n(a) - F_1(a)) = O_p(1)$ under F_1, $\sqrt{n}(F_1(a) - F_0(a)) \to \infty$ and, as stated above, $G_n^{-1}(1-\alpha) = O(1)$.

Remark. The same argument establishes the consistency of the other EDF (empirical distribution function)-based tests against all alternatives. In contrast, we will later see that chi-square goodness-of-fit tests cannot be consistent against all alternatives.

The invariance principle argument that we used to derive the limit distributions under H_0 also produces the limit distributions under F, a specified alternative. The limit distributions are still the distributions of appropriate functionals of suitable Gaussian processes (see Raghavachari (1973)). First we need some notation. Let F be a specified CDF different from F_0. Without loss of generality, we assume $F(0) = 0$ and $F(1) = 1$ and $F_0(t) = t$. Let

$$\begin{aligned}
\alpha &= \sup_{0 \le t \le 1} |F(t) - F_0(t)| = \sup_{0 \le t \le 1} |F(t) - t|, \\
\alpha^+ &= \sup_{0 \le t \le 1} (F(t) - F_0(t)) = \sup_{0 \le t \le 1} (F(t) - t), \\
\alpha^- &= \inf_{0 \le t \le 1} (F(t) - F_0(t)) = \inf_{0 \le t \le 1} (F(t) - t), \\
K_1 &= \{0 \le t \le 1 : F(t) - t = \alpha\}, \\
K_2 &= \{0 \le t \le 1 : t - F(t) = \alpha\}, \\
K^+ &= \{0 \le t \le 1 : F(t) - t = \alpha^+\}, \\
K^- &= \{0 \le t \le 1 : F(t) - t = \alpha^-\}.
\end{aligned}$$

Also, let W_F denote a Gaussian process on $[0, 1]$ with $W_F(0) = 0$, $E(W_F(t)) = 0$, and $\text{cov}(W_F(s), W_F(t)) = F(s) \wedge F(t) - F(s)F(t), 0 \le s \le t \le 1$.

Theorem 26.2 $P_F(\sqrt{n}(D_n - \alpha) \le \lambda) \to P\left(\sup_{t \in K_1} W_F(t) \le \lambda, \inf_{t \in K_2} W_F(t) \ge -\lambda\right)$.

Remark. This result also gives a proof of the consistency of the test based on D_n. For given $0 < \gamma < 1$, $P_F(\sqrt{n} D_n < G_n^{-1}(1 - \gamma)) = P_F(\sqrt{n}(D_n - \alpha) < G_n^{-1}(1 - \gamma) - \alpha\sqrt{n}) \to 0$ from the theorem above as $n \to \infty$ since $G_n^{-1}(1 - \gamma) = O(1)$ and $\alpha > 0$.

One can likewise find the limiting distributions of the other EDF-based statistics (e.g., D_n^+ and D_n^-) under an alternative. For example, $P_F(\sqrt{n}(D_n^+ - \alpha^+) \le \lambda) \to P(\sup_{t \in K^+} W_F(t) \le \lambda)$. As regards the Kuiper statistic V_n, $P_F(\sqrt{n}(V_n - (\alpha^+ + \alpha^-))) \le \lambda \to P(\sup_{t \in K^+} W_F(t) - \inf_{t \in K^-} W_F(t) \le \lambda)$.

26.6 Finite Sample Distributions and Other EDF-Based Tests

Kolmogorov himself studied the problem of the finite sample distribution of D_n under H_0 (Kolmogorov (1933)). He gave recurrence relations for finding the pmf of D_n. Wald and Wolfowitz (1940, 1941) gave exact formulas easy to use for small n. Since then, the exact percentiles and exact CDFs have been numerically evaluated and extensively tabulated. For the reader's convenience, we report a short table of exact percentiles of D_n for some selected values of n.

n	95th Percentile	99th Percentile
20	.294	.352
21	.287	.344
22	.281	.337
23	.275	.330
24	.269	.323
25	.264	.317
26	.259	.311
27	.254	.305
28	.250	.300
29	.246	.295
30	.242	.290
35	.224	.269
40	.210	.252
>40	$\frac{1.36}{\sqrt{n}}$	$\frac{1.63}{\sqrt{n}}$

Smirnov (1941) found the exact distribution of the one-sided statistic D_n^+. Indeed, he found the limiting distribution of $\sqrt{n}D_n^+$ under the null by taking the exact distribution for given n and then finding the pointwise limit. Smirnov's formula given n is

$$P_{H_0}(D_n^+ > \epsilon) = (1-\epsilon)^n + \epsilon \sum_{j=1}^{[n(1-\epsilon)]} \binom{j}{n} \left(1 - \epsilon - \frac{j}{n}\right)^{n-j} \left(\epsilon + \frac{j}{n}\right)^{j-1}.$$

By symmetry, therefore, one also knows the exact distribution of D_n^- for any given n. Weighted versions of D_n, C_n, and A_n are also sometimes used. In particular, the weighted Kolmogorov-Smirnov statistic is $D_n(g) = \sup_{-\infty<x<\infty} \frac{|F_n(t)-F_0(t)|}{g(F_0(t))}$, and the weighted Anderson-Darling statistic is $w_n(g) = \int \frac{(F_n(t)-F_0(t))^2}{g(F_0(t))} dF_0(t)$ for some suitable function g. For specific types of alternatives, the weighted versions provide greater power than the original unweighted versions, if the weighting function g is properly chosen. It is *not true* that the weighted versions converge in law to what would seem to be the obvious limit for arbitrary g. In fact, the question of weak convergence of the weighted versions is surprisingly delicate. Here is the precise theorem; see del Barrio, Deheuvels, and van de Geer (2007) for a proof.

Theorem 26.3 (a) Let g be a strictly positive function on (0, 1), nondecreasing in a neighborhood of $t = 0$ and nonincreasing in a neighborhood of $t = 1$. Assume that, for some $c > 0$, $\int_0^1 \frac{1}{t(1-t)} e^{-c\frac{g^2(t)}{t(1-t)}} dt < \infty$. Then, $\sqrt{n} D_n(g) \stackrel{\mathcal{L}}{\Rightarrow} \sup_{0<t<1} \frac{|B(t)|}{g(t)}$, where $B(t)$ is a Brownian bridge on [0, 1].

(b) Let g be a strictly positive function on (0, 1). Assume that $\int_0^1 \frac{t(1-t)}{g(t)} dt < \infty$. Then, $n w_n(g) \stackrel{\mathcal{L}}{\Rightarrow} \int_0^1 \frac{B^2(t)}{g(t)} dt$, where $B(t)$ is a Brownian bridge on [0, 1].

Remark. See Csörgo (1986) for part (a) and Araujo and Giné (1980) for part (b).

26.7 The Berk-Jones Procedure

Berk and Jones (1979) proposed an intuitively appealing method of testing the simple goodness-of-fit null hypothesis $F = F_0$ for some specified continuous F_0 in the one-dimensional iid situation. It is also based on the empirical CDF, and quite a bit of useful work has been done on finite sample distributions of the Berk-Jones test statistic. It has also led to subsequent developments of other tests for the simple goodness-of-fit problem as generalizations of the Berk-Jones idea. On the other hand, there are some unusual aspects about the asymptotic behavior of the Berk-Jones test statistic and the statistics corresponding to its generalizations. We discuss the Berk-Jones test in this section and certain generalizations in the next section.

The Berk-Jones method is to transform the simple goodness-of-fit problem into a family of binomial testing problems. More specifically, if the true underlying CDF is F, then for any given x, $nF_n(x) \sim \text{Bin}(n, F(x))$. Suppressing the x and writing p for $F(x)$ and p_0 for $F_0(x)$, for the given x, we want to test $p = p_0$. Since F_0 is specified, by the usual quantile transform method, we may assume that the observations take values in [0, 1] and that F_0 is the CDF of the $U[0, 1]$ distribution. We can use a likelihood ratio test corresponding to a two-sided alternative to test this hypothesis. It will require maximization of the binomial likelihood function over all values of p that corresponds to maximization over $F(x)$, with x being fixed, while F is an arbitrary CDF. The likelihood is maximized at $F(x) = F_n(x)$, resulting in the likelihood ratio statistic

$$\lambda_n(x) = \frac{F_n(x)^{nF_n(x)}(1 - F_n(x))^{n-nF_n(x)}}{F_0(x)^{nF_n(x)}(1 - F_0(x))^{n-nF_n(x)}}$$

$$= \left(\frac{F_n(x)}{F_0(x)}\right)^{nF_n(x)} \left(\frac{1-F_n(x)}{1-F_0(x)}\right)^{n-nF_n(x)}.$$

But, of course, the original problem is to test that $F(x) = F_0(x) \forall x$. So, it would make sense to take a supremum of the log-likelihood ratio statistics over x. The Berk-Jones statistic is

$$R_n = n^{-1} \sup_{0 \le x \le 1} \log \lambda_n(x).$$

As always, the questions of interest are the asymptotic and fixed sample distributions of R_n under the null and, if possible, under suitable alternatives. We present some key available results on these questions below. The principal references are Berk and Jones (1979), Wellner and Koltchinskii (2003), and Jager and Wellner (2006).

To study the asymptotics of the Berk-Jones statistic, first it is useful to draw a connection between it and the Kullback-Leibler distance between Bernoulli distributions. Let $K(p, \theta) = p(\log p - \log \theta) + (1-p)(\log(1-p) - \log(1-\theta))$ be the Kullback-Leibler distance between the Bernoulli distributions with parameters p and θ. Then, it is easily seen that $\log \lambda_n(x) = K(F_n(x), F_0(x))$ and hence $R_n = \sup_{0 \le x \le 1} K(F_n(x), F_0(x))$. Properties of the Kullback-Leibler distance and empirical process theory are now brought together, in an entirely nontrivial way, to derive the limiting distribution of R_n under the null hypothesis. See Jager and Wellner (2006) for a proof of the next theorem.

Theorem 26.4 Let $c_n = 2 \log \log n + \frac{1}{2} \log \log \log n - \frac{1}{2} \log(4\pi)$, $b_n = \sqrt{2 \log \log n}$. Under $H_0 : F = F_0$, $nR_n - \frac{c_n^2}{2b_n^2} \stackrel{\mathcal{L}}{\Rightarrow} V$, where V has the CDF $e^{-4e^{-x}}, -\infty < x < \infty$.

Approximations to finite sample percentiles of nR_n are presented in Owen (1995). Somewhat simpler, but almost as accurate, approximations for the 95th percentile of nR_n are $\frac{c_n^2}{2b_n^2} - \log(-.25 \log .95)$.

26.8 φ-Divergences and the Jager-Wellner Tests

The Kullback-Leibler connection to the Berk-Jones statistic is usefully exploited to produce a more general family of tests for the simple goodness-of-fit problem in Jager and Wellner (2006). We describe these tests and the asymptotic distribution theory below. Some remarks about the efficiency of these tests are made at the end of this section.

The generalizations are obtained by considering generalizations of the $K(p, \theta)$ function above. The $K(p, \theta)$ function arises from the Kullback-Leibler distance, as we explained. The more general functions are obtained from distances more general than the Kullback-Leibler distance. In the information theory literature, these more general distances are known as φ-divergences. Let P_1, P_2 be two probability measures on some space, absolutely continuous with respect to a common measure λ (such an λ always exists). Let p_1, p_2 be the densities of P_1, P_2 with respect to λ, and let $g = \frac{p_2}{p_1} I_{p_1 > 0}$. Given a nonnegative convex function φ on the nonnegative reals, let $K_\varphi(P_1, P_2) = E_{P_1}[\varphi(g)]$. These are known as φ-divergence measures between a pair of probability distributions. See Csizár (1963) for the apparently first introduction of these divergences. Divergence measures have also been used usefully in estimation, and particularly robust estimation; one reference for an overview is Basu et al. (1998).

Some examples of the φ-function in common use are

$$\begin{aligned}\varphi(x) =& (x-1)\log x; \varphi(x) = -\log x; \varphi(x) = (\sqrt{x}-1)^2; \varphi(x)\\ =& |x-1|; \varphi(x) = -x^{1-t}, 0 < t < 1.\end{aligned}$$

For the purpose of writing tests for the goodness-of-fit problem, Jager and Wellner (2006) use the following one-parameter family of φ-functions:

$$\begin{aligned}\varphi_s(x) =& x - \log x - 1, \ s = 0;\\ \varphi_s(x) =& x \log x - x + 1, \ s = 1;\\ \varphi_s(x) =& \frac{1 - s + sx - x^s}{s(1-s)}, \ s \neq 0, 1.\end{aligned}$$

These functions result in the corresponding divergence measures

$$K_s(p, \theta) = \theta \varphi_s(p/\theta) + (1-\theta)\varphi_s((1-p)/(1-\theta)).$$

Accordingly, one has the family of test statistics

$$S_n(s) = \sup_{0 \le x \le 1} K_s(F_n(x), F_0(x)),$$

or, instead of taking suprema, one can take averages and get the test statistics

$$T_n(s) = \int_0^1 K_s(F_n(x), F_0(x))dx.$$

It is interesting to note that S_n and T_n generalize well-known tests for the simple goodness-of-fit problem; for example, in particular, $S_n(1)$ is the Berk-Jones statistic and $T_n(2)$ the integral form of the Anderson-Darling statistic.

The central limit theorem under the null for the families of test statistics $S_n(s)$, $T_n(s)$ is described in the following theorem; see Jager and Wellner (2006) for a proof.

Theorem 26.5 (a) Let b_n, c_n, V be as in the previous theorem. For $s \in [-1, 2]$, $nS_n(s) - \frac{c_n^2}{2b_n^2} \stackrel{\mathcal{L}}{\Rightarrow} V$ under H_0 as $n \to \infty$.

(b) For $s \in (-\infty, 2]$, $nT_n(s) \stackrel{\mathcal{L}}{\Rightarrow} \int_0^1 \frac{B^2(t)}{2t(1-t)} dt$ under H_0 as $n \to \infty$, where $B(t)$ is a Brownian bridge on $[0, 1]$.

The question of comparison naturally arises. We now have a large family of possible tests, all for the simple goodness-of-fit problem. Which one should one use? The natural comparison would be in terms of power. This can be done theoretically or by large-scale simulations. The theoretical study focuses on comparing Bahadur slopes of these various statistics. However, this is considerably more subtle than one would first imagine. The problem is that sometimes, depending on the exact alternative, when one intuitively expects the sequence of statistics to have an obvious almost sure limit, in reality it converges in law to a nondegenerate random variable. There is a boundary phenomenon going on. On one side of the boundary, there is an almost sure constant limit, while on the other side there is a nondegenerate weak limit. This would make comparing Bahadur slopes essentially meaningless. However, some qualitative understanding of power comparison has been achieved; see Berk and Jones (1979), Groeneboom and Shorack (1981), and Jager and Wellner (2006). Simulations are available in Jager (2006). Berk-Jones type supremum statistics appear to come out well in these theoretical studies and the simulations, but perhaps with a truncated supremum, over $x \in [X_{(1)}, X_{(n)}]$.

26.9 The Two-Sample Case

Suppose $X_i, i = 1, 2, \cdots, n$ are iid samples from some continuous CDF F_0 and $Y_i, i = 1, 2, \cdots, m$ are iid samples from some continuous CDF F, and all random variables are mutually independent. Without loss of generality, assume $F_0(t) = t, 0 \leq t \leq 1$, and that F is a CDF on $[0, 1]$. Let F_n and G_m denote the empirical CDFs of the X_i''s and the Y_i''s, respectively. Analogous to the one-sample case, one can define two- and one-sided Kolmogorov-Smirnov and Kuiper test statistics

$$
\begin{aligned}
D_{m,n} &= \sup_{0\le t\le 1} |F_n - G_m|, \\
D^+_{m,n} &= \sup_{0\le t\le 1} (F_n - G_m), \\
D^-_{m,n} &= \sup_{0\le t\le 1} (G_m - F_n) = -\inf_{0\le t\le 1} (F_n - G_m), \\
V_{m,n} &= D^+_{m,n} + D^-_{m,n}.
\end{aligned}
$$

There is substantial literature on the two-sample equality-of-distributions problem. In particular, see Kiefer (1959), Anderson (1962), and Hodges (1958). As in the one-sample case, the multivariate problem is a lot harder, it being difficult to come up with distribution-free simple and intuitive tests, even in the continuous case. See Bickel (1968), and Weiss (1960) for some results.

The limiting distribution of the two-sided Kolmogorov-Smirnov (KS) statistic is as follows.

Theorem 26.6 Let $X_i, 1 \le i \le n \overset{\text{iid}}{\sim} F_0, Y_j, 1 \le j \le m \overset{\text{iid}}{\sim} F$, where F_0, F are continuous CDFs. Consider testing $H_0 : F = F_0$. Then,

$$
\begin{aligned}
\lim_{m,n\to\infty} P_{H_0}\left(\sqrt{\frac{mn}{m+n}} D_{m,n} \le \lambda\right) &= P(\sup_{0\le t\le 1} |B(t)| \le \lambda) \\
&= 1 - 2\sum_{k=1}^{\infty} (-1)^{k-1} e^{-2k^2\lambda^2},
\end{aligned}
$$

provided that, for some $0 < \gamma < 1$, $\frac{m}{m+n} \to \gamma$.

Remark. Notice that the limiting distribution of the two-sample two-sided K-S statistic under H_0 is the same as that of the one-sample two-sided K-S statistic. The reason is $\sqrt{\frac{mn}{m+n}}(F_n(t) - G_m(t)) = \sqrt{\frac{mn}{m+n}}(F_n(t) - t - (G_m(t) - t)) = \sqrt{\frac{mn}{m+n}}(F_n(t) - t) - \sqrt{\frac{mn}{m+n}}(G_m(t) - t) \overset{\mathcal{L}}{\Rightarrow} \sqrt{1-\gamma} B_1(t) - \sqrt{\gamma} B_2(t)$, where $B_1(.)$ and $B_2(.)$ are two independent Brownian bridges. But $\sqrt{1-\gamma} B_1(t) - \sqrt{\gamma} B_2(t)$ is another Brownian bridge. Therefore, by our usual continuous-mapping argument, $\sqrt{\frac{mn}{m+n}} D_{m,n} \overset{\mathcal{L}}{\Rightarrow} \sup_{0\le t\le 1} |B(t)|$. The asymptotic null distribution of the two-sample Kuiper statistic is also easily found and is stated next.

Theorem 26.7 Assume the same conditions as in the previous theorem. Let $B(t)$ be a standard Brownian bridge on [0,1]. Then

$$\lim_{m,n\to\infty} P_{F_0}\left(\sqrt{\frac{mn}{m+n}}V_{m,n} \le \lambda\right) = P(\sup_{0\le t\le 1} B(t) - \inf_{0\le t\le 1} B(t) \le \lambda).$$

Remark. Notice that again, in the two-sample case, the asymptotic null distribution of the Kuiper statistic is the same as that in the one-sample case. The reason is the same as the explanation given above for the case of the K-S statistic.

The asymptotic distributions of $D_{m,n}$ and $V_{m,n}$ under an alternative are also known, and we describe them below.

Theorem 26.8 Suppose $\frac{m}{m+n} \to \gamma, 0 < \gamma < 1$. Then

$$\begin{aligned}
&\lim_{m,n\to\infty} P_F\left(\sqrt{\frac{mn}{m+n}}(D_{m,n} - \alpha) \le \lambda\right) \\
&= P\left(\sup_{t\in K_1}(\sqrt{\gamma}W_F(t) - \sqrt{1-\gamma}B(t)) \le \lambda,\right. \\
&\qquad \left.\inf_{t\in K_2}(\sqrt{1-\gamma}W_F(t) - \sqrt{\gamma}B(t)) \ge -\lambda\right),
\end{aligned}$$

where α, K_1, K_2, $W_F(t)$ are as in Section 26.5 and $W_F(t)$ and $B(t)$ are independent.

Theorem 26.9 Suppose $\frac{m}{m+n} \to \gamma, 0 < \gamma < 1$. Then

$$\begin{aligned}
&\lim_{m,n\to\infty} P_F\left(\sqrt{\frac{mn}{m+n}}(V_{m,n} - \alpha^+ + \alpha^-) \le \lambda\right) \\
&= P\left(\sup_{t\in K^+}(\sqrt{\gamma}W_F(t) - \sqrt{1-\gamma}B(t))\right. \\
&\qquad \left.- \inf_{t\in K^-}(\sqrt{\gamma}W_F(t) - \sqrt{1-\gamma}B(t)) \le \lambda\right),
\end{aligned}$$

where $K^{\pm}$ are as in Section 26.5 and $W_F(t)$ and $B(t)$ are again independent.

Remark. See Raghavachari (1973) for details and proofs of the last two theorems in this section. In examples, one of the sets K_1 and K_2 and one of the sets K^+, K^- may be empty and the other one a singleton set. This will facilitate analytical calculations of the asymptotic CDFs in Theorems 26.8 and 26.9. In general, analytical calculation may be cumbersome. For discrete cases, it is wrong to use the statistics; however, if our "wrong" p-value is very small, the true p-value is even smaller than the wrong one, and so it is probably safe to reject the hypothesis in such a case.

26.10 Tests for Normality

Because of its obvious practical importance, there has been a substantial amount of work on devising tests for normality. Thus, suppose $X_1, X_2, \ldots, X_n$ are iid observations from a CDF F on the real line. The problem is to test that F belongs to the family of normal distributions. Although we will discuss modifications of the Kolmogorov-Smirnov and the chi-square tests for testing this hypothesis in Chapters 27 and 28, we will describe some other fairly popular tests for normality here due to the practical interest in the problem.

The Q-Q Plot The Q-Q plot is a hugely popular graphical method for testing for normality. It appears to have been invented by the research group at Bell Labs. See Gnanadesikan (1997). The simple rationale is that the quantile function of a general normal distribution satisfies $Q(\alpha) = \mu + \sigma z_\alpha$, with obvious notation. So a plot of $Q(\alpha)$ against z_α would be linear, with an intercept of μ and a slope σ. With given data, the order statistics of the sample are plotted against the standard normal percentiles. This plot should be roughly linear if the data are truly normal. A visual assessment of linearity is then made. To avoid singularities, the plot consists of the pairs $(\Phi^{-1}(\frac{i}{n+1}), X_{(i)}), i = 1, 2, \ldots, n$ (or some such modification of $\Phi^{-1}(\frac{i}{n+1})$).

In the hands of a skilled analyst, the Q-Q plot can provide useful information about the nature of the true CDF from which the observations are coming. For example, it can give information about the tail and skewness of the distribution and about its unimodality. See Marden (1998, 2004). Brown et al. (2004) show that the basic assessment of linearity, however, is fundamentally unreliable, as most types of data would produce remarkably linear Q-Q plots, except for a detour only in the extreme tails. If this detour in the tails is brushed aside as unimportant, then the Q-Q plot becomes a worthless tool. Brown et al. give the following theorem.

Theorem 26.10 Let r_n denote the correlation coefficient computed from the bivariate pairs $(\Phi^{-1}(\frac{i}{n+1}), X_{(i)}), i = 1, 2, \ldots, n$. Let F denote the true CDF with finite variance σ^2. Then

$$\lim_{n \to \infty} r_n = \rho_F = \frac{\int_0^1 F^{-1}(\alpha)\Phi^{-1}(\alpha)d\alpha}{\sigma}, \text{ a.s.}$$

The a.s. limit is very close to 1 for all kinds of distributions, as can be seen in the following table.

F	$\lim r_n$
normal	1
uniform	.9772
double exponential	.9811
t_3	.9008
t_5	.9832
χ_5^2	.9577
exponential	.9032
tukey	.9706
logistic	.9663

They also show that if just 5% of the points from each tail are deleted, then the corresponding a.s. limits are virtually equal to 1 even for skewed data such as χ_5^2. Having said that, a formal test that uses essentially the correlation coefficient r_n above is an omnibus consistent test for normality. This test is described next.

The Shapiro-Francia-Wilk Test The Shapiro-Francia test is a slight modification of the wildly popular Shapiro-Wilk test but is easier to describe; see Shapiro and Wilk (1965) and de Wet and Ventner (1972). It rejects the hypothesis of normality when r_n is small. If the true CDF is nonnormal, then r_n on centering and norming by $\sqrt{n}$ has a limiting normal distribution. If the true CDF is normal, $r_n \to 1$ faster than $\sqrt{n}$. Against all nonnormal alternatives with a finite variance, the test is consistent; see Sarkadi (1985).

Theorem 26.11 (de Wet and Ventner (1972), Sarkadi (1985))

(a) If F does not belong to the family of normal distributions, then $\sqrt{n}(r_n - \rho_F) \stackrel{\mathcal{L}}{\Rightarrow} N(0, \tau_F^2)$, where ρ_F is as above and τ_F is a suitable functional of F.

(b) If F belongs to the family of normal distributions, then $n(1 - r_n) \stackrel{\mathcal{L}}{\Rightarrow} \sum_{i=1}^{\infty} c_i(W_i - 1)$, where W_i are iid χ_1^2 variables and c_i are suitable constants.

(c) If F does not belong to the family of normal distributions, then $\lim_{n\to\infty} P(r_n < 1 - \frac{c}{n}) = 1$ for any $c > 0$.

Among many other tests for normality available in the literature, tests based on the skewness and the kurtosis of the sample are quite popular; see the exercises. There is no clear-cut comparison between these tests without focusing on the type of alternative for which good power is desired.

26.11 Exercises

Exercise 26.1 * For the dataset −1.88, −1.71, −1.40, −0.95, 0.22, 1.18, 1.25, 1.41, 1.70, 1.97, calculate the values of D_n^+, D_n^-, D_n, C_n, A_n, and V_n when the null distribution is $F_0 = N(0, 1)$, $F_0 =$ Double Exp(0, 1), and $F_0 = C(0, 1)$. Do the computed values make sense intuitively?

Exercise 26.2 * For $n = 15, 25, 50, 100$, simulate $U[0, 1]$ data and then plot the normalized uniform empirical process $\sqrt{n}(U_n(t) - t)$. Comment on the behavior of your simulated trajectory.

Exercise 26.3 * By repeatedly simulating $U[0, 1]$ data for $n = 50, 80, 125$, compare the simulated 95th percentile of D_n with the approximation $\frac{1.358}{\sqrt{n}}$ quoted in the text.

Exercise 26.4 Show that on $D[0, 1]$, the functionals $x \to \sup_t |x(t)|$, $x \to \int x^2(t)dt$ are continuous with respect to sup norm (see Chapter 12).

Exercise 26.5 * Find the mean and the variance of the asymptotic null distribution of the two-sided K-S statistic.

Exercise 26.6 * By using Corollary 26.1, find the mean and the variance of the asymptotic null distribution of nC_n.

Exercise 26.7 * By careful computing, plot and superimpose the CDFs of the asymptotic null distributions of $\sqrt{n}D_n$ and nC_n.

Exercise 26.8 * By careful computing, plot the density of the asymptotic null distribution of the Kuiper statistic.

Exercise 26.9 * Prove that each of the Kuiper, Cramér-von Mises, and Anderson-Darling tests is consistent against any alternative.

Exercise 26.10 * By using Theorem 26.4, approximate the power of the two-sided K-S test for testing $H_0 : F = U[0, 1]$ against a Beta(α, α) alternative, with $\alpha = 0.5, 1.5, 2, 5$ and $n = 25, 50, 100$.

Exercise 26.11 * Consider the statistic N_n, the number of times F_n crosses F_0. Simulate the expected value of N_n under the null and compare this to its known limiting value $\sqrt{\frac{\pi}{2}}$.

Exercise 26.12 * Simulate a sample of size $m = 20$ from the $N(0, 25)$ distribution and a sample of size $n = 10$ from the $C(0, 1)$ distribution. Test the hypothesis that the data come from the same distribution by using the two-sample K-S and the two-sample Kuiper statistics. Find the p-values.

Exercise 26.13 * Simulate a sample of size $n = 20$ from the $N(0, 25)$ distribution. Test the hypothesis that the data come from a standard Cauchy distribution by using the Berk-Jones statistic. Use the percentile approximation given in the text to approximately compute a p-value. Repeat the exercise for testing that the simulated data come from an $N(0, 100)$ distribution.

Exercise 26.14 * Simulate a sample of size $n = 20$ from the $N(0, 25)$ distribution and compute the values of the $S_n(s)$ and $T_n(s)$ statistics for $s = -1, 0, \frac{1}{2}, 1, 2$.

Exercise 26.15 * **(Geary's "a")** Given iid observations $X_1, X_2, \ldots, X_n$ from a distribution with finite variance, let $a = \frac{\frac{1}{n}\sum_{i=1}^{n}|X_i - \bar{X}|}{s}$.

(a) Derive the asymptotic distribution of a in general and use it to derive the asymptotic distribution of a when the underlying distribution is normal.
(b) Hence suggest a test for normality based on Geary's a.
(c) Find the *exact* mean and variance of a in finite samples under normality.

Exercise 26.16 (Q-Q Plot for Exponentiality) Let $X_1, X_2, \ldots, X_n$ be iid observations from a distribution on $(0, \infty)$ and let F_n be the empirical CDF. Let $t_{(i)}$ denote the order statistics of the sample and let $Q(\alpha) = -\log(1-\alpha)$ be the quantile function of the standard exponential. Justify why a plot of the pairs $(t_{(i)}, Q(F_n(t_{(i)}) - .5n))$ can be used to test that $X_1, X_2, \ldots, X_n$ are samples from some exponential distribution.

Exercise 26.17 * **(Test for Normality)** Let $X_1, X_2, \ldots, X_n$ be iid observations from a distribution F on the real line, and let b_1, b_2 denote the usual sample skewness and sample kurtosis coefficients; i.e., $b_1 = \frac{\frac{1}{n}\sum(X_i-\bar{X})^3}{s^3}$ and $b_2 = \frac{\frac{1}{n}\sum(X_i-\bar{X})^4}{s^4}$.

(a) Show that if F is a normal distribution, then (i) $\sqrt{n}b_1 \stackrel{\mathcal{L}}{\Rightarrow} N(0, 6)$, and (ii) $\sqrt{n}(b_2 - 3) \stackrel{\mathcal{L}}{\Rightarrow} N(0, 24)$.
(b) Suggest a test for normality based on b_1 and one based on b_2.

(c) Are these tests consistent against all alternatives or only certain ones?
(d) Can you suggest a test based jointly on (b_1, b_2)?

References

Anderson, T.W. (1962). On the distribution of the two sample Cramér-von Mises criterion, Ann. Math. Stat., 33, 1148–1159.

Araujo, A. and Giné, E. (1980). *The Central Limit Theorem for Real and Banach Valued Random Variables*, Wiley, New York.

Basu, A., Harris, I., Hjort, N., and Jones, M. (1998). Robust and efficient estimation by minimizing density power divergence, Biometrika, 85, 549–559.

Berk, R. and Jones, D. (1979). Goodness of fit statistics that dominate the Kolmogorov statistics, Z. Wahr. Verw. Geb., 47, 47–59.

Bickel, P.J. (1968). A distribution free version of the Smirnov two sample test in the p-variate case, Ann. Math. Stat., 40, 1–23.

Billingsley, P. (1999). *Convergence of Probability Measures*, 2nd ed., John Wiley, New York.

Brown, L., DasGupta, A., Marden, J., and Politis, D.(2004). *Characterizations, Sub and Resampling, and Goodness of Fit*, IMS Lecture Notes Monograph Series, Vol. 45, Institute of Mathematical Statistics, Beachwood, OH, 180–206.

Csizár, I. (1963). Magy. Tud. Akad. Mat. Kutató Int. Közl, 8, 85–108 (in German).

Csörgo, M., Csörgo, S., Horváth, L., and Mason, D. (1986). Weighted empirical and quantile process, Ann. Prob., 14, 31–85.

D'Agostino, R. and Stephens, M. (1986). *Goodness of Fit Techniques*, Marcel Dekker, New York.

de Wet, T. and Ventner, J. (1972). Asymptotic distributions of certain test criteria of normality, S. Afr. Stat. J., 6, 135–149.

del Barrio, E., Deheuvels, P., and van de Geer, S. (2007). *Lectures on Empirical Processes*, European Mathematical Society, Zurich.

Gnanadesikan, R. (1997). *Methods for Statistical Data Analysis of Multivariate Observations*, John Wiley, New York.

Groeneboom, P. and Shorack, G. (1981). Large deviations of goodness of fit statistics and linear combinations of order statistics, Ann. Prob., 9, 971–987.

Hodges, J.L. (1958). The significance probability of the Smirnov two sample test, Ark. Mat., 3, 469–486.

Jager, L. (2006). Goodness of fit tests based on phi-divergences, Technical Report, University of Washington.

Jager, L. and Wellner, J. (2006). Goodness of fit tests via phi-divergences, in Press.

Kiefer, J. (1959). k sample analogues of the Kolmogorov-Smirnov and the Cramér-von Mises tests, Ann. Math. Stat., 30, 420–447.

Kolmogorov, A. (1933). Sulla determinazione empirica di una legge di distribuzione, Giorn. Ist. Ital. Attuari, 4, 83–91.

Lehmann, E.L. (1999). *Elements of Large Sample Theory*, Springer, New York.

Marden, J. (1998). Bivariate qq and spider-web plots, Stat. Sinica, 8(3), 813–816.

Marden, J. (2004). Positions and QQ plots, Stat. Sci., 19(4), 606–614.

Martynov, G.V. (1992). Statistical tests based on Empirical processes and related problems, Sov. J. Math., 61(4), 2195–2271.

Owen, A. (1995). Nonparametric likelihood confidence hands for a distribution function, J. Amer. Stat., Assoc., 90, 430, 516–521.

Pollard, D. (1989). Asymptotics via Empirical processes, Stat. Sci., 4(4), 341–366.

Raghavachari, M. (1973). Limiting distributions of the Kolmogorov-Smirnov type statistics under the alternative, Ann. Stat., 1, 67–73.

Sarkadi, K. (1985). On the asymptotic behavior of the Shapiro-Wilk test, in *Proceedings of the 7th Conference on Probability* Theory, Brasov, Romania, IVNU Science Press, Utrecht, 325–331.

Shapiro, S.S. and Wilk, M.B. (1965). An analysis of variance test for normality, Biometrika, 52, 591–611.

Shorack, G. and Wellner, J. (1986). *Empirical Processes, with Applications to Statistics*, John Wiley, New York.

Smirnov, N. (1941). Approximate laws of distribution of random variables from empirical data, Usp. Mat. Nauk., 10, 179–206 (in Russian).

Stephens, M. (1993). *Aspects of Goodness of Fit:Statistical Sciences and Data Analysis*, VSP, Utrecht.

Stuart, A. and Ord, K. (1991). *Kendall's Advanced Theory of Statistics*, Vol. II, Clarendon Press, New York.

Wald, A. and Wolfowitz, J. (1940). On a test whether two samples are from the same population, Ann. Math. Stat., 11, 147–162.

Wald, A. and Wolfowitz, J. (1941). Note on confidence limits for continuous distribution functions, Ann. Math. Stat., 12, 118–119.

Weiss, L. (1960). Two sample tests for multivariate distributions, Ann. Math. Stat., 31, 159–164.

Wellner, J. and Koltchinskii, V. (2003). A note on the asymptotic distribution of the Berk-Jones type statistics under the null distribution, in *High Dimensional Probability III*, Progress in Probability, J. Hoffman Jørgensen, M. Marcus and J. Wellner (eds.), Vol. 55, Birkhäuser, Basel, 321–332.

Chapter 27
Chi-square Tests for Goodness of Fit

Chi-square tests are well-known competitors to EDF-based statistics. They discretize the null distribution in some way and assess the agreement of observed counts to the postulated counts, so there is obviously some loss of information and hence a loss in power. But they are versatile. Unlike EDF-based tests, a chi-square test can be used for continuous as well as discrete data and in one dimension as well as many dimensions. Thus, a loss of information is being exchanged for versatility of the principle and ease of computation.

27.1 The Pearson χ^2 Test

Suppose $X_1, \ldots, X_n$ are IID observations from some distribution F in a Euclidean space and that we want to test $H_0 : F = F_0$, F_0 being a completely specified distribution. Let S be the support of F_0 and, for some given $k \geq 1$, $A_{k,i}$, $i = 1, 2, \cdots, k$ form a partition of S. Let $p_{0,i} = P_{F_0}(A_{k,i})$ and $n_i = \#\{j : x_j \in A_{k,i}\} =$ the observed frequency of the partition set $A_{k,i}$. Therefore, under H_0, $E(n_i) = np_{0,i}$. Karl Pearson suggested that as a measure of discrepancy between the observed sample and the null hypothesis, one compare $(n_1, ..., n_k)$ with $(np_{0i}, ..., np_{0k})$. The Pearson chi-square statistic is defined as

$$\chi^2 = \sum_{i=1}^{k} \frac{(n_i - np_{0i})^2}{np_{0i}}.$$

For fixed n, certainly χ^2 is not distributed as a chi-square, for it is just a quadratic form in a multinomial random vector. However, the asymptotic distribution of χ^2 is χ^2_{k-1} if H_0 holds, hence the name Pearson chi-square for this test.

As hard as it is to believe, Pearson's chi-square test is actually more than a century old (Pearson (1900)). Cox (2000) and Rao (2000) give well-written accounts. Serfling (1980) and Ferguson (1996) contain theoretical developments. Greenwood and Nikulin (1996) is a masterly treatment. Modifications of Pearson's chi-square have been suggested; see, among others, Rao and Robson (1974).

27.2 Asymptotic Distribution of Pearson's Chi-square

Theorem 27.1 Suppose $X_1, X_2, \ldots, X_n$ are iid observations from some distribution F in a finite-dimensional Euclidean space. Consider testing $H_0 : F = F_0$ (specified). Let χ^2 be the Pearson χ^2 statistic defined above. Then $\chi^2 \overset{\mathcal{L}}{\Rightarrow} \chi^2_{k-1}$ under H_0.

Proof. It is easy to see why the asymptotic null distribution of χ^2 should be χ^2_{k-1}. Define $Y = (Y_1, ..., Y_k) = (\frac{n_1 - np_{01}}{\sqrt{np_{01}}}, ..., \frac{n_k - np_{0k}}{\sqrt{np_{0k}}})$. By the multinomial CLT (see Chapter 1 exercises), $Y \overset{\mathcal{L}}{\Rightarrow} N_k(0, \Sigma)$, where $\Sigma = I - \mu\mu^{'}$, where $\mu^{'} = (\sqrt{p_{01}}, ...\sqrt{p_{0k}})$, trace $(\Sigma) = k - 1$. The eigenvalues of Σ are 0 with multiplicity 1 and 1 with multiplicity $(k - 1)$. Notice now that Pearson's $\chi^2 = Y^{'}Y$, and if $Y \backsim N_k(0, \Sigma)$ for any general Σ, then $Y^{'}Y = X^{'}P^{'}PX = X^{'}X \overset{\mathcal{L}}{=} \sum_{i=1}^{k} \lambda_i w_i$, where $w_i \overset{\text{iid}}{\sim} \chi^2_1$, λ_i are the eigenvalues of Σ, and $P^{'}\Sigma P = \text{diag}(\lambda_1, ..., \lambda_k)$ is the spectral decomposition of Σ. So $X \sim N_k(0,$ $\text{diag}(\lambda_1, ..., \lambda_k))$, and it follows that $X^{'}X = \sum_{i=1}^{k} X_i^2 \overset{\mathcal{L}}{=} \sum_{i=1}^{k} \lambda_i w_i$. For our Σ, $k - 1$ of $\lambda_i^{'}$s are 1 and the remaining one is zero. Since a sum of independent chi-squares is again a chi-square, it follows from the multinomial CLT that $\chi^2 \overset{\mathcal{L}}{\Rightarrow} \chi^2_{k-1}$ under H_0.

Remark. The so-called Freeman-Tukey statistic is a kind of symmetrization of Pearson χ^2 with respect to the vector of observed and expected frequencies. It is defined as $\text{FT} = 4\sum_{i=1}^{k}(\sqrt{n_i} - \sqrt{np_{0i}})^2$, and it turns out that FT also converges to χ^2_{k-1} under H_0, which follows by an easy application of the delta theorem. The Freeman-Tukey statistic is sometimes preferred to Pearson's chi-square. See Stuart and Ord (1991) for some additional information.

27.3 Asymptotic Distribution under Alternatives and Consistency

Let F_1 be a distribution different from F_0 and let $p_{1i} = P_{F_1}(A_{k,i})$. Clearly, if by chance $p_{1i} = p_{0i}\ \forall i = 1, .., k$ (which is certainly possible), then a test based on the empirical frequencies of $A_{k,i}$ cannot distinguish F_0 from F_1, even asymptotically. In such a case, the χ^2 test cannot be consistent against F_1. However, otherwise it will be consistent, as can be seen easily from the following result.

Theorem 27.2 $\frac{\chi^2}{n} \stackrel{P}{\Rightarrow} \sum_{i=1}^{k} \frac{(p_{1i}-p_{0i})^2}{p_{0i}}$ under F_1.

This is evident as $\chi^2 = \sum_{i=1}^{k} \frac{(n_i-np_{0i})^2}{np_{0i}} = n\sum_{i=1}^{k} \frac{(\frac{n_i}{n}-p_{0i})^2}{p_{0i}}$. But $(\frac{n_1}{n}, \ldots, \frac{n_k}{n}) \stackrel{P}{\Rightarrow} (p_{11}, \ldots, p_{1k})$ under F_1. Therefore, by the continuous mapping theorem, $\frac{\chi^2}{n} \stackrel{P}{\Rightarrow} \sum_{i=1}^{k} \frac{(p_{1i}-p_{0i})^2}{p_{0i}}$.

Corollary 27.1 If $\sum_{i=1}^{k} \frac{(p_{1i}-p_{0i})^2}{p_{0i}} > 0$, then $\chi^2 \stackrel{\mathcal{P}}{\longrightarrow} \infty$ under F_1 and hence the χ^2 test is consistent against F_1.

Remark. Thus, for a fixed alternative F_1 such that the vector $(p_{11}, \ldots, p_{1k}) \neq (p_{01}, \ldots, p_{0k})$, Pearson's χ^2 cannot have a nondegenerate limit distribution under F_1. However, if the alternative is very close to the null, in the sense of being a Pitman alternative, there is a nondegenerate limit distribution. We have seen this phenomenon occur previously in other testing problems.

Theorem 27.3 Consider an alternative $F_1 = F_{1,n} = F_0 + \frac{1}{\sqrt{n}}G$, where the total mass of G is 0. Let $p_{1i} = p_{0i} + \frac{1}{\sqrt{n}}c_i$, where $c_i = \int_{A_{k,i}} dG, \sum_{i=1}^{k} c_i = 0$. Then $\chi^2 \stackrel{\mathcal{L}}{\Rightarrow} NC\chi^2(k-1, \delta^2)$, where $\delta^2 = \sum_{i=1}^{k} \frac{c_i^2}{p_{0i}}$.

Remark. This result can be used to approximate the power of the χ^2 test at a close alternative by using the noncentral χ^2 CDF as an approximation to the exact CDF of χ^2 under the alternative.

27.4 Choice of k

A key practical question in the implementation of χ^2 tests is the choice of k and the actual partitioning sets $A_{k,i}$. Both are hard problems, and despite a huge body of literature on the topic, there are no clear-cut solutions. Some major references on this hard problem are Mann and Wald (1942), Oosterhoff (1985), and Stuart and Ord (1991).

A common assumption in much of the theoretical work is to take some suitable value of k and use the partition sets $A_{k,i}$, which make $p_{0i} \equiv \frac{1}{k}$. In other words, the cells are equiprobable under H_0. Note that generally this will make the cells of unequal size (e.g., of unequal width if they are intervals). The problem then is to seek the optimum value of k. The crux of the problem in optimizing k is that a large k may or may not be a good choice, depending on the alternative. One can see this by simple moment calculations, and further comments on this are made below.

Theorem 27.4

(a) $E_{H_0}(\chi^2) = k - 1$.

(b) $\mathrm{Var}_{H_0}(\chi^2) = \frac{1}{n}\left(2(n-1)(k-1) - k^2 + \sum_{i=1}^{k} \frac{1}{p_{0i}}\right)$.

(c) $E_{F_1}(\chi^2) = \sum_{i=1}^{k} \frac{p_{1i}(1-p_{1i})}{p_{0i}} + n\left(\sum_{i=1}^{k} \frac{p_{1i}^2}{p_{0i}} - 1\right)$.

Remark. See Sen and Singer (1993) or Serfling (1980) for simple derivations of the moments of Pearson's χ^2. The variance under the alternative has a somewhat messy expression. The formula for $\mathrm{Var}_{H_0}(\chi^2)$ indicates the problem one will have with many cells. If k is very large, then some value of p_{0i} would be small, making $\sum_{i=1}^{k} \frac{1}{p_{0i}}$ a large number and $\mathrm{Var}_{H_0}(\chi^2)$ quite a bit larger than $2(k-1)$. This would indicate that the χ^2_{k-1} approximation to the null distribution of χ^2 is not accurate. So even the size of the test may differ significantly from the nominal value if k is too large. Clearly, the choice of k is a rather subtle issue.

Example 27.1 Here are some values of the power of the Pearson χ^2 test when $F_0 = N(0,1)$ and F_1 is a Cauchy or another normal and when $F_0 = U[0,1]$ and F_1 is a Beta distribution. The numbers in Table 27.1 are quite illuminating.

For the case $F_0 = N(0,1)$, the power increases monotonically in k when the alternative is Cauchy, which is thick tailed, but actually deteriorates for the larger k, when the alternative is another normal, which is thin tailed. Similarly, when $F_0 = U[0,1]$, the power increases monotonically in k when F_1 is a U-shaped Beta distribution but deteriorates for the larger k when F_1 is a unimodal Beta distribution. We shall later see that some general results

Table 27.1 ($n = 50, \alpha = 0.05, p_{0i} = \frac{1}{k}$)

		k			
F_0	F_1	4	6	8	15
$N(0,1)$	$C(0,\sigma), \sigma = \frac{1}{2}$	0.18	0.25	0.28	0.40
$N(0,1)$	$N(0,\sigma^2), \sigma = \frac{4}{3}$	0.32	0.32	0.30	0.24
$U[0,1]$	$\mathrm{Beta}(\frac{2}{3}, \frac{2}{3})$	0.20	0.23	0.25	0.26
$U[0,1]$	$\mathrm{Beta}(\frac{5}{3}, \frac{5}{3})$	0.39	0.34	0.32	0.24

can be given that justify such an empirical finding. We now present some results on selecting the number of cells k.

27.5 Recommendation of Mann and Wald

Mann and Wald (1942) formulated the problem of selecting the value of k in a (somewhat complicated) paradigm and came out with an optimal rate of growth of k as $n \to \infty$. The formulation of Mann and Wald was along the following lines.

Fix a number $0 < \Delta < 1$. Let F_0 be the null distribution and F_1 a plausible alternative. Consider the class of alternatives $\mathcal{F} = \mathcal{F}_\Delta = \{F_1 : d_K(F_0, F_1) \geq \Delta\}$, where $d_K(F_0, F_1)$ is the Kolmogorov distance between F_0 and F_1. Let $\beta(F_1, n, k, \alpha) = P_{F_1}(\chi^2 > \chi^2_{k-1}(\alpha))$. Mann and Wald (1942) consider $\inf_{F_1 \in \mathcal{F}_\Delta} \beta(F_1, n, k, \alpha)$ and suggest $k_n = k_n(\alpha, \Delta) = \operatorname{argmax}_k \inf_{F_1 \in \mathcal{F}_\Delta} \beta(F_1, n, k, \alpha)$ as the value of k. Actually, the criterion is a bit more complex than that; see Mann and Wald (1942) for the exact criterion. They prove that k_n grows at the rate $n^{\frac{2}{5}}$; i.e., $k_n \sim n^{\frac{2}{5}}$. Actually, they also produce a constant in this rate result. Later empirical experience has suggested that the theoretical constant is a bit too large.

A common practical recommendation influenced by the Mann-Wald result is $k = 2n^{\frac{2}{5}}$. The recommendation seems to produce values of k that agree well with practical choices of k; see Table 27.2.

These values seem to be close to common practice. The important points are that k should be larger when n is large. But it is not recommended that one use a very large value for k, and a choice in the range 5–15 seems right.

27.6 Power at Local Alternatives and Choice of k

Suppose we wish to test that $X_1, X_2, \ldots, X_n$ are iid H with density h. Thus the null density is h. For another density g and $0 \leq \theta \leq 1$, consider alternatives

Table 27.2 Integer nearest to $2n^{2/5}$ for various n

n	Integer nearest to $2n^{2/5}$
25	7
50	10
80	12
100	13

$$g_\theta = (1-\theta)h + \theta g.$$

If $0 < \theta < 1$ is fixed, then the Pearson $\chi^2 \xrightarrow{\mathcal{P}} \infty$ under g_θ, as we saw previously (provided the cell probabilities are not the same under g and h). But if $\theta = \theta_n$ and θ_n converges to zero at the rate $\frac{1}{\sqrt{n}}$, then the Pearson χ^2 has a noncentral χ^2 limit distribution and the power under the alternative g_{θ_n} has a finite limit for any fixed k. The question is, if we let $k \to \infty$, then what happens to the power? If it converges to 1, letting k grow would be a good idea. If it converges to the level α, then letting k grow arbitrarily would be a bad idea.

To describe the results, we first need some notation. We suppress k and n in the notation. Let

$$\begin{aligned} p_{0i} &= \int_{A_{k,i}} h(x)dx, \\ p_i &= \int_{A_{k,i}} g_{\theta_n}(x)dx, \\ p_i^* &= \int_{A_{k,i}} g(x)dx, \\ \Delta_k &= \sum_{i=1}^{k} \frac{(p_i^* - p_{0i})^2}{p_{0i}}, \\ f &= \frac{g}{h} - 1. \end{aligned}$$

Then one has the following results (Kallenberg, Oosterhoff, and Schriever (1985)).

Theorem 27.5 Suppose

(i) $k = k(n) \to \infty$ such that $k = o(n)$,
(ii) $\liminf_n \min_i (kp_{0i}) > 0$,
(iii) $\lim_n n\theta_n^2$ exists and is nonzero and finite.

Then

$$\begin{aligned} \lim_n \beta(g_{\theta_n}, n, k, \alpha) &= 1, \text{ iff } \lim \tfrac{\Delta_k}{\sqrt{k}} = \infty, \\ &= \alpha, \text{ iff } \lim \tfrac{\Delta_k}{\sqrt{k}} = 0, \end{aligned}$$

where as before $\beta(.)$ denotes the power of the test.

Remark. If $0 < \lim \frac{\Delta_k}{\sqrt{k}} < \infty$, then the power would typically converge to a number between α and 1, but a general characterization is lacking. The issue about letting k grow is that the approximate noncentral χ^2 distribution for Pearson χ^2 under the alternative has a noncentrality parameter increasing in k, which would make the distribution stochastically larger. On the other hand, by increasing k, the degree of freedom also increases, which would increase the variance.

Thus, there are two conflicting effects of increasing k, and it is not clear which one will win. For certain alternatives, the increase in Δ_k beats the effect of the increase in the variance and the power converges to 1. For certain other alternatives, it does not. The tail of g relative to h is the key factor. The next result makes this precise.
The following result (Kallenberg, Oosterhoff, and Schriever (1985)) connects the condition $\lim \frac{\Delta_k}{\sqrt{k}} = \infty$ (0) to the thickness of the tail of the fixed alternative g.

Theorem 27.6 (a) Suppose $\limsup_n \min_i \; kp_{0i} > 0$. If, for some $r > \frac{4}{3}$, $\int |f|^r dH < \infty$, then $\lim \frac{\Delta_k}{\sqrt{k}} = 0$.
(b) Suppose $\liminf_n \min_i kp_{0i} > 0$ and $\limsup_n \max_i kp_{0i} < \infty$. If, for some $0 < r < \frac{4}{3}$, $\int |f|^r dH = \infty$, then $\lim \frac{\Delta_k}{\sqrt{k}} = \infty$.

Remark. The assumption $\liminf_n \min_i kp_{0i} > 0$ says that none of the cells $A_{k,i}$ should have very small probabilities under h. Assumption (b), that $\limsup \max_i \; kp_{0i} < \infty$, likewise says that none of the cells should have a high probability under h. The two assumptions are both satisfied if $p_{0i} \sim \frac{1}{k}$ for all i and k.

If g has a thick tail relative to h, then, for small r, $\int |f|^r dH$ would typically diverge. On the contrary, if g has a thin tail relative to h, then $\int |f|^r dH$ would typically converge even for large r. So the combined qualitative conclusion of Theorem 27.5 and Theorem 27.6 is that if g has thick tails relative to h, then we can afford to choose a large number of cells, and if g has thin tails relative to h, then we should not use a large number of cells. These are useful general principles.

The next two examples illustrate the phenomenon.

Example 27.2 Let $h(x) = \frac{1}{\sqrt{2\pi}} e^{-\frac{x^2}{2}}$ and $g(x) = \frac{1}{\pi(1+x^2)}$. Note that g has thick tails relative to h. Therefore,

$$f(x) = \frac{g(x)}{h(x)} - 1 = \frac{ce^{\frac{x^2}{2}}}{1+x^2} - 1 \text{ for some } 0 < c < \infty$$

$$\Rightarrow \int |f|^r dH = \frac{1}{\sqrt{2\pi}} \int \left| \frac{ce^{\frac{x^2}{2}}}{1+x^2} - 1 \right|^r e^{-\frac{x^2}{2}} dx.$$

For any $r > 1$, this integral diverges. So, from Theorem 27.5 and Theorem 27.6, $\lim \beta(g_{\theta_n}, n, k_n, \alpha) = 1$.

Example 27.3 Let $h(x) = \frac{1}{\sqrt{2\pi}} e^{-\frac{x^2}{2}}$ and $g(x) = \frac{1}{\sigma\sqrt{2\pi}} e^{-\frac{x^2}{2\sigma^2}}$. The larger the σ is, the thicker is the tail of g relative to h. Now, $f(x) = ce^{\frac{x^2}{2}(1-\frac{1}{\sigma^2})} - 1$ for some $0 < c < \infty$. Therefore

$$\int |f|^r dH = \frac{1}{\sqrt{2\pi}} \int \left| ce^{\frac{x^2}{2}(1-\frac{1}{\sigma^2})} - 1 \right|^r e^{-\frac{x^2}{2}} dx$$

$$\sim \int e^{\frac{x^2}{2}\left[r(1-\frac{1}{\sigma^2})-1\right]} dx.$$

If $r = \frac{4}{3}$ and $\sigma^2 = 4$, $r(1 - \frac{1}{\sigma^2}) - 1 = 0$. Also, $r(1 - \frac{1}{\sigma^2}) - 1 < 0$ and the integral $\int e^{\frac{x^2}{2}\left[r(1-\frac{1}{\sigma^2})-1\right]} dx$ converges for some $r < \frac{4}{3}$ iff $\sigma^2 < 4$. On the other hand, $r(1 - \frac{1}{\sigma^2}) - 1 > 0$ and the integral $\int e^{\frac{x^2}{2}\left[r(1-\frac{1}{\sigma^2})-1\right]} dx$ diverges for some $r < \frac{4}{3}$ iff $\sigma^2 > 4$. So, if g has a "small" variance, then letting $k \to \infty$ is not a good idea, while if g has a "large" variance, then one can let $k \to \infty$. Note that the conclusion is similar to that in the previous example.

27.7 Exercises

Exercise 27.1 * For testing that $F = N(0, 1)$, and with the cells as $(-\infty, -3), (-3, -2), \ldots, (2, 3), (3, \infty)$, find explicitly an alternative F_1 such that Pearson's chi-square is not consistent.

Exercise 27.2 For testing that a p-dimensional distribution is $N(0, I)$, find $k = 10$ spherical shells with equal probability under the null.

Exercise 27.3 * For testing $H_0 : F = N(0, 1)$ vs. $H_1 : F = C(0, 1)$, and with the cells as $(-\infty, -4a), (-4a, -3a), (-3a, -2a), \ldots, (3a, 4a)$,

$(4a, \infty)$, find a that maximizes $\sum_{i=1}^{k} \frac{(p_{1i}-p_{0i})^2}{p_{0i}}$. Why would you want to maximize it?

Exercise 27.4 For $k = 6, 8, 10, 12$, $n = 15, 25, 40$, and with the equiprobable cells, find the approximate power of the chi-square test for testing $F = \text{Exp}(1)$ vs. $F = \text{Gamma}(2, 1); \text{Gamma}(5, 1)$. Do more cells help?

Exercise 27.5 For $k = 6, 8, 10, 12$, $n = 15, 25, 40$, and with the equiprobable cells, find the approximate power of the chi-square test for testing $F = N(0, 1)$ vs. $F = \text{DoubleExp}(0, 1)$. Do more cells help?

Exercise 27.6 For $k = 6, 8, 10, 12$, $n = 15, 25, 40$, and with the equiprobable cells, find the approximate power of the chi-square test for testing $F = C(0, 1)$ vs. $F = t(m)$, $m = 2, 5, 10$. Do more cells help?

Exercise 27.7 Prove that the Freeman-Tukey statistic defined in the text is asymptotically a chi-square.

Exercise 27.8 * Prove or disprove that $E_{F_1}\chi^2 \geq E_{F_0}\chi^2 \,\forall F_1 \neq F_0$.

Exercise 27.9 * Find a formula for $\text{Var}_{F_1}\chi^2$.

Exercise 27.10 * Find the limiting distribution under the null of $\frac{\chi^2-k}{\sqrt{k}}$, where $k = k(n) \to \infty$. Does a weak limit always exist?

Exercise 27.11 * With $h = N(0, 1)$, $g = \text{DoubleExp}(0, 1)$, in the notation of Section 27.6, does $\beta(g_{\theta_n}, n, k, \alpha)$ converge to 1, 0, or something in between?

Exercise 27.12 * With $h = \text{Gamma}(2, 1)$, $g = \text{lognormal}(0, 1)$, in the notation of Section 27.6, does $\beta(g_{\theta_n}, n, k, \alpha)$ converge to 1, 0, or something in between?

References

Cox, D.R. (2000). Karl Pearson and the chisquared test, in *Goodness of Fit Tests and Model Validity*, Birkhäuser, Boston, 3–8.

Ferguson, T.S. (1996). *A Course in Large Sample Theory*, Chapman and Hall, London.

Greenwood, P.E. and Nikulin, M.S. (1996). *A Guide to Chi-squared Testing*, John Wiley, New York.

Kallenberg, W.C.M., Oosterhoff, J. and Schriever, B.F. (1985). The number of classes in chi-squared goodness of fit tests, J. Am. Stat. Assoc., 80(392), 959–968.

Mann, H.B. and Wald, A. (1942). On the choice of the number of class intervals in the application of the chisquare test, Ann. Math. Stat., 13, 306–317.

Oosterhoff, J. (1985). Choice of cells in chisquare tests, Stat. Neerlandica, 39(2), 115–128.

Pearson, K. (1900). On a criterion that a given system of deviations from the probable in the case of a correlated system of variables is such that it can be reasonably supposed to have arisen from random sampling, Philos. Mag., Ser. 5, 50, 157–175.

Rao, C.R. (2000). Pearson chisquare test, the dawn of statistical inference, in *Goodness of Fit Tests and Model Validity*, C. Huber-Carol, N. Balakrishnan, M. Nikulin and M. Mesbah. Birkhäuser, Boston, 9–24.

Rao, K.C. and Robson, D.S. (1974). A chi-square statistic for goodness of fit tests within the Exponential family, Commun. Stat., 3, 1139–1153.

Serfling, R. (1980). *Approximation Theorems of Mathematical Statistics*, John Wiley, New York.

Stuart, A. and Ord, K. (1991). *Kendall's Theory of Statistics*, 4th ed., Vol. II, Clarendon Press, New York.

Chapter 28
Goodness of Fit with Estimated Parameters

The tests described in the previous chapter correspond to the null hypothesis $H_0 : F = F_0$ (specified). Usually, however, one is interested in testing that F belongs to a family $\mathcal{F}$, say the family of all $N(\mu, \sigma^2)$ distributions. In such a case, the EDF-based test statistics as well as chi-square tests would have to be redefined, as there is no longer any fixed null CDF F_0. If the true value of the parameter θ that indexes the family $\mathcal{F}$ is θ_0 and $\hat{\theta}_n = \hat{\theta}_n(X_1, \ldots, X_n)$ is an estimate of θ_0, then the specified F_0 of the previous chapter may be replaced by $F(t, \hat{\theta}_n) = P_{\theta=\hat{\theta}_n}(X_1 \leq t)$. So the adjusted two-sided Kolmogorov-Smirnov statistic would be $\widetilde{D}_n = \sup_t |F_n(t) - F(t, \hat{\theta}_n)|$. The asymptotic distribution of $\sqrt{n}\widetilde{D}_n$ under $F(t, \theta)$ will depend on θ, the exact CDF F, and the exact choice of $\hat{\theta}_n$; $\sqrt{n}\widetilde{D}_n$ does not have a fixed asymptotic null distribution as did $\sqrt{n}D_n$. The same holds for the correspondingly adjusted versions of V_n, C_n, and A_n. This problem is known as the goodness-of-fit problem with estimated parameters or the composite goodness-of-fit problem. There is wide misunderstanding in parts of the statistics community about the invalidity of the simple null procedures, such as the Kolmogorov-Smirnov test, in the composite situation.

Some major references on goodness-of-fit tests for composite nulls are Chernoff and Lehmann (1954), Stephens (1976), Neuhaus (1979), Boos (1981), and Beran and Millar (1989). A Bayesian approach is presented in Verdinelli and Wasserman (1998). Csörgo (1974), Mudholkar and Lin (1987), and Brown (2004) use characterization results to test for composite nulls. Hall and Welsh (1983) and Csörgo (1984) discuss goodness of fit by using the empirical characteristic function. At a textbook level, van der Vaart (1998) gives a rigorous treatment of basic material. Also see Stuart and Ord (1991).

A. DasGupta, *Asymptotic Theory of Statistics and Probability*,

28.1 Preliminary Analysis by Stochastic Expansion

Let us run through a quick explanation of why $\sqrt{n}\widetilde{D}_n$ has an asymptotic null distribution that depends on F, θ, and the choice of $\hat{\theta}_n$.
Fix t. Then

$$\begin{aligned}
&F_n(t) - F(t, \hat{\theta}_n) \\
&\quad = F_n(t) - F(t, \theta_0) - \big(F(t, \hat{\theta}_n) - F(t, \theta_0)\big) \\
&\quad = F_n(t) - F(t, \theta_0) - \left((\hat{\theta}_n - \theta_0)\frac{\partial}{\partial\theta}F(t, \theta)|_{\theta=\theta_0} + o_p(\hat{\theta}_n - \theta_0)\right).
\end{aligned}$$

Therefore

$$\begin{aligned}
&\sqrt{n}\left(F_n(t) - F(t, \hat{\theta}_n)\right) \\
&= \sqrt{n}\big(F_n(t) - F(t, \theta_0)\big) - \sqrt{n}(\hat{\theta}_n - \theta_0)\frac{\partial}{\partial\theta}F(t, \theta)|_{\theta=\theta_0} + o_p\big(\sqrt{n}(\hat{\theta}_n - \theta_0)\big).
\end{aligned}$$

Now suppose $\sqrt{n}(\hat{\theta}_n - \theta_0) \stackrel{\mathcal{L}}{\Rightarrow} Z$, some random variable, usually a mean zero normal. Then

$$\begin{aligned}
\sqrt{n}\left(F_n(t) - F(t, \hat{\theta}_n)\right) = {} & \sqrt{n}\Big(F_n(t) - F(t, \theta_0)\Big) \\
& - \sqrt{n}(\hat{\theta}_n - \theta_0)\frac{\partial}{\partial\theta}F(t, \theta)|_{\theta=\theta_0} + o_p(1).
\end{aligned}$$

Suppose also that $\hat{\theta}_n$ is an asymptotically linear statistic (i.e., $\hat{\theta}_n$ admits a Bahadur expansion)

$$\hat{\theta}_n = \theta_0 + \frac{1}{n}\sum_{i=1}^{n} Y_i + o_p\left(\frac{1}{\sqrt{n}}\right).$$

Then

$$\sqrt{n}\left(F_n(t) - F(t, \hat{\theta}_n)\right) = \sqrt{n}\Big(F_n(t) - F(t, \theta_0)\Big) - \frac{\partial}{\partial\theta}F(t, \theta)|_{\theta=\theta_0}\sqrt{n}\bar{Y} + o_p(1).$$

Now define

$$\underset{\sim}{U_i} = \begin{pmatrix} I_{X_i \le t} - F(t, \theta_0) \\ Y_i \end{pmatrix},$$

where $X_1, \ldots, X_n$ are the original iid observations from F_{θ_0}. By the multivariate CLT, $\sqrt{n}\underset{\sim}{\bar{U}}$ converges under F_{θ_0} to a bivariate normal with mean vector $\underset{\sim}{0}$ and some covariance matrix Σ_{θ_0}. Therefore, the limit distribution of $\sqrt{n}(F_n(t) - F(t, \hat{\theta}_n))$ is a normal but with a variance that depends on F, θ_0, and the choice of $\hat{\theta}_n$. Likewise, for any fixed $t_1 < t_2 < \ldots < t_k$, the joint distribution of

$$\sqrt{n}\begin{pmatrix} F_n(t_1) - F(t_1, \hat{\theta}_n) \\ \vdots \\ F_n(t_k) - F(t_k, \hat{\theta}_n) \end{pmatrix}$$

converges to a k-dimensional normal with mean $\underset{\sim}{0}$ and some covariance matrix $\Sigma(\theta_0, t_1, \ldots, t_k)$. The covariance elements σ_{ij} in $\Sigma(\theta_0, t_1, \ldots, t_k)$ correspond to the covariance $\rho(t_i, t_j)$ between X_{t_i} and X_{t_j} of some mean zero Gaussian process $X(t)$. The process $\sqrt{n}(F_n(t) - F(t, \hat{\theta}_n))$ weakly converges to this Gaussian process $X(t)$. By a continuous-mapping argument, $\sqrt{n}\sup_t |F_n(t) - F(t, \hat{\theta}_n)|$ converges in law under F_{θ_0} to the distribution of $\sup_t |X(t)|$, but this distribution, in general, depends on everything: the value θ_0, F, and the choice of the estimate $\hat{\theta}_n$.

Remark. In general, therefore, the covariance kernel $\rho(s, t)$ of the limiting Gaussian process depends on θ_0, F, and the choice of $\hat{\theta}_n$. If there is a group structure in the distribution of $X_1, \ldots, X_n$, and if $\hat{\theta}_n$ obeys the group structure (i.e., $\hat{\theta}_n$ is equivariant with respect to the particular group), then $\rho(s, t)$ will be free of θ but will still depend on the exact form of F and the choice of $\hat{\theta}_n$.

For example, if $X_1, \ldots, X_n$ are iid $N(\mu, \sigma^2)$ and $\hat{\mu} = \bar{X}$, $\hat{\sigma}^2 = s^2$, then $\rho(s, t)$ will not depend on μ and σ^2. So, subject to fixing $\hat{\theta}_n$, and having a fixed functional form for F, $\rho(s, t)$ will be a fixed kernel and in principle can be found.

28.2 Asymptotic Distribution of EDF-Based Statistics for Composite Nulls

For the case of the simple null $H_0 : F = F_0$, we discussed a few other EDF-based statistics besides D_n, namely

$$D_n^+ = \sup_t (F_n(t) - F_0(t)),$$

$$\begin{aligned}
D_n^- &= \sup_t (F_0(t) - F_n(t)), \\
V_n &= D_n^+ + D_n^-, \\
C_n &= \int (F_n(t) - F_0(t))^2 \, dF_0(t), \\
A_n &= \int \frac{(F_n(t) - F_0(t))^2}{F_0(t)(1 - F_0(t))} dF_0(t).
\end{aligned}$$

For the case of estimated parameters, the adjusted versions are, for example,

$$\begin{aligned}
\widetilde{C}_n &= \int \left(F_n(t) - F(t, \hat{\theta}_n)\right)^2 dF(t, \hat{\theta}_n), \\
\widetilde{A}_n &= \int \frac{\left(F_n(t) - F(t, \hat{\theta}_n)\right)^2}{F(t, \hat{\theta}_n)\left(1 - F(t, \hat{\theta}_n)\right)} dF(t, \hat{\theta}_n).
\end{aligned}$$

We consider these two statistics in particular. Both $n\widetilde{C}_n$ and $n\widetilde{A}_n$ have non-degenerate limit distributions, provided $\hat{\theta}_n$ is $\sqrt{n}$-consistent; i.e., $\hat{\theta}_n - \theta_0 = o_p(n^{-\frac{1}{2}})$ under F_{θ_0}.

Theorem 28.1 Suppose $X_1, \ldots, X_n$ are iid F_{θ_0} for some $\theta_0 \in \Theta$. Let $\hat{\theta}_n = \hat{\theta}_n(X_1, \ldots, X_n)$ be $\sqrt{n}$-consistent for θ_0 and be asymptotically linear. Then each of $n\widetilde{C}_n$ and $n\widetilde{A}_n$ converges in law to the distribution of $\sum_{i=1}^{\infty} \lambda_i W_i$ for suitable $\{\lambda_i\}$, where $W_1, W_2, \ldots$ are iid χ_1^2. The coefficients $\{\lambda_i\}$ are the eigenvalues corresponding to the eigenfunctions $\{f_i\}$ of the covariance kernel $\rho(s,t)$ of a suitable zero-mean Gaussian process $X(t)$; i.e., λ_i satisfies

$$\lambda_i f_i(s) = \int \rho(s,t) f_i(t) dt.$$

Remark. See del Bario, Deheuvels, and van de Geer (2007) for a proof. The $\{\lambda_i\}$ can be found numerically by solving certain partial differential equations.

Example 28.1 Suppose $X_1, \ldots, X_n$ are iid $N(\mu, \sigma^2)$, $\hat{\mu} = \bar{X}$, and $\hat{\sigma}^2 = s^2$. The values of the largest ten $\lambda_i's$ for $\widetilde{C}_n$ are 10^{-2}(1.834, 1.344, 0.539, 0.436, 0.252, 0.216, 0.146, 0.129, 0.095, 0.085). The percentage points of $\sum_{i=1}^{\infty} \lambda_i W_i$ can be approximated by Monte Carlo simulation. For $\widetilde{A}_n$, the ten largest eigenvalues are 10^{-2}(9.386, 7.206, 3.593, 2.879, 1.810, 1.584, 1.148, 1.002, 0.777, 0.692).

Table 28.1 Percentage points

	$n\widetilde{C}_n$		$n\widetilde{A}_n$	
Hypothesis	$\alpha = 0.05$	$\alpha = 0.01$	$\alpha = 0.05$	$\alpha = 0.01$
Simple	0.461	0.743	2.492	3.857
Composite	0.126	0.178	0.751	1.029

Example 28.2 Table 28.1 below gives the 95th and the 99th percentiles of the limiting distributions of $n\widetilde{C}_n$ and $n\widetilde{A}_n$ when $X_1, \ldots, X_n$ are iid $N(\mu, \sigma^2)$ and $\hat{\mu} = \bar{X}$, $\hat{\sigma}^2 = s^2$. For comparison, the percentiles for the case of the simple null are also provided. The change in the percentage points from the simple to the composite case is radical. Using the simple percentage points would cause a very significant error in the size of the test. Unfortunately, in applied work, it is routinely done.

28.3 Chi-square Tests with Estimated Parameters and the Chernoff-Lehmann Result

Suppose $X_1, \ldots, X_n$ are iid observations from a CDF F and we wish to test $H_0 : F \in \mathcal{F} = \{F_\theta : \theta \in \Theta\}$. If $A_{ki}, i = 1, ..., k$ form a partition of S, the common support of F_θ, then the null probabilities $p_{0i} = P_{F_\theta}(A_{ki})$ will depend on θ. So the Pearson χ^2 statistic $\sum_{i=1}^{k} \frac{(n_i - np_{0i}(\theta))^2}{np_{0i}(\theta)}$ will not be computable. An adjusted version is $\widetilde{\chi}^2 = \sum_{i=1}^{k} \frac{(n_i - np_{0i}(\hat{\theta}))^2}{np_{0i}(\hat{\theta})}$, where $\hat{\theta}$ is some estimate of θ. Replacing θ by $\hat{\theta}$ will cause a change in the asymptotic distribution of $\widetilde{\chi}^2$ under F_θ, and it may depend on θ, the choice of $\hat{\theta}$, and the exact function F_θ. Depending on the choice of $\hat{\theta}$, $\widetilde{\chi}^2$ may not even be asymptotically χ^2 under F_θ.

Fix an i. Then, suppressing θ,

$$\begin{aligned}
\Delta_i &= \frac{n_i - n\hat{p}_{0i}}{\sqrt{n\hat{p}_{0i}}} \\
&= \frac{n_i - np_{0i} + np_{0i} - n\hat{p}_{0i}}{\sqrt{n\hat{p}_{0i}}} \\
&= \frac{n_i - np_{0i}}{\sqrt{n\hat{p}_{0i}}} + \frac{\sqrt{n}(p_{0i} - \hat{p}_{0i})}{\sqrt{\hat{p}_{0i}}}.
\end{aligned}$$

The vector $\left(\frac{n_i - np_{0i}}{\sqrt{n\hat{p}_{0i}}}, \frac{\sqrt{n}(p_{0i} - \hat{p}_{0i})}{\sqrt{\hat{p}_{0i}}}\right)$ will typically be asymptotically bivariate normal, and so Δ_i is asymptotically normal with a zero mean. Even $\underset{\sim}{\Delta} = (\Delta_1, ..., \Delta_k)$ is asymptotically k-dimensional normal, perhaps with a singular covariance matrix $V = V_\theta$. From general theory, therefore, $\widetilde{\chi}^2 = \|\Delta\|^2 \stackrel{\mathcal{L}}{\Rightarrow} \sum_{i=1}^k \lambda_i(\theta) w_i$, where w_i are iid χ_1^2 and $\lambda_i = \lambda_i(\theta)$ are the eigenvalues of V. If V is idempotent with rank t, then t of the eigenvalues are 1 and the rest are 0. In that case, $\widetilde{\chi}^2 \stackrel{\mathcal{L}}{\Rightarrow} \chi_t^2$. If V is not idempotent, then $\tilde{\chi}^2$ will not be distributed as χ^2 asymptotically.

Case 1 Suppose θ is known and so there is no need to estimate θ. Thus, $\hat{p}_{0i}$ is in fact the known p_{0i} itself. Writing $\sqrt{\underset{\sim}{p_0}} = (\sqrt{p_{01}}, \ldots \sqrt{p_{0k}})$, the matrix

$V = I - \sqrt{\underset{\sim}{p_0}}\left(\sqrt{\underset{\sim}{p_0}}\right)'$. It is idempotent with rank $k-1$ and $\widetilde{\chi}^2 \stackrel{\mathcal{L}}{\Rightarrow} \chi_{k-1}^2$. This is the simple case.

Case 2 Suppose we do not know θ and estimate it from the frequencies $n_1, n_2, \ldots, n_k$ of the predetermined cells $A_{k1}, A_{k2}, \ldots, A_{kk}$ by using the corresponding multinomial likelihood MLE for θ and then using $\hat{p}_{0i} = p_{0i}(\hat{\theta})$. In this case, the matrix $V = I - B(B'B)^{-1}B' - \sqrt{\underset{\sim}{p_0}}\left(\sqrt{\underset{\sim}{p_0}}\right)'$, where $B = (b_{ij})$ with $b_{ij} = \frac{1}{\sqrt{p_{0i}(\theta)}} \frac{\partial}{\partial \theta_j} p_{0i}(\theta)$. If θ is s-dimensional, then B is of the order $k \times s$. It is still the case that V is idempotent, with rank $k - s - 1$. Therefore, still $\widetilde{\chi}^2 \stackrel{\mathcal{L}}{\Rightarrow} \chi_{k-s-1}^2$. Because s unknown parameters $\theta_1, \theta_2, \ldots, \theta_s$ have to be estimated, we lose s degrees of freedom in the asymptotic χ^2 distribution.

Case 3 The raw observations $X_1, \ldots, X_n$ are iid F_θ for some unknown θ. We can estimate θ from the original likelihood $\prod_{i=1}^n f(x_i|\theta)$. Denoting $\hat{\hat{\theta}}$ as the MLE from this likelihood and using $\hat{p}_{0i} = p_{0i}(\hat{\hat{\theta}})$, the matrix V now is

$$V = I - \sqrt{\underset{\sim}{p_0}}\left(\sqrt{\underset{\sim}{p_0}}\right)' - BCB' + B(C - D)B',$$

where B is as in case 2, $C = (B'B)^{-1}$, and D is the inverse of the Fisher information matrix $I_f(\theta)$. Since C and $C - D$ are both nonnegative definite matrices, it follows that V is in between the V in case 1 and the V in case 2 in the sense of Loewner ordering of matrices. So, in case 3, the limiting distribution of $\widetilde{\chi}^2$ under F_θ is stochastically larger than χ_{k-s-1}^2 but stochastically smaller than χ_{k-1}^2. The matrix V is no longer idempotent. There are some eigenvalues that are strictly between 0 and 1. Actually, $k - s - 1$ eigenvalues are 1, one is 0, and the rest are between 0 and 1, so $\widetilde{\chi}^2 \stackrel{\mathcal{L}}{\Rightarrow} \sum_{i=1}^{k-s-1} w_i + \sum_{i=k-s+1}^k \lambda_i(\theta) w_i$, where $0 < \lambda_i(\theta) < 1, i = k - s + 1, \ldots, k$. This is the classic Chernoff (1954) result.

Remark. The Chernoff-Lehmann result leads to the intriguing question of whether semioptimal likelihoods should be used in preference to the complete likelihood in order to simplify distribution theory and obtain directly implementable procedures. The result is regarded as a pioneering result in the goodness-of-fit literature.

28.4 Chi-square Tests with Random Cells

The problem of encountering unknown parameters can also be addressed by choosing the cells $A_{k1}, A_{k2}, ..., A_{kk}$ as random sets. For example, in the one-dimensional case, each A_{ki} may be of the form $(\bar{X} - k_2 s, \bar{X} + k_1 s)$ for some $k_1, k_2 > 0$. The cell limits would act like $\mu - k_2\sigma$ and $\mu + k_1\sigma$ if μ, σ are the mean and the standard deviation in the population. Furthermore, if the population distribution is a location-scale parameter distribution, then the cell probabilities are free of μ and σ. This is a major simplification obtained by approximate choice of the random cells. A good first-generation reference is Moore and Spruill (1975). Chi-square tests with random cells have undergone massive advances since the development of modern empirical process theory. These tests probably have better potential than is appreciated in multivariate problems; see del Bario, Deheuvels, and van de Geer (2007).

28.5 Exercises

Exercise 28.1 Simulate a sample of size $n = 20$ from an $N(0, 1)$ distribution and compute the value of the adjusted K-S statistic using $\bar{X}$, s as the estimates of μ, σ. Next, compute it without using the estimated values but using the true values $\mu = 0, \sigma = 1$. Compare them and repeat the simulation.

Exercise 28.2 * Derive the limiting distribution of $\sqrt{n}(F_n(t) - F(t, \hat{\theta_n}))$ for samples from an $N(\mu, 1)$ distribution when $\theta = \mu$ is estimated by the MLE $\bar{X}$.

Exercise 28.3 * Derive the limiting distribution of $\sqrt{n}(F_n(t) - F(t, \hat{\theta_n}))$ for samples from an Exp(λ) distribution when $\theta = \lambda$ is estimated by the MLE $\bar{X}$.

Exercise 28.4 * Derive the limiting distribution of $\sqrt{n}(F_n(t) - F(t, \hat{\theta_n}))$ for samples from an $N(\mu, \sigma^2)$ distribution when $\theta = (\mu, \sigma)$ is estimated by the MLE $(\bar{X}, \sqrt{\frac{1}{n}\sum_{i=1}^{n}(X_i - \bar{X})^2})$.

Exercise 28.5 * Derive the limiting distribution of $\sqrt{n}(F_n(t) - F(t, \hat{\theta_n}))$ for samples from a Double Exp(μ, 1) distribution when $\theta = \mu$ is estimated by the MLE M_n, the sample median.

Exercise 28.6 * Derive the limiting distribution of $\sqrt{n}(F_n(t) - F(t, \hat{\theta_n}))$ for samples from an $N(\mu, 1)$ distribution when $\theta = \mu$ is estimated by M_n, the sample median. Compare the result with Exercise 28.2.

Exercise 28.7 * Derive the limiting distribution of $\sqrt{n}(F_n(t) - F(t, \hat{\theta_n}))$ for samples from an $N(\mu, 1)$ distribution when $\theta = \mu$ is estimated by the posterior mean with respect to a general normal prior. Is it the same as in Exercise 28.2?

Can you make any general statements about irrelevance of the prior for this limiting distribution?

Exercise 28.8 * Derive the limiting distribution of $\sqrt{n}(F_n(t) - F(t, \hat{\theta_n}))$ for samples from a Double Exp(μ, 1) distribution when $\theta = \mu$ is estimated by $\bar{X}$. Compare the result with Exercise 28.5.

Exercise 28.9 * For iid samples from a general location-scale density, derive the joint asymptotic distribution of the number of observations in $(\bar{X} - s, \bar{X} + s), (\bar{X} + s, \bar{X} + 2s), (\bar{X} + 2s, \infty), (\bar{X} - 2s, \bar{X} - s), (-\infty, \bar{X} - 2s)$.

Hence, suggest a test for normality.

Exercise 28.10 * Compute the matrix B in the notation of the text when samples are from an Exp(λ) distribution and the MLE is calculated from grouped data, the groups being intervals. Should it be idempotent?

Exercise 28.11 Verify by simulation the entries in Table 28.1.

Exercise 28.12 * Approximately calculate the eigenvalues $\lambda_i(\theta)$ of the Chernoff-Lehmann result in the exponential case if the true $\theta = \lambda = 1$.

References

Beran, R. and Millar, P.W. (1989). A stochastic minimum distance test for multivariate parametric models, Ann. Stat., 17(1), 125–140.

Boos, D. (1981). Minimum distance estimators for location and goodness of fit, J. Am. Stat. Assoc., 76(375), 663–670.

Brown, L., DasGupta, A., Marden, J.I., and Politis, D. (2004). Characterizations, sub and resampling, and goodness of fit, in Festschrift for Herman Rubin *IMS Lecture*

Notes Monograph Series, A. Dasgupta (eds.) Vol. 45, Institute of Mathematical Statistics, Beachwood, OH, 180–206.

Chernoff, H. and Lehmann, E.L. (1954). The use of maximum likelihood estimates in chisquare tests for goodness of fit, Ann. Math. Stat., 25, 579–586.

Csörgo, M. (1974). *On the Problem of Replacing Composite Hypotheses by Equivalent Simple Ones*, Colloquia of Mathematical Society Janos Bolyai, Vol. 9, North-Holland, Amsterdam.

Csörgo, S. (1984). *Testing by Empirical Characteristic Functions: A Survey*, Asymptotic Statistics, Vol. 2, Elsevier, Amsterdam.

del Bario, E., Deheuvels, P., and van de Geer, S. (2007). *Lectures on Empirical Processes*, European Mathematical Society, Zurich.

Hall, P. and Welsh, A. (1983). A test for normality based on the empirical characteristic function, Biometrika, 70(2), 485–489.

Moore, D.S. and Spruill, M.C. (1975). Unified large sample theory of general chi-square statistics for tests of fit, Ann. Stat., 3, 599–616.

Mudholkar, G. and Lin, C.T. (1987). *On Two Applications of Characterization Theorems to Goodness-of-Fit*, Colloquia of Mathematical Society Janos Bolyai, Vol. 45, North-Holland, Amsterdam.

Neuhaus, G. (1979). Asymptotic theory of goodness of fit tests when parameters are present: a survey, Math. Operationforsch. Stat. Ser. Stat.,10(3), 479–494.

Stephens, M.A. (1976). Asymptotic results for goodness-of-fit statistics with unknown parameters, Ann. Stat., 4(2), 357–369.

Stuart, A. and Ord, K. (1991). *Kendall's Theory of Statistics*, 4th ed.,Vol. II, Clarendon Press, New York.

van der Vaart, A. (1998). *Asymptotic Statistics*, Cambridge University Press, Cambridge.

Verdinelli, I. and Wasserman, L. (1998). Bayesian goodness-of-fit testing using infinite dimensional Exponential families, Ann. Stat., 26(4), 1215–1241.

Chapter 29
The Bootstrap

The bootstrap is a resampling mechanism designed to provide information about the sampling distribution of a functional $T(X_1, X_2, ..., X_n, F)$, where $X_1, X_2, ..., X_n$ are sample observations and F is the CDF from which $X_1, X_2, ..., X_n$ are independent observations. The bootstrap is not limited to the iid situation. It has been studied for various kinds of dependent data and complex situations. In fact, this versatile nature of the bootstrap is the principal reason for its popularity. There are numerous texts and reviews of bootstrap theory and methodology at various technical levels. We recommend Efron and Tibshirani (1993) and Davison and Hinkley (1997) for applications-oriented broad expositions and Hall (1992) and Shao and Tu (1995) for detailed theoretical development. Modern reviews include Hall (2003), Beran (2003), Bickel (2003), and Efron (2003). Bose and Politis (1992) is a well-written nontechnical account, and Lahiri (2003) is a rigorous treatment of the bootstrap for various kinds of dependent data.

Suppose $X_1, X_2, \ldots, X_n \stackrel{\text{iid}}{\sim} F$ and $T(X_1, X_2, ..., X_n, F)$ is a functional; e.g., $T(X_1, X_2, ..., X_n, F) = \frac{\sqrt{n}(\bar{X}-\mu)}{\sigma}$, where $\mu = E_F(X_1)$ and $\sigma^2 = \mathrm{Var}_F(X_1)$. In statistical problems, we frequently need to know something about the sampling distribution of T; e.g., $P_F(T(X_1, X_2, ..., X_n, F) \leq t)$. If we had replicated samples from the population, resulting in a series of values for the statistic T, then we could form estimates of $P_F(T \leq t)$ by counting how many of the T_i's are $\leq t$. But statistical sampling is not done that way. We do not usually obtain replicated samples; we obtain just one set of data of some size n. However, let us think for a moment of a finite population. A large sample from a finite population should be well representative of the full population itself, so replicated samples (with replacement) from the original sample, which would just be an iid sample from the empirical CDF F_n, could be regarded as proxies for replicated samples from the population itself, provided n is large. Suppose that for some number B we draw B resamples of size n from the original sample. Denoting the resamples from the original

A. DasGupta, *Asymptotic Theory of Statistics and Probability*,

sample as $(X_{11}^*, X_{12}^*, ..., X_{1n}^*), (X_{21}^*, X_{22}^*, ..., X_{2n}^*), ..., (X_{B1}^*, X_{B2}^*, ..., X_{Bn}^*)$, with corresponding values $T_1^*, T_2^*, ..., T_B^*$ for the functional T, one can use simple frequency-based estimates such as $\frac{\#\{j:T_j^* \leq t\}}{B}$ to estimate $P_F(T \leq t)$. This is the basic idea of the bootstrap. Over time, the bootstrap has found its use in estimating other quantities, e.g., $\text{Var}_F(T)$ or quantiles of T. The bootstrap is thus an omnibus mechanism for approximating sampling distributions or functionals of sampling distributions of statistics. Since frequentist inference is mostly about sampling distributions of suitable statistics, the bootstrap is viewed as an immensely useful and versatile tool, further popularized by its automatic nature. However, it is also frequently used in situations where it should not be used. In this chapter, we give a broad methodological introduction to various types of bootstraps, explain their theoretical underpinnings, discuss their successes and limitations, and try them out in some trial cases.

29.1 Bootstrap Distribution and the Meaning of Consistency

The formal definition of the bootstrap distribution of a functional is the following.

Definition 29.1 Let $X_1, X_2, \ldots, X_n \overset{\text{iid}}{\sim} F$ and $T(X_1, X_2, ..., X_n, F)$ be a given functional. The *ordinary bootstrap distribution* of T is defined as

$$H_{\text{Boot}}(x) = P_{F_n}(T(X_1^*, ..., X_n^*, F_n) \leq x),$$

where $(X_1^*, ..., X_n^*)$ is an iid sample of size n from the empirical CDF F_n.

It is common to use the notation P_* to denote probabilities under the bootstrap distribution.

Remark. $P_{F_n}(\cdot)$ corresponds to probability statements corresponding to all the n^n possible resamples with replacement from the original sample $(X_1, \ldots, X_n)$. Since recalculating T from all n^n resamples is basically impossible unless n is very small, one uses a smaller number of B resamples and recalculates T only B times. Thus $H_{\text{Boot}}(x)$ itself is estimated by a Monte Carlo, known as the *bootstrap Monte Carlo*, so the final estimate for $P_F(T(X_1, X_2, ..., X_n, F_n) \leq x)$ absorbs errors from two sources: (i) pretending $(X_{i1}^*, X_{i2}^*, ..., X_{in}^*)$ to be bona fide resamples from F; (ii) estimating the true $H_{\text{Boot}}(x)$ by a Monte Carlo. By choosing B adequately large, the Monte Carlo error is generally ignored. The choice of B that would let one ignore the Monte Carlo error is a hard mathematical problem; Hall (1986, 1989a) are two key references. It is customary to choose $B \approx 300$ for variance estimation and a somewhat larger value for estimating quantiles. It is hard to give any general reliable prescriptions on B.

It is important to note that the resampled data need not necessarily be obtained from the empirical CDF F_n. Indeed, it is a natural question whether resampling from a smoothed nonparametric distribution estimator can result in better performance. Examples of such smoothed distribution estimators are integrated kernel density estimates. It turns out that, in some problems, smoothing does lead to greater accuracy, typically in the second order. See Silverman and Young (1987) and Hall, DiCiccio, and Romano (1989) for practical questions and theoretical analysis of the benefits of using a smoothed bootstrap. Meanwhile, bootstrapping from F_n is often called the *naive or orthodox bootstrap*, and we will sometimes use this terminology.

Remark. At first glance, the idea appears to be a bit too simple to actually work. But one has to have a definition for what one means by the bootstrap working in a given situation. It depends on what one wants the bootstrap to do. For estimating the CDF of a statistic, one should want $H_{\text{Boot}}(x)$ to be numerically close to the true CDF $H_n(x)$ of T. This would require consideration of metrics on CDFs. For a general metric ρ, the definition of "the bootstrap working" is the following.

Definition 29.2 Let F and G be two CDFs on a sample space $\mathcal{X}$. Let $\rho(F, G)$ be a metric on the space of CDFs on $\mathcal{X}$. For $X_1, X_2, \ldots, X_n \stackrel{\text{iid}}{\sim} F$, and a given functional $T(X_1, X_2, ..., X_n, F)$, let

$$H_n(x) = P_F(T(X_1, X_2, ..., X_n, F) \leq x),$$
$$H_{\text{Boot}}(x) = P_*(T(X_1^*, X_2^*, ..., X_n^*, F_n) \leq x).$$

We say that the bootstrap is weakly consistent under ρ for T if $\rho(H_n, H_{\text{Boot}}) \stackrel{P}{\Rightarrow} 0$ as $n \to \infty$. We say that the bootstrap is strongly consistent under ρ for T if $\rho(H_n, H_{\text{Boot}}) \stackrel{\text{a.s.}}{\Rightarrow} 0$.

Remark. Note that the need for mentioning convergence to zero in probability or a.s. in this definition is due to the fact that the bootstrap distribution H_{Boot} is a random CDF. That H_{Boot} is a random CDF has nothing to do with bootstrap Monte Carlo; it is a random CDF because as a function it depends on the original sample $(X_1, X_2, ..., X_n)$. Thus, the bootstrap uses a random CDF to approximate a deterministic but unknown CDF, namely the true CDF H_n of the functional T.

Example 29.1 How does one apply the bootstrap in practice? Suppose, for example, $T(X_1, \ldots, X_n, F) = \frac{\sqrt{n}(\bar{X}-\mu)}{\sigma}$. In the orthodox bootstrap scheme, we take iid samples from F_n. The mean and the variance of the empirical distribution F_n are $\bar{X}$ and $s^2 = \frac{1}{n}\sum_{i=1}^{n}(X_i - \bar{X})^2$ (note the n rather than $n-1$ in the denominator). The bootstrap is a device for estimating $P_F(\frac{\sqrt{n}(\bar{X}-\mu(F))}{\sigma} \leq x)$ by $P_{F_n}(\frac{\sqrt{n}(\bar{X}_n^*-\bar{X})}{s} \leq x)$. We will further approximate $P_{F_n}(\frac{\sqrt{n}(\bar{X}_n^*-\bar{X})}{s} \leq x)$ by resampling only B times from the original sample set $\{X_1, \ldots, X_n\}$. In other words, finally we will report as our estimate for $P_F(\frac{\sqrt{n}(\bar{X}-\mu)}{\sigma} \leq x)$ the number $\#\{j : \frac{\sqrt{n}(\bar{X}_{n,j}^*-\bar{X})}{s} \leq x\}/B$.

29.2 Consistency in the Kolmogorov and Wasserstein Metrics

We start with the case of the sample mean of iid random variables. If $X_1, \ldots, X_n \overset{\text{iid}}{\sim} F$ and if $\text{Var}_F(X_i) < \infty$, then $\sqrt{n}(\bar{X} - \mu)$ has a limiting normal distribution by the CLT. So a probability such as $P_F(\sqrt{n}(\bar{X}-\mu) \leq x)$ could be approximated for example by $\Phi(\frac{x}{s})$, where s is the sample standard deviation. An interesting property of the bootstrap approximation is that, even when the CLT approximation $\Phi(\frac{x}{s})$ is available, the bootstrap approximation may be more accurate. We will later describe theoretical results in this regard. But first we present two consistency results corresponding to the following two specific metrics that have earned a special status in this literature:

(i) Kolmogorov metric

$$K(F, G) = \sup_{-\infty<x<\infty} |F(x) - G(x)|;$$

(ii) Mallows-Wasserstein metric

$$\ell_2(F, G) = \inf_{\Gamma_{2,F,G}} (E|Y - X|^2)^{\frac{1}{2}},$$

where $X \sim F$, $Y \sim G$, and $\Gamma_{2,F,G}$ is the class of all joint distributions of (X, Y) with marginals F and G, each with a finite second moment.

ℓ_2 is a special case of the more general metric

$$\ell_p(F, G) = \inf_{\Gamma_{p,F,G}} (E|Y - X|^p)^{\frac{1}{p}},$$

with the infimum being taken over the class of joint distributions with marginals as F, G, and the pth moment of F, G being finite.

Of these, the Kolmogorov metric is universally regarded as a natural one. But how about ℓ_2? ℓ_2 is a natural metric for many statistical problems because of its interesting property that $\ell_2(F_n, F) \to 0$ iff $F_n \stackrel{\mathcal{L}}{\Rightarrow} F$ and $E_{F_n}(X^i) \to E_F(X^i)$ for $i = 1, 2$. Since one might want to use the bootstrap primarily for estimating the CDF, mean, and variance of a statistic, consistency in ℓ_2 is just the right result for that purpose.

Theorem 29.1 Suppose $X_1, X_2, \ldots, X_n \stackrel{\text{iid}}{\sim} F$ and that $E_F(X_1^2) < \infty$. Let $T(X_1, \ldots, X_n, F) = \sqrt{n}(\bar{X} - \mu)$. Then $K(H_n, H_{\text{Boot}})$ and $\ell_2(H_n, H_{\text{Boot}}) \stackrel{\text{a.s.}}{\longrightarrow} 0$ as $n \to \infty$.

Remark. Strong consistency in K is proved in Singh (1981), and that for ℓ_2 is proved in Bickel and Freedman (1981). Notice that $E_F(X_1^2) < \infty$ guarantees that $\sqrt{n}(\bar{X} - \mu)$ admits a CLT. And Theorem 29.1 says that the bootstrap is strongly consistent (w.r.t. K and ℓ_2) under that assumption. This is in fact a very good rule of thumb: if a functional $T(X_1, X_2, ..., X_n, F)$ admits a CLT, then the bootstrap would be at least weakly consistent for T. Strong consistency might require a little more assumption.

We sketch a proof of the strong consistency in K. The proof requires use of the Berry-Esseen inequality, Polya's theorem (see Chapter 1 or Chapter 2), and a strong law known as the Zygmund-Marcinkiewicz strong law, which we state below.

Lemma 29.1 **(Zygmund-Marcinkiewicz SLLN)** Let $Y_1, Y_2, \ldots$ be iid random variables with CDF F and suppose, for some $0 < \delta < 1$, $E_F|Y_1|^\delta < \infty$. Then $n^{-1/\delta} \sum_{i=1}^n Y_i \stackrel{\text{a.s.}}{\Rightarrow} 0$.

We are now ready to sketch the proof of strong consistency of H_{Boot} under K. Using the definition of K, we can write $K(H_n, H_{\text{Boot}}) = \sup_x \left| P_F \{T_n \le x\} - P_* \{T_n^* \le x\} \right|$

$$
\begin{aligned}
&= \sup_x \left| P_F \left\{ \frac{T_n}{\sigma} \le \frac{x}{\sigma} \right\} - P_* \left\{ \frac{T_n^*}{s} \le \frac{x}{s} \right\} \right| \\
&= \sup_x \left| P_F \left\{ \frac{T_n}{\sigma} \le \frac{x}{\sigma} \right\} - \Phi\left(\frac{x}{\sigma}\right) + \Phi\left(\frac{x}{\sigma}\right) - \Phi\left(\frac{x}{s}\right) + \Phi\left(\frac{x}{s}\right) \right. \\
&\quad \left. - P_* \left\{ \frac{T_n^*}{s} \le \frac{x}{s} \right\} \right| \\
&\le \sup_x \left| P_F \left\{ \frac{T_n}{\sigma} \le \frac{x}{\sigma} \right\} - \Phi\left(\frac{x}{\sigma}\right) \right| + \sup_x \left| \Phi\left(\frac{x}{\sigma}\right) - \Phi\left(\frac{x}{s}\right) \right| \\
&\quad + \sup_x \left| \Phi\left(\frac{x}{s}\right) - P_* \left\{ \frac{T_n^*}{s} \le \frac{x}{s} \right\} \right| \\
&= A_n + B_n + C_n, \quad \text{say.}
\end{aligned}
$$

That $A_n \to 0$ is a direct consequence of Polya's theorem. Also, s^2 converges almost surely to σ^2 and so, by the continuous mapping theorem, s converges almost surely to σ. Then $B_n \Rightarrow 0$ almost surely by the fact that $\Phi(\cdot)$ is a uniformly continuous function. Finally, we can apply the Berry-Esseen theorem to show that C_n goes to zero:

$$\begin{aligned} C_n &\leq \frac{4}{5\sqrt{n}} \cdot \frac{E_{F_n}|X_1^* - \overline{X}_n|^3}{[\text{var}_{F_n}(X_1^*)]^{3/2}} = \frac{4}{5\sqrt{n}} \cdot \frac{\sum_{i=1}^n |X_i - \overline{X}_n|^3}{ns^3} \\ &\leq \frac{4}{5n^{3/2}s^3} \cdot 2^3 \left[\sum_{i=1}^n |X_i - \mu|^3 + n|\mu - \overline{X}_n|^3 \right] \\ &= \frac{M}{s^3} \left[\frac{1}{n^{3/2}} \sum_{i=1}^n |X_i - \mu|^3 + \frac{|\overline{X}_n - \mu|^3}{\sqrt{n}} \right], \end{aligned}$$

where $M = \frac{32}{5}$.

Since $s \Rightarrow \sigma > 0$ and $\overline{X}_n \Rightarrow \mu$, it is clear that $|\overline{X}_n - \mu|^3/(\sqrt{n}s^3) \Rightarrow 0$ almost surely. As regards the first term, let $Y_i = |X_i - \mu|^3$ and $\delta = 2/3$. Then the $\{Y_i\}$ are iid and

$$E|Y_i|^\delta = E_F|X_i - \mu|^{3 \cdot 2/3} = \text{Var}_F(X_1) < \infty.$$

It now follows from the Zygmund-Marcinkiewicz SLLN that

$$\frac{1}{n^{3/2}} \sum_{i=1}^n |X_i - \mu|^3 = n^{-1/\delta} \sum_{i=1}^n Y_i \Rightarrow 0 \text{ a.s.} \quad \text{as } n \to \infty.$$

Thus, $A_n + B_n + C_n \Rightarrow 0$ almost surely, and hence $K(H_n, H_{\text{Boot}}) \Rightarrow 0$.

We now proceed to a proof of convergence under the Wasserstein-Kantorovich-Mallows metric ℓ_2. Recall that convergence in ℓ_2 allows us to conclude more than weak convergence. We start with a sequence of results that enumerate useful properties of the ℓ_2 metric.

These facts (see Bickel and Freedman (1981)) are needed to prove consistency of H_{Boot} in the ℓ_2 metric.

Lemma 29.2 Let $G_n, G \in \Gamma_2$. Then $\ell_2(G_n, G) \to 0$ if and only if

$$G_n \stackrel{\mathcal{L}}{\Rightarrow} G \quad \text{and} \quad \lim_{n\to\infty} \int x^k dG_n(x) = \int x^k \, dG(x), \quad k = 1, 2.$$

Lemma 29.3 Let $G, H \in \Gamma_2$, and suppose $Y_1, \ldots, Y_n$ are iid G and $Z_1, \ldots, Z_n$ are iid H. If $G^{(n)}$ is the CDF of $\sqrt{n}(\bar{Y} - \mu_G)$ and $H^{(n)}$ is the CDF of $\sqrt{n}(\bar{Z} - \mu_H)$, then $\ell_2(G^{(n)}, H^{(n)}) \leq \ell_2(G, H), \quad \forall\, n \geq 1$.

Lemma 29.4 (Glivenko-Cantelli) Let $X_1, X_2, \ldots, X_n$ be iid F and let F_n be the empirical CDF. Then $F_n(x) \to F(x)$ almost surely, uniformly in x.

Lemma 29.5 Let $X_1, X_2, \ldots, X_n$ be iid F and let F_n be the empirical CDF. Then $\ell_2(F_n, F) \Rightarrow 0$ almost surely.

The proof that $\ell_2(H_n, H_{\text{Boot}})$ converges to zero almost surely follows on simply putting together the lemmas 29.2–29.5. We omit this easy verification.

It is natural to ask if the bootstrap is consistent for $\sqrt{n}(\bar{X} - \mu)$ even when $E_F(X_1^2) = \infty$. If we insist on strong consistency, then the answer is negative. The point is that the sequence of bootstrap distributions is a sequence of random CDFs and so it cannot be expected a priori that it will converge to a fixed CDF. It may very well converge to a random CDF, depending on the particular realization $X_1, X_2, \ldots$. One runs into this problem if $E_F(X_1^2)$ does not exist. We state the result below.

Theorem 29.2 Suppose $X_1, X_2, \ldots$ are iid random variables. There exist $\mu_n(X_1, X_2, ..., X_n)$, an increasing sequence c_n, and a fixed CDF $G(x)$ such that

$$P_* \left(\frac{\sum_{i=1}^{n} (X_i^* - \mu(X_1, \ldots, X_n))}{c_n} \leq x \right) \xrightarrow{\text{a.s.}} G(x)$$

if and only if $E_F(X_1^2) < \infty$, in which case $\frac{c_n}{\sqrt{n}} \longrightarrow 1$.

Remark. The moral of Theorem 29.2 is that the existence of a nonrandom limit itself would be a problem if $E_F(X_1^2) = \infty$. See Athreya (1987), Giné and Zinn (1989), and Hall (1990) for proofs and additional examples.

The consistency of the bootstrap for the sample mean under finite second moments is also true for the multivariate case. We record consistency under the Kolmogorov metric next; see Shao and Tu (1995) for a proof.

Theorem 29.3 Let $\underset{\sim}{X}_1, \cdots, \underset{\sim}{X}_n, \cdots$ be iid F with $\text{cov}_F(\underset{\sim}{X}_1) = \Sigma$, Σ finite. Let $T(\underset{\sim}{X}_1, \underset{\sim}{X}_2, ..., \underset{\sim}{X}_n, F) = \sqrt{n}(\underset{\sim}{\bar{X}} - \underset{\sim}{\mu})$. Then $K(H_{\text{Boot}}, H_n) \xrightarrow{\text{a.s.}} 0$ as $n \to \infty$.

29.3 Delta Theorem for the Bootstrap

We know from the ordinary delta theorem that if T admits a CLT and $g(\cdot)$ is a smooth transformation, then $g(T)$ also admits a CLT. If we were to believe in our rule of thumb, then this would suggest that the bootstrap should be consistent for $g(T)$ if it is already consistent for T. For the case of sample mean vectors, the following result holds; again, see Shao and Tu (1995) for a proof.

Theorem 29.4 Let $\underset{\sim}{X}_1, \underset{\sim}{X}_2, \ldots, \underset{\sim}{X}_n \overset{\text{iid}}{\sim} F$ and let $\Sigma_{p\times p} = \text{cov}_F(\underset{\sim}{X}_1)$ be finite. Let $T(\underset{\sim}{X}_1, \underset{\sim}{X}_2, \ldots, \underset{\sim}{X}_n, F) = \sqrt{n}(\bar{\underset{\sim}{X}} - \underset{\sim}{\mu})$ and, for some $m \geq 1$, let $g : \mathbb{R}^p \to \mathbb{R}^m$. If $\nabla g(\cdot)$ exists in a neighborhood of $\underset{\sim}{\mu}, \nabla g(\underset{\sim}{\mu}) \neq \underset{\sim}{0}$, and if $\nabla g(\cdot)$ is continuous at $\underset{\sim}{\mu}$, then the bootstrap is strongly consistent w.r.t. K for $\sqrt{n}(g(\bar{\underset{\sim}{X}}) - g(\underset{\sim}{\mu}))$.

Example 29.2 Let $X_1, X_2, \ldots, X_n \overset{\text{iid}}{\sim} F$, and suppose $E_F(X_1^4) < \infty$. Let $\underset{\sim}{Y}_i = \binom{X_i}{X_i^2}$. Then, with $p = 2$, $\underset{\sim}{Y}_1, \underset{\sim}{Y}_2, \ldots, \underset{\sim}{Y}_n$ are iid p-dimensional vectors with $\text{cov}(\underset{\sim}{Y}_1)$ finite. Note that $\bar{\underset{\sim}{Y}} = \binom{\bar{X}}{\frac{1}{n}\sum_{i=1}^n X_i^2}$. Consider the transformation $g : \mathbb{R}^2 \to \mathbb{R}^1$ defined as $g(u, v) = v - u^2$. Then $\frac{1}{n}\sum_{i=1}^n (X_i - \bar{X})^2 = \frac{1}{n}\sum_{i=1}^n X_i^2 - (\bar{X})^2 = g(\bar{\underset{\sim}{Y}})$. If we let $\underset{\sim}{\mu} = E(\underset{\sim}{Y}_1)$, then $g(\underset{\sim}{\mu}) = \sigma^2 = \text{Var}(X_1)$. Since $g(\cdot)$ satisfies the conditions of the Theorem 29.4, it follows that the bootstrap is strongly consistent w.r.t. K for $\sqrt{n}(\frac{1}{n}\sum_{i=1}^n (X_i - \bar{X})^2 - \sigma^2)$.

29.4 Second-Order Accuracy of the Bootstrap

One philosophical question about the use of the bootstrap is whether the bootstrap has any advantages at all when a CLT is already available. To be specific, suppose $T(X_1, \ldots, X_n, F) = \sqrt{n}(\bar{X} - \mu)$. If $\sigma^2 = \text{Var}_F(X) < \infty$, then $\sqrt{n}(\bar{X} - \mu) \overset{\mathcal{L}}{\Rightarrow} N(0, \sigma^2)$ and $K(H_{\text{Boot}}, H_n) \overset{\text{a.s.}}{\to} 0$. So two competitive approximations to $P_F(T(X_1, \ldots, X_n, F) \leq x)$ are $\Phi(\frac{x}{\hat{\sigma}})$ and $P_{F_n}(\sqrt{n}(\bar{X}^* - \bar{X}) \leq x)$. It turns out that, for certain types of statistics, the bootstrap approximation is (theoretically) more accurate than the approximation provided by the CLT. Because any normal distribution is symmetric, the CLT cannot capture information about the skewness in the finite sample distribution of T. The bootstrap approximation does so. So the bootstrap succeeds in correcting for skewness, just as an Edgeworth expansion would do. This is called Edgeworth correction by the bootstrap, and the property is called second-order accuracy of the bootstrap. It is important to remember that second-order accuracy is not automatic; it holds for certain types of T but not for others. It is also important to understand that practical accuracy and theoretical higher-order accuracy can be different things. The following heuristic calculation will illustrate when second-order accuracy can be anticipated. The first result on higher-order accuracy of the bootstrap is due to Singh (1981). In addition to the references we provided in the beginning, Lehmann (1999) gives a very readable treatment of higher-order accuracy of the bootstrap.

Suppose $X_1, X_2, \ldots, X_n \stackrel{\text{iid}}{\sim} F$ and $T(X_1, \ldots, X_n, F) = \frac{\sqrt{n}(\bar{X}-\mu)}{\sigma}$; here $\sigma^2 = \mathrm{Var}_F(X_1) < \infty$. We know that T admits the Edgeworth expansion

$$
\begin{aligned}
P_F(T \leq x) &= \Phi(x) + \frac{p_1(x|F)}{\sqrt{n}}\varphi(x) + \frac{p_2(x|F)}{n}\varphi(x) \\
&\quad + \text{smaller order terms}, \\
P_*(T^* \leq x) &= \Phi(x) + \frac{p_1(x|F_n)}{\sqrt{n}}\varphi(x) + \frac{p_2(x|F_n)}{n}\varphi(x) \\
&\quad + \text{smaller order terms}, \\
H_n(x) - H_{\text{Boot}}(x) &= \frac{p_1(x|F) - p_1(x|F_n)}{\sqrt{n}} + \frac{p_2(x|F) - p_2(x|F_n)}{n} \\
&\quad + \text{smaller order terms}.
\end{aligned}
$$

Recall now that the polynomials p_1, p_2 are given as

$$
\begin{aligned}
p_1(x|F) &= \frac{\gamma}{6}(1 - x^2), \\
p_2(x|F) &= x\left[\frac{\kappa - 3}{24}(3 - x^2) - \frac{\kappa^2}{72}(x^4 - 10x^2 + 15)\right],
\end{aligned}
$$

where $\gamma = \frac{E_F(X_1-\mu)^3}{\sigma^3}$ and $\kappa = \frac{E_F(X_1-\mu)^4}{\sigma^4}$. Since $\gamma_{F_n} - \gamma = O_p(\frac{1}{\sqrt{n}})$ and $\kappa_{F_n} - \kappa = O_p(\frac{1}{\sqrt{n}})$, just from the CLT for γ_{F_n} and κ_{F_n} under finiteness of four moments, one obtains $H_n(x) - H_{\text{Boot}}(x) = O_p(\frac{1}{n})$. If we contrast this with the CLT approximation, in general, the error in the CLT is $O(\frac{1}{\sqrt{n}})$, as is known from the Berry-Esseen theorem. The $\frac{1}{\sqrt{n}}$ rate cannot be improved in general even if there are four moments. Thus, by looking at the standardized statistic $\frac{\sqrt{n}(\bar{X}-\mu)}{\sigma}$, we have succeeded in making the bootstrap one order more accurate than the CLT. This is called second-order accuracy of the bootstrap. If one does not standardize, then

$$
P_F(\sqrt{n}(\bar{X} - \mu) \leq x) = P_F\left(\frac{\sqrt{n}(\bar{X} - \mu)}{\sigma} \leq \frac{x}{\sigma}\right) \to \Phi\left(\frac{x}{\sigma}\right),
$$

and the leading term in the bootstrap approximation in this unstandardized case would be $\Phi(\frac{x}{\hat{\sigma}})$. So the bootstrap approximates the true CDF $H_n(x)$ also at the rate $\frac{1}{\sqrt{n}}$; i.e., if one does not standardize, then $H_n(x) - H_{\text{Boot}}(x) = O_p(\frac{1}{\sqrt{n}})$. We have now lost the second-order accuracy. The following second rule of thumb often applies.

Rule of Thumb Let $X_1, X_2, \ldots, X_n \overset{\text{iid}}{\sim} F$ and $T(X_1, \ldots, X_n, F)$ a functional. If $T(X_1, \ldots, X_n, F) \overset{\mathcal{L}}{\Rightarrow} N(0, \tau^2)$, where τ is independent of F, then second-order accuracy is likely. Proving it will depend on the availability of an Edgeworth expansion for T. If τ depends on F (i.e., $\tau = \tau(F)$), then the bootstrap should be just first-order accurate.

Thus, as we will now see, the orthodox bootstrap is second-order accurate for the standardized mean $\frac{\sqrt{n}(\bar{X}-\mu)}{\sigma}$, although from an inferential point of view it is not particularly useful to have an accurate approximation to the distribution of $\frac{\sqrt{n}(\bar{X}-\mu)}{\sigma}$ because σ would usually be unknown, and the accurate approximation could not really be used to construct a confidence interval for μ. Still, the second-order accuracy result is theoretically insightful.

We state a specific result below for the case of standardized and nonstandardized sample means. Let $H_n(x) = P_F(\sqrt{n}(\bar{X} - \mu) \le x)$, $H_{n,0}(x) = P_F(\frac{\sqrt{n}(\bar{X}-\mu)}{\sigma} \le x)$, $H_{\text{Boot}}(x) = P_*(\sqrt{n}(\bar{X}^* - \bar{X}) \le x)$, $H_{\text{Boot},0}(x) = P_{F_n}(\frac{\sqrt{n}(\bar{X}^*-\bar{X})}{s} \le x)$.

Theorem 29.5 Let $X_1, X_2, \ldots, X_n \overset{\text{iid}}{\sim} F$.

(a) If $E_F|X_1|^3 < \infty$ and F is nonlattice, then $K(H_{n,0}, H_{\text{Boot},0}) = o_p(\frac{1}{\sqrt{n}})$.

(b) If $E_F|X_1|^3 < \infty$ and F is lattice, then $\sqrt{n}K(H_{n,0}, H_{\text{Boot},0}) \overset{P}{\to} c$, $0 < c < \infty$.

Remark. See Lahiri (2003) for a proof. The constant c in the lattice case equals $\frac{h}{\sigma\sqrt{2\pi}}$, where h is the span of the lattice $\{a + kh, k = 0, \pm 1, \pm 2, \ldots\}$ on which the X_i are supported. Note also that part (a) says that higher-order accuracy for the standardized case obtains with three moments; Hall (1988) showed that finiteness of three absolute moments is in fact necessary and sufficient for higher-order accuracy of the bootstrap in the standardized case. Bose and Babu (1991) investigate the *unconditional probability* that the Kolmogorov distance between H_{Boot} and H_n exceeds a quantity of the order $o(n^{-\frac{1}{2}})$ for a variety of statistics and show that, with various assumptions, this probability goes to zero at a rate faster than $O(n^{-1})$.

Example 29.3 How does the bootstrap compare with the CLT approximation in actual applications? The question can only be answered by case-by-case simulation. The results are mixed in the following numerical table. The X_i are iid Exp(1) in this example and $T = \sqrt{n}(\bar{X} - 1)$ with $n = 20$. For the bootstrap approximation, $B = 250$ was used.

t	$H_n(t)$	CLT approximation	$H_{\text{Boot}}(t)$
−2	0.0098	0.0228	0.0080
−1	0.1563	0.1587	0.1160
0	0.5297	0.5000	0.4840
1	0.8431	0.8413	0.8760
2	0.9667	0.9772	0.9700

29.5 Other Statistics

The ordinary bootstrap that resamples with replacement from the empirical CDF F_n is consistent for many other natural statistics besides the sample mean and even higher-order accurate for some, but under additional conditions. We mention a few such results below; see Shao and Tu (1995) for further details on the theorems in this section.

Theorem 29.6 (Sample Percentiles)

Let $X_1, \ldots, X_n$ be $\stackrel{\text{iid}}{\sim} F$ and let $0 < p < 1$. Let $\xi_p = F^{-1}(p)$ and suppose F has a positive derivative $f(\xi_p)$ at ξ_p. Let $T_n = T(X_1, \ldots, X_n, F) = \sqrt{n}(F_n^{-1}(p) - \xi_p)$ and $T_n^* = T(X_1^*, \ldots, X_n^*, F_n) = \sqrt{n}(F_n^{*-1}(p) - F_n^{-1}(p))$, where F_n^* is the empirical CDF of $X_1^*, \ldots, X_n^*$. Let $H_n(x) = P_F(T_n \le x)$ and $H_{\text{Boot}}(x) = P_*(T_n^* \le x)$. Then, $K(H_{\text{Boot}}, H_n) = O(n^{-1/4}\sqrt{\log \log n})$ almost surely.

Remark. So again we see that, under certain conditions that ensure the existence of a CLT, the bootstrap is consistent.

Next we consider the class of one-sample U-statistics.

Theorem 29.7 (U-statistics)

Let $U_n = U_n(X_1, \ldots, X_n)$ be a U-statistic with a kernel h of order 2. Let $\theta = E_F(U_n) = E_F[h(X_1, X_2)]$, where $X_1, X_2 \stackrel{\text{iid}}{\sim} F$. Assume:

(i) $E_F\left(h^2(X_1, X_2)\right) < \infty$.
(ii) $\tau^2 = \text{Var}_F\left(\tilde{h}(X)\right) > 0$, where $\tilde{h}(x) = E_F[h(X_1, X_2)|X_2 = x]$.
(iii) $E_F|h(X_1, X_1)| < \infty$.

Let $T_n = \sqrt{n}(U_n - \theta)$ and $T_n^* = \sqrt{n}(U_n^* - U_n)$, where $U_n^* = U_n(X_1^*, \ldots, X_n^*)$, $H_n(x) = P_F(T_n \le x)$, and $H_{\text{Boot}}(x) = P_*(T_n^* \le x)$. Then $K(H_n, H_{\text{Boot}}) \xrightarrow{\text{a.s}} 0$.

Remark. Under conditions (i) and (ii), $\sqrt{n}(U_n - \theta)$ has a limiting normal distribution. Condition (iii) is a new additional condition and actually cannot be relaxed. Condition (iii) is vacuous if the kernel h is bounded or a function of $|X_1 - X_2|$. Under additional moment conditions on the kernel h, there is also a higher-order accuracy result; see Helmers (1991).

Previously, we observed that the bootstrap is consistent for smooth functions of a sample mean vector. That lets us handle statistics such as the sample variance. Under some more conditions, even higher-order accuracy obtains. Here is a result in that direction.

Theorem 29.8 (Higher-Order Accuracy for Functions of Means)

Let $X_1, \ldots, X_n \overset{\text{iid}}{\sim} F$ with $E_F(X_1) = \mu$ and $\text{cov}_F(X_1) = \Sigma_{p \times p}$. Let $g : \mathbb{R}^p \to \mathbb{R}$ be such that $g(\cdot)$ is twice continuously differentiable in some neighborhood of μ and $\nabla g(\mu) \neq 0$. Assume also:

(i) $E_F ||X_1 - \mu||^3 < \infty$.
(ii) $\limsup_{||t|| \to \infty} \left| E_F \left(e^{it' X_1} \right) \right| < 1$.

Let $T_n = \frac{\sqrt{n}(g(\bar{X}) - g(\mu))}{\sqrt{(\nabla g(\mu))' \Sigma (\nabla g(\mu))}}$ and $T_n^* = \frac{\sqrt{n}(g(\bar{X}^*) - g(\bar{X}))}{\sqrt{(\nabla g(\bar{X}))' S (\nabla g(\bar{X}))}}$, where $S = S(X_1, \ldots, X_n)$ is the sample variance-covariance matrix. Also let $H_n(x) = P_F(T_n \le x)$ and $H_{\text{Boot}}(x) = P_*(T_n^* \le x)$. Then $\sqrt{n} K(H_n, H_{\text{Boot}}) \xrightarrow{\text{a.s}} 0$.

Finally, let us describe the case of the t-statistic. By our previous rule of thumb, we would expect the bootstrap to be higher-order accurate simply because the t-statistic is already studentized and has an asymptotic variance function independent of the underlying F.

Theorem 29.9 (Higher-Order Accuracy for the t-statistic)

Let $X_1, \ldots, X_n \overset{\text{iid}}{\sim} F$. Suppose F is nonlattice and that $E_F(X^6) < \infty$. Let $T_n = \frac{\sqrt{n}(\bar{X} - \mu)}{s}$ and $T_n^* = \frac{\sqrt{n}(\bar{X}^* - \bar{X})}{s^*}$, where s^* is the standard deviation of $X_1^*, \ldots, X_n^*$. Let $H_n(x) = P_F(T_n \le x)$ and $H_{\text{Boot}}(x) = P_*(T_n^* \le x)$. Then $\sqrt{n} K(H_n, H_{\text{Boot}}) \xrightarrow{\text{a.s}} 0$.

29.6 Some Numerical Examples

The bootstrap is used in practice for a variety of purposes. It is used to estimate a CDF, a percentile, or the bias or variance of a statistic T_n. For example, if T_n is an estimate for some parameter θ, and if $E_F(T_n - \theta)$ is the bias of T_n, the bootstrap estimate $E_{F_n}(T_n^* - T_n)$ can be used to estimate the bias. Likewise, variance estimates can be formed by estimating $\text{Var}_F(T_n)$ by $\text{Var}_{F_n}(T_n^*)$. How accurate are the bootstrap-based estimates in reality?

This can only be answered on the basis of case-by-case simulation. Some overall qualitative phenomena have emerged from these simulations. They are:

(a) The bootstrap captures information about skewness that the CLT will miss.

(b) The bootstrap tends to underestimate the variance of a statistic T_n.

Here are a few numerical examples.

Example 29.4 Let $X_1, \ldots, X_n \overset{\text{iid}}{\sim} \text{Cauchy}(\mu, 1)$. Let M_n be the sample median and $T_n = \sqrt{n}(M_n - \mu)$. If n is odd, say $n = 2k + 1$, then there is an exact variance formula for M_n. Indeed

$$\text{Var}(M_n) = \frac{2n!}{(k!)^2 \pi^n} \int_0^{\pi/2} x^k (\pi - x)^k (\cot x)^2 dx;$$

see David (1981). Because of this exact formula, we can easily gauge the accuracy of the bootstrap variance estimate. In this example, $n = 21$ and $B = 200$. For comparison, the CLT-based variance estimate is also used, which is

$$\widehat{\text{Var}(M_n)} = \frac{\pi^2}{4n}.$$

The exact variance, the CLT-based estimate, and the bootstrap estimate for the specific simulation are 0.1367, 0.1175, and 0.0517, respectively. Note the obvious underestimation of variance by the bootstrap. Of course, one cannot be sure if it is the idiosyncrasy of the specific simulation.

A general useful result on consistency of the bootstrap variance estimate for medians under very mild conditions is in Ghosh et al. (1984).

Example 29.5 Suppose $X_1, \ldots, X_n$ are iid Poi(μ), and let T_n be the t-statistic $T_n = \sqrt{n}(\bar{X} - \mu)/s$. In this example, $n = 20$ and $B = 200$, and for the actual data, μ was chosen to be 1. Apart from the bias and the variance of T_n, in this example we also report percentile estimates for T_n. The bootstrap percentile estimates are found by calculating T_n^* for the B resamples and calculating the corresponding percentile value of the B values of T_n^*. The bias and the variance are estimated to be −0.18 and 1.614, respectively. The estimated percentiles are reported in the following table.

α	Estimated 100αPercentile
0.05	−2.45
0.10	−1.73
0.25	−0.76
0.50	−0.17
0.75	0.49
0.90	1.25
0.95	1.58

On observing the $100(1 - \alpha)\%$ estimated percentiles, it is clear that there seems to be substantial skewness in the distribution of T. Whether the skewness is truly as serious can be assessed by a large-scale simulation.

Example 29.6 Suppose $(X_i, Y_i),\ i = 1, 2, \cdots, n$ are iid $BVN(0, 0, 1, 1, \rho)$, and let r be the sample correlation coefficient. Let $T_n = \sqrt{n}(r - \rho)$. We know that $T_n \overset{\mathcal{L}}{\Rightarrow} N(0, (1 - \rho^2)^2)$; see Chapter 3. Convergence to normality is very slow. There is also an exact formula for the density of r. For $n \geq 4$, the exact density is

$$f(r|\rho) = \frac{2^{n-3}(1 - \rho^2)^{(n-1)/2}}{\pi(n-3)!}(1 - r^2)^{(n-4)/2} \sum_{k=0}^{\infty} \Gamma\left(\frac{n+k-1}{2}\right)^2 \frac{(2\rho r)^k}{k!};$$

see Tong (1990). In the following table, we give simulation averages of the estimated standard deviation of r by using the bootstrap. We used $n = 20$ and $B = 200$. The bootstrap estimate was calculated for 1000 independent simulations, and the table reports the average of the standard deviation estimates over the 1000 simulations.

n	True ρ	True s.d. of r	CLT estimate	Bootstrap estimate
	0.0	0.230	0.232	0.217
20	0.5	0.182	0.175	0.160
	0.9	0.053	0.046	0.046

Again, except when ρ is large, the bootstrap underestimates the variance and the CLT estimate is better.

29.7 Failure of the Bootstrap

In spite of the many consistency theorems in the previous sections, there are instances where the ordinary bootstrap based on sampling with replacement from F_n actually does not work. Typically, these are instances where the functional T_n fails to admit a CLT. Before seeing a few examples, we list a few situations where the ordinary bootstrap fails to estimate the CDF of T_n consistently:

(a) $T_n = \sqrt{n}(\bar{X} - \mu)$ when $\mathrm{Var}_F(X_1) = \infty$.
(b) $T_n = \sqrt{n}(g(\bar{X}) - g(\mu))$ and $\nabla g(\mu) = 0$.
(c) $T_n = \sqrt{n}(g(\bar{X}) - g(\mu))$ and g is not differentiable at μ.
(d) $T_n = \sqrt{n}(F_n^{-1}(p) - F^{-1}(p))$ and $f(F^{-1}(p)) = 0$ or F has unequal right and left derivatives at $F^{-1}(p)$.
(e) The underlying population F_θ is indexed by a parameter θ, and the support of F_θ depends on the value of θ.
(f) The underlying population F_θ is indexed by a parameter θ, and the true value θ_0 belongs to the boundary of the parameter space Θ.

Example 29.7 Let $X_1, X_2, \ldots, X_n \overset{\text{iid}}{\sim} F$ and $\sigma^2 = \mathrm{Var}_F(X) = 1$. Let $g(x) = |x|$ and $T_n = \sqrt{n}(g(\bar{X}) - g(\mu))$. If the true value of μ is 0, then by the CLT for $\bar{X}$ and the continuous mapping theorem, $T_n \overset{\mathcal{L}}{\Rightarrow} |Z|$ with $Z \sim N(0, \sigma^2)$. To show that the bootstrap does not work in this case, we first need to observe a few subsidiary facts.

(a) For almost all sequences $\{X_1, X_2, \cdots\}$, the conditional distribution of $\sqrt{n}(\overline{X}_n^* - \overline{X}_n)$, given $\overline{X}_n$, converges in law to $N(0, \sigma^2)$ by the triangular array CLT (see van der Vaart (1998).
(b) The joint asymptotic distribution of $(\sqrt{n}(\overline{X}_n - \mu), \sqrt{n}(\overline{X}_n^* - \overline{X}_n)) \overset{\mathcal{L}}{\Rightarrow} (Z_1, Z_2)$, where Z_1, Z_2 are iid $N(0, \sigma^2)$.

In fact, a more general version of part (b) is true. Suppose (X_n, Y_n) is a sequence of random vectors such that $X_n \overset{\mathcal{L}}{\Rightarrow} Z \sim H$ (some Z) and $Y_n|X_n \overset{\mathcal{L}}{\Rightarrow} Z$ (the same Z) almost surely. Then $(X_n, Y_n) \overset{\mathcal{L}}{\Rightarrow} (Z_1, Z_2)$, where Z_1, Z_2 are iid $\sim H$.

Therefore, returning to the example, when the true μ is 0,

$$\begin{aligned} T_n^* &= \sqrt{n}(|\overline{X}_n^*| - |\overline{X}_n|) \\ &= |\sqrt{n}(\overline{X}_n^* - \overline{X}_n) + \sqrt{n}\,\overline{X}_n| - |\sqrt{n}\,\overline{X}_n| \\ &\stackrel{\mathcal{L}}{\Rightarrow} |Z_1 + Z_2| - |Z_1|, \end{aligned}$$

where Z_1, Z_2 are iid $N(0, \sigma^2)$. But this is not distributed as the absolute value of $N(0, \sigma^2)$. The sequence of bootstrap CDFs is therefore not consistent when $\mu = 0$.

Example 29.8 Let $X_1, X_2, \ldots, X_n \stackrel{\text{iid}}{\sim} U(0, \theta)$ and let $T_n = n(\theta - X_{(n)})$, $T_n^* = n(X_{(n)} - X_{(n)}^*)$. The ordinary bootstrap will fail in this example in the sense that the conditional distribution of T_n^* given $X_{(n)}$ does not converge to the Exp(θ) a.s. Let us assume $\theta = 1$. Then, for $t \geq 0$,

$$\begin{aligned} P_{F_n}(T_n^* \leq t) &\geq P_{F_n}(T_n^* = 0) \\ &= P_{F_n}(X_{(n)}^* = X_{(n)}) \\ &= 1 - P_{F_n}(X_{(n)}^* < X_{(n)}) \\ &= 1 - \left(\frac{n-1}{n}\right)^n \\ &\stackrel{n\to\infty}{\longrightarrow} 1 - e^{-1}. \end{aligned}$$

For example, take $t = 0.0001$. Then $\lim_n P_{F_n}(T_n^* \leq t) \geq 1 - e^{-1}$, while $\lim_n P_F(T_n \leq t) = 1 - e^{-0.0001} \approx 0$. So $P_{F_n}(T_n^* \leq t) \not\to P_F(T_n \leq t)$.

The phenomenon of this example can be generalized essentially to any CDF F with a compact support $[\underline{\omega}(F), \overline{\omega}(F)]$ with some conditions on F, such as existence of a smooth and positive density. This is one of the earliest examples of the failure of the ordinary bootstrap. We will revisit this issue in the next section.

29.8 *m* out of *n* Bootstrap

In the particular problems presented above and several other problems where the ordinary bootstrap fails to be consistent, resampling fewer than n observations from F_n, say m observations, cures the inconsistency problem. This is called the m out of n bootstrap. Typically, consistency will be regained if $m = o(n)$; in some general theorems in this regard, one requires $m^2 = o(n)$ or some similar stronger condition than $m = o(n)$. If the n out of n ordinary bootstrap is already consistent, then there can still be m out of n schemes with m going to ∞ slower than n that are also consistent, but the m out of n scheme will perform somewhat worse than the n out of n. See Bickel, Göetze, and van Zwet (1997) for an overall review.

We will now present a collection of results that show that the m out of n bootstrap, written as the m/n bootstrap, solves the orthodox bootstrap's inconsistency problem in a number of cases; see Shao and Tu (1995) for proofs and details on all of the theorems in this section.

Theorem 29.10 Let $X_1, X_2, \ldots$ be iid F, where F is a CDF on $\mathbb{R}^d$, $d \geq 1$. Suppose $\mu = E_F(X_1)$ and $\Sigma = \text{cov}_F(X_1)$ exist, and suppose Σ is positive definite. Let $g : \mathbb{R}^d \to \mathbb{R}$ be such that $\nabla g(\mu) = 0$ and the Hessian matrix $\nabla^2 g(\mu)$ is not the zero matrix. Let $T_n = n(g(\bar{X}_n) - g(\mu))$ and $T^*_{m,n} = m(g(\bar{X}_m{}^*) - g(\bar{X}_n))$ and define $H_n(x) = P_F\{T_n \leq x\}$ and $H_{\text{Boot},m,n}(x) = P_*\{T^*_{m,n} \leq x\}$. Here $\bar{X}_m{}^*$ denotes the mean of an iid sample of size $m = m(n)$ from F_n, where $m \to \infty$ with n.

(a) If $m = o(n)$, then $K(H_{\text{Boot},m,n}, H_n) \overset{\mathcal{P}}{\Rightarrow} 0$.
(b) If $m = o(\frac{n}{\log\log n})$, then $K(H_{\text{Boot},m,n}, H_n) \overset{\text{a.s.}}{\Rightarrow} 0$.

Theorem 29.11 Let $X_1, X_2, \ldots$ be iid F, where F is a CDF on $\mathbb{R}$. For $0 < p < 1$, let $\xi_p = F^{-1}(p)$. Suppose F has finite and positive left and right derivatives $f(\xi_p+)$, $f(\xi_p-)$ and that $f(\xi_p+) \neq f(\xi_p-)$. Let $T_n = \sqrt{n}(F_n^{-1}(p) - \xi_p)$ and $T^*_{m,n} = \sqrt{m}(F_m^{*-1}(p) - F_n^{-1}(p))$, and define $H_n(x) = P_F\{T_n \leq x\}$and $H_{\text{Boot},m,n}(x) = P_*\{T^*_{m,n} \leq x\}$. Here, $F_m^{*-1}(p)$ denotes the pth quantile of an iid sample of size m from F_n.

(a) If $m = o(n)$, then $K(H_{\text{Boot},m,n}, H_n) \overset{\mathcal{P}}{\Rightarrow} 0$.
(b) If $m = o(\frac{n}{\log\log n})$, then $K(H_{\text{Boot},m,n}, H_n) \overset{a.s.}{\Rightarrow} 0$.

Theorem 29.12 Suppose F is a CDF on $\mathbb{R}$, and let $X_1, X_2, \ldots$ be iid F. Suppose $\theta = \theta(F)$ is such that $F(\theta) = 1$ and $F(x) < 1$ for all $x < \theta$. Suppose, for some $\delta > 0$, $P_F\left\{n^{1/\delta}(\theta - X_{(n)}) > x\right\} \longrightarrow e^{-(x/\theta)^\delta}, \quad \forall\, x$. Let $T_n = n^{1/\delta}(\theta - X_{(n)})$ and $T^*_{m,n} = m^{1/\delta}(X_{(n)} - X^*_{(m)})$, and define $H_n(x) = P_F\{T_n \leq x\}$ and $H_{\text{Boot},m,n}(x) = P_*\{T^*_{m,n} \leq x\}$.

(a) If $m = o(n)$, then $K(H_{\text{Boot},m,n}, H_n) \overset{\mathcal{P}}{\Rightarrow} 0$.
(b) If $m = o(\frac{n}{\log\log n})$, then $K(H_{\text{Boot},m,n}, H_n) \overset{a.s.}{\Rightarrow} 0$.

Remark. Clearly an important practical question is the choice of the bootstrap resample size m. This is a difficult question to answer, and no precise prescriptions that have any sort of general optimality are possible. A rule of thumb is to take $m \approx 2\sqrt{n}$.

29.9 Bootstrap Confidence Intervals

The standard method to find a confidence interval for a parameter θ is to find a studentized statistic, sometimes called a pivot, say $T_n = \frac{\hat{\theta}_n - \theta}{\hat{\sigma}_n}$, such that $T_n \stackrel{\mathcal{L}}{\Rightarrow} T$, with T having some known CDF G. An equal-tailed confidence interval for θ, asymptotically correct, is constructed as

$$\hat{\theta}_n - G^{-1}(1 - \alpha/2)\hat{\sigma}_n \leq \theta \leq \hat{\theta}_n - G^{-1}(\alpha/2)\hat{\sigma}_n.$$

This agenda requires the use of a standard deviation estimate $\hat{\sigma}_n$ for the standard deviation of $\hat{\theta}_n$ and the knowledge of the function $G(x)$. Furthermore, in many cases, the limiting CDF G may depend on some unknown parameters, too, that will have to be estimated in turn to construct the confidence interval. The bootstrap methodology offers an omnibus, sometimes easy to implement, and often more accurate method of constructing confidence intervals. Bootstrap confidence intervals and lower and upper one-sided confidence limits of various types have been proposed in great generality. Although, as a matter of methodology, they can be used in an automatic manner, a theoretical evaluation of their performance requires specific structural assumptions. The theoretical evaluation involves an Edgeworth expansion for the relevant statistic and an expansion for their quantiles, called Cornish-Fisher expansions. Necessarily, we are limited to the cases where the underlying statistic admits a known Edgeworth and Cornish-Fisher expansion. The main reference is Hall (1988), but see also Göetze (1989), Hall and Martin (1989), Bickel (1992), Konishi (1991), DiCiccio and Efron (1996), and Lee (1999), of which the article by DiCiccio and Efron is a survey article and Lee (1999) discusses m/n bootstrap confidence intervals. There are also confidence intervals based on more general *subsampling* methods, which work asymptotically under the mildest conditions. These intervals and their extensions to higher dimensions are discussed in Politis, Romano, and Wolf (1999).

Over time, various bootstrap confidence limits have been proposed. Generally, the evolution is from the algebraically simplest to progressively more complicated and computer-intensive formulas for the limits. Many of these limits have, however, now been incorporated into standard statistical software. We present below a selection of these different bootstrap confidence limits and bounds. Let $\hat{\theta}_n = \hat{\theta}_n(X_1, \ldots, X_n)$ be a specific estimate of the underlying parameter of interest θ.

(a) *The bootstrap percentile lower bound* (BP). Let $G(x) = G_n(x) = P_F\{\hat{\theta}_n \leq x\}$ be the exact distribution and let $\hat{G}(x) = P_*\{\hat{\theta}_n^* \leq x\}$ be the bootstrap distribution. The lower $1 - \alpha$ bootstrap percentile confidence bound would be $\hat{G}^{-1}(\alpha)$, so the reported interval would be $[\hat{G}^{-1}(\alpha), \infty)$. This was present in Efron (1979) itself, but it is seldom used because it tends to have a significant coverage bias.

(b) *Transformation-based bootstrap percentile confidence bound.* Suppose there is a suitable 1-1 transformation $\varphi = \varphi_n$ of $\hat{\theta}_n$ such that $P_F\{\varphi(\hat{\theta}_n) - \varphi(\theta) \leq x\} = \psi(x)$, with ψ being a known continuous, strictly increasing, and symmetric CDF (e.g., the $N(0, 1)$ CDF). Then a transformation-based bootstrap percentile lower confidence bound for θ is $\varphi^{-1}(\hat{\varphi}_n + z_\alpha)$, where $\hat{\varphi}_n = \varphi(\hat{\theta}_n)$ and $z_\alpha = \psi^{-1}(\alpha)$. Transforming may enhance the quality of the confidence bound in some problems. But, on the other hand, it is rare that one can find such a 1-1 transformation with a known ψ.

(c) *Bootstrap-t* (BT). Let $t_n = \frac{\hat{\theta}_n - \theta}{\hat{\sigma}_n}$, where $\hat{\sigma}_n$ is an estimate of the standard error of $\hat{\theta}_n$, and let $t_n^* = \frac{\hat{\theta}_n^* - \hat{\theta}_n}{\hat{\sigma}_n^*}$ be its bootstrap counterpart. As usual, let $H_{\text{Boot}}(x) = P_*\{t_n^* \leq x\}$. The bootstrap-*t* lower bound is $\hat{\theta}_n - H_{\text{Boot}}^{-1}(1 - \alpha)\hat{\sigma}_n$, and the two-sided BT confidence limits are $\hat{\theta}_n - H_{\text{Boot}}^{-1}(1 - \alpha_1)\hat{\sigma}_n$ and $\hat{\theta}_n - H_{\text{Boot}}^{-1}(\alpha_2)\hat{\sigma}_n$, where $\alpha_1 + \alpha_2 = \alpha$, the nominal confidence level.

(d) *Bias-corrected bootstrap percentile bound* (BC). The derivation of the BC bound involves quite a lot of calculation; see Efron (1981) and Shao and Tu (1995). The BC lower confidence bound is given by $\underline{\theta}_{\text{BC}} = \hat{G}^{-1}[\psi(z_\alpha + 2\psi^{-1}(\hat{G}(\hat{\theta}_n)))]$, where $\hat{G}$ is the bootstrap distribution of $\hat{\theta}_n^*$, ψ is as above, and $z_\alpha = \psi^{-1}(\alpha)$.

(e) *Hybrid bootstrap confidence bound* (BH). Suppose for some deterministic sequence $\{c_n\}$, $c_n(\hat{\theta}_n - \theta) \sim H_n$ and let H_{Boot} be the bootstrap distribution; i.e., the distribution of $c_n(\hat{\theta}_n^* - \hat{\theta}_n)$ under F_n. We know that $P_F\{c_n(\hat{\theta}_n - \theta) \leq H_n^{-1}(1 - \alpha)\} = 1 - \alpha$.
If we knew H_n, then we could turn this into a $100(1 - \alpha)\%$ lower confidence bound, $\theta \geq \hat{\theta}_n - \frac{1}{c_n} H_n^{-1}(1 - \alpha)$. But H_n is, in general, not known, so we approximate it by H_{Boot}. That is, the hybrid bootstrap lower confidence bound is defined as $\underline{\theta}_{\text{BH}} = \hat{\theta}_n - \frac{1}{c_n} H_{\text{Boot}}^{-1}(1 - \alpha)$.

(f) *Accelerated bias-corrected bootstrap percentile bound* (BC_a). The ordinary bias-corrected bootstrap bound is based on the assumption that we can find $z_0 = z_0(F, n)$ and ψ (for known ψ) such that

$$P_F\{\hat{\varphi}_n - \varphi + z_0 \le x\} = \psi(x).$$

The accelerated bias-corrected bound comes from the modified assumption that there exists a constant $a = a(F, n)$ such that $P_F\left\{\frac{\hat{\varphi}_n - \varphi}{1 + a\varphi} + z_0 \le x\right\} = \psi(x)$. In applications, it is rare that even this modification holds exactly for any given F and n. Manipulation of this probability statement results in a lower bound, $\underline{\theta}_{\mathrm{BC_a}} = \hat{G}^{-1}\left(\psi\left(z_0 + \frac{z_\alpha + z_0}{1 - a(z_\alpha - z_0)}\right)\right)$, where $z_\alpha = \psi^{-1}(\alpha)$, a is the acceleration parameter, and $\hat{G}$ is as before. We repeat that, of these, z_0 and a both depend on F and n. They will have to be estimated. Moreover, the CDF ψ will generally have to be replaced by an asymptotic version; e.g., an asymptotic normal CDF of $(\hat{\varphi}_n - \varphi)/(1 + a\varphi)$. The exact manner in which z_0 and a depend on F and n is a function of the specific problem. For example, suppose that the problem to begin with is a parametric problem, $F = F_\theta$. In such a case, $z_0 = z_0(\theta, n)$ and $a = a(\theta, n)$. The exact form of $z_0(\theta, n)$ and $a(\theta, n)$ depends on F_θ, $\hat{\theta}_n$, and φ.

Remark. As regards computational simplicity, BP, BT, and BH are the simplest to apply; BC and BC_a are harder to apply and, in addition, are based on assumptions that will rarely hold exactly for finite n. Furthermore, BC_a involves estimation of a very problem-specific acceleration constant a. The bootstrap-*t* intervals are popular in practice, provided an estimate $\hat{\sigma}_n$ is readily available. The BP method usually suffers from a large bias in coverage and is seldom used.

Remark. If the model is parametric, $F = F_\theta$, and $\hat{\theta}_n$ is the MLE, then one can show the following general and useful formula: $a = z_0 = \frac{1}{6} \times$ skewness coefficient of $\dot{\ell}(\theta)$, where $\dot{\ell}(\theta)$ is the score function, $\dot{\ell}(\theta) = \frac{d}{d\theta} \log f(x_1, \ldots, x_n|\theta)$. This expression allows for estimation of a and z_0 by plug-in estimates. Nonparametric estimates of a and z_0 have also been suggested; see Efron (1987) and Loh and Wu (1987).

We now state the theoretical coverage properties of the various one-sided bounds and two-sided intervals.

Definition 29.3 Let $0 < \alpha < 1$ and $I_n = I_n(X_1, \ldots, X_n)$ be a confidence set for the functional $\theta(F^{(n)})$, where $F^{(n)}$ is the joint distribution of $(X_1, \ldots, X_n)$. Then I_n is called kth-order accurate if $P_{F^{(n)}}\left\{I_n \ni \theta(F^{(n)})\right\} = 1 - \alpha + O(n^{-k/2})$.

The theoretical coverage properties below are derived by using Edgeworth expansions as well as Cornish-Fisher expansions for the underlying estimate $\hat{\theta}_n$. If $X_1, X_2, \ldots$ are iid F on $\mathbb{R}^d$, $1 \le d < \infty$, and if $\theta = \varphi(\mu)$, $\hat{\theta} = \varphi(\bar{X})$, for a sufficiently smooth map $\varphi : \mathbb{R}^d \to \mathbb{R}$, then such Edgeworth and Cornish-Fisher expansions are available. In the results below, it is assumed that θ and $\hat{\theta}$ are the images of μ and $\bar{X}$, respectively, under such a smooth mapping φ. See Hall (1988) for the exact details.

Theorem 29.13 The CLT, BP, BH and BC one-sided confidence bounds are first-order accurate. The BT and BC$_a$ one-sided bounds are second-order accurate. The CLT, BP, BH, BT, and BC$_a$ two-sided intervals are all second-order accurate.

Remark. For two-sided intervals, the higher-order accuracy result is expected because the coverage bias for the two tails cancels in the $n^{-1/2}$ term, as can be seen from the Edgeworth expansion. The striking part of the result is that the BT and BC$_a$ can achieve higher-order accuracy even for one-sided bounds.

The second-order accuracy of the BT lower bound is driven by an Edgeworth expansion for H_n and an analogous one for H_{Boot}. One can invert these expansions for the CDFs to get expansions for their quantiles; i.e., to obtain Cornish-Fisher expansions. Under suitable conditions on F, H_n^{-1} and H_{Boot}^{-1} admit expansions of the forms

$$H_n^{-1}(t) = z_t + \frac{q_{11}(z_t, F)}{\sqrt{n}} + \frac{q_{12}(z_t, F)}{n} + o\left(\frac{1}{n}\right)$$

and

$$H_{\text{Boot}}^{-1}(t) = z_t + \frac{q_{11}(z_t, F_n)}{\sqrt{n}} + \frac{q_{12}(z_t, F_n)}{n} + o\left(\frac{1}{n}\right) \text{ (a.s.)},$$

where $q_{11}(\cdot, F)$ and $q_{12}(\cdot, F)$ are polynomials with coefficients that depend on the moments of F. The exact polynomials depend on what the statistic $\hat{\theta}_n$ is. For example, if $\hat{\theta}_n = \bar{X}$ and $\hat{\sigma} = \sqrt{\frac{1}{n-1}\sum(X_i - \bar{X})^2}$, then $q_{11}(x, F) = -\frac{\gamma}{6}(1 + 2x^2)$, $q_{12} = x[\frac{x^2+3}{4} - \frac{\kappa(x^2-3)}{12} + \frac{5\gamma^2}{72}(4x^2 - 1)]$, where $\gamma = E_F\frac{(X-\mu)^3}{\sigma^3}$ and $\kappa = E_F\frac{(X-\mu)^4}{\sigma^4} - 3$. For a given t, $0 < t < 1$, on subtraction,

$$\begin{aligned} H_n^{-1}(t) - H_{\text{Boot}}^{-1}(t) &= \frac{1}{\sqrt{n}}[q_{11}(z_t, F) - q_{11}(z_t, F_n)] \\ &\quad + \frac{1}{n}[q_{12}(z_t, F) - q_{12}(z_t, F_n)] + o\left(\frac{1}{n}\right) \text{ (a.s.)} \end{aligned}$$

$$= \frac{1}{\sqrt{n}} O_p\left(\frac{1}{\sqrt{n}}\right) + \frac{1}{n} O_p\left(\frac{1}{\sqrt{n}}\right) + o\left(\frac{1}{n}\right) \text{(a.s.)}$$
$$= O_p\left(\frac{1}{n}\right).$$

The actual confidence bounds obtained from H_n, H_{Boot} are $\underline{\theta}_{H_n} = \hat{\theta}_n - \hat{\sigma}_n H_n^{-1}(1-\alpha)$ and $\underline{\theta}_{\text{BT}} = \hat{\theta}_n - \hat{\sigma}_n H_{\text{Boot}}^{-1}(1-\alpha)$. On subtraction,

$$|\underline{\theta}_{H_n} - \underline{\theta}_{\text{BT}}| = \hat{\sigma}_n O_p\left(\frac{1}{n}\right) \overset{\text{typically}}{=} O_p(n^{-\frac{3}{2}}).$$

Thus, the bootstrap-t lower bound is approximating the idealized lower bound with third-order accuracy. In addition, it can be shown that $P(\theta \geq \underline{\theta}_{\text{BT}}) = 1 - \alpha + \frac{p(z_\alpha)\varphi(z_\alpha)}{n} + o\left(\frac{1}{n}\right)$, where $p(\cdot)$ is again a polynomial depending on the specific statistic and F. For the case of $\bar{X}$, as an example, $p(x) = \frac{x}{6}(1 + 2x^2)(\kappa - \frac{3}{2}\gamma^2)$. Notice the second-order accuracy in this coverage statement in spite of the fact that the confidence bound is one sided. Again, see Hall (1988) for full details.

29.10 Some Numerical Examples

How accurate are the bootstrap confidence intervals in practice? Only case-by-case numerical investigation can give an answer to that question. We report in the following table results of simulation averages of coverage and length in two problems. The sample size in each case is $n = 20$, in each case $B = 200$, the simulation size is 500, and the nominal coverage $1 - \alpha = .9$.

$\theta(F)$	Type of CI	F					
		$N(0,1)$		$t(5)$		Weibull	
		coverage	length	coverage	length	coverage	length
μ	Regular t	.9	0.76	.91	1.8	.75	2.8
	BP	.91	0.71	.84	1.7	.73	2.6
	BT	.92	0.77	.83	2.7	.83	5.5
σ^2	BP	.79	0.86	.68	1.1	.65	1.3
	BT	.88	1.5	.85	3.2	.83	5.5

From the table, the bootstrap-t interval seems to buy more accuracy (i.e., a smaller bias in coverage) with a larger length than the BP interval. But the BP interval has such a serious bias in coverage that the bootstrap-t may be preferable. To kill the bias, modifications of the BP method have been suggested, such as the bias-corrected BP and the accelerated bias-corrected BP intervals. Extensive numerical comparisons are reported in Shao and Tu (1995).

29.11 Bootstrap Confidence Intervals for Quantiles

Another interesting problem is the estimation of quantiles of a CDF F on $\mathbb{R}$. We know, for example, that if $X_1, X_2, \ldots$ are iid F, if $0 < p < 1$, and if $f = F'$ exists and is strictly positive at $\xi_p = F^{-1}(p)$, then $\sqrt{n}(F_n^{-1}(p) - \xi_p) \stackrel{\mathcal{L}}{\Rightarrow} N(0, p(1-p)[f(\xi_p)]^{-2})$. So, a standard CLT-based interval is

$$F_n^{-1}(p) \pm \frac{z_{\alpha/2}}{\sqrt{n}} \cdot \frac{\sqrt{p(1-p)}}{\widehat{f(\xi_p)}},$$

where $\widehat{f(\xi_p)}$ is some estimate of the unknown $f = F'$ at the unknown ξ_p.

For a bootstrap interval, let H_n be the CDF of $\sqrt{n}(F^{-1}(p) - \xi_p)$ and H_{Boot} its bootstrap counterpart. Using the terminology from before, a hybrid bootstrap two-sided confidence interval for ξ_p is

$$\left[F_n^{-1}(p) - H_{\text{Boot}}^{-1}(1 - \tfrac{\alpha}{2})/\sqrt{n},\, F_n^{-1}(p) - H_{\text{Boot}}^{-1}(\tfrac{\alpha}{2})/\sqrt{n}\right].$$

It turns out that this interval is not only asymptotically correct but also comes with a surprising asymptotic accuracy. The main references are Hall, DiCiccio, and Romano (1989) and Falk and Kaufman (1991).

Theorem 29.14 Let $X_1, X_2, \ldots$ be iid and F a CDF on $\mathbb{R}$. For $0 < p < 1$, let $\xi_p = F^{-1}(p)$, and suppose $0 < f(\xi_p) = F'(\xi_p) < \infty$. If I_n is the two-sided hybrid bootstrap interval, then $P_F\{I_n \ni \xi_p\} = 1 - \alpha + O(n^{-1/2})$.

Remark. Actually, the best result available is stronger and says that $P_F\{I_n \ni \xi_p\} = 1 - \alpha + \frac{c(F,\alpha,p)}{\sqrt{n}} + o(n^{-1/2})$, where $c(F, \alpha, p)$ has an explicit but complicated formula. That the bias of the hybrid interval is $O(n^{-1/2})$ is *still* a surprise in view of the fact that the bootstrap distribution of $F_n^{-1}(p)$ is consistent at a very slow rate; see Singh (1981).

29.12 Bootstrap in Regression

Regression models are among the key ones that differ from the iid setup and are also among the most widely used. Bootstrap for regression cannot be model-free; the particular choice of the bootstrap scheme depends on whether the errors are iid or not. We will only talk about the linear model with deterministic X and iid errors. Additional moment conditions will be necessary depending on the specific problem to which the bootstrap will be applied. The results here are available in Freedman (1981). First let us introduce some notation.

Model: $y_i = \beta' x_i + \epsilon_i$, where β is a $p \times 1$ vector and so is x_i, and ϵ_i are iid with mean 0 and variance $\sigma^2 < \infty$.

X is the $n \times p$ design matrix with ith row equal to x_i'; $H = X(X'X)^{-1}X'$ and $h_i = H_{ii} = x_i'(X'X)^{-1}x_i$.

$\hat{\beta} = \hat{\beta}_{LS} = (X'X)^{-1}X'y$ is the least squares estimate of β, where $y = (y_1, \cdots, y_n)'$ and $(X'X)^{-1}$ is assumed to be nonsingular.

The bootstrap scheme is defined below.

29.13 Residual Bootstrap

Let $e_1, e_2, \cdots, e_n$ denote the residuals obtained from fitting the model (i.e., $e_i = y_i - x_i'\hat{\beta}$); $\bar{e} = 0$ if $x_i = (1, x_{i1}, \cdots, x_{i,p-1})'$ but not otherwise. Define $\tilde{e}_i = e_i - \bar{e}$, and let $e_1^*, \cdots, e_n^*$ be a sample with replacement of size n from $\{\tilde{e}_1, \cdots, \tilde{e}_n\}$. Let $y_i^* = x_i'\hat{\beta} + e_i^*$ and let β^* be the LSE of β computed from (x_i, y_i^*), $i = 1, \cdots, n$. This is the bootstrapped version of $\hat{\beta}$, and the scheme is called the residual bootstrap (RB).

Remark. The more direct approach of resampling the pairs (x_i, y_i) is known as the paired bootstrap and is necessary when the errors are not iid; for example, the case where the errors are still independent but their variances depend on the corresponding covariate values (called the heteroscedastic case). In such a case, the residual bootstrap scheme would not work.

By simple matrix algebra, it can be shown that

$$E_*(\beta^*) = \hat{\beta},$$
$$\text{cov}_*(\beta^*) = \hat{\sigma}^2(X'X)^{-1},$$

where $\hat{\sigma}^2 = (1/n)\sum_{i=1}^n (e_i - \bar{e})^2$. Note that $E(\hat{\sigma}^2) < \sigma^2$. So on average the bootstrap covariance matrix estimate will somewhat underestimate $\text{cov}(\hat{\beta})$. However, $\text{cov}_*(\beta^*)$ is still consistent under some mild conditions. See Shao and Tu (1995) or Freedman (1981) for the following result.

Theorem 29.15 Suppose $|X'X| \to \infty$ and $\max_{1\le i\le n} h_i \to 0$ as $n \to \infty$. Then $[\mathrm{cov}_*(\beta^*)]^{-1}\mathrm{cov}(\hat{\beta}) \Rightarrow I_{p\times p}$ almost surely.

Example 29.9 The only question is, when do the conditions $|X'X| \to \infty$, $\max_{1\le i\le n} h_i \to 0$ hold? As an example, take the basic regression model $y_i = \beta_0 + \beta_1 x_i + \epsilon_i$ with one covariate. Then, $|X'X| = n\sum_i (x_i - \bar{x})^2$ and $h_i = (\sum_j x_j^2 - 2x_i \sum_j x_j + n x_i^2)/(n\sum_j (x_j - \bar{x})^2)$.

$$\therefore h_i \le \frac{4n \max_j x_j^2}{n\sum_j (x_j - \bar{x})^2} = \frac{4 \max_j x_j^2}{\sum_j (x_j - \bar{x})^2}.$$

Therefore, for the theorem to apply, it is enough to have $\max |x_j| / \sqrt{\sum (x_j - \bar{x})^2} \to 0$ and $n\sum (x_i - \bar{x})^2 \to \infty$.

29.14 Confidence Intervals

We present some results on bootstrap confidence intervals for a linear combination $\theta = c'\beta_1$, where $\beta' = (\beta_0, \beta_1')$; i.e., there is an intercept term in the model. Correspondingly, $x_i' = (1, t_i')$. The confidence interval for θ or confidence bounds (lower or upper) are going to be in terms of the studentized version of the LSE of θ, namely $\hat{\theta} = c'\hat{\beta}_1$. In fact, $\hat{\beta}_1 = S_{tt}^{-1} S_{ty}$, where $S_{tt} = \sum_i (t_i - \bar{t})(t_i - \bar{t})'$ and $S_{ty} = \sum_i (t_i - \bar{t})(y_i - \bar{y})'$. The bootstrapped version of $\hat{\theta}$ is $\theta^* = c'\beta_1^*$, where $\beta^{*'} = (\beta_0^*, \beta_1^{*'})$ as before. Since the variance of $\hat{\theta}$ is $\sigma^2 c' S_{tt}^{-1} c$, the bootstrapped version of the studentized $\hat{\theta}$ is

$$\theta_s^* = \frac{\theta^* - \hat{\theta}}{\sqrt{\frac{1}{n}\sum_i (y_i - x_i'\beta^*)^2 c' S_{tt}^{-1} c}}.$$

The bootstrap distribution is defined as $H_{\mathrm{Boot}}(x) = P_*(\theta_s^* \le x)$. For given α, let $H_{\mathrm{Boot}}^{-1}(\alpha)$ be the αth quantile of H_{Boot}. We consider the bootstrap-t (BT) confidence bounds and intervals for θ. They are obtained as

$$\underline{\theta}_{\mathrm{BT}}^{(\alpha)} = \hat{\theta} - H_{\mathrm{Boot}}^{-1}(1-\alpha)\sqrt{\hat{\sigma}^2 c' S_{tt}^{-1} c},$$

$$\bar{\theta}_{\mathrm{BT}}^{(\alpha)} = \hat{\theta} - H_{\mathrm{Boot}}^{-1}(\alpha)\sqrt{\hat{\sigma}^2 c' S_{tt}^{-1} c},$$

and the intervals $\theta_{L,\mathrm{BT}} = \underline{\theta}_{\mathrm{BT}}^{(\alpha/2)}$ and $\theta_{U,\mathrm{BT}} = \bar{\theta}_{\mathrm{BT}}^{(\alpha/2)}$.

There are some remarkable results on the accuracy in coverage of the BT one-sided bounds and confidence intervals. We state one key result below.

Theorem 29.16 (a) $P(\theta \geq \underline{\theta}_{\mathrm{BT}}) = (1-\alpha) + O(n^{-3/2})$.
(b) $P(\theta \leq \bar{\theta}_{\mathrm{BT}}) = (1-\alpha) + O(n^{-3/2})$.
(c) $P(\theta_{L,\mathrm{BT}} \leq \theta \leq \theta_{U,\mathrm{BT}}) = (1-\alpha) + O(n^{-2})$.

These results are derived in Hall (1989).

Remark. It is remarkable that one already gets third-order accuracy for the one-sided confidence bounds and fourth-order accuracy for the two-sided bounds. There seems to be no intuitive explanation for this phenomenon. It just happens that certain terms cancel in the Cornish-Fisher expansions used in the proof for the regression case.

29.15 Distribution Estimates in Regression

The residual bootstrap is also consistent for estimating the distribution of the least squares estimate $\hat{\beta}$ of the full vector β. The metric chosen is the Mallows-Wasserstein metric we used earlier for sample means of iid data. See Freedman (1981) for the result below. We first state the model and the required assumptions below.

Let $y_i = x_i'\beta + \varepsilon_i$, where x_i is the p-vector of covariates for the ith sample unit. Write the design matrix as X_n. We assume that the ε_i's are iid with mean 0 and variance $\sigma^2 < \infty$ and that $\{X_n\}$ is a sequence of nonstochastic matrices. We assume that, for every n ($n > p$), $X_n'X_n$ is positive definite. Let $h_i = x_i'(X'X)^{-1}x_i$ and let $h_{\max} = \max\{h_i\}$. We assume, for the consistency theorem below, that:

(C1) **Stability**: $\frac{1}{n}X_n'X_n \to V$, where V is a $p \times p$ positive definite matrix.
(C2) **Uniform asymptotic negligibility**: $h_{\max} \to 0$.

Under these conditions, we have the following theorem of Freedman (1981) for RB.

Theorem 29.17 Under conditions C1 and C2 above, we have the following:

(a) $\sqrt{n}(\hat{\beta} - \beta) \stackrel{\mathcal{L}}{\Rightarrow} N_p(0, \sigma^2 V^{-1})$.

(b) For almost all $\{\varepsilon_i : i \geq 1\}$, $\sqrt{n}(\beta^* - \hat{\beta}) \stackrel{\mathcal{L}}{\Rightarrow} N_p(0, \sigma^2 V^{-1})$.

(c) $\frac{1}{\sigma}(X_n'X_n)^{1/2}(\hat{\beta} - \beta) \stackrel{\mathcal{L}}{\Rightarrow} N_p(0, I_p)$.

(d) For almost all $\{\varepsilon_i : i \geq 1\}$, $\frac{1}{\hat{\sigma}}(X_n'X_n)^{1/2}(\beta^* - \hat{\beta}) \stackrel{\mathcal{L}}{\Rightarrow} N_p(0, I_p)$.

(e) If H_n and H_{Boot} are the true and bootstrap distributions of $\sqrt{n}(\hat{\beta} - \beta)$ and $\sqrt{n}(\beta^* - \hat{\beta})$, respectively, then for almost all $\{\varepsilon_i : i \geq 1\}$, $\ell_2(H_n, H_{\text{Boot}}) \to 0$.

Remark. This theorem gives a complete picture of the consistency issue for the case of a nonstochastic design matrix and iid errors using the residual bootstrap. If the errors are iid but the design matrices are random, the same results hold as long as the conditions of stability and uniform asymptotic negligibility stated earlier hold with probability 1. See Shao and Tu (1995) for the case of independent but not iid errors (for example, the heteroscedastic case).

29.16 Bootstrap for Dependent Data

The orthodox bootstrap does not work when the sample observations are dependent. This was already pointed out in Singh (1981). It took some time before consistent bootstrap schemes were offered for dependent data. There are consistent schemes that are meant for specific dependence structures (e.g., stationary autoregression of a known order) and also general bootstrap schemes that work for large classes of stationary time series without requiring any particular dependence structure. The model-based schemes are better for the specific models but can completely fall apart if some assumption about the specific model does not hold.

We start with examples of some standard short-range dependence time series models. As opposed to these models, there are some that have a long memory or long-range dependence. The bootstrap runs into problems for long-memory data; see Lahiri (2006).

Standard time series models for short-range dependent processes include:

(a) *Autoregressive processes*. The observations y_t are assumed to satisfy

$$y_t = \mu + \theta_1 y_{t-1} + \theta_2 y_{t-2} + \ldots \theta_p y_{t-p} + \varepsilon_t,$$

where $1 \leq p < \infty$ and the ε_t's are iid white noise with mean 0 and variance $\sigma^2 < \infty$. The $\{y_t\}$ process is stationary if the solutions of the polynomial equation

$$1 + \theta_1 z + \theta_2 z^2 + \ldots + \theta_p z^p = 0$$

lie strictly *outside* the unit circle in the complex plane. This process is called autoregression of order p and is denoted by AR(p).

(b) *Moving average processes.* Given a white noise process $\{\varepsilon_t\}$ with mean 0 and variance $\sigma^2 < \infty$, the observations are assumed to satisfy

$$y_t = \mu + \varepsilon_t - \varphi_1 \varepsilon_{t-1} - \varphi_2 \varepsilon_{t-2} - \ldots - \varphi_q \varepsilon_{t-q},$$

where $1 \leq q < \infty$. The process $\{y_t\}$ is stationary if the roots of

$$1 - \varphi_1 z - \varphi_2 z^2 - \ldots - \varphi_q z^q = 0$$

lie strictly *outside* the unit circle. This process is called a moving average process of order q and is denoted by MA(q).

(c) *Autoregressive moving average processes.* This combines the two previously mentioned models. The observations are assumed to satisfy

$$y_t = \mu + \theta_1 y_{t-1} + \ldots \theta_p y_{t-p} + \varepsilon_t - \varphi_1 \varepsilon_{t-1} - \ldots - \varphi_q \varepsilon_{t-q}.$$

The process $\{y_t\}$ is called an autoregressive moving average process of order (p, q) and is denoted by ARMA(p, q).

For all of these processes, the autocorrelation sequence dies off quickly; in particular, if ρ_k is the autocorrelation of lag k, then $\sum_k |\rho_k| < \infty$.

29.17 Consistent Bootstrap for Stationary Autoregression

A version of the residual bootstrap (RB) was offered in Bose (1988) and shown to be consistent and even higher-order accurate for the least squares estimate (LSE) of the vector of regression coefficients in the stationary AR(p) case. For ease of presentation, we assume $\mu = 0$ and $\sigma = 1$. In this case, the LSE of $\theta = (\theta_1, \ldots, \theta_p)'$ is defined as $\hat{\theta} = \arg\min_\theta \sum_{t=1}^n \left[y_t - \sum_{j=1}^p \theta_j y_{t-j}\right]^2$, where $y_{1-p}, \ldots, y_0, y_1, \ldots, y_n$ is the observed data sequence. There is a closed-form expression of $\hat{\theta}$; specifically, $\hat{\theta} = S_{nn}^{-1}\big(\sum_{t=1}^n y_t y_{t-1}, \sum_{t=1}^n y_t y_{t-2}, \ldots, \sum_{t=1}^n y_t y_{t-p}\big)$, where $S_{nn} = ((S_{nn}^{ij}))_{p\times p}$ and $S_{nn}^{ij} = \sum_{t=1}^n y_{t-i} y_{t-j}$. Let $\sigma_k = \mathrm{cov}(y_i, y_{i+k})$ and let

$$\Sigma = \begin{vmatrix} \sigma_0 & \sigma_1 & \ldots \sigma_{p-1} \\ \sigma_1 & \sigma_0 & \ldots \sigma_{p-2} \\ \vdots & & \ddots \\ \sigma_{p-1} & \sigma_{p-2} & \ldots \sigma_0 \end{vmatrix}.$$

Assume Σ is positive definite. It is known that under this condition $\sqrt{n}\Sigma^{-1/2}(\hat{\theta}-\theta) \stackrel{\mathcal{L}}{\Rightarrow} N(0, I)$. So we may expect that with a suitable bootstrap scheme $\sqrt{n}\hat{\Sigma}^{-1/2}(\theta^* - \hat{\theta})$ converges a.s. in law to $N(0, I)$. Here $\hat{\Sigma}$ denotes the sample autocovariance matrix. We now describe the bootstrap scheme given in Bose (1988).

Let $\hat{y}_t = \sum_{j=1}^{p} \hat{\theta}_j y_{t-j}$ and let the residuals be $e_t = y_t - \hat{y}_t$. To obtain the bootstrap data, define $\{y^*_{1-2p}, y^*_{2-2p}, \ldots, y^*_{-p}\} \equiv \{y_{1-p}, y_{2-p}, \ldots . y_0\}$. Obtain bootstrap residuals by taking a random sample with replacement from $\{e_t - \bar{e}\}$. Then obtain the "starred" data by using the equation $y^*_t = \sum_{j=1}^{p} \hat{\theta}_j y^*_{t-j} + e^*_t$. Then θ^* is the LSE obtained by using $\{y^*_t\}$. Bose (1988) proves the following result.

Theorem 29.18 Assume that ε_1 has a density with respect to Lebesgue measure and that $E(\varepsilon_1^8) < \infty$. If $H_n(x) = P\{\sqrt{n}\Sigma^{-1/2}(\hat{\theta} - \theta) \leq x\}$ and $H_{\text{Boot}}(x) = P_*\{\sqrt{n}\hat{\Sigma}^{-1/2}(\theta^* - \hat{\theta}) \leq x\}$, then $\|H_n - H_{\text{Boot}}\|_\infty = o(n^{-1/2})$, almost surely.

Remark. This was the first result on higher-order accuracy of a suitable form of the bootstrap for dependent data. One possible criticism of the otherwise important result is that it assumes a specific dependence structure and that it assumes the order p is known. More flexible consistent bootstrap schemes involve some form of block resampling, which we describe next.

29.18 Block Bootstrap Methods

The basic idea of the block bootstrap method is that if the underlying series is a stationary process with short-range dependence, then blocks of observations of suitable lengths should be approximately independent and the joint distribution of the variables in different blocks would be (about) the same due to stationarity. So, if we resample blocks of observations rather than observations one at a time, then that should bring us back to the nearly iid situation, a situation in which the bootstrap is known to succeed. The block bootstrap was first suggested in Carlstein (1986) and Künsch (1989). Various block bootstrap schemes are now available. We only present three such schemes, for which the block length is nonrandom. A small problem with some of the blocking schemes is that the "starred" time series is not stationary, although the original series is, by hypothesis, stationary. A version of the block bootstrap that resamples blocks of *random* length allows the "starred" series to be provably stationary. This is called the *stationary bootstrap*, proposed in Politis and Romano (1994), and Politis, Romano, and Wolf (1999). However, later theoretical studies have established that the auxiliary randomization to determine the block lengths can make the stationary bootstrap less accurate. For this reason, we only discuss three blocking methods with nonrandom block lengths.

(a) *Nonoverlapping block bootstrap (NBB).* In this scheme, one splits the observed series $\{y_1, \dots, y_n\}$ into nonoverlapping blocks

$$B_1 = \{y_1, \dots, y_h\}, \; B_2 = \{y_{h+1}, \dots, y_{2h}\}, \dots, \\ B_m = \{y_{(m-1)h+1}, \dots, y_{mh}\},$$

where it is assumed that $n = mh$. The common block length is h. One then resamples $B_1^*, B_2^*, \dots, B_m^*$ at random, with replacement, from $\{B_1, \dots, B_m\}$. Finally, the B_i^*'s are pasted together to obtain the "starred" series $y_1^*, \dots, y_n^*$.

(b) *Moving block bootstrap (MBB).* In this scheme, the blocks are

$$B_1 = \{y_1, \dots, y_h\}, \; B_2 = \{y_2, \dots, y_{h+1}\}, \dots, B_N = \{y_{n-h+1}, \dots, y_n\},$$

where $N = n - h + 1$. One then resamples $B_1^*, \dots, B_m^*$ from $B_1, \dots, B_N$, where still $n = mh$.

(c) *Circular block bootstrap (CBB).* In this scheme, one periodically extends the observed series as $y_1, y_2, \dots, y_n, y_1, y_2, \dots, y_n, \dots$. Suppose we let z_i be the members of this new series, $i = 1, 2, \dots$. The blocks are defined as

$$B_1 = \{z_1, \dots, z_h\}, \; B_2 = \{z_{h+1}, \dots, z_{2h}\}, \dots, B_n = \{z_n, \dots, z_{n+h-1}\}.$$

One then resamples $B_1^*, \dots, B_m^*$ from $B_1, \dots, B_n$.

Next we give some theoretical properties of the three block bootstrap methods described above. The results below are due to Lahiri (1999).

Suppose $\{y_i : -\infty < i < \infty\}$ is a d-dimensional stationary process with a finite mean μ and spectral density f. Let $h : \mathbb{R}^d \to \mathbb{R}^1$ be a sufficiently smooth function. Let $\theta = h(\mu)$ and $\hat{\theta}_n = h(\bar{y}_n)$, where $\bar{y}_n$ is the mean of the realized series. We propose to use the block bootstrap schemes to estimate the bias and variance of $\hat{\theta}_n$. Precisely, let $b_n = E(\hat{\theta}_n - \theta)$ be the bias and let $\sigma_n^2 = \text{Var}(\hat{\theta}_n)$ be the variance. We use the block bootstrap-based estimates of b_n and σ_n^2, denoted by $\hat{b}_n$ and $\hat{\sigma_n^2}$, respectively.

Next, let $T_n = \hat{\theta}_n - \theta = h(\bar{y}_n) - h(\mu)$, and let $T_n^* = h(\bar{y}_n^*) - h(E_* \bar{y}_n^*)$. The estimates $\hat{b}_n$ and $\hat{\sigma_n^2}$ are defined as $\hat{b}_n = E_* T_n^*$ and $\hat{\sigma_n^2} = \text{Var}_*(T_n^*)$. Then the following asymptotic expansions hold; see Lahiri (1999).

Theorem 29.19 Let $h : \mathbb{R}^d \to \mathbb{R}^1$ be a sufficiently smooth function.

(a) For each of the NBB, MBB, and CBB, there exists $c_1 = c_1(f)$ such that

$$E\hat{b}_n = b_n + \frac{c_1}{nh} + o((nh)^{-1}), \quad n \to \infty.$$

(b) For the NBB, there exists $c_2 = c_2(f)$ such that

$$\text{Var}(\hat{b}_n) = \frac{2\pi^2 c_2 h}{n^3} + o(hn^{-3}), \quad n \to \infty,$$

and for the MBB and CBB,

$$\text{Var}(\hat{b}_n) = \frac{4\pi^2 c_2 h}{3n^3} + o(hn^{-3}), \quad n \to \infty.$$

(c) For each of NBB, MBB, and CBB, there exists $c_3 = c_3(f)$ such that $E(\hat{\sigma_n^2}) = \sigma_n^2 + \frac{c_3}{nh} + o((nh)^{-1}), \quad n \to \infty$.

(d) For NBB, there exists $c_4 = c_4(f)$ such that $\text{Var}(\hat{\sigma_n^2}) = \frac{2\pi^2 c_4 h}{n^3} + o(hn^{-3}), \quad n \to \infty$, and for the MBB and CBB, $\text{Var}(\hat{\sigma_n^2}) = \frac{4\pi^2 c_4 h}{3n^3} + o(hn^{-3}), \quad n \to \infty$.

These expansions are used in the next section.

29.19 Optimal Block Length

The asymptotic expansions for the bias and variance of the block bootstrap estimates, given in Theorem 29.19, can be combined to produce MSE-optimal block lengths. For example, for estimating b_n by $\hat{b}_n$, the leading term in the expansion for the MSE is

$$m(h) = \frac{4\pi^2 c_2 h}{3n^3} + \frac{c_1^2}{n^2 h^2}.$$

To minimize $m(\cdot)$, we solve $m'(h) = 0$ to get

$$h_{\text{opt}} = \left(\frac{3c_1^2}{2\pi^2 c_2}\right)^{1/3} n^{1/3}.$$

Similarly, an MSE-optimal block length can be derived for estimating σ_n^2 by $\hat{\sigma_n^2}$. We state the following optimal block-length result of Lahiri (1999) below.

Theorem 29.20 For the MBB and the CBB, the MSE-optimal block length for estimating b_n by $\hat{b}_n$ satisfies

$$h_{\text{opt}} = \left(\frac{3c_1^2}{2\pi^2 c_2} \right)^{1/3} n^{1/3}(1 + o(1)),$$

and the MSE-optimal block length for estimating σ_n^2 by $\hat{\sigma_n^2}$ satisfies

$$h_{\text{opt}} = \left(\frac{3c_3^2}{2\pi^2 c_4} \right)^{1/3} n^{1/3}(1 + o(1)).$$

Remark. Recall that the constants c_i depend on the spectral density f of the process. So, the optimal block lengths cannot be used directly. Plug-in estimates for the c_i may be substituted, or the formulas can be used to try block lengths proportional to $n^{1/3}$ with flexible proportionality constants. There are also other methods in the literature on selection of block lengths; see Hall, Horowitz, and Jing (1995) and Politis and White (2004).

29.20 Exercises

Exercise 29.1 For $n = 10, 20, 50$, take a random sample from an $N(0, 1)$ distribution and bootstrap the sample mean $\bar{X}$ using a bootstrap Monte Carlo size $B = 200$. Construct a histogram and superimpose on it the exact density of $\bar{X}$. Compare the two.

Exercise 29.2 For $n = 5, 25, 50$, take a random sample from an Exp(1) density and bootstrap the sample mean $\bar{X}$ using a bootstrap Monte Carlo size $B = 200$. Construct a histogram and superimpose on it the exact density of $\bar{X}$ and the CLT approximation. Compare the two and discuss if the bootstrap is doing something that the CLT answer does not.

Exercise 29.3 * By using combinatorial coefficient matching cleverly, derive a formula for the number of distinct orthodox bootstrap samples with a general value of n.

Exercise 29.4 * For which, if any, of the sample mean, the sample median, and the sample variance is it possible to explicitly obtain the bootstrap distribution $H_{\text{Boot}}(x)$?

Exercise 29.5 * For $n = 3$, write an expression for the exact Kolmogorov distance between H_n and H_{Boot} when the statistic is $\bar{X}$ and $F = N(0, 1)$.

Exercise 29.6 For $n = 5, 25, 50$, take a random sample from an Exp(1) density and bootstrap the sample mean $\bar{X}$ using a bootstrap Monte Carlo size $B = 200$ using both the canonical bootstrap and the natural parametric bootstrap. Construct the corresponding histograms and superimpose them on the exact density. Is the parametric bootstrap more accurate?

Exercise 29.7 * Prove that under appropriate moment conditions, the bootstrap is consistent for the sample correlation coefficient r between two jointly distributed variables X, Y.

Exercise 29.8 * Give examples of three statistics for which the condition in the rule of thumb on second-order accuracy of the bootstrap does not hold.

Exercise 29.9 * By gradually increasing the value of n, numerically approximate the constant c in the limit theorem for the Kolmogorov distance for the Poisson(1) case (see the text for the definition of c).

Exercise 29.10 * For samples from a uniform distribution, is the bootstrap consistent for the second-largest order statistic? Prove your assertion.

Exercise 29.11 For $n = 5, 25, 50$, take a random sample from an Exp(1) density and compute the bootstrap-t, bootstrap percentile, and the usual t 95% lower confidence bounds on the population mean. Use $B = 300$. Compare them meaningfully.

Exercise 29.12 * Give an example of:

(a) a density such that the bootstrap is not consistent for the median;
(b) a density such that the bootstrap is not consistent for the mean;
(c) a density such that the bootstrap is consistent but not second-order accurate for the mean.

Exercise 29.13 For simulated independent samples from the $U[0, \theta)$ density, let $T_n = n(\theta - X_{(n)})$. For $n = 20, 40, 60$, numerically approximate

$K(H_{\text{Boot},m,n}, H_n)$ with varying choices of m and investigate the choice of an optimal m.

Exercise 29.14 * Suppose (X_i, Y_i) are iid samples from a bivariate normal distribution. Simulate $n = 25$ observations taking $\rho = .5$, and compute:

(a) the usual 95% confidence interval;
(b) the interval based on the variance stabilizing transformation (Fisher's z) (see Chapter 4);
(c) the bootstrap percentile interval;
(d) the bootstrap hybrid percentile interval;
(e) the bootstrap-t interval with $\hat{\sigma_n}$ as the usual estimate;
(f) the accelerated bias-corrected bootstrap interval using φ as Fisher's z, $z_0 = \frac{r}{2\sqrt{n}}$ (the choice coming from theory), and three different values of a near zero.
Discuss your findings.

Exercise 29.15 * In which of the following cases are the results in Hall (1988) not applicable and why?

(a) estimating the 80th percentile of a density on $\mathcal{R}$;
(b) estimating the variance of a Gamma density with known scale and unknown shape parameter;
(c) estimating θ in the $U[0, \theta]$ density;
(d) estimating $P(X > 0)$ in a location-parameter Cauchy density;
(e) estimating the variance of the t-statistic for Weibull data;
(f) estimating a binomial success probability.

Exercise 29.16 Using simulated data, compute a standard CLT-based 95% confidence interval and the hybrid bootstrap interval for the 90th percentile of a (i) standard Cauchy distribution and (ii) a Gamma distribution with scale parameter 1 and shape parameter 3. Compare them and comment. Use $n = 20, 40$.

Exercise 29.17 * Are the centers of the CLT-based interval and the hybrid bootstrap interval for a population quantile always the same? Sometimes the same?

Exercise 29.18 * Simulate a series of length 50 from a stationary $AR(p)$ process with $p = 2$ and then obtain the starred series by using the scheme in Bose (1988).

Exercise 29.19 * For the simulated data in Exercise 29.18, obtain the actual blocks in the NBB and the MBB schemes with $h = 5$. Hence, generate the starred series by pasting the resampled blocks.

Exercise 29.20 For $n = 25$, take a random sample from a bivariate normal distribution with zero means, unit variances, and correlation .6. Implement the residual bootstrap using $B = 150$. Compute a bootstrap estimate of the variance of the LSE of the regression slope parameter. Comment on the accuracy of this estimate.

Exercise 29.21 For $n = 25$, take a random sample from a bivariate normal distri-bution with zero means, unit variances, and correlation .6. Implement the paired bootstrap using $B = 150$. Compute a bootstrap estimate of the variance of the LSE of the regression slope parameter. Compare your results with the preceding exercise.

Exercise 29.22 * Give an example of two design matrices that do not satisfy the conditions C1 and C2 in the text.

Exercise 29.23 * Suppose the values of the covariates are $x_i = \frac{1}{i}$, $i = 1, 2, \cdots, n$ in a simple linear regression setup. Prove or disprove that the residual bootstrap consistently estimates the distribution of the LSE of the slope parameter if the errors are (i) iid $N(0, \sigma^2)$, (ii) iid $t(m, 0, \sigma^2)$, where m denotes the degree of freedom.

Exercise 29.24 * Suppose $\bar{X}_n$ is the sample mean of an iid sample from a CDF F with a finite variance and $\bar{X}_n{}^*$ is the mean of a bootstrap sample. Consistency of the bootstrap is a statement about the bootstrap distribution, conditional on the observed data. What can you say about the unconditional limit distribution of $\sqrt{n}(\bar{X}_n{}^* - \mu)$, where μ is the mean of F?

References

Athreya, K. (1987). Bootstrap of the mean in the infinite variance case, Ann. Stat.,15(2), 724–731.

Beran, R. (2003). The impact of the bootstrap on statistical algorithms and theory, Stat. Sci., 18(2), 175–184.

Bickel, P.J. (1992). Theoretical comparison of different bootstrap t confidence bounds, in *Exploring the Limits of Bootstrap*, R. LePage and L. Billard (eds.) John Wiley, New York, 65–76.

Bickel, P.J. (2003). Unorthodox bootstraps, Invited paper, J. Korean Stat. Soc., 32(3), 213–224.

Bickel, P.J. and Freedman, D. (1981). Some asymptotic theory for the bootstrap, Ann. Stat., 9(6), 1196–1217.

Bickel, P.J., Göetze, F., and van Zwet, W. (1997). Resampling fewer than n observations: gains, losses, and remedies for losses, Stat. Sinica, 1, 1–31.

Bose, A. (1988). Edgeworth correction by bootstrap in autoregressions, Ann. Stat., 16(4), 1709–1722.

Bose, A. and Babu, G. (1991). Accuracy of the bootstrap approximation, Prob. Theory Related Fields, 90(3), 301–316.

Bose, A. and Politis, D. (1992). A review of the bootstrap for dependent samples, in *Stochastic Processes and Statistical Inference*, B.L.S.P Rao and B.R. Bhat, (eds.), New Age, New Delhi.

Carlstein, E. (1986). The use of subseries values for estimating the variance of a general statistic from a stationary sequence, Ann. Stat., 14(3), 1171–1179.

David, H.A. (1981). *Order Statistics*, Wiley, New York.

Davison, A.C. and Hinkley, D. (1997). *Bootstrap Methods and Their Application*, Cambridge University Press, Cambridge.

DiCiccio, T. and Efron, B. (1996). Bootstrap confidence intervals, with discussion, Stat. Sci., 11(3), 189–228.

Efron, B. (1979). Bootstrap methods: another look at the Jackknife, Ann. Stat., 7(1), 1–26.

Efron, B. (1981). Nonparametric standard errors and confidence intervals, with discussion, Can. J. Stat., 9(2), 139–172.

Efron, B. (1987). Better bootstrap confidence intervals, with comments, J. Am. Stat. Assoc., 82(397), 171–200.

Efron, B. (2003). Second thoughts on the bootstrap, Stat. Sci., 18(2), 135–140.

Efron, B. and Tibshirani, R. (1993). *An Introduction to the Bootstrap*, Chapman and Hall, New York.

Falk, M. and Kaufman, E. (1991). Coverage probabilities of bootstrap confidence intervals for quantiles, Ann. Stat., 19(1), 485–495.

Freedman, D. (1981). Bootstrapping regression models, Ann. Stat., 9(6), 1218–1228.

Ghosh, M., Parr, W., Singh, K., and Babu, G. (1984). A note on bootstrapping the sample median, Ann. Stat., 12, 1130–1135.

Giné, E. and Zinn, J. (1989). Necessary conditions for bootstrap of the mean, Ann. Stat., 17(2), 684–691.

Göetze, F. (1989). Edgeworth expansions in functional limit theorems, Ann. Prob., 17, 1602–1634.

Hall, P. (1986). On the number of bootstrap simulations required to construct a confidence interval, Ann. Stat., 14(4), 1453–1462.

Hall, P. (1988). Rate of convergence in bootstrap approximations, Ann. Prob., 16(4), 1665–1684.

Hall, P. (1989a). On efficient bootstrap simulation, Biometrika, 76(3), 613–617.

Hall, P. (1989b). Unusual properties of bootstrap confidence intervals in regression problems, Prob. Theory Related Fields, 81(2), 247–273.

Hall, P. (1990). Asymptotic properties of the bootstrap for heavy-tailed distributions, Ann. Prob., 18(3), 1342–1360.
Hall, P. (1992). *Bootstrap and Edgeworth Expansion*, Springer-Verlag, New York.
Hall, P. (2003). A short prehistory of the bootstrap, Stat. Sci., 18(2), 158–167.
Hall, P., DiCiccio, T., and Romano, J. (1989). On smoothing and the bootstrap, Ann. Stat., 17(2), 692–704.
Hall, P. and Martin, M.A. (1989). A note on the accuracy of bootstrap percentile method confidence intervals for a quantile, Stat. Prob. Lett., 8(3), 197–200.
Hall, P., Horowitz, J., and Jing, B. (1995). On blocking rules for the bootstrap with dependent data, Biometrika, 82(3), 561–574.
Helmers, R. (1991). On the Edgeworth expansion and bootstrap approximation for a studentized U-statistic, Ann. Stat., 19(1), 470–484.
Konishi, S. (1991). Normalizing transformations and bootstrap confidence intervals, Ann. Stat., 19(4), 2209–2225.
Künsch, H.R. (1989). The Jackknife and the bootstrap for general stationary observations, Ann. Stat., 17(3), 1217–1241.
Lahiri, S.N. (1999). Theoretical comparisons of block bootstrap methods, Ann. Stat., 27(1), 386–404.
Lahiri, S.N. (2003). *Resampling Methods for Dependent Data*, Springer-Verlag, New York.
Lahiri, S.N. (2006). Bootstrap methods, a review, in *Frontiers in Statistics*, J. Fan and H. Koul (eds.), Imperial College Press, London, 231–256.
Lee, S. (1999). On a class of m out of n bootstrap confidence intervals, J.R. Stat. Soc. B, 61(4), 901–911.
Lehmann, E.L. (1999). *Elements of Large Sample Theory*, Springer, New York.
Loh, W. and Wu, C.F.J. (1987). Discussion of "Better bootstrap confidence intervals" by Efron, B., J. Amer. Statist. Assoc., 82, 188–190.
Politis, D. and Romano, J. (1994). The stationary bootstrap, J. Am. Stat. Assoc., 89(428), 1303–1313.
Politis, D., Romano, J., and Wolf, M. (1999). *Subsampling*, Springer, New York.
Politis, D. and White, A. (2004). Automatic block length selection for the dependent bootstrap, Econ. Rev., 23(1), 53–70.
Shao, J. and Tu, D. (1995). *The Jackknife and Bootstrap*, Springer-Verlag, New York.
Silverman, B. and Young, G. (1987). The bootstrap: to smooth or not to smooth?, Biometrika, 74, 469–479.
Singh, K. (1981). On the asymptotic accuracy of Efron's bootstrap, Ann. Stat., 9(6), 1187–1195.
Tong, Y.L. (1990). *The Multivariate Normal Distribution*, Springer, New York.
van der Vaart, A. (1998). *Asymptotic Statistics*, Cambridge University Press, Cambridge.

Chapter 30
Jackknife

The jackknife was initially suggested as a bias reduction tool (Quenouille (1949)). Gradually, it has started to be used for other purposes, such as variance estimation, estimation of distributions of statistics, and interval estimation. In many ways, it is a complement to the bootstrap. There are problems in which the jackknife would work better than the bootstrap and vice versa. It is more appropriate to call the jackknife a subsampling method, as it recomputes the statistic by leaving out a fixed number of the original observations. We recommend Miller (1974), Shao and Tu (1995), and Politis, Romano, and Wolf (1999) for detailed expositions of the jackknife and general subsampling methods, the latter two references being at a higher technical level. First we introduce some standard notation.

30.1 Notation and Motivating Examples

Notation. Let $X_1, \cdots, X_n \stackrel{\text{iid}}{\sim} F$, $T_n = T_n(X_1, \ldots, X_n)$ a statistic. Let

$$T_{n-1,i} = T_{n-1}(x_1, \ldots, x_{i-1}, x_{i+1}, \ldots, x_n),\ \bar{T}_n = \frac{1}{n}\sum_{i=1}^{n} T_{n-1,i},$$

$$\widetilde{T}_{n,i} = nT_n - (n-1)T_{n-1,i},\ \overline{\widetilde{T}}_n = \frac{1}{n}\sum_{i=1}^{n} \widetilde{T}_{n,i},$$

$$b_{\text{Jack}} = (n-1)(\bar{T}_n - T_n),\ V_{\text{Jack}} = \frac{1}{n-1}\sum_{i=1}^{n}(\widetilde{T}_{n,i} - \overline{\widetilde{T}}_n)^2.$$

Remark. $T_{n-1,i}$ is called the delete-1 recomputed T_n, and $\{\widetilde{T}_{n.i}\}$ are known as Tukey pseudo-values (see Tukey (1958)) and are proxies for iid copies of $\sqrt{n}T_n$. b_{Jack} is the jackknife estimate of the bias of T_n and V_{Jack} the jackknife estimate of its variance. T_{Jack} is the bias-corrected version of T_n when the bias is estimated by b_{Jack}.

Example 30.1 Let $T_n = \bar{X}$. Then $T_{n-1,i} = \frac{1}{n-1}(x_1 + \ldots + x_{i-1} + x_{i+1} + \ldots + x_n) = \frac{n\bar{x}-x_i}{n-1} = \frac{n}{n-1}\bar{x} - \frac{x_i}{n-1}$, so $\bar{T}_n = \frac{n}{n-1}\bar{x} - \frac{\bar{x}}{n-1} = \bar{x}$. Therefore, $b_{\text{Jack}} = (n-1)(\bar{T}_n - T_n) = 0$. Thus, the jackknife estimates the bias to be zero, which is the correct answer.

Example 30.2 Let $T_n = \bar{X}_n^2$. Then,

$$
\begin{aligned}
b_{\text{Jack}} &= (n-1)(\bar{T}_n - T_n) \\
&= (n-1)\left[\frac{1}{n}\sum_{i=1}^{n} \bar{x}_{n-1,i}^2 - \bar{x}_n^2\right] \\
&= \frac{n-1}{n}\sum_{i=1}^{n}[\bar{x}_{n-1,i}^2 - \bar{x}_n^2] \\
&= \frac{n-1}{n}\sum_{i=1}^{n}[\bar{x}_{n-1,i} + \bar{x}_n][\bar{x}_{n-1,i} - \bar{x}_n] \\
&= \frac{1}{n(n-1)}\sum_{i=1}^{n}[(n-1)\bar{x}_{n-1,i} - (n-1)\bar{x}_n][(n-1)\bar{x}_{n-1,i} + (n-1)\bar{x}_n] \\
&= \frac{1}{n(n-1)}\sum_{i=1}^{n}[\bar{x}_n - x_i][(2n-1)\bar{x}_n - x_i] \\
&= \frac{1}{n(n-1)}\sum_{i=1}^{n}[\bar{x}_n - x_i]^2.
\end{aligned}
$$

Hence, $T_{\text{Jack}} = T_n - b_{\text{Jack}} = \bar{x}_n^2 - \frac{1}{n(n-1)}\sum_{i=1}^{n}(x_i - \bar{x}_n)^2$, and $ET_{\text{Jack}} = \mu^2 + \frac{\sigma^2}{n} - \frac{\sigma^2}{n} = \mu^2$. So, in this example, jackknife removes the bias of T_n altogether.

In fact, it is true in general that the jackknife would reduce the bias of a statistic T_n in some precise sense. First, let us see a heuristic calculation for why it does so. For a function $\Theta(F)$, a statistic T_n of interest would often have a bias of the form $\text{bias}(T_n) = \frac{a}{n} + \frac{b}{n^2} + o(n^{-2})$, so, for each i, $\text{bias}(T_{n-1,i}) = \frac{a}{n-1} + \frac{b}{(n-1)^2} + o(n^{-2})$. Since $b_{\text{Jack}} = (n-1)(\bar{T}_n - T_n)$, one has

$$
\begin{aligned}
E(b_{\text{Jack}}) &= (n-1)(E(\bar{T}_n - E(T_n))) \\
&= (n-1)(\text{bias}(\bar{T}_n) - \text{bias}(T_n))
\end{aligned}
$$

$$= (n-1)\left(\frac{a}{n-1} + \frac{b}{(n-1)^2} - \frac{a}{n} - \frac{b}{n^2} + o(n^{-2})\right)$$
$$= \frac{a}{n} + \frac{2n-1}{n^2(n-1)}b + o(n^{-1}).$$

Thus,

$$\begin{aligned}
\text{bias}(T_{\text{Jack}}) &= \text{bias}(T_n) - E(b_{\text{Jack}}) \\
&= \frac{a}{n} + \frac{b}{n^2} - \left(\frac{a}{n} + \frac{2n-1}{n^2(n-1)}b\right) + o(n^{-1}) \\
&= o(n^{-1})
\end{aligned}$$

because the $\frac{a}{n}$ term cancels.

The original T_n has bias $O(\frac{1}{n})$, while the bias-corrected T_{Jack} has bias $o(\frac{1}{n})$. In this sense, the jackknife reduces the bias of a statistic. It is not true that the jackknife necessarily reduces bias at a fixed sample size n.

The next two examples are on the jackknife estimate of the variance of a statistic.

Example 30.3 Let $T_n = \bar{X}_n$. By simple algebra,

$$\begin{aligned}
V_{\text{Jack}} &= \frac{1}{n(n-1)} \sum_{i=1}^{n} (\widetilde{T}_{n,i} - \bar{\widetilde{T}}_n)^2 \\
&= \frac{n-1}{n} \sum_{i=1}^{n} (T_{n-1,i} - \bar{T}_n)^2 \\
&= \frac{n-1}{n} \sum_{i=1}^{n} \left(\frac{n}{n-1}\bar{x} - \frac{x_i}{n-1} - \bar{x}\right)^2 \\
&= \frac{1}{n(n-1)} \sum_{i=1}^{n} (x_i - \bar{x})^2 \\
&= \frac{s^2}{n}.
\end{aligned}$$

So, in this example, the jackknife variance estimate for $\text{Var}_F(T_n)$ is just the unbiased estimate $\frac{s^2}{n}$.

Example 30.4 Let $T_n = \bar{X}_n^2$. The exact variance of T_n is $\text{Var}T_n = \frac{4\mu^2\sigma^2}{n} + \frac{4\mu\mu_3}{n^2} + \frac{\mu_4}{n^3}$, where $\mu_j = E(X-\mu)^j$. So a traditional plug-in estimate for $\text{Var}T_n$ would be $\frac{4\bar{x}^2 s^2}{n} + \frac{4\bar{x}\hat{\mu}_3}{n^2} + \frac{\hat{\mu}_4}{n^3}$, where $\hat{\mu}_j = \frac{1}{n}\sum_{i=1}^n (x_i - \bar{x})^j$. On calculation, the jackknife variance estimate turns out to be $V_{\text{Jack}} = \frac{4\bar{x}^2 s^2}{n} - \frac{4\bar{x}\hat{\mu}_3}{n^2} + \frac{\hat{\mu}_4}{n(n-1)^2} - \frac{s^2}{n^2(n-1)}$. Thus the traditional and the jackknife variance estimates differ at the $\frac{1}{n^2}$ term.

30.2 Bias Correction by the Jackknife

There are an array of theorems on asymptotic bias reduction by the jackknife for various classes of statistics. Here is a result for smooth functions of sample means, $T_n = g(\bar{X}_n)$; we give an informal proof below. See Shao and Tu (1995) for a formal proof.

Theorem 30.1 Let $X_1, \ldots, X_n$ be iid p-vectors with mean μ and $\Sigma_{p\times p} = \text{cov}(X_1)$ finite. Let $g : \mathbb{R}^p \to \mathbb{R}$ be such that $\nabla^2 g(\cdot) = ((\frac{\partial^2}{\partial x_i \partial x_j} g(\cdot)))$ exists in a neighborhood of μ and is continuous at μ and let $T_n = g(\bar{X})$. Let $a = \frac{1}{2}\text{tr}(\nabla^2 g(\mu)\Sigma)$. Then $nb_{Jack} \overset{\text{a.s.}}{\to} a$. If, in addition, $\frac{\partial^3}{\partial x_i \partial x_j \partial x_k} g(\cdot)$ is uniformly bounded $\forall i, j, k$ in a neighborhood of μ, then $b_{Jack} = \frac{a}{n} + O_p(n^{-2})$.

Proof. The consistency result can be seen by putting together the Taylor expansions

$$g(\bar{X}) - g(\mu) = (\bar{X} - \mu)'\nabla g(\mu) + \frac{1}{2}(\bar{X} - \mu)'\nabla^2 g(\mu)(\bar{X} - \mu) + R_n$$

and

$$\begin{aligned} g(\bar{X}) - g(\bar{X}_{n-1,i}) &= (\bar{X} - \bar{X}_{n-1,i})'\nabla g(\bar{X}) \\ &\quad + \frac{1}{2}(\bar{X} - \bar{X}_{n-1,i})'\nabla^2 g(X_{n,i}^*)(\bar{X} - \bar{X}_{n-1,i}), \end{aligned}$$

where $X_{n,i}^*$ is a point on the line segment $\lambda\bar{X} + (1-\lambda)\bar{X}_{n-1,i}$, $0 \le \lambda \le 1$. Therefore,

$$b_{\text{Jack}} = (n-1)(T_n - \bar{T}_n) = \frac{n-1}{2n}\text{tr}\sum_{i=1}^n \nabla^2 g(X_{n,i}^*)(\bar{X} - \bar{X}_{n-1,i})(\bar{X} - \bar{X}_{n-1,i})'.$$

But $\bar{X}_{n-1,i} = \frac{n\bar{X} - X_i}{n-1} = \frac{n}{n-1}\bar{X} - \frac{X_i}{n-1}$. Substituting,

$$
\begin{aligned}
b_{\text{Jack}} &= \frac{n-1}{2n(n-1)^2} \sum_{i=1}^{n} \text{tr}\nabla^2 g(X_{n,i}^*)(\bar{X} - X_i)(\bar{X} - X_i)' \\
&= \frac{1}{n-1} \frac{1}{2n} \sum_{i=1}^{n} \text{tr}\nabla^2 g(X_{n,i}^*)(\bar{X} - X_i)(\bar{X} - X_i)'
\end{aligned}
$$

and

$$
\begin{aligned}
(n-1)b_{\text{Jack}} &= \frac{1}{2n} \sum_{i=1}^{n} \text{tr}\nabla^2 g(X_{n,i}^*)(\bar{X} - X_i)(\bar{X} - X_i)' \\
&= \frac{1}{2n} \sum_{i=1}^{n} \text{tr}[\nabla^2 g(X_{n,i}^*) - \nabla^2 g(\mu) + \nabla^2 g(\mu)](\bar{X} - X_i)(\bar{X} - X_i)' \\
&\overset{\text{a.s.}}{\to} a
\end{aligned}
$$

under the assumed derivative conditions.

30.3 Variance Estimation

As we remarked before, the jackknife is also used to estimate variances of statistics. Analogous to the theorem on estimation of bias, there are also results available on consistent estimation of the variance by using the jackknife. We present a few key results on consistency of the jackknife variance estimate.

Theorem 30.2 Let $X_i \overset{iid}{\sim} (\mu, \Sigma_{p\times p})$. Let $g : \mathbb{R}^p \to \mathbb{R}$ be such that $\nabla g(\cdot)$ exists in a neighborhood of μ, is continuous at μ, and $\nabla g(\mu) \neq 0$. Let $T_n = g(\bar{X})$. Then

$$
\frac{n V_{\text{Jack}}}{(\nabla g(\mu))' \Sigma (\nabla g(\mu))} \overset{\text{a.s.}}{\to} 1.
$$

Remark. A proof can be seen in Shao and Tu (1995). The utility of this result is that if we want an estimate of the standard error of $g(\bar{X})$, then the delta theorem estimate would call for plug-in estimates for μ and Σ and analytic calculation for $\nabla g(\cdot)$. The jackknife variance estimate, on the other hand, is an automated estimate requiring only programming and no further analytic calculations or plug-in estimates. Note, however, that for the consistency of

V_{Jack}, we need the continuity of $\nabla g(\cdot)$ at μ, which is not needed for the CLT for $g(\bar{X})$.

Theorem 30.3 Let $U_n = U_n(X_1, \ldots, X_n)$ be a U-statistic with a kernel h of order $r \geq 1$. Assume $\tau^2 = \text{cov}_F(h(X_1, X_2, \ldots, X_r), h(X_1, Y_2, \ldots, Y_r)) > 0$, where $X_1, X_2, Y_2, \ldots, X_r, Y_r$ are iid F. Let $g : \mathbb{R} \to \mathbb{R}$ and suppose g' exists in a neighborhood of $\theta = E_F(h)$, is continuous at θ, and $g'(\theta) \neq 0$. Let $T_n = g(U_n)$. If $E_F(h^2) < \infty$, then

$$\frac{nV_{\text{Jack}}}{(g'(\theta))^2 r^2 \tau^2} \stackrel{\text{a.s.}}{\to} 1.$$

Remark. Again, see Shao and Tu (1995) for additional details on this theorem. An example of a U-statistic for which τ^2 is hard to calculate is the Gini mean difference $G = \frac{1}{\binom{n}{2}} \sum\sum_{i<j} |X_i - X_j|$. Therefore, the delta theorem variance estimate will be cumbersome to implement. The jackknife variance estimate would be useful to estimate the standard error of G.

Remark. There are some statistics whose variance is not consistently estimated by the jackknife. Lack of smoothness of the statistic would typically be the problem. An example of a statistic that is not sufficiently smooth in the observations is the sample median. Efron (1982) showed that the jackknife does not consistently estimate the variance of sample percentiles. Modification of the basic jackknife is needed to address these situations. In fact, Efron showed that if $T_n = F_n^{-1}(\frac{1}{2})$, then $\frac{nV_{\text{Jack}}}{4f^2(F^{-1}(\frac{1}{2}))} \stackrel{\mathcal{L}}{\Rightarrow} w^2$, where $w \sim \text{Exp}(1)$. So evidently $\frac{nV_{\text{Jack}}}{4f^2(F^{-1}(\frac{1}{2}))}$ cannot converge to 1 even in probability. For example, $\lim_n P(\frac{nV_{\text{Jack}}}{\sigma_F^2} < 1) = P(w^2 < 1) = 1 - e^{-1} > 0.5$. So there is a propensity to underestimate the variance.

In fact, there are some general results, too, in this direction. The required modification to cure the inconsistency problem is that the statistic T_n is recomputed by deleting many more than just one observation at a time. This is called the delete-d jackknife. We discuss the delete-d jackknife in the next section. See Shao and Wu (1989) and Wu (1990) for further detailed analysis of this fundamental idea.

30.4 Delete-*d* Jackknife and von Mises Functionals

First we need some notation and definitions. Let $\mathfrak{X} = \{x_1, \ldots, x_n\}$, $1 \leq d \leq n$, $r = n - d$, S be a generic subset of $\mathfrak{X}$ of cardinality d, $T_{r,S}$ be the statistic T_n recomputed with the observations belonging to S deleted, and $N = \binom{n}{d}$.

Definition 30.1 The delete-d jackknife variance estimate is defined as

$$V_{\text{Jack}.d} = \frac{r}{Nd} \sum_{S} \left(T_{r,S} - \frac{1}{N} \sum_{S} T_{r,S} \right)^2 .$$

For an appropriate choice of d, $V_{\text{Jack}.d}$ consistently estimates the variance of a variety of statistics T_n, including sample percentiles, for which the basic jackknife fails.

There is a meta-theorem on consistency of $V_{\text{Jack}.d}$ covering a broad class of statistics T_n, collectively known as von Mises functionals. We give a quick treatment of von Mises functionals first in order to be able to state our meta-theorem. We recommend Serfling (1980) at a textbook level for further examples and proofs of basic theorems on von Mises functionals. Fernholz (1983), and Sen (1996) are expositions on the same topic at a somewhat higher technical level.

Definition 30.2 Let $X_1, \ldots, X_n \overset{iid}{\sim} F$ and F_n the empirical CDF. Let $\mathfrak{F}$ be the class of all CDFs on the marginal sample space of X_1. Let $h : \mathfrak{F} \to \mathcal{R}$ be an operator on $\mathfrak{F}$. The statistic $T_n = h(F_n)$ is called a von Mises functional.

Examples of von Mises functions are $T_n = \bar{X} = \int x dF_n(x)$, $T_n = F_n^{-1}(p)$ for any $0 < p < 1$, any sample moment $T_n = \frac{1}{n} \sum_{i=1}^{n} X_i^k = \int x^k dF_n(x)$, and the Cramér-von Mises statistic $T_n = n \int (F_n(x) - F(x))^2 dF(x)$. Von Mises functionals that are sufficiently smooth in the sense of being smooth operators admit a Bahadur type expansion. Such an expansion guarantees a CLT for them; with a little more assumption, the jackknife variance estimate $V_{\text{Jack}.d}$ consistently estimates the variance of a von Mises functional T_n. Here is a fundamental result. See Shao (1993) for further details on the technical issues.

Theorem 30.4 Let $T_n = h(F_n)$ be a von Mises functional, and suppose T_n admits the Bahadur expansion,

$$T_n = h(F) + \frac{1}{n} \sum_{i=1}^{n} \varphi(X_i, F) + R_n,$$

for some function φ with zero mean and where $\sqrt{n} R_n \xrightarrow{\mathcal{P}} 0$.

Let $\sigma_F^2 = E_F \varphi^2(X, F)$. Suppose, in addition, that either

$$n \text{Var}_F(T_n) - \sigma_F^2 \to 0$$

or $\{n(T_n - h(F))^2\}$ is uniformly integrable.

Then, with $d \sim n\delta$ for $0 < \delta < 1$, $\frac{V_{\text{Jack.}d}}{\text{Var}_F(T_n)}$ and $\frac{nV_{\text{Jack.}d}}{E_F\varphi^2(X,F)}$ each converge in probability to 1.

Remark. A more general version of Theorem 30.4 is proved in Shao and Tu (1995). The function $\varphi(x, F)$ is called the *influence function* of T_n. Typically, one would take $\delta \approx \frac{1}{2}$. So the delete-$d$ jackknife can consistently estimate variances of many statistics if about half the observations are thrown away in recomputing it each time. Remember, however, that not all subsets S of size d can be exhausted in practice and some J randomly selected ones have to be used. This leads to interesting consistency issues about the associated Monte Carlo; see Shao and Tu (1995).

Remark. The exact form of the function $\varphi(x, F)$ is essential to know for actual application. The function $\varphi(x, F)$ is what is known in operator theory as the Gateaux derivative of the operator $h(F)$ at F in the direction $\delta_{\{x\}} - F$, where $\delta_{\{x\}}$ denotes the CDF of the distribution degenerate at x. Precisely, $\varphi(x, F) = \lim_{\delta\to 0} \frac{h(F+\delta(\delta_{\{x\}}-F))-h(F)}{\delta}$, which works out to the operationally useful formula

$$\varphi(x, F) = \lim_{\varepsilon\to 0} \frac{d}{d\varepsilon} h((1-\varepsilon)F + \varepsilon(\delta_{\{x\}})).$$

Let us see some examples on the calculation of $\varphi(x, F)$.

Example 30.5 Let $X_1, \ldots, X_n \stackrel{\text{iid}}{\sim} F$ and let $T_n = h(F_n) = \bar{X}$, so that $h(F) = \int x dF = \mu$. Therefore,

$$\begin{aligned} & h((1-\epsilon)F + \epsilon\delta_{\{x\}}) = (1-\epsilon)\mu + \epsilon x \\ \Rightarrow\ & \frac{d}{d\epsilon} h((1-\epsilon)F + \epsilon\delta_{\{x\}}) = x - \mu \\ \Rightarrow\ & \varphi(x, F) = \lim_{\epsilon\to 0} \frac{d}{d\epsilon} h((1-\epsilon)F + \epsilon\delta_{\{x\}}) = x - \mu. \end{aligned}$$

Therefore,

$$\sqrt{n}(h(F_n) - h(F)) = \sqrt{n}(\bar{X} - \mu) \stackrel{\mathcal{L}}{\Rightarrow} N(0, E_F(\varphi(X, F))^2) = N(0, \sigma^2),$$

which is the CLT.

Example 30.6 Consider the Cramér-von Mises goodness-of-fit statistic for testing $H_0 : F = F_0$ on the basis of $X_1, \ldots, X_n \stackrel{\text{iid}}{\sim} F$. The statistic is defined as $C_n = \int (F_n - F_0)^2 dF_0$ (see Chapter 26). This is a von Mises functional $h(F_n)$ with $h(F) = \int (F - F_0)^2 dF_0$. Therefore,

$$h((1-\epsilon)F+\epsilon\delta_{\{x\}})$$
$$=\int_{t<x}((1-\epsilon)F(t)-F_0(t))^2dF_0(t)+\int_{t\geq x}((1-\epsilon)F(t)+\epsilon-F_0(t))^2dF_0(t)$$
$$=(1-\epsilon)^2\int F^2dF_0+\int F_0^2dF_0-2(1-\epsilon)\int FF_0dF_0+\epsilon^2\int_{t\geq x}dF_0$$
$$+2\epsilon(1-\epsilon)\int_{t\geq x}FdF_0-2\epsilon\int_{t\geq x}F_0dF_0.$$

By straight differentiation,

$$\varphi(x,F)=\lim_{\epsilon\to 0}\frac{d}{d\epsilon}h((1-\epsilon)F+\epsilon\delta_{\{x\}})$$
$$=-2\int F^2dF_0+2\int FF_0dF_0+2\int_{t\geq x}FdF_0-2\int_{t\geq x}F_0dF_0$$
$$=2\int(F(t)-F_0(t))(I_{t\geq x}-F(t))dF_0(t).$$

Therefore,

$$\sqrt{n}(C_n-h(F))$$
$$\stackrel{\mathcal{L}}{\Rightarrow}N\left(0,4\int\left[\int(F(t)-F_0(t))(I_{t\geq x}-F(t))dF_0(t)\right]^2dF(x)\right).$$

Clearly, the asymptotic variance is very complicated as a function of F. The delete-d jackknife is routinely used to approximate the limiting variance. For completeness, note that under the null F_0, the limit distribution is not normal and in fact $nC_n\xrightarrow[H_0]{\mathcal{L}}\sup_{0\leq t\leq 1}|B(t)|$, where $B(\cdot)$ is a Brownian bridge (see Chapter 26). Therefore, under the null, there is no issue of estimating the variance of C_n.

30.5 A Numerical Example

Example 30.7 How does $V_{\text{Jack}.d}$ work in practice? It can only be answered by case-by-case simulation. We report the result of a simulation in two specific examples. In each case, $n = 40$, and the results are simulation averages over 2000 simulations.

Remark. The sample median $F_n^{-1}(\frac{1}{2})$ is a nonsmooth statistic, and we know that V_{Jack} itself is inconsistent in that case. It is clear from the table above

	$\frac{\text{Bias}(V_{\text{Jack},d})}{\sigma_F^2/n}$			
	$T_n = \bar{X}_n^2$		$T_n = F_n^{-1}(\frac{1}{2})$	
d	$F = N(\frac{5}{4}, 1)$	$F = \text{Exp}(\frac{1}{2})$	$F = N(\frac{5}{4}, 1)$	$F = C(\frac{5}{2}, 2)$
1	1.2%	7.2%	92.2%	106.8%
10	1.2%	9.8%	21.8%	42.1%
20	1.6%	13.0%	8.9%	29.0%

that the delete-d jackknife improves the quality of the variance estimation. On the other hand, $\bar{X}_n^2$ is a smooth statistic. We know that V_{Jack} is already consistent if F has four moments. The delete-d jackknife is unnecessary in this case and in fact even seems to worsen the quality of the variance approximation. This is akin to the m out of n bootstrap, which leads to some loss of efficiency if the ordinary bootstrap is already consistent.

30.6 Jackknife Histogram

Just like the bootstrap, the jackknife is also used to estimate distributions of statistics. It will frequently be consistent and sometimes even strongly consistent, but it does not have the higher-order accuracy properties comparable to the bootstrap. However, it is an important aspect of jackknife methodology. Let us first define the jackknife distribution estimator.
For simplicity of exposition, consider the case $T_n = \bar{X}_n$, where $X_1, \ldots, X_n \stackrel{\text{iid}}{\sim} F$, $E_F(X_1) = \mu$, $\text{Var}_F(X_1) = \sigma^2$. Let

$$H_n(x) = P_F(\sqrt{n}(\bar{X} - \mu) \le x) = P_F\left(\frac{\sqrt{n}(\bar{X} - \mu)}{\sigma} \le \frac{x}{\sigma}\right).$$

An analog of $\frac{\sqrt{n}(\bar{X}-\mu)}{\sigma}$ based on the delete-d jackknife (for some specific d) is $\frac{\sqrt{r}(\bar{X}_{r,S}-\bar{X})}{\sqrt{(1-\frac{r}{n})\hat{\sigma}^2}}$, where $(1 - \frac{r}{n})\hat{\sigma}^2$ is an estimate of $(1 - \frac{r}{n})\sigma^2$, the latter being the variance of a mean of a sample without replacement of size r from the finite population $\{X_1, \ldots, X_n\}$. The jackknife distribution estimator is

$$H_J(x) = P_J\left(\frac{\sqrt{r}(\bar{X}_{r,S} - \bar{X})}{\sqrt{(1 - \frac{r}{n})\hat{\sigma}^2}} \le \frac{x}{\hat{\sigma}}\right)$$

$$= P_J\left(\sqrt{\frac{rn}{d}}(\bar{X}_{r,S} - \bar{X}) \le x\right),$$

where $P_J(\cdot)$ means the proportion of subsets S of size d for which the stated event $\{\sqrt{\frac{rn}{d}}(\bar{X}_{r,S} - \bar{X}) \le x\}$ occurs. This has to be approximated by a further Monte Carlo estimate, as all subsets of size d cannot be exhausted in the computation.

This is the jackknife distribution estimator for $\sqrt{n}(\bar{X} - \mu)$ and is called the jackknife histogram. It is a misnomer to call it a histogram, as it estimates the CDF rather than the PDF. Here is a consistency theorem; see Shao and Tu (1995) for proofs and more details on each of the next three theorems.

Theorem 30.5 Let $X_1, \ldots, X_n \overset{\text{iid}}{\sim} F$, and $\mu = E_F(X_1)$, $\sigma^2 = \text{Var}_F(X_1) < \infty$.

(a) If $r, d \to \infty$, then $\sup_x |H_J(x) - H_n(x)| \overset{\text{a.s.}}{\to} 0$.

(b) If $E_F(|X - \mu|^3) < \infty$, then $\sup_x |H_J(x) - H_n(x)| = O\left(\frac{1}{\sqrt{\min(d,r)}}\right)$ a.s.

A similar theorem is known for the case of standardized as well as studentized sample means. For the standardized mean $T_n = \sqrt{n}(\overline{X}_n - \mu)/\sigma$, the jackknife distribution estimator, still called a jackknife histogram, is defined as $P_J(\sqrt{\frac{nr}{d}}\frac{\bar{X}_{r,S} - \overline{X}_n}{\hat{\sigma}} \le x)$, where $\hat{\sigma} = s$ is the sample standard deviation. For the studentized mean, the jackknife estimator is $P_J(\sqrt{\frac{nr}{d}}\frac{\bar{X}_{r,S} - \overline{X}_n}{s_{r,S}} \le x)$, where $s_{r,S}$ is the standard deviation of the delete-d jackknife sample.

Consistency theorems for each case are available. However, in each case, the straight delete-d jackknife can, in general, provide *only* first-order accuracy. If the underlying CDF has zero skewness, then, of course, second-order accuracy will not be in question. But the bootstrap is second-order accurate in probability for the standardized and the studentized samples regardless of whether the underlying CDF has zero skewness or not. In this sense, the delete-d jackknife falls short of the bootstrap.

Theorem 30.6 Let $X_1, X_2, \ldots, X_n \overset{iid}{\sim} F$ and let $T_n = \frac{\sqrt{n}(\overline{X}_n - \mu)}{\sigma}$ or $T_n = \frac{\sqrt{n}(\overline{X}_n - \mu)}{\hat{\sigma}}$. Suppose $d/n \to 0 < \delta < 1$. Then,

(a) $\sup_x |H_J(x) - H_n(x)| \overset{\text{a.s.}}{\to} 0$.
(b) If $E_F|X - \mu|^3 < \infty$, then $\sup_x |H_J(x) - H_n(x)| = O(\frac{1}{\sqrt{n}})$.

The consistency of jackknife histograms for much more general statistics is also known. The assumption needed is that the statistic should admit a Bahadur expansion as follows:

Assumption. For some functional $\theta(F)$ and a function $\varphi(x, F)$, $T_n = \theta + \frac{1}{n}\sum \varphi(X_i, F) + R_n$, where $EF(\varphi(X, F)) = 0$, $E_F(\varphi^2(X, F)) = \sigma^2 = \sigma^2(F) < \infty$ and $\sqrt{n}R_n \xrightarrow{\mathcal{P}} 0$.

Examples of statistics that admit such an expansion include von Mises functionals, U-statistics, and M-estimates. We have the following consistency theorem for such statistics.

Theorem 30.7 Let $X_1, X_2, \ldots, X_n \stackrel{iid}{\sim} F$, and suppose T_n satisfies the assumption above. Let $H_n(x) = P_F(\frac{\sqrt{n}(T_n-\theta)}{\sigma} \leq x)$ and $H_J(x) = P_J(\sqrt{\frac{nr}{d}} \times \frac{T_{r,S}-T_n}{\hat{\sigma}} \leq x)$, where $\hat{\sigma}$ is a consistent estimate of $\sigma = \sigma(F)$. Let $d/n \to 0 < \delta < 1$. Then,

$$\sup_x |H_n(x) - H_J(x)| \xrightarrow{\mathcal{P}} 0.$$

Remark. Note that the consistency is only in probability. Also note that we do not have any result on the rate of convergence of $|H_n(x) - H_J(x)|$ either a.s. or in probability.

Remark. Also notice that the assumption in each of these theorems is that $d \asymp n\delta$, while $r \to \infty$; $d = [n\delta]$ for $0 < \delta < 1$ satisfies these requirements. The parameter δ has to be chosen in a given problem. The optimal choice is obviously case-specific. It is recommended that one take $1/4 \leq \delta \leq 3/4$ and, in particular, $\delta = 1/2$ is a popular choice, the corresponding jackknife being known as the *half-sample jackknife*.

Remark. The bootstrap is an iid sample from F_n. On the contrary, the jackknife is a resampling scheme without replacement from the observations $\{x_1, x_2, \ldots, x_n\}$. The original $\{x_1, x_2, \ldots, x_n\}$ being a sample with replacement from F, the bootstrap mimics the original sampling design better than the jackknife. This is at the heart of why the bootstrap can provide second-order accuracy for standardized and studentized statistics, whereas the jackknife in general cannot. Subsampling methods are also successfully used to construct confidence intervals and sets, although we do not discuss them. See Politis and Romano (1994) for a very general construction applicable to very generic situations.

30.7 Exercises

Exercise 30.1 Let $T_n = \frac{1}{n}\sum_{i=1}^{n}(X_i - \bar{X})^2$. Find the jackknife bias estimate and hence the bias-corrected estimate T_{Jack}.

Exercise 30.2 * Let T_n be the posterior mean with respect to the Beta(α, β) prior for a binomial success probability p. Find the jackknife bias estimate and hence the bias-corrected estimate T_{Jack}.

Exercise 30.3 * Let T_n be the sample standard deviation s. Find the coefficients a, b in the expansion for the bias of T_n in the $N(0, \sigma^2)$ case.

Exercise 30.4 * Is the jackknife bias estimate necessarily zero for an unbiased estimate? If not, give a counterexample.

Exercise 30.5 Simulate a sample of size $n = 25$ from the $N(0, 1)$ distribution, and compute the jackknife variance estimate V_{Jack} for $T_n = |\bar{X}_n|$. Check its accuracy by comparing it with the exact variance.

Exercise 30.6 * Simulate a sample of size $n = 10, 25, 40, 60$ from the $N(0, 1)$ distribution, and compute the jackknife variance estimate V_{Jack} for $T_n = F_n^{-1}(\frac{1}{2})$. Compare it with the estimate obtained from the CLT for the median. Is the ratio getting close to 1? Should it?

Exercise 30.7 Find the influence function for the sample second moment. Hence find a CLT for it.

Exercise 30.8 Simulate a sample of size $n = 40$ from the $N(0, 1)$ distribution, and compute the delete-d jackknife variance estimate for $T_n = |\bar{X}_n|$ with $d = 1, 5, 10, 20, 30$. Repeat this. Is $d = 1$ the worst choice for accurate estimation of the variance of T_n? What is the best value of d?

Exercise 30.9 * Let $T_n = \bar{X}_n$. Formulate and prove a CLT for $V_{\text{Jack}} - \text{Var}(T_n)$.

Exercise 30.10 * Simulate a sample of size $n = 50$ from an exponential distribution with mean 1, and for the standardized sample mean compute

(a) the bootstrap CDF estimate $H_{\text{Boot}}(x)$;
(b) the jackknife histogram with $d = 1, 10, 25$.

According to theory, H_{Boot} is second-order accurate, but the jackknife histogram is only consistent. Is H_{Boot} actually more accurate than the jackknife histogram for each value of d?

References

Efron, B. (1982). *The Jackknife,the Bootstrap, and other Resampling Plans*, SIAM, Philadelphia.

Fernholz, L. (1983). *Von Mises Calculus for Statistical Functionals*, Lecture Notes in Statistics, Vol. 19, Springer, New York.

Miller, R.G. (1974). The jackknife—a review, Biometrika, 61, 1–15.

Politis, D. and Romano, J. (1994). Large sample confidence regions based on subsamples under minimal assumptions, Ann. Stat., 22(4), 2031–2050.

Politis, D., Romano, J. and Wolf, M. (1999). *Subsampling*, Springer, New York.

Quenouille, M.H. (1949). Approximate tests of correlation in time series, J.R.Stat.Soc.B, 11, 68–84.

Sen, P.K. (1996). Statistical functionals, Hadamard differentiability, and martingales, in *Probability Models and Statistics*, A. Borthakur and H. Choudhury (eds.), New Age, New Delhi, 29–47.

Serfling, R. (1980). *Approximation Theorems of Mathematical Statistics*, John Wiley, New York.

Shao, J. (1993). Differentiability of statistical functionals and consistency of the jackknife, Ann. Stat., 21(1), 61–75.

Shao, J. and Tu, D. (1995). *The Jackknife and Bootstrap*, Springer-Verlag, New York.

Shao, J. and Wu, C.J. (1989). A general theory for jackknife variance estimation, Ann. Stat., 17(3), 1176–1197.

Tukey, J. (1958). Bias and confidence in not quite large samples, Ann. Math. Stat., 29(2), 614.

Wu, C.J. (1990). On the asymptotic properties of the jackknife histogram, Ann. Stat., 18(3), 1438–1452.

Chapter 31
Permutation Tests

Permutation tests were first described by Fisher. In Fisher (1935), permutation tests were derived for testing the absence of a treatment effect in a randomized block experiment situation. The idea is rather simple and intuitive.

Suppose one group of observations is $X_1, \ldots, X_m \stackrel{\text{iid}}{\sim} f(x)$ and another group is $Y_1, \ldots, Y_n \stackrel{\text{iid}}{\sim} f(x - \Delta)$. Suppose that $T = T_{m,n}$ is some statistic for testing $H_0 : \Delta = 0$; e.g., the two-sample t-statistic.

If indeed $\Delta = 0$, then we have one set of $m + n$ iid observations; i.e., $X_1, \ldots, X_m, Y_1, \ldots, Y_n \stackrel{\text{iid}}{\sim} f(x)$. Then any m of the $N = m+n$ observations could be treated as the iid sample from the first group. So, under any permutation of $(X_1, \ldots, X_m, Y_1, \ldots, Y_n)$, say S, the value $T_{m,n}(S)$ that would be obtained for our statistic $T_{m,n}$ should be similar to the actual observed value we get for the actual observed data. Therefore, if the realized value of $T_{m,n}$ is in an extreme tail of the permutation distribution of $T_{m,n}(S)$, then this would cast suspicion on the hypothesis $\Delta = 0$.

A permutation test rejects $H_0 : \Delta = 0$ when $T_{m,n} > c$, where c is the $100(1 - \alpha)$th percentile of the permutation distribution of $T_{m,n}(S)$. Clearly c is a random variable. That is, the permutation test uses a random cutoff value rather than the fixed value that a parametric test would use. In exchange for this extra complexity, an exact size α is ensured unconditionally for all m, n and for all f (in some appropriate nonparametric class).

One difficulty with permutation tests is that it is impossible to find the exact value of c unless the sample sizes are small. Usually c is estimated by Monte Carlo permutation. No theoretical accuracy results for the Monte Carlo permutation are known. An alternative is to calculate the p-value $P_S(T_{m,n} \geq T_{m,n,OBS})$, where P_S stands for the permutation distribution of the statistic. This, of course, also cannot be found exactly, but theoretical approximations are available (although often difficult to use).

We recommend Lehmann (1986), Gebhard and Schmitz (1998), and Good (2005) for overall expositions of permutation tests. A rigorous theory of best permutation tests of a specified size with several applications is

A. DasGupta, *Asymptotic Theory of Statistics and Probability*,

presented in Runger and Eaton (1992). Permutation tests fall under the category of more general tests called *randomization tests*. See Basu (1980) and Romano (1989) for discussion and analytical properties of randomization tests.

31.1 General Permutation Tests and Basic Group Theory

We start with a simple illustrative example.

Example 31.1 Suppose $m = 3, n = 4$, and the data are

$$(X_1, X_2, X_3) = (3, 4, 5) \quad \text{and} \quad (Y_1, Y_2, Y_3, Y_4) = (1, 1.5, 2, 2.5).$$

We want to use the statistic $T_{m,n} = |\overline{X} - \overline{Y}|$. From the observed data, $T_{m,n} = 2.25$. Because the sample sizes are small, the permutation distribution can be written exactly. The pmf of the permutation distribution is easy and given below.

Therefore, the permutation p-value is $1/35 = 0.0286$. In contrast, if we did this test with the two-sample t-test, the p-value would be 0.015. Therefore, the permutation test is less sensitive than the parametric test.

So far, we have discussed permutation tests only in the context of a two-sample test of equality of locations. However, permutation tests are useful in much more general settings. In this most general context, the name "permutation" test is slightly misleading, but this is the name they were given.

All of these tests have the following nonparametric unconditional size property. Suppose a null hypothesis is of the form $H_0 : F \in \mathcal{F}_0$ for some appropriate (large nonparametric) family $\mathcal{F}_0$. If $\varphi(\cdot)$ is the test function of such a permutation test, then $E_F\varphi(X) = \alpha$ for all $F \in \mathcal{F}_0$. Such tests are called "similar" tests of size α.

Table 31.1 Permutation distribution of $T_{m,n}$

t	$35 \times p(t)$	t	$35 \times p(t)$
0.083	3	1.25	2
0.208	4	1.375	2
0.375	4	1.542	2
0.5	3	1.667	1
0.667	2	1.833	1
0.792	3	1.958	1
0.958	3	2.125	1
1.083	2	2.25	1

These general permutation tests are described using the terminology of *group theory*. A brief description of basic group theory terminology is given below. For those seeking a modern and more detailed exposition of group theory, we recommend Rotman (1994).

Definition 31.1 A set G together with a binary operation $*$ is called a group if

(i) $(a * b) * c = a * (b * c)$ for all $a, b, c \in G$.
(ii) There exists a unique $e \in G$ such that $e * a = a * e = a$, $\forall a \in G$.
(iii) For all $a \in G$, there exists $a^{-1} \in G$ such that $a * a^{-1} = a^{-1} * a = e$.

Remark. To be precise, a group G is defined by the pair $(G, *)$, but often the binary operation is implicit. If properties (i) and (ii) hold but not (iii), then G is called a semigroup.

- We define $a^n = a * \ldots * a$ if $n \geq 1$ and $a^0 = e$. Also $a^{-n} = a^{-1} * \ldots * a^{-1}$ for $n \geq 1$. It follows that, for all $m, n \in \mathbb{Z}$ and for all $a \in G$,

$$a^{m+n} = a^m * a^n \quad \text{and} \quad a^{mn} = (a^m)^n.$$

- If G has m elements, where $m < \infty$, then G is a finite group of order $|G| = m$. If G has infinitely many elements, then it is said to be of an infinite order.

Example 31.2

(1) If $G = \mathbb{Z}$ and $* = +$, then G is a group.
(2) If $G = \mathbb{R}$ and $* = \times$, then property (iii) does not hold and G is only a semigroup. However, if $G = \mathbb{R} \setminus \{0\}$, then G is a group.
(3) Let G be the set of all $n \times n$ nonsingular matrices and let $*$ be matrix multiplication. Then G is a group, denoted by $GL(n, \mathbb{R})$, the *general linear group*.
(4) If H is the set of all $n \times n$ orthogonal matrices, then H is a *subgroup* (see below) of $G = GL(n, \mathbb{R})$ and H is called the *orthogonal group*, denoted O_n.
(5) The set of all permutation matrices under matrix multiplication is a group.
(6) Let $\mathcal{X}$ be a finite alphabet of n elements, written $\mathcal{X} = \{1, 2, \ldots, n\}$. Let G be the set of all $n!$ permutations of the elements of $\mathcal{X}$. Under composition, G is a group written as S_n and called the *symmetric group*.

Definition 31.2 Let $G = (G, *)$ be a group and $H \subset G$. If $(H, *)$ is itself a group, then H is called a subgroup of G.

Theorem 31.1 Let $G = (G, *)$ be a group and $\mathcal{X}$ an arbitrary subset of G. Then there exists a smallest set $\mathcal{Y} \supset X$ such that $(\mathcal{Y}, *)$ is a group.

$\mathcal{Y}$ is called the subgroup generated by $\mathcal{X}$ ($\mathcal{Y}$ may be G itself). If $\mathcal{Y} = G$, then $\mathcal{X}$ is called a generator of G. In general, $\mathcal{Y}$ consists of all finite products of powers of x and x^{-1} for $x \in \mathcal{X}$.

Example 31.3 The following are examples of generators of two important groups.

(1) Let $G = S_n$ and let $\mathcal{X}$ be the subset of all transpositions. Then $\mathcal{X}$ is a generator of G.
(2) Let $G = O_n$ and let $\mathcal{X}$ be the subset of all reflections; i.e., $\mathcal{X}$ is the collection of all matrices of the form

$$M = I_n - 2\frac{\xi\xi'}{\|\xi\|^2}, \quad \xi \in \mathbb{R}^n.$$

Then $\mathcal{X}$ generates G.

Definition 31.3 A group G is Abelian if $a * b = b * a$ for all $a, b \in G$.

For example, $(\mathbb{R}, +)$ is Abelian but $GL(n, \mathbb{R})$ is non-Abelian. However, the subset $H \subset GL(n, \mathbb{R})$ of all *diagonal* matrices is an Abelian group under matrix multiplication.

31.2 Exact Similarity of Permutation Tests

As mentioned above, permutations have the similarity property at any level α and for any sample size. This property is described below in the context of a general finite group.

Let X, taking values in some space $\mathcal{X}$ (usually some Euclidean space), be distributed according to $P = P_n$. Suppose G is some finite group, $|G| = M$, of transformations acting on $\mathcal{X}$. Let H_0 be some specified null hypothesis, say $H_0 : P \in \mathcal{F}_0$. We assume that, under H_0, gX and X have the same distribution for all $g \in G$. That is, P is invariant under transformations in G under H_0. For example, if $X \sim N_p(0, I_p)$ and G is the orthogonal group, the UX is still $N_p(0, I_p)$ for all $U \in G$.

Let $t(\cdot)$ be a specified function on $\mathcal{X}$. This function plays the role of the "parent" statistic, always required for a permutation test.

For $0 < \alpha < 1$, let $k = M - [M\alpha]$. Let $t_{(1)}(x) \leq \ldots \leq t_{(M)}(x)$ be the ordered values of $t(gx)$ based on all M elements $g \in G$, where x is the observed $x \in \mathcal{X}$. Also define

$$\gamma(x) = \frac{M\alpha - M^+(x)}{M_0(x)}, \quad \text{where} \begin{cases} M_0(x) = \#\{i : t_{(i)}(x) = t_{(k)}(x)\} \\ M^+(x) = \#\{i : t_{(i)}(x) > t_{(k)}(x)\} \end{cases}.$$

Then the permutation test with respect to G is defined by

$$\varphi(x) = \begin{cases} 1, & t(x) > t_{(k)}(x) \\ \gamma(x), & t(x) = t_{(k)}(x). \\ 0, & t(x) < t_{(k)}(x) \end{cases}$$

Theorem 31.2 The permutation test φ is similar of size α if the distribution of X is invariant under the group operation in G for each P in the null.

Proof. From its definition, for all $x \in \mathcal{X}$,

$$\sum_{g \in G} \varphi(gx) = M^+(x) + \gamma(x) M_0(x) = M\alpha,$$

which implies, for all $P \in \mathcal{F}_0$,

$$\begin{aligned} E_P\left[\sum_{g \in G} \varphi(gX)\right] &= \sum_{g \in G} E_P \varphi(gX) = \sum_{g \in G} E_P \varphi(X) \\ &= E_P \varphi(X) \sum_{g \in G} 1 = M E_P \varphi(X). \end{aligned}$$

But we previously showed that $\sum_{g \in G} \varphi(gx) = M\alpha$ for all $x \in \mathcal{X}$. So, it follows that

$$M\alpha = E_P\left[\sum_{g \in G} \varphi(gX)\right] = M E_P \varphi(X),$$

which implies $E_P \varphi(X) = \alpha$ for all $P \in \mathcal{F}_0$. That is, the permutation test $\varphi(\cdot)$ has the claimed similarity property.

Remark. To execute the permutation test, one must compute the statistic $t(gx)$ for all $g \in G$. This computation can be difficult, an issue we discuss later. Notice that the permutation test is essentially a parametric test with a *random* cutoff $t_{(k)}(x)$.

Because computation of $t(gx)$ for every g can be difficult or costly, theoretical asymptotic approximations for $t_{(k)}(x)$ are practically useful. Under a general sequence of distributions P_n, it is often the case that

$$t_{(k)}(x) = t_{(k),n}(X^{(n)}) \approx \lambda,$$

where λ is a suitable deterministic constant. Also, $E_{P_n}\varphi_n(X^{(n)})$ is approximately equal to $1 - H(\lambda)$ for a suitable function $H(\cdot)$, where $0 \leq H \leq 1$. For alternatives $P = P_n$ asymptotically separated from $\mathcal{F}_0$, usually $H \equiv 0$ and the power of the permutation test converges to 1. That is, for large classes of alternatives but with some assumptions, permutation tests have the consistency property.

There is a way to identify the number λ by consideration of G and also to identify the function H. This statement is in complete generality; in specific examples, however, this may require a lot of calculations. Here is a familiar problem where the constant λ can be explicitly identified.

Example 31.4 Consider the p-group ANOVA problem where

$$X_{ij} = Z_{ij} + b_i + t_j, \quad \text{where} \begin{cases} i = 1, \ldots, n \\ j = 1, \ldots, p \end{cases}.$$

One wishes to test $H_0 : t_1 = \ldots = t_p = 0$. We take the Z_{ij} to be white noise; i.e., $Z_{ij} \overset{\text{iid}}{\sim} \Delta(0, \sigma^2)$ for some distribution Δ. Suppose we choose t to be the usual ANOVA F-statistic. Note that we are not assuming the Z_{ij} to be normal. This is the attraction of permutation tests; we get the similarity property without making parametric assumptions.

The underlying group G is the set of all permutations of the treatment labels and $|G| = (p!)^n$. More precisely, $G = S_p \times \ldots S_P$, the n-product of the symmetric group S_p. Since $M = |G|$ is rather large, identification of λ would be useful. Under any distribution in the null,

$$t_{(k),n}(X^{(n,p)}) \longrightarrow \lambda = \chi^2_{p-1,\alpha} \text{ in probability.}$$

Notice that, if we use $\chi^2_{p-1,\alpha}$ as our approximation to $t_{(k)}(x)$, then the test is essentially exactly the parametric F-test. Therefore, as $n \to \infty$, the limiting power of the permutation test is the same as that of the F-test.

31.3 Power of Permutation Tests

Because the random cutoff point of the permutation test "converges" to the cutoff point of the corresponding parametric test, the permutation test and the parametric test are asymptotically equivalent in power in some suitable sense. Approximations and expansions for power are available. Here we consider only the location-shift problem, using the sample mean as our basic statistic. See Bickel and van Zwet (1978) for derivations of the asymptotic power expansions in this section.

Theorem 31.3 Let $X_1, \ldots, X_m \overset{\text{iid}}{\sim} F(x)$ and $Y_1, \ldots, Y_n \overset{\text{iid}}{\sim} F(x-\theta)$. Define $N = m+n$ and $\lambda = \frac{n}{N}$. Fix $\alpha > 0$ and let $0 \leq \theta \leq \frac{D}{\sqrt{N}}$ and $\varepsilon \leq \lambda \leq 1-\varepsilon$, and suppose there is a number $r > 8$ such that

$$\int |x|^k \, dF(x) \leq C, \quad \forall k \leq r.$$

Let $\Pi_p(\theta, m, n)$ and $\Pi_t(\theta, m, n)$ denote the power functions of the permutation test based on $n\bar{X}$ and the test based on the two-sample t-statistic, respectively. Then there exists a number $B = B(\alpha, C, D, \varepsilon) < \infty$ such that

$$\left|\Pi_p(\theta, m, n) - \Pi_t(\theta, m, n)\right| \leq BN^{-(1+\beta)}$$

uniformly in θ, λ, and F, where

$$\beta = \min\left\{\frac{r-8}{2r+8}, \frac{1}{4}\right\}.$$

Remark.

(1) Π_p and Π_t admit expansions of the form

$$c_0 + \frac{c_1}{\sqrt{N}} + \frac{c_2}{N} + o\left(\frac{1}{N}\right).$$

The coefficients c_0, c_1, c_2 happen to be the same for both the permutation test and the parametric test. The difference is therefore $o\left(\frac{1}{N}\right)$. Theorem 31.3 states that it is in fact $O(N^{-1-\beta})$ in general, with the assumption of more than eight moments.

(2) For known functional forms of F, it may be possible to strengthen the rate. For example, if F is some normal CDF, one can strengthen the rate

to $O(N^{-3/2})$. So, in this case, the permutation test based on the mean is third-order accurate.

(3) It is also possible to give rough approximations to Π_p, and these approximations can be made fully rigorous with additional notation. Let

$$\eta = \frac{\sqrt{\lambda(1-\lambda)N}\theta}{\sigma_F} \quad \text{and} \quad \kappa_3 = \frac{E_F(X - E_F X)^3}{\sigma_F^3}.$$

Then

$$\Pi_p(F, \theta, m, n) \approx 1 - \Phi(z_\alpha - \eta) + \frac{(1-2\lambda)(\eta - 2z_\alpha)\theta\kappa_3\varphi(z_\alpha - \eta)}{6\sigma_F}.$$

31.4 Exercises

Exercise 31.1 Suppose the X data are $-1, 0, 1$ and the Y data are $-2, -1, 1, 2$. Exactly compute the permutation p-value by using $|\bar{X} - \bar{Y}|$ and then $|\text{med}(X) - \text{med}(Y)|$ as the test statistic.

Exercise 31.2 * Suppose the X data are $-1, 0, 1$ and the Y data are $-2, -1, 1, 2$. Think of a permutation test for testing that the two populations have an equal variance if they are known to have an equal mean.

Exercise 31.3 Can a permutation test be nonrandomized if you insist on the exact similarity property? Give a precise answer.

Exercise 31.4 * Give a precise statement for what it means to say that *permutation Monte Carlo methods work.*

Exercise 31.5 * For testing the equality of two means, simulate samples of sizes $m = n = 20$ from $N(0, 1)$ and $N(4, 1)$ distributions and approximate the random cutoff of the 5% permutation t-test by a Monte Carlo estimation. Compare the result with the nonrandom cutoff of a t-test.

Exercise 31.6 Compare, by a simulation, the power of the permutation t-test with the likelihood ratio test for testing that the means of two exponential populations are equal. Use $m = n = 20$.

Exercise 31.7 * Derive the Bickel-van Zwet power approximation for the permutation test based on $n\bar{X}$ when the null is a standard exponential and the alternative is a shift of the null. Roughly plot it. Is it always between 0 and 1? Use selected values for m, n.

Exercise 31.8 * Can you think of a way to construct a permutation test in the Behrens-Fisher problem?

References

Basu, D. (1980). Randomization analysis of experimental data: The Fisher randomization test, J. Am. Stat. Assoc., 371, 575–595.

Bickel, P.J. and van Zwet, W. (1978). Asymptotic expansions for the power of distribution-free tests in the two sample problem, Ann. Stat., 6(5), 937–1004.

Fisher, R.A. (1935). *Design of Experiments*, Oliver and Boyd,Edinburgh.

Gebhard, J. and Schmitz, N. (1998). Permutation tests - a revival?, Stat. Papers, 39(1), 75–85.

Good, P.I. (2005). *Permutation, Parametric and Bootstrap Tests of Hypotheses*, Springer-Verlag, New York.

Lehmann, E.L. (1986). *Testing Statistical Hypotheses*, John Wiley, New York.

Romano, J. (1989). Bootstrap and randomization tests of some nonparametric hypotheses, Ann. Stat., 17(1), 141–159.

Rotman, J. (1994). *An Introduction to the Theory of Groups*, Springer-Verlag, New York.

Runger, G.C. and Eaton, M.L. (1992). Most powerful invariant permutation tests, J. Multivar.Anal., 42(2), 202–209.

Chapter 32
Density Estimation

Nonparametric density estimation is one of the most researched and still active areas in statistical theory, and the techniques and the theory are highly sophisticated and elegant. A lot of development in statistics has taken place around the themes, methods, and mathematics of density estimation. For example, research in nonparametric regression and nonparametric function estimation has been heavily influenced by the density estimation literature. High-dimensional density estimation problems still remain a major challenge in statistics. Even if the impact on practical statistical methodology is not proportional to the research volume, density estimation is a topic very much worth studying. There are a number of very good texts on density estimation at various technical levels. We recommend Scott (1992), Silverman (1986), Thompson and Tapia (1990), and Hardle et al. (2004), of which Scott (1992) is excellent for technical material and Hardle et al. (2004) treats density estimation in the larger context of nonparametric statistics. We also recommend the review article of Revesz (1984) for a quick exposition on the most basic theorems in density estimation up to that time. Although we discuss primarily $\mathcal{L}_2$ methods, Devroye (1987) and Devroye and Lugosi (2000) contain a wealth of material, ideas, and derivations on density estimation in general and the latest technical tools in particular. Rosenblatt (1956) and Parzen (1962) are the two principal references on kernel density estimates, which form the backbone of this chapter.

32.1 Basic Terminology and Some Popular Methods

The basic problem is the following. There is an unknown density f with respect to Lebesgue measure on $\mathbb{R}^d$, where $d < \infty$. We take an iid sample $X_1, \ldots, X_n$ from f and want to use the data to estimate the infinite-dimensional parameter f. In some particular cases, we may be interested in estimating only $f(x_0)$ for some fixed and prespecified x_0.

A. DasGupta, *Asymptotic Theory of Statistics and Probability*,

Given the infinite-dimensional nature of the parameter, we cannot expect parametric convergence rates in our inference; the rates will be slower. The exact rates depend the assumptions we are willing to make and the number of dimensions.

Several standard types of density estimators are briefly discussed below. We will revisit some of them later to discuss their properties in more detail.

(1) *Plug-in estimates*. This is a parametric reduction of the problem. Assume $f \in \mathcal{F}_\Theta = \{f_\theta : \theta \in \Theta \subset \mathbb{R}^k, k < \infty\}$. Let $\hat{\theta}$ be any reasonable estimator of θ. Then the plug-in estimate of f is $\hat{f} = f_{\hat{\theta}}$.

(2) *Histograms*. Let the support of f be $[a, b]$ and consider the partition $a = t_0 < t_1 < \ldots < t_m = b$. Let $c_0, \ldots, c_{m-1}$ be constants. The histogram density estimate is

$$\hat{f}(x) = \begin{cases} c_i, \; t_i \leq x < t_{i+1} \\ c_{m-1}, \; x = b \\ 0, \; x \notin [a, b] \end{cases}.$$

The restriction $c_i \geq 0$ and $\sum_i c_i(t_{i+1} - t_i) = 1$ makes $\hat{f}$ a true density; i.e., $\hat{f} \geq 0$ and $\int \hat{f}(x)\,dx = 1$. The canonical histogram estimator is

$$\hat{f}_0(x) = \frac{n_i}{n(t_{i+1} - t_i)}, \quad t_i \leq x < t_{i+1},$$

where $n_i = \#\{i : t_i \leq X_i < t_{i+1}\}$. Then $\hat{f}_0$ is in fact the nonparametric maximum likelihood estimator of f. For a small interval $[t_i, t_{i+1})$ containing a given point x,

$$E_f n_i = n P_f(t_i \leq X < t_{i+1}) \approx n(t_{i+1} - t_i) f(x).$$

Then it follows that

$$E_f \hat{f}_0(x) = E_f \left(\frac{n_i}{n(t_{i+1} - t_i)} \right) \approx f(x).$$

(3) *Shifted histograms*. Fix x and take an interval $x \pm h_n$. The shifted histogram estimator is

$$\hat{f}(x) = \frac{\#\{i : x - h_n < X_i \leq x + h_n\}}{2nh_n}$$

$$= \frac{F_n(x+h_n) - F_n(x-h_n)}{2nh_n},$$

where F_n is the empirical CDF. The reason for choosing x to be the center of the interval is that it helps to reduce the bias of the density estimator. As above, $E_f \hat{f}(x) \approx f(x)$. It is interesting to note that $\hat{f}(x)$ may be represented in the form

$$\hat{f}(x) = \frac{1}{nh_n} \sum_{i=1}^{n} K\left(\frac{x - X_i}{h_n}\right), \quad \text{where } K(z) = \frac{1}{2}\mathbb{I}_{(-1,1]}(z).$$

(4) *Kernel estimates*. Let $K(z) \geq 0$ be a general nonnegative function, usually integrating to 1, and let

$$\hat{f}_K(x) = \frac{1}{nh_n} \sum_{i=1}^{n} K\left(\frac{x - X_i}{h_n}\right).$$

The function $K(\cdot)$ is called the kernel and $\hat{f}_K$ is called the kernel density estimator. A common kernel function is a Gaussian kernel. In such a case, $\hat{f}_K$ is a mixture of $N(X_i, h_n^2)$ densities. Both the shifted histogram and kernel estimates were proposed in Rosenblatt (1956).

(5) *Series estimates*. Suppose that $f \in \mathcal{L}^2(\lambda)$, where λ is the Lebesgue measure on $\mathbb{R}$. Let $\{\varphi_k : k \geq 0\}$ be a collection of orthonormal basis functions. Then f admits a Fourier expansion

$$f(x) = \sum_{k=0}^{\infty} c_k \varphi_k(x), \quad \text{where } c_k = \int \varphi_k(x) f(x)\, dx.$$

Since $c_k = E_f \varphi_k(X)$, as an estimate one may use

$$\hat{f}(x) = \sum_{k=0}^{l_n} \left[\frac{1}{n} \sum_{i=1}^{n} \varphi_k(X_i)\right] \varphi_k(x)$$

for a suitably large cutoff l_n. This is the orthonormal series estimator.

(6) *Nearest neighbor estimates*. Fix $k = k_n$ and fix x. Identify the k closest sample values to x. Suppose $r = r_{k,n}$ is the distance between x and the kth closest one. The nearest neighbor estimate is $\hat{f}(x) = k/2nr$.

32.2 Measures of the Quality of Density Estimates

As in parametric estimation, different reasonable criteria have been studied. The criteria could be local or global. Several such indices are mentioned below, where $\hat{f}$ denotes a generic estimate of the unknown density f.
Local Indices:

(i) $\xi_n(x) = |\hat{f}(x) - f(x)|$.
(ii) $L(\hat{f}(x), f(x))$ for some general loss function $L(\cdot, \cdot)$.

Global Indices:

(i) Expected $\mathcal{L}^1$ error: $E_f\left[\int \xi_n(x)\,dx\right]$.
(ii) Expected $\mathcal{L}^2$ error: $E_f\left[\int \xi_n^2(x)\,dx\right]$.
(iii) $\mathbb{E}_f\left[\sup_{x\in\mathbb{R}} \xi_n(x)\right]$.
(iv) The distribution of $\sup_x \xi_n(x)$, $\int \xi_n(x)\,dx$, or $\int \xi_n^2(x)\,dx$.
(v) If t_n is one of the quantities mentioned in (iv) above, then we can consider the limiting distribution of t_n; i.e., we could identify $\{a_n\}, \{b_n\}$, and G such that

$$b_n(t_n - a_n) \stackrel{\mathcal{L}}{\Rightarrow} G$$

and use some or all of a_n, b_n, G as the performance criteria.
(vi) Strong laws: Identify $\{a_n\}, \{b_n\}$ such that $b_n(t_n - a_n) \stackrel{\text{a.s.}}{\longrightarrow} c$ for some c, and again use some or all of a_n, b_n, c as the performance criteria.

32.3 Certain Negative Results

We mentioned above that we cannot expect to have convergence rates similar to parametric problems in the nonparametric density estimation problem. Here we give several particular examples to show how bad or good the rates can be. We also present a number of other fundamental results, including a famous result on the lack of unbiased estimates. These results and examples illustrate what can and cannot be done in the density estimation problem.

A. Nonexistence of Unbiased Estimates.

Proposition 32.1 *Suppose $X_1, \ldots, X_n \stackrel{\text{iid}}{\sim} f$, where $f \in \mathcal{F}$, the family of all continuous densities on $\mathbb{R}$. Then there does not exist an estimator*

$\hat{f}_n$ *such that* $\hat{f}_n$ *is a density and*

$$E_f[\hat{f}_n(x)] = f(x) \;\; \forall x \in \mathbb{R}, \; \forall f \in \mathcal{F}.$$

Proof. Suppose there is such an estimator $\hat{f}_n$. Then $\hat{I}(a,b) \equiv \int_a^b \hat{f}_n(x)dx$ is an unbiased estimator of $I(a,b) \equiv \int_a^b f(x)\,dx = F(b) - F(a)$. To prove this, we use the fact that $\hat{f}_n \geq 0$ and apply Fubini's theorem.

But an unbiased estimate of $F(b) - F(a)$ is $F_n(b) - F_n(a)$, where F_n is the empirical CDF. Since $F_n(\cdot)$ is a permutation-invariant function of $X_1, \ldots, X_n$, it must be a function of the order statistics $X_{(1)}, \ldots, X_{(n)}$ alone. Now, if $\hat{f}_n$ is not already permutation invariant, we may make it so by averaging over all $n!$ permutations of $X_1, \ldots, X_n$ and it will still be unbiased. Therefore, we may assume that $\hat{I}(a,b)$ is permutation invariant and hence a function of the order statistics alone. However, it is known that the order statistics form a complete and sufficient statistic when the parameter space is $\mathcal{F}$. This implies that

$$\hat{I}(a,b) = \int_a^b \hat{f}_n(x)\,dx = F_n(b) - F_n(a) \;\; \text{a.s.}$$

But the left-hand side, as a function of b alone, is a smooth function, while the right-hand side is a step function. Therefore, they cannot be equal as functions, which is a contradiction. Therefore, $\hat{f}_n$ cannot be unbiased. □

This result is due to Rosenblatt (1956).

B. An example due to Devroye (1983) shows that with just smoothness of f, even a lot of smoothness, convergence rates of $\hat{f}_n$ can be arbitrarily slow.

Let $\{a_n\}$ be any positive sequence. Then, given essentially any sequence of estimates $\hat{f}_n$, one would be able to find a density f such that

$$E_f\left[\int |\hat{f}_n(x) - f(x)|\,dx\right] > a_n, \;\; \text{i.o.}$$

Another way to think about this is as follows. Take any $\{c_n\}$, perhaps rapidly decreasing to 0. Then we could find f such that

$$\limsup_n c_n \times E_f\left[\int |\hat{f}_n(x) - f(x)|\,dx\right] = \infty.$$

C. The phenomenon in (B) reemerges even if we put in a lot of shape restrictions. For example, let $\mathcal{F}$ be the family of densities on $[0, \infty)$ that are monotone decreasing and $\mathcal{C}^{\infty}$-smooth. Then the result of (B) still holds (Devroye and Lugosi (2000, p. 91).

D. Farrell (1972) showed that, in smooth nonparametric classes, there is actually a demonstrable best attainable rate for estimation of f at a point, and this rate is less than the parametric rate. Precisely, let $\mathcal{F}$ be the family of densities with $(k-1)$ continuous derivatives, for some $k \geq 1$, and the kth derivative absolutely uniformly bounded by some $\alpha < \infty$. Fix $x \in \mathbb{R}$. If

$$P_f(|\hat{f}_n(x) - f(x)| \leq a_n) \longrightarrow 1$$

uniformly in $f \in \mathcal{F}$, then

$$\liminf_n \left(a_n^2 \times n^{\frac{2k}{2k+1}}\right) > 0.$$

E. This example shows that even pointwise consistency can be problematic if f acts erratically indefinitely in the tails; e.g., if the density has spikes of fixed height on infinitely many shrinking intervals. In such a case, no standard density estimate can catch the spikes, even asymptotically. In other words,

$$\limsup_{x \to \infty} f(x) = C > 0$$

but, for every standard density estimate $\hat{f}_n$,

$$\lim_{x \to \infty} \hat{f}_n(x) = 0 \ \ \forall n.$$

That is, there exists $x \in \mathbb{R}$ such that $E_f(\hat{f}_n(x) - f(x)) \not\to 0$.

F. Can we let f be completely arbitrary and hope to achieve some sort of consistency? The answer is *no*. The following proposition illustrates such a result; see Devroye (2000, p. 91).

Proposition 32.2 *For any kernel density estimator $\hat{f}_{n,h}$ and for all n,*

$$\sup_f \inf_h E_f\left[\int |\hat{f}_{n,h}(x) - f(x)|\, dx\right] = 2.$$

32.4 Minimaxity Criterion

Minimaxity is a well-established paradigm in inference. Exact minimaxity is not regarded as so important nowadays. But minimaxity still serves the purpose of being a touchstone, some default with which to compare a procedure. Furthermore, the idea of minimaxity has led to many other developments in statistical theory and practice. Brown (1992, 2000) gives two lucid accounts of the history, impact, and inferential importance of minimaxity or near minimaxity.

In the density estimation context, let $X_1, \ldots, X_n \stackrel{\text{iid}}{\sim} f$, where $f \in \mathcal{F}$, a suitable family of densities. Let $\hat{f}_n(x)$ be an estimate of f at $x \in \mathbb{R}$, a function of $X_1, \ldots, X_n$. We let $\hat{f}_n$ be arbitrary and, for a specific loss function $L(f, \hat{f}_n)$, define the risk as $R(f, \hat{f}_n) \equiv E_f L(f, \hat{f}_n)$. The minimax estimate $\hat{f}_n^*$ satisfies

$$\sup_f R(f, \hat{f}_n^*) = \inf_{\hat{f}_n} \sup_f R(f, \hat{f}_n) \equiv \mathcal{R}_n(\mathcal{F}).$$

If $\mathcal{R}_n(\mathcal{F})$ is of the order of some sequence b_n, written $\mathcal{R}_n(\mathcal{F}) \asymp b_n$, we say that the minimax rate of the density estimation problem is b_n. Devroye and Lugosi (2000) contains the following interesting collection of examples and results.

Example 32.1

(1) Let $\mathcal{F} = \{N(\mu, 1) : \mu \in \mathbb{R}\}$, and define $L(f, \hat{f}_n) = \int |\hat{f}_n(x) - f(x)|\, dx$. If we use the plug-in estimator $N(\overline{X}_n, 1)$, then, by calculating the $\mathcal{L}^1$ error directly, we find that $\mathcal{R}_n(\mathcal{F}) \asymp n^{-1/2}$. However, interestingly, the minimax density estimate for any given n is not a normal density. It is a mixture of a discrete number of normal densities, and it is very difficult to find the components in the mixture and the mixing proportions.

(2) Let $\mathcal{F}$ be the family of all Beta densities. Again, with the $\mathcal{L}^1$ loss, we achieve the parametric rate: $\mathcal{R}_n(\mathcal{F}) \asymp n^{-1/2}$. This is expected because the $\mathcal{F}$ is a finitely parametrized class, and simple plug-in estimates will already provide the $n^{-1/2}$ rate.

(3) Let $\mathcal{F}$ be the family of $U[0, \theta]$ densities. Under $\mathcal{L}^1$ loss, the minimax estimate $\hat{f}^*$, restricted to $\mathcal{F}$, based on one observation X from f, is the $U[0, \sqrt{e}X]$ density; this is verified by simple calculus. If we do not restrict the estimator to $\mathcal{F}$, then, for all n, the minimax estimator is not in $\mathcal{F}$.

(4) Let $\mathcal{F}$ be the family of all densities f with a unique mode at 0, $f(0) = c$, and $\log(f)$ a concave function. Under $\mathcal{L}^1$ loss, it can be shown that $\mathcal{R}_n(\mathcal{F}) \asymp n^{-2/5}$.

The examples show that, in density estimation, convergence rates depend very much on exactly what assumptions one is making.

32.5 Performance of Some Popular Methods: A Preview

Before we can compare the methods of density estimation, we must first define a metric for comparison. As in the usual parametric inference, there are two issues: systematic error (in the form of bias) and random error (in the form of variance). To include both of these aspects, we consider the mean squared error (MSE). If $\hat{f}_n$ is some estimate of the unknown density f, we want to consider

$$\text{MSE}[\hat{f}_n] \equiv E[\hat{f}_n(x) - f(x)]^2 = \text{Var}[\hat{f}_n(x)] + E^2[\hat{f}_n(x) - f(x)].$$

As usual, there is a bias–variance trade-off. The standard density estimates require specification of a suitable tuning parameter. The tuning parameter is optimally chosen after weighing the bias–variance trade-off. With the optimal choices, some of the methods may perform better than others, depending on the assumptions we are willing to make about f.

Below is a preview of some of the results we will later derive. Notice that no assumptions are given here. This is not to imply that the assumptions are not important. In fact, the performance of a particular method depends heavily on these assumptions. The general approach taken below is to determine a rate of convergence of the MSE to zero.

A. *Histograms.* If h_n is the width of the equally spaced mesh, then, as we later show,

$$\begin{aligned} \sup \mathbb{E}|\hat{f}_n(x) - f(x)| &= O(h_n), \\ \text{Var}[\hat{f}_n(x)] &= O(1/nh_n), \\ \text{MSE}[\hat{f}_n(x)] &= O(h_n^2 + 1/nh_n). \end{aligned}$$

The term $h_n^2 + \frac{1}{nh_n}$ is minimized by taking $h_n \asymp n^{-1/3}$. Plugging back into the formula for MSE, the rate corresponding to the optimal choice of h_n is

$$\text{MSE}[\hat{f}_n(x)] = O(n^{-2/3}).$$

B. *Smoothed histograms*. With smoothing, we get

$$\sup_x E|\hat{f}_n(x) - f(x)| = O(h_n^2),$$
$$\mathrm{Var}[\hat{f}_n(x)] = O(1/nh_n),$$
$$\mathrm{MSE}[\hat{f}_n(x)] = O(h_n^4 + 1/nh_n).$$

Then the optimal bin width is $h_n \asymp n^{-1/5}$, and hence the rate corresponding to the optimal choice of h_n is $n^{-4/5}$. Note that smoothing improves the convergence rate of the histogram.

C. *Kernel estimates*. Just as in (B), the optimal bandwidth is $h_n \asymp n^{-1/5}$, yielding the rate $n^{-4/5}$ for the MSE. We discuss this in detail later.

D. *Orthogonal series estimates*. Let l_n denote the number of terms kept in the series expansion for the estimate. Then

$$\sup_x E|\hat{f}_n(x) - f(x)| = O(l_n^{-3/2}),$$
$$\mathrm{Var}[\hat{f}_n(x)] = O(l_n/n),$$
$$\mathrm{MSE}[\hat{f}_n(x)] = O(l_n^{-3} + l_n/n).$$

Then, it follows that the optimal cutoff is $l_n \asymp n^{1/4}$ and the corresponding rate is $n^{-3/4}$. Particularly for series estimates, the rate of convergence depends heavily on the assumptions (e.g., smoothness) we make on f.

32.6 Rate of Convergence of Histograms

Let h_n be the common length of the m interior intervals of the histogram. For fixed x, there is a unique interval capturing x; call the midpoint of this interval $c = c(x, h_n)$. The idea is to do a Taylor series expansion of f about c:

$$f(x) = f(c) + f'(c)(x - c) + \frac{f''(c)}{2}(x - c)^2 + O((x - c)^3).$$

This yields an expansion for the probability p_i of the ith interval:

$$p_i = 2h_n f(c) + \frac{h_n^3}{3} f''(c) + O(h_n^4).$$

From this, we can get an expansion for the variance of $\hat{f}_n(c)$:

$$\mathrm{Var}[\hat{f}_n(c)] = \frac{1}{2nh_n}[f(c) - 2h_n f^2(c)] + \text{lower-order terms.}$$

Also,

$$E^2[\hat{f}_n(c) - f(c)] = \frac{h_n^4}{36}[f''(c)]^2 + \text{lower-order terms.}$$

Adding the variance and squared bias yields

$$\mathrm{MSE}[\hat{f}_n(c)] = \frac{f(c)}{2nh_n} + \frac{h_n^4}{36}[f''(c)]^2 + O(n^{-1}) + O(h_n^5).$$

But we want an expansion for the MSE about x, not c. However, from the way we defined c, we know that $\hat{f}_n(x) = \hat{f}_n(c)$. Therefore,

$$\begin{aligned}\mathrm{MSE}[\hat{f}_n(x)] &= E[\hat{f}_n(c) - f(x)]^2 \\ &= E[\hat{f}_n(c) - f(c) + f(c) - f(x)]^2 \\ &\leq 2\left\{E[\hat{f}_n(c) - f(c)]^2 + [f(c) - f(x)]^2\right\}.\end{aligned}$$

We already have expansions for both of these terms. Plugging these in, we get

$$\mathrm{MSE}[\hat{f}_n(x)] \leq \frac{f(c)}{nh_n} + 2h_n^2[f'(c)]^2 + O(\max\{n^{-1}, h_n^3\}),$$

which goes to zero if $h_n \to 0$ and $nh_n \to \infty$. In that case, the histogram estimator is consistent.

Optimization of the upper bound leads to the optimal bandwidth

$$h_{\text{loc,opt}} = n^{-1/3}\left[\frac{f(c)}{4(f'(c))^2}\right]^{1/3}.$$

Plugging back into the MSE formula, we find that

$$\mathrm{MSE}[\hat{f}_n(x)] = O(n^{-2/3}).$$

This is the result we had previously mentioned in our preview.

Sometimes it is of greater interest to globally estimate f. In that case, the standard criterion is the integrated mean squared error (IMSE) given by

$$\text{IMSE}[\hat{f}_n] = \int E[\hat{f}_n(x) - f(x)]^2 \, dx.$$

Based on the IMSE criterion, the optimal global bandwidth is

$$h_{\text{opt}} = n^{-1/3} \left[\frac{1}{4\|f'\|_2^2} \right]^{1/3}$$

so that, optimally, IMSE $= O(n^{-2/3})$. Notice that the local $n^{-2/3}$ rate is maintained globally.

32.7 Consistency of Kernel Estimates

Kernel density estimates have become by far the most common method for nonparametric density estimation. They generally provide the best rates of convergence, as well as provide a great deal of flexibility through the choice of the kernel. Actually, in an overall asymptotic sense, the choice of the kernel is of much less importance than the choice of bandwidth. The calculations below provide some guidelines as to how the bandwidth should be chosen. Bandwidth choice is by far the hardest part of density estimation, and we will discuss it in detail later.

As mentioned above for general density estimates, there will always be a variance –bias trade-off. To study the performance of kernel estimates, we will consider both the variance and the bias, starting with the bias.

Define the kernel estimate by

$$\hat{f}_n(x) = \hat{f}_{n,h}(x) = \frac{1}{nh} \sum_{i=1}^{n} K\left(\frac{x - X_i}{h}\right),$$

where $K(\cdot)$ is the kernel function. Often $K(\cdot)$ is assumed to be a density, but for now we assume only that $K \geq 0$ and $K \in \mathcal{L}^1$. We have already proved that the density estimation problem cannot have nonparametric unbiased estimates. A natural question to ask is if $\hat{f}_n$ is at least asymptotically unbiased.

Under various sets of assumptions on f and K, asymptotic unbiasedness indeed holds. One such result is given below; see Rosenblatt (1956).

Theorem 32.1 Assume f is uniformly bounded by some $M < \infty$. Also assume that $K \in \mathcal{L}^1$. Then, for any x that is a continuity point of f,

$$E[\hat{f}_n(x)] \longrightarrow f(x) \int K(z) \, dz \quad \text{as } n \to \infty.$$

Proof. Since the X_i's are iid and $\hat{f}_n(x) = \frac{1}{nh}\sum K\left(\frac{x-X_i}{h}\right)$,

$$E[\hat{f}_n(x)] = \frac{1}{h}\int_{-\infty}^{\infty} K\left(\frac{x-z}{h}\right) f(z)\,dz = \int_{-\infty}^{\infty} K(z)f(x-hz)\,dz.$$

But f is continuous at x and uniformly bounded, so $K(z)f(x-hz) \leq MK(z)$ and $f(x-hz) \to f(x)$ as $h \to 0$. Since $K(\cdot) \in \mathcal{L}^1$, we can apply the Lebesgue dominated convergence theorem to interchange the limit and the integral. That is, if we let $h = h_n$ and $h_n \to 0$ as $n \to \infty$, then

$$\begin{aligned}
\lim_{n\to\infty} E[\hat{f}_n(x)] &= \int \lim_{n\to\infty} K(z)f(x-hz)\,dz \\
&= \int K(z)f(x)\,dz \\
&= f(x)\int K(z)\,dz.
\end{aligned}$$

Remark. In particular, if $\int K(z)\,dz = 1$, then $\hat{f}_n(x)$ is asymptotically unbiased at all continuity points of f, provided that $h = h_n \to 0$.

Next, we consider the variance of $\hat{f}_n$. Consistency of the kernel estimate does not follow from asymptotic unbiasedness alone; we need something more. To get a stronger result, we need to assume more than simply $h_n \to 0$. Loosely stated, we want to drive the variance of $\hat{f}_n$ to zero.

Obviously, since $\hat{f}_n(x)$ is essentially a sample mean,

$$\text{Var}[\hat{f}_n(x)] = \frac{1}{n}\text{Var}\left[\frac{1}{h}K\left(\frac{x-X}{h}\right)\right],$$

which implies

$$nh\text{Var}[\hat{f}_n(x)] = h\text{Var}\left[\frac{1}{h}K\left(\frac{x-X}{h}\right)\right].$$

By an application of the Lebesgue dominated convergence theorem, as in the proof of Theorem 32.1, $E[\hat{f}_n(x)]^2$ converges, at continuity points x, to $f(x)\|K\|_2^2$, which is finite if $K \in \mathcal{L}^2$. We already know that, for all continuity points x, $E[\hat{f}_n(x)]$ converges to $f(x)\|K\|_1$, so it follows that

$$h\left(E[\hat{f}_n(x)]\right)^2 \longrightarrow 0, \;\; h \to 0.$$

Combining these results, we get, for continuity points x,

$$nh\text{Var}[\hat{f}_n(x)] = h\text{Var}\left[\frac{1}{h}K\left(\frac{x-X}{h}\right)\right] \longrightarrow f(x)\|K\|_2^2 < \infty,$$

provided $K \in \mathcal{L}^2$. Consequently, if $h \to 0$, $nh \to \infty$, $f \le M$, and $K \in \mathcal{L}^2$, then, at continuity points x,

$$\text{Var}[\hat{f}_n(x)] \longrightarrow 0.$$

We summarize the derivation above in the following theorem.

Theorem 32.2 Suppose f is uniformly bounded by $M < \infty$, and let $K \in \mathcal{L}^2(\mathbb{R})$ with $\|K\|_1 = 1$. At any continuity point x of f,

$$\hat{f}_n(x) \underset{\rightarrow}{p} f(x), \text{ provided } h \to 0 \text{ and } nh \to \infty.$$

32.8 Order of Optimal Bandwidth and Superkernels

Next, we consider the choice of the bandwidth and the corresponding convergence rates. It turns out that the optimal choice of bandwidth and the corresponding convergence rate for the MSE can be figured out by studying the properties of K.

Assume that $K \in \mathcal{L}^2$, $\|K\|_1 = 1$, and $\int zK(z)\,dz = 0$. Let $k(u)$ be the characteristic function of K; i.e.,

$$k(u) = E_K(e^{iuX}) = \int_{-\infty}^{\infty} e^{iux}K(x)\,dx.$$

Let r be the smallest integer such that

$$\lim_{u\to 0} \frac{1-k(u)}{u^r} = k_r, \quad 0 < |k_r| < \infty.$$

It should be noted that k_r is a *complex* number. Typically, the value of r is 2; in fact, if

$$\int zK(z)\,dz = \ldots = \int z^{r_0-1}K(z)\,dz = 0 \text{ and } \int z^{r_0}K(z)\,dz \neq 0,$$

then $r = r_0$. Hence, if K is nonnegative, we must have $r = 2$. The number r is called the *exponent* of $k(\cdot)$, and k_r is called the *characteristic coefficient* of k or K. It will be useful later to note that

$$k_r = \frac{1}{r!} \int z^r K(z)\, dz.$$

By decomposing the MSE into variance and squared bias, we get

$$E[\hat{f}_n(x) - f(x)]^2 \asymp \frac{f(x)}{nh} \|K\|_2^2 + h^{2r} |k_r f^{(r)}(x)|^2,$$

provided $K \in \mathcal{L}^2$ and f has $(r+1)$ continuous derivatives at x. Minimizing the right-hand side of the expansion above with respect to h, we have the asymptotically optimal local bandwidth

$$h_{\text{loc,opt}} = [f(x)\|K\|_2^2]^{\frac{1}{2r+1}} \left[2nr|k_r f^{(r)}(x)|^2\right]^{-\frac{1}{2r+1}},$$

and on plugging this in, the convergence rate of the MSE is MSE $\sim n^{-2r/(2r+1)}$. Since usually $r = 2$, the pointwise convergence rate is MSE $= O(n^{-4/5})$, slower than the parametric n^{-1} rate.

The corresponding global optimal bandwidth found by considering the IMSE is given by

$$h_{\text{opt}} = \frac{C(K)}{[n \int |f^{(r)}|^2]^{\frac{1}{2r+1}}},$$

where $C(K) = \frac{[||K||_2]^{\frac{2}{2r+1}}}{[2rk_r^2]^{\frac{1}{2r+1}}}$.

This is a famous formula in density estimation. If we do not insist on using a nonnegative kernel K, then it is possible to make the characteristic exponent r as large as one wants; i.e., given any r, one can find kernels K that satisfy all the conditions given above but have $\int z^s K(z) = 0\, \forall s < r$. Our calculations above show that for such a kernel the IMSE would converge to zero at the rate $n^{-2r/(2r+1)}$. By choosing r very large, we can therefore not only improve on the $n^{-4/5}$ rate but also come *close* to the parametric n^{-1} rate. Kernels with a large characteristic exponent are sometimes called *superkernels*. The problematic aspect of density estimation with superkernels is that one can get estimates that take negative values. As a matter of practicality, one could replace the negative values by zero. Of course, the theoretical nearly parametric convergence rate is then no longer true. Also, a density estimate that keeps taking zero values in parts of the sample space could be aesthetically unattractive. The exact pros and cons of superkernels in density estimation from a practical perspective need further investigation. Formally, kernels with a characteristic exponent $= \infty$ are called *infinite-order kernels*. Devroye (1992) is a good reference for theoretical aspects of superkernels.

Example 32.2 Recall that the shifted histogram is a special type of kernel estimate. Therefore, we can specialize the results above by choosing the kernel that corresponds to the shifted histogram. That is, if we take the kernel function to be

$$K(z) = \frac{1}{2} I_{[-1,1]}(z),$$

it turns out that the optimal bandwidth and corresponding MSE are

$$h^* = h_{\text{loc,opt}} = n^{-1/5} \left[\frac{9 f(x)}{2\{f''(x)\}^2} \right]^{1/5},$$
$$\text{MSE}[\hat{f}_{n,h^*}(x)] \asymp \frac{5}{4} 9^{-1/5} 2^{-4/5} [f(x)]^{1/5} |f''(x)|^{2/5} n^{-4/5}.$$

The corresponding global bandwidth and IMSE satisfy

$$h^{**} = h_{\text{opt}} = \left[\frac{9}{2\|f''\|_2^2} \right] n^{-1/5},$$
$$\text{IMSE}[\hat{f}_{n,h^{**}}] \asymp \frac{5}{4} 9^{-1/5} 2^{4/5} \|f''\|_2^{2/5} n^{-4/5}.$$

Thus, both locally and globally, the optimal bandwidth is of the order $n^{-1/5}$ and the convergence rate is $n^{-4/5}$.

Example 32.3 The results above clearly say that the choice of the kernel does *not* affect the order of the bandwidth or the rate of mean square convergence. Any kernel from a large class satisfying the assumptions stated above can be used.

The following table gives several common kernels and some associated important quantities.

Table 32.1 Kernels and quantities needed for bandwidth and rate calculations

K	$\int K^2(z)\,dz$	$\int z^2 K(z)\,dz$
$\frac{1}{2} I_{[-1,1]}(z)$	1/2	1/3
$(1-\|z\|) I_{[-1,1]}(z)$	2/3	1/6
$(2\pi)^{-1/2}\text{Exp}\{-z^2/2\}$	$1/\sqrt{2\pi}$	1
$\frac{1}{2}\text{Exp}\{-\|z\|\}$	1/2	2
$\frac{1}{2\pi}\left[\frac{\sin(z/2)}{z/2}\right]^2$	$1/3\pi$	$2(1-\sin 1)/\pi$

These are known as the uniform, the triangular, the Gaussian, the double exponential, and the Fejer kernels. There are plenty of other kernels in use. Specifically, the *Epanechnikov kernel*, with a certain optimality property, is introduced below.

Example 32.4 If one does not mind using kernels that take negative values, then theoretically better convergence rates can be obtained by using superkernels. Here are some common superkernels with characteristic exponent $r = 4$.

Müller kernel. $K(z) = 105/64\,(1-5z^2+7z^4-3z^6)I_{|z|\leq 1}$; (Müller (1984))

Gasser, Müller, and Mammitzsch kernel. $K(z) = 75/16\,(1 - z^2)_+ - 105/32(1 - z^4)_+$; (Gasser, Müller and Mammitzsch (1985))

Stuetzle-Mittal kernel. $K(z) = 2L(z) - (L * L)(z)$ for any kernel L with characteristic exponent $r = 2$. For example, L can be the standard normal kernel or the Epanechnikov kernel (Stuetzle and Mittal (1979))

$K(z) = 1/8\,(9 - 15z^2)I_{|z|\leq 1}$.

A broad family of superkernels were proposed and analyzed in Hall and Wand (1988); these are kernels with Fourier transforms $e^{-|t|^s}$, $s > 0$. For $s > 2$, the corresponding kernels cannot be probability density functions; thus, the family contains both traditional kernels and superkernels.

32.9 The Epanechnikov Kernel

As stated above, the order of the bandwidth and the rate of convergence do not depend on the choice of kernel. But still it is legitimate to look for an optimal kernel for a given n and f. In the mean square formulation, it was shown in Epanechnikov (1969) that there is an optimal kernel. To describe the sense in which the Epanechnikov kernel is optimal, we must first make several assumptions. Assume that we only consider kernels of the following type.

(a) $K(z) \geq 0$ and uniformly bounded.
(b) $K(z) = K(-z)$.
(c) $\int K(z)\,dz = 1$.
(d) $\int z^2 K(z)\,dz = 1$.
(e) $\int |z|^k K(z)\,dz < \infty$ for all k.

Now, recall from above that, under these assumptions,

$$\text{IMSE}[\hat{f}_{h,n}] \asymp \frac{1}{nh}\int K^2(z)\,dz + \frac{h^4}{4}\int [f''(x)]^2\,dx.$$

Then the optimal kernel may be formulated as the one that minimizes the IMSE subject to the assumptions (a)–(e). Assumption (b) lets us reduce to the interval $(0,\infty)$. If the minimizer of $\int K^2(z)dz$ subject to $\int K(z)dz = \int z^2 K(z)dz = \frac{1}{2}$ satisfies (a) and (e), then the solution to our problem has been found. By calculus of variation methods, Epanechnikov found the solution to be

$$K_0(z) = \frac{3}{20\sqrt{5}}(5 - z^2)I_{[-\sqrt{5},\sqrt{5}]}(z),$$

which is well known as the Epanechnikov kernel. Note that K_0 has a compact support. However, in practice, Gaussian kernels appear to be used more than the Epanechnikov kernel.

32.10 Choice of Bandwidth by Cross Validation

The formulas above show that bandwidth (local or global) depends on the true f. In particular, we have seen that, under assumptions,

$$h_{\text{opt}} = C(K)\left[n\|f''\|_2^2\right]^{-1/5}.$$

Such a circularity phenomenon is true for almost any density estimation scheme. There is always a tuning parameter, whose optimal value depends on the unknown f. Therefore, the actual choice of h is a critical issue. Earlier suggestions considered estimating the norm $\|f''\|_2^2$ from the data and using a plug-in estimate of h. Generally, this plug-in estimate of h_{opt} is consistent. However, currently popular methods for choosing h are different from first-generation plug-in methods.

We consider here the method of *cross-validation* (CV). Cross-validation was not first suggested for density estimation. But once it was established as a method of choosing smoothing parameters, it was naturally considered for density estimation. See Stone (1974) and the discussions in that article for an introduction to the general concept of cross-validation. For the discussions and theorems presented below, we refer to Bowman (1984), Chow, Geman, and Wu (1983), Stone (1984), Hall (1982, 1983), and Schuster and

Gregory (1981). Cross-validation is still undergoing refinements. See Wang and Zidek (2005) for a recent innovative method of cross-validation.

Cross-validation is not uniquely defined; there can be many methods of cross-validation in a given problem. It turns out that CV in density estimation is subtle in the sense that certain natural types of CV actually do *not* work; modifications are needed. We will discuss several methods of CV in density estimation and their consequences.

The basic idea of CV is very intuitive. Fix h, and select a part of the data to fit the model. Then, apply the fitted model to the rest of the data to assess goodness of fit. Typically, if the entire sample is of size n, then $(n-1)$ values would be used to fit the model and would then be applied to the one that was left out. The value left out would be taken to be each unit, one at a time, and an average measure of goodness of fit would be computed. By varying h, a function $CV(h)$ will be formed and then maximized. The maximum is the CV bandwidth.

Of course, there are many measures of goodness of fit. Also, it is very rare that $CV(h)$ can be maximized analytically; numerical methods must be used. If h is a vector tuning parameter, numerical maximization of $CV(h)$ is a very hard problem.

We discuss below two specific methods of cross-validation for kernel density estimation.

32.10.1 Maximum Likelihood CV

As remarked above, the basic idea is to build the procedure on a part of the sample and try it on the rest of the sample to assess the goodness of fit.

Let $\hat{f}_{n,h}$ denote a specific density estimate with generic bandwidth h. Let

$$\hat{f}_{n,h}^{(j)}(x) = \frac{1}{(n-1)h} \sum_{i \neq j} K\left(\frac{x - X_i}{h}\right)$$

be the estimated density based on the sample values except X_j. We apply the estimate $\hat{f}_{n,h}^{(j)}(x)$ to $x = X_j$ to obtain $\hat{f}_{n,h}^{(j)}(X_j)$. Since X_j was actually observed, a good choice of h should give large values of $\hat{f}_{n,h}^{(j)}(X_j)$.

Define the CV likelihood as

$$\hat{L} = \hat{L}(h) = \prod_{j=1}^{n} \hat{f}_{n,h}^{(j)}(X_j).$$

Note that if we used the ordinary kernel estimates $\hat{f}_{n,h}(X_j)$ in the likelihood, then the likelihood would be unbounded. The maximum likelihood CV (MLCV) bandwidth is the value of h that maximizes this CV likelihood; i.e.,

$$h^* = \text{argmax}_h \hat{L}(h).$$

It turns out that although the method seems reasonable on its surface, it can cause serious problems, both practical and theoretical. These problems can appear in very innocuous situations. The general problem is that h^* often produces estimates $\hat{f}_{n,h^*}$ that behave badly when the support of the true f is unbounded and especially when the kernel K has unbounded support. Thus, MLCV is dangerous to use to estimate densities with unbounded support. Even when f has bounded support, h^* can lead to estimates with suboptimal performance.

That is, while $\hat{f}_{n,h^*}$ may be consistent, even strongly consistent, it may have a convergence rate strictly worse than the optimal convergence rate using $h = h_{\text{opt}}$. The primary reason for this phenomenon is that h^* does not go to zero at the correct rate. In fact, it can converge *faster* than h_{opt}. This will result in an undersmoothed estimate and an order of magnitude increase in the variance. It can also happen that h^* oversmoothes, in which case the bias goes up and again the performance is suboptimal.

Example 32.5 We report results of a simulation. Consider the case where f is the $N(0, 1)$ density and we want to estimate it on $[-1, 1]$. Let $K(u)$ be the $N(0, 1)$ kernel. First, consider $n = 500$. In this case, it turns out that $h_{\text{opt}} = 0.24$. On the other hand, over simulations, the range of h^* was found to be 0.09 to 0.11, with an average of 0.10. Clearly, $\hat{f}_{n,h^*}$ will undersmooth. Next, take $n = 2000$. In this case, $h_{\text{opt}} = 0.29$, while the simulation average of h^* is 0.07. Again, $\hat{f}_{n,h^*}$ drastically undersmoothes.

The following theorem shows the undersmoothing phenomenon.

Theorem 32.3 Let $-\infty < a < b < \infty$ and let f have support (a, b). Suppose $f'(b) < f'(a)$. Then $h^* \asymp n^{-1/3}$.

Remark. Suboptimal performance by h^* does not, however, imply that there will be an inconsistency problem. In fact, the following elegant consistency result holds; see Chow, Geman, and Wu (1983), Bowman (1984), and Devroye and Penrod (1984) for more details on this theorem and the delicate consistency issue for cross-validation in density estimation.

Theorem 32.4 Assume the following:

(a) f is uniformly bounded with compact support.
(b) K is uniformly bounded with compact support.
(c) K is symmetric and unimodal about zero.
(d) K is strictly positive in some neighborhood of zero.

Then

$$\int_{-\infty}^{\infty} |\hat{f}_{n,h^*}(x) - f(x)|\, dx \stackrel{\text{a.s}}{\longrightarrow} 0.$$

Remark. There are known examples of inconsistency of $\hat{f}_{n,h^*}$ if f has unbounded support. Thus, the conditions of Theorem 32.4 cannot be relaxed.

32.10.2 Least Squares CV

The problems mentioned above provide motivation for an alternative CV procedure that results in asymptotically optimal performance. It turns out that CV using least squares will provide asymptotically optimal performance if IMSE is the criterion.

We therefore redefine the CV index from $\hat{L}(h)$ to something else. Here is the motivation. Clearly,

$$\int |\hat{f}_{n,h}(x) - f(x)|^2\, dx = \int \hat{f}_{n,h}(x)^2\, dx - 2\int \hat{f}_{n,h}(x) f(x)\, dx + \int f(x)^2\, dx,$$

provided that $f, K \in \mathcal{L}^2$. But $\hat{f}_{n,h} \approx \hat{f}_{n,h}^{(j)}$ and

$$\int \hat{f}_{n,h}(x) f(x)\, dx = E_f[\hat{f}_{n,h}(X)].$$

The least squares CV (LSCV) criterion is defined as

$$J(h) = J_n(h) = \frac{1}{n}\sum_{j=1}^{n}\int [\hat{f}_{n,h}^{(j)}(x)]^2\, dx - \frac{2}{n}\sum_{j=1}^{n} \hat{f}_{n,h}^{(j)}(X_j).$$

The quantity $\int f(x)^2 dx$ is ignored because it does not involve h. The LSCV bandwidth is defined as

$$h^+ = \text{argmin}_{h \in H(n,\varepsilon,\gamma)}\, J(h),$$

where $H(n, \varepsilon, \gamma) = \{h : \varepsilon n^{-1/5} \leq h \leq \gamma n^{-1/5}\}$ for specified $0 < \varepsilon < \gamma < \infty$ such that the constant C in $h_{\text{opt}} = Cn^{-1/5}$, the theoretical optimal bandwidth (see Section 32.8), is covered inside the interval (ε, γ). Hall (1983) shows that h^+ is well defined and asymptotically optimal in a sense made formal in the theorem below. The intuition in minimizing $J(h)$ over $H(n, \varepsilon, \gamma)$ is that to begin one would want the chosen bandwidth to be of the order $n^{-1/5}$.

Theorem 32.5 Assume the following:

(a) f'' exists and is uniformly continuous on $(-\infty, \infty)$.
(b) $f' \in \mathcal{L}^1$.
(c) For some $\delta > 0$, $E_f[X^2|\log|X||^{2+\delta}] < \infty$.
(d) K is a nonnegative, symmetric density with two uniformly continuous derivatives.
(e) $\int z^2[K(z) + |K'(z)| + |K''(z)|]\,dz < \infty$.

Let h_{opt} be the theoretical optimal bandwidth. Then

$$\frac{\int |\hat{f}_{n,h^+}(x) - f(x)|\,dx}{\int |\hat{f}_{n,h_{\text{opt}}}(x) - f(x)|\,dx} - 1 = o_p(1).$$

Remarks.

(i) The conditions imposed on f and K, which are only slightly stronger than those in Hall (1983), imply constraints on the tails. Obviously, f must have two moments. So, for example, a Cauchy density is not covered by this theorem. In addition, simple calculus shows that, under the conditions imposed on K, $K(z) = o(z^{-2})$ as $|z| \to \infty$. So, again, a Cauchy kernel is not covered by the theorem.
(ii) It is also true that the LSCV bandwidth h^+ itself is approximately equal to h_{opt}. More precisely,

$$\frac{h^+}{h_{\text{opt}}} - 1 = o_p(1).$$

This is a very positive property of the LSCV bandwidth. A rate of convergence of h^+/h_{opt} to 1 was later established. This rate, however, is quite slow. Indeed,

$$n^{1/10}\left(\frac{h^+}{h_{\text{opt}}}-1\right)$$

has a nondegenerate limiting distribution.

32.10.3 Stone's Result

Stone (1984) writes a slight modification of $J(h)$ (call it $M(h)$) and shows that an unconstrained minimum of $M(h)$ is asymptotically optimal under strikingly mild conditions on f at the cost of stronger conditions on K.

Recall that $J(h)$ was the CV-based estimate of

$$\int[\hat{f}_{n,h}(x)-f(x)]^2\,dx-\int f(x)^2\,dx=\int \hat{f}_{n,h}(x)^2\,dx-2\int \hat{f}_{n,h}(x)f(x)\,dx.$$

Stone (1984) seeks an unbiased estimate for

$$E_f[\hat{f}_{n,h}(X)]=\int \hat{f}_{n,h}(x)f(x)\,dx$$

and exactly evaluates the other term, $\int \hat{f}_{n,h}(x)^2\,dx$. On calculation, the unbiased estimate of the expectation above works out to

$$\frac{1}{n(n-1)h}\sum_{i\neq j}K\left(\frac{X_i-X_j}{h}\right).$$

Also, on further calculation, it turns out that

$$\frac{1}{n^2h}\sum_{i\neq j}K\left(\frac{X_i-X_j}{h}\right)=\frac{2}{n}\sum_j \hat{f}_{n,h}^{(j)}(X_j).$$

This, as we recall, is exactly the second term in the expression for $J(h)$. Stone shows that the other term is

$$\int \hat{f}_{n,h}(x)^2\,dx=\frac{1}{n^2}\sum_i\sum_j K_h^{(2)}(X_i-X_j),$$

where $K_h(u)=K(u/h)/h$ and $K_h^{(2)}=K_h*K_h$, the twofold convolution of K_h with itself. Finally, then, Stone's modified criterion is

$$M(h) = M_n(h) = \frac{1}{n^2}\sum_i \sum_j K_h^{(2)}(X_i - X_j) + \frac{2}{n}\sum_{i \neq j} \hat{f}_{n,h}^{(j)}(X_j).$$

Remark. Stone's modified $M(h)$ is particularly suitable for the proof he gives of his consistency theorem. Asymptotically, $M(h)$ and $J(h)$ are approximately equal. Here is the principal result in Stone (1984). This result is regarded as a landmark in the cross-validation literature.

Theorem 32.6 Assume the following:

(a) f is uniformly bounded.
(b) K is nonnegative, symmetric, and unimodal around zero.
(c) $\int K(z)\,dz = 1$.
(d) K is compactly supported.
(e) K is Holder continuous of order β; i.e., for some $C, \beta \in \mathbb{R}^+$,

$$|K(z_1) - K(z_2)| \leq C|z_1 - z_2|^\beta, \; \forall\, z_1, z_2 \in \mathcal{R}.$$

Let $L_{n,h} = \int [\hat{f}_{n,h}(x) - f(x)]^2\,dx$ be the incurred loss. Let $\tilde{h} = \text{argmin}_h M(h)$. Then

$$\frac{L_{n,\tilde{h}}}{L_{n,h_{\text{opt}}}} - 1 \stackrel{\text{a.s.}}{\longrightarrow} 0.$$

Remark. Notice that the theorem asserts optimal performance of the modified LSCV without practically any conditions on f.

32.11 Comparison of Bandwidth Selectors and Recommendations

(a) *Straight plug-in.* Recall that the asymptotic optimal bandwidth is of the form

$$h_{\text{opt}} = n^{-1/5}\frac{C(K)}{\|f''\|_2^2}.$$

In the straight plug-in method, a pilot estimate of f, say $\hat{g}_n$, is found, and f'' is estimated by $\hat{g}_n''$. Then the estimated bandwidth is

$$h_{PI} = n^{-1/5} \frac{C(K)}{\|\hat{g}_n''\|_2^2}.$$

This is very basic and also naive because $\hat{g}_n''$ is usually not a good estimate of f'' even if $\hat{g}_n$ is a good estimate of f.

(b) *Plug-in equation solving*. Instead of using a pilot estimate $\hat{g}_n$, consider for general h the corresponding $\hat{f}_{n,h}$ itself. The bandwidth h_{PI}^* is defined as a suitably selected root of the equation

$$h = n^{-1/5} \frac{C(K)}{\|\hat{f}_{n,h}''\|_2^2}.$$

This plug-in bandwidth is regarded as a first-generation plug-in bandwidth.

(c) *Reference bandwidth*. The reference bandwidth, due to Silverman (1986), assumes that f is the density of some $N(0, \sigma^2)$, where the mean can be assumed to be zero, without loss of generality. The model-based estimate of σ is s, the sample standard deviation. Since we really do not believe that f is normal, we consider a more robust estimator of σ. The particular robust estimate that is used is Q/β, where Q is the interquartile range and β is the unique number such that $Q/\beta \stackrel{\text{a.s.}}{\longrightarrow} \sigma$ if f is $N(0, \sigma^2)$; actually, $\beta \approx 1.34$.
Pretending that f is normal would already cause some oversmoothing. To avoid further escalation, σ is estimated by $\min\{s, Q/\beta\}$. The corresponding bandwidth is

$$h_R = 1.06 n^{-1/5} \times \min\{s, Q/\beta\}.$$

Even if one does not ultimately use h_R, it is possibly a good idea to know in a particular problem the value of h_R.

(d) *Second-generation plug-in*. One starts with the formula for h_{opt} once again. Instead of using the same bandwidth h for forming an estimate of f and an estimate of f'', a different bandwidth $g = g(h)$ is used for f''. The details about the choice of $g(h)$ are complex and there are some variations in the methods different people propose. The important point is that g is *not* of the order $n^{-1/5}$. It can be shown that the best g for estimating $\|f''\|_2^2$ is of the order of $n^{-1/7}$. The resulting bandwidth is

$$h_{\text{SGPI}} = n^{-1/5} \frac{C(K)}{\|\hat{f}_{n,g}''\|_2^2},$$

where the final selection of g is interactive and numerical. The exact details can be found in Jones and Sheather (1991); also see Rice (1986) and Hall et al. (1991) for other efficient second-generation plug-in bandwidths.

(e) *Bootstrap.* Contrary to the developments so far in which h_{opt} optimizes only an asymptotic approximation of IMSE, in the bootstrap approach, in this problem, one can attack the IMSE directly. Thus, the bootstrap estimate of $E_f[\int(\hat{f}_{n,h} - f)^2]$ can be formed exactly and the bootstrap Monte Carlo method is not needed. The bootstrap bandwidth h_B minimizes the bootstrap estimate of IMSE. Usually, however, the standard bootstrap is not used. A suitable "smoothed" bootstrap is used instead. Analytical expressions for the bootstrap estimate are available in the literature; see Jones, Marron, and Sheather (1996).

(f) *Cross-validation.* There are many CV-based methods. We have seen that maximum likelihood CV is dangerous; in particular, when f is not compactly supported. Stone's least squares CV is asymptotically sound, as we have seen. There are other CV-based methods besides MLCV and LSCV. On balance, LSCV is a relatively easy-to-execute scheme with sound consistency properties.

The current favorite is h_{SGPI}, although h_{LSCV} is also popular.

32.12 $\mathcal{L}^1$ Optimal Bandwidths

The literature on the selection of bandwidths is dominated by $\mathcal{L}^2$ loss. The main reason is that the results are much more structured than, for example, the $\mathcal{L}^1$ loss, although conceptually $\mathcal{L}^1$ is arguably more attractive. Does it matter ultimately which loss function is used? Asymptotically, the answer is "yes" and "no." The tail of the unknown f has an influence on whether the choice between $\mathcal{L}^1$ or $\mathcal{L}^2$ matters.

First, we present a result in Hall and Wand (1988) that characterizes the optimal bandwidth under $\mathcal{L}^1$ loss. One key difference is that, unlike the case of $\mathcal{L}^2$, where an asymptotically optimal bandwidth can be written analytically, it is a numerical problem in the case of the $\mathcal{L}^1$ loss. However, the result below (see Hall and Wand (1988)) simplifies the numerical effort.

Theorem 32.7 Let $E_f|X|^{1+\delta} < \infty$ for some $\delta > 0$. Let f'' be continuous, bounded, and square integrable. Let K be a symmetric, square integrable density with finite second moment and let h_{opt} be the optimal bandwidth under $\mathcal{L}^2$ loss. Then

(a) $E\left[\int |\hat{f}_{n,h_{\text{opt}}} - f|\right] \asymp \inf_h E\left[\int |\hat{f}_{n,h} - f|\right]$.

(b) Let h^1_{opt} denote the optimal bandwidth under $\mathcal{L}^1$. Let $c_1 = \int z^2 K(z)\,dz$ and $c_2 = \int K^2(z)\,dz$. Define

$$\lambda(u) = \int \int \left| c_1 u^4 f''(x) - \frac{2c_2}{u} z \sqrt{f(x)} \right| \varphi(z)\,dz\,dx.$$

Then

$$h^1_{\text{opt}} = n^{-1/5} \left[\operatorname{arginf}_u \lambda(u)\right]^2.$$

Remark. This says that as long as a little more than one moment of f exists, the order of the bandwidth under $\mathcal{L}^1$ and $\mathcal{L}^2$ is the same.

Let $h_{\text{opt}} = n^{-1/5} C_2(K, f)$ and $h^1_{\text{opt}} = n^{-1/5} C_1(K, f)$. Hall and Wand (1988) show through some examples that $C_2 \approx C_1$. So, from a practical point of view, there is some evidence that using the $\mathcal{L}^2$ bandwidth, which has an analytical formula, for $\mathcal{L}^1$ is acceptable.

Remark. The moment condition $E_f|X|^{1+\delta} < \infty$ cannot be relaxed. Indeed, if f is a Cauchy density, then

$$\frac{E\left[\int |\hat{f}_{n,h_{\text{opt}}} - f|\right]}{\inf_h E\left[\int |\hat{f}_{n,h} - f|\right]} \longrightarrow \infty \quad \text{as } n \to \infty,$$

and furthermore, h^1_{opt} is not of the order of $n^{-1/5}$ (Devroye and Lugosi (2000)). Because spacings for Cauchy data will tend to be large, the optimal bandwidth is larger than $O(n^{-1/5})$. It is interesting that this tail-heaviness affects the $\mathcal{L}^1$ bandwidth but not the $\mathcal{L}^2$ bandwidth, as, under $\mathcal{L}^2$, the optimal bandwidth is still $O(n^{-1/5})$, even for the Cauchy case.

32.13 Variable Bandwidths

In all of the discussions thus far, the bandwidth has been taken as a deterministic and constant sequence $\{h_n\}$. In particular, $\{h_n\}$ does not depend on x. There is something to be said about choosing h differently in different parts of the sample space. This is because, in sparse regions of the sample space, one would want a large window. On the other hand, in dense regions, one can afford a smaller window. The greatest generalization is to let $h = h(x; X_1, \ldots, X_n)$. The variable kernel estimate is of the form

$$\hat{f}_{n,h,\mathrm{Var}}(x) = \frac{1}{n}\sum_{i=1}^{n} \frac{1}{h(x;X_i)} K\left(\frac{x - X_i}{h(x;X_i)}\right).$$

We have the immediate problem that, in general, $\hat{f}_{n,h,\mathrm{Var}}$ is *not* a density. So usually h is allowed to be a function of the data but not of x. Still, it is done in such a way that it automatically adapts to a large or a small window in sparse or dense regions. Such bandwidths are known as variable bandwidths and were first proposed by Breiman, Meisel, and Purcell (1977). We mention the original suggestion by Breiman, Meisel, and Purcell and then a more sophisticated version.

(a) Breiman, Meisel, and Purcell proposed the following. Fix $k = k(n) \in \mathbb{N}$, the set of natural numbers. Let d_{jk} denote the distance from X_j to its kth nearest neighbor. Let $\alpha = \{\alpha_k\}$ be a sequence of constants. Then the estimate is defined as

$$\hat{f}_{n,k,\alpha}(x) = \frac{1}{n}\sum_{j=1}^{n} \frac{1}{\alpha_k d_{jk}} K\left(\frac{x - X_j}{\alpha_k d_{jk}}\right).$$

The intuition is that for x near a "lonely" X_j, the important component has the bandwidth $\alpha_k d_{jk}$, which would be large. On the other hand, for x in a region densely occupied by the data, the relevant components all have a small bandwidth. Breiman, Meisel, and Purcell give information on the choice of k and α_k. Typically, k grows with n but at a slower rate than n.

(b) A similar idea leads to estimates of the following form, with variable bandwidths. Fix $r \in \mathbb{N}$ and a partition $\mathcal{A} = \{A_1, \ldots, A_r\}$ of the sample space. Then take r corresponding bandwidths $h = \{h_1, \ldots, h_r\}$. Let $n_j = \#\{i : X_i \in A_j\}$. The estimate is defined as

$$\hat{f}_{n,h,\mathcal{A}}(x) = \sum_{j=1}^{r} \frac{n_j}{n} \sum_{i=1}^{n} \frac{1}{n_j h_j} K\left(\frac{x - X_i}{h_j}\right) I_{\{X_i \in A_j\}}.$$

The intuition of the estimate is that we form separate kernel estimates based on data in different subgroups and then pool them through a weighted average.

These variable bandwidths are conceptually attractive but obviously much more computationally intensive and also much harder to deal with analytically. However, the idea itself is appealing and there continues to be some research activity on variable bandwidths. Devroye and Lugosi (1999) is an interesting and informative modern reference on variable bandwidths.

32.14 Strong Uniform Consistency and Confidence Bands

Another interesting and useful direction is a study of the magnitude of the actual error in estimation. For example, we could ask about the behavior of $\int(\hat{f}_n - f)^2$ or $\sup_x |\hat{f}_n - f|$ as random variables. There is some literature on these two, but perhaps it is not completely satisfactory.

Suppose f is the true unknown density and $\hat{f}_n$ is an estimate. The error committed can be considered locally or globally. Global measures of the error include the previously introduced quantities

$$w_n = \int |\hat{f}_n(x) - f(x)|\,dx, \quad v_n = \int [\hat{f}_n(x) - f(x)]^2\,dx,$$
$$t_n = \sup_x |\hat{f}_n(x) - f(x)|.$$

Each of w_n, v_n, and t_n are random variables, and we want to study their distributions. As always, we have to resort to asymptotics, as the finite sample distributional questions are basically unanswerable. Below is an important property of the global index t_n due to Devroye and Penrod (1984).

Theorem 32.8 Assume the following:

(a) f is uniformly continuous.
(b) K is uniformly continuous.
(c) K is a density.
(d) $K(z) \to 0$ as $|z| \to \infty$.
(e) K' exists.
(f) For some $\delta > 0$, $\int |z|^{\delta+1/2}|K'(z)|\,dz < \infty$.
(g) Let $\omega_K(u)$ be the modulus of continuity of K; i.e.,

$$\omega_K(u) = \sup_{|x-y|<u} |K(x) - K(y)|.$$

Let $\gamma(u) = \sqrt{\omega_K(u)}$. We assume that $\int_0^1 \sqrt{\log(u^{-1})}\,d\gamma(u) < \infty$.

Let $\hat{f}_{n,h}$ be a kernel estimate based on K, where $h \to 0$ and $\log(n)/nh \to 0$. Then $t_n \xrightarrow{\text{a.s.}} 0$.

Remark. The result says that strong uniform consistency of the usual kernel estimates holds with innocuous assumptions on f as long as we choose the kernel K carefully. Also, there is no assumption on the support or the tails of f. As regards condition (g), which looks complicated, if $|K'|$ is bounded and if K'' exists, then $\omega_K(u) = O(u)$ and condition (g) holds because in this case

$$(g) \iff \int_0^\infty e^{-u/2}\sqrt{u}\, du < \infty.$$

There are numerous choices of K that satisfy conditions (b)–(g); for example, a Gaussian kernel.

The next result, due to Bickel and Rosenblatt (1973), is on the asymptotic distribution of the $\mathcal{L}^2$ error, which we have called v_n.

Theorem 32.9 Assume the following:

(a) f vanishes outside $-\infty < C < D < \infty$.
(b) f'' is uniformly bounded.
(c) K vanishes outside $-\infty < A < B < \infty$.
(d) K is a density with $\int_A^B zK(z)\,dz = 0$.
(e) K is of bounded variation.

Let $\Lambda^2 = \int K^2(z)\,dz$ and

$$\sigma^2 = \left(\int f^2(x)\,dx\right)\int\left[\int K(x+y)K(x)\,dx\right]^2 dy.$$

Suppose $h_n = n^{-\gamma}$ and $\frac{2}{9} < \gamma < \frac{2}{3}$. Then

$$\frac{1}{\sigma\sqrt{h_n}}(nh_n v_n - \Lambda^2) \stackrel{\mathcal{L}}{\Rightarrow} N(0, 1).$$

Remarks.

(i) From elementary arguments, it is intuitively clear that the kernel estimate is pointwise asymptotically normal since it is "like" a sample mean. However, to handle global indices, we have to consider asymptotics for the entire process, with x being the time parameter of the processes. The proof uses embedding of the uniform empirical process into sequences of Brownian bridges; see Chapter 12 for these embeddings.

(ii) There are two aspects of Theorem 32.9 that are not fully satisfactory. They are that f is assumed to have bounded support and that the optimal case $h_n \asymp n^{-1/5}$ is not covered.

(iii) Theorem 32.9 is useful for approximating $P(v_n \geq c)$. Similar results are available for w_n, the $\mathcal{L}^1$ error.

The final result presented here, also due to Bickel and Rosenblatt (1975), concerns the $\mathcal{L}^\infty$ error, t_n.

Theorem 32.10 Fix $0 < \alpha < 1$. Assume that $h_n = n^{-\gamma}$, where $\frac{1}{5} < \gamma < \frac{1}{2}$. For a suitable (explicit) constant $c = c(n, \gamma, \alpha, K)$,

$$\lim_{n\to\infty} P\left(|\hat{f}_{n,h}(x) - f(x)| \leq c\sqrt{\frac{\hat{f}_{n,h}(x)}{nh}},\ \forall x\right) = 1 - \alpha$$

under appropriate regularity conditions on f and K.

Remark. The main application of this result is in the construction of an explicit asymptotic $100(1-\alpha)\%$ confidence band for f. Such a confidence band can be used for goodness-of-fit tests, for insight into where the modes of the unknown density are located, etc. We discuss the mode estimation problem in detail later in this chapter.

32.15 Multivariate Density Estimation and Curse of Dimensionality

Akin to the one-dimensional case, nonparametric estimation of a density in a higher dimensional space $\mathbb{R}^d$ is also an interesting problem. Most of the methods (e.g., kernel estimates, histograms, estimates based on orthogonal series, etc.) extend, without any difficulty, to the case of $\mathbb{R}^d$. Kernel estimates remain the most popular, just as in the case of $d = 1$.

The basic problem of density estimation in high-dimensional spaces is, however, very difficult. The reason for this is that, in high dimensions, local neighborhoods tend to be empty of sample observations unless the sample size is very large. For example, kernel estimates with a given bandwidth are just local averages, and yet, in high dimensions, there will be no local averages to take unless the bandwidth is very large. However, large bandwidths are not good for error properties of the density estimate. Large biases will plague estimates formed from large bandwidths. The practical requirements are contradictory to the requirements for good accuracy. This general problem was termed the *curse of dimensionality* in Huber (1985). We start with a sequence of results and examples that illustrate the curse of dimensionality. Most of the material below in the next two theorems is taken from Scott (1992) and DasGupta (2000). Scott and Wand (1991) discuss the feasibility and difficulties of high-dimensional density estimation with examples and computation.

Theorem 32.11 Let $C = C_d \subset \mathbb{R}^d$ be the d-dimensional unit cube

$$C = \left\{ x \in \mathbb{R}^d : \max_{1 \le i \le d} |x_i| \le 1 \right\}.$$

For $p > 0$, let $B_p \subset C$ be the unit $\mathcal{L}^p$ ball

$$B_p = \{x \in \mathbb{R}^d : |x_1|^p + \ldots + |x_d|^p \le 1\}.$$

Let $\varepsilon > 0$ and define

$$a = a_p = \left(\frac{e}{p}\right)^{1/p} \Gamma\left(1 + \frac{1}{p}\right) \quad \text{and} \quad b = b_p = \frac{\varepsilon\sqrt{2\pi}}{\sqrt{p}}.$$

Then, $\text{vol}(B_p)/\text{vol}(C) < \varepsilon$ provided

$$\log d > p\left(\log a - \frac{\log b}{d} - \frac{\log d}{2d}\right).$$

The following table gives the value of d_0 such that, for $d \ge d_0$, 90% of the volume of the cube is outside the ball B_p.

p	1/2	1	2	3	4
d_0	3	4	5	7	11

For example, for $d \ge 5$, 90% of the volume of the cube is outside the inscribed sphere. That is, most of the volume is in the corners.

The next theorem shows that the problem of neighborhoods devoid of sample observations is typical in much more generality, and the uniform distribution on the cube is nothing special. First we need a definition.

Definition 32.1 Let $f : \mathbb{R}^d \rightarrow \mathbb{R}$. The least radial majorant of f is the function

$$g(x) = g(\|x\|_2) = \sup_{y:\|y\|=\|x\|} f(y).$$

Theorem 32.12 Let $f = f_d$ be a density on $\mathbb{R}^d$ and let $g = g_d$ be its least radial majorant. Let

$$M = M_d = \sup_{r>0} \frac{g(r) - g(1)}{r - 1}.$$

Assume that, as $d \rightarrow \infty$,

$$\frac{\sqrt[d]{g(1)}}{\sqrt{d}} = o(1) \quad \text{and} \quad \frac{\sqrt[d]{M_d}}{\sqrt{d}} = o(1).$$

Then, for any fixed $k > 0$,

$$P_f\left(\|X\|_2 \leq k\right) \longrightarrow 0 \quad \text{as } d \rightarrow \infty.$$

Remark. Theorem 32.12 says that under frequently satisfied conditions, fixed neighborhoods, however large, have very small probabilities in high dimensions. In particular, the normal case is reported below.

Example 32.6 Let $X \sim N_d(0, \Sigma)$. On calculation,

$$g(r) = (2\pi)^{-d/2}|\Sigma|^{-1/2}\text{Exp}\left\{-\frac{r^2}{2\lambda_{\max}}\right\},$$

where $\lambda_{\max}$ is the largest eigenvalue of Σ. This implies that

$$\frac{\sqrt[d]{g(1)}}{\sqrt{d}} \leq \frac{1}{\sqrt{2\pi d\lambda_{\min}}}$$

(where $\lambda_{\min}$ is the minimum eigenvalue of Σ), which goes to zero if $d\lambda_{\min} \rightarrow \infty$. Again, on calculation,

$$M = M_d \leq C\left[(2\pi)^d \lambda_{\min}^d \lambda_{\max}\right]^{-1/2}, \quad \text{for some } C > 0.$$

But this implies

$$\frac{\sqrt[d]{M_d}}{\sqrt{d}} \leq \frac{\sqrt[d]{C}}{\sqrt{2\pi d\lambda_{\min}}\sqrt[d]{\lambda_{\max}}},$$

which goes to zero if

$$d\lambda_{\min} \to \infty \quad \text{and} \quad \liminf_{d\to\infty} \sqrt[d]{\lambda_{\max}} > 0.$$

In other words, if $\lambda_{\min}$ and $\lambda_{\max}$ satisfy the conditions

$$d\lambda_{\min} \to \infty \quad \text{and} \quad \liminf_{d\to\infty} \sqrt[d]{\lambda_{\max}} > 0,$$

then Theorem 32.12 implies that

$$P_{N(0,\Sigma)}(\|X\|_2 \leq k) \longrightarrow 0, \quad \forall k > 0.$$

Note, however, that the second condition is automatically satisfied if the first condition holds, so all that is required is that $d\lambda_{\min} \to \infty$ as $d \to \infty$. If $\Sigma = \sigma_d^2 I_d$, then the condition is satisfied if $d\sigma_d^2 \to \infty$; i.e., unless $\sigma_d^2 \to 0$ faster than d^{-1} as $d \to \infty$. Similar results can be proved about multivariate t-distributions. The following table gives actual values of $P(\|X\|_2 \leq 1)$ in the multivariate t case with scale matrix I_d.

$d\downarrow$, df $\to$	1	3	5	∞
1	0.5	0.6090	0.6368	0.6822
6	0.0498	0.0308	0.0249	0.0145
10	0.0101	0.0024	0.0012	0.0002

Example 32.7 This example further illustrates the curse of dimensionality phenomenon. Suppose we take n iid observations from a uniform distribution in $C = C_d$. Suppose we plant a set at random within the cube. Take

$$B = \{x \in \mathbb{R}^d : \|x - \mu\|_2 \leq r\},$$

where $\mu \sim U(C_d)$ and $r|\mu \sim U[0, 1 - \max|\mu_i|]$. The following table gives the minimum necessary sample size n^* such that P (there is at least one observation inside B) $\geq \frac{1}{2}$.

d	2	3	4	5
n^*	70	2400	145,000	1.2×10^6

It is striking that, even in Five dimensions, a randomly planted sphere will probably be empty unless *extremely* large samples are available. In spite of these well-understood difficulties, there is a good body of literature on multivariate density estimation. We discuss the kernel method briefly.

32.15.1 Kernel Estimates and Optimal Bandwidths

The technical results in this subsection are derived in Scott (1992). Let $f : \mathbb{R}^d \to [0, \infty)$ be a d-dimensional density and $K : \mathbb{R}^d \to \mathbb{R}$ be a kernel function. Let $h > 0$ be a bandwidth. Then a kernel estimate of f is

$$\hat{f}_{n,h}(x) = \frac{1}{nh^d} \sum_{i=1}^{n} K\left(\frac{x - X_i}{h}\right), \quad x \in \mathbb{R}^d.$$

In practice, K is often taken to be a product kernel or an ellipsoidal kernel; i.e.,

(i) $K(z) = \prod_{i=1}^{d} K_0(z_i)$ for $z \in \mathbb{R}^d$.
(ii) $K(z) = K_0(z'\Sigma^{-1}z)$, where $z \in \mathbb{R}^d$ and $\Sigma_{d\times d}$ is a prespecified matrix.

For example, the multivariate Gaussian kernel corresponds to (ii), where K_0 is essentially the $N(0, 1)$ density. More general kernel estimates in high dimensions that allow different amounts of smoothing along different directions can be seen in Hardle et al. (2004).

In the following, we assume that K is permutation invariant. Before we talk about optimal bandwidths, we need some notation. For a function $h : \mathbb{R}^d \to \mathbb{R}$, let $\nabla^2 h$ denote the Laplacian of h; i.e.,

$$\nabla^2 h(x) = \sum_{i=1}^{d} \frac{\partial^2}{\partial x_i^2} h(x).$$

Also, define the additional notation

- $\kappa = \int z_1^2 K(z)\, dz$;
- $b_d(x) = \frac{1}{2}\kappa \nabla^2 f(x)$;
- $\sigma_d^2(x) = f(x) \int K^2(z)\, dz$;
- $r_d(x) = b_d(x)/\sigma_d(x)$.

Theorem 32.13 The asymptotically optimal global $\mathcal{L}^1$ and $\mathcal{L}^2$ bandwidths are as follows.

$$h_{\text{opt}}(\mathcal{L}^1) = c^*_{1,d} n^{-1/(d+4)},$$
$$h_{\text{opt}}(\mathcal{L}^2) = c^*_{2,d} n^{-1/(d+4)},$$

where

$$c^*_{1,d} = (v^*)^{2/(d+4)},$$
$$c^*_{2,d} = \left[d \int K^2(z)\, dz \div 4 \int b_d^2(x)\, dx \right]^{1/(d+4)},$$

v^* being the unique positive root of the equation,

$$\int \sigma_d(x) \left[4vr_d(x) \left\{ \Phi(vr_d(x)) - \frac{1}{2} \right\} - d\varphi(vr_d(x)) \right] dx = 0,$$

where, as usual, φ, Φ denote the standard normal density and CDF.

Remark. It turns out that, in relatively low dimensions, $c^*_{1,d}$ and $c^*_{2,d}$ are often quite close. This may save cumbersome exact computation of $c^*_{1,d}$.

Example 32.8 Let f be the $N_d(0, I_d)$ density. Then $c^*_{1,d}$ and $c^*_{2,d}$ are given for various values of d in the following table.

d	$c^*_{1,d}$	$c^*_{2,d}$
1	1.030	1.059
2	1.043	1.000
3	1.078	0.969
5	1.151	0.940
10	1.249	0.925

Remark. Despite these demands for astronomical sample sizes in high dimensions for accurate estimation of f, limited success is sometimes attainable with relatively small sample sizes. A famous example is due to Scott and Wand (1991), where, with $n = 225$, they succeed in identifying all three modes in a two-dimensional projection of a ten-dimensional density, where the projection was a three-component mixture of three distinct bivariate normals.

So, pronounced shape-related structure present in f may sometimes be visible with relatively small n, but function estimation itself cannot be done well except with huge and unrealistic sample sizes.

32.16 Estimating a Unimodal Density and the Grenander Estimate

In common statistical inference, we are often willing to make the assumption that the underlying density f is unimodal. Indeed, unimodality comes more naturally than symmetry; real data usually show some skewness, but unimodality is more common. The natural question is, if we add in the unimodality assumption, what would be our appropriate density estimate and how would that estimate perform? For example, what will be the convergence rate? The results have a flavor of being paradoxical, as we show below.

There are two ideas that come quickly to mind:

- Enforce the unimodality constraint and select an appropriate unimodal density estimate.
- Ignore the constraint and form a usual density estimate and hope that it turns out to be unimodal or at least close to unimodal.

Both ideas have been explored, and the results are very interesting. This is an example of a rare infinite-dimensional problem where maximum likelihood actually works! For example, with just derivative assumptions on f, an MLE for f cannot exist for any n. So, maximum likelihood does not even apply there. Indeed, maximum likelihood is notorious for being problematic in nonparametric problems. But that is not so in the case we are considering now.

32.16.1 The Grenander Estimate

We mainly consider the case of a unimodal density with *known* mode ξ. In fact, we assume f to be strictly unimodal; i.e., f is strictly increasing on $(-\infty, \xi]$ and strictly decreasing on $[\xi, \infty)$. Since f is the "union" of two strictly monotone pieces, it is sufficient to consider the problem of estimating a decreasing density on $[\xi, \infty)$ for some known ξ. The form of the MLE of a decreasing density was first derived by Grenander (1956), and the estimate is the famous *Grenander estimate*.

Definition 32.2 Given iid observations $X_1, \ldots, X_n$, let F_n be the empirical CDF. Let $_n$ be its smallest concave majorant; i.e., the smallest concave function $\omega_n(x)$ such that $\omega_n(x) \geq F_n(x)\, \forall x$. Let $_n$ be the *left* derivative of $_n$, which exists, is a piecewise constant function, and is called the Grenander estimate.

Remark. In fact, $_n$ can be written a little more explicitly as

$$_n(x) = \begin{cases} m_i, & X_{(i)} \leq x < X_{(i+1)} \\ 0, & \text{otherwise} \end{cases},$$

where m_i is the slope of $_n$ on $[X_{(i)}, X_{(i+1)})$.

We now present a fairly large collection of results on the performance of the Grenander estimate. The first is due to Groeneboom (1985). Results of this kind are called *cube root asymptotics*. Cube root and more generally slower rate asymptotics are common in shape-restricted inference. For example, see Rao (1969), Banerjee and Wellner (2001), and references in the latter.

Theorem 32.14 Assume f is strictly decreasing, bounded, twice continuously differentiable, and compactly supported. Let $_n$ be the Grenander estimate and Z_t an independent Brownian motion on $(-\infty, \infty)$ with $Z_0 = 0$. Then:

(a) For fixed $x \in \mathbb{R}$,

$$n^{1/3}[_n(x) - f(x)] \overset{\mathcal{L}}{\Rightarrow} |4f(x)f'(x)|^{1/3} \operatorname{argmax}_t(Z_t - t^2).$$

(b) Let $M = E[\arg\max(Z_t - t^2)]$. Then,

$$n^{1/3} \int |_n(x) - f(x)|\, dx \overset{P}{\Rightarrow} M \int |4f(x)f'(x)|^{1/3}\, dx.$$

(c) For a suitable σ^2 completely independent of f,

$$n^{1/6} \left\{ n^{1/3} \int |_n(x) - f(x)|\, dx - M \int |4f(x)f'(x)|^{1/3}\, dx \right\} \overset{\mathcal{L}}{\Rightarrow} N(0, \sigma^2).$$

Remark. At first glance, it seems paradoxical that with the extra information that we have, namely unimodality, we do not have even the $n^{2/5}$ asymptotics for ordinary kernel estimates. However, that sort of intuition is parametric. Parametric intuition is usually not very helpful in nonparametric problems. In nonparametric problems, it is actually quite common to improve one's performance by using procedures outside the known parameter space. For example, it is known that if we construct kernel estimates using a kernel that can assume negative values, we can improve on the convergence rate of the usual kernel estimates.

Complementary results can also be established in a decision theory framework using a risk function. Both fixed n and asymptotic results are available. The next two theorems are due to Groeneboom (1985) and Birge (1989).

Let $\mathcal{F}(\xi, H, L)$ be the family of all decreasing f that are bounded by H with support $[\xi, \xi + L]$, where $L, H < \infty$. We need some notation.

$$U = \log(1 + LH),$$
$$\kappa = \frac{1}{2}\left(1 + \sqrt{\frac{\pi}{2}}\right),$$
$$I(f) = \int \left|\frac{1}{2} f(x) f'(x)\right|^{1/3} dx,$$
$$R_n(f, \hat{f}_n) = E\left[\int |\hat{f}_n(x) - f(x)|\, dx\right].$$

Theorem 32.15 (a) For a suitable functional $\Lambda(f, z)$, $z \geq 0$,

$$R_n(f,_n) \leq 2\Lambda(f, \kappa/\sqrt{n}), \quad \forall n \geq 1.$$

(b) $\lim_{n\to\infty} n^{1/3} R_n(f,_n) = \tau I(f)$, where $\tau \approx 0.82$.

(c) For all $n \geq 40U$,

$$\sup_{f\in\mathcal{F}(\xi,H,L)} n^{1/3} R_n(f,_n) \leq 4.75 U^{1/3}.$$

(d) For all $n \geq 40U$,

$$0.195 U^{1/3} \leq \inf_{\hat{f}_n} \sup_{f\in\mathcal{F}(\xi,H,L)} n^{1/3} R_n(f, \hat{f}_n) \leq 1.95 U^{1/3}.$$

Remark. Parts (a), (c), and (d) are exceptional in the sense that they give fixed n bounds on $R_n(f,_n)$ and the minimax risks. Usually, fixed n bounds with constants are very hard to obtain in nonparametrics. Also, interestingly, parts (c) and (d) show that the Grenander estimate is within a constant factor of being minimax for all large n.

The functional $\Lambda(f, z)$ collapses at the rate $z^{2/3}$, and even explicit bounds $\Lambda(f, z) \leq Az^{2/3} + Bz^{4/3}$ are known; see Birge (1989).

Remark. As regards the idea of ignoring the unimodality constraint, a remarkably nice result was proved in Hall, Minnotte, and Zhang (2004). Basically, it says that, for certain types of kernels K, the kernel estimate is going to be unimodal for all datasets unless the bandwidth is too small. Here is a special result from their article.

Theorem 32.16

(a) Given $X_1, \ldots, X_n$, there exists a unique $\tilde{h} = \tilde{h}(X_1, \ldots, X_n)$ such that, for all $h > \tilde{h}$, $\hat{f}_{n,h}$ is unimodal with probability 1, provided the kernel K is strictly unimodal, concave in some neighborhood of the origin, and compactly supported.

(b) $\tilde{h} \asymp n^{-1/5}$ in the sense that $n^{1/5}\tilde{h}$ has a nondegenerate limiting distribution.

32.17 Mode Estimation and Chernoff's Distribution

Closely related to the problem of density estimation is that of estimation of the mode of a density. There can be several reasons for wanting to estimate the mode; for example, as a part of graphical or exploratory data analysis or for secondary confirmation of estimates of other parameters of location, such as the median, which should be close to the mode for approximately symmetric unimodal densities. It is also a hard and therefore intrinsically interesting problem.

Two main approaches have been used in mode estimation. *Direct estimates* are those that address mode estimation on its own; *indirect estimates* are those that first estimate the whole density function and then locate a global maximum or an approximate global maximum. For example, one could look at the maxima of a suitable kernel estimate of the density. Such an estimate uses all of the data. But, there are also mode estimates that use only a part of the data. Both the univariate case and the multivariate case have been addressed in the literature. The most important references are Chernoff (1964), Parzen (1962), Venter (1967), Wegman (1971), and Sager (1975). With respect to Chernoff's estimate specifically, there are some hard distribution theory problems, and Groeneboom and Wellner (2001) is the latest reference on that problem.

We assume that the sample observations $X_1, X_2, \cdots$ are iid from a density f on the real line with support (a, b), $-\infty \leq a < b \leq \infty$, f achieves a unique global maximum at some $\theta \in (a, b)$, which we call the *mode* of f.

The problem we consider is estimation of θ. Three specific estimates and their properties are discussed below.

Parzen's estimate. Let $\hat{f}_n(x)$ be a kernel density estimate corresponding to a kernel $K(.)$. Parzen's estimate is defined to be the maximizer of $\hat{f}_n(x)$ over $-\infty < x < \infty$. We denote it as $\hat{\theta}_P$.

Chernoff's estimate. Let a_n be a positive sequence going to zero as $n \to \infty$. Chernoff's estimate is defined as the midpoint of any interval of length $2a_n$ containing the maximum number of observations among $X_1, \cdots, X_n$. We denote it as $\hat{\theta}_C$.

Venter's estimate. Let k_n be a positive integral sequence going to ∞ as $n \to \infty$. Venter's estimate is the midpoint of the shortest interval containing k_n of the sample observations $X_1, \cdots, X_n$. We denote it as $\hat{\theta}_V$.

First, we address consistency of these estimates; the parts of Theorem 32.17 below are proved in Parzen (1962), Chernoff (1964), and Venter (1967), respectively.

Theorem 32.17 (a) Suppose $f(x)$ is uniformly continuous on (a, b), that the kernel function $K(.)$ is twice continuously differentiable on $(-\infty, \infty)$, and that $\int_{-\infty}^{\infty} |K''(x)|dx < \infty$. Then $\hat{\theta}_P \stackrel{P}{\Rightarrow} \theta$ as $n \to \infty$.

(b) Suppose $f(x)$ is eventually monotone decreasing and converges to zero as $|x| \to \infty$. Also suppose that $\sqrt{\frac{n}{\log \log n}} a_n \to \infty$ as $n \to \infty$. Then $\hat{\theta}_C \stackrel{\text{a.s.}}{\to} \theta$ as $n \to \infty$.

(c) Let $\alpha_1(\delta) = \inf\{f(x) : |x - \theta| \leq \delta\}$ and $\alpha_2(\delta) = \sup\{f(x) : |x - \theta| > 2\delta\}$.
Suppose that for $0 < \delta \leq \delta_0$ for some $\delta_0, \alpha_1(\delta) > \alpha_2(\delta)$. Also assume that f is either left continuous or right continuous at every point of jump discontinuity, if any exists, that $k_n \to \infty$, $\frac{k_n}{n} \to 0$, and $\sum_{n=1}^{\infty} n\lambda^{k_n} < \infty \, \forall \, 0 < \lambda < 1$. Then, $\hat{\theta}_V \stackrel{\text{a.s.}}{\to} \theta$ as $n \to \infty$.

Thus, the conditions for consistency are mild. The Chernoff estimate is consistent if the density is not erratic at the tails and if the window length a_n is not too small. The Venter estimate is consistent if the density has some degree of smoothness at any potential jump discontinuities and if the density is continuous in some neighborhood of the mode θ. Somewhat stronger conditions are needed for strong consistency of the Parzen estimate.

Asymptotic distribution theory, however, is very sophisticated, so much so that until quite recently (Groeneboom (1989), Groeneboom and Wellner (2001)) computing percentiles of the limiting distributions was considered a nearly impossible problem. The distribution theory is described next;

again, see Parzen (1962), Chernoff (1964), and Venter (1967) respectively, for proofs of the three parts

Theorem 32.18 (a) Assume the following conditions on the kernel function $K(.)$, the density function f, and the bandwidths h_n:

i. $K(.)$ has zero mean and a finite variance and is four times continuously differentiable with an absolutely integrable fourth derivative.
ii. f is uniformly continuous on (a, b) and is also four times continuously differentiable with an absolutely integrable fourth derivative.
iii. $\lim_{n\to\infty} nh_n^6 = \infty$, and for some $0 < \delta < 1$, $\lim_{n\to\infty} nh_n^{5+2\delta} = 0$.
Then, $\sqrt{nh_n^3}(\hat{\theta}_P - \theta) \stackrel{\mathcal{L}}{\Rightarrow} N(0, \frac{f(\theta)}{[f''(\theta)]^2}J)$, where $J = \int_{-\infty}^{\infty}[K'(x)]^2 dx$.

(b) Let $\alpha_i(\delta), i = 1, 2$ be as defined previously, and assume that for $0 < \delta \leq \delta_0$ for some δ_0, $\alpha_1(\delta) > \alpha_2(\delta)$. Suppose f is once continuously differentiable, and with $\gamma_2 > 0$, $f(x) = f(\theta) - \frac{\gamma_2}{2}(x-\theta)^2 + \frac{\gamma_3}{6}(x-\theta)^3 + o(|x-\theta|^3)$ as $x \to \theta$, that $a_n \to 0$, $na_n^2 \to \infty$ as $n \to \infty$. Then, $(na_n^2)^{1/3}(\hat{\theta}_C - \theta) \stackrel{\mathcal{L}}{\Rightarrow} (\frac{2f(\theta)}{\gamma_2^2})^{1/3}T$, where $T = \text{argmax}\{W(t) - t^2, -\infty < t < \infty\}$ and $W(t)$ is a two-sided standard Wiener process with $W(0) = 0$.

(c) Suppose f is continuous, and with $\gamma_2 > 0$, $f(x) = f(\theta) - \frac{\gamma_2}{2}(x - \theta)^2 + \frac{\gamma_3}{6}(x - \theta)^3 + o(|x - \theta|^3)$ as $x \to \theta$ that $k_n \sim An^\nu$, $\frac{4}{5} \leq \nu < \frac{7}{8}$. Then, $n^{1/5}(\hat{\theta}_V - \theta) \stackrel{\mathcal{L}}{\Rightarrow} 2A^{-2/3}\gamma_2^{-2/3} f(\theta)T$, where $T = \text{argmax}\{W(t) - t^2, -\infty < t < \infty\}$, and $W(t)$ is a two-sided standard Wiener process with $W(0) = 0$.

Remark. At first sight, it seems odd that the normalizing constants for the three estimates are different. This is because the different parts of Theorem 32.18 assume varying degrees of smoothness of the underlying density f. The smoother the density, the better would be our ability to estimate the location of the mode. Thus, Parzen's estimate assumes the strongest smoothness assumptions and in return gets the best rate. Note that the optimal bandwidths for the problem of estimating the whole density function are of the order $n^{-1/5}$ and do not satisfy the conditions of part (a) of the theorem.

The random variable T referred to in parts (b) and (c) is indeed uniquely defined with probability 1, and its distribution is known as *Chernoff's distribution*. It is absolutely continuous, and its density function was derived in Groeneboom (1989). The density function is obviously symmetric around zero and satisfies

$$p(t) \sim \frac{2^{5/3}}{Ai'(w)} e^{-\frac{2}{3}|t|^3 + 2^{1/3}w|t|}, \; t \to \infty,$$

where $Ai(.)$ is the Airy function and $w \approx -2.338$ its largest root. Also, $Ai'(w) \approx 0.702$. The density function $p(t)$ is characterized as a limiting solution of a partial differential equation with certain boundary value conditions; see Chernoff (1964). Computing the density function by numerically solving the boundary value problem has met with partial success. However, Groeneboom and Wellner (2001) provide another representation, which is essentially analytic and is amenable to numerical computations. The following percentiles of Chernoff's distribution are taken from Groeneboom and Wellner (2001).

p	pth percentile of T
.5	0
.75	0.3533
.8	0.4398
.9	0.6642
.95	0.8451
.975	1.0015
.99	1.1715
.995	1.2867

Example 32.9 Suppose the underlying density was $N(\theta, 1)$. It is uniformly continuous and infinitely differentiable, and all derivatives are absolutely integrable. Thus the conditions on f for asymptotic normality of the Parzen estimate of θ are satisfied. Take the kernel function K to be the standard normal density, which satisfies all the conditions required on the kernel. For the bandwidth, take $h_n = n^{-1/6} \log n$. Then $nh_n^6 \to \infty$. Furthermore, $nh_n^{5+2\delta} = n^{(1-2\delta)/6}(\log n)^{(5+2\delta)/6} \to 0$ for any $\frac{1}{2} < \delta < 1$. It follows from Theorem 32.18 that $n^{1/4}(\log n)^{3/2}(\hat{\theta}_P - \theta)$ has a limiting normal distribution. Note that the normalizing constant $n^{1/4}(\log n)^{3/2}$ is much slower than $\sqrt{n}$, the rate one would obtain by using the sample mean as an estimate of θ.

32.18 Exercises

Exercise 32.1 * Consider the $.5N(-2, 1) + .5N(2, 1)$ density, which is bimodal. Simulate a sample of size $n = 40$ from this density, and compute and plot each of the following density estimates. Compare visually, and give, with reasons, a ranking for your preference of the estimates: (a) a plug-in parametric estimate that uses a $.5N(\theta, 1) + .5N(-\theta, 1)$ parametric model; (b) a histogram that uses $m = 11$ equal width cells; (c) an orthogonal series estimate that uses the Hermite polynomials and keeps six terms; (d) a kernel estimate that uses a Gaussian kernel and Silverman's reference bandwidth.

Exercise 32.2 * Suppose the true model is $N(\theta, \sigma^2)$, and a parametric plug-in estimate using MLEs of the parameters is used. Derive an expression for the global error index $E_f \int |f - \hat{f}|dx$. At what rate does this converge to zero?

Exercise 32.3 * Suppose the true model is a double exponential location parameter density but you thought it was $N(\theta, 1)$ and used a parametric plug-in estimate with an MLE. Does $E_f \int |f - \hat{f}|dx$ still converge to zero? Is it the same rate as in the previous exercise?

Exercise 32.4 * Suppose the true model is $N(\theta, 1)$. Does there exist an estimate of f that is unbiased under all θ? Prove or disprove.

Exercise 32.5 * Suppose the true model is $N(\theta, 1)$. Using $\mathcal{L}_1$, $\mathcal{L}_2$ loss, what are the parametric minimax estimates?

Exercise 32.6 * Suppose the true model is $U[-\theta, \theta]$. Using $\mathcal{L}_1$ loss, what is the parametric minimax estimate?

Exercise 32.7 For what sort of binwidths does the bias part dominate the MSE in pointwise estimation of f with histogram estimates?

Exercise 32.8 * Derive a reasonable *reference global binwidth* for histogram estimates of a density. Follow Silverman's idea for kernel estimates.

Exercise 32.9 * If the kernel is a bounded density, is it true that the kernel density estimate is consistent a.e.? What conditions on f, h would ensure this?

Exercise 32.10 * Suppose the true density is $N(0, 1)$. Find the full range of values for the optimal global bandwidth if you use the known f but vary the kernel K in the family of all densities of the form $e^{-|x|^\alpha}$, $\alpha \geq 1$. Apply it to $n = 30$ and compute a numerical range. Is it too wide?

Exercise 32.11 Give a proof of Epanechnikov's theorem.

Exercise 32.12 * Suppose the true density is a Beta and a kernel estimate with the Epanechnikov kernel is used. Is maximum likelihood CV consistent under $\mathcal{L}_1$ loss? Prove your answer.

Exercise 32.13 * Simulate a sample of size $n = 30$ from $N(0, 1)$. For a Gaussian kernel estimate, plot the CV likelihood carefully. Repeat the simulation. Examine the shapes of the CV likelihoods and comment.

Exercise 32.14 * Simulate a sample of size $n = 30$ from $N(0, 1)$. For a Gaussian kernel and the triangular kernel, plot Stone's modified CV likelihood. Repeat the simulation. Examine the shapes of the CV likelihoods and comment. Compare the results with the previous exercise.

Exercise 32.15 Give examples of three different kernels that satisfy conditions in Stone's consistency theorem.

Exercise 32.16 * How does Silverman's reference bandwidth change if you take the double exponential as your reference family? Does this cause more smoothing, less, or neither in general?

Exercise 32.17 * Suppose the true f is the $N(0, 1)$ density and a Gaussian kernel is used. Find the theoretical global optimum bandwidths under $\mathcal{L}_1$, $\mathcal{L}_2$ loss. Plot the ratio of the two optimal bandwidths against n and comment.

Exercise 32.18 * Take your simulated data of Exercise 32.1. Compute and plot the least squares CV estimate, still using the standard Gaussian kernel. Compare it with the kernel estimate you previously found using the reference bandwidth.

Exercise 32.19 * Find, by using the Bickel-Rosenblatt theorem, an approximation to the probability that the IMSE exceeds .1 if the true density is a standard normal truncated to $[-3, 3]$ and the kernel is the triangular kernel. Use only those bandwidths that the theorem covers.

Exercise 32.20 * Suppose the true density is a normal. Give three distinct kernels for which the kernel estimate is strongly uniformly consistent using a bandwidth of the $n^{-1/5}$ rate.

Exercise 32.21 Write an analysis of the two Bickel-Rosenblatt theorems, discussing their pros and cons.

Exercise 32.22 Suppose a d-dimensional density is the standard multivariate normal. Give a direct proof that as $d \to \infty$, the probability of any compact set converges to zero. What is the rate of convergence?

Exercise 32.23 * Suppose $d = 2$ and the joint density is that of two independent exponentials. What is the least radial majorant? Characterize it.

Exercise 32.24 Prove that the least radial majorant of a general multivariate normal density is the one stated in the text.

Exercise 32.25 * Suppose we plant a random sphere inside the d-dimensional cube in the way described in the text. For $d = 2, 3, 5$, find the smallest number of uniform observations from the cube necessary to ensure with an 80% probability that the randomly planted sphere is not observation-void.

Exercise 32.26 * Suppose the true density is a d-dimensional standard multinormal. Find the global optimum bandwidth under $\mathcal{L}_2$ loss using a multivariate standard Gaussian kernel. Compute it for $d = 2, 3, 5, 8, 10$ with $n = 100, 250, 1000$. Comment on the magnitudes of the bandwidths and what you like and dislike about them.

Exercise 32.27 * Suppose the true density is a ten-dimensional standard multinormal. Use the global optimum bandwidth under $\mathcal{L}_2$ loss and a standard Gaussian kernel. Plot some selected contours of equal density of the true and the estimated densities and compare them. Use suitable values of n of your choice.

Exercise 32.28 * Explicitly compute and plot the Grenander estimate for a simulated $N(0, 1)$ sample with $n = 20$ observations. Find a reasonably smoothed version of it and plot that one, too.

Exercise 32.29 * Do you consider it paradoxical that the convergence rate of the Grenander estimate is slower than that of kernel density estimates? Discuss.

Exercise 32.30 * Suggest a method to approximately calculate the constant M in Groeneboom's theorem on the weak limit of the $\mathcal{L}_1$ error of the Grenander estimate.

Exercise 32.31 Suppose the underlying density is a chi-square with 12 degrees of freedom. Using the standard normal density as the kernel function, explicitly evaluate the limiting distribution of the Chernoff estimate of the mode if $a_n = \frac{1}{n^\delta}$ for suitable δ.

Exercise 32.32 * Simulate a standard normal sample of size $n = 40$. Compute the Parzen estimate, the Chernoff estimate, and the Venter estimate by using the standard normal kernel and $h = n^{-1/6} \log n$ for the Parzen estimate, $a_n = n^{-1/4}$ for the Chernoff estimate, and $k_n = n^{4/5}$ for the Venter estimate. Repeat the simulation. Comment on the relative accuracies; which estimate does the best?

Exercise 32.33 * Write an essay on how you may want to use a density estimate.

References

Banerjee, M. and Wellner, J. (2001). Likelihood ratio tests for monotone functions, Ann. Stat., 29(6), 1699–1731.
Bickel, P.J. and Rosenblatt, M. (1973). On some global measures of the deviations of density function estimates, Ann. Stat., 1, 1071–1095.
Birge, L. (1989). The Grenander estimator: a nonasymptotic approach, Ann. Stat., 17(4), 1532–1549.
Bowman, A.W. (1984). An alternative method of cross-validation for the smoothing of density estimates, Biometrika, 71(2), 353–360.
Breiman, L., Meisel, W., and Purcell, E. (1977). Variable kernel estimates of multivariate densities, Technometrics, 19, 135–144.
Brown, L. (1992). Minimaxity, more or less, *in Statistical Decision Theory and Related Topics*, Vol. V, J. Berger and S.S. Gupta (eds.), Springer-Verlag, New York, 1–18.
Brown, L. (2000). An essay on statistical decision theory, J. Am. Stat. Assoc., 95, 1277–1281.
Chernoff, H. (1964). Estimation of the mode, Ann. Inst. Stat. Math., 16, 31–41.
Chow, Y.S., Geman, S. and Wu, L.D. (1983). Consistent cross-validated density estimation, Ann. Stat., 11(1), 25–38.
DasGupta, A. (2000). Some results on the curse of dimensionality and sample size recommendations, Cal. Stat. Assoc. Bull., 50(199/200), 157–177.
Devroye, L. (1983). On arbitrarily slow rates of convergence in density estimation, Z. Wahr. Verw. Geb., 62(4), 475–483.
Devroye, L. (1987). *A Course in Density Estimation*, Birkhäuser, Boston.
Devroye, L. (1992). A note on the usefulness of superkernels in density estimation, Ann. Stat., 20, 2037–2056.
Devroye, L. and Penrod, C.S. (1984). The consistency of automatic kernel density estimates, Ann. Stat., 12(4), 1231–1249.
Devroye, L. and Lugosi, G. (1999). Variable kernel estimates: on the impossibility of tuning the parameters, in *High Dimensional Probability*, Giné, E. and D. Mason (eds.) Vol. II, Birkhäuser, Boston, 405–424.
Devroye, L. and Lugosi, G. (2000). *Combinatorial Methods in Density Estimation*, Springer, New York.
Epanechnikov, V.A. (1969). Nonparametric estimates of a multivariate probability density, Theory Prob. Appl., 14, 153–158.
Farrell, R.H. (1972). On the best attainable asymptotic rates of convergence in estimation of a density at a point, Ann. Math. Stat., 43, 170–180.

Gasser, T., Müller, H., and Mammitzsch, V. (1985). Kernels for nonparametric curve estimations, J.R. Stat. Soc. B, 47, 238–252.

Grenander, U. (1956). On the theory of mortality measurement, Skand. Aktuarietidskr, 39, 70–96.

Groeneboom, P. (1985). Estimating a monotone density, *in Proceedings of the Berkeley Conference in Honor of Jerzy Neyman and Jack Kiefer*, L. Le cam and R. A. O eslen (eds.) Wadsworth, Belmont, CA, 539–555.

Groeneboom, P. (1989). Brownian motion with a parabolic drift and Airy functions, Prob. Theory Related Fields, 81, 79–109.

Groeneboom, P. and Wellner, J. (2001). Computing Chernoff's distribution, J. Comput. Graph. Stat., 10, 388–400.

Hall, P. (1982). Cross-validation in density estimation, Biometrika, 69(2), 383–390.

Hall, P. (1983). Large sample optimality of least squares cross-validation in density estimation, Ann. Stat., 11(4), 1156–1174.

Hall, P. and Marron, J.S. (1988). Choice of kernel order in density estimation, Ann. Stat., 16, 161–173.

Hall, P., Minnotte, M.C., and Zhang, C., (2004). Bump hunting with non-Gaussian kernels, Ann. Stat., 32(5), 2124–2141.

Hall, P., Sheather, S.J., Jones, M.C., and Marron, J.S. (1991). On optimal data-based bandwidth selection in kernel density estimation, Biometrika, 78(2), 263–269.

Hall, P. and Wand, M.P. (1988). Minimizing $\mathcal{L}_1$ distance in nonparametric density estimation, J. Multivar. Anal., 26(1), 59–88.

Hardle, W., Muller, M., Sperlich, S., and Werwatz, A. (2004). *Nonparametric and Semiparametric Models*, Springer, New York.

Huber, P.J. (1985). Projection pursuit, with discussion, Ann. Stat., 13(2), 435–525.

Jones, M.C., Marron, J.S., and Sheather, S.J. (1996). A brief survey of bandwidth selection for density estimation, J. Am. Stat. Assoc., 91(433), 401–407.

Jones, M.C. and Sheather, S.J. (1991). Using nonstochastic terms to advantage in kernel-based estimation of integrated squared density derivatives, Stat. Prob. Lett., 11(6), 511–514.

Müller, H. (1984). Smooth optimum kernel estimators of densities, regression curves, and modes, Ann. Stat., 12, 766–774.

Parzen, E. (1962). On estimation of a probability density function and mode, Ann. Math. Stat., 33, 1065–1076.

Rao, B.L.S.P. (1969). Estimation of a unimodal density, Sankhya, Ser. A, 31, 23–36.

Revesz, P. (1984). Density estimation, *in Handbook of Statistics*, Vol. 4, P.R. Krishnaiah and P.K. Sen (eds.), North-Holland, Amsterdam, 531–549.

Rice, J. (1986). Bandwidth choice for differentiation, J. Multivar. Anal., 19(2), 251–264.

Rosenblatt, M. (1956). Remarks on some nonparametric estimates of a density function, Ann. Math. Stat., 27, 832–835.

Sager, T. (1975). Consistency in nonparametric estimation of the mode, Ann. Stat., 3, 698–706.

Schuster, E.F. and Gregory, G.G. (1981). On the inconsistency of maximum likelihood density estimators, in *Computer Science and Statistics*, Proceedings of the 13th Symposium on Interface, W.F. Eddy Springer, New York, 295–298.

Scott, D.W. (1992). *Multivariate Density Estimation: Theory, Practice and Visualization*, John Wiley, New York.

Scott, D.W. and Wand, M.P. (1991). Feasibility of multivariate density estimates, Biometrika, 78(1), 197–205.

Silverman, B.W. (1986). *Density Estimation for Statistics and Data Analysis*, Chapman and Hall, London.

Stone, C.J. (1984). An asymptotically optimal window selection rule for kernel density estimates, Ann. Stat., 12(4), 1285–1297.

Stone, M. (1974). Cross-validatory choice and assessment of statistical predictions, J.R. Stat. Soc. Ser. B, 36, 111–147.

Stuetzle, W. and Mittal, Y. (1979). Some comments on the asymptotic behavior of robust smoothers, *in Proceedings of the Heidelberg Workshop*, T. Gasser and M. Rosenblatt (eds.), Lecture Notes in Mathematics, Springer, Berlin, 191–195.

Thompson, J. and Tapia, R. (1990). *Nonparametric Function Estimation, Modelling and Simulation*, SIAM, Philadelphia.

Venter, J. (1967). On estimation of the mode, Ann. Math. Stat., 38, 1446–1455.

Wang, X. and Zidek, J.V. (2005). Selecting likelihood weights by cross-validation, Ann. Stat., 33(2), 463–500.

Wegman, E. (1971). A note on the estimation of the mode, Ann. Math. Stat., 42, 1909–1915.

Chapter 33
Mixture Models and Nonparametric Deconvolution

Mixture models are popular in statistical practice. They can arise naturally as representations of how the data are generated. They can also be viewed as mathematical artifacts, justified by elegant theoretical results that say that mixtures can well approximate essentially any density in a Euclidean space. Application areas where mixtures are extensively used include reliability, survival analysis, population studies, genetics, and astronomy, among others. The mixtures can be finite mixtures (e.g., a mixture of two normal distributions) or infinite mixtures (e.g., a variance mixture of normals). In spite of their inherent appeal as statistical models, mixtures are also notoriously difficult models from an inference perspective. Mixture problems are riddled with difficulties such as nonidentifiability, unbounded or hard-to-compute likelihoods, and slow convergence rates in infinite mixture models. Mixture models thus provide a very intriguing mix of appeal and difficulty to a statistician. There is a fantastically huge body of literature on almost all aspects of mixture models, namely methods, theory, and computing. We recommend Lindsay (1983a,b, 1995), Teicher (1961), Redner (1981), Pfanzagl (1988), Roeder (1992), Cutler and Cordero-Brana (1996), and Hall and Zhou (2003) for general theory and consistency questions, Carroll and Hall (1988), Fan (1991), and Chen (1995) for nonparametric convergence rates, and Titterington (1985) and McLachlan and Peel (2000) for book-length treatments. Computational and practical issues are discussed in Dempster, Laird, and Rubin (1977), Redner, and Walker (1984), Kabir (1968), Bowman and Shenton (1973), Hosmer (1973), Peters and Walker (1978), and Roeder and Wasserman (1997), among many others. Certain nonstandard asymptotics about the likelihood ratio tests for homogeneity of a mixture distribution are given in Hartigan (1985), Ghosal and Sen (1985), and Hall and Stewart (2005). Approximation-theoretic properties are available in Cheney and Light (1999).

A. DasGupta, *Asymptotic Theory of Statistics and Probability*,

33.1 Mixtures as Dense Families

We start with a general mathematical motivation for using mixtures as probability models in any finite-dimensional Euclidean space. The available results are very strong; for example, mean-variance mixtures of normal densities are dense in the class of *all* densities in a suitable sense. In fact, there is nothing special about taking mixtures of normals; mixtures of almost any type of density have the same denseness properties.

Theorem 33.1 Let f be a density function in $\mathcal{R}^d$, $1 \le d < \infty$. Let g be any fixed density in $\mathcal{R}^d$. Let $\mathcal{F}_g$ be the class of location-scale mixtures of g,

$$\mathcal{F}_g = \left\{ h : h(x) = \int_{[0,\infty)} \int_{\mathcal{R}^d} \frac{1}{\sigma^d} g\left(\frac{x-\mu}{\sigma}\right) p(\mu)q(\sigma)d\mu d\sigma \right\},$$

where p, q are densities on $\mathcal{R}^d$ and $\mathcal{R}^+$. Then, given $\varepsilon > 0$, there exists $h \in \mathcal{F}_g$ such that $d_{\text{TV}}(f, h) < \varepsilon$, where d_{TV} denotes total variation distance.

Remark. This result says that location-scale mixtures of *anything* provide strong approximations to any density in a finite-dimensional Euclidean space. The proof is not hard; it only needs the use of standard facts, such as Scheffe's theorem and the dominated convergence theorem. Cheney and Light (1999) give a clean proof. We can think of this result as providing a mathematical justification for considering mixture models, statistical motivations aside. The intuition for the result is that if we write $X = \mu + \sigma Z$, then, for small σ, the distribution of X should be close to the distribution of μ, whatever the distribution of Z. That is why in Theorem 33.1 g can be taken to be any density.

Example 33.1 Suppose we take g to be the $U[-1, 1]$ density and μ to be degenerate at zero (thus, strictly speaking, we are not assigning μ a density p), and suppose σ has the density $q(\sigma) = \sigma e^{-\sigma}$. Then, the mixture density works out to the double exponential density $h(x) = \frac{1}{2}e^{-|x|}$. In fact, any symmetric unimodal density can be generated as such a mixture, with g being the $U[-1, 1]$ density.

If we were to take a fixed small σ, then the approximating density h in the notation above would simply be the convolution $p * g_\sigma$, where $g_\sigma(x) = \frac{1}{\sigma^d} g(\frac{x}{\sigma})$. In turn, the convolution should be approximately equal to a Riemann sum, which would be just a linear combination of translates of g_σ. This would suggest that *finite mixtures* should provide accurate approximations to a given density f in any finite-dimensional Euclidean space. The following result holds; once again, see Cheney and Light (1999) for a proof.

Theorem 33.2 Let $1 \leq d < \infty$ and f be a density on $\mathcal{R}^d$. Assume that f is continuous. If g is any density on $\mathcal{R}^d$ and is also continuous, then, given $\varepsilon > 0$ and a compact set $K \subset \mathcal{R}^d$, there exists a finite mixture of the form $\hat{f}(x) = \sum_{i=1}^{m} c_i \frac{1}{\sigma_i^d} g(\frac{x-\mu_i}{\sigma_i})$ such that $\sup_{x \in K} |f(x) - \hat{f}(x)| < \varepsilon$.

Remark. The assumption that g is continuous is not necessary. But continuity of g would ensure that the approximating finite mixture is a continuous function, which may be considered an aesthetic necessity. Note that the theorem says that any smooth density can be well approximated, except possibly at the extreme tails, by finite mixtures. However, the theorem does not say anything about how many components are needed in the mixture. Still, the result does provide a solid mathematical reason for considering finite mixtures of *one fixed type* of density as a model in finite-dimensional problems.

33.2 z Distributions and Other Gaussian Mixtures as Useful Models

Although the theorems in the previous section make it clear that location-scale mixtures of almost any type of density can well approximate an arbitrary density in Euclidean spaces, Gaussian mixtures have historically played a special role. We present a discussion of normal variance mixtures and certain specific normal mean-variance mixtures below. The mean-variance mixtures are the so-called *z distributions* and *generalized hyperbolic distributions*. Principal references for the material below are Fisher (1921), Prentice (1975), Barndorff-Nielsen (1978), and Barndorff-Nielsen, Kent and Sorensen (1982).

First we consider normal variance mixtures. Here is a formal definition.

Definition 33.1 Suppose, conditional on $u \geq 0$, $X \sim N(\mu, u\Delta)$, where Δ is a fixed p.d. matrix of order $d \times d$ with determinant 1, μ is a fixed vector in d dimensions, and $u \sim F$, a CDF on $[0, \infty)$. Then the unconditional distribution of X is called a variance mixture of normals.

Normal variance mixtures are popular and useful models to describe symmetric data that are heavier tailed than any normal. Familiar symmetric distributions such as t (and thus, Cauchy distributions) and the double exponential and the logistic in the case of one dimension are normal variance mixtures. So are all symmetric stable laws; see Feller (1966). We give a result that characterizes a normal variance mixture; this is essentially a restatement of the theorem in Feller (1966, Vol. II) on completely monotone densities.

Theorem 33.3 (Characterization of a Variance Mixture) Let $h(x) = g((x-\mu)'\Delta^{-1}(x-\mu))$ be a density function in $\mathcal{R}^d$.

h is a variance mixture of normals if and only if $g(y) : \mathcal{R}^+ \to \mathcal{R}^+$ is a completely monotone function; i.e., $\forall k \geq 1, (-1)^k g^{(k)}(y) \geq 0$ for all y.

Example 33.2 Let $d = 1$, and suppose $g(y) = \frac{1}{\pi(1+y)}$, which gives $h(x)$ as the standard Cauchy density. By inspection, $\int \sqrt{v} e^{-\frac{vx^2}{2}} \frac{e^{-\frac{v}{2}}}{\sqrt{v}} dv = \text{constant} \times \frac{1}{1+x^2}$, and thus the standard Cauchy density is a normal variance mixture with the mixing distribution for the reciprocal of the variance having a density $p(v) = \text{constant} \times \frac{e^{-\frac{v}{2}}}{\sqrt{v}}$, which is a $\chi^2(1)$ density.

Example 33.3 Again, let $d = 1$, and suppose $h(x) = \frac{1}{2}e^{-|x|}$, the standard double exponential density. This is also a variance mixture of normals, and the mixing distribution for the variance has the exponential density $f(u) = \frac{1}{2}e^{-u/2}$. This can be seen directly from the integration formula $\int \frac{e^{-x^2/(2u)-u/2}}{\sqrt{u}} du = \sqrt{2\pi} e^{-|x|}$ or from the values of the even moments of a standard double exponential. The $(2k)$th moment of a standard double exponential equals $(2k)!$, while the $(2k)$th moment of a standard normal equals $\frac{(2k)!}{2^k k!}$. The ratio of the two moments of order $(2k)$ is therefore $2^k k!$, which is the kth moment of an exponential with mean 2. Since the sequence of even moments determines the distribution (because $\sum_{n=1}^{\infty} \frac{1}{2n} = \infty$, a generalization of a criterion in the theory of moments known as Muntz's theorem), it follows that the mixing density for the variance is an exponential with mean 2.

Example 33.4 Let $d = 1$, and suppose $X|u \sim N(0, \frac{1}{2}\kappa^2 u)$. If $u \sim$ Beta$(a, 1)$, then the unconditional density of X is $\frac{a}{\kappa\sqrt{\pi}}(\frac{|x|}{\kappa})^{2a-1}\Gamma(\frac{1}{2} - a, \frac{x^2}{\kappa^2})$, which is the *type I modulated normal density*. In the above, $\Gamma(.,.)$ denotes the incomplete Gamma function.

We next discuss certain special mean-variance mixtures of normals. Prominent among them are the z-distributions with density $\frac{1}{\tau B(\alpha,\beta)} \frac{e^{\alpha(\frac{x-c}{\tau})}}{(1+e^{\frac{x-c}{\tau}})^{\alpha+\beta}}$, $\alpha, \beta, \tau > 0, x, c \in \mathcal{R}$. Note that the logistic density is a special case with $\alpha = \beta = 1$. Fisher (1921) introduced z-distributions into the statistical literature. A formal definition of mean-variance mixtures of normals is given below. Note that the definition is an unusual one.

Definition 33.2 Suppose $X|u \sim N(\mu + u\beta, u\Delta)$, where $u \geq 0, \mu, \beta$ are d-dimensional vectors and Δ is a $d \times d$ p.d. matrix of determinant 1, and suppose $u \sim F$, a CDF on $[0, \infty)$. Then the unconditional distribution of X is called a mean-variance mixture of normals.

Example 33.5 Suppose X has the hyperbolic cosine density $\frac{1}{\sqrt{2\pi}} \frac{1}{\cosh(\sqrt{\pi/2}x)}$. Then, on manipulation, this is seen to be a z-distribution with $\alpha = \beta = \frac{1}{2}, c = 0, \tau = \frac{1}{\sqrt{2\pi}}$.

Suppose F has the central F-distribution with m, n degrees of freedom. Then $\log X$ has a z-distribution with $\alpha = \frac{m}{2}, \beta = \frac{n}{2}, c = \log n - \log m$, $\tau = 1$.

In general, all z-distributions are mean-variance mixtures of normals, as the following result says; see Barndorff-Nielsen, Kent and Sorensen (1982) for more details and a proof.

Theorem 33.4 (Mixing Distribution for z-distributions) Every z-distribution is a mean-variance mixture of normals, and the mixing CDF F has the mgf

$$\psi(s) = e^{cs} \prod_{k=0}^{\infty} \left[1 - \frac{s\tau}{(\delta + k)^2/2 - \gamma} \right]^{-1},$$

where $\delta = (\alpha + \beta)/2, \gamma = (\alpha - \beta)^2/8$.

Remark. The mixing distribution for the z-distributions is absolutely continuous. But no simple formula for the mixing density is possible. The mixing density is an infinite mixture of exponentials; see Barndorff-Nielsen, Kent and Sorensen (1982) for further details. It can also be shown, by considering of the mixing distribution, that all z-distributions are infinitely divisible. They provide a rich variety of distributions with some attractive properties.

Next, we describe the generalized hyperbolic distributions, which are also mean-variance mixtures of normals, with the mixing density being a generalized inverse Gaussian. The formal definition is as follows.

Definition 33.3 Let F be a CDF on $[0, \infty)$ with the generalized inverse Gaussian density $f(u) = \frac{(\kappa/\delta)^\lambda}{2K_\lambda(\kappa\delta)} u^{\lambda-1} e^{-\frac{1}{2}(\kappa^2 u + \frac{\delta^2}{u})}$, where K_λ is the Bessel K function, and

(a) $\kappa > 0, \delta \geq 0$, if $\lambda > 0$,

(b) $\kappa > 0, \delta > 0$, if $\lambda = 0$, and

(c) $\kappa \geq 0, \delta > 0$, if $\lambda < 0$.

Let X be a mean-variance mixture of normals with a generalized inverse Gaussian F as the mixing distribution. Then the distribution of X is called a generalized hyperbolic distribution and has the density

$$h(x) = \frac{(\kappa/\delta)^{\lambda}}{(2\pi)^{d/2} K_{\lambda}(\kappa\delta)} \frac{K_{\lambda-d/2}(\alpha[\delta^2 + (x-\mu)'\Delta^{-1}(x-\mu)]^{\frac{1}{2}})e^{\beta'(x-\mu)}}{([\delta^2 + (x-\mu)'\Delta^{-1}(x-\mu)]^{\frac{1}{2}}/\alpha)^{d/2-\lambda}},$$

where $\alpha = \kappa^2 + \beta'\Delta\beta$.

Example 33.6 Suppose $\kappa = 0, \lambda < 0, \beta = 0$. Then, formally, $\alpha = 0$. However, on considering the power series expansion for the Bessel K function, the formula for $h(x)$ reduces to the density of a d-dimensional t-distribution. That is, a t-density is obtained as a limit by letting $\kappa \downarrow 0$.

Example 33.7 Suppose $d = 1, \delta = 0, \lambda = 1$, and $\beta = 0$. Then the expression for h simplifies to the double exponential density.

The reason for calling these distributions generalized hyperbolic is that the logarithm of the density function plots to a hyperboloid. Moreover, if $\lambda = \frac{d+1}{2}$, the d-dimensional hyperbolic density is exactly obtained; that is, the hyperbolic densities are special cases. Like the z-distributions, the generalized hyperbolic distributions are also all infinitely divisible. Examples 33.6 and 33.7 show that a wide variety of tails is contained in the class, and in fact the normal densities themselves are also obtained as limits.

For purposes of practical modeling, it is worth noting that the tail of a mean-variance mixture of normals will typically be polynomial-like or Gamma-like and thus heavier than normal. We end this section with a more precise description of the tail behavior below. The key point to note is that the tail of the mixture is tied up with the tail of the mixing distribution. Here is a Tauberian theorem from Barndorff-Nielsen, Kent and Sorensen (1982).

Theorem 33.5 (**Tail of a Mean-Variance Mixture of Normals**) Suppose $d = 1$, and $\mu = \beta = 0$. Let $\psi(s)$ denote the mgf of the mixing distribution F, and suppose F has a density f. Let $\psi_+ = \sup\{s \in \mathcal{R} : \psi(s) < \infty\} \geq 0$. Assume that f satisfies the tail property $f(u) \sim e^{-\psi_+ u} u^{\lambda-1} L(u), u \to \infty$, for some λ and some slowly varying function $L(u)$.

(a) If $\psi_+ = 0$, then the density h of the mean-variance mixture satisfies $h(x) \sim |x|^{2\lambda-1} L(x^2), |x| \to \infty$.

(b) If $\psi_+ > 0$, h satisfies $h(x) \sim e^{-\sqrt{2\psi_+}|x|} |x|^{\lambda-1} L(|x|), |x| \to \infty$.

33.3 Estimation Methods and Their Properties: Finite Mixtures

A variety of methods for estimating the component densities and the mixture proportions in finite mixtures and estimating the mixing distribution in a general infinite mixture have been proposed in the literature. It should be emphasized that the mixture problem is inherently a difficult one, and there is no single method that outperforms the others. Even more, computing the estimates is never easy. We treat here principally the maximum likelihood and minimum distance estimates, with only a brief discussion of moment estimates. Only a small selection of a huge body of results is presented.

33.3.1 Maximum Likelihood

Maximum likelihood, of course, is always a candidate. But, in the mixture problem, the term is misleading. An overall finite maximum of the likelihood function does not exist in the finite mixture case. So, traditionally, maximum likelihood meants a root of the likelihood equation. The root used is only a local maximum. No optimality properties of an arbitrary local maximum of the likelihood can be guaranteed. Some positive properties can be guaranteed for specific local maxima (e.g., the largest local maximum). In any case, computing either a root of the likelihood equation or the largest local maximum involves nontrivial computing, such as iterative root finding or the EM algorithm. Because of the special status of finite parametric mixtures, we treat that case first. We consider the statistically most common case of all unlabeled observations (i.e., no samples from the component densities are available). The results in the next theorem are available, for example, in Peters and Walker (1978) and Redner, and Walker (1984).

Theorem 33.6 (Likelihood Roots) Let $X_1, X_2, \cdots \overset{\text{iid}}{\sim} f(x) = f(x|\xi) = \sum_{i=1}^m p_i f_i(x|\theta_i)$, where $\xi = (p_1, \cdots, p_m, \theta_1, \cdots, \theta_m)$. Let

$$\Omega = \{\xi : p_i \geq 0, \sum p_i = 1, \theta_i \in \Omega_i \subset \mathcal{R}^{d_i}\}.$$

Let the components of ξ be denoted as $\xi_1, \cdots, \xi_\nu$, where $\nu = m - 1 + \sum d_i$, and let ξ_0 denote the true value of ξ. Assume that:

(C1) There exist functions g_i, g_{ij}, g_{ijk} such that $|\frac{\partial}{\partial \xi_i} f(x|\xi)| \leq g_i(x)$, $|\frac{\partial^2}{\partial \xi_i \partial \xi_j} \times f(x|\xi)| \leq g_{ij}(x)$, $|\frac{\partial^3}{\partial \xi_i \partial \xi_j \partial \xi_k} \log f(x|\xi)| \leq g_{ijk}(x)$, with $\int g_i(x)$, $\int g_{ij}(x)$, $\int g_{ijk}(x) f(x|\xi_0) < \infty$.

(C2) The Fisher information matrix $I(\xi)$ exists in a neighborhood of ξ_0 and is positive definite at ξ_0.

Then, given any neighborhood $\mathcal{U}$ of ξ_0, for all large n with probability 1 under ξ_0, there is a unique root $\hat{\xi}_n$ of the likelihood equation in $\mathcal{U}$ and $\sqrt{n}(\hat{\xi}_n - \xi_0) \stackrel{\mathcal{L}}{\Rightarrow} N_\nu(0, I^{-1}(\xi_0))$.

Remark. The result says that in finite mixture problems with smooth parametric component densities and a *known* number of components, there is a consistent, asymptotically efficient, and asymptotically normal sequence of roots of the likelihood equation. Because the problem is still parametric, we get the $\sqrt{n}$ rate. However, it does not say anything about the choice of the root among the possibly multiple roots of the likelihood equation. Simplification of the likelihood equations when the component densities are normal is provided in Behboodian (1970). Solution by EM methods is discussed in Dempster, Laird, and Rubin (1977) and Redner, and Walker (1984).

33.3.2 Minimum Distance Method

Next, we consider minimum distance methods. Minimum distance methods minimize the distance of an omnibus distribution estimate from a given parametric family. The distance can be chosen in many ways. Thus, minimum distance estimates are not uniquely defined. Basu, Harris, and Basu (1997) is a nice overview. The material below is taken from Choi and Bulgren (1967), Beran (1977), Hall and Titterington (1984), Tamura and Boos (1986), Chen and Kalbfleisch (1996), and Cutler and Cordero-Brana (1996).

Thus, let $X_1, \cdots, X_n$ be iid observations from some density g on $\mathcal{R}^d$ and let $\mathcal{F}_\theta = \{f_\theta : \theta \in \Theta \subset \mathcal{R}^\nu\}, d, \nu < \infty$, be a parametric family of densities on $\mathcal{R}^d$. Let $d(f, g)$ denote a distance on densities and $\hat{g}_n$ a suitable (sequence of) estimate of g. Then the minimum distance estimate of θ with respect to the distance d is defined as $T(\hat{g}_n) = \operatorname{arginf}_{\theta\in\Theta} d(f_\theta, \hat{g}_n)$. Note that $T(\hat{g}_n)$ need not exist, and if it does, it may be multivalued.

We consider the Hellinger distance in particular. Recall that it is defined as $H(f, g) = (\int(\sqrt{f} - \sqrt{g})^2)^{\frac{1}{2}}$. Beran (1977) explains that minimum distance estimates corresponding to the Hellinger distance are *similar* to maximum likelihood estimates. Thus, it is a distinguished distance. First, we note a result (see Beran (1977)) on existence of a minimum Hellinger distance estimate.

Theorem 33.7 (Minimum Hellinger Distance Estimates) Suppose the mixture density $f(x|\xi)$ is of the form $f(x|\xi) = \sum_{i=1}^m p_i \frac{1}{\sigma_i^d} f_i(\frac{x-\mu_i}{\sigma_i})$, where

m is considered fixed. Assume that the component densities $\frac{1}{\sigma_i^d} f_i(\frac{x-\mu_i}{\sigma_i})$ are distinct (as functions), that $f(x|\xi)$ is identifiable subject to a permutation of $(p_i, \mu_i, \sigma_i, f_i)$, and that each component density $\frac{1}{\sigma_i^d} f_i(\frac{x-\mu_i}{\sigma_i})$ is continuous in (μ_i, σ_i) for almost all values of x with respect to the Lebesgue measure. Then a minimum density estimate (MDE) with respect to the Hellinger distance exists if $\hat{g}_n$ is a probability density function.

Remark. The identifiability and continuity assumptions hold if the component densities are $N_d(\mu_i, \sigma_i^2 I_d)$. The continuity assumption will usually hold; see Teicher (1961) for general results on identifiability of the mixture and distinctness of the components.

We now give a result on consistent estimation by using the minimum Hellinger distance estimate. Results are available on the asymptotic distribution of the minimum Hellinger distance estimate. However, the conditions needed are quite technical; see Tamura and Boos (1986) for the exact conditions.

Theorem 33.8 Suppose $X_1, X_2, \cdots$ are iid observations from the mixture model in the previous theorem. Let ξ_0 denote the true value. Assume that the conditions of the previous theorem hold. Also assume, in addition, that:

(a) Each f_i, $1 \leq i \leq m$, is uniformly continuous.
(b) $\hat{g}_n$ is a kernel density estimate, with the kernel K being of compact support, and absolutely continuous with a bounded first derivative.
(c) The bandwidth $h = h_n$ is such that $h \to 0$, $\sqrt{n}h \to \infty$, as $n \to \infty$.

Then the minimum Hellinger distance estimate $\hat{\xi}_n \stackrel{P}{\Rightarrow} \xi_0$.

Remark. Note that the mixture model is supposed to be the *true* model in Theorems 33.7 and 33.8. If the mixture model is only an approximation to a correct model that is not in the postulated mixture family, then the results are not valid.

33.3.3 Moment Estimates

Moment estimates have also been considered for the mixture problem. The positive aspect of moment estimates is that, given the choice of functions, the solutions of the moment equations are less intensive than for some other methods. Also, to a certain extent, the estimates can be given an analytical form. On the negative side, the derivations are entirely case by case, and more than three components in the mixture or higher dimensions are hard to handle. Also, if the problem is not univariate, then moment estimates perform relatively poorly in comparison with likelihood roots and minimum distance estimates. The first work on moment estimates of all five parameters (two means, two standard deviations, and a proportion) of a mixture of two univariate normals was Pearson (1894). Pearson equated the first five moments of the mixture density to the corresponding moments of the unlabeled sample. Impressive algebraic manipulation enabled him to reduce the primary calculation to solving for a root of a nonic polynomial. The estimates of the five parameters are then found in terms of this root and the sample moments. The exact equations were later simplified in Cohen (1967). A problem with Pearson's estimates is that the nonic equation can (and often does) have more than one root. So, as in the case of likelihood-based estimates, a choice issue arises. Pearson gave some suggestions on which root one should use.

Lindsay and Basak (1993) provide moment estimates for multivariate normal mixtures and give useful computing tips. Mixtures of nonnormal distributions and parameter estimates for them have been considered in Blischke (1962, binomial mixtures), Hasselblad (1966, general exponential family), Rider (1962, Poisson mixtures), and John (1970, mixtures of Gamma), among others. Generally, likelihood roots are the common choice in these papers, with some consideration of moment estimates.

33.4 Estimation in General Mixtures

General mixtures are of the form $f(x) = f_Q(x) = \int_\Omega f_\theta(x)dQ(\theta)$, where Q is a probability measure on the parameter space Ω of the component densities f_θ. The problem now is to estimate Q based on samples from the mixture density f_Q. We focus here on the maximum likelihood estimate of Q. The mixing distribution Q being arbitrary, we refer to the MLE as the NPMLE (nonparametric maximum likelihood estimate).

Definition 33.4 Let $X_1, \cdots, X_n \stackrel{\text{iid}}{\sim} f_Q(x)$. Any $\hat{Q}_n$ that maximizes the nonparametric likelihood function $l(Q) = \prod_{i=1}^n f_Q(x_i)$ is called an NPMLE of Q. The important questions are when an NPMLE exists, when it is unique, and how to find it. Methods of convex geometry and Choquet-type representations of elements of the closed convex hull of $\Gamma = \{f_\theta : \theta \in \Omega\}$ are used to answer these questions. To simplify the topological issues, we assume below that Ω is a subset of a finite-dimensional Euclidean space. We follow the results in Lindsay (1983a,b); earlier literature on similar questions in special setups includes Simar (1976) and Laird (1978).

We first introduce some notation. Let $x_1, \cdots, x_n$ denote an iid sample of size n from f_Q. Let m denote the number of distinct values in the set $\{x_1, \cdots, x_n\}$, and let the distinct values be denoted as $y_1 < \cdots < y_m$. Suppose $n_j = \#\{i : x_i = y_j\}, 1 \leq j \leq m$. A characterization of an NPMLE of Q is conveniently described through the operator $D(\theta, Q) = \sum_{j=1}^{m} n_j (\frac{f_\theta(y_j)}{f_Q(y_j)} - 1)$. The idea behind characterizing an NPMLE is that the directional derivative of the loglikelihood at an NPMLE along any direction should be ≤ 0 and should be zero along a direction f_θ if the particular θ is in the support of this NPMLE. For discussions of uniqueness of an NPMLE, it is useful to define the polynomials $p_k(x) = \prod_{i=1}^{k}(x - y_i), 1 \leq k \leq m$. Here is a collection of results from Lindsay (1983a,b).

Theorem 33.9 (a) $\forall Q, \sup_\theta D(\theta, Q) \leq 0$.

(b) $\sup_\theta D(\theta, \hat{Q}_n) = 0$.

(c) At any point θ in the support of $\hat{Q}_n$, $D(\theta, \hat{Q}_n) = 0$.

(d) The cardinality of the support of $\hat{Q}_n$ is at most m.

(e) In the special case where each f_θ is a member of the one-parameter exponential family, $f_\theta(x) = e^{\theta x - \psi(\theta)} d\mu(x), \theta \in \Omega$, support$(\mu) = \mathcal{X} \subseteq \mathcal{R}$, $\hat{Q}_n$ is unique if the μ-measure of every cofinite subset of $\mathcal{R}$ is strictly positive and if $\mathcal{X} \cap \partial\Omega = \varphi$ (the empty set).

Remark. The most interesting aspect of Theorem 33.9 is that an NPMLE of f_Q will thus turn out to be a finite mixture. Actual computation of the NPMLE is not trivial; Lindsay (1983a) provides an explicit algorithm and cites other algorithms found in the optimal design literature. As regards the uniqueness part for mixtures of an exponential family density, if μ is Lebesgue, then the assumption on measures of cofinite sets is obviously true. The assumption on the nonoverlap between $\mathcal{X}$ and the boundary of Ω, which would consist of at most two points, the natural parameter space Ω being necessarily an interval, also holds, if necessary by removing the (at most) two points from $\mathcal{X}$. As regards the discrete cases, the boundary of the natural parameter space in the Poisson case is $\{\pm\infty\}$, and so the Poisson case is covered by this theorem.

The situation is not so simple in some other discrete cases. For example, in the geometric case, the support set $\mathcal{X}$ includes the value 0, which is also a boundary point of the natural parameter space in the geometric case.

33.5 Strong Consistency and Weak Convergence of the MLE

The asymptotics of the NPMLE are much harder than in the parametric case. Since identifiability itself can be a problem for mixture models, clearly even consistency can be an issue because the problem is ill posed without identifiability. Broadly speaking, subject to identifiability, the NPMLE in the mixture model would be consistent, although additional technical conditions are needed. Kiefer and Wolfowitz (1956) gave sufficient conditions in addition to identifiability that guarantee consistency of the NPMLE in the mixture problem. We will limit ourselves to the case of exponential family mixtures, treated in Pfanzagl (1988). Beyond consistency, results are known in some special cases. For example, for mixtures of exponential densities, Jewell (1982) gives results on weak convergence of the NPMLE to the true mixing distribution. It is also interesting to consider asymptotic behavior of the NPMLE of f_Q itself. Actually, asymptotics of the NPMLE of f_Q are of more direct inferential relevance than the asymptotics of the NPMLE of Q. We address both topics below. Establishing convergence rates of the NPMLE has proved to be extremely hard, and results are sporadic. Carroll and Hall (1988), Fan (1991), Chen (1995), and van de Geer (1996) give results on the mixture and deconvolution problems, and Ghosal and van der Vaart (2001) consider certain special types of normal mixtures.

We define strong consistency below.

Definition 33.5 Suppose the mixture f_Q is identifiable (i.e., if f_{Q_1}, f_{Q_2} result in the same probability measure). Then $Q_1 = Q_2$ as probability measures. An estimator sequence $\hat{Q}_n$ is called strongly consistent if for almost all sequences $X_1, X_2, \cdots$ with respect to the true f_Q, $\hat{Q}_n \stackrel{\mathcal{L}}{\Rightarrow} Q$.

Theorem 33.10 (Consistency of the NPMLE) Suppose each f_θ is a member of the one-parameter exponential family and, in the notation of the previous theorem, the support $\mathcal{X}$ of the dominating measure μ has a nonempty interior or has at least an accumulation point in the interior of the closed convex hull of $\mathcal{X}$. Then f_Q is identifiable and a unique NPMLE $\hat{Q}_n$ exists with probability 1 that is strongly consistent.

Remark. This result was first observed in Kiefer and Wolfowitz (1956). Once again, as was the case in deciding uniqueness, we see that consistency is assured if each f_θ is one of the common exponential family densities, such as normal or Gamma. Theorem 33.10 also ensures strong consistency for truncated versions of such densities as long as the truncation contains a nonempty interval. For cases such as Poisson mixtures, this theorem does not apply. So, we state the Poisson mixture case separately, it being an important case in applications; the Poisson case was treated in Simar (1976).

Theorem 33.11 (Poisson Mixtures) Let $f_\theta(x) = \frac{e^{-\theta}\theta^x}{x!}, x = 0, 1, \cdots$. Then f_Q is identifiable and a unique NPMLE $\hat{Q}_n$ of Q exists with probability 1, that is strongly consistent.

We now give some examples of distributions that are mixtures of some exponential family density. Here we are using the word density loosely, and we will give some discrete examples also.

Example 33.8 A famous theorem in analysis says that a PDF on the real line f is a mixture of exponential densities; i.e., $f(x) = \int_0^\infty \frac{1}{\lambda} e^{-\frac{x}{\lambda}} dQ(\lambda)$ iff f is completely monotone on $(0, \infty)$. That is, $(-1)^k f^{(k)}(x) \geq 0\, \forall k \geq 1, \forall x > 0$; for example, see Cheney and Light (1999). One example of such a density is the Weibull density with shape parameter $\gamma < 1$; i.e., $f(x) = \gamma x^{\gamma-1} \frac{e^{-x^\gamma/\lambda}}{\lambda}, x, \lambda > 0, 0 < \gamma < 1$. Another example is $f(x) = c(a, b, \lambda) \frac{e^{-x/\lambda}}{(x+a\lambda)^b}, x, \lambda, a, b > 0$, where $c(a, b, \lambda)$ is the normalization constant (and actually can be written explicitly). A limiting case worth mentioning separately is the Pareto density $f(x) = \frac{c}{(1+x)^{(c+1)}}, x, c > 0$. This is also a mixture of exponential densities.

Example 33.9 Unlike the case of exponential mixtures, there is no simple characterization of Poisson mixtures in terms of the pmf itself. Characterizations in terms of the characteristic function or the probability generating function are available; see Johnson and Kotz (1969). Poisson mixtures are quite popular models in actuarial sciences and queueing theory. In many applications, the mixing distribution has been chosen to be a uniform or a Beta on some finite interval, a Gamma or an inverse Gamma, and an inverse Gaussian. Of these, a clean and familiar result obtains with the Gamma mixing density; i.e., if Q has the Gamma density $\frac{e^{-\lambda/\beta}\lambda^{\alpha-1}}{\beta^\alpha \Gamma(\alpha)}$, then the mixture pmf works out to $P(X = x) = \frac{\Gamma(\alpha+x)}{\Gamma(\alpha)x!}(1 + \beta)^{-(\alpha+x)}\beta^x, x = 0, 1, 2, \cdots$. This is a negative binomial distribution with parameters α and $\frac{\beta}{1+\beta}$.

A practically interesting question is how the estimated mixture density itself performs if we use the estimate $\hat{f}_Q(x) = \int f(x|\theta) d\hat{Q}_n(\theta)$. We will refer to it as the MLE of f_Q. Certainly, consistency is the first question. But even more important practically is to form asymptotic estimates of standard errors and to obtain results on the asymptotic distribution, so that approximate confidence intervals can be computed. We present results on all of these questions in the Poisson mixture case below. These are available in Lambert and Tierney (1984).

Theorem 33.12 (Asymptotic Normality) Let $X_1, X_2, \cdots$ be iid observations from the Poisson mixture $f_Q(x) = \int \frac{e^{-\theta}\theta^x}{x!} dQ(\theta)$. Let $\hat{f}_n(x) = \int \frac{e^{-\theta}\theta^x}{x!} d\hat{Q}_n(\theta)$ and $\hat{p}_n(x) = \frac{\#\{j: X_j = x, 1 \leq j \leq n\}}{n}$. Suppose the true Q is continuously differentiable and $\geq c > 0$ in some right neighborhood of $\theta = 0$. Then the finite-dimensional limiting distributions of $\sqrt{n}(\hat{f}_n - f_Q)$ and $\sqrt{n}(\hat{p}_n - f_Q)$ are the same, and hence the finite-dimensional limiting distributions of $\sqrt{n}(\hat{f}_n - f_Q)$ are multivariate normal.

Remark. Theorem 33.12 says that estimates of the class probabilities constructed by using the nonparametric mixture model in large samples are close in their numerical values and in their sampling distributions to the fully nonparametric estimates based on the sample proportions. A heuristic explanation for this asymptotic agreement is that the nonparametric mixture class is sufficiently rich. Asymptotic agreement cannot be expected if the mixing distribution Q had heavy restrictions. Note that each $\hat{f}_n(x)$ is strictly positive, although no more than n of the proportion-based estimates can be strictly positive. Thus, as estimates of asymptotic standard error, it is perhaps preferable to use $\sqrt{\frac{\hat{f}_n(x)(1-\hat{f}_n(x))}{n}}$.

33.6 Convergence Rates for Finite Mixtures and Nonparametric Deconvolution

As we remarked before, establishing convergence rates of the NPMLE has proved to be extremely hard, and results are sporadic. If the mixing distribution Q is nonparametric, then we cannot expect parametric rates of convergence for the MLE. The exact rate would typically depend on fine aspects of the problem, and a single simple answer is no longer possible. However, quite typically, the rate would be logarithmic, or at best a polynomial rate slower than the parametric $n^{-\frac{1}{2}}$ rate. A sample of the available results is presented here. Again, we start with the finite mixture case. We let $\mathcal{G}_j$ denote the set of distributions on a compact set $\Theta \subset \mathcal{R}$, supported on exactly j points. Given two CDFs G_1, G_2 on Θ, define the distance $d(G_1, G_2) = \int_\Theta |G_1(\theta) - G_2(\theta)| d\theta$. The component densities $f(x|\theta)$ in the result below must satisfy some regularity conditions; see Chen (1995) for a listing of these conditions.

Theorem 33.13 (Finite Mixtures) Let $X_1, X_2, \cdots,$ be iid observations from $f_Q(x) = \int_\Theta f(x|\theta) dQ(\theta)$, where $Q \in \cup_{j=1}^m \mathcal{G}_j$ for a fixed known m. Let $\hat{Q}_n$ be any consistent estimate of $Q, Q \in \cup_{j=1}^m \mathcal{G}_j$ and let a_n be a sequence such that $\frac{a_n}{n^{\frac{1}{4}}} \to \infty$. Then $a_n \sup_{\cup_{j=1}^m \mathcal{G}_j} d(\hat{Q}_n, Q) \stackrel{P}{\Rightarrow} \infty$.

Remark. The result says that, even in the finite mixture case, if we know the number of components *only* up to an upper bound, the convergence rate (in the d-distance) of no estimate can be better than $n^{-\frac{1}{4}}$. The reason that a better rate than $n^{-\frac{1}{4}}$ cannot be obtained is that there is a parametric subproblem in which the rate is exactly $n^{-\frac{1}{4}}$. It was originally Charles Stein's idea to get a nonparametric rate of convergence by considering the hardest parametric subproblem embedded in it. It takes clarity of understanding of the nonparametric problem to extract out of it the hardest parametric subproblem. If it can be done, parametric methods, which are simpler in nature, can be used to obtain the nonparametric rate. Interestingly, if we know the number of components is exactly m for some fixed m, the $n^{-\frac{1}{2}}$ rate can be obtained. The $n^{-\frac{1}{4}}$ best possible rate in the formulation above, however, is sharp in the sense that explicit estimates $\hat{Q}_n$ can be constructed that attain this rate; see Chen (1995).

33.6.1 Nonparametric Deconvolution

Next, we consider the deconvolution problem. Formally, an observable random variable X has the convolution form $X = Y + Z$, where Y is the latent unobservable variable of interest and Z is another random variable independent of Y. The distribution of Z is assumed to be known, while the distribution of Y is unknown and we would like to estimate it. However, the estimation has to be done on the basis of samples on the X variable. This model is of practical interest in any situation where a signal cannot be directly observed and is always contaminated by background noise. The noise distribution can be reasonably estimated by physically turning off any potential signals, and so we consider it as known. The question is how well we can infer the distribution of signals or its functionals; for example, quantiles and means. It turns out that the denoising problem, known as deconvolution, is hard if we do not make parametric assumptions about the signal distribution. Moreover, the smoother the noise density is, assuming that the noise distribution is continuous, the harder is the deconvolution. The smoothness of a density corresponds, mathematically, to the tail of its characteristic function and vice versa. The smoother the density, the thinner is the tail of the characteristic function; see, e.g., Champeney (2003). Thus, it seems natural that the convergence rate in the deconvolution problem would depend on how fast the characteristic function of the noise goes to zero. We consider two different types of tails of the noise characteristic function: polynomial tails and exponential-type tails. Formally, we define smooth and supersmooth densities as follows, following terminology in Fan (1991).

Definition 33.6 A density function on $\mathcal{R}$ is called smooth if for $j = 0, 1, 2$, and for some positive constants c_j, d_j, β, the characteristic function $\psi(t)$ of the density satisfies $\liminf_{t\to\infty} |\psi^{(j)}(t)||t^{\beta+j}| = c_j$, $\limsup_{t\to\infty} |\psi^{(j)}(t)||t^{\beta+j}| = d_j$.

A density function on $\mathcal{R}$ is called supersmooth if for some positive constants c, d, α, γ, and finite constants β_1, β_2, $\liminf_{t\to\infty} |\psi(t)||t|^{\beta_1} e^{|t|^\alpha/\gamma} = c$, $\limsup_{t\to\infty} |\psi(t)||t^{\beta_2}|e^{|t|^\alpha/\gamma} = d$.

Note that because of the assumption that $\alpha, \gamma > 0$, the supersmooth densities have a more rapidly decaying characteristic function. Examples of such supersmooth densities are (any) normal, Cauchy, t, logistic, finite mixtures of normals or t, etc., and their finite convolutions. Examples of ordinary smooth densities include Gamma, double exponential, and triangular.

We also need some notation. Let

$$\mathcal{C}_{k,B} = \{f : |f^{(k)}(x)| \leq B < \infty\, \forall x\},$$

where f is a density on $\mathcal{R}$ and k is a fixed positive integer. Principal results on this very interesting problem are due to Carroll and Hall (1988) and Fan (1991), and some of their results are discussed below.

Theorem 33.14 (**Attainable Optimal Convergence Rates**) For the model $X = Y + Z, Y \sim f \in \mathcal{C}_{k,B}, Z \sim h, Y, Z$ independent, let $X_1, X_2, \cdots$ be iid observations from the distribution of X and let $\hat{f}_n(x_0)$ be any estimate of $f(x_0)$ based on $X_1, X_2, \cdots, X_n$, where x_0 is any fixed real number. Assume that h has a nonvanishing characteristic function.

(a) If h is supersmooth, and if an estimate $\hat{f}_n(x_0)$ satisfies $\lim_{n\to\infty} \sup_{f\in\mathcal{C}_{k,B}} P_f(|\hat{f}_n(x_0) - f(x_0)| > a_n) = 0$, for any $0 < B < \infty$, then the sequence a_n must satisfy $\limsup_{n\to\infty} a_n(\log n)^{\frac{k}{\alpha}} > 0$, where the quantity α is as in the definition of a supersmooth density. Furthermore, the rate $(\log n)^{-\frac{k}{\alpha}}$ is attainable.

(b) If h is smooth, and if an estimate $\hat{f}_n(x_0)$ satisfies $\lim_{n\to\infty} \sup_{f\in\mathcal{C}_{k,B}} P_f(|\hat{f}_n(x_0) - f(x_0)| > a_n) = 0$, for any $0 < B < \infty$, then the sequence a_n must satisfy $\limsup_{n\to\infty} a_n n^{\frac{k}{2k+2\beta+1}} > 0$, where the quantity β is as in the definition of a smooth density. Furthermore, the rate $n^{-\frac{k}{2k+2\beta+1}}$ is attainable.

Carroll and Hall (1988) and Fan (1991) provide explicit kernel estimates $\hat{f}_n$ based on kernels with characteristic functions of compact support that attain the best convergence rates for each part in Theorem 33.14. These estimates are in the spirit of the Stefanski-Carroll estimate based on a Fourier inversion of the quotient of the characteristic functions of X and Z; see Stefanski and Carroll (1990). The literature also includes results on estimating CDFs and other functionals, and for some of these problems, the best attainable rates are better.

Example 33.10 Suppose $h(x)$ is the standard normal density. Its characteristic function is $e^{-\frac{t^2}{2}}$, and so, according to our definition, h is supersmooth with $\alpha = 2$. The best convergence rate attainable over $\mathcal{C}_{k,B}$ is therefore $(\log n)^{-\frac{k}{2}}$. On the other hand, if $h(x)$ is the standard Cauchy density, then its characteristic function is $e^{-|t|}$, and so h is supersmooth with $\alpha = 1$. The best convergence rate attainable over $\mathcal{C}_{k,B}$ is therefore $(\log n)^{-k}$.

Example 33.11 Suppose next that $h(x)$ is a t-density with 3 degrees of freedom. Then its characteristic function can be shown to be $\frac{1+\sqrt{3}|t|}{e^{\sqrt{3}|t|}}$, so h is supersmooth with $\alpha = 1$, and hence the best convergence rate attainable over $\mathcal{C}_{k,B}$ is $(\log n)^{-k}$.

Example 33.12 Suppose $h(x)$ is the density of a double exponential density. Its characteristic function is $\frac{1}{1+t^2}$, and the first two derivatives are $-\frac{2t}{(1+t^2)^2}$ and $\frac{6t^2-2}{(1+t^2)^3}$. Thus, h is smooth with $\beta = 2$, and the two derivative conditions on the characteristic function are satisfied. Thus, the best convergence rate attainable over $\mathcal{C}_{k,B}$ is $n^{-\frac{k}{2k+5}}$.

33.7 Exercises

Exercise 33.1 Find the mixture normal density h when the basic density g is $N(0, 1)$, the mixing density p is also $N(0, 1)$, and the mixing density q is such that $\frac{1}{\sigma^2}$ is a chi-square with a general degree of freedom.

Exercise 33.2 Find the mixture density h when the basic density g is $U[-1, 1]$, μ is taken to be degenerate at 0, and the mixing density q is $q(\sigma) = \text{constant} \times \sigma^2 e^{-\sigma^2/2}$.

Exercise 33.3 * Prove that every density on $\mathcal{R}$ that is symmetric and unimodal around zero can be written as a mixture in which the basic density g is $U[-1, 1]$, μ is taken to be degenerate at 0, and σ has a suitable CDF G. That is, every symmetric and unimodal density on the real line is a mixture of symmetric uniforms.

Exercise 33.4 * Find, by a numerical search, a finite mixture of normals that approximates the standard double exponential density to a uniform error of at most .01 in the compact interval $[-4, 4]$.

Exercise 33.5 * Find, by a numerical search, a finite mixture of normals that approximates the standard Cauchy density to a uniform error of at most .01 in the compact interval $[-4, 4]$.

Exercise 33.6 Give examples of half a dozen densities on the real line that *are* normal variance mixtures.

Exercise 33.7 Give examples of half a dozen densities on the real line that *are not* normal variance mixtures.

Exercise 33.8 * Give an example of a density on the real line that is a mean-variance mixture of normals but *not* a normal variance mixture.

Exercise 33.9 * Show that neither the double exponential nor the t-densities can be normal mean mixtures.

Exercise 33.10 * Give an example of a density on the real line that is *both* a normal variance mixture and a normal mean mixture.

Exercise 33.11 Plot the type I modulated normal density for $\kappa = 1, 2, 5$, $a = .5, 1, 2, 5$. Comment on the shapes.

Exercise 33.12 Suppose $X|u \sim N(0, \frac{1}{2}\kappa^2 u)$, $u \sim \frac{a}{u^{a+1}}$, $u \geq 1$. Find the unconditional density of X. This is known as type II modulated normal.

Exercise 33.13 * Write the mgf of the mixing distribution while writing the standard logistic distribution as a z-distribution.

Exercise 33.14 Is the standard logistic distribution infinitely divisible?

Exercise 33.15 Verify in a few selected cases in dimension $d = 1$ that the logarithm of the hyperbolic density plots to a hyperboloid.

Exercise 33.16 Is the standard double exponential distribution infinitely divisible?

Exercise 33.17 * Simulate a sample of size $n = 25$ from the mixture $.9N(0, 1) + .1N(1, 9)$, and try to identify the local maxima of the likelihood function, in which all five relevant parameters are treated as unknown.

Exercise 33.18 * Prove that the conditions assumed for the existence of a minimum Hellinger distance estimate in the finite mixture model are satisfied if the component densities are normal.

Exercise 33.19 * Suppose the component densities in a finite mixture are Cauchy and the kernel K for the construction of the nonparametric density estimate is the Epanechnikov kernel. Show that the conditions assumed for consistency of the mini- mum Hellinger distance estimate are satisfied.

Exercise 33.20 * Find the Poisson mixtures corresponding to a $U[0, a]$, Beta$(a, 1)$, Gamma(α, θ) mixing density, where α, θ are the shape and the scale parameters of Gamma.

Exercise 33.21 * Find the best attainable convergence rate in the nonparametric deconvolution problem if the error Z has the density constant $\times e^{-z^4}$.

Exercise 33.22 * Find the best attainable convergence rate in the nonparametric deconvolution problem if the error Z has the density constant $\times e^{-z^2} z^2, z \geq 0$.

References

Barndorff-Nielsen, O. (1978). Hyperbolic distributions and distributions on hyperbolae, Scand. J. Stat., 5, 151–157.

Barndorff-Nielsen, O., Kent, J., and Sorensen, M. (1982). Normal variance-mean mixtures and z distributions, Int. Stat. Rev., 50, 145–159.

Basu, A., Harris, I., and Basu, S. (1997). Minimum distance estimation: the approach using density based divergences, in *Handbook of Statistics*, Vol. 15, Maddala, G. and Rao, C.R. (eds.), North-Holland, Amsterdam, 21–48.

Behboodian, J. (1970). On a mixture of normal distributions, Biometrika, 57 (1), 215–217.

Beran, R. (1977). Minimum Hellinger distance estimates for parametric models, Ann. Stat., 5(3), 445–463.

Blischke, W. (1962). Moment estimators for parameters of mixtures of two Binomial distributions, Ann. Math. Stat., 33, 444–454.

Bowman, K. and Shenton, L. (1973). Space of solutions for a normal mixture, Biometrika, 60, 629–636.

Carroll, R. and Hall, P. (1988). Optimal rates of convergence for deconvolving a density, J. Am. Stat. Assoc., 83(404), 1184–1186.

Champeney, D. (2003). *Handbook of Fourier Theorems*, Cambridge University Press, Cambridge.
Chen, J. (1995). Optimal rate of convergence for finite mixture models, Ann. Stat., 23(1), 221–233.
Chen, J. and Kalbfleisch, J. (1996). Penalized minimum distance estimates in finite mixture models, Can. J. Stat., 24 (2), 167–175.
Cheney, W. and Light, L. (1999). *A Course in Approximation Theory*, Brooks and Cole, Boston CA.
Choi, K. and Bulgren, W. (1967). An estimation procedure for mixtures of distributions, J.R. Stat. Soc. B, 30, 444–460.
Cohen, A.C. (1967). Estimation in mixtures of two normal distributions, Technometrics, 9(1), 15–28.
Cutler, A. and Cordero-Brana, O. (1996). Minimum Hellinger distance estimation for finite mixture models, J. Am. Stat. Assoc., 91 (436), 1716–1723.
Dempster, A., Laird, N., and Rubin, D. (1977). Maximum likelihood from incomplete data via the EM algorithm, with discussion, J. R. Stat. Soc. B, 39 (1), 1–38.
Fan, J. (1991). On the optimal rates of convergence for nonparametric deconvolution problems, Ann. Stat., 19 (3), 1257–1272.
Feller, W. (1966). An Introduction to Probability Theory and its Applications, Vol. II, Wiley, New York.
Fisher, R.A. (1921). On the probable error of a coefficient of correlation deduced from a small sample, Metron, 1, 3–32.
Ghosal, S. and van der Vaart, A. (2001). Entropies and rates of convergence for maximum likelihood and Bayes estimation for mixtures of normal densities, Ann. Stat., 29 (5), 1233–1263.
Ghosh, J. and Sen, P.K. (1985). On the asymptotic performance of the log-likelihood ratio statistic for the mixture model, in *Proceedings of the Berkeley Conference in Honor of Jerzy Neyman and Jack Kiefer*, L. Le Cam and R.Olshen (eds.), Wadsworth, Belmont, CA, 789–806.
Hall, P. and Stewart, M. (2005). Theoretical analysis of power in a two component normal mixture model, J. Stat. Planning Infer., 134 (1), 158–179.
Hall, P. and Titterington, D. (1984). Efficient nonparametric estimation of mixture proportions, J. R. Stat. Soc. B., 46 (3), 465–473.
Hall, P. and Zhou, X. (2003). Nonparametric estimation of component distributions in a multivariate mixture, Ann. Stat., 31 (1), 201–224.
Hartigan, J. (1985). A failure of likelihood asymptotics for normal mixtures, in *Proceedings of the Berkeley Conference in Honor of Jerzy Neyman and Jack Kiefer*, L. Le Cam and R.Olshen (eds.), Wadsworth, Belmont, CA, 807–810.
Hasselblad, V. (1966). Estimation of parameters for a mixture of normal distributions, Technometrics, 8, 431–444.
Hosmer, D. (1973). On MLE of the parameters of the mixture of two normal distributions when the sample size is small, Commun. Stat., 1, 217–227.
Jewell, N. (1982). Mixtures of exponential distributions, Ann. Stat., 10, 479–484.
John, S. (1970). On identifying the population of origin of each observation in a mixture of observations from two gamma populations, Technometrics, 12, 565–568.
Johnson, N. and Kotz, S. (1969). *Distributions in Statistics: Continuous Univariate Distributions*, Vol. 2, Houghton Mifflin, Boston.
Kabir, A. (1968). Estimation of parameters of a finite mixture of distributions, J. R. Stat. Soc. B, 30, 472–482.

Kiefer, J. and Wolfowitz, J. (1956). Consistency of the maximum likelihood estimator in the presence of infinitely many nuisance parameters, Ann. Math. Stat., 27, 887–906.

Laird, N. (1978). Nonparametric maximum likelihood estimation of a mixing distribution, J. Am. Stat. Assoc., 73, 805–811.

Lambert, D. and Tierney, L. (1984). Asymptotic properties of maximum likelihood estimates in the mixed Poisson model, Ann. Stat., 12(4), 1388–1399.

Lindsay, B. (1983a). The geometry of mixture likelihoods: a general theory, Ann. Stat., 11 (1), 86–94.

Lindsay, B. (1983b). The geometry of mixture likelihoods II: the Exponential family, Ann. Stat., 11(3), 783–792.

Lindsay, B. (1995). *Mixture Models: Theory, Geometry, and Applications*, NSF-CBMS Series in Probability and Statistics, Institute of Mathematical Statistics, Hayward, CA.

Lindsay, B. and Basak, P. (1993). Multivariate normal mixtures: a fast consistent method of moments, J. Am. Stat. Assoc., 88 (422), 468–476.

McLachlan, G. and Peel, D. (2000). *Finite Mixture Models*, John Wiley, New York.

Pearson, K. (1894). Contributions to the mathematical theory of evolution, Philos. Trans. R. Soc. A, 185, 71–110.

Peters, B. and Walker, H. (1978). An iterative procedure for obtaining maximum likelihood estimates of the parameters for a mixture of normal distributions, SIAM J. Appl. Math., 35, 362–378.

Pfanzagl, J. (1988). Consistency of maximum likelihood estimators for certain nonparametric families, in particular mixtures, J. Stat. Planning Infer., 19(2), 137–158.

Prentice, R. (1975). Discrimination among some parametric models, Biometrika, 62, 607–614.

Redner, R. (1981). Consistency of the maximum likelihood estimate for nonidentifiable distributions, Ann. Stat., 9(1), 225–228.

Redner, R. and Walker, H. (1984). Mixture densities, maximum likelihood, and the EM algorithm, SIAM Rev., 26 (2), 195–240.

Rider, P. (1962). Estimating the parameters of mixed Poisson, binomial, and Weibull distributions by the method of moments, Bull. Int. Stat. Inst., 39(2), 225–232.

Roeder, K. (1992). Semiparametric estimation of normal mixture densities, Ann. Stat., 20(2), 929–943.

Roeder, K. and Wasserman, L. (1997). Practical Bayesian density estimation using mixtures of normals, J. Am. Stat. Assoc., 92 (439), 894–902.

Simar, L. (1976). Maximum likelihood estimation of a compound Poisson process, Ann. Stat., 4, 1200–1209.

Stefanski, L. and Carroll, R. (1990). Deconvoluting kernel density estimators, Statistics, 21 (2), 169–184.

Tamura, R. and Boos, D. (1986). Minimum Hellinger distance estimation for multivariate location and covariance, J. Am. Stat. Assoc., 81 (393), 223–229.

Teicher, H. (1961). Identifiability of mixtures, Ann. Math. Stat., 32, 244–248.

Titterington, D., Smith, A., and Makov, U. (1985). *Statistical Analysis of Finite Mixture Distributions*, John Wiley, New York.

van de Geer, S. (1996). Rates of convergence for the maximum likelihood estimator in mixture models, J. Nonparamet. Stat., 6, 293–310.

Chapter 34
High-Dimensional Inference and False Discovery

Advances in technology, engineering, and computing power, as well as real problems in diverse areas such as genomics, cosmology, clinical trials, finance, and climate studies, have recently given rise to new types of inference problems that somehow involve a very large number of unknown parameters. Blind use of traditional asymptotics in these problems leads to unacceptable errors (e.g., wildly inaccurate p-values, stated significance levels, or coverage statements). A popular mathematical paradigm for these problems is to have a sequence of finite-dimensional parameter spaces with dimension p that one assumes to depend on n and lets $p = p_n \to \infty$ as $n \to \infty$, n being the sample size. Sometimes one *knows* from the nature of the underlying physical problem that many or even most of the components of the large-dimensional parameter are small, or even zero, and interest is on the nonzero ones. Collectively, these new families of problems are known as *high-dimensional inference* or *sparse inference* problems. The problems overlap, and so does the literature. One abiding feature of these interesting new problems seems to be that meaningful asymptotics exist only for cleverly chosen configurations and dependence of p on n, and when such meaningful asymptotics exist, they are different from the fixed-dimensional case and typically the results tend to lead to a whole array of new and generally more difficult questions. The entire area is quite new, with a seemingly endless supply of problems with both real contemporary practical relevance and aesthetic mathematical appeal. In this chapter, we give an account of some of the historical development that centers around these high-dimensional problems and also an account of some of the most important and innovative recent results and tools that apply to these problems. It is interesting that we are seeing a revival of parametric models in the most well-known literature on high-dimensional inference and false-discovery control. It may be difficult to treat these problems fully nonparametrically and provide practical procedures with nearly exact theoretical properties.

A. DasGupta, *Asymptotic Theory of Statistics and Probability*,

Examples of the types of scenarios where such high-dimensional problems arise include, among others, goodness-of-fit tests, regression, multiple comparisons, and more generally simultaneous or multiple testing, false discovery, and classification. In goodness-of-fit situations, they arise when the number of cells in a chi-square test is large, so that some cells have small expected frequencies. High-dimensional problems arise in regression when an experimenter has a large collection of explanatory variables, most of which contribute a small power to the regression model, but it is not a priori clear which ones are relatively more important. Similarly, in classification one could have a classification rule that is based on a large number of measurements, each coming with one or more unknown parameters. Multiple testing problems easily get high dimensional even if originally the number of parameters was not very large. For example, in an ANOVA type situation, paired comparisons would call for 45 t-tests if there are ten treatments to begin with. Careless use of off-the shelfe methods has almost no validity in any of these problems.

For historical developments, we recommend Tukey (1953, 1991, 1993), Miller (1966), Shaffer (1995), Bernhard, Klein, and Hommel (2004), andl Hochberg and Tamhane (1987). For regression problems with many parameters, we recommend Portnoy (1984, 1985, 1988). For goodness-of-fit tests with many cells, we recommend Holst (1972) and Morris (1975). For modern sparse inference problems, some key references are Donoho and Jin (2004), Abramovich et al.(2006), Bickel and Levina (2004), and Hall and Jin (2007). The literature on false discovery is huge and growing; a few references are Sóric (1989), Benjamini and Hochberg (1995), Storey (2002, 2003), Storey (2003), Storey, Taylor, and Siegmund (2004), Efron (2007), Genovese and Wasserman (2002, 2004), Genovese (2006), Benjamini and Yekutieli (2001), Finner and Roters (2002), Sarkar (2006), Meinshausen and Rice (2006), and DasGupta and Zhang (2006). Scott and Berger (2006) give Bayesian perspectives on the false-discovery problem; scopes of resampling for multiple testing are discussed in Westfall and Young (1993). Additional references are provided in the following sections.

34.1 Chi-square Tests with Many Cells and Sparse Multinomials

The standard paradigm for the chi-square test of Pearson assumes that one has a sample of size n from a multinomial distribution with k cells, where k is considered fixed and $n \to \infty$. The operating characteristics of Pearson's chi-square test are not close to the nominal levels if k is so large that the expected frequencies of some cells are small. Indeed, this has led to the classic rules of thumb such as five counts or more per cell for using the chi-square test.

In practice, k is often not small compared with n. For example, the cells may be generated by cross-classifying a set of individuals according to a number of categorical variables. Just four variables with three categories each lead to 81 multinomial cells. The standard chi-square asymptotics cannot be expected to perform well under such circumstances unless n is quite large.

To handle this problem, one lets k depend on n and allows k to go to ∞ with n. The question is what configurations of k, n and the multinomial probability vector allow meaningful asymptotics and what these asymptotics are. Once the asymptotics are established, one can use them in the usual manner to find critical values and to approximate the power. Holst (1972) and Morris (1975) obtained the asymptotics for Pearson's chi-square statistic as well as the likelihood ratio statistic in this new paradigm by using the Poissonization technique, which makes the cell counts independent on dropping a Poisson number of balls into the cells. Here is their result.

Theorem 34.1 For $k = k(n), n \geq 1, \mathbf{p}_k = (p_1, \cdots, p_k)$ a probability vector, let $N = (N_1, \cdots, N_k)$ have a multinomial distribution with parameters $n, \mathbf{p}_k$. Suppose $X_i, i = 1, \cdots, k$ are independent Poisson(np_i) variables and $f_i, i = 1, \cdots, k$ are polynomials of a bounded degree r such that $E(f_i(X_i)) = 0$ for all i and $\sum_{i=1}^{k} \text{cov}(X_i, f_i(X_i)) = 0$. Denote $\sigma_i^2 = \text{Var}(f_i(X_i)), \sigma^2 = \sum_{i=1}^{k} \sigma_i^2$. Assume the following conditions:

(a) $\min_i np_i$ is bounded away from zero.

(b) $\max_i p_i = o(1)$.

(c) $\frac{\max_i \sigma_i^2}{\sigma^2} = o(1)$.

Then $\frac{\sum_{i=1}^{k} f_i(N_i)}{\sigma} \stackrel{\mathcal{L}}{\Rightarrow} N(0, 1)$ as $n \to \infty$.

Two major consequences of this general theorem are the following results on the asymptotic distribution of Pearson's chi-square and the likelihood ratio statistic for the multinomial hypothesis $H_0 : \mathbf{p} = \mathbf{p}_0$. We give the asymptotic distributions under the null only, although Theorem 34.1 above covers the alternative also under suitable conditions on the alternative. A key aspect of these new asymptotics in the new paradigm is that Pearson's chi-square and the likelihood ratio statistic no longer have the *same* asymptotic distribution under the null. Recall from Chapter 27 that, when k is fixed, they do have the same asymptotic distributions under the null. The next two theorems are from Morris (1975) and Holst (1972).

Theorem 34.2 Let $\chi^2 = \sum_{i=1}^{k} \frac{(N_i - np_{0i})^2}{np_{0i}}$ be the chi-square statistic for $H_0 : \mathbf{p} = \mathbf{p}_0$. Let $\sigma_i^2 = 2 + \frac{(kp_{0i}-1)^2}{np_{0i}}$ and $\sigma^2 = \sum_{i=1}^{k} \sigma_i^2$. Assume that $\max_i p_{0i} = o(1)$ and $\min_i np_{0i}$ is bounded away from zero. Then, under H_0, $\frac{\chi^2 - k}{\sigma} \stackrel{\mathcal{L}}{\Rightarrow} N(0, 1)$ as $n \to \infty$.

The next result is about the likelihood ratio statistic. The likelihood ratio test rejects the null hypothesis for large values of $T_n = \sum_{i=1}^{k} N_i \log(\frac{N_i}{np_{0i}})$.

Theorem 34.3 Let $I(x, y) = x \log(\frac{x}{y}) - (x - y), x, y > 0$, and $I(0, y) = y, y > 0$. Under the assumptions of the preceding theorem, $\frac{T_n - \sum_{i=1}^{k} E[I(X_i, \lambda_i)]}{\sigma} \stackrel{\mathcal{L}}{\Rightarrow} N(0, 1)$ as $n \to \infty$ under H_0.

Remark. In an actual problem with data, one would have the option of using the chi-square approximation or the normal approximation presented here. The chi-square approximation is omnibus in the sense that the expectation would not have to be computed. But the normal approximation may be more accurate if the smallest expected cell frequencies are small. Koehler and Larntz (1980) have reported some simulations on accuracy of the two approximations and also some power comparisons between the chi-square and the likelihood ratio test when k is large. There is no uniform preference according to the simulations.

Example 34.1 Suppose that, based on $n = 60$ observations from a CDF F on the real line, we wish to test that F is the $C(0, 1)$ CDF (the standard Cauchy). Suppose we use Pearson's chi-square to test this hypothesis by using the observed frequencies of the intervals $(0, .5], (.5, 1], \cdots, (4.5, 5], (5, \infty)$ and their corresponding counterparts on the negative half line. This gives $k = 22$ with the null probability vector $\mathbf{p}_0 = (.0628, .0068, .0084, .0106, .0138, .0187, .0265, .0396, .0628, .1024, .1476, .1476, .1024, \cdots, .0628)$. The minimum of np_{0i} is $60 \times .0068 = .41$, which is not very small, but the maximum of p_{0i} is .1476, which is not small. This is a situation where neither the chi-square nor the normal approximation is likely to be very accurate. By direct computation, the normal approximation in the Theorem 34.2 will approximate Pearson's χ^2 as $N(22, 55.86)$, while the chi-square approximation of course approximates it as a chi-square with 21 degrees of freedom, which has a variance of 42. We see that the normal approximation entails a larger variance and may approximate small-tail areas better. But it can only be verified by a good simulation.

34.2 Regression Models with Many Parameters: The Portnoy Paradigm

Another important high-dimensional problem is regression with a large number of explanatory variables. We will only discuss linear models here. Huber (1973) and Yohai and Maronna (1979) treated problems of consistency and asymptotic normality of least squares and general M estimates of the regression parameters when the number of parameters goes to infinity with n. Portnoy (1984, 1985) weakened the conditions and streamlined the results, and we mostly follow Portnoy's results here.

Consider the usual linear model $Y = X\beta + R$, where Y is $n \times 1$, $X = X_n$ is $n \times p$, β is $p \times 1$, and R is $n \times 1$. We let p, the dimension of the regression parameter, depend on n; i.e., $p = p_n$. But the dependence on n will be suppressed for notational convenience. We let x_i' denote the ith row of the design matrix X. For a suitable function $\psi : \mathcal{R} \to \mathcal{R}$, an *M estimate* of β is defined to be a root of the equation $\sum_{i=1}^{n} x_i \psi(Y_i - x_i'\beta) = 0$. The question of uniqueness of the root does arise; we will comment on it a bit later. The results are about consistency and weak convergence of a general M estimate of β. As one would expect, consistency can be achieved with weaker conditions than weak convergence. The next two theorems present some convenient adaptations of results from Portnoy (1984, 1985). We have chosen not to present the best possible results in order to avoid clumsy conditions.

Theorem 34.4 Assume that:

(a) $R_1, \cdots, R_n$ are iid.

(b) $\psi(.)$ is an odd function differentiable with a bounded derivative and with $E[\psi'(R)] = 0$, $E[\psi^2(R)] \leq B < \infty$.

(c) $x_1, x_2, \cdots$ are iid, distributed according to a mixture of $N_p(0, \sigma^2 I)$ distributions.

(d) $E[x_{ij}^2] \leq B_0 < \infty$.

(e) $\frac{p \log p}{n} \to 0$ as $n \to \infty$.

Then there is a sequence of roots $\hat{\beta}_n$ of the equation $\sum_{i=1}^{n} x_i \psi(Y_i - x_i'\beta) = 0$ such that $\hat{\beta}_n \stackrel{P}{\Rightarrow} \beta$.

From the point of view of applications, the result is not useful unless something can be said about uniqueness of the root or how to choose the desired root when there are multiple roots. Portnoy (1984) gives the following useful result.

Proposition 34.1 *If, in addition to the conditions of the preceding theorem, $\psi'(.) > 0$ everywhere, then with probability tending to 1 as $n \to \infty$, there is a unique root of the equation $\sum_{i=1}^{n} x_i \psi(Y_i - x_i'\beta) = 0$.*

Remark. Proposition 34.1 and Theorem 34.4 above show that if p does not grow too fast, then the regression coefficients can be estimated consistently by a broad variety of M estimates. See Chapter 17 for examples of ψ functions that satisfy all of the conditions here.

Weak convergence with an explicit weak limit is more useful for statistical purposes than mere consistency. We now provide a theorem due to Portnoy (1985) that shows that asymptotic normality, too, holds for suitable M estimates under conditions somewhat stronger than those needed for consistency.

Theorem 34.5 Assume, in addition to all the conditions in the previous theorem and the previous proposition, that:

(i) $\psi(.)$ is three times continuously differentiable with each of the first three derivatives being bounded or R has a density function g such that $g^{(4)}$ is continuous, bounded, and integrable.
(ii) $\psi(R)$ has a finite mgf in some neighborhood of zero.
(iii) $\frac{(p \log p)^{3/2}}{n} \to 0$ as $n \to \infty$.

Let $\mathbf{c}_n$ be any sequence of p-vectors with bounded L_2 norm. Then,

$$\frac{\mathbf{c}_n'(\hat{\beta}_n - \beta)}{\sqrt{\mathrm{Var}(\psi(R))\mathbf{c}_n'(X'X)^{-1}\mathbf{c}_n}} \stackrel{\mathcal{L}}{\Rightarrow} N\left(0, \frac{1}{(E\psi'(R))^2}\right).$$

Remark. It is useful that the stringent smoothness conditions on ψ in (i) can be shuffled to g, the density of R. As a result, the asymptotic normality results hold for many of the examples of ψ functions provided in Chapter 17.

Example 34.2 Consider a multiple linear regression model with observations $Y_i = \beta_0 + \beta_1 x_{i1} + \cdots + \beta_9 x_{i9} + R_i$, so that $p = 10$ and $(p \log p)^{3/2} = 110.5$. The asymptotic normality theorem (Theorem 34.5) above assumes that $\frac{(p \log p)^{3/2}}{n} \to 0$. Thus, for reasonably accurate asymptotics, we need n to be quite large. Suppose also that the R_i are iid $C(0, 1)$, and we want an asymptotic confidence interval for $\sum_{i=0}^{9} \beta_i$. Note that by renormalization we can make the corresponding $\mathbf{c}_n$-vector have norm 1.

First of all, because of the tails of the Cauchy distribution, we need to carefully choose our ψ function such that Theorem 34.5 applies. As a specific choice, take ψ to be Tukey's biweight function (see Chapter 17) defined as $\psi(z) = z(r^2 - z^2)^2 I_{|z| \leq r}$, where $r > 0$ is fixed but arbitrarily finite. Interestingly, the required moments and variances can be found in closed form. The expressions (found in Mathematica) are

$$E\psi'(R) = \frac{6(5 + 6r^2 + r^4)\arctan r - 2r(15 + 13r^2)}{3\pi},$$

$$\mathrm{Var}(\psi(R)) = 2\frac{(315r + 1155r^3 + 1533r^5 + 837r^7 + 128r^9) - 315(1 + Lr^2)^4 \arctan r}{315\pi}.$$

Denoting $E\psi'(R)$ by d and $\mathrm{Var}(\psi(R))$ by v^2, provided the estimating equation has a unique root, an asymptotic confidence interval is found as $1'\hat{\beta}_n \pm z_{\alpha/2}\frac{v\sqrt{1'(X'X)^{-1}1}}{d}$. Note that Tukey's biweight ψ function is not monotone and so a unique root is not guaranteed.

34.3 Multiple Testing and False Discovery: Early Developments

The terms *multiple testing* and *simultaneous testing* refer to either many tests of statistical significance performed using a common dataset or repeated independent tests based on separate datasets. By virtue of the Neyman-Pearson formulation, it is unavoidable that in a large collection of tests (or, equivalently, confidence intervals) some statistical significance would be found by the laws of probability alone, although the effects actually do not exist. If an experimenter is trying to find nonzero effects, then sooner or later, by chance alone, she or he will discover an effect; the colorful term *false discovery* has been coined for this phenomenon.

Multiple testing and attempts to protect against such false discoveries are quite old topics. But new directions and important new developments have taken place in the area, primarily influenced by new types of problems in genomics and more generally in what is called data mining. With the advances in engineering and technology, collection and storage of a massive amount of data have become possible. It is a natural temptation to do a large number of statistical tests with a massive amount of data in hand. Although each test may have a low potential of registering a false positive, collectively some and even many false positives would be discovered. The general problem is to devise methods that provably prevent large-scale false discoveries. For example, in genomic studies, if numerous loci are simultaneously tested for association with a serious disease, and if a number of innocuous loci are discovered to be associated, the consequences obviously can be hazardous. Unfortunately, however, with the present state of knowledge, methods to control propensities for discovering nonexistent effects are generally problem-specific. The probability theory associated with devising such methods for realistic types of situations is very hard, and the area is likely to need and see substantial new developments. We provide an account of the key new ideas and results available now and also provide an account of some of the classic material that goes back to the works of John Tukey and Rupert Miller.

The two most influential early expositions on multiple testing are Tukey (1953) and Miller (1966). A lot of the early work was on simultaneous inference in the ANOVA setting. They make the normality assumption. A key development in these early works is a procedure suggested by Tukey and independently rediscovered in Kramer (1956). It is a method for forming a family of confidence intervals for all pairwise differences of means in one-way ANOVA, with the mathematical goal of ensuring a prescribed probability for the intersection of all the coverage statements. The proof of this mathematical property, only conjectured by Tukey, is due to Brown (1979, 1984) and Hayter (1984) and is highly involved and long. Here is their result.

Theorem 34.6 Let $X_{ij} = \mu_i + \epsilon_{ij}, 1 \leq i \leq k, 1 \leq j \leq n_i, \epsilon_{ij} \overset{\text{iid}}{\sim} N(0, \sigma^2)$, σ^2 unknown. Let s^2 denote the usual ANOVA estimate of σ^2, $\bar{X}_i$ the group sample means, and T_{α,ν_1,ν_2} the $100(1-\alpha)$th percentile of the studentized range distribution. Then,

$$P\left(\cap_{i,j=1}^{k}\left\{\bar{X}_i - \bar{X}_j \pm T_{\alpha,k,N-k}s\sqrt{\frac{n_i+n_j}{2n_in_j}} \ni \mu_i - \mu_j\right\}\right) \geq 1-\alpha,$$

$\forall \mu_i, \mu_j, \sigma, n_i, n_j, k$, where $N = \sum_{i=1}^{k} n_i$.

Remark. Some limited results are available when the samples from the different groups are not independent. See Brown (1984) for the case of three treatments. Extensions to the multivariate case have been sporadic, and the problem remains largely open; see Seo, Mano, and Fujikoshi (1994).

Example 34.3 It is instructive to know how much longer the Tukey-Kramer intervals are compared with the one-at-a-time t confidence intervals that ignore multiplicity altogether. We consider the balanced case when all $n_i = n$. The percentage increase in the length is found from the percentiles of the T and the t-distribution. Here is an illustration; the nominal confidence is 95% in the example.

(Length of Tukey-Kramer CI − Length of ordinary t-CI)/(Length of ordinary t-CI)$\times$ 100

n	k	k	k
–	3	5	10
5	73	104	138
10	72	102	135
30	71	99	134

In the balanced case, the Tukey-Kramer procedure in fact has simultaneous coverage exactly equal to $1-\alpha$. Therefore, the entries in the table above show how much less the simultaneous coverage of the ordinary t-intervals is likely to be. For example, when there are five groups and 30 observations per group, the Tukey-Kramer intervals are 99% longer, i.e., basically they are *twice* as long and yet *only exact, not conservative*. This illustrates the seriousness of the multiplicity issue.

Simultaneous 95% coverage is considered to be too expensive and unnecessary by many practitioners. Tukey (1993) suggests simultaneous 50% intervals if a large number of intervals are to be computed. We will readdress the issue of stringent simultaneous control of error rates in the next section.

Alternatives to the simple Bonferroni adjustment wherein α is replaced with $\frac{\alpha}{m}$ for a family of m confidence intervals have been suggested. These are related to the corresponding problem of testing a family of m null hypotheses, with the goal of controlling the probability of falsely rejecting any true null hypotheses. We discuss some of these in the next section.

34.4 False Discovery: Definitions, Control, and the Benjamini-Hochberg Rule

A relatively large number of definitions of what controlling a false discovery means are now available in the literature. We will treat a few of these approaches. In this section, we present the first approaches to the problem. More modern versions are treated in the next sections. The first attempts to give a definition for a false discovery and procedures to control them are due to Simes (1986), Hommel (1988), Sóric (1989), Wright (1992), and

Benjamini and Hochberg (1995). We need some notation to describe some of these developments.

For given $m \geq 1$, let $\Theta_i \subseteq \mathcal{R}^{d_i}, 1 \leq d_i < \infty$, and $A_i \subset \Theta_i$, $i = 1, \cdots, m$. Let $X \sim P_{\theta_1, \cdots, \theta_m}, \theta_i \in \Theta_i, i = 1, \cdots, m$, and $X \in \mathcal{R}^d$ for some $d \geq 1$. Let $H_{0i} : \theta_i \in A_i, i = 1, \cdots, m$ be given null hypotheses. Let $T_i, = T_i(X), i = 1, \cdots, m$ be specified test statistics, and suppose H_{0i} is rejected if $T_i > c_i$; i.e., H_{0i} is rejected for large values of T_i. Given the data, let p_i be the p-value based on the statistic T_i. Here it is understood that the hypotheses H_{0i} are such that a p-value is actually defined (it is well known that p-values cannot be defined for many testing problems) and is (marginally) uniformly distributed under the corresponding null. For some $m_0, 0 \leq m_0 \leq m$, suppose m_0 of the null hypotheses happen to be true. Let R denote the total number of rejections out of the m tests in a generic application of the given sequence of tests and define $W = m - R$; $V = \sum_{i=1}^{m} I_{T_i > c_i \cap \{H_{0i} \text{ is true}\}}$; $S = R - V; U = m_0 - V; T = m - m_0 - S$.

Thus, V is the number of occasions in the m cases where a true null hypothesis is rejected (a *false discovery*), T is the number of occasions where a false null has not been rejected (a *false nondiscovery*), and U and S are the respective frequencies of correct actions. It seems reasonable to want to keep U and S large or, equivalently, V and T small. Typically, critical regions of statistical tests are nested, and so the desires to keep both V and T small are conflicting. One can try to keep V small (in some well-formulated sense) or try to strike a balance between keeping V small and keeping T small. Both approaches have been suggested.

Some possible criteria for control of false discovery are:

Familywise error rate (FWER) $P_{\theta_1, \cdots, \theta_m}(V \geq 1)$;
False-discovery rate (FDR) $E_{\theta_1, \cdots, \theta_m}(Q)$, $Q = \frac{V}{R} I_{R>0} + 0 I_{R=0} = \frac{V}{R \vee 1}$;
Marginal false-discovery rate (MFDR) $\frac{E_{\theta_1, \cdots, \theta_m}(V)}{E_{\theta_1, \cdots, \theta_m}(R)}$;
Marginal realized false-discovery rate (MRFDR) $\frac{E_{\theta_1, \cdots, \theta_m}(V)}{r} I_{r>0} + 0 I_{r=0}$,

where r is the realized value of R;

False-discovery proportion distribution (FDPD) $P_{\theta_1, \cdots, \theta_m}(Q > q)$;
Positive false-discovery proportion distribution (PFDPD) $P_{\theta_1, \cdots, \theta_m}(Q > q | R > 0)$;
Positive false-discovery rate (PFDR) $E_{\theta_1, \cdots, \theta_m}(Q | R > 0)$.

There are pros and cons to each of these indices; see the discussions in Sóric (1989), Benjamini and Hochberg (1995), and Storey (2003). Besides these indices, obviously many other indices are also reasonable; for example, indices that take account of not only the mean but also the variance of Q or indices that use the median or quantiles of Q rather than its mean. In any case, having selected an index, the goal is to devise the test statistics in such a way that the selected index is *small*. Here are three specific results on particular procedures with specific multiplicity properties.

Theorem 34.7 Let $p_{(i)}, i = 1, \cdots, m$ denote the ordered p-values of tests based on the statistics $T_i, i = 1, \cdots, m$.

(a) Consider the procedure φ_1 that rejects $\cap_{i=1}^m H_{0i}$ if $\min_{1 \le j \le m} \frac{p_{(j)}}{j} \le \frac{\alpha}{m}$. If the null distributions of $T_1, \cdots, T_m$ are independent, then φ_1 satisfies $P_{\cap_{i=1}^m H_{0i}}(\varphi_1 \text{ rejects } \cap_{i=1}^m H_{0i}) = \alpha$. In words, if every null hypothesis is true, the probability that φ_1 fails to recognize it as such is equal to α.

(b) Suppose hypothesis $H_{0(i)}$ corresponds to the ith ordered p-value $p_{(i)}$. Consider the procedure φ_2 that rejects the hypotheses $H_{0(1)}, \cdots, H_{0(k)}$, where $k = \max\{j : (m - j + 1)p_{(j)} \le \alpha\}$. If $T_1, \cdots, T_m$ are independent under all configurations of $\theta_1, \cdots, \theta_m$, then φ_2 satisfies FWER $\le \alpha$, with the definition of FWER as given above.

(c) Consider the procedure φ_3 that rejects the hypotheses $H_{0(1)}, \cdots, H_{0(k)}$, where $k = \max\{j : \frac{mp_{(j)}}{j} \le \eta\}$. If $T_1, \cdots, T_m$ are independent under all configurations of $\theta_1, \cdots, \theta_m$, then φ_3 satisfies FDR $\le \eta$, with the definition of FDR as given above.

Remark. See Simes (1986), Hommel (1988), Hochberg (1988) and Benjamini and Hochberg (1995) for proofs of the parts of Theorem 34.7 (part (a) is almost obvious). Note that φ_1 does not give a prescription for which of the individual hypotheses are to be rejected; it only says if the entire collection is true or not. φ_2 and φ_3 reject those hypotheses that correspond to sufficiently small p-values, but the threshold is decided jointly. It is clear that whenever φ_2 rejects a particular null, so does φ_3, and therefore φ_3 has a larger power function than φ_2. φ_3, due to Benjamini and Hochberg (1995), is an example of an *FDR controlling multiple testing procedure*. It is a *step-wise procedure* in the sense that if the largest p-value $p_{(m)}$ is not $\le \alpha$, then we keep going down until we meet a p-value that is $\le \frac{j}{m}\alpha$ and reject all those nulls whose p-values fall below the line through the origin with slope $\frac{\alpha}{m}$. Power simulations of $\varphi_1, \varphi_2, \varphi_3$ in some selected cases can be seen in Simes (1986) and Benjamini and Hochberg (1995). The assumption on the mutual independence of $T_1, \cdots, T_m$ is restrictive, and the stated properties do not hold in general without the assumption.

Storey (2003) gives motivations for considering the PFDR instead of the FDR and draws some Bayesian connections. In many practical multiple testing situations, the number of hypotheses tested is very large (in the thousands), and in these cases FDR and PFDR are the same because one or the other of the hypotheses would end up getting rejected.

34.5 Distribution Theory for False Discoveries and Poisson and First-Passage Asymptotics

The FDR is the expectation of the proportion of rejections that are false, and the overall level is $P(V \geq 1)$. Expectation is one summary feature of a distribution, and similarly $P(V \geq 1) = 1 - P(V = 0)$ is essentially the probability of one value of V. In principle, the whole distribution should be much more informative. Distribution theory for $\frac{V}{R \vee 1}$ is complicated. But quite a bit has been done on distribution theory for V, the number of false rejections. Under the intersection null hypothesis, there would be a small probability of falsely rejecting any individual one. The decisions to reject or retain the individual nulls could be weakly dependent, depending on the exact multiple testing procedure. So, one might expect that the total number of false rejections, V, is a sum of weakly dependent Bernoullis and could be asymptotically Poisson. This intuition does work for some multiple testing procedures, but not always. We give an account of the fixed sample and asymptotic distribution theory for V in this section.

Theorem 34.8 (a) For the procedures φ_1 and φ_2, $V = V_m$ is uniformly integrable under the sequence of intersection nulls $\cap_{i=1}^m H_{0i}$, $V_m \stackrel{\mathcal{L}}{\Rightarrow}$ Poisson(α), and $E_{\cap_{i=1}^m H_{0i}}(V_m) \to \alpha$ as $m \to \infty$.

(b) For the procedure φ_3, using $\eta = \alpha$,

$$P_{\cap_{i=1}^m H_{0i}}(V_m = i) = (1-\alpha)\binom{m}{i}\left(\frac{i}{m}\alpha\right)^i\left(1 - \frac{i}{m}\alpha\right)^{m-1-i}, i = 0, 1, \cdots, m;$$

$$E_{\cap_{i=1}^m H_{0i}}(V_m) = \alpha \sum_{i=0}^{m-1} \binom{m-1}{i}(i+1)!\left(\frac{\alpha}{m}\right)^i;$$

$$P_{\cap_{i=1}^m H_{0i}}(V_m = i) \to (1-\alpha)\frac{i^i}{i!}\alpha^i e^{-i\alpha};$$

$$E_{\cap_{i=1}^m H_{0i}}(V_m) \to \frac{\alpha}{(1-\alpha)^2}.$$

Proofs of these four results can be seen in Finner and Roters (2002).

Remark. Note that the number of false discoveries made by φ_3 is not asymptotically Poisson. Its finite sample distribution has a nice barrier-crossing interpretation. Consider a plot of the empirical CDF of a $U[0, 1]$ random sample of size m. Now consider an infinite set of straight lines joining the points $(0, \alpha)$ to $(1 - \alpha, 1), 0 \leq \alpha \leq 1$. The curve $y = F_m(x)$ will be crossed by a random subset of these lines. Consider among these the line farthest from the diagonal that just touches the curve $y = F_m(x)$. The x coordinate of the point of touch is a data point, and the y coordinate is a value of the empirical CDF. Denote the point of touch as $(U^*, \frac{i^*}{m})$. The distribution of i^* is basically the finite sample distribution of V_m for the procedure φ_3. Birnbaum and Pyke (1958) and Dempster (1959) study this barrier-crossing problem. It is interesting that nearly 50 years after these papers were published, a connection has been found to a concrete and important statistical problem. We have not been able to find a name for the interesting distribution $(1 - \alpha)\frac{i^i}{i!}\alpha^i e^{-i\alpha}$ on the nonnegative integers.

Example 34.4 Suppose $m = 100$ nulls are to be tested and we want to keep the FDR controlled at $\alpha = .05$ while using the procedure φ_3. We use the formulas for the exact distribution of V to obtain $P(V = i)$ for some values of i.

i	$P(V = i)$
0	.95
1	.0452
2	.0043
3	.0004

With a virtually 100% probability, the number of false rejections is 0 or 1 if all the nulls are true. Of course, if some of the nulls are false, V could be larger with a high probability, depending on how close to the null the alternatives are. Distribution theory when some nulls are false is much more complex and is quite an open area.

Now increase m to 1000. Then the probabilities change as follows.

i	$P(V = i)$
0	.95
1	.04518
2	.00429
3	.00046

The numbers are remarkably stable. The asymptotics under the intersection null hypothesis are accurate at quite small m. Note that these results are all *distribution-free* because φ_3 itself is a distribution-free procedure.

34.6 Newer FDR Controlling Procedures

An alternative equivalent way to think of the procedure φ_3 is that it rejects those hypotheses H_{0i} for which $p_i \leq p_{(k)}$, k being as defined in the definition of φ_3. In other words, hypotheses for which the p-values fall below a data-dependent threshold, namely $p_{(k)}$, are rejected. With this data-dependent threshold, FDR control is achieved. An alternative approach is to consider a fixed but arbitrary threshold t. Use of any specific t will result in a corresponding FDR value, say FDR(t). This, of course, is a parametric function. One could try to estimate it and then use a judiciously selected t so that the corresponding estimated FDR is below some level, say α. One would intuitively expect that such a procedure will control FDR, too. Storey, Taylor, and Siegmund (2004) proved that this method leads to FDR controlling procedures. Furthermore, their estimates of FDR(t) are upwardly biased, and in that sense *honest estimates*, and in a suitable asymptotic sense, their procedures have greater power than φ_3 when both are subjected to the same level of FDR control value, say α. To describe a few of their results, we first need some notation.

34.6.1 Storey-Taylor-Siegmund Rule

Let φ_t denote the procedure that rejects all those nulls for which $p_i \leq t$. Let π_0 be the proportion of nulls that are true (i.e., $\pi_0 = \frac{m_0}{m}$). It is convenient to think of π_0 as the probability that a particular null is true. Let FDR(t) denote the FDR of φ_t. Apart from t, we need another constant λ in the unit interval; this constant λ is ultimately going to be chosen using some means involving some subjective judgment and some help from the data. It is important not to confuse t with λ; there is a role for each in the development of the final procedure. For any given t, there will be a family of estimates $\widehat{\text{FDR}(t)}$ using different λ. So as long as both t and λ are allowed to vary freely, there is a double array of estimated FDRs. But eventually t would be chosen to give a concrete multiple testing procedure φ_t; the constant λ is still free at this point, but practically it, too, will have to be chosen at the end, as we remarked before.

Given a monotone nondecreasing function G taking values between 0 and 1 and a constant α between 0 and 1, let $t_\alpha(G) = \sup\{t \in [0, 1] : G(t) \leq \alpha\}$. Corresponding to every fixed threshold level λ for retaining or rejecting the null hypotheses, we will have variables $R(\lambda)$, $W(\lambda)$, $V(\lambda)$, $S(\lambda)$, $U(\lambda)$, and $T(\lambda)$. These are defined exactly as before, when a hypothesis is rejected if and only if the p-value is $\leq \lambda$. The Storey-Taylor-Siegmund multiple testing procedure is now described and motivated. Its important properties will be presented following the description below.

For a given t, FDR(t) is estimated by a plug-in rule. Since FDR is defined as $E(\frac{V}{R \vee 1})$, FDR(t) is estimated as $\frac{E(\hat{V}(t))}{R(t) \vee 1}$; note that $R(t) \vee 1$ is observable, but $V(t)$ is not. Now, interpreting π_0 as the probability that a particular null is true, elementary algebra gives $E(V(t)) = mt\pi_0$. So, to estimate $E(V(t))$, we need a data-based estimate of π_0. In fact, a family of estimates of π_0 is given as $\hat{\pi_{0,\lambda}} = \frac{W(\lambda)}{m(1-\lambda)}$; these estimates are easily motivated by using the uniform distribution of the p-value when the null is true and the fact that the large p-values should mostly correspond to the true nulls. Now, plugging in, a family of estimates of FDR(t) are obtained:

$$\widehat{\mathrm{FDR}_\lambda(t)} = \frac{tW(\lambda)}{(1-\lambda)(R(t) \vee 1)}.$$

A slight modification of this is necessary for certain technical reasons, which we make precise later. But this is basically the Storey-Taylor-Siegmund estimate of the FDR of φ_t. The key point is that π_0 has been estimated instead of using the conservative estimate that π_0 is 1. One hopes that this leads to better power. These FDR estimates are honest (i.e., they are upwardly biased). Here is a result from Storey, Taylor, and Siegmund (2004).

Theorem 34.9 Suppose the null distributions of $T_1, \cdots, T_m$ are independent. Then, for any $t, \lambda \in [0, 1)$, $E(\widehat{\mathrm{FDR}_\lambda(t)}) \geq \mathrm{FDR}(t)$.

The primary goal is to develop concrete procedures that demonstrably control FDR in finite samples. If we can force $\widehat{\mathrm{FDR}_\lambda(t)} \leq \alpha$, then the honest nature of the FDR estimates will lead to FDR control. Among the infinitely many possible values of t that may lead to $\widehat{\mathrm{FDR}_\lambda(t)} \leq \alpha$, we should choose the largest value of t. This will allow us to reject the maximum number of nulls while controlling FDR at the same time. Hence, we should use the specific t given by $t = t_\alpha(\widehat{\mathrm{FDR}_\lambda})$. One minor technical point is that we modify the estimate of π_0 as $\hat{\pi_{0,\lambda}} = \frac{1+W(\lambda)}{m(1-\lambda)}$ to ensure that we do not estimate π_0 to be zero. The correspondingly modified FDR estimate is

$$\widehat{\text{FDR}_{*,\lambda}}(t) = \frac{t(1 + W(\lambda))}{(1 - \lambda)(R(t) \vee 1)} I_{t \leq \lambda} + 1 I_{t > \lambda}.$$

This gives a whole new family of verifiably FDR controlling procedures. Here is another result proved in Storey, Taylor, and Siegmund (2004).

Theorem 34.10 Let $\lambda, \alpha \in (0, 1)$. Then the procedure $\varphi_* := \varphi_{t_\alpha(\widehat{\text{FDR}_{*,\lambda}})}$, which rejects every null with a p-value $\leq t_\alpha(\widehat{\text{FDR}_{*,\lambda}})$, satisfies $\text{FDR}(\varphi_*) \leq (1 - \lambda^{m\pi_0})\alpha \leq \alpha$.

Remark. Although the procedure above gives a specific choice of t, λ has been left arbitrary. One could choose λ as a suitable value near 1 that minimizes the MSE $E(\hat{\pi_{0,\lambda}} - \pi_0)^2$. λ should preferably be chosen near 1, as that is the case when the intuition that p-values larger than λ should correspond to true nulls is most likely to be valid; this was used in writing the estimate $\hat{\pi_{0,\lambda}}$. Note, however, that the MSE itself is unknown. So instead, λ may be chosen by using an estimated MSE, which can be formed by the bootstrap, cross-validation, or other subsampling methods. A default choice is $\lambda = .5$. Other possibilities are suggested in Storey (2002) and Genovese and Wasserman (2002).

34.7 Higher Criticism and the Donoho-Jin Developments

Recently, there has been a spurt of statistical interest in devising tests for multiple testing situations where most of the component problems correspond to the null, or *nothing going on*, and there *may be* a small fraction of component problems where something small might be going on. A principal example is the so-called quantitative trait loci search, with the intention of identifying parts of the DNA that are *associated with* a given specific trait (e.g., a disease). Since numerous segments are being searched as candidates, one can expect to find some statistically significant results. Because a false-discovery in a problem of this nature has obvious negative consequences, control of the false discovery rate in the overall family of tests is considered important. The three problems of testing the global null hypothesis that there are no effects or signals at all, of estimating the number of nonzero signals, and of identifying precisely which ones are nonzero are progressively more difficult. Donoho and Jin (2004) give an elegant analysis connecting control of false discovery to asymptotic detectability of existence of an effect when the fraction and magnitude of the nonzero effects are small in a suitable sense. They do this in a specific mixture model setting that is supposed to model the marginal distribution in the population of the various test results. We give a description of the Donoho-Jin results in this section. We first need some notation and definitions.

As in the previous section, $H_{0i}, i = 1, \cdots, m$ denote m null hypotheses, and based on some test statistics, T_i, p_i denote the observed p-values and $p_{(i)}$ the ordered p-values. We assume that the null distributions of T_i are independent for the discussion below. Let $U_m(t) = \frac{1}{m}\sum_{i=1}^{m} I_{p_i \le t}$ be the empirical CDF of the p-values; information about our problem is carried by $U_m(t)$, and we will use it shortly. If each test was performed at a fixed level α, then under the intersection null hypothesis $\cap_{i=1}^{m} H_{0i}$, the number N_m of hypotheses rejected would be distributed as $\text{Bin}(m\alpha, m\alpha(1-\alpha))$. The standardized version of it is $Z_{m,\alpha} = \frac{\sqrt{m}(\frac{N_m}{m}-\alpha)}{\sqrt{\alpha(1-\alpha)}}$. One can view this as a two-parameter stochastic process and consider the asymptotics of $\sup_{\alpha \le \alpha_0} Z_{m,\alpha}$. It is supposed to measure the maximum deviation of the observed proportion of rejections from what one would expect it to be purely by chance as the type I error level changes. Presumably, although depending on the application, α_0 is a number like .1 or .2 because testing at higher α levels would not be practically done. There is also a technical necessity to stay away from the two boundary values 0, 1.

Writing $p_{(i)}$ in place of α, $\frac{i}{m}$ in place of $\frac{N_m}{m}$, and now taking a maximum over i, one gets a possible test statistic for assessing the strength of deviation of the proportion of significant results from what is expected just by chance; it is

$$\hat{Z}_m = \max_{i \le m\alpha_0} \frac{\sqrt{m}(\frac{i}{m} - p_{(i)})}{\sqrt{p_{(i)}(1 - p_{(i)})}}.$$

Under the intersection null hypothesis, the $p_{(i)}$ are jointly distributed as the order statistics of a $U[0, 1]$ random sample of size m. As a consequence, asymptotically, under the intersection null hypothesis, $\hat{Z}_m$ behaves like the supremum of the weighted normalized uniform empirical process, $Z_m^* = \sup_{0<t\le\alpha_0} \frac{\sqrt{m}(U_m(t)-t)}{\sqrt{t(1-t)}}$. The statistic Z_m^* is called the *higher criticism statistic* and was possibly known to John Tukey.

Theoretically, it is possible to use the higher-criticism statistic for any multiple testing problem, but we discuss here a particular normal mixture problem, for which meaningful asymptotics exist, and one that is emphasized in Donoho and Jin (2004). Other similar models have been discussed in Donoho and Jin (2004). The normal mixture model is the following: let $\epsilon_m = m^{-\beta}, \frac{1}{2} < \beta < 1; \mu_m = \sqrt{2r \log m}, 0 < r < 1, H_{0i} : X_i \sim N(0, 1), H_{1i} : X_i \sim (1 - \epsilon_m)N(0, 1) + \epsilon_m N(\mu_m, 1)$, X_i independent, regardless of whether H_{0i} or H_{1i} holds. The choices of ϵ_m, μ_m are made such that meaningful asymptotics exist in the specific normal mixture setting, and the mixture model itself is supposed to epitomize the situation where most probably there are no effects in any of the m cases at all, and if there are any, then they are not very large. It is helpful to understand that the model is an abstraction of an interesting practical question, but it may not necessarily reflect reality in an application. Unlike the discussions in our previous section, where the procedures tell us specifically which nulls should be rejected, here we only discuss whether the intersection null hypothesis should be rejected. The question is whether the higher-criticism statistic will eventually (that is, after many trials) know if there are any effects at all or not. A principal result given in Donoho and Jin (2004) is the following.

Theorem 34.11 Let $\rho_{\text{HC}}(\beta) = \beta - \frac{1}{2}, \frac{1}{2} < \beta \leq \frac{3}{4}, \rho_{\text{HC}}(\beta) = (1 - \sqrt{1-\beta})^2, \frac{3}{4} < \beta < 1$. Given α, let $c(\alpha, m)$ be such that $P_{\cap_{i=1}^m H_{0i}}(Z_m^* > c(\alpha, m)) = \alpha$. Assume that α_m is a sequence such that $c(\alpha_m, m) = \sqrt{2 \log \log m}(1 + o(1))$ as $m \to \infty$. Let φ_{HC} denote the procedure that rejects $\cap_{i=1}^m H_{0i}$ if $Z_m^* > c(\alpha_m, m)$. Let $\beta_m = P_{\cap_{i=1}^m H_{1i}}(Z_m^* \leq c(\alpha_m, m))$ and let $\gamma_m = \alpha_m + \beta_m$. Then, $\gamma_m \to 0$ as $m \to \infty$ if $r > \rho_{\text{HC}}(\beta)$, and $\gamma_m \to 1$ as $m \to \infty$ if $r < \rho_{\text{HC}}(\beta)$.

Remark. The message of the result is that for a given fraction of small nonzero effects, the higher-criticism statistic will make the right decision with a high probability in the long run if the magnitude of the nonzero effect is sufficiently large relative to the fraction of signals present. While this looks like an expected outcome, it is interesting that the boundary between detectability and nondetectability is analytically so simple. Donoho and Jin (2004) call the curve $r = \rho_{\text{HC}}(\beta)$ *the detection boundary*. Note that the result does not assert whether $\gamma_m \to 0$, or 1, or may be a function of r and β on the boundary $r = \rho_{\text{HC}}(\beta)$.

Interestingly, other procedures, and not just the higher criticism, also come with their own detection boundaries. Several other procedures were studied in Donoho and Jin (2004). We report the detection boundary for φ_3 of the previous section, namely the Benjamini-Hochberg (1995) procedure. The result says higher criticism works whenever φ_3 works and sometimes it works, even when φ_3 does not work. In that specific sense, φ_3 is inadmissible; the result below is also from Donoho and Jin (2004).

Theorem 34.12 Let γ_m^* denote the sum of the type I and type II error probabilities of φ_3 for testing the intersection null hypothesis in the given normal mixture model. Then, $\gamma_m^* \to 0$ as $m \to \infty$ if $r > \rho^*(\beta)$, and $\gamma_m^* \to 1$ as $m \to \infty$ if $r < \rho^*(\beta)$, where $\rho^*(\beta) = (1 - \sqrt{1-\beta})^2, \frac{1}{2} < \beta < 1$.

Remark. Thus, the detection boundaries of higher criticism and φ_3 coincide if $\frac{3}{4} < \beta < 1$ (which means the boundaries coincide when the fraction of nonzero effects is smaller). However, for $\frac{1}{2} < \beta < \frac{3}{4}$, $\beta - \frac{1}{2} < (1-\sqrt{1-\beta})^2$, and so in that case higher criticism works for more values of r than does φ_3.

It is not clear for what m the asymptotics start to give reasonably accurate descriptions of the actual finite sample performance and actual finite sample comparison. An exact theory for the finite sample case would be impossible. So, such a question can only be answered by large-scale simulations taking values of m, ϵ, and a small nonzero value for μ_m. One also has to simulate when the null is true. Simulation would be informative and even necessary. But the range in which m has to be in order that the procedures work well when the distance between the null and the alternative is so small would make the necessary simulations time consuming.

34.8 False Nondiscovery and Decision Theory Formulation

Just as false discoveries can have costly consequences, a false null that ends up getting accepted also constitutes a wrong decision; Tukey had termed it a *missed opportunity*. Genovese and Wasserman (2002) call T (using our previous notation) the number of false nondiscoveries; see also DasGupta and Zhang (2006). Genovese and Wasserman (2002) give a formulation for assessing of a multiple testing procedure by combining the two components, false discovery and false nondiscovery. They provide a loss function and obtain some results on comparison of a few common procedures based on the risk computed from this loss function. An account of the Genovese and Wasserman (2002) results is given below.

As before, we assume that the null hypotheses are such that p-values are defined and iid $U[0, 1]$ under the m nulls. We now make the *important additional assumption* that the distributions of T_i under H_{1i} are the same. Consequently, the distributions of the p-values p_i under H_{1i} are also the same. This common p-value distribution will be denoted as F. An example of such a scenario is the presence of a common normal mean, say θ, and independent testing of $\theta = 0$ by m experimenters, or the different experimenters could have different parameters θ_i but their alternate values happen to be the same. One could make either interpretation. The common distribution F of the p-values will have a precipitious role in the results below.

Define $\beta = \frac{\frac{1}{\alpha}-\pi_0}{1-\pi_0}$ and u^* the unique root of $F(u) = \beta u$, assuming a unique root exists. Suitable assumptions made below ensure the existence of such a unique root. Consider procedures φ of the form that rejects the hypotheses $H_{(0i)}, i = 1, \cdots, k$ if $k = \max\{j : p_{(j)} \leq l_j\}, l_j$ being a prescribed sequence. Note that procedures φ_2, φ_3 are both of this form, and so are the procedures that do not correct for multiplicity at all and also the procedure based on the naive Bonferroni correction. In these last two cases, the l_j sequences are α and $\frac{\alpha}{m}$, respectively.

Finally, consider the loss function $L(F, \varphi) = \frac{V+T}{m}$, the fraction of wrong actions out of the total m actions taken. The risk function is defined as $R_m(F, \varphi) = E(L(F, \varphi))$. Genovese and Wasserman (2002) give the following result.

34.8.1 Genovese-Wasserman Procedure

Theorem 34.13 Suppose $F(.)$ is strictly concave and $F'(0+)$ exists and is $> \beta$. Suppose also that π_0 is a fixed constant and $0 < \pi_0 < 1$. Then, $R_m(F, \varphi_3) \to R_\infty(F) := \pi_0 u^* + (1-\pi_0)[1-F(u^*)] = \pi_0 u^* + (1-\pi_0)[1-\beta u^*]$ as $m \to \infty$.

This result gives a basis for asymptotic comparison of φ_3 with other procedures for which the limiting risk also can be found. Here is an example.

Example 34.5 Consider two other procedures, φ_U, φ_B, the uncorrected and the ordinary Bonferroni corrected procedure, respectively. They admit the risk functions

$$R_m(F, \varphi_U) = \alpha\pi_0 + (1 - F(\alpha))(1 - \pi_0),$$

$$R_m(F, \varphi_B) = \pi_0\frac{\alpha}{m} + (1 - \pi_0)\left[1 - F\left(\frac{\alpha}{m}\right)\right].$$

Therefore, directly, as $m \to \infty$,

$$\lim[R_m(F, \varphi_3) - R_m(F, \varphi_U)] < 0 \Leftrightarrow \frac{\pi_0}{1-\pi_0} > \frac{F(\alpha) - F(u^*)}{\alpha - u^*}$$

and

$$\lim[R_m(F, \varphi_3) - R_m(F, \varphi_B)] < 0 \Leftrightarrow 2\alpha < \frac{1}{\pi_0}.$$

Thus, for any conventional α level, φ_3 dominates the Bonferroni correction in asymptotic risk because 2α would be < 1. But, curiously, the uncorrected version need not be dominated by φ_3. This may be a manifestation of the intuitive fact that correction is essentially superfluous when the false nulls are abundantly false, and in such a situation, correction leads to false nondiscoveries, which is a part of the Genovese-Wasserman loss function.

Extending this analysis to the case where there is no common distribution of the p-values under the alternatives is an important open problem.

Genovese and Wasserman (2002) give another formulation for the choice of a multiple testing procedure. Let FNR (false nondiscovery rate) denote the expectation of the fraction of acceptances that are false; i.e., FNR $= E_{\theta_1,\cdots,\theta_m}(\frac{T}{U+T})$. Asymptotically, at least one hypothesis is certain to be accepted unless $\pi_0 = 0$, and so we do not write $\frac{T}{(U+T)\vee 1}$ in the definition. The goal is to minimize the FNR subject to the constraint FDR $\le \alpha$. For very general multiple testing procedures, we cannot hope to characterize the solution. However, Genovese and Wasserman (2002) give the following result in the subclass of *single-step tests*.

Theorem 34.14 Assume the conditions of the previous theorem. Let φ_c denote the procedure that rejects H_{0i} iff $p_i \le c$. Let

$$I(c, F) = \frac{1}{1 + \frac{(1-\pi_0)F(c)}{\pi_0 c}},$$

$$J(c, F) = \frac{1}{1 + \frac{\pi_0(1-c)}{(1-\pi_0)(1-F(c))}}.$$

Then,

(a) FDR$(\varphi_c) = I(c, F) + O(m^{-1/2})$.
(b) FNR$(\varphi_c) = J(c, F) + O(m^{-1/2})$.
(c) The procedure φ_{c^*} that minimizes $J(c, F)$ subject to $I(c, F) \le \alpha$ is characterized by $\frac{F(u^*)}{u^*} - \frac{F(c^*)}{c^*} = \frac{1}{\alpha}$.

Remark. It is shown in Genovese and Wasserman (2002) that asymptotically the procedure φ_3 behaves like the single-step procedure φ_{u^*}; see Genovese and Wasserman (2002) for an exact meaning of this statement. Therefore, the result above says that up to the $m^{-1/2}$ order, the Benjamini-Hochberg procedure φ_3 does not minimize FNR subject to FDR control at a given level α. The study of minimization of FNR subject to FDR control is a very important question and merits a lot of further work in practical settings. It is likely to be a hard optimization problem and quite possibly challenging if stepwise procedures are allowed. Sun and Cai (2007) have recently given a formulation for selection of a multiple testing rule based on essentially empirical Bayes ideas, which they use to provide asymptotically adaptive multiple testing rules. Their procedure is based on Efron's *z values* (Efron (2003), Efron et al. (2001)) instead of the p-values, and they claim that using the z-values leads to asymptotically better procedures. Abramovich et al. (2006) and Donoho and Jin (2006) consider traditional minimax estimation of sparse signals in some specific parametric setups and show that estimates obtained from certain FDR considerations are approximately asymptotically minimax. This can be an interesting theoretical connection between formal decision theory and FDR methodologies, which constitute a less formal inference. The entire area is likely to see swift developments in many directions in the very near future.

34.9 Asymptotic Expansions

There is no Bayesian component in the formulations and the results on false discovery that we have presented so far. Interestingly, there is in fact a Bayesian connection. To understand this connection, suppose the parameters $\theta_1, \cdots, \theta_m$ of our m null hypotheses are iid realizations from a distribution G on some set Θ. To keep matters simple, assume that $\Theta \subseteq \mathcal{R}$. Assume that the m experimenters are testing the same null on their respective parameters, $H_{0i} : \theta_i \in S$, for some specified $S \subset \Theta$, and their procedures are the same (i.e., H_{0i} is rejected if $T_i \in C$ for some common C). Consider now the fraction of false discoveries, which, in our notation, is $\frac{V}{R \vee 1}$. Note that this is simultaneously a function of the data and the parameters $\theta_1, \cdots, \theta_m$. DasGupta and Zhang (2006) show that the sequence of fractions of false discoveries converge almost surely to an intrinsically Bayesian quantity, namely $P(\theta_1 \in S | T_1 \in C)$, which is the posterior probability of a false discovery from the point of view of a Bayesian. DasGupta and Zhang (2006) prove this by using a particular strong law for sequences of independent but not iid random variables; the details are available there. Here is the formal result.

Theorem 34.15 For $m, n \geq 1$, let $X_{i1}, \cdots, X_{in}, i = 1, \cdots, m$ be independent, with $X_{i1}, \cdots, X_{in} \stackrel{\text{iid}}{\sim} f(x|\theta_i) << \mu$, μ a σ-finite measure on some set $\mathcal{X}$. Suppose $\theta_i \in \Theta \subseteq \mathcal{R}$, and let $S \subset \Theta$. Let $H_{0i} : \theta_i \in S$ be m hypotheses, and suppose H_{0i} is rejected if $T_{ni} = T_n(X_{i1}, \cdots, X_{in}) \in C$ for some real functional T_n and some $C \subset \mathcal{R}$. Assume that $\theta_1, \cdots, \theta_m \stackrel{\text{iid}}{\sim} G$, for some distribution G on Θ. Define $\delta_n = P(\theta_1 \in S | T_{n1} \in C), n \geq 1$. Then, for each given n, for almost all sequences $\theta_1, \theta_2, \cdots$ with respect to the measure G,

(a) $\frac{V_m}{R_m \vee 1} \overset{\text{a.s.}}{\Rightarrow} \delta_n$ as $m \to \infty$, and

(b) $\text{FDR}_m = E_{\theta_1,\cdots,\theta_m}(\frac{V_m}{R_m \vee 1}) \overset{\text{a.s.}}{\Rightarrow} \delta_n$ as $m \to \infty$.

Remark. In general, FDR is a function of the unknown values $\theta_1, \cdots, \theta_m$. We have previously seen estimates of the FDR due to Storey, Taylor, and Siegmund (2004) based on single-step testing procedures. Those estimates are distribution-free. Theorem 34.15 gives a purely Bayesian method of approximating the FDR. This estimate, too, is based on single-step procedures and requires a model, namely $f(.)$. But the theorem says that whatever G is, the posterior probability that a false discovery has been made is a strongly consistent approximation of the purely frequentist quantity, the FDR. It is well known that, for complicated models or nonconventional priors, computing a posterior probability in closed analytical form can be impossible. With that in mind, DasGupta and Zhang (2006) provide the following asymptotic expansion for the posterior probability δ_n. Note that now the word *asymptotic* refers to $n \to \infty$. The combined consequence of Theorem 34.15 and the asymptotic expansion below is that the expansion can be used as an estimate of FDR when sample sizes are large and the total number of tests being done is *also* large; *both m* and n need to be large. It is shown in DasGupta and Zhang (2006) that an asymptotic expansion also holds for the nondiscovery rate, namely the FNR. Here is the asymptotic expansion result when the underlying f is a member of the one-parameter continuous exponential family.

Theorem 34.16 Let $f(x|\nu) = e^{\nu x - a(\nu)} b(x), x \in \mathcal{X} \subseteq \mathcal{R}$. Let $S = (-\infty, \nu_0)$ and suppose $T_n = \bar{X}_n$, the minimal sufficient statistic, and $C = (k_{\nu_0,n}, \infty)$ for some suitable constant $k_{\nu_0,n}$. Suppose $\nu \sim G$, an absolutely continuous probability measure with density g. Let $\delta_n = P(\nu \in S|T_n \in C)$ and $\epsilon_n = P(\nu \in S^c|T_n \in C^c)$. If g is three times continuously differentiable with a bounded third derivative, then for explicit constants $c_1, c_2, c_3, d_1, d_2, d_3, \delta_n, \epsilon_n$ admit the asymptotic expansions

$$\delta_n = \frac{c_1}{\sqrt{n}} + \frac{c_2}{n} + \frac{c_3}{n^{3/2}} + O(n^{-2}),$$

$$\epsilon_n = \frac{d_1}{\sqrt{n}} + \frac{d_2}{n} + \frac{d_3}{n^{3/2}} + O(n^{-2}).$$

The constants $c_1, c_2, c_3, d_1, d_2, d_3$ are explicitly provided in DasGupta and Zhang (2006), but the expressions for a general member of the continuous exponential family are complicated, and so we do not write the formulas here. Because the asymptotic expansion is concrete and computable, and it is extremely accurate, they can be useful model-based approximations of the FDR and the FNR based on purely Bayesian methods.

Example 34.6 Suppose the model density is $N(\theta, 1)$, and the hypotheses are $H_{0i} : \theta_i \leq 0$. Suppose the testing procedure is the single-step procedure that rejects H_{0i} when $\sqrt{n}\bar{X}n, i > z_\alpha$, which is the UMP level α test. We let $\theta_1, \cdots, \theta_m \stackrel{\text{iid}}{\sim} g$, a density on $(-\infty, \infty)$. In this case, the formulas for c_1, c_2, c_3 work out to the following expressions (see DasGupta and Zhang (2006)):

$$c_1 = 2g(0)[\varphi(z_\alpha) - \alpha z_\alpha]; c_2 = 4z_\alpha g^2(0)[\varphi(z_\alpha) - \alpha z_\alpha];$$

$$c_3 = 2\varphi(z_\alpha)[4z_\alpha^2 g^3(0) + g''(0)(z_\alpha^2 + 2)/6] - \alpha[g''(0)(z_\alpha^3 + 3z_\alpha)/3 + 8z_\alpha^3 g^3(0)].$$

Now, as an example, taking g to be the standard normal density and plugging into the expressions above, we can get approximations to the FDR for any value of n with the implicit understanding that n and m both need to be large. Because the asymptotic expansion for δ_n is a three-term expansion, even for $n = 4$, the expansion is nearly exact. For $n = 4$ itself, the expansion with the values above for c_1, c_2, c_3 gives a value of about .02 for δ_n when $\alpha = .05$. The interpretation is that if a large number of normal means were tested, say $m = 1000$ of them, using a single-step 5% procedure, then about 2% of the rejections are false rejections and the other 98% are correct. Note that the standard normal prior density is only a step in the middle; it is not an assumption in the basic FDR problem itself. The basic FDR problem is frequentist, but Bayesian methods were used to give a solution to the frequentist question. DasGupta and Zhang (2006) give many other examples using their asymptotic expansions to produce FDR and FNR approximations based on Bayesian methods.

34.10 Lower Bounds on the Number of False Hypotheses

A very interesting quantity in many types of multiple testing problems is $m_1 = m - m_0$, the total number of false null hypotheses. The corresponding proportion is $\pi_1 = \frac{m_1}{m}$. Meinshausen and Bühlmann (2005) and Meinshausen and Rice (2006) give many examples of problems where inference on m_1 would be useful. For example, a problem of substantial interest to astronomers is to estimate the number of objects of size greater than some threshold in the Kuiper Belt, an extreme outlying area in our solar system.

Direct observation or counting is simply impossible. Indirect methods, one of which is *occultation*, are used to try to estimate the number of Kuiper Belt objects (KBOs). When such an object, in its travel along its orbit, passes in front of a known star, it causes a momentary reduction in the flux related to the star, from which one concludes that a KBO was responsible for this reduction. But obviously one cannot conclude with certainty that the observed occultation was real. As a consequence, one would want to know how many were real, which is the same as estimating m_1 in a large number of tests. Meinshausen and Bühlmann (2005) give examples of problems in microarray analysis where estimation of m_1 would be useful. We present some ideas and results from Meinshausen and Rice (2006) and Meinshausen and Bühlmann (2005) on this problem. We first need some notation and a definition.

Any strictly positive real-valued function on $(0, 1)$ would be called a *bounding function* below. Given iid $U[0, 1]$ observations $U_1, U_2, \cdots$, the uniform empirical process $U_m(t) = \frac{1}{m}\sum_{i=1}^{m} I_{U_i \le t}$, and a bounding function $\delta(t)$, let $W_{m,\delta}(t) = \frac{U_m(t)-t}{\delta(t)}$ and $V_{m,\delta} = \sup_{0<t<1} W_{m,\delta}(t)$. Note that the distribution of the whole $W_{m,\delta}(t)$ process, and therefore that of $V_{m,\delta}$, is a fixed distribution depending only on the bounding function $\delta(t)$ and m. We also need the definition of a so-called *bounding sequence*.

Definition 34.1 A sequence of reals $\{h(\alpha, m)\}$ is called a bounding sequence (with respect to the bounding function $\delta(t)$) if $mh(\alpha, m) \to \infty$ as $m \to \infty$ and $P(V_{m,\delta} > h(\alpha, m)) \le \alpha$ for all $m \ge 1$.

The choice of a bounding function $\delta(t)$ and a bounding sequence $h(\alpha, m)$ would be crucial in ultimately finding good estimates of m_1; we discuss this later. The estimate analyzed below uses the observed sequence of p-values; more precisely, the estimate is based on the empirical CDF of the p-values $F_m(t) = \frac{1}{m}\sum_{i=1}^{m} I_{p_i \le t}$.

34.10.1 Bühlmann-Meinshausen-Rice Method

Meinshausen and Rice (2006) present the following point estimator of m_1:

$$\hat{m}_1 = \hat{m}_1(\delta, h) = \sup_{0<t<1} \frac{F_m(t) - t - h(\alpha, m)\delta(t)}{1 - t}.$$

This estimate has a number of good properties and in addition has interesting connections to other methods we have discussed in this chapter, such as higher criticism. We describe some of the most important properties of the Meinshausen-Rice estimate $\hat{m}_1$ now.

Theorem 34.17 For any bounding function $\delta(t)$ and a bounding sequence $\{h(\alpha, m)\}$, $P(\hat{m}_1 \leq m_1) \geq 1 - \alpha$ for any $m \geq 1$.

The result says that with $100(1-\alpha)\%$ confidence, we can assert that the true total number of false nulls is at least as large as the estimated value $\hat{m}_1$, regardless of the sample size. But note that this result does not say anything about the choice of $\delta(t)$ or $\{h(\alpha, m)\}$. Meinshausen and Rice (2006) give an extensive discussion of this issue. Among their results, the following result is a particularly practically useful one.

Proposition 34.2 *Suppose $\delta(t)$ is a given bounding function and that for suitable sequences a_m, b_m, the sequence of random variables $a_m V_{m,\delta} - b_m \stackrel{\mathcal{L}}{\Rightarrow} Z$ for some random variable Z. Let Q be the CDF of Z. Then $h(\alpha, m) = \frac{b_m + Q^{-1}(1-\alpha)}{a_m}$ is a bounding sequence with respect to $\delta(t)$.*

Example 34.7 Let $\delta(t) = \sqrt{t(1-t)}$. Then, with $a_m = \sqrt{2m \log\log m}$ and $b_m = 2\log\log m + \frac{1}{2}\log\log\log m - \frac{1}{2}\log \pi$, the weak convergence assumption of Proposition 34.2 holds, Q being the CDF of the so-called Gumbel law; i.e., $Q(x) = e^{-e^{-x}}, -\infty < x < \infty$. Note that $Q^{-1}(1-\alpha) = -\log\log\frac{1}{1-\alpha}$. This results in a concrete pair (δ, h) that can be plugged into the formula for $\hat{m}_1$. Of course, a concrete α has to be chosen in a real application. The theorems assume that $\alpha = \alpha_m \to 0$, which means that one needs to select small values of α. As a rule, in any real application using these estimates, a sensitivity analysis with different combinations of $\delta(t)$, α, and $h(m, \alpha)$ is desirable to ensure that the component inputs are not giving drastically different answers for different reasonable choices of the inputs.

For example, in the Kuiper Belt object (KBO) survey, about 3000 individual stars are being monitored for occultation, almost continuously, over several years. This results in measurements up to $m = 10^{12}$, a real example of large-scale multiple testing. With $m = 10^{12}$, the formulas above lead to $a_m = 2.5764 \times 10^6$, $b_m = 6.66534$. If we select $\alpha = .01$, then $h(m, \alpha)$ of Proposition 34.2 works out to 4.37257×10^{-6}. Finally, then, the estimated value of the true total number of occultations is found from the formula $\hat{m}_1 = \sup_{t\in(0,1)} \frac{F_m(t) - t - h(m,\alpha)\delta(t)}{1-t}$, $F_m(t)$ being the empirical distribution of the observed p-values. The observed p-values wll come from the flux measurements of the survey. What does $\hat{m}_1$ have to do with estimating the number of KBO objects? The answer is that if we pretend that each true occultation is caused by a *separate* KBO, then $\hat{m}_1$ will give us a rough lower bound for the total number of KBOs. This is evidently a hard problem, and sharp estimates cannot and should not be expected. Rough ideas are useful in this problem.

Theorem 34.17 given above says that, with a large probability, the estimate $\hat{m}_1$ is smaller than the true m_1. It would be an undesirable property, however, if $\hat{m}_1$ could be much smaller than m_1 with a substantial probability. Now the story gets interesting. If the true value of m_1 is too small, then, relatively speaking, $\hat{m}_1$ is *not* a good estimate of m_1; it underestimates m_1 drastically in that case. But if the true value of m_1 is only moderately small, then $\hat{m}_1$ estimates m_1 well; specifically, under well-formulated assumptions, $\hat{m}_1$ is consistent for m_1 in a *relative* sense. It is the relative error that is perhaps more meaningful here. To give the precise result, we need some assumptions.

Condition A (i) $\delta(1-t) \geq \delta(t)$ for all $t \in (0, \frac{1}{2})$.
(i) For all $c > 0$, $\lim_{t \to 0} \frac{\delta(ct)}{\delta(t)} = c^{\nu}$ for some $\nu \in [0, \frac{1}{2})$.
(ii) α_m and $h(\alpha_m, m)$ are such that

$$\frac{V_{m,\delta}}{h(\alpha_m, m)} \overset{P}{\Rightarrow} 0$$

and

$$\limsup_{m} \left(\frac{m}{\log m}\right)^{\frac{1}{2}} h(\alpha_m, m) < \infty.$$

Remark. It is shown in Meinshausen and Rice (2006) that condition (iii) above is not vacuous; i.e., one can choose a sequence $\alpha_m \to 0$ and a bounding sequence $h(\alpha_m, m)$ satisfying condition (iii) as long as $\delta(t)$ satisfies (i) and (ii) above.

Here is their consistency result on $\hat{m}_1$.

Theorem 34.18 Suppose that Condition A holds. Also assume that the p-values have a common distribution F with density f under the respective alternatives and that m_1 is of the form $m_1 \sim m^{1-\beta}$ for some $\beta \in [0, \frac{1}{2})$. Then, provided $0 \leq \nu \leq \frac{1}{2}$ and $\inf_t f(t) = 0$, $\frac{\hat{m}_1}{m_1} \overset{P}{\Rightarrow} 1$ as $m \to \infty$.

Remark. Notice the partial similarity with the results in Donoho and Jin (2004). If the proportion of false null hypotheses is of the order of $m^{-\beta}$ for $\beta < \frac{1}{2}$, procedures can be devised that will pick up all the false nulls in a relative sense.

It is shown in Meinshausen and Rice (2006) that if $\beta > \frac{1}{2}$, then a relative consistency result such as the one above cannot hold if the p-values have a fixed distribution F, independent of m, under the respective alternatives. One has to make the problem more detectable by letting F depend on m, say

$F = F_m$, in a way such that F_m diverges away from the null distribution, which is $U[0, 1]$, as m increases. Because the alternatives are now more markedly distinct from the nulls, there is still a way to pick up the false nulls in a relative sense. This was also the message in Donoho and Jin (2004), who achieved the divergent difference of the alternative from the null by letting the alternative mean, which was $\sqrt{2r \log n}$ in their model, diverge from 0, the null mean as the number of tests increased. Of course, these theoretical developments are not intended to mean that real problems fit these models. That is a separate exercise for someone to verify.

34.11 The Dependent Case and the Hall-Jin Results

In many, and perhaps even most, applications, the statistics T_i used to test the individual hypotheses are not independent, even under the nulls. For example, in comparing many treatments with a common control, clearly the T_i are not independent. In microarray analysis, it is well understood that the expression levels of the different genes are not independent. In climate studies, temperature measurements at different stations cannot be independent due to the mutual geographic proximity of subsets of stations. However, all the results we have presented so far, and indeed an overwhelming majority of the available literature, are based on the assumption of independence in some magnitude or the other. We are beginning to see some theoretical developments on multiple testing and false-discovery control for dependent test statistics. The advances so far are limited. This is partly because of the considerable theoretical difficulty in handling the relevant quantities under the realistic kinds of dependence, and also because developing the right new procedures needs new insight. We present here some results due to Benjamini and Yekutieli (2001), Sarkar (2006), Sarkar and Chang (1997) Chang, Rom, and Sarkar (1996), and Hall and Jin (2007). The first three papers deal with the general issue of developing single-step or Benjamini-Hochberg type stepwise procedures for traditional dependence structures, while the work of Hall and Jin (2007) studies higher criticism under strong dependence.

34.11.1 Increasing and Multivariate Totally Positive Distributions

Definition 34.2 A set $C \subset \mathcal{R}^m$ is called an increasing set if $x \in C, x \preceq y \Rightarrow y \in C$, where $\preceq$ denotes the partial ordering with respect to coordinatewise monotonicity.

Definition 34.3 A family of distributions $P_\theta, \theta \in \Theta \subseteq \mathcal{R}^m$, is called increasing in θ if $P_\theta(C)$ is increasing in θ for every increasing set C, where increasing in θ means coordinatewise monotonicity in each θ_i.

Here is a result on FDR control by single-step procedures when the test statistics have a joint distribution increasing in the parameters (see Sarkar (2006)).

Theorem 34.19 Suppose $H_{0i} : \theta_i \leq c_i, H_{1i} : \theta_i > c_i, i = 1, \cdots, m$. Suppose that $(T_1, \cdots, T_m)$ have a joint distribution increasing in $\theta = (\theta_1, \cdots, \theta_m)$ and that their joint distribution under $\theta = (c_1, \cdots, c_m)$ is permutation invariant. Let φ_t denote the procedure that rejects H_{0i} iff $p_i \leq t$. Then, $\text{FDR}(\varphi_t) \leq P_{\theta=(c_1,\cdots,c_m)}(p_{(1)} \leq t)$, provided $m_0 \neq 0$.

Example 34.8 Suppose $(T_1, \cdots, T_m) \sim N_m(\theta, \Sigma)$ for a given p.d. matrix Σ. If H_{0i} is rejected for large values of T_i, then p_i has the expression $p_i = 1 - \Phi(\frac{T_i - c_i}{\sigma_i})$, and so $P_{\theta=(c_1,\cdots,c_m)}(p_{(1)} \leq t) = P(Z_{(m)} \geq \Phi^{-1}(1-t))$, where $Z_{(m)}$ is the maximum of $Z_i = \frac{T_i - c_i}{\sigma_i}$. Therefore, by an application of Theorem 34.19, FDR of φ_t is controlled at a given α if $P(Z_{(m)} \geq \Phi^{-1}(1-t)) \leq \alpha$. Note that the Z_i are not independent unless Σ is diagonal. So, the distribution of $Z_{(m)}$ in general is not something simple. The required t in the last sentence can be found, in principle, when Σ is explicitly provided. Alternatively, a conservative t can be used; one obvious choice for a conservative value of t is $\frac{\alpha}{m}$ by the simple inclusion-exclusion inequality, and of course this is the naive Bonferroni correction.

Remark. The *increasing in* θ property is satisfied by many other parametric families of joint distributions. If the final joint distribution is amenable to a conditionally independent representation given some ancillary variable, and if these component conditional densities have a suitable stochastic monotonicity property, then the increasing in θ property holds. The chapter exercises give some precise instances.

Note that practitioners are often more interested in stepwise procedures rather than single-step procedures. Thus, a natural question is what sorts of FDR controls can be guaranteed by using stepwise procedures when the test statistics are not independent. The following result is a principal one in that direction; stronger versions of Theorem 34.20 below can be seen in Benjamini and Yekutieli (2001). We first need an important definition.

Definition 34.4 A joint density or a joint pmf $f(x_1, \cdots, x_m)$ is called multivariate totally positive of order two (MTP_2) if $f(x \wedge y) f(x \vee y) \geq f(x) f(y)$

for all x, y, where $\wedge, \vee$ are defined as the vectors of coordinatewise minima and maxima.

Multivariate total positivity was introduced by Karlin and Rinott (1980). Some families of multivariate distributions encountered in practice satisfy the MTP_2 property. Here is a useful collection of results; most of the parts in the next theorem are available in Karlin and Rinott (1980).

Theorem 34.20 (a) Let $X \sim N_m(0, \Sigma)$ and let $B = \Sigma^{-1}$. X is MTP_2 iff $b_{ij} \leq 0$ for all $i, j, i \neq j$.

(b) Let $f_i(.)$ be densities on the real line such that $f_i(x - \mu)$ is MLR (monotone likelihood ratio) for each $i = 1, \cdots, m$. Let $g(y_1, \cdots, y_m)$ be MTP_2. Then $f(x_1, \cdots, x_m) = \int \prod_{i=1}^m f_i(x_i - \mu_i) g(\mu_1, \cdots, \mu_m) d(\mu_1 \cdots d\mu_m)$ is MTP_2.

(c) Let $f_i(.)$ be densities on the real line such that $\frac{1}{\sigma} f_i(x/\sigma)$ is MLR for each $i = 1, \cdots, m$. Then, $f(x_1, \cdots, x_m) = \int_{[0,\infty)} [\prod_{i=1}^m \frac{1}{\sigma} f_i(x_i/\sigma)] dG(\sigma)$ is MTP_2 for any probability measure G on $[0, \infty)$.

(d) **General Composition Formula**. Suppose $g(x, y)$ is MTP_2 on $\mathcal{R}^{m_1+m_2}$ and $h(y, z)$ is MTP_2 on $\mathcal{R}^{m_2+m_3}$. Then $f(x, z) = \int g(x, y) h(y, z) dy$ is MTP_2 on $\mathcal{R}^{m_1+m_3}$.

(e) If f, g are MTP_2, then, subject to integrability, $\frac{fg}{\int fg}$ is MTP_2.

(f) Suppose $Z_i, i = 1, \cdots, m$ are iid from a density g on the real line. Let $X_i = Z_{(i)}$, the ith order statistic. Then the joint density of $(X_1, \cdots, X_m)$ is MTP_2.

Example 34.9 For $0 \leq i \leq m$, let Z_i be independent $\chi^2(\nu_i)$ variables. Let $X_i = \frac{\nu_0 Z_i}{\nu_i Z_0}, i = 1, \cdots, m$. Then, it follows that the joint density of $(X_1, \cdots, X_m)$ is MTP_2. This distribution is commonly known as a multivariate F-distribution.

Similarly, for $0 \leq i \leq m$, let Z_i be independent Gamma(α_i, λ) variables and let $X_i = Z_i + Z_0, i = 1, \cdots, m$. Then, it follows that the joint density of $(X_1, \cdots, X_m)$ is MTP_2. This distribution is commonly known as a multivariate Gamma distribution.

Now we state the result on FDR control; again, see Sarkar (2006).

Theorem 34.21 Suppose the joint distribution of the test statistics T_i corresponding to the *true* null hypotheses is MTP_2. Then the procedure φ_3 (the Benjamini-Hochberg procedure) with $\eta = \alpha$ controls FDR at α.

This leads to a whole array of useful parametric families of distributions for which φ_3 controls FDR at a specified level when the test statistics are not assumed to be independent. Note the particular feature of the assumption in Theorem 34.21 that the MTP_2 property is demanded of only those T_i involved with the true nulls. So, for example, if one is testing for the means of m normal distributions to be zero, then our result above on the MTP_2 property of suitable multivariate normals with mean vector zero comes in very handy in applying this FDR control result.

These results we have presented are by no means all that are known in this direction. We have presented a selection of basic results. Now, we deal with another question of great contemporary interest: How good or bad is higher criticism when the test statistics are not independent? This was addressed very recently in Hall and Jin (2007), and it turns out that a very neat dichotomy of phenomena occurs. We discuss this next.

34.11.2 Higher Criticism under Dependence: Hall-Jin Results

For independent normally distributed observations, above the detection boundary the higher-criticism statistic correctly chooses between the null and the alternative with probability tending to 1 as the number of tests (observations) tends to ∞ in the mixture model described there. It was also described how higher criticism is more sensitive than using the series maximum in choosing between the null and the alternative. In this section, we discuss some results in Hall and Jin (2007) on the performance of higher criticism in detecting possible sparse signals in a stationary Gaussian time series when the autocorrelation function decays at a suitable polynomial rate. The most important findings of these results include the fact that the phenomenon of the presence of a detection boundary persists and that testing based on the series maximum has certain uniformity (robustness) properties that the higher criticism is not known to possess. Performance of the higher-criticism statistic for some traditional short-range dependent processes has been treated in Delaigle and Hall (2006). The description of the exact model and the choice of various relevant sequences, such as the critical values for using the higher-criticism statistic, under which meaningful asymptotics exist are a bit complicated. Hall and Jin (2007) give quite detailed discussions on the specification of the model and the various sequences of constants.

For $n \geq 1$, let $X_{n1}, \cdots, X_{nn}$ be the first n values of a stationary zero-mean Gaussian time series with autocorrelation function $\rho_k = \rho_{k,n} = \max\{0, 1 - k^\delta n^{-\tau\delta}\}$, where $\delta > 0, 0 < \tau < 1$. The stationary variance will be set equal to 1. The null hypothesis is that the observed series follows this model. Under the alternative, the null series would be contaminated, in a suitable random sense, by a signal. But the signal is added to a smaller order of the observations than the full length n of the series. A useful way to think of it is to imagine that the data have been partitioned into blocks, a suitable subset of all the blocks are chosen, and members of the chosen blocks are endowed with a signal. The blocking argument is in fact also technically useful in deriving the intricate results in the sense that on partitioning the series into blocks, approximate independence is restored and then one can borrow intuition from the independent case, keeping in mind that the effective sample size is now the number of blocks rather than the original n. This is used very cleverly in Hall and Jin (2007). Here is the model for the alternative hypothesis.

Let $N = n^{1-\tau}, 0 < r < 1, \frac{1}{2} < \beta < 1, \mu = \mu_n = \sqrt{2r \log N}$, and $Y_{ni} = X_{ni} + \mu I_{ni}$, where $I_{ni} = 1$ for a subset of $m = m_n = N^{1-\beta} n^{\tau}$ indices and zero for the rest. The choice of the m indices can be done deterministically, although in practice it could be sensible (and cannot hurt) to choose them at random. The same detection boundary as in the Donoho and Jin (2004) case will be used: $p_{\mathrm{HC}}(\beta) = \beta - \frac{1}{2}, \frac{1}{2} < \beta \leq \frac{3}{4}, p_{\mathrm{HC}}(\beta) = (1 - \sqrt{1-\beta})^2, \frac{3}{4} \leq \beta < 1$.

It remains to exactly define the higher-criticism statistic and the exact testing rule; i.e., what is the cutoff value for rejecting the null. We define the test statistic as

$$Z_n^* = \sup_{t:|t| \leq t_n} \frac{\sum_{i=1}^n [I_{X_{ni} > t} - P_{H_0}(X_{ni} > t)]}{\sqrt{n\Phi(t)(1 - \Phi(t))}},$$

where the window width t_n is chosen so that

$$t_n \to \infty, \liminf n^{1-\tau}(1 - \Phi(t_n)) > 0.$$

The level is held at a constant level $\alpha, 0 < \alpha < 1$, and the cutoff value for Z_n^* is a sequence $c(\alpha, n)$ satisfying

$$c(\alpha, n) = n^{\tau/2} d(\alpha, n), d(\alpha, n) \to \infty, d(\alpha, n) = O(n^{\epsilon})$$

for any $\epsilon > 0$. Choosing the cutoff in this manner ensures a limiting type I error probability equal to the specified α; see Hall and Jin (2007).

Here is the first key result from Hall and Jin (2007).

Theorem 34.22 With the notation and the assumptions given above, let $\varphi_{\mathrm{HC},dep}$ denote the procedure that rejects the null hypothesis H_0 that the observed series equals the series $\{X_{ni}\}$ in distribution if $Z_n^* > c(\alpha, n)$. Then, $P_{H_1}(Z_n^* \leq c(\alpha, n)) \to 0$ as $n \to \infty$ if $r > p_{\mathrm{HC}}(\beta)$; i.e., $\varphi_{\mathrm{HC},dep}$ correctly detects the alternative with probability tending to 1 as $n \to \infty$.

Remark. This is a positive result. There is no effect of the polynomially strong dependence as long as we are above the detection boundary. The behavior of the higher-criticism statistic below the detection boundary (i.e., for $r < \rho_{HC}(\beta)$) is a little involved to describe. An informal description is that we cannot expect that it will work. A discussion is given in Hall and Jin (2007). They also present a robustness property of the test that rejects the null model for large values of the series maximum; thus, for a broad variety of autocorrelation functions, the maximum-based test will give some basic protection, which so far has not been established for other procedures.

34.12 Exercises

Exercise 34.1 Suppose participants in a poll of US residents are cross-classified according to sex, age (18–30, 30–45, 45–60, $>$ 60), income group ($<$ 30,000; 30,000–50,000; 50,000–75,000; 75,000–100,000; $>$ 100,000), geographic location of residence (northeast, mid-atlantic, midwest, south, southwest, mountain, west), and political affiliation (independent, Democrat, Republican). To test that the multinomial cell probability vector is equal to the uniform vector, how large a sample size would make you personally comfortable that the CLT for Pearson's chi-square statistic would be applicable?

Exercise 34.2 * Give a proof of the Poissonization theorem that if a Poisson number of balls are dropped into the cells of a multinomial distribution, then, marginally, the cell counts are independent Poisson.

Exercise 34.3 * Suppose the number of cells in a multinomial is $k = 100$, the null probability vector is the uniform vector, and $n = 300$. Conduct a careful simulation to investigate whether the chi-square or the normal approximation is better for the chi-square statistic as well as the likelihood ratio statistic.

Exercise 34.4 Simulate $n = 50$ observations from a Beta distribution with both parameters equal to 2. Using 20 cells of equal length, conduct a test of the hypothesis that the sampling distribution is $U[0, 1]$ by using Pearson's chi-square statistic. Use first the chi-square approximation and then the normal approximation. Comment on your finding; for example, comment on the p-values you get.

Exercise 34.5 Give examples of two ψ functions that satisfy the assumptions of Theorem 34.5 and another two ψ functions that do not.

Exercise 34.6 * Theorems 34.4 and 34.5 do not say anything about the case of nonrandom covariates. Does the nonrandom case follow under suitable conditions on them?

Exercise 34.7 * Suppose the error R in Theorem 34.4 has an $N(0, 1)$ distribution. Using Tukey's biweight ψ function, find the moments necessary for applying Theorem 34.5.

Exercise 34.8 For each of the following cases, simulate the *simultaneous* coverage probability of the Tukey-Kramer intervals:

(a) $k = 5, n_i = 15, \alpha = .05$, sampling distribution $= N(0, 1)$;
(b) $k = 5, n_i = 15, \alpha = .05$, sampling distribution $= t(5)$;
(c) $k = 10, n_i = 15, \alpha = .05$, sampling distribution $= t(5)$;
(d) $k = 5, n_i = 15, \alpha = .05$, sampling distribution $= \chi^2(5)$.

Exercise 34.9 * Suppose you want to test that the mean of each of two normal distributions with unit variance is zero. Suppose the simple Bonferroni corrected procedure is used. For a general configuration of false and true nulls, find expressions for the FWER, FDR, MFDR, and PFDR in the best closed form that you can. Assume that the groups are independently sampled.

Exercise 34.10 * Suppose you want to test the hypotheses $\theta_i = 1$ against $\theta_i > 1$ with $X_{ij} \sim U[0, \theta_i], i = 1, \cdots, m, j = 1, \cdots, n$. The test statistics are $T_i = X_{i,(n)}$, the group maxima. Find $E_{\theta_1, \cdots, \theta_m}(R)$, $E_{\theta_1, \cdots, \theta_m}(R \vee 1)$, and $E_{\theta_1, \cdots, \theta_m}(V)$ for a general vector $(\theta_1, \cdots, \theta_m)$. Specialize it to the case of the intersection null hypothesis and comment on what happens as $m \to \infty$.

Exercise 34.11 * Suppose $m = 3, 5, 10$, the group samples are independent, and the group sample sizes are all 20. Simulate the FWER of the Benjamini-Hochberg procedure under the intersection null hypothesis that the means of m normal distributions are all zero; use $\eta = .01, .05, .1$.

Exercise 34.12 Why is the result of Theorem 34.8 completely free of any assumptions about the underlying continuous distributions from which the samples are obtained?

Exercise 34.13 * Find the exact distribution of the number of false discoveries made by the Benjamini-Hochberg procedure when $m = 10, 25, 50, 500$

and the nominal FDR control level is $\alpha = .10$. Compare it with the asymptotic distribution.

Exercise 34.14 * Plot the set of (π_0, α) for which the Bonferroni-Hochberg procedure beats the uncorrected testing procedure asymptotically under the Genovese-Wasserman risk function when the common alternative is $N(\theta, 1), \theta = 1, 3, 5, 10$. Comment on the size and the shape of your plots.

Exercise 34.15 * Fix $\pi_0 = .01, \alpha = .05$, and let the common alternative be $N(\theta, 1)$. Find and plot the asymptotic risk function of the Benjamini-Hochberg procedure under the Genovese-Wasserman formulation.

Exercise 34.16 * Find the procedure φ_c that makes the FDR and the FNR equal up to the leading term when the common alternative is $N(\theta, 1)$ in the Genovese-Wasserman formulation. From this, suggest a multiple testing procedure by estimating the common θ and π_0.

Exercise 34.17 Compute the asymptotic expansion δ_n of Theorem 34.16 for the problem of testing that m normal means are ≤ 0 using a standard double exponential and a standard Cauchy prior and using $n = 15, \alpha = .05$. Do you get similar answers for both priors?

Exercise 34.18 * Can the coefficient c_1 in the asymptotic expansion of Theorem 34.16 ever be zero when the underlying distributions are normal?

Exercise 34.19 Show that the coefficient c_1 in the asymptotic expansion of Theorem 34.16 is strictly monotone in α when the underlying distributions are normal. Interpret this heuristically.

Exercise 34.20 * Simulate $n = 20$ observations from each of $m_0 = 100$ standard normal densities and $m_1 = 5$ $N(.25, 1)$ densities. By using the explicit $h(\alpha, m), \delta(t)$ as in Example 34.7, compute the Meinshausen-Rice estimate of m_1. Use $\alpha = .05$. Repeat the simulation a few times to study the sensitivity of the Meinshausen-Rice estimate.

Exercise 34.21 * Suppose X_i are conditionally independent $N(\theta_i, \tau)$ given (some) τ and $\tau \sim G$. Under what conditions on G do the X_i have a joint distribution with the property of increasing in θ?

Exercise 34.22 * Suppose X_i are conditionally independent $\text{Exp}(\tau\theta_i)$ given (some) τ and $\tau \sim G$. Under what conditions on G do the X_i have a joint distribution with the property of increasing in θ?

Exercise 34.23 Give examples of three parametric families of distributions that are *not* MTP_2.

Exercise 34.24 Let $Z_i, i = 0, \cdots, m$ be independent Poisson(λ_i), and let $X_i = Z_0 + Z_i, i = 1, \cdots, m$. Prove or disprove that the joint distribution of the X_i is MTP_2; this is called a multivariate Poisson distribution.

Exercise 34.25 Suppose f is $N(\theta, 1)$ and $g = \text{Exp}(\lambda)$. Prove or disprove that the normalized density fg is MTP_2; find this density in closed form. What does MTP_2 mean in this case?

Exercise 34.26 * Suppose f is $N(\theta, 1)$ and $g = C(0, 1)$, the standard Cauchy. Prove or disprove that the normalized density fg is MTP_2.

Exercise 34.27 * Find the areas (i.e., the integrals) in the (β, r) plane that are below the Donoho-Jin detection boundary $r = p_{\text{HC}}(\beta)$ between $p_{\text{HC}}(\beta)$ and $p^*(\beta)$ and above $p^*(\beta)$. Interpret the results heuristically.

Exercise 34.28 * Take $\beta = .75$ and $r = p_{\text{HC}}(\beta)$ to get a point on exactly the Donoho-Jin detection boundary. Simulate the higher-criticism statistic by generating data from the mixture model in their setup, and draw a histogram of the higher-criticism values. Use $m = 100, 500, 1000$. Comment on your results.

Exercise 34.29 If $\tau = .5$, explicitly characterize all those window widths t_n that satisfy the Hall-Jin condition in the text.

Exercise 34.30 * Take $\delta = 1, \tau = .05$, and simulate the Hall-Jin higher-criticism statistic by using a series of length $n = 100, 500$ from the null case. Use a window width as in the previous exercise.

References

Abramovich, F., Benjamini, Y., Donoho, D., and Johnstone, I. (2006). Adapting to unknown sparsity by controlling the FDR, Ann. Stat., 34, 584–653.

Benjamini, Y. and Hochberg, Y. (1995). Controlling the false discovery rate: a practical and powerful approach to multiple testing, J.R. Stat. Soc. B, 57, 289–300.

Benjamini, Y. and Yekutieli, D. (2001). The control of the false discovery rate in multiple testing under dependency, Ann. Stat., 29, 1165–1188.

Bernhard, G., Klein, M., and Hommel, G. (2004). Global and multiple test procedures using ordered P-values—a review, Stat. Papers, 45, 1–14.

Bickel, P. and Levina, E. (2004). Some theory of Fisher's linear discriminant function, 'naive Bayes', and some alternatives when there are many more variables than observations, Bernoulli, 10, 989–1010.

Birnbaum, Z. and Pyke, R. (1958). On some distributions related to the statistic D_n^+, Ann. Math. Stat., 29, 179–187.

Brown, L. (1979). A proof that the Tukey-Kramer multiple comparison procedure is level α for 3, 4, or 5 treatments, Technical Report, Cornell University.

Brown, L. (1984). A note on the Tukey-Kramer procedure for pairwise comparison of correlated means, in *Design of Experiments: Ranking and Selection*, T.J. Santner and A. Tamhane (eds.), Marcel Dekker, New York, 1–6.

Chang, C., Rom, D., and Sarkar, S. (1996). A modified Bonferroni procedure for repeated significance testing, Technical Report, Temple University.

DasGupta, A. and Zhang, T. (2006). On the false discovery rates of a frequentist: asymptotic expansions, Lecture Notes and Monograph Series, Vol. 50, Institute of Mathematical Statistics, Beachwood, OH, 190–212.

Delaigle, A. and Hall, P. (2006). Using thresholding methods to extend higher criticism classification to non-normal dependent vector components, manuscript.

Dempster, A. (1959). Generalized D_n^+ statistic, Ann. Math. Stat., 30, 593–597.

Donoho, D. and Jin, J. (2004). Higher criticism for detecting sparse heterogenous mixtures, Ann. Stat., 32, 962–994.

Donoho, D. and Jin, J. (2006). Asymptotic minimaxity of FDR thresholding for sparse Exponential data, Ann. Stat., 34, 2980–3018.

Efron, B. (2003). Robbins, Empirical Bayes, and microarrays, Ann. Stat., 31, 366–378.

Efron, B. (2007). Correlation and large scale simultaneous significance testing, J. Am. Stat. Assoc., 102, 93–103.

Efron, B., Tibshirani, R., Storey, J. and Tusher, V. (2001). Empirical Bayes analysis of a microarray experiment, J. Am. Stat. Assoc., 96, 1151–1160.

Finner, H. and Roters, M. (2002). Multiple hypothesis testing and expected number of type I errors, Ann. Stat., 30, 220–238.

Genovese, C. and Wasserman, L. (2002). Operating characteristics and extensions of the false discovery rate procedure, J. R. Stat. Soc. B, 64, 499–517.

Genovese, C. and Wasserman, L. (2004). A stochastic process approach to false discovery control, Ann. Stat., 32, 1035–1061.

Genovese, C., Roeder, K. and Wasserman, L. (2006). False discovery control with P-value weighting, Biometrika, 93, 509–524.

Hall, P. and Jin, J. (2007). Performance of higher criticism under strong dependence (in press).

Hayter, A. (1984). A proof of the conjecture that the Tukey-Kramer multiple comparisons procedure is conservative, Ann. Stat., 12, 61–75.

Hochberg, Y. and Tamhane, A. (1987). *Multiple Comparisons Procedures*, John Wiley, New York.

Hochberg, Y. (1988). A sharper Bonferroni procedure for multiple tests of significance, Biometrika, 75, 800–803.

Holst, L. (1972). Asymptotic normality and efficiency for certain goodness of fit tests, Biometrika, 59, 137–145.

Hommel, G. (1988). A stage-wise rejective multiple test procedure based on a modified Bonferroni test, Biometrika, 75, 383–386.

Huber, P. (1973). Robust regression: asymptotics, conjectures, and Monte Carlo, Ann. Stat., 1, 799–821.

Karlin, S. and Rinott, Y. (1980). Classes of orderings of measures and related correlation inequalities I: multivariate totally positive distributions, J. Multivar. Anal., 10, 467–498.

Koehler, K. and Larntz, K. (1980). An empirical investigation of goodness of fit statistics for sparse multinomials, J. Am. Stat. Assoc., 75, 336–344.

Kramer, C. (1956). Extension of multiple range tests to group means with unequal numbers of replications, Biometrics, 12, 307–310.

Meinshausen, M. and Bühlmann, P. (2005). Lower bounds for the number of false null hypotheses for multiple testing of association under general dependence structures, Biometrika, 92, 893–907.

Meinshausen, M. and Rice, J. (2006). Estimating the proportion of false null hypotheses among a large number of independent hypotheses, Ann. Stat., 34, 373–393.

Miller, R. (1966). *Simultaneous Statistical Inference*, McGraw-Hill, New York.

Morris, C. (1975). Central limit theorems for multinomial sums, Ann. Stat., 3, 165–188.

Portnoy, S. (1984). Asymptotic behavior of M-estimates of p regression parameters when p^2/n is large I: consistency, Ann. Stat., 12, 1298–1309.

Portnoy, S. (1985). Asymptotic behavior of M-estimates of p regression parameters when p^2/n is large II: normal approximation, Ann. Stat., 13, 1403–1417.

Portnoy, S. (1988). Asymptotic behavior of likelihood methods for Exponential families when the number of parameters tends to infinity, Ann. Stat., 16, 356–366.

Sarkar, S. and Chang, C. (1997). The Simes method for multiple hypotheses testing with positively dependent test statistics, J. Am. Stat. Assoc., 92, 1601–1608.

Sarkar, S. (2006). False discovery and false nondiscovery rates in single step multiple testing procedures, Ann. Stat., 34, 394–415.

Scott, J. and Berger, J. (2006). An exploration of aspects of Bayesian multiple testing, J. Stat. Planning Infer, 136, 2144–2162.

Seo, T., Mano, S., and Fujikoshi, Y. (1994). A generalized Tukey conjecture for multiple comparison among mean vectors, J. Am. Stat. Assoc., 89, 676–679.

Shaffer, J. (1995). Multiple hypothesis testing: a review, Annu. Rev. Psychol., 46, 561–584.

Simes, R. (1986). An improved Bonferroni procedure for multiple tests of significance, Biometrika, 73, 751–754.

Sóric, B. (1989). Statistical 'discoveries' and effect-size estimation, J. Am. Stat. Assoc., 84, 608–610.

Storey, J. (2002). A direct approach to false discovery rates, J. R. Stat. Soc. B, 64, 479–498.

Storey, J. (2003). The positive false discovery rate: a Bayesian interpretation and the q-value, Ann. Stat., 31, 2013–2035.

Storey, J. and Tibshirani, R. (2003). Statistical significance for genomewide studies, Proc. Natl. Acad. Sci. USA, 16, 9440–9445.

Storey, J., Taylor, J., and Siegmund, D. (2004). Strong control, conservative point estimation and simultaneous conservative consistency of false discovery rates: a unified approach, J. R. Stat. Soc. B, 66, 187–205.

Sun, W. and Cai, T. (2007). Oracle and adaptive compound decision rules for false discovery rate control, in Press.

Tukey, J. (1953). The problem of multiple comparisons, in *The Collected Works of John Tukey*, Vol. VIII, Chapman and Hall, New York, 1–300.

Tukey, J. (1991). The philosophy of multiple comparison, Stat. Sci., 6, 100–116.

Tukey, J. (1993). Where should multiple comparisons go next? In *Multiple Comparisons, Selections, and Applications in Biometry*, F. Hoppe (ed.), Marcel Dekker, New York.

Westfall, P. and Young, S. (1993). *Resampling Based Multiple Testing*, John Wiley, New York.

Wright, S. (1992). Adjusted P-values for simultaneous inference, Biometrics, 48, 1005–1013.

Yohai, V. and Maronna, R. (1979). Asymptotic behavior of M-estimators for the linear model, Ann. Stat., 7, 258–268.

Chapter 35
A Collection of Inequalities in Probability, Linear Algebra, and Analysis

35.1 Probability Inequalities

35.1.1 Improved Bonferroni Inequalities

Galambos Inequality Given $n \geq 3$, events $A_1, \cdots, A_n$, $S_{1,n} = \sum P(A_i)$, $S_{2,n} = \sum_{1 \leq i < j \leq n} P(A_i \cap A_j)$, $S_{3,n} = \sum_{1 \leq i < j < k \leq n} P(A_i \cap A_j \cap A_k)$,

$$S_{1,n} - S_{2,n} + \frac{2}{n-1} S_{3,n} \leq P(\cup A_i) \leq S_{1,n} - \frac{2}{n} S_{2,n}.$$

Galambos, J. and Simonelli, I. (1996). *Bonferroni-Type Inequalities with Applications*, Springer-Verlag, New York.

Móri-Székely Inequality Given events $A_1, \cdots, A_n$, $S_{1,n} = \sum P(A_i)$, $S_{2,n} = \sum_{1 \leq i < j \leq n} P(A_i \cap A_j)$, $1 \leq m \leq n-1$, and M_n the number of events among $A_1, \cdots, A_n$ that occur,

$$P(M_n = m) \geq (2m-1)S_{1,n} - 2S_{2,n} + 1 - m^2.$$

Móri, T. and Székely, G. (1985). J. Appl. Prob., 22, 836–843.

Chung-Erdös Inequality Given events $A_1, \cdots, A_n$, $P_i = P(A_i)$, $P_{ij} = P(A_i \cap A_j)$, $B = \sum P_i$, $C = \sum_{j > i = 1}^{n-1} P_{ij}$,

$$P(\cup_{i=1}^n A_i) \geq \frac{B^2}{2C + B}.$$

Rao, C.R. (1973). *Linear Statistical Inference and Applications*, Wiley, New York.

Dawson-Sankoff Inequality

Given events $A_1, \cdots, A_n, P_i = P(A_i), P_{ij} = P(A_i \cap A_j), B = \sum P_i, C = \sum_{j>i=1}^{n-1} P_{ij}, \rho = [\frac{2C}{B}]$,

$$P(\cup_{i=1}^n A_i) \geq \frac{2(B+C)}{2+\rho} - \frac{2C}{1+\rho}.$$

Dawson, D. and Sankoff, D. (1967). Proc. Am. Math. Soc., 18, 504–507.

Kounias Inequality Given events $A_1, \cdots, A_n, P_i = P(A_i), P_{ij} = P(A_i \cap A_j), P' = (P_1, \cdots, P_n), Q = ((P_{ij}))$, and Q^- a generalized inverse such that $QQ^-Q = Q$,

$$P(\cup_{i=1}^n A_i) \geq P'Q^-P.$$

Kounias, E. (1968). Ann. Math. Stat., 39, 2154–2158.

35.1.2 Concentration Inequalities

Cantelli's Inequality If $E(X) = \mu$, Var$(X) = \sigma^2 < \infty$, then

$$P(X - \mu \geq \lambda) \leq \frac{\sigma^2}{\sigma^2 + \lambda^2}, \lambda > 0,$$

$$P(X - \mu \leq \lambda) \leq \frac{\sigma^2}{\sigma^2 + \lambda^2}, \lambda < 0.$$

Rao, C.R. (1973). *Linear Statistical Inference and Applications*, Wiley, New York.

Paley-Zygmund Inequality Given a nonnegative random variable X with a finite mean μ and any $c > 0$,

$$P(X > c\mu) \geq (1-c)_+^2 \frac{\mu^2}{EX^2}.$$

Paley, R. and Zugmund, A. (1932). Proc. Cambridge Philos. Soc., 28, 266–272.

Alon-Spencer Inequality Given a nonnegative integer-valued random variable X with a finite variance,

$$P(X = 0) \leq \frac{\text{Var}(X)}{E(X^2)}.$$

Alon, N. and Spencer, J. (2000). *The Probabilistic Method*, Wiley, New York.

Carlson Inequality Given any random variable X with a density f on the real line,

$$\text{Var}(X) \geq \frac{1}{4\pi^2}(\int \sqrt{f})^4.$$

Bullen, P. (1998). *A Dictionary of Inequalities*, Addison-Wesley Longman, Harlow.

Reverse Carlson Inequality Given any random variable X with a density f on the real line,

$$E|X| \leq \sup(|x|\sqrt{f(x)}) \int \sqrt{f}.$$

Anderson Inequality I Given $X_{n\times 1} \sim f(x)$, f symmetric around the origin, and such that $\{x : f(x) \geq c\}$ is convex for any $c > 0$,

$$P(X + s\theta \in C) \geq P(X + t\theta \in C)$$

for all $0 \leq s < t < \infty$, all $\theta \in \mathcal{R}^n$, and any symmetric convex set C.

Tong, Y. (1980). *Probability Inequalities in Multivariate Distributions*, Academic Press, New York.

Anderson Inequality II Given $X \sim N(0, \Sigma_1)$, $Y \sim N(0, \Sigma_2)$, $\Sigma_2 - \Sigma_1$ nnd, and C any symmetric convex set,

$$P(X \in C) \geq P(Y \in C).$$

Tong, Y. (1980). *Probability Inequalities in Multivariate Distributions*, Academic Press, New York.

Prékopa-Leindler Convolution Inequality Given density functions f, g on the real line and $p, q \geq 1$, $\frac{1}{p} + \frac{1}{q} - 1 = \frac{1}{r}$,

$$\int_{\mathcal{R}} \left(\int_{\mathcal{R}} (f(y)g(x-y))^r dy \right)^{1/r} dx \geq ||f||_p ||g||_q,$$

provided all the integrals exist.

Bullen, P. (1998). *A Dictionary of Inequalities*, Addison-Wesley Longman, New York.

Loève's Inequality Given independent random variables $X_1, \cdots, X_n$, $E(X_i) = \mu_i$, $\text{Var}(X_i) = \sigma_i^2$, $X_i \leq \mu_i + M$, for any $t > 0$,

$$P(\sum(X_i - \mu_i) \geq tB_n) \leq e^{-\frac{t^2}{2}(1-\frac{tM}{2B_n})}$$

if $\frac{tM}{B_n} \leq 1$ and

$$P(\sum(X_i - \mu_i) \geq tB_n) \leq e^{-\frac{tB_n}{4M}}$$

if $\frac{tM}{B_n} \geq 1$, where $B_n^2 = \sum \sigma_i^2$.

Loève, M. (1955). *Probability Theory*, Van Nostrand, New York.

Berry's Inequality Given independent random variables $X_1, \cdots, X_n$, $E(X_i) = \mu_i$, $\text{Var}(X_i) = \sigma_i^2$, $X_i \leq \mu_i + M$, for any $t > 0$,

$$P(\sum(X_i - \mu_i) \geq tB_n) \leq 1 - \Phi(t) + 2\frac{M}{B_n},$$

where Φ is the standard normal CDF.

Berry, A. (1941). Trans. Am. Math. Soc., 49, 122–136.

Bernstein Inequality Given a nonnegative random variable X with a finite mgf $\psi(a) = E(e^{aX})$,

$$P(X \geq t) \leq \inf_{a>0} e^{-at}\psi(a).$$

Hoeffding Inequality Given constants $a_1, \cdots, a_n$, and iid *Rademacher variables* $\varepsilon_1, \cdots, \varepsilon_n$, $P(\varepsilon_i = \pm 1) = \frac{1}{2}$, for any $t > 0$,

$$P(\sum a_i \varepsilon_i \geq t) \leq e^{-\frac{t^2}{2\sum a_i^2}}.$$

Hoeffding, W. (1963). J. Am. Stat. Assoc., 58, 13–30.

Hoeffding-Chernoff Inequality Given iid random variables $X_1, \cdots, X_n$, $0 \leq X_i \leq 1$, $E(X_i) = \mu$, for any $t > 0, t + \mu < 1$,

$$P\left(\sum(X_i - \mu) > nt\right) \leq e^{-nD(\mu+t,\mu)},$$

where $D(x, y) = x \log \frac{x}{y} + (1 - x) \log \frac{1-x}{1-y}$.

Chernoff, H. (1952). Ann. Math. Stat., 23, 493–507.

Prohorov Inequality Given independent random variables $X_1, \cdots, X_n$, $E(X_i) = \mu_i$, $\text{Var}(X_i) = \sigma_i^2$, $|X_i - \mu_i| \le M$,

$$P\left(\sum (X_i - \mu_i) \ge t B_n\right) \le e^{-\frac{tB_n}{2M} \text{arcsinh} \frac{tM}{2B_n}},$$

where $B_n^2 = \sum \sigma_i^2$.

Prohorov, Y. (1959). Theory Prob. Appl., 4, 201–204.

Eaton Inequality Given independent random variables $X_1, \cdots, X_n$, $E(X_i) = \mu_i$, $|X_i - \mu_i| \le 1$, constants $a_1, \cdots, a_n$, such that $\sum_{i=1}^n a_i^2 = 1$, and $t > 0$,

$$P\left(|\sum_{i=1}^n a_i (X_i - \mu_i)| \ge t\right) \le 2 \inf_{0 \le u \le t} \int_u^\infty \frac{(x-u)^3}{(t-u)^3} \varphi(x) dx$$

$$\le \frac{4e^3}{9} \frac{\varphi(t) e^{-\frac{9}{2t^2}}}{t(1 - 3/t^2)^4}$$

if $t > \sqrt{3}$.

Eaton, M. (1974). Ann. Stat., 2, 609–613.

Pinelis Inequality Given iid *Rademacher variables* $\varepsilon_1, \cdots, \varepsilon_n$, with $P(\varepsilon_1 = \pm 1) = \frac{1}{2}$, constants $a_1, \cdots, a_n$, such that $\sum_{i=1}^n a_i^2 = 1$, and $t > 0$,

$$P\left(|\sum_{i=1}^n a_i \varepsilon_i| \ge t\right) \le \frac{4e^3}{9}(1 - \Phi(t)) \le \frac{4e^3}{9} \frac{\varphi(t)}{t}.$$

van der Vaart, A. and Wellner, J. (1996). *Weak Convergence and Empirical Processes*, Springer, New York.

Reformulated Bennett Inequality of Shorack Given independent random variables $X_1, \cdots, X_n$, $E(X_i) = \mu_i$, $\text{Var}(X_i) = \sigma_i^2$, $X_i \le \mu_i + M$, for any $t > 0$,

$$P(\sqrt{n}(\bar{X} - \bar{\mu}) \ge t) \le e^{-\frac{t^2}{2\bar{\sigma}^2} \psi\left(\frac{tM}{\sqrt{n}\bar{\sigma}^2}\right)},$$

where $\bar{\mu} = \frac{1}{n}\sum_{i=1}^n \mu_i$, $\bar{\sigma}^2 = \frac{1}{n}\sum_{i=1}^n \sigma_i^2$, and $\psi(\lambda) = \frac{2}{\lambda^2}[1 + (1+\lambda)(\log(1+\lambda) - 1)] \ge \frac{3}{3+\lambda}$ for all $\lambda > 0$.

Shorack, G. (1980). Aust. J. Stat., 22, 50–59.

Robbins Inequality Given independent random variables $X_1, \cdots, X_n$,

$$P(\cap_{k=1}^n \{X_1 + \cdots + X_k \leq a_k\}) \geq \prod_{k=1}^n P(X_1 + \cdots + X_k \leq a_k).$$

Robbins, H. (1954). Ann. Math. Stat., 25, 614–616.

Kolmogorov's Maximal Inequality For independent random variables $X_1, \cdots, X_n$, $E(X_i) = 0$, $\text{Var}(X_i) < \infty$, for any $t > 0$,

$$P\left(\max_{1 \leq k \leq n} |X_1 + \cdots + X_k| \geq t\right) \leq \frac{\sum_{k=1}^n \text{Var}(X_k)}{t^2}.$$

Bullen, P. (1998). *A Dictionary of Inequalities*, Addison-Wesley Longman, Harlow.

Doob Submartingale Inequality Given a nonnegative submartingale X_t on a probability space $(\Omega, \mathcal{A}, P)$, $0 \leq t \leq T$, $M_t = \sup_{0 \leq s \leq t} X_s$, and any constant $c > 0$,

$$P(M_T \geq c) \leq \frac{EX_T}{c},$$

$$||X_T||_p \leq ||M_T||_p \leq \frac{p}{p-1}||X_T||_p, \; p > 1.$$

Bullen, P. (1998). *A Dictionary of Inequalities*, Addison-Wesley Longman, Harlow.

Hájek-Rényi Inequality For independent random variables $X_1, \cdots, X_n$, $E(X_i) = 0$, $\text{Var}(X_i) = \sigma_i^2$, and a nonincreasing positive sequence c_k,

$$P\left(\max_{m \leq k \leq n} c_k |(X_1 + \cdots + X_k)| \geq \varepsilon\right) \leq \frac{1}{\varepsilon^2}\left(c_m^2 \sum_{k=1}^m \sigma_k^2 + \sum_{k=m+1}^n c_k^2 \sigma_k^2\right).$$

Hájek, J. and Rényi, A. (1955). Acta Math. Acad. Sci. Hung., 6, 281–283.

Lévy Inequality Given independent random variables $X_1, \cdots, X_n$, each symmetric about zero,

$$P\left(\max_{1 \leq k \leq n} |X_k| \geq x\right) \leq 2P(|X_1 + \cdots + X_n| \geq x),$$

$$P\left(\max_{1\le k\le n}|X_1+\cdots+X_k|\ge x\right)\le 2P(|X_1+\cdots+X_n|\ge x).$$

Bullen, P. (1998). *A Dictionary of Inequalities*, Addison-Wesley Longman, Harlow.

Improved Lévy Inequality of Bhattacharya Given independent random variables $X_1, \cdots, X_n$, each symmetric about zero,

$$P\left(\max_{1\le k\le n}(X_1+\cdots+X_k)\ge x\right)\le 2P(X_1+\cdots+X_n\ge x)$$
$$-P(X_1+\cdots+X_n=x).$$

Doob Inequality Given independent random variables $X_1, \cdots, X_n$, each symmetric about zero, for any $\varepsilon > 0$,

$$P\left(\max_{1\le k\le n}(X_1+\cdots+X_k)\ge x\right)\ge 2P(X_1+\cdots+X_n\ge x+2\varepsilon)$$
$$-2\sum_{k=1}^{n}P(X_k\ge\varepsilon).$$

Doob, J. (1953). *Stochastic Processes*, Wiley, New York.

Bickel Inequality Given independent random variables $X_1, \cdots, X_n$, each symmetric about zero, a nonincreasing positive sequence c_k, and a nonnegative convex function g,

$$P\left(\max_{1\le k\le n}c_k g(X_1+\cdots+X_k)\ge x\right)$$
$$\le 2P\left(\sum_{k=1}^{n-1}(c_k-c_{k+1})g(X_1+\cdots+X_k)+c_n g(X_1+\cdots+X_n)\ge x\right).$$

Bickel, P. (1970). Acta Math. Acad. Sci. Hung., 21, 199–206.

35.1.3 Tail Inequalities for Specific Distributions

Normal Distribution Given $X \sim N(\mu, \sigma^2)$, for any $k > 0$,

$$P(|X-\mu|>k\sigma)<\frac{1}{3k^2}.$$

DasGupta, A. (2000). Metrika, 51, 185–200.

Gamma Distribution Given $X \sim \text{Gamma}(\alpha, \lambda)$ with density $f(x) = \frac{\lambda^\alpha}{\Gamma(\alpha)} e^{-\lambda x} x^{\alpha-1}$, for any $k > E(X) = \frac{\alpha}{\lambda}$,

$$P(X > k) \le \frac{k}{k\lambda - \alpha} f(k).$$

DasGupta, A. (2000). Metrika, 51, 185–200.

Beta Distribution Given $X \sim \text{Beta}(\alpha, \beta), \beta > 1$ with density $f(x) = \frac{x^{\alpha-1}(1-x)^{\beta-1}}{B(\alpha,\beta)}$, for all $k > \frac{\alpha}{\alpha+\beta-1}$,

$$P(X > k) \le \frac{k(1-k)}{k(\alpha+\beta-1)-\alpha} f(k).$$

DasGupta, A. (2000). Metrika, 51, 185–200.

t-distribution Given $X \sim t_\alpha$, the standard t-distribution with α degrees of freedom, for any $k > 1$,

$$P(X > k) \le \frac{k(\alpha + k^2)}{\alpha(k^2 - 1)} f(k),$$

where f denotes the t_α density.

DasGupta, A. (2000). Metrika, 51, 185–200.

Bhattacharya-Rao Inequality Given a random variable X with CDF F, mean 0, and variance 1,

$$\sup_{-\infty < x < \infty} |F(x) - \Phi(x)| \le \frac{11}{20},$$

where Φ is the $N(0, 1)$ CDF.

Bhattacharya, R. and Rao, R. (1976). *Normal Approximation and Asymptotic Expansions*, Wiley, New York.

Cirel'son et al. Concentration Inequality for Gaussian Measures I Given $X_{n\times 1} \sim N(0, I_n)$ and $f : \mathcal{R}^n \to \mathcal{R}$ a Lipschitz function with Lipschitz norm $= \sup_{x,y} \frac{|f(x)-f(y)|}{||x-y||} \le 1$, for any $t > 0$,

$$P(f(X) - Ef(X) \ge t) \le e^{-\frac{t^2}{2}}.$$

van der Vaart, A. and Wellner, J. (1996). *Weak Convergence and Empirical Processes*, Springer, New York.

Borell Concentration Inequality for Gaussian Measures II Given $X_{n\times 1} \sim N(0, I_n)$ and $f : \mathcal{R}^n \to \mathcal{R}$ a Lipschitz function with Lipschitz norm $= \sup_{x,y} \frac{|f(x)-f(y)|}{||x-y||} \leq 1$, for any $t > 0$,

$$P(f(X) - M_f \geq t) \leq e^{-\frac{t^2}{2}},$$

where M_f denotes the median of $f(X)$.

van der Vaart, A. and Wellner, J. (1996). *Weak Convergence and Empirical Processes*, Springer, New York.

35.1.4 Inequalities under Unimodality

3σ Rule for Unimodal Variables If X is unimodal with an absolutely continuous distribution, $\mu = E(X)$, $\sigma^2 = \text{Var}(X) < \infty$, then

$$P(|X - \mu| \geq 3\sigma) \leq \frac{4}{81} < .05.$$

Dharmadhikari, S. and Joag-Dev, K. (1988). *Unimodality, Convexity and Applications*, Academic Press, New York.

Vysochanskiĭ-Petunin Inequality for Unimodal Variables If X is unimodal with an absolutely continuous distribution, $\alpha \in \mathcal{R}$, $\tau^2 = E(X - \alpha)^2$, then

$$P(|X - \alpha| \geq k) \leq \frac{4\tau^2}{9k^2}, k \geq \frac{\sqrt{8}\tau}{\sqrt{3}},$$

$$P(|X - \alpha| \geq k) \leq \frac{4\tau^2}{3k^2} - \frac{1}{3}, k \leq \frac{\sqrt{8}\tau}{\sqrt{3}}.$$

Pukelsheim, F. (1994). Am. Stat., 48, 88–91.

Johnson-Rogers Inequality for Unimodal Variables If X is unimodal around M with an absolutely continuous distribution, $E(X) = \mu$, $\text{Var}(X) = \sigma^2$, then

$$|\mu - M| \leq \sigma\sqrt{3}.$$

Dharmadhikari, S. and Joag-Dev, K. (1988). *Unimodality, Convexity and Applications*, Academic Press, New York.

Dictionary Ordering for Unimodal Variables If X is unimodal around a unique mode M with median m and finite mean μ, and if the density of $(X - m)^-$ crosses the density of $(X - m)^+$ only once and from above, then $M \le m \le \mu$.

Dharmadhikari, S. and Joag-Dev, K. (1988). *Unimodality, Convexity and Applications*, Academic Press, New York.

Berge Inequality for Bivariate Distributions If $X = (X_1, X_2)$ is a two-dimensional variable with coordinate means μ_1, μ_2, variances σ_1^2, σ_2^2, and correlation ρ, then

$$P\left(\max\left(\frac{|X_1 - \mu_1|}{\sigma_1}, \frac{|X_2 - \mu_2|}{\sigma_2}\right) \ge k\right) \le \frac{1 + \sqrt{1 - \rho^2}}{k^2}.$$

Dharmadhikari, S. and Joag-Dev, K. (1988). *Unimodality, Convexity and Applications*, Academic Press, New York.

Dharmadhikari-Joagdev Inequality for Bivariate Unimodal Variables If $X = (X_1, X_2)$ is a two-dimensional variable with density f, coordinate means μ_1, μ_2, variances σ_1^2, σ_2^2, and correlation ρ, and such that $f(t(x_1 - \mu_1, x_2 - \mu_2))$ is nonincreasing in $t, t > 0$ for all (x_1, x_2), then

$$P\left(\max\left(\frac{|X_1 - \mu_1|}{\sigma_1}, \frac{|X_2 - \mu_2|}{\sigma_2}\right) \ge k\right) \le \frac{1 + \sqrt{1 - \rho^2}}{2k^2}, k > 0.$$

Dharmadhikari, S. and Joag-Dev, K. (1988). *Unimodality, Convexity and Applications*, Academic Press, New York.

Multivariate Chebyshev Inequality of Olkin and Pratt I

Given $X = (X_1, \cdots, X_n)$, $E(X_i) = \mu_i$, $\text{Var}(X_i) = \sigma_i^2$, $\text{corr}(X_i, X_j) = 0$ for $i \ne j$,

$$P\left(\cup_{i=1}^n \frac{|X_i - \mu_i|}{\sigma_i} \ge k_i\right) \le \sum_{i=1}^n k_i^{-2}.$$

Olkin, I. and Pratt, J. (1958). Ann. Math. Stat., 29, 226–234.

Multivariate Chebyshev Inequality of Olkin and Pratt II

Given $X = (X_1, \cdots, X_n)$, $E(X_i) = \mu_i$, $\text{Var}(X_i) = \sigma_i^2$, $\text{corr}(X_i, X_j) = \rho$ for $i \ne j$,

$$P\left(\cup_{i=1}^n \frac{|X_i - \mu_i|}{\sigma_i} \ge k\right) \le \frac{[(n-1)\sqrt{1-\rho} + \sqrt{1+(n-1)\rho}]^2}{nk^2}.$$

Olkin, I. and Pratt, J. (1958). Ann. Math. Stat., 29, 226–234.

35.1.5 *Moment and Monotonicity Inequalities*

Hotelling-Solomons Inequality Given a random variable X with $E(X) = \mu$, Var$(X) = \sigma^2 < \infty$, and median m,

$$|\mu - m| \leq \sigma.$$

Hotelling, H. and Solomons, L. (1932). Ann. Math. Stat., 3, 141–142.

Lyapounov Inequality Given a nonnegative random variable X and $0 < \alpha < \beta$,

$$(EX^{\alpha})^{\frac{1}{\alpha}} \leq (EX^{\beta})^{\frac{1}{\beta}}.$$

Improved Lyapounov Inequality of Petrov Given a nonnegative random variable X, $0 < \alpha < \beta, and q = P(X = 0)$,

$$EX^{\alpha} \leq (1-q)^{1-\alpha/\beta}(EX^{\beta})^{\frac{\alpha}{\beta}}.$$

Petrov, V. (1995). *Limit Theorems of Probability Theory*, Oxford University Press, Oxford.

Reciprocal Moment Inequality of Mark Brown Given positive independent random variables X, Y,

$$\frac{E\frac{1}{X+Y}}{E\frac{1}{(X+Y)^2}} \geq \frac{E\frac{1}{X}}{E\frac{1}{X^2}} + \frac{E\frac{1}{Y}}{E\frac{1}{Y^2}}.$$

Olkin, I. and Shepp, L. (2006). Am. Math. Mon., 113, 817–822.

Generalized Mark Brown Inequality of Olkin and Shepp Given positive independent random variables X, Y, and a positive logconvex function g,

$$\frac{E[(X+Y)g(X+Y)]}{E[g(X+Y)]} \geq \frac{E[Xg(X)]}{E[g(X)]} + \frac{E[Yg(Y)]}{E[g(Y)]}.$$

Olkin, I. and Shepp, L. (2006). Am. Math. Mon., 113, 817–822.

Littlewood Inequality Given $r \geq s \geq t \geq 0$ and any random variable X such that $E|X|^r < \infty$,

$$(E|X|^r)^{s-t}(E|X|^t)^{r-s} \geq (E|X|^s)^{r-t},$$

and, in particular, for any variable X with a finite fourth moment,

$$E|X| \geq \frac{(EX^2)^{\frac{3}{2}}}{\sqrt{EX^4}}.$$

de la Peña, V. and Giné, E. (1999). *Decoupling: From Dependence to Independence*, Springer, New York.

Log Convexity Inequality of Lyapounov Given a nonnegative random variable X and $0 \leq \alpha_1 < \alpha_2 \leq \frac{\beta}{2}$,

$$EX^{\alpha_1} EX^{\beta-\alpha_1} \geq EX^{\alpha_2} EX^{\beta-\alpha_2}.$$

Arnold Inequality Given a nonnegative random variable X and $0 < \alpha < \beta, q = P(X = 0), p = P(X = 1)$,

$$EX^\alpha \leq p + (1 - p - q)^{1-\alpha/\beta}(EX^\beta - p)^{\frac{\alpha}{\beta}}.$$

Arnold, B. (1978). SIAM J. Appl. Math., 35, 117–118.

von Mises Inequality Given an integer-valued random variable X with $\mu = E(X), \nu_k = E|X - \mu|^k$,

$$\nu_k \leq 2\nu_{k+1}.$$

von Mises, R. (1939). Skand. Aktuarietidskr., 32–36.

Chow-Studden Inequality Given a random variable X and constants a, b, with $U = \min(X, a), V = \max(X, b), and W = \max(b, U)$,

$$\text{Var}(U, V, W) \leq \text{Var}(X).$$

Chow, Y. and Studden, W. (1969). Ann. Math. Stat., 40, 1106–1108.

Efron-Stein Inequality Given iid random variables $X_1, \cdots, X_n$, a permutation-invariant function $S(X_1, \cdots, X_n)$, $S_i = S(X_1, \cdots, X_{i-1}, X_{i+1}, \cdots, X_n)$, $\bar{S} = \frac{1}{n}\sum_{i=1}^n S_i, and V_{\text{Jack}} = \sum_{i=1}^n (S_i - \bar{S})^2$,

$$E[V_{\text{Jack}}] \geq \text{Var}[S(X_1, \cdots, X_{n-1})], n \geq 2.$$

Efron, B. and Stein, C. (1981). Ann. Stat., 9, 586–596.

Feller Inequality on Poisson Mixtures Given a random variable $Y \sim F_Q$ that is a Q-mixture of Poisson(λ) distributions, Var(Y) $\geq$ Var(X), where $X \sim$ Poisson with mean $= E(Y)$.

Feller, W. (1943). Ann. Math. Stat., 14 389–400.

Generalized Feller Inequality of Molenaar and van Zwet Given a family of densities $\{f(x|\theta)\}, x, \theta \in \mathcal{R}$, if $f(x|\theta)$ is TP3 (totally positive of order 3), then $\text{Var}(Y) \geq \text{Var}(X)$, where Y has a mixture density $\int f(x|\theta)dQ(\theta)$ and $X \sim f(x|\theta_0)$, where θ_0 is such that X and Y have the same expectation.

Molenaar, W. and van Zwet, W. (1966). Ann. Math. Stat., 37, 281–283.

Kemperman Inequality Given $X_i \overset{\text{indep.}}{\sim} \text{Ber}(p_i)$, N a Poisson random variable independent of the X_i, and $Y = X_1 + \cdots X_N$,

$$\text{Var}(Y) \leq E(Y)$$

if p_i are nonincreasing and

$$\text{Var}(Y) \geq E(Y)$$

if p_i are nondecreasing.

Kemperman, J. (1979). J. Appl. Prob., 16, 220–225.

Keilson-Steutel Inequality on Scale Mixtures Given random variables $X, Y, W, X \overset{\mathcal{L}}{=} YW$, where Y, W are independent, and $W \geq 0$,

$$\text{cv}(X) \geq \text{cv}(Y),$$

where cv denotes the coefficient of variation, $\text{cv}^2(U) = \frac{\text{Var}(U)}{(E|U|)^2}$.

Keilson, J. and Steutel, F. (1974). Ann. Prob., 2, 112–130.

Keilson-Steutel Inequality on Powers Given a random variable $W \geq 0$, $E(W^{2r}) < \infty$, for some given $r > 0$, and $0 < \alpha < \beta < r$,

$$\text{cv}(W^{\alpha}) \leq \text{cv}(W^{\beta}).$$

Keilson, J. and Steutel, F. (1974). Ann. Prob., 2, 112–130.

Keilson-Steutel Inequality on Power Mixtures Given an infinitely divisible random variable Y with characteristic function $\psi_Y(t)$ and X defined as the random variable with characteristic function $\psi_X(t) = \int_{[0,\infty)}(\psi_Y(t))^w dQ(w)$, Q a probability measure,

$$\text{cv}(X) \geq \text{cv}(Y).$$

Keilson, J. and Steutel, F. (1974). Ann. Prob., 2, 112–130.

Bronk Inequality Given nonnegative random variables $X, Y, E(X) = E(Y)$, with density functions f, g, such that f, g cross exactly twice, and $g(x) > f(x)$ for all x after the second crossing, $\text{Var}(X) \leq \text{Var}(Y)$.

Bronk, B. (1979). J. Appl. Prob., 16, 665–670.

Brown et al. Inequality I Given $Z \sim N(0, 1)$, $g(z) = f(z^2)$ a symmetric convex function, with $f'(t) \geq a \geq 0$, $f''(t) \geq b \geq 0$ for all t, then

$$E[g(Z)] \geq g(0) + \frac{a}{2} + \frac{3}{4}.$$

Brown, L., DasGupta, A., Haff, L., and Strawderman, W. (2006). J. Stat. Planning Infer., 136, 2254–2278.

Brown et al. Inequality II Given $X \sim N(\mu, \sigma^2)$, $g(x, \mu)$ a twice continuously differentiable function in x for any μ, such that $|g(x, \mu)|$, $|g_x(x, \mu)| \leq ke^{c|x|}$ for some $0 < c, k < \infty$,

$$E[g(X, \mu)] \leq g(\mu, \mu) + \sigma^2 E[|g_{xx}(X, \mu)|],$$

$$E[g(X, \mu)] \geq g(\mu, \mu) + \sigma^2 E\left[\frac{g_{xx}(X, \mu)}{\frac{(X-\mu)^2}{\sigma^2} + 3}\right],$$

if $g(x, \mu)$ is convex in x.

Brown, L., DasGupta, A., Haff, L., and Strawderman, W. (2006). J. Stat. Planning Infer., 136, 2254–2278.

Generalized Jensen Inequality on $\mathcal{R}$ Given a real-valued random variable X, with a finite mean, a continuous convex function f, a constant c, and any function τ with a finite mean such that $\tau(x)$ and $x - \tau(x)$ are both nondecreasing,

$$Ef(\tau(X) - E\tau(X) + c) \leq Ef(X - EX + c).$$

Pečaric, J., Proschan, F., and Tong, Y. (1992). *Convex Functions, Partial Orderings and Statistical Applications*, Academic Press, New York.

Chernoff Inequality Given $X \sim N(\mu, \sigma^2)$ and an absolutely continuous function $g(z)$ such that $\text{Var}(g(X))$, $E[|g'(X)|] < \infty$,

$$\sigma^2(Eg'(X))^2 \leq \text{Var}(g(X)) \leq \sigma^2 E[(g'(X))^2].$$

Chernoff, H. (1981). Ann. Prob., 9, 533–535.

Chen Inequality Given $X \sim MVN(0, \Sigma)$ and a real-valued function $g(x)$ having all partial derivatives g_i, $E(|g_i(X)|^2) < \infty$,

$$E(\nabla g(X))'\Sigma E(\nabla g(X)) \leq \text{Var}(g(X)) \leq E[(\nabla g(X))'\Sigma(\nabla g(X))].$$

Chen, L. (1982). J. Multivar. Anal., 12, 306–315.

Interpolated Chernoff Inequality of Beckner Given $X \sim MVN(0, I)$, $1 \leq p \leq 2$, and a real-valued function $g(x)$ having all partial derivatives g_i, $E(|g_i(X)|^2) < \infty$,

$$E(g^2(X)) - (E|g(X)|^p)^{2/p} \leq (2-p)E(||\nabla g(X)||^2).$$

Beckner, W. (1989). Proc. Am. Math. Soc., 105, 397–400.

log-Sobolev Inequality of Beckner Given $X \sim MVN(0, I)$ and a real-valued function $g(x)$ having all partial derivatives g_i, $E(|g_i(X)|^2) < \infty$,

$$E(f^2(X)\log|f(X)|) - \frac{1}{2}E(f^2(X))\log[E(f^2(X))] \leq E(||\nabla g(X)||^2).$$

Beckner, W. (1998). Bull. London Math. Soc., 30, 80–84.

Chernoff Inequality of Cacoullos for Exponential and Poisson Given $X \sim f(x) = \theta e^{-\theta x}$, $x > 0$, and a continuously differentiable function g,

$$(E[Xg'(X)])^2 \leq \text{Var}(g(X)) \leq \frac{\text{Var}(g'(X))}{\theta^2} + \frac{1}{\theta}E[X(g'(X))^2].$$

Given $X \sim \text{Poisson}(\lambda)$,

$$\lambda(E[g(X+1) - g(X)])^2 \leq \text{Var}(g(X)) \leq \lambda E[g(X+1) - g(X)]^2$$

$$+\lambda\int_0^\lambda (E_\theta[g(X+1) - g(X)]^2)d\theta.$$

Cacoullos, T. (1982). Ann. Prob., 10, 799–809.

Cacoullos, T. and Papathanasiou, V. (1997). J. Stat. Planning Infer., 63, 157–171.

Minkowski's Inequality Given random variables X, Y and $p \geq 1$,

$$(E|X+Y|^p)^{\frac{1}{p}} \leq (E|X|^p)^{\frac{1}{p}} + (E|Y|^p)^{\frac{1}{p}}.$$

Improved Minkowski Inequality of Petrov Given random variables X, Y and $p \geq 1$,

$$(E|X+Y|^p)^{\frac{1}{p}} \leq (E|X|^p I_{X+Y\neq 0})^{\frac{1}{p}} + (E|Y|^p I_{X+Y\neq 0})^{\frac{1}{p}}.$$

Petrov, V. (1995). *Limit Theorems of Probability Theory*, Oxford University Press, Oxford.

Positive Dependence Inequality for Multivariate Normal Given $X_{n\times 1} \sim \text{MV}N(\mu, \Sigma)$, Σ p.d. and such that $\sigma^{ij} \leq 0$ for all $i, j, i \neq j$,

$$P(X_i \geq c_i, i = 1, \cdots, n) \geq \prod_{i=1}^{n} P(X_i \geq c_i)$$

for any constants $c_1, \cdots, c_n$.

Tong, Y. (1980). *Probability Inequalities in Multivariate Distributions*, Academic Press, New York.

Sidak Inequality for Multivariate Normal Given $X_{n\times 1} \sim MVN(0, \Sigma)$, Σ p.d.,

$$P(|X_i| \leq c_i, i = 1, \cdots, n) \geq \prod_{i=1}^{n} P(|X_i| \leq c_i)$$

for any constants $c_1, \cdots, c_n$.

Tong, Y. (1980). *Probability Inequalities in Multivariate Distributions*, Academic Press, New York.

Wilks Inequality Given $X_{n\times 1} \sim F$ with marginal CDFs F_i,

$$F(x_1, \cdots, x_n) \leq (\prod_{i=1}^{n} F_i(x_i))^{\frac{1}{n}}.$$

Mallows Inequality for Multinomial Given a multinomial vector $(f_1, \cdots, f_k)$,

$$P(f_1 \leq a_1, \cdots, f_k \leq a_k) \leq \prod_{i=1}^{k} P(f_i \leq a_i).$$

Mallows, C. (1968). Biometrika, 55, 422–424.

Karlin-Rinott-Tong Inequality I Given $X_{n\times 1} \sim f(x)$, where f is permutation invariant and logconcave,

$$P(\cap\{a_j \le X_j \le b_j\}) \le P(\cap\{\bar{a} \le X_j \le \bar{b}\}).$$

Karlin, S. and Rinott, Y. (1980a). J. Multivar. Anal., 10, 467–498.

Karlin, S. and Rinott, Y. (1980b). J. Multivar. Anal., 10, 499–516.

Tong, Y. (1980). *Probability Inequalities in Multivariate Distributions*, Academic Press, New York.

Karlin-Rinott-Tong Inequality II Given $X_{n\times 1} \sim f(x)$, where f is *schur-concave*,

$$P(\cap\{a_j \le X_j \le b_j\}) \le P(\cap\{\bar{a} \le X_j \le \bar{b}\}).$$

Karlin, S. and Rinott, Y. (1980a). J. Multivar. Anal., 10, 467–498.

Karlin, S. and Rinott, Y. (1980b). J. Multivar. Anal., 10, 499–516.

Tong, Y. (1980). *Probability Inequalities in Multivariate Distributions*, Academic Press, New York.

Prekopa-Brunn-Minkowski Inequality Given $X_{n\times 1} \sim f(x)$, a density such that f is logconcave, a constant $0 < \alpha < 1$, and general Borel sets B_1, B_2,

$$P(X \in \alpha B_1 + (1-\alpha)B_2) \ge (P(X \in B_1))^{\alpha}(P(X \in B_2))^{1-\alpha}.$$

Tong, Y. (1980). *Probability Inequalities in Multivariate Distributions*, Academic Press, New York.

FKG Inequality on $\mathcal{R}^n$ Given $X_{n\times 1} \sim p(x)$, p such that $p(x)p(y) \le p(x \wedge y)p(x \vee y)$ for all x, y, and functions f, g, each nonnegative bounded coordinatewise increasing in all n coordinates,

$$\text{Cov}(f(X), g(X)) \ge 0.$$

Pečaric, J., Proschan, F., and Tong, Y. (1992). *Convex Functions, Partial Orderings and Statistical Applications*, Academic Press, New York.

Tukey Inequality Given independent random variables $X_1, \cdots, X_n$, each with median zero,

$$E\left|\sum X_i\right| \ge \frac{w_n}{n}\sum E|X_i|,$$

where $w_{2k+1} = w_{2k+2} = \frac{1\cdot 3\cdot 5\cdots(2k+1)}{2\cdot 4\cdot 6\cdots 2k}$.

Tukey, J. (1946). Ann. Math. Stat., 17, 75–78.

Devroye-Györfi Inequality Given iid random variables $X_1, \cdots, X_n$, each with mean zero,

$$E\left|\sum X_i\right| \geq \sqrt{\frac{n}{8}} E|X_1|.$$

Devroye, L. and Györfi, L. (1985). *Nonparametric Density Estimation: The L_1 View*, Wiley, New York.

Partial Sums Moment Inequality Given n random variables $X_1, \cdots, X_n$, $p > 1$,

$$E\left|\sum X_i\right|^p \leq n^{p-1} \sum E|X_k|^p.$$

Given n independent random variables $X_1, \cdots, X_n$, $p \geq 2$,

$$E\left|\sum X_i\right|^p \leq c(p) n^{p/2-1} \sum E|X_k|^p$$

for some finite universal constant $c(p)$.

Petrov, V. (1995). *Limit Theorems of Probability Theory*, Oxford University Press, Oxford.

von Bahr-Esseen Inequality I Given independent random variables $X_1, \cdots, X_n$ with mean zero, $1 \leq p \leq 2$,

$$E\left|\sum X_i\right|^p \leq \left(2 - \frac{1}{n}\right) \sum E|X_k|^p.$$

von Bahr, B. and Esseen, C. (1965). Ann. Math. Stat., 36, 299–303.

von Bahr-Esseen Inequality II Given independent random variables $X_1, \cdots, X_n$, symmetric about zero, $1 \leq p \leq 2$,

$$E\left|\sum X_i\right|^p \leq \sum E|X_k|^p.$$

von Bahr, B. and Esseen, C. (1965). Ann. Math. Stat., 36, 299–303.

Rosenthal Inequality I Given independent random variables $X_1, \cdots, X_n$, $p > 1$,

$$E\left|\sum X_i\right|^p \leq 2^{p^2} \max\left\{\sum E|X_k|^p, \left(\sum E|X_k|\right)^p\right\}.$$

Rosenthal, H. (1970). Isr. J. Math., 8, 273–303.

Rosenthal Inequality II Given independent random variables $X_1, \cdots, X_n$, symmetric about zero,

$$\left(E|\sum X_i|^p\right)^{1/p} \leq \left(1 + \frac{2^{P/2}}{\sqrt{\pi}}\Gamma\left(\frac{p+1}{2}\right)\right)^{1/p}$$

$$\max\left\{\left(E|\sum X_i|^2\right)^{1/2}, \left(\sum E|X_i|^p\right)^{1/p}\right\}, 2 < p < 4.$$

Ibragimov, R. and Sharakhmetov, Sh. (1998). Theory Prob. Appl., 42, 294–302.

Doob-Klass Prophet Inequality Given iid mean-zero random variables $X_1, \cdots, X_n$,

$$E\left[\max_{1\leq k\leq n}(X_1 + \cdots + X_k)\right] \leq 3E|X_1 + \cdots + X_n|.$$

Klass, M. (1988). Ann. Prob., 16, 840–853.

Klass Prophet Inequality Given iid mean-zero random variables $X_1, \cdots, X_n$,

$$E\left[\max_{1\leq k\leq n}(X_1 + \cdots + X_k)^+\right] \leq \left(2 - \frac{1}{n}\right)E[(X_1 + \cdots + X_n)^+].$$

Klass, M. (1989). Ann. Prob., 17, 1243–1247.

Bickel Inequality Given independent random variables $X_1, \cdots, X_n$, each symmetric about $E(X_k) = 0$, and a nonnegative convex function g,

$$E\left[\max_{1\leq k\leq n} g(X_1 + \cdots + X_k)\right] \leq 2E[g(X_1 + \cdots + X_n)].$$

Bickel, P. (1970). Acta Math. Acad. Sci. Hung., 21, 199–206.

General Prophet Inequality of Bickel Given independent random variables $X_1, \cdots, X_n$, each with mean zero,

$$E\left[\max_{1\leq k\leq n}(X_1 + \cdots + X_k)\right] \leq 4E[X_1 + \cdots + X_n].$$

35.1.6 Inequalities in Order Statistics

Thomson Inequality Given any set of numbers $x_1, \cdots, x_n$, the order statistics $x_{(1)} \leq \cdots \leq x_{(n)}$, and $s_2^2 = \frac{1}{n} \sum (x_i - \bar{x})^2$,

$$x_{(n)} - x_{(1)} \geq \frac{2\sqrt{2}n}{\sqrt{2n^2 - 1 + (-1)^n}} s_2.$$

Thomson, G. (1955). Biometrika, 42, 268–269.

Bickel-Lehmann Inequality Given independent random variables $X_1, \cdots, X_n$, and the order statistics $X_{(1)} \leq \cdots \leq X_{(n)}$, always

$$P(\cap_{k=1}^{n} \{X_{(k)} \leq a_k\}) \geq \prod_{k=1}^{n} P(X_{(k)} \leq a_k).$$

Lehmann, E. (1966). Ann. Math. Stat., 37, 1137–1153.

Bickel, P. (1967). *Proceedings of the Fifth Berkeley Symposium*, L. Le Cam and J. Neyman (eds.), Vol. I, University of California Press, Berkeley, 575–591.

Positive Dependence Inequality Given independent random variables $X_1, \cdots, X_n$ and the order statistics $X_{(1)} \leq \cdots \leq X_{(n)}$, for any i, j,

$$\text{cov}(X_{(i)}, X_{(j)}) \geq 0.$$

Tong, Y. (1980). *Probability Inequalities in Multivariate Distributions*, Academic Press, New York.

Reiss Inequality I Given iid random variables $X_1, \cdots, X_n$, the order statistics $X_{(1)} \leq \cdots \leq X_{(n)}$, and $m \geq 1$,

$$E|X_{(r)}|^m \leq \frac{n!}{(r-1)!(n-r)!} E|X_1|^m$$

for any $1 \leq r \leq n$.

Reiss, R. (1989). *Approximate Distributions of Order Statistics*, Springer, New York.

Reiss Inequality II Given iid random variables $X_1, \cdots, X_n$, the order statistics $X_{(1)} \leq \cdots \leq X_{(n)}$, $r_0 = 0 < r_1 < \cdots < r_k < n+1 = r_{k+1}$, and g a general nonnegative function of k variables,

$$Eg(X_{(r_1)}, \cdots, X_{(r_k)}) \leq \frac{n!}{\prod_{i=1}^{k+1}(r_i - r_{i-1} - 1)!} Eg(X_1, \cdots, X_k).$$

Reiss, R. (1989). *Approximate Distributions of Order Statistics*, Springer, New York.

Hartley-David Inequality Given absolutely continuous iid random variables $X_1, \cdots, X_n$, with mean μ and variance σ^2,

$$EX_{(n)} \leq \mu + \frac{\sigma\sqrt{n-1}}{\sqrt{2}}.$$

Hartley, H. and David, H. (1954). Ann. Math. Stat., 25, 85–99.

Caraux-Gascuel Inequality Given absolutely continuous iid random variables $X_1, \cdots, X_n$, with mean μ and variance σ^2, and $1 \leq m \leq n$,

$$\mu - \sqrt{\frac{n-m}{m}}\sigma \leq EX_{(m)} \leq \mu + \sqrt{\frac{m-1}{n-m+1}}\sigma.$$

Caraux, G. and Gascuel, O. (1992a). Stat. Prob. Lett., 14, 103–105.

Gascuel, O. and Caraux, G. (1992b). Stat. Prob. Lett., 15, 143–148.

Gilstein Inequality Given iid random variables $X_1, \cdots, X_n$, $1 < p < \infty$, $c_p = (E|X_1|^p)^{1/p}$,

$$EX_{(n)} \leq n \left(\frac{p-1}{np-1}\right)^{\frac{p-1}{p}} c_p.$$

Gilstein, C. (1981). Preprint.

Gilstein, C. (1983). Ann. Stat., 11, 913–920.

de la Cal-Cárcamo Inequality Given iid random variables $X_1, \cdots, X_n$ from a CDF F with a finite mean μ,

$$EX_{(n)} \geq \sup_{a \geq \mu}(a + (\mu - a)F^{n-1}(a)),$$

$$EX_{(1)} \leq \inf_{a \leq \mu}(a + (\mu - a)F^{n-1}(a)).$$

de la Cal, J. and Cárcamo, J. (2005). Stat. Prob. Lett., 73, 219–231.

Gajek-Rychlik Inequality Given absolutely continuous symmetric unimodal random variables $X_1, \cdots, X_n$, with mean μ and variance σ^2,

$$EX_{(m)} \leq \mu + \sqrt{3}\frac{m-1}{n}\sigma$$

if $\frac{1}{3} \leq \frac{m-1}{n} \leq \frac{2}{3}$ and

$$EX_{(n)} \leq \mu + \frac{2}{3}\frac{\sigma}{\sqrt{n}}$$

for $n \geq 4$.

Gajek, L. and Rychlik, T. (1996). J. Multivar. Anal., 57, 156–174.

Gajek, L. and Rychlik, T. (1998). J. Multivar. Anal., 64, 156–182.

David Inequality for Normal Distribution Given $X_1, \cdots, X_n \stackrel{\text{iid}}{\sim} N(\mu, \sigma^2)$, for $r \geq \frac{n+1}{2}$,

$$\mu + \Phi^{-1}\left(\frac{r}{n+1}\right)\sigma \leq EX_{(r)} \leq \mu + \min\left\{\Phi^{-1}\left(\frac{r}{n+1/2}\right), \Phi^{-1}\left(\frac{r-1/2}{n}\right)\right\}\sigma.$$

Patel, J. and Read, C. (1996). *Handbook of the Normal Distribution*, CRC Press, Boca Raton, FL.

Esary-Proschan-Walkup Inequality for Normal Distribution Given $X_{n\times 1} \sim N(\mu, \Sigma)$ such that $\sigma_{ij} \geq 0$ for any i, j and $X_{(1)}, \cdots, X_{(n)}$ the ordered coordinates of $X_{n\times 1}$,

$$\text{corr}(X_{(i)}, X_{(j)}) \geq 0$$

for all i, j, and

$$P(\cap_{k=1}^{n}\{X_{(k)} \leq a_k\}) \geq \prod_{k=1}^{n} P(X_{(k)} \leq a_k).$$

Esary, J., Proschan, F., and Walkup, D. (1967). Ann. Math. Stat., 38, 1466–1474.

Large Deviation Inequality for Normal Distribution Given $X_{n\times 1} \sim N(0, \Sigma)$, $\text{Var}(X_i) = \sigma_i^2$, $\sigma^2 = \max_{1\leq i\leq n}\sigma_i^2$, for any $t > 0$,

$$P(X_{(n)} \geq EX_{(n)} + \sigma t) \leq e^{-\frac{t^2}{2}}.$$

Patel, J. and Read, C. (1996). *Handbook of the Normal Distribution*, CRC Press, Boca Raton, FL.

35.1.7 Inequalities for Normal Distributions

With $\Phi(x)$ and $\varphi(x)$ as the $N(0, 1)$ CDF and PDF, $R(x) = \frac{1-\Phi(x)}{\phi(x)}$ the *Mill's Ratio*.

Polýa Inequality

$$\Phi(x) < \frac{1}{2}\left[1 + \sqrt{1 - e^{-\frac{2}{\pi}x^2}}\right].$$

Patel, J. and Read, C. (1996). *Handbook of the Normal Distribution*, CRC Press, Boca Raton, FL.

Chu Inequality

$$\Phi(x) \geq \frac{1}{2}\left[1 + \sqrt{1 - e^{-\frac{x^2}{2}}}\right].$$

Patel, J. and Read, C. (1996). *Handbook of the Normal Distribution*, CRC Press, Boca Raton, FL.

Gordon Inequality For $x > 0$,

$$\frac{x}{x^2 + 1} \leq R(x) \leq \frac{1}{x}.$$

Patel, J. and Read, C. (1996). *Handbook of the Normal Distribution*, CRC Press, Boca Raton, FL.

Mitrinovic Inequality For $x > 0$,

$$\frac{2}{x + \sqrt{x^2 + 4}} < R(x) < \frac{2}{x + \sqrt{x^2 + \frac{8}{\pi}}}.$$

Patel, J. and Read, C. (1996). *Handbook of the Normal Distribution*, CRC Press, Boca Raton, FL.

Szarek-Werner Inequality For $x > -1$,

$$\frac{2}{x + \sqrt{x^2 + 4}} < R(x) < \frac{4}{3x + \sqrt{x^2 + 8}}.$$

Szarek, S. and Werner, E. (1999). J. Multivar. Anal., 68, 193–211.

Boyd Inequality For $x > 0$,

$$R(x) < \frac{\pi}{2x + \sqrt{(\pi - 2)^2 x^2 + 2\pi}}.$$

Patel, J. and Read, C. (1996). *Handbook of the Normal Distribution*, CRC Press, Boca Raton, FL.

Savage Inequality for Multivariate Mill's Ratio Given $X_{n\times 1} \sim N(0, \Sigma)$, where $\sigma_{ii} = 1$,

$$\frac{1 - \sum_{i=1}^{n} m_{ii}/\Delta_i^2 - \sum_{1\le i<j\le n} m_{ij}/(\Delta_i \Delta_j)}{\prod_{i=1}^{n} \Delta_i} \le R(a, M) \le \frac{1}{\prod_{i=1}^{n} \Delta_i},$$

where $M = \Sigma^{-1}$, $R(a, M) = P(X_{n\times 1} \ge a)/f(a, M)$, the n-dimensional Mill's ratio, and $\Delta = Ma$, assuming that each coordinate of Δ is positive.

Savage, I. (1962). J. Res. Nat. Bur. Standards Sect. B, 66, 93–96.

35.1.8 Inequalities for Binomial and Poisson Distributions

Bohman Inequality Given $X \sim \text{Poisson}(\lambda)$,

$$P(X \ge k) \ge 1 - \Phi((k - \lambda)/\sqrt{\lambda}).$$

Bohman, H. (1963). Skand. Actuarietidskr., 46, 47–52.

Alon-Spencer Inequality Given $X \sim \text{Poisson}(\lambda)$,

$$P(X \le \lambda(1 - \varepsilon)) \le e^{-\varepsilon^2 \lambda/2}$$

for any $\varepsilon > 0$.

Alon, N. and Spencer, J. (2000). *The Probabilistic Method*, Wiley, New York.

Anderson-Samuels Inequality Given $X \sim \text{Bin}(n, p)$, $Y \sim \text{Poisson}(np)$,

$$P(Y \ge k) \ge \max\{P(X \ge k), 1 - \Phi((k - np)/\sqrt{np(1 - p)})\}$$

if $k \ge np + 1$ and

$$P(X \ge k) \ge P(Y \ge k) \ge 1 - \Phi((k - np)/\sqrt{np})$$

if $k \leq np$.

Slud, E. (1977). Ann. Prob., 5, 404–412.

Slud Inequality Given $X \sim \text{Bin}(n, p)$,

$$P(X = k) \geq \Phi((k - np + 1)/\sqrt{np(1-p)}) - \Phi((k - np)/\sqrt{np(1-p)})$$

if $\max\{2, np\} \leq k \leq n(1-p)$.

Slud, E. (1977). Ann. Prob., 5, 404–412.

Giné-Zinn Inequality Given $X \sim \text{Bin}(n, p)$,

$$P(X \geq k) \leq \binom{n}{k} p^k \leq \left(\frac{enp}{k}\right)^k.$$

Shorack, G. and Wellner, J. (1986). *Empirical Processes with Applications to Statistics*, Wiley, New York.

McKay Inequality Given $X \sim \text{Bin}(n, p)$, $Y \sim \text{Bin}(n-1, p)$,

$$\frac{P(X \geq k)}{P(Y = k-1)} = \sqrt{np(1-p)} R(x) e^{E(k,n,p)/\sqrt{np(1-p)}},$$

where $k \geq np, x = (k - np)/\sqrt{np(1-p)}$, $R(x)$ is the univariate Mill's ratio, and $0 \leq E(k, n, p) \leq \min\{\sqrt{\pi/8}, \frac{1}{x}\}$.

McKay, B. (1989). Adv. Appl. Prob., 21, 475–478.

Szarek Inequality Given $X \sim \text{Bin}(n, \frac{1}{2})$,

$$E|X - \frac{n}{2}| \geq \sqrt{\frac{n}{8}}.$$

Szarek, S. (1976). Studia Math., 58(2), 197–208.

Sheu Inequality Given $X \sim \text{Bin}(n, p)$, $Y \sim \text{Poisson}(np)$,

$$\sum_{k \geq 0} |P(X = k) - P(Y = k)| \leq \min\{2np^2, 3p\}.$$

Sheu, S. (1984). Am. Stat., 38, 206–207.

LeCam Inequality Given $X_i \overset{\text{indep.}}{\sim} \text{Ber}(p_i)$, $1 \leq i \leq n$, $\lambda = \sum p_i$, $X = \sum X_i$, $Y \sim \text{Poisson}(\lambda)$,

$$\sum_{k \geq 0} |P(X = k) - P(Y = k)| \leq 2 \sum p_i^2.$$

Neyman, J. and Le Cam, L. (eds.) (1965)., On the distribution of sums of independent random variables. In Proceedings of the Bernoulli-Bayes-Laplace Seminar, University of California, Berkeley, 179–202.

Steele Inequality Given $X_i \overset{\text{indep.}}{\sim} \text{Ber}(p_i), 1 \le i \le n, \lambda = \sum p_i, X = \sum X_i, Y \sim \text{Poisson}(\lambda)$,

$$\sum_{k\ge 0} |P(X=k) - P(Y=k)| \le 2\frac{1-e^{-\lambda}}{\lambda} \sum p_i^2.$$

Steele, J. M. (1994). Am. Math Mon., 101, 48–54.

35.1.9 Inequalities in the Central Limit Theorem

Berry-Esseen Inequality Given independent random variables $X_1, \cdots, X_n$ with mean zero such that $E|X_1|^{2+\delta} < \infty$ for some $0 < \delta \le 1$, $B_n^2 = \sum_{i=1}^n \text{Var}(X_i)$, and $F_n(x) = P\left(\frac{\sum_{i=1}^n X_i}{B_n} \le x\right)$,

$$\sup_{-\infty<x<\infty} |F_n(x) - \Phi(x)| \le A \frac{\sum_{i=1}^n E|X_i|^{2+\delta}}{B_n^{2+\delta}}$$

for a finite universal constant A.

Petrov, V. (1995). *Limit Theorems of Probability Theory*, Oxford University Press, Oxford.

Generalized Berry-Esseen Inequality Given independent random variables $X_1, \cdots, X_n$ with mean zero and finite variance and a function g such that g is nonnegative, even, nondecreasing for $x > 0$, $\frac{g(x)}{x}$ is nonincreasing for $x > 0$, and such that $EX_i^2 g(X_i) < \infty$ for $i = 1, \cdots, n$,

$$\sup_{-\infty<x<\infty} |F_n(x) - \Phi(x)| \le A \frac{\sum_{i=1}^n EX_i^2 g(X_i)}{B_n^2 g(B_n)}.$$

Petrov, V. (1995). *Limit Theorems of Probability Theory*, Oxford University Press, Oxford.

Sazonov Inequality Given iid random variables $X_1, \cdots, X_n$ with mean zero, variance 1, and common CDF F,

$$\sup_{-\infty<x<\infty} |F_n(x) - \Phi(x)| \le \frac{A}{\sqrt{n}} \int_{-\infty}^{\infty} \max\{1, |x|^3\} |F(x) - \Phi(x)| dx.$$

Petrov, V. (1995). *Limit Theorems of Probability Theory*, Oxford University Press, Oxford.

Bikelis Nonuniform Inequality Given independent random variables $X_1, \cdots, X_n$ with mean zero and finite third absolute moments,

$$|F_n(x) - \Phi(x)| \leq A \frac{\sum_{i=1}^n E|X_i|^3}{B_n^3} \frac{1}{(1 + |x|)^3}$$

for all real x.

Petrov, V. (1995). *Limit Theorems of Probability Theory*, Oxford University Press, Oxford.

Osipov Nonuniform Inequality Given iid random variables $X_1, \cdots, X_n$ with mean zero, variance σ^2, and $E|X_1|^\alpha < \infty$ for some $\alpha \geq 3$,

$$|F_n(x) - \Phi(x)| \leq A(\alpha) \left[\frac{E|X_1|^3}{\sigma^3 \sqrt{n}} + \frac{E|X_1|^\alpha}{\sigma^\alpha n^{(\alpha-2)/2}} \right] \frac{1}{(1 + |x|)^\alpha}$$

for all real x.

Petrov, V. (1995). *Limit Theorems of Probability Theory*, Oxford University Press, Oxford.

Petrov Nonuniform Inequality Given independent random variables $X_1, \cdots, X_n$ with mean zero and finite variance and a function g such that g is nonnegative, even, nondecreasing for $x > 0$, $\frac{g(x)}{x}$ is nonincreasing for $x > 0$, and such that $EX_i^2 g(X_i) < \infty$ for $i = 1, \cdots, n$,

$$|F_n(x) - \Phi(x)| \leq A \frac{\sum_{i=1}^n EX_i^2 g(X_i)}{B_n^2} \frac{1}{(1 + |x|)^2 g((1 + |x|)B_n)}$$

for all real x. Petrov, V. (1995). *Limit Theorems of Probability Theory*, Oxford University Press, Oxford.

Reverse Berry-Esseen Inequality of Hall and Barbour Given independent random variables $X_1, \cdots, X_n$ with mean zero, variances σ_i^2 scaled so that $\sum_{i=1}^n \sigma_i^2 = 1$, and $\delta_n = \sum_{i=1}^n E[(X_i^3 + X_i^4) I_{|X_i| \leq 1}] + \sum_{i=1}^n E[X_i^2 I_{|X_i| > 1}]$,

$$\sup_{-\infty < x < \infty} |F_n(x) - \Phi(x)| \geq \frac{1}{392} \left(\delta_n - 121 \sum_{i=1}^n \sigma_i^4 \right).$$

Hall, P. and Barbour, A. (1984). Proc. Am. Math. Soc., 90, 107–110.

Multidimensional Berry-Esseen Inequality of Bentkus Given d-dimensional iid random variables $X_1, \cdots, X_n$, with mean zero and identity covariance matrix, and $S_n = \sqrt{n}\bar{X}$,

$$\sup_{A \in \mathcal{A}_1} |P(S_n \in A) - \Phi(A)| \leq \frac{400 d^{1/4} \rho_3}{\sqrt{n}}$$

and

$$\sup_{A \in \mathcal{A}_2} |P(S_n \in A) - \Phi(A)| \leq \frac{C d^{1/4} \rho_3}{\sqrt{n}}$$

where $\rho_3 = E||X_1||^3$, Φ is the d-dimensional standard normal measure, $\mathcal{A}_1$ is the class of all measurable convex sets in $\mathcal{R}^d$, $\mathcal{A}_2$ is the class of all closed balls in $\mathcal{R}^d$, and C is a finite universal constant.

Bentkus, V. (2003). J. Stat. Planning Infer., 113, 385–402.

Multidimensional Berry-Esseen Inequality of Bhattacharya Given d-dimensional independent random variables $X_1, \cdots, X_n$, with mean zero and with covariance matrices Σ_i scaled so that $\frac{1}{n}\sum_{i=1}^{n} \Sigma_i = I_d$,

$$\sup_{A \in \mathcal{A}_{K,\alpha}} |P(S_n \in A) - \Phi(A)| \leq \frac{C_1 \rho_3}{\sqrt{n}} + \frac{C_2 \rho_3^{\alpha}}{n^{\alpha/2}},$$

where $\rho_3 = \frac{1}{n}\sum_{i=1}^{n} E||X_i||^3$, $\mathcal{A}_{K,\alpha}$ is any class of measurable sets of $\mathcal{R}^d$ satisfying $\sup_{y \in \mathcal{R}^d} \Phi((\partial A)^{\varepsilon} + y) \leq K\varepsilon^{\alpha}$ for each $\varepsilon > 0$, $(\partial A)^{\varepsilon}$ being the ε-parallel body of ∂A (the topological boundary of A), C_1 is a universal constant, and C_2 is a constant depending on K and α.

Bhattacharya, R. (1968). Bull. Am. Math. Soc., 74, 285–287.

Bhattacharya, R. (1971). Ann. Math. Stat., 42, 241–259.

Bhattacharya, R. (1977). Ann. Prob., 5, 1–27.

Arak Inequality on Approximation by id Laws Given a CDF F on the real line, $F^{(n)}$ its n-fold convolution, $\mathcal{F}$ the class of all distributions on the real line, and $\mathcal{I}$ the class of all id (infinitely divisible) laws on the real line,

$$C_1 n^{-2/3} \leq \sup_{F \in \mathcal{F}} \inf_{G \in \mathcal{I}} \sup_{-\infty < x < \infty} |F^{(n)}(x) - G(x)| \leq C_2 n^{-2/3},$$

where C_1, C_2 are strictly positive finite universal constants.

Arak, T. (1981a). Theory Prob. Appl., 26, 219–239.

Arak, T. (1981b). Theory Prob. Appl., 26, 437–451.

Arak, T. (1982). Theory Prob. Appl., 27, 826–832.

35.1.10 Martingale Inequalities

Doob Submartingale Inequality Given a nonnegative submartingale X_t on a probability space $(\Omega, \mathcal{A}, P), 0 \leq t \leq T, M_t = \sup_{0\leq s\leq t} X_s$, and any constant $c > 0$,

$$P(M_T \geq c) \leq \frac{EX_T}{c},$$

$$||X_T||_p \leq ||M_T||_p \leq \frac{p}{p-1}||X_T||_p, \; p > 1.$$

Bullen, P. (1998). *A Dictionary of Inequalities*, Addison-Wesley Longman, Harlow.

Kolmogorov Inequality for Square Integrable Martingales Given a discrete-time martingale $(X_n, \mathcal{F}_n, n \geq 0)$, $E[X_n^2] < \infty$, and any constant $c > 0$,

$$P\left(\max_{0\leq k\leq n} |X_k| \geq c\right) \leq \frac{E[X_n^2]}{c^2}.$$

Burkholder Inequality Given a discrete-time martingale $(X_n, \mathcal{F}_n, n \geq 0)$, $X_0 = 0$, $\Delta(X_i) = X_i - X_{i-1}$, $D_n = \sum_{i=1}^n (\Delta(X_i))^2$, $M_n = \max_{1\leq i\leq n} |X_i|$, for any $p > 1$,

$$\frac{p-1}{18p^{3/2}}||\sqrt{D_n}||_p \leq ||M_n||_p \leq \frac{18p^{3/2}}{(p-1)^{3/2}}||\sqrt{D_n}||_p.$$

Bullen, P. (1998). *A Dictionary of Inequalities*, Addison-Wesley Longman, Harlow.

Davis Inequality Given a discrete-time martingale $(X_n, \mathcal{F}_n, n \geq 0)$, $X_0 = 0$, $\Delta(X_i) = X_i - X_{i-1}$, $D_n = \sum_{i=1}^n (\Delta(X_i))^2$, $M_n = \max_{1\leq i\leq n} |X_i|$,

$$c_1||\sqrt{D_n}||_1 \leq ||M_n||_1 \leq c_2||\sqrt{D_n}||_1,$$

where c_1, c_2 are universal finite positive constants and c_2 may be taken to be $\sqrt{3}$ (Burkholder).

Davis, B. (1970). Isr. J. Math., 8, 187–190.

Burkholder-Davis-Gundy Inequality Given a discrete-time martingale $(X_n, \mathcal{F}_n, n \geq 0)$, $X_0 = 0$, $\Delta(X_i) = X_i - X_{i-1}$, $D_n = \sum_{i=1}^n (\Delta(X_i))^2$, $M_n = \max_{1 \leq i \leq n} |X_i|$, and φ a general convex function,

$$c_\varphi E[\varphi(\sqrt{D_n})] \leq E[\varphi(M_n)] \leq C_\varphi E[\varphi(\sqrt{D_n})],$$

where c_φ, C_φ are universal finite positive constants depending only on φ.

Burkholder, D., Davis, B., and Gundy, R. (1972). In *Proceedings of the Sixth Berkeley Symposium*, L. Le Cam, J. Neyman and E. Scott (eds.), Vol. II, University of California Press, Berkeley, 223–240.

Doob Upcrossing Inequality Given a discrete-time submartingale $(X_n, \mathcal{F}_n, n \geq 0)$, constants $a < b$, and $U_n^{[a,b]}$ the *number of upcrossings* of $[a, b]$ in $\{X_0, \cdots, X_n\}$,

$$E[U_n^{[a,b]}] \leq \frac{E[(X_n - a)^+ - (X_0 - a)^+]}{b - a}.$$

Breiman, L. (1968). *Probability*, Addison-Wesley, Reading, MA.

Optional Stopping Inequality Given a discrete-time submartingale $(X_n, \mathcal{F}_n, n \geq 0)$ and stopping times S, T such that

(a) $P(T < \infty) = 1$,
(b) $E|X_T| < \infty$,
(c) $\lim inf_n E[|X_n| I_{T>n}] = 0$,
(d) $P(S \leq T) = 1$,

one has $E[X_S] \leq E[X_T]$.

Breiman, L. (1968). *Probability*, Addison-Wesley, Reading, MA.

Williams, D. (1991). *Probability with Martingales*, Cambridge University Press, Cambridge.

Azuma Inequality Given a discrete-time martingale $(X_n, \mathcal{F}_n, n \geq 0)$ such that $|X_k - X_{k-1}| \leq c_k$ a.s. and any constant $c > 0$,

$$P(|X_n - X_0| \geq c) \leq 2e^{-\frac{c^2}{2\sum_{k=1}^n c_k^2}}.$$

Azuma, K. (1967). Tõhoku Math. J., 19, 357–367.

35.2 Matrix Inequalities

35.2.1 Rank, Determinant, and Trace Inequalities

(1) **Fröbenius Rank Inequality** Given real matrices A, B, C such that ABC is defined,

$$\text{Rank}(AB) + \text{Rank}(BC) \leq \text{Rank}(B) + \text{Rank}(ABC).$$

Hogben, L. (2007). *Handbook of Linear Algebra*, Chapman and Hall, London.

(2) **Sylvester Rank Inequality** Given real matrices $A_{m \times n}$, $B_{n \times p}$,

$$\text{Rank}(A) + \text{Rank}(B) - n \leq \text{Rank}(AB) \leq \min\{\text{Rank}(A), \text{Rank}(B)\}.$$

Hogben, L. (2007). *Handbook of Linear Algebra*, Chapman and Hall, London.

(3) Given any real matrices A, B,

$$\text{Rank}(A + B) \leq \text{Rank}(A) + \text{Rank}(B).$$

Hogben, L. (2007). *Handbook of Linear Algebra*, Chapman and Hall, London.

(4) Given a p.d. matrix $A_{n \times n}$ and any general matrix $B_{n \times n}$,

$$|A| \leq \prod_{i=1}^{n} a_{ii}$$

and

$$|B| \leq (M\sqrt{n})^n,$$

where $M = \max_{i,j} |a_{ij}|$.

Hogben, L. (2007). *Handbook of Linear Algebra*, Chapman and Hall, London.

Zhang, F. (1999). *Matrix Theory*, Springer, New York.

(5) Given any real $n \times n$ nnd matrix A,

$$|A| \leq \text{per}(A),$$

where per(A) is the *permanent* of A.

Bhatia, R. (1997). *Matrix Analysis*, Springer, New York.

(6) **Hadamard Inequality** Given any real $n \times n$ matrix A,

$$|A|^2 \leq \prod_{i=1}^{n} \left(\sum_{j=1}^{n} a_{ij}^2 \right).$$

Hogben, L. (2007). *Handbook of Linear Algebra*, Chapman and Hall, London.

(7) **Fischer's Inequality** Given a p.d. matrix A partitioned as $\begin{bmatrix} Y & X \\ X' & Z \end{bmatrix}$,

$$|A| \leq |Y||Z|,$$

where Y, Z are square nonnull matrices.

Hogben, L. (2007). *Handbook of Linear Algebra*, Chapman and Hall, London.

(8) Given real $n \times n$ matrices A, B,

$$(|A + B|)^2 \leq |I + AA'||I + B'B|.$$

Zhang, F. (1999). *Matrix Theory*, Springer, New York.

(9) **Minkowski's Inequality** Given any real $n \times n$ nnd matrices A, B,

$$|A + B|^{\frac{1}{n}} \geq |A|^{\frac{1}{n}} + |B|^{\frac{1}{n}}.$$

Bhatia, R. (1997). *Matrix Analysis*, Springer, New York.

(10) **Ky Fan Inequality** Given nnd real matrices A, B, and $0 < p < 1$,

$$|pA + (1 - p)B| \geq |A|^p |B|^{1-p}.$$

Zhang, F. (1999). *Matrix Theory*, Springer, New York.

(11) Given any real $n \times n$ symmetric matrices A, B, with eigenvalues α_i and β_j,

$$\min_{\sigma} \prod_{i=1}^{n} (\alpha_i + \beta_{\sigma(i)}) \leq |A + B| \leq \max_{\sigma} \prod_{i=1}^{n} (\alpha_i + \beta_{\sigma(i)}),$$

where σ denotes a permutation of $1, 2, \cdots, n$.

Beckenbach, E. and Bellman, R. (1961). *Inequalities*, Springer, New York.

(12) **Schur Inequality** Given any real $n \times n$ nnd matrices A, B,

$$\text{tr}(A'B)^2 \leq \text{tr}(AA')(BB').$$

Hogben, L. (2007). *Handbook of Linear Algebra*, Chapman and Hall, London.

Beckenbach, E. and Bellman, R. (1961). *Inequalities*, Springer, New York.

(13) Given p.d. real matrices A, B,

$$\text{tr}(A^{-1}B) \geq \frac{\text{tr}(B)}{\lambda_{\max}(A)} \geq \frac{\text{tr}(B)}{\text{tr}A},$$

where $\lambda_{\max}(A)$ is the largest eigenvalue of A.

Zhang, F. (1999). *Matrix Theory*, Springer, New York.

(14) **Ostrowski Inequality** Given any real matrices A, B such that AB is defined and $1 \leq p \leq 2$,

$$||AB||_p \leq ||A||_p ||B||_p,$$

where $||C||_p$ denotes $(\sum_{i,j} |c_{ij}|^p)^{1/p}$.

Beckenbach, E. and Bellman, R. (1961). *Inequalities*, Springer, New York.

(15) Given any real $n \times n$ nnd matrices A, B, and $0 \leq p \leq 1$,

$$\text{tr}(A^p B^{1-p}) \leq \text{tr}[pA + (1-p)B].$$

Beckenbach, E. and Bellman, R. (1961). *Inequalities*, Springer, New York.

(16) Given any real $n \times n$ nnd matrices A, B, and $0 \leq p \leq 1$,

$$\text{tr}(A^p B^{1-p}) \leq (\text{tr}A)^p (\text{tr}B)^{1-p}.$$

Beckenbach, E. and Bellman, R. (1961). *Inequalities*, Springer, New York.

(17) **Lieb-Thirring Inequality I** Given any real $n \times n$ nnd matrices A, B, and $m \geq 1$,

$$\text{tr}(AB)^m \leq \text{tr}(A^m B^m).$$

Bhatia, R. (1997). *Matrix Analysis*, Springer, New York.

(18) **Lieb-Thirring Inequality II** Given any real $n \times n$ nnd matrices A, B, and $m \geq k \geq 1$,

$$\text{tr}(A^k B^k)^m \leq \text{tr}(A^m B^m)^k.$$

Bhatia, R. (1997). *Matrix Analysis*, Springer, New York.

(19) Given any real $n \times n$ nnd matrices A, B, and $p > 1$,

$$[\text{tr}(A+B)^p]^{1/p} \leq (\text{tr}A^p)^{1/p} + (\text{tr}B^p)^{1/p}.$$

(20) Given any real $n \times n$ nnd matrices A, B,

$$\text{tr}(AB) \geq n|A|^{1/n}|B|^{1/n}.$$

Beckenbach, E. and Bellman, R. (1961). *Inequalities*, Springer, New York.

(21) Given nnd real matrices A, B,

$$(\text{tr}(A+B))^{1/2} \leq (\text{tr}A)^{1/2} + (\text{tr}B)^{1/2}.$$

Zhang, F. (1999). *Matrix Theory*, Springer, New York.
Given real symmetric matrices A, B,

$$\text{tr}(AB) \leq \sqrt{\text{tr}(A^2)\text{tr}(B^2)}.$$

Bullen, P. (1998). *A Dictionary of Inequalities*, Addison-Wesley Longman.

(22) Given p.d. real matrices A, B,

$$\text{tr}((A-B)(A^{-1}-B^{-1})) \leq 0.$$

Zhang, F. (1999). *Matrix Theory*, Springer, New York.

(23) **Golden-Thompson Inequality** Given any real $n \times n$ symmetric matrices A, B,

$$\text{tr}(e^{A+B}) \leq \text{tr}(e^A e^B).$$

Bhatia, R. (1997). *Matrix Analysis*, Springer, New York.

35.2.2 Eigenvalue and Quadratic Form Inequalities

(24) Given any real symmetric matrix $A_{n\times n}$ and λ any eigenvalue of A,

$$|\lambda| \leq n \max |a_{ij}|.$$

Hogben, L. (2007). *Handbook of Linear Algebra*, Chapman and Hall, London.

(25) Given any real symmetric matrix $A_{n\times n}$, $R_i = \sum_j |a_{ij}|$, $C_j = \sum_i |a_{ij}|$, $R = \max(R_i)$, $C = \max(C_j)$, and λ any eigenvalue of A,

$$|\lambda| \leq \min(R, C).$$

Hogben, L. (2007). *Handbook of Linear Algebra*, Chapman and Hall, London.

(26) **Interlacing Inequality** Given any real symmetric matrix $A_{n\times n}$, and $B_{k\times k}$ any principal submatrix of A,

$$\lambda_1(A) \leq \lambda_1(B) \leq \lambda_k(B) \leq \lambda_n(A),$$

where λ_1 indicates the minimum eigenvalue and λ_k, λ_n denote the corresponding maximum eigenvalues.

Zhang, F. (1999). *Matrix Theory*, Springer, New York.

(27) **Poincaré Inequality I** Given any real symmetric matrix $A_{n\times n}$ and the ordered eigenvalues $\lambda_1 \leq \cdots \leq \lambda_n$,

$$\sum_{i=1}^{k} \lambda_i \leq \sum_{i=1}^{k} a_{ii} \leq \sum_{i=1}^{k} \lambda_{n-k+i}$$

for any $1 \leq k \leq n$.

Hogben, L. (2007). *Handbook of Linear Algebra*, Chapman and Hall, London.

(28) **Poincaré Inequality II** Given any real symmetric matrix $A_{n\times n}$ and the ordered eigenvalues $\lambda_1 \leq \cdots \leq \lambda_n$,

$$\text{tr}(X'AX) \geq \sum_{i=1}^{k} \lambda_i$$

and

$$\mathrm{tr}(X'AX) \leq \sum_{i=1}^{k} \lambda_{n-k+i},$$

where X is any matrix such that $X'X = I_k$, $1 \leq k \leq n$.

Bhatia, R. (1997). *Matrix Analysis*, Springer, New York.

(29) **Poincaré Inequality III** Given any real p.d. matrix $A_{n\times n}$ and the ordered eigenvalues $\lambda_1 \leq \cdots \leq \lambda_n$,

$$|X'AX| \geq \prod_{i=1}^{k} \lambda_i$$

and

$$|X'AX| \leq \prod_{i=1}^{k} \lambda_{n-k+i},$$

where X is any matrix such that $X'X = I_k$, $1 \leq k \leq n$.

Bhatia, R. (1997). *Matrix Analysis*, Springer, New York.

(30) **Poincaré Inequality IV** Given any real p.d. matrix $A_{n\times n}$, the ordered eigenvalues $\lambda_1 \leq \cdots \leq \lambda_n$, and A_k the kth principal submatrix,

$$\prod_{i=1}^{k} \lambda_i \leq |A_k| \leq \prod_{i=1}^{k} \lambda_{n-k+i}.$$

Bhatia, R. (1997). *Matrix Analysis*, Springer, New York.

(31) **Karamata Inequality** Given any real symmetric nondiagonal matrix $A_{n\times n}$, the ordered eigenvalues $\lambda_1 \leq \cdots \leq \lambda_n$, and any strictly convex function $\varphi : \mathcal{R} \to \mathcal{R}$,

$$\sum_{i=1}^{n} \varphi(\lambda_i) > \sum_{i=1}^{n} \varphi(a_{ii}).$$

Beckenbach, E. and Bellman, R. (1961). *Inequalities*, Springer, New York.

(32) **Weyl Inequality I** Given any real symmetric matrices A, B,

$$\lambda_1(A+B) \geq \lambda_1(A) + \lambda_1(B),$$

$$\lambda_n(A+B) \leq \lambda_n(A) + \lambda_n(B).$$

Hogben, L. (2007). *Handbook of Linear Algebra*, Chapman and Hall, London.

Bhatia, R. (1997). *Matrix Analysis*, Springer, New York.

(33) **Weyl Inequality II** Given any real symmetric matrices A, B,

$$|\lambda_k(A) - \lambda_k(B)| \leq \text{spr}(A-B),$$

where $1 \leq k \leq n$ and spr denotes the *spectral radius* (the largest absolute eigenvalue) of a matrix.

Hogben, L. (2007). *Handbook of Linear Algebra*, Chapman and Hall, London.

Bhatia, R. (1997). *Matrix Analysis*, Springer, New York.

(34) Given any real symmetric matrix $A_{n\times n}$ and an nnd matrix $B_{n\times n}$, for any $1 \leq k \leq n$,

$$\lambda_k(A+B) \geq \lambda_k(A),$$

where λ_k indicates the kth smallest eigenvalue.

(35) Given any real symmetric matrices A, B, and $0 \leq p \leq 1$,

$$\lambda_1(pA+(1-p)B) \geq p\lambda_1(A) + (1-p)\lambda_1(B),$$

$$\lambda_n(pA+(1-p)B) \leq p\lambda_n(A) + (1-p)\lambda_n(B).$$

Bhatia, R. (1997). *Matrix Analysis*, Springer, New York.

(36) Given any real p.d. matrices A, B, and $0 \leq p \leq 1$,

$$(pA+(1-p)B)^{-1} \leq pA^{-1} + (1-p)B^{-1},$$

where the $\leq$ notation means the difference of the right- and the left-hand sides is nnd.

Bhatia, R. (1997). *Matrix Analysis*, Springer, New York.

(37) **Hoffman-Wielandt Inequality** Given any real symmetric matrices A, B, with eigenvalues α_i and β_j,

$$\min_{\sigma} \sum_{i=1}^{n} (\alpha_i - \beta_{\sigma(i)})^2 \leq ||A-B||_2^2 \leq \max_{\sigma} \sum_{i=1}^{n} (\alpha_i - \beta_{\sigma(i)})^2,$$

where σ is a permutation of $1, 2, \cdots, n$, and $||C||_2^2$ denotes $\sum_{i,j} c_{ij}^2$.
Bhatia, R. (1997). *Matrix Analysis*, Springer, New York.

(38) Given any real nnd matrices A, B, and $t \geq 1$,

$$\text{spr}(A^t B^t) \geq (\text{spr}(AB))^t.$$

Bhatia, R. (1997). *Matrix Analysis*, Springer, New York.

(39) **Fiedler Inequality** Given any real p.d. matrices A, B of the same order,

$$\sum_{i,j}(a_{ij} - a^{ij})(b_{ij} - b^{ij}) \leq 0.$$

Hogben, L. (2007). *Handbook of Linear Algebra*, Chapman and Hall, London.

(40) Given any real symmetric matrices A, B,

$$\text{spr}(e^{A+B}) \leq \text{spr}(e^A e^B).$$

Bhatia, R. (1997). *Matrix Analysis*, Springer, New York.

(41) **Kantorovich Inequality** Given a p.d. matrix A with all eigenvalues in the interval $[a, b]$ and any vector x,

$$\frac{(x'x)^2}{(x'Ax)(x'A^{-1}x)} \geq \frac{4ab}{(a+b)^2}.$$

Zhang, F. (1999). *Matrix Theory*, Springer, New York.

(42) **Khatri-Rao Inequality** Given a p.d. matrix $A_{n\times n}$ with ordered eigenvalues $\lambda_1 \leq \cdots \leq \lambda_n$, $n \times k$ matrices X, Y of rank k such that $X'X = Y'Y = I_k$, for $n \geq 2k$,

$$\text{tr}(X'AYY'A^{-1}X) \leq \sum_{i=1}^{k} \frac{(\lambda_i + \lambda_{n-i+1})^2}{4\lambda_i \lambda_{n-i+1}}.$$

Khatri, C. and Rao, C.R. (1982). Sankhya Ser. A, 44, 91–102.

(43) **Inequality of the Gershgorin Disks** Given a complex square matrix $A_{n\times n}$, $r_i = \sum_{j\neq i} |a_{ij}|$, for each $i = 1, 2, \cdots, n$, $B(a_{ii}, r_i)$ being the ball with center at a_{ii} and of radius r_i, and λ any eigenvalue of A,

$$N(\lambda) \geq 1,$$

where $N(\lambda)$ denotes the number of balls $B(a_{ii}, r_i)$ to which λ belongs.
Bhatia, R. (1997). *Matrix Analysis*, Springer, New York.

(44) **Stronger Gershgorin Disks Characterization** Given a complex symmetric matrix $A_{n\times n}$, $\delta(A) = \min\{|a_{ii} - a_{jj}|, 1 \le i < j \le n\}$, $r(A) = \max\{r_i, 1 \le i \le n\}$ such that $r(A) < \frac{\delta(A)}{2}$, $n_i = 1$ for each $i = 1, 2, \cdots, n$, where n_i denotes the total number of eigenvalues of A in the ith ball $B(a_{ii}, r_i)$.
Bhatia, R. (1997). *Matrix Analysis*, Springer, New York.

35.3 Series and Polynomial Inequalities

Arithmetic vs. Geometric Mean Progressive Inequality Given nonnegative numbers $x_1, x_2, \cdots, x_n$, and $A_k = \frac{\sum_{i=1}^{k} x_i}{k}$, $G_k = (\prod_{i=1}^{k} x_i)^{1/k}$,

$$A_n - G_n \ge \frac{n-1}{n}(A_{n-1} - G_{n-1}).$$

Beckenbach, E. and Bellman, R. (1961). *Inequalities*, Springer, New York.

Elementary Symmetric Function Inequality I Given nonnegative numbers $x_1, x_2, \cdots, x_n$, not all equal, $S_r = S_r(x_1, x_2, \cdots, x_n)$ the rth *elementary symmetric function*, and $p_r = \frac{S_r}{\binom{n}{r}}$,

$$p_1 > p_2^{1/2} > p_3^{1/3} > \cdots > p_n^{1/n}.$$

Beckenbach, E. and Bellman, R. (1961). *Inequalities*, Springer, New York.

Elementary Symmetric Function Inequality II

$$S_{r-1}S_{r+1} \le S_r^2,$$
$$p_{r-1}p_{r+1} \le p_r^2,$$
$$p_{r+s} \le p_r p_s,$$
$$p_s \ge (p_r)^{\frac{t-s}{t-r}}(p_t)^{\frac{s-r}{t-r}},$$

for $1 \le r < s < t \le n$.

Beckenbach, E. and Bellman, R. (1961). *Inequalities*, Springer, New York.

Elementary Symmetric Function Inequality III Given two sets of nonnegative numbers $x_1, x_2, \cdots, x_n, y_1, y_2, \cdots, y_n$,

$$[S_r(x+y)]^{1/r} \geq [S_r(x)]^{1/r} + [S_r(y)]^{1/r}.$$

Beckenbach, E. and Bellman, R. (1961). *Inequalities*, Springer, New York.

Reverse Cauchy-Schwartz Inequality for Sums Given $0 < a \leq x_i < A, 0 < b \leq y_i < B, i = 1, 2, \cdots, n,$

$$(\sum x_i y_i)^2 \geq \sum x_i^2 \sum y_i^2 \frac{4}{[\sqrt{\frac{AB}{ab}} + \sqrt{\frac{ab}{AB}}]^2}.$$

Bullen, P. (1998). *A Dictionary of Inequalities*, Addison-Wesley Longman, Harlow.

Special Sequences and Ratio and Spacings Inequalities For $p > 0$,

$$\frac{n^{p+1}-1}{p+1} < \sum_{i=1}^{n} i^p < \frac{n^{p+1}-1}{p+1} + n^p,$$

$$2(\sqrt{n+1}-1) < \sum_{i=1}^{n} \frac{1}{\sqrt{i}} < 2\sqrt{n} - 1,$$

$$\frac{1}{n} + \log n < \sum_{i=1}^{n} \frac{1}{i} < 1 + \log n, n \geq 2,$$

$$\gamma + \frac{1}{2n} - \frac{1}{8n^2} < \sum_{i=1}^{n} \frac{1}{i} - \log n < \gamma + \frac{1}{2n},$$

where γ is Euler's constant,

$$\frac{1}{n} - \frac{1}{2n+1} < \sum_{i=1}^{n} \frac{1}{i^2} < \frac{1}{n-1} - \frac{1}{2n}, n \geq 2,$$

$$\frac{1}{n[n^{s-1}-(n-1)^{s-1}]} < \sum_{i \geq n} \frac{1}{i^s} < \frac{(n+1)^{s-1}}{n^s[(n+1)^{s-1}-n^{s-1}]}, s > 1,$$

$$\sqrt{2\pi} n^{n+\frac{1}{2}} e^{-n+\frac{1}{12n+1/4}} < n! < \sqrt{2\pi} n^{n+\frac{1}{2}} e^{-n+\frac{1}{12n}}.$$

Given any positive constants $a_1, a_2, \cdots, a_n$,

$$\frac{a_1}{a_2} + \frac{a_2}{a_3} + \cdots + \frac{a_{n-1}}{a_n} + \frac{a_n}{a_1} \geq n.$$

Given any constants $a_1, a_2, \cdots, a_n$, and $\Delta a_i = a_{i+1} - a_i$, $a_0 = a_{n+1} = 0$,

$$\sum_{i=0}^{n} (\Delta a_i)^2 \geq 4 \sin^2 \left(\frac{\pi}{2(n+1)} \right) \sum_{i=1}^{n} a_i^2.$$

For the entire collection, see: Bullen, P. (1998). *A Dictionary of Inequalities*, Addison-Wesley Longman.

Binomial and Bernstein Polynomial Inequalities

$$\frac{4^n}{n+1} < \binom{2n}{n} < \frac{(2n+2)^n}{(n+1)!},$$

$$\binom{2n}{n} > \frac{4^n}{2\sqrt{n}},$$

$$\binom{n}{k} p^k (1-p)^{n-k} < e^{-2n(x - \frac{k}{n})^2}, 0 \leq k \leq n, 0 < p < 1,$$

$$\binom{n}{k} p^k (1-p)^{n-k} < \sqrt{\frac{1}{2\pi k (1 - \frac{k}{n})}} e^{-2n(x - \frac{k}{n})^2}, 1 \leq k \leq n-1, 0 < p < 1,$$

$$B_n^2(f, x) \leq B_n(f^2, x),$$

for all $x \in [0, 1]$, where $B_n(f, x)$ is the nth Bernstein polynomial $B_n(f, x) = \sum_{k=0}^{n} \binom{n}{k} f(k/n) x^k (1-x)^{n-k}$, $0 \leq x \leq 1$,

$f : [0, 1] \to \mathcal{R}$ is convex iff

$$B_{n+1}(f, x) \leq B_n(f, x)$$

for all $n \geq 1$, $x \in [0, 1]$, or

$$f(x) \leq B_n(f, x)$$

for all $n \geq 1$, $x \in [0, 1]$.

For the entire collection, see: Bullen, P. (1998). *A Dictionary of Inequalities*, Addison-Wesley Longman.

Descartes' Sign Change Inequality Given a power series with radius of convergence ρ, Z its number of zeros in $(0, \rho)$, and C the number of sign changes in its coefficients,

$$Z \leq C.$$

Pólya, G. and Szegö, G. (1998). *Problems and Theorems in Analysis II*, Springer, New York.

Sign Change Inequality for Polynomials Given $P_n(x) = a_0 + a_1 x + \cdots + a_n x^n$, C^+ its total number of changes of sign, and C^- the total number of changes of sign of $P_n(-x)$,

$$C^+ + C^- \leq n.$$

Pólya, G. and Szegö, G. (1998). *Problems and Theorems in Analysis II*, Springer, New York.

Number of Roots Inequality Given a polynomial $P(x)$, C = its total number of complex roots, and C' = the total number of complex roots of $P'(x)$,

$$C' \leq C.$$

Pólya, G. and Szegö, G. (1998). *Problems and Theorems in Analysis II*, Springer, New York.

Location of Roots Inequality I Given $P(x) = x^n + a_1 x^{n-1} + \cdots + a_n$ and z any root of $P(x)$,

$$|z| \leq \max\{(n|a_i|)^{1/i}, 1 \leq i \leq n\}.$$

Pólya, G. and Szegö, G. (1998). *Problems and Theorems in Analysis II*, Springer, New York.

Location of Roots Inequality II Given $P(x) = a_0 + a_1 x + \cdots a_n x^n$, $a_0 > a_1 > \cdots > a_n > 0$, and z any root of $P(x)$,

$$|z| > 1.$$

Pólya, G. and Szegö, G. (1998). *Problems and Theorems in Analysis II*, Springer, New York.

Location of Roots Inequality III Given $P(x) = a_0x^n + a_1x^{n-1} + \cdots + a_{n-1}x + a_n$, $a_0, a_1, a_2, \cdots, a_n > 0$, and z any root of $P(x)$,

$$\alpha \le |z| \le \beta,$$

where $\alpha = \min\{\frac{a_{i+1}}{a_i}, 0 \le i \le n-1\}$, $\beta = \max\{\frac{a_{i+1}}{a_i}, 0 \le i \le n-1\}$.

Pólya, G. and Szegö, G. (1998). *Problems and Theorems in Analysis II*, Springer, New York.

Markov Polynomial Growth Inequality I Given a polynomial $P(x)$ of degree at most n,

$$||P'||_{\infty,[a,b]} \le \frac{2n^2}{b-a}||P||_{\infty,[a,b]}.$$

Bullen, P. (1998). *A Dictionary of Inequalities*, Addison-Wesley Longman, Harlow.

Markov Polynomial Growth Inequality II Given a polynomial $P(x)$ of degree at most n on $[-1, 1]$ such that $||P||_\infty \le 1$,

$$||P^{(k)}||_\infty \le \frac{n^2(n^2-1)(n^2-4)\cdots(n^2-(k-1)^2)}{(2k-1)!!}, k \ge 1.$$

Bullen, P. (1998). *A Dictionary of Inequalities*, Addison-Wesley Longman, Harlow.

35.4 Integral and Derivative Inequalities

Wirtinger Inequality Given $f(0) = f(1) = 0$, $f' \in L^2$,

$$\int_0^1 f^2 \le \frac{1}{\pi^2}\int_0^1 (f')^2.$$

Zwillinger, D. (1992). *Handbook of Integration*, Jones and Bartlett, Boston.

Generalized Wirtinger Inequality I Given $1 \le k < \infty$, f such that $f \in L^k[0, 1]$, and $f' \in L^2[0, 1]$,

$$\left(\int_0^1 |f|^k\right)^{1/k} \le \sqrt{\frac{k}{\pi}}2^{(1-k)/k}(k+2)^{(k-2)/(2k)}\frac{\Gamma((k+2)/(2k))}{\Gamma(1/k)}\left(\int_0^1 |f'|^2\right)^{1/2}.$$

Zwillinger, D. (1992). *Handbook of Integration*, Jones and Bartlett, Boston.

Generalized Wirtinger Inequality II Given $k \geq 1$, f such that $f(0) = 0$, and $f, f' \in L^{2k}[0, 1]$,

$$\int_0^1 f^{2k} \leq \frac{1}{2k-1}\left(\frac{2k}{\pi}\sin\left(\frac{\pi}{2k}\right)\right)^{2k}\int_0^1 (f')^{2k}.$$

Hardy, G., Littlewood, J., and Pólya, G. (1952). *Inequalities*, Cambridge University Press, Cambridge.

Talenti Inequality Given f such that $f(0) = f(1) = 0, p > 1, q \geq 1, r = \frac{p}{p-1}$,

$$\left(\int_0^1 |f|^q\right)^{1/q} \leq \frac{q}{2}\left(1+\frac{r}{q}\right)^{1/p}\left(1+\frac{q}{r}\right)^{-1/q}\frac{\Gamma(\frac{1}{q}+\frac{1}{r})}{\Gamma(\frac{1}{q})\Gamma(\frac{1}{r})}\left(\int_0^1 |f'|^p\right)^{1/p}.$$

Milovanović, G. (1998). *Recent Progress in Inequalities*, Kluwer, Dordrecht.

Finch, S. (2003). *Mathematical Constants*, Cambridge University Press, Cambridge.

Friedrich's Inequality Given $f \in C^1[0, 1]$,

$$\int_0^1 (f^2 + (f')^2) \leq \beta[f^2(0) + f^2(1) + \int_0^1 (f')^2]$$

$$\Leftrightarrow \int_0^1 f^2 \leq \beta[f^2(0) + f^2(1)] + (\beta - 1)\int_0^1 (f')^2,$$

where $\beta = 1 + \theta^{-2}, \theta = 2.472548$ is the unique solution of $\cos\theta - \frac{\theta}{\theta^2+1}\sin\theta = -1, 0 < \theta < \pi$.

Finch, S. (2003). *Mathematical Constants*, Cambridge University Press, Cambridge.

Weighted Poincaré Inequality Given $f \in C^1[0, 1]$ such that $f(0) = f(1) = 0$, and given $k > 0$,

$$\int_0^1 e^{-kx} f^2 \leq \frac{4}{k^2 + 4\pi^2}\int_0^1 e^{-kx}(f')^2.$$

Flavin, J. (1996). *Qualitative Estimates for Partial Differential Equations*, CRC Press, Boca Raton, FL.

Rod Inequality Given $f \in C^2[0,1]$ such that $f(0) = f'(0) = f(1) = f'(1) = 0$,

$$\int_0^1 f^2 \leq \mu \int_0^1 (f'')^2,$$

where $\mu = \theta^{-4}, \theta = 4.730041$ is the smallest positive root of $\cos\theta \cosh\theta = 1$.

Finch, S. (2003). *Mathematical Constants*, Cambridge University Press, Cambridge.

Flavin-Rionero Inequality Given $f \in C^2[0,h]$ such that $f(0) = f'(0) = f(h) = f'(h) = 0$,

$$\int_0^h (f')^2 \leq \frac{h^2}{4\pi^2} \int_0^h (f'')^2.$$

Flavin, J. (1996). *Qualitative Estimates for Partial Differential Equations*, CRC Press, Boca Raton, FL.

Hardy-Littlewood-Polya Inequality I Given f such that $f(0) = f(1) = 0$,

$$\int_0^1 \frac{f^2(x)}{x(1-x)} dx \leq \frac{1}{2} \int_0^1 (f'(x))^2 dx.$$

Hardy, G., Littlewood, J., and Pólya, G. (1952). *Inequalities*, Cambridge University Press, Cambridge.

Opial Inequality Given f absolutely continuous on $[0,h]$, $f(0) = 0$,

$$\int_0^h |ff'| \leq \frac{h}{2} \int_0^h (f')^2.$$

Bullen, P. (1998). *A Dictionary of Inequalities*, Addison-Wesley Longman, Harlow.

Opial-Yang Inequality Given f absolutely continuous on $[a,b]$, $f(a) = 0, r \geq 0, s \geq 1$,

$$\int_a^b |f|^r |f'|^s \leq \frac{s}{r+s}(b-a)^r \int_a^b |f'|^{r+s}.$$

Bullen, P. (1998). *A Dictionary of Inequalities*, Addison-Wesley Longman, Harlow.

Young Inequality Given f strictly increasing continuous on $[0, c]$, $f[0] = 0, 0 \le a, b \le c$,

$$\int_0^a f + \int_0^b f^{-1} \ge ab.$$

Bullen, P. (1998). *A Dictionary of Inequalities*, Addison-Wesley Longman.

Hardy-Littlewood-Polya Inequality II Given $1 \neq a \ge 0, b \ge 0, f$ nonnegative and strictly decreasing,

$$\left(\int_0^1 x^{a+b} f\right)^2 \le \left[1 - \left(\frac{a-b}{a+b+1}\right)^2\right] \int_0^1 x^{2a} f \int_0^1 x^{2b} f.$$

Hardy, G., Littlewood, J., and Pólya, G. (1952). *Inequalities*, Cambridge University Press, Cambridge.

Hardy-Littlewood-Polya Inequality III Given f, g such that $0 \le f' \le 1, 0 \le g(x) < x$,

$$\int_0^1 \frac{f(x) - f(g(x))}{x - g(x)} \le f(1)(1 - \log f(1)).$$

Hardy, G., Littlewood, J., and Pólya, G. (1952). *Inequalities*, Cambridge University Press, Cambridge.

Hardy-Littlewood-Polya Inequality IV Given f, g such that $0 \le f' \le 1, 0 \le g(x) < x, k > 1$,

$$\int_0^1 \left(\frac{f(x) - f(g(x))}{x - g(x)}\right)^k \le \frac{kf(1) - f^k(1)}{k-1}.$$

Hardy, G., Littlewood, J., and Pólya, G. (1952), *Inequalities*, Cambridge University Press, Cambridge.

Reverse Holder Inequality of Brown-Shepp Given f, g of compact support, $p > 0, \frac{1}{p} + \frac{1}{q} = 1$,

$$\left(\int |f|^p\right)^{1/p} \left(\int |g|^q\right)^{1/q} \le \int \sup_y [f(x-y)g(y)]dx.$$

Zwillinger, D. (1992). *Handbook of Integration*, Jones and Bartlett, Boston.

Ostrowski Inequality I Given f monotone decreasing and nonnegative and g integrable on $[a, b]$,

$$\left|\int_a^b fg\right| \le f(a) \sup_{a \le \xi \le b} \left|\int_a^\xi g\right|.$$

Zwillinger, D. (1992). *Handbook of Integration*, Jones and Bartlett, Boston.

Ostrowski Inequality II Given $f \in C^1[a, b]$ and $\bar{f} = \frac{1}{b-a}\int_a^b f$,

$$|f(x) - \bar{f}| \le \frac{(x-a)^2 + (b-x)^2}{2(b-a)} ||f'||_{\infty,[a,b]}.$$

Bullen, P. (1998), *A Dictionary of Inequalities*, Addison-Wesley Longman, Harlow.

Steffensen's Inequality Given f, g nonnegative, f nonincreasing on $[a, b]$, and $g(x) \le 1$, $\int_a^b g = k$,

$$\int_{b-k}^b f \le \int_a^b fg \le \int_a^{a+k} f.$$

Bullen, P. (1998). *A Dictionary of Inequalities*, Addison-Wesley Longman, Harlow.

Convex Function Inequality I Given f nonnegative and convex on $[0, 1]$,

$$\frac{1}{(n+1)(n+2)}\int_0^1 f \le \int_0^1 t^n f \le \frac{2}{n+2}\int_0^1 f.$$

Bullen, P. (1998). *A Dictionary of Inequalities*, Addison-Wesley Longman, Harlow.

Ting Convex Function Inequality II Given $f \in C[0, a]$ nonnegative and convex and $\alpha > 2$,

$$\frac{\alpha(\alpha-1)}{\alpha+1} \le \frac{\int_0^a x^{\alpha-1} f}{\int_0^a x^{\alpha-2} f} \le \frac{\alpha a}{\alpha+1}.$$

Bullen, P. (1998). *A Dictionary of Inequalities*, Addison-Wesley Longman, Harlow.

Hermite-Hadamard Inequality for Convex Functions Given f convex on $[a, b]$, and $\bar{f} = \frac{1}{b-a}\int_a^b f$,

$$\bar{f} \geq \frac{1}{n}\sum_{i=0}^{n-1} f\left(a + \frac{i}{n-1}(b-a)\right).$$

Bullen, P. (1998). *A Dictionary of Inequalities*, Addison-Wesley Longman, Harlow.

Thunsdorff Inequality Given f nonnegative and concave on $[a, b]$, and $0 < r < s$,

$$\left(\frac{s+1}{b-a}\int_a^b f^s\right)^{1/s} \leq \left(\frac{r+1}{b-a}\int_a^b f^r\right)^{1/r}.$$

Bullen, P. (1998). *A Dictionary of Inequalities*, Addison-Wesley Longman, Harlow.

Grüss-Barnes Inequality Given $p, q \geq 1$, f, g nonnegative and concave on $[0, 1]$,

$$\int_0^1 fg \geq \frac{(p+1)^{1/p}(q+1)^{1/q}}{6}||f||_p||g||_q.$$

Bullen, P. (1998). *A Dictionary of Inequalities*, Addison-Wesley Longman, Harlow.

Reverse Cauchy-Schwartz Inequality Given f, g defined on an arbitrary interval such that $0 < a \leq f(x) \leq A, 0 < b \leq g(x) \leq B$,

$$\left(\int fg\right)^2 \geq \frac{4}{\left(\sqrt{\frac{AB}{ab}} + \sqrt{\frac{ab}{AB}}\right)^2}\int f^2 \int g^2.$$

Zwillinger, D. (1992). *Handbook of Integration*, Jones and Bartlett, Boston.

Carlson Inequality Given f nonnegative such that $f, |x|f \in L^2[0, \infty)$,

$$\int_0^\infty f \leq \sqrt{\pi}\left(\int_0^\infty f^2\right)^{1/4}\left(\int_0^\infty x^2 f^2\right)^{1/4}.$$

Bullen, P. (1998). *A Dictionary of Inequalities*, Addison-Wesley Longman, Harlow.

Generalized Carlson Inequality of Levin Given f nonnegative, and $p, q > 1, \lambda, \mu > 0$, such that both integrals on the right-hand side exist,

$$\int_0^\infty f \leq \frac{1}{(ps)^s(qt)^t}\left[\frac{\Gamma(\frac{s}{r})\Gamma(\frac{t}{r})}{(\lambda+\mu)\Gamma(\frac{s+t}{r})}\right]^r \left(\int_0^\infty x^{p-1-\lambda} f^p\right)^s \left(\int_0^\infty x^{q-1+\mu} f^q\right)^t,$$

where $s = \dfrac{\mu}{p\mu + q\lambda}, t = \dfrac{\lambda}{p\mu + q\lambda}, r = 1 - s - t.$

Bullen, P. (1998). *A Dictionary of Inequalities*, Addison-Wesley Longman, Harlow.

Landau Inequality Given f twice differentiable such that f, f'' are both uniformly bounded,

$$||f'||_{\infty,[0,\infty)} \leq 2||f||^{1/2}_{\infty,[0,\infty)}||f''||^{1/2}_{\infty,[0,\infty)}.$$

Milovanović, G. (1998). *Recent Progress in Inequalities*, Kluwer, Dordrecht.

Schoenberg-Cavaretta Extension of Landau Inequality Given f that is n times differentiable and such that f, $f^{(n)}$ are both uniformly bounded, for $1 \leq k < n$,

$$||f^{(k)}||_{\infty,[0,\infty)} \leq C(n,k)(||f||_{\infty,[0,\infty)})^{1-k/n}(||f^{(n)}||_{\infty,[0,\infty)})^{k/n},$$

where $C(3,1) = (\frac{243}{8})^{1/3} = 4.35622$, $C(4,1) = 4.288$, $C(3,2) = 24^{1/3} = 2.88449$, $C(4,2) = 5.750$, $C(4,3) = 3.708$. (No general explicit value for a general $C(n,k)$ is known.)

Milovanović, G. (1998). *Recent Progress in Inequalities*, Kluwer, Dordrecht.

Hadamard-Landau Inequality on $(-\infty, \infty)$ Given f twice differentiable and such that f, f'' are both uniformly bounded,

$$||f'||_{\infty,(-\infty,\infty)} \leq \sqrt{2}||f||^{1/2}_{\infty,(-\infty,\infty)}||f''||^{1/2}_{\infty,(-\infty,\infty)}.$$

Milovanović, G. (1998). *Recent Progress in Inequalities*, Kluwer, Dordrecht.

Kolmogorov Inequality on $(-\infty, \infty)$ Given f that is n times differentiable and such that f, $f^{(n)}$ are both uniformly bounded, for $1 \leq k < n$,

$$||f^{(k)}||_{\infty,(-\infty,\infty)} \leq C(n,k)(||f||_{\infty,(-\infty,\infty)})^{1-k/n}(||f^{(n)}||_{\infty,(-\infty,\infty)})^{k/n},$$

where $C(n,k) = a_{n-k}a_n^{k/n-1}$ with $a_m = \frac{4}{\pi}\sum_{j=0}^{\infty}\left[\frac{(-1)^j}{2j+1}\right]^{m+1}$ and, in particular,

$$C(3,1) = \left(\frac{9}{8}\right)^{1/3}, C(4,1) = \left(\frac{512}{375}\right)^{1/4},$$

$$C(3,2) = 3^{1/3}, C(4,2) = \sqrt{\frac{6}{5}}, C(4,3) = \left(\frac{24}{5}\right)^{1/4}.$$

Milovanović, G. (1998). *Recent Progress in Inequalities*, Kluwer, Dordrecht.

Finch, S. (2003). *Mathematical Constants*, Cambridge University Press, Cambridge.

Hardy-Littlewood-Polya Inequality V Given f nonnegative and decreasing on $[0,\infty)$ and $a, b \geq 0, a \neq b$,

$$\left(\int_0^\infty x^{a+b} f\right)^2 \leq \left[1 - \left(\frac{a-b}{a+b+1}\right)^2\right]\int_0^\infty x^{2a} f \int_0^\infty x^{2b} f.$$

Hardy, G., Littlewood, J., and Pólya, G. (1952). *Inequalities*, Cambridge University Press, Cambridge.

Zagier Inequality Given $f, g : [0,\infty) \to [0,1]$ decreasing integrable functions,

$$\int_0^\infty fg \geq \frac{\int_0^\infty f^2 \int_0^\infty g^2}{\max\{\int_0^\infty f, \int_0^\infty g\}}.$$

Bullen, P. (1998). *A Dictionary of Inequalities*, Addison-Wesley Longman, Harlow.

Hardy Inequality I Given $p > 1$, f such that $f \in L^p[0,\infty)$ and is nonnegative, and with $F(x) = \int_0^x f(t)dt$,

$$\int_0^\infty (F(x)/x)^p \leq \left(\frac{p}{p-1}\right)^p \int_0^\infty f^p.$$

Bullen, P. (1998). *A Dictionary of Inequalities*, Addison-Wesley Longman, Harlow.

Hardy Inequality II Given $p > 1$, f such that $xf \in L^p[0, \infty)$ and f nonnegative, with $F(x) = \int_0^x f(t)dt$,

$$\int_0^\infty \left(\int_x^\infty f \right)^p dx \leq p^p \int_0^\infty (xf)^p.$$

Bullen, P. (1998). *A Dictionary of Inequalities*, Addison-Wesley Longman, Harlow.

Hardy Inequality III Given $p, m > 1$, f nonnegative, $F(x) = \int_0^x f(t)dt$, and the integrals below existing,

$$\int_0^\infty x^{-m} F^p \leq \left(\frac{p}{m-1} \right)^p \int_0^\infty x^{p-m} f^p.$$

Bullen, P. (1998). *A Dictionary of Inequalities*, Addison-Wesley Longman, Harlow.

Hardy Inequality in Higher Dimensions Given $f : \Omega \to \mathcal{R}^1$ in the Sobolev space $H_0^{1,p}(\Omega) = \{f : \sum_{|\alpha| \leq p} \int_\Omega |D^\alpha f| < \infty, \}$, Ω a bounded convex domain, and $f = 0$ on $\partial\Omega$,

$$\int_\Omega \left| \frac{f}{d} \right|^p \leq (p/(p-1))^p \int_\Omega ||\nabla f||^p,$$

where $d = d(x)$ denotes the distance of x from $\partial\Omega$.

Finch, S. (2003). *Mathematical Constants*, Cambridge University Press, Cambridge.

Heisenberg-Weyl Uncertainty Inequality Given $f \in C^1(0, \infty)$, $p > 1, q = \frac{p}{p-1}, r > -1$,

$$\int_0^\infty x^r |f|^p \leq \frac{p}{r+1} \left(\int_0^\infty x^{q(r+1)} |f|^p \right)^{1/q} \left(\int_0^\infty |f'|^p \right)^{1/p}.$$

Zwillinger, D. (1992). *Handbook of Integration*, Jones and Bartlett, Boston.

***L^1* Version of Uncertainty Inequality** Given f such that $f, f' \in L^2(-\infty, \infty)$, f is absolutely continuous, and $x^2 f \in L^1(-\infty, \infty)$,

$$\int |f| \int f^2 \leq C \int |x^2 f| \int (f')^2,$$

where C may be taken as 2.35.

Laeng, E. and Morpurgo, C. (1999). Proc. Am. Math. Soc., 127, 3565–3572.

Hardy-Littlewood-Polya Inequality VI Given f such that $f, f'' \in L^2(0, \infty)$, it follows that $f' \in L^2(0, \infty)$ and

$$\left(\int_0^\infty (f')^2\right)^2 \leq 2\int_0^\infty f^2 \int_0^\infty (f'')^2.$$

Hardy, G., Littlewood, J., and Pólya, G. (1952). *Inequalities*, Cambridge University Press, Cambridge.

Ljubic-Kupcov Extension of Hardy-Littlewood-Polya Inequality VI Given f such that $f, f^{(n)} \in L^2(0, \infty)$, and $1 \leq k < n$, it follows that $f^{(k)} \in L^2(0, \infty)$ and

$$||f^{(k)}||_{2,[0,\infty)} \leq C(n,k)(||f||_{2,[0,\infty)})^{1-k/n}(||f^{(n)}||_{2,[0,\infty)})^{k/n},$$

where $C(3,1) = C(3,2) = 1.84420, C(4,1) = C(4,3) = 2.27432$, $C(4,2) = 2.97963$.

Milovanović, G. (1998). *Recent Progress in Inequalities*, Kluwer, Dordrecht.

Hardy-Littlewood-Polya Inequality VII Given f such that $f, f'' \in L^2(0, \infty)$, it follows that $f' \in L^2(0, \infty)$ and

$$\int_0^\infty [f^2 - (f')^2 + (f'')^2] \geq 0.$$

Hardy, G., Littlewood, J., and Pólya, G. (1952). *Inequalities*, Cambridge University Press, Cambridge.

Pointwise Inequality I Given any function f such that $f, f' \in L^2[0, \infty)$,

$$(f(0))^4 \leq 4\int_0^\infty f^2 \int_0^\infty (f')^2.$$

Hardy, G., Littlewood, J., and Pólya, G. (1952). *Inequalities*, Cambridge University Press, Cambridge.

Pointwise Inequality II Given f such that $f, f'' \in L^2(0, \infty)$, it follows that $f' \in L^2(0, \infty)$ and

$$(f(0))^2 \leq \frac{2}{3}\int_0^\infty [f^2 + 2(f')^2 + (f'')^2].$$

Hardy, G., Littlewood, J., and Pólya, G. (1952). *Inequalities*, Cambridge University Press, Cambridge.

Pointwise Inequality III Given f such that $f \in C^1(-\infty, \infty)$, $|xf|$, $f' \in L^2(-\infty, \infty)$, for any $x > 0$,

$$xf^2(x) \leq 4\sqrt{\int_x^\infty t^2 f^2(t)}\sqrt{\int_x^\infty (f'(t))^2}.$$

Zwillinger, D. (1992). *Handbook of Integration*, Jones and Bartlett, Boston.

Kolmogorov-Landau Inequality Given $f : \mathcal{R} \rightarrow \mathcal{R}$ such that $f \in L^r, f^{(n)} \in L^p$ for some $n, p, r, 1 \leq n < \infty, 1 \leq p, r \leq \infty$, for any $1 \leq k < n, 1 \leq q \leq \infty$,

$$||f^{(k)}||_q \leq C(n, p, q, r, k)||f||_r^\nu ||f^{(n)}||_p^{1-\nu},$$

where C is a universal positive finite constant and $\nu = \dfrac{n - k - 1/p + 1/q}{n - 1/p + 1/r}$.

Milovanović, G. (1998). *Recent Progress in Inequalities*, Kluwer, Dordrecht.

Planar Sobolev Inequality Given $f \in C^2(\mathcal{R}^2)$,

$$||f||_{\infty,\mathcal{R}^2} \leq \alpha \left[\int_{\mathcal{R}^2} (f^2 + f_x^2 + f_y^2 + f_{xx}^2 + f_{yy}^2 + f_{xy}^2)\right]^{1/2},$$

where $\alpha = \sqrt{\frac{1}{2\pi} \int_1^\infty \frac{1}{\sqrt{t^4+5t^2+6}} dt} = .3187591.$

Finch, S. (2003). *Mathematical Constants*, Cambridge University Press, Cambridge.

Three-dimensional Sobolev Inequality Given $f \in C^1(\Omega), with \Omega$ a bounded domain in $\mathcal{R}^3$ with a smooth boundary and such that $f = 0$ on $\partial\Omega$,

$$\int_\Omega f^2 \leq \frac{4}{\sqrt{3}\pi^2} \left(\int_\Omega ||\nabla f||^2\right)^3.$$

Finch, S. (2003). *Mathematical Constants*, Cambridge University Press, Cambridge.

Higher-Dimensional Sobolev Inequality Given $f : \mathcal{R}^n \rightarrow \mathcal{R}$ in the Sobolev space $\{f : \sum_{\alpha \leq 1} \int_{\mathcal{R}^n} |D^\alpha f|^p < \infty\}$ and k such that $1 \leq p < \frac{n}{k}$,

$$||f||_{np/(n-kp)} \leq C(||D^k f||_p + ||f||_p),$$

where C is a finite positive constant.

Finch, S. (2003). *Mathematical Constants*, Cambridge University Press, Cambridge.

One-dimensional Nash Inequality Given $f \in C^1(\mathcal{R})$,

$$\left(\int_{\mathcal{R}} f^2\right)^3 \leq \frac{27}{16\pi^2} \int_{\mathcal{R}} (f')^2 \left(\int_{\mathcal{R}} |f|\right)^4.$$

Laeng, E. and Morpurgo, C. (1999). Proc. Am. Math. Soc., 127, 3565–3572.

***n*-dimensional Nash Inequality** Given $f \in C^1(\mathcal{R}^n)$,

$$\left(\int_{\mathcal{R}^n} f^2\right)^{(n+2)/n} \leq C_n \int_{\mathcal{R}^n} |\nabla f|^2 \left(\int_{\mathcal{R}^n} |f|\right)^{4/n},$$

where C_n is a universal positive finite constant and may be taken to be $\dfrac{2(n/2+1)^{(n+2)/n}}{n\omega_n^{2/n} k_n^2}$, with $\omega_n = \dfrac{\pi^{n/2}}{\Gamma(\frac{n}{2}+1)}$ the volume of the n-dimensional unit ball, and k_n the first positive root of the Bessel function $J_{n/2}$.

Carlen, E. and Loss, M. (1993). Int. Mat. Res. Notices, 7 301–305.

***n*-dimensional Isoperimetric Inequality** Given $f \in C^1(\mathcal{R}^n)$ of compact support,

$$\left(\int_{\mathcal{R}^n} |f|^{\frac{n}{n-1}}\right)^{\frac{n-1}{n}} \leq \frac{1}{n\omega_n^{1/n}} \int_{\mathcal{R}^n} ||\nabla f||,$$

where $\omega_n = \frac{\pi^{n/2}}{\Gamma(\frac{n}{2}+1)}$ is the volume of the n-dimensional unit ball.

Finch, S. (2003). *Mathematical Constants*, Cambridge University Press, Cambridge.

***n*-dimensional Poincaré Inequality I** Given $f, g \in C^1[0, a]^n$, $1 \leq n < \infty$,

$$\int_{[0,a]^n} fg \leq \frac{1}{a^n} \int_{[0,a]^n} f \int_{[0,a]^n} g + \frac{na^2}{4} \int_{[0,a]^n} [||\nabla f||^2 + ||\nabla g||^2].$$

Bullen, P. (1998). *A Dictionary of Inequalities*, Addison-Wesley Longman.

***n*-dimensional Poincaré Inequality II** Given $f : \Omega \to \mathcal{R}$ such that $f = 0$ on $\partial\Omega$, Ω a bounded convex domain, and $\sum_{|\alpha|\leq 1} \int_\Omega |D^\alpha f|^2 < \infty$,

$$\int_\Omega |f|^2 \leq c(\Omega) \int_\Omega ||\nabla f||^2,$$

where $c(\Omega)$ is a finite positive constant.

Bullen, P. (1998). *A Dictionary of Inequalities*, Addison-Wesley Longman, Harlow.

Glossary of Symbols

$\stackrel{P}{\Rightarrow}, \stackrel{P}{\rightarrow}$	convergence in probability
$o_p(1)$	convergence in probability to zero
$O_p(1)$	bounded in probability
$a_n \sim b_n$	$0 < \liminf \frac{a_n}{b_n} \leq \limsup \frac{a_n}{b_n} < \infty$
$a_n \asymp b_n, a_n \approx b_n$	$\lim \frac{a_n}{b_n} = 1$
$\stackrel{\text{a.s.}}{\Rightarrow}, \stackrel{\text{a.s.}}{\rightarrow}$	almost sure convergence
with probability 1	almost surely
a.e.	almost everywhere
i.o.	infinitely often
$\stackrel{\mathcal{L}}{\Rightarrow}, \stackrel{\mathcal{L}}{\rightarrow}$	convergence in distribution
$\stackrel{r}{\Rightarrow}, \stackrel{r}{\rightarrow}$	convergence in rth mean
F_n	empirical CDF
F_n^{-1}	sample quantile function
$X^{(n)}$	sample observation vector $(X_1, \ldots, X_n)$
M_n, med	sample median
δ_x	point mass at x
$P(\{x\})$	probability of the point x
λ	Lebesgue measure
$*$	convolution
$\ll$	absolutely continuous
$\frac{dP}{d\mu}$	Radon-Nikodym derivative
$\bigotimes$	product measure; Kronecker product
Var	variance
μ_k	$E(X-\mu)^k$
κ_r	rth cumulant
ρ_r	rth standardized cumulant; correlation of lag r
$\rho(A, B)$	Hausdorff distance between A and B
$I(\theta)$	Fisher information function or matrix
$\pi(\theta \mid X^{(n)})$	posterior density of θ

Π_n	posterior measure
$\theta_{n,\alpha}$	αth percentile of posterior of θ
ARE	asymptotic relative efficiency
$e_P(T_1, T_2)$	Pitman efficiency of T_1 w.r.t. T_2
$e_B(T_1, T_2)$	Bahadur efficiency of T_1 w.r.t. T_2
LRT	likelihood ratio test
Λ_n	likelihood ratio
$W(t), B(t)$	Brownian motion and Brownian bridge
$W^d(t)$	d-dimensional Brownian motion
$\mathcal{R}$	real line
$\mathcal{R}^d$	d-dimensional Euclidean space
$C(X)$	real-valued continuous functions on X
$C_k(\mathcal{R})$	k times continuously differentiable functions
$C_0(\mathcal{R})$	real continuous functions f on $\mathcal{R}$ such that $f(x) \to 0$ as $\lvert x \rvert \to \infty$
$\mathcal{F}$	family of functions
$\nabla, \triangle$	gradient vector and Laplacian
$f^{(m)}$	mth derivative
$D^k f(x_1, \ldots, x_n)$	$\sum_{m_1, m_2, \ldots, m_n \geq 0, m_1 + \ldots m_n = k} \frac{\partial^{m_1} f}{\partial x_1^{m_1}} \cdots \frac{\partial^{m_n} f}{\partial x_n^{m_n}}$
$\lvert\lvert . \rvert\rvert$	Euclidean norm
$\lvert\lvert . \rvert\rvert_\infty$	supnorm
tr	trace of a matrix
$\lvert A \rvert$	determinant of a matrix
$B(x, r)$	sphere with center at x and radius r
I_A	indicator function of A
$I(t)$	large-deviation rate function
$\{\}$	fractional part
$[.]$	integer part
sgn	signum function
x_+, x^+	$\max\{x, 0\}$
max, min	maximum, minimum
sup, inf	supremum, infimum
K_ν, I_ν	Bessel functions
$L_p(\mu), L^p(\mu)$	set of functions such that $\int \lvert f \rvert^p d\mu < \infty$
d_K, d_L, d_{TV}	Kolmogorov, Levy, and total variation distances
H, K	Hellinger and Kullback-Leibler distances
ℓ_2	Wasserstein distance

$N_p(\mu, \Sigma)$, $MVN(\mu, \Sigma)$	multivariate normal distribution
BVN	bivariate normal distribution
$MN(n, p_1, \ldots, p_k)$	multinomial distribution with these parameters
$t_n, t_{(n)}$	t-distribution with n degrees of freedom
$\text{Ber}(p)$, $\text{Bin}(n, p)$	Bernoulli and binomial distributions
$\text{Poi}(\lambda)$	Poisson distribution
$\text{Geo}(p)$	geometric distribution
$\text{NB}(r, p)$	negative binomial distribution
$\text{Exp}(\lambda)$	exponential distribution with mean λ
$\text{Gamma}(\alpha, \lambda)$	Gamma density with shape parameter α and scale parameter λ
$\chi^2_n, \chi^2_{(n)}$	chi-square distribution with n degrees of freedom
χ^2	Pearson's chi-square statistic
$C(\mu, \sigma)$	Cauchy distribution
$\text{DoubleExp}(\mu, \lambda)$	double exponential with parameters μ, λ
$X_{(k)}, X_{k:n}$	kth order statistic
H_{Boot}	bootstrap distribution of a statistic
H_{Jack}	jackknife histogram
P_*	bootstrap measure
$b_{\text{Jack}}, V_{\text{Jack}}$	jackknife bias and variance estimate
$V_{\text{Jack},d}$	delete-d jackknife variance estimate
$\varphi(x, F)$	influence function
$\hat{f}_{n,h}$	kernel density estimate with bandwidth h
h_{opt}	optimal bandwidth
$h_{\text{opt}}(\mathcal{L}_1)$	optimal bandwidth under $\mathcal{L}_1$ loss
NPMLE	nonparametric maximum likelihood estimate
LSE	least squares estimate
Z^*_m	higher-criticism statistic
FDR	false-discovery rate
FNR	false-nondiscovery rate
MTP_2	multivariate totally positive of order 2

Index

D

I

Q

R

Springer Texts in Statistics

(continued from p. ii)

Robert: The Bayesian Choice: From Decision-Theoretic Foundations to Computational Implementation, Second Edition
Robert/Casella: Monte Carlo Statistical Methods, Second Edition
Rose/Smith: Mathematical Statistics with *Mathematica*
Ruppert: Statistics and Finance: An Introduction
Sen/Srivastava: Regression Analysis: Theory, Methods, and Applications
Shao: Mathematical Statistics, Second Edition
Shorack: Probability for Statisticians
Shumway/Stoffer: Time Series Analysis and Its Applications, Second Edition
Simonoff: Analyzing Categorical Data
Terrell: Mathematical Statistics: A Unified Introduction
Timm: Applied Multivariate Analysis
Toutenberg: Statistical Analysis of Designed Experiments, Second Edition
Wasserman: All of Nonparametric Statistics
Wasserman: All of Statistics: A Concise Course in Statistical Inference
Weiss: Modeling Longitudinal Data
Whittle: Probability via Expectation, Fourth Edition

Printed in the United States of America